Reliability Engineering

Singiresu S. Rao
Department of Mechanical and Aerospace Engineering
University of Miami Coral Gables, Florida

PEARSON

Boston Columbus Indianapolis New York San Francisco Upper Saddle River
Amsterdam Cape Town Dubai London Madrid Milan Munich Paris Montréal Toronto
Delhi Mexico City São Paulo Sydney Hong Kong Seoul Singapore Taipei Tokyo

Vice President and Editorial Director, ECS: *Marcia J. Horton*
Acquisitions Editor: *Norrin Dias*
Senior Managing Editor: *Scott Disanno*
Project Manager: *Irwin Zucker*
Full-Service Project Management: *Abinaya Rajendran*
Operations Specialist: *Linda Sager*
Senior Marketing Manager: *Tim Galligan*
Art Director: *Jayne Conte*
Cover Designer: *Suzanne Behnke*

Library of Congress Cataloging-in-Publication Data on File

PEARSON

ISBN-10: 0-13-601572-7
ISBN-13: 978-0-13-601572-7

ScoutAutomatedPrintCode

To Lord Sri Venkateswara

Contents

CHAPTER 3

Random Variables and Probability Distributions 87

CHAPTER 4

Extremal Distributions 161

CHAPTER 5

Functions of Random Variables 191

CHAPTER 6

Time-Dependent Reliability of Components and Systems 232

CHAPTER 7

Modeling of Geometry, Material Strength, and Loads 301

CHAPTER 14

Quality Control and Reliability 642

CHAPTER 15

Maintainability and Availability 706

CHAPTER 16

Warrantiesv 739

APPENDIX A

APPENDIX B

APPENDIX C

APPENDIX D

Preface

In today's competitive world, every device and system is expected to perform satisfactorily over the duration of its expected life span. No device or system will perform reliably unless it is designed specifically for *reliability*. The field of reliability became more prominent after several of the high-profile system failures such as the space shuttle disasters of *Challenger* and *Columbia*, Fukushima nuclear power plant accident, and the collapse of the Mississippi River Bridge.

This book is intended for use as an introduction to reliability engineering, including the aspects of analysis, design, testing, production, and quality control of engineering components and systems. The material can be used for senior or dual-level courses on reliability. The mathematical background necessary in the area of probability and statistics is covered briefly to make the presentation complete and self-contained. Expanded explanations of the fundamental concepts are given throughout the book, with emphasis on the physical significance of the ideas. Numerous analytical and numerical examples and problems are used to illustrate the principles and concepts. The solution of probability and reliability problems using the commonly used software systems, such as MATLAB and Excel, is also presented.

Features

All the concepts are explained fully and the derivations are presented with complete details for the benefit of the reader. The computational aspects of reliability analysis and design are adequately emphasized. Specific features include:

- More than 230 illustrative examples follow the presentation of most of the topics.
- More than 800 review questions to help readers in reviewing and testing their understanding of the text material.
- Nearly 50 examples are given to illustrate the use of MATLAB and Excel for solving probability and reliability problems.
- Over 550 references to lead the reader to specialized and advanced literature.
- More than 500 problems to help readers in the application of the theory to practical problems.
- Biographical information about the mathematicians and scientists who contributed to the development of the theories of probability and reliability on chapter-opening pages.
- Answers to review questions are available on the companion website at www.pearsonhighered.com/rao.
- Computer programs for the reliability analysis and design of engineering systems are provided on the companion website.
- Solutions to selected problems are available for Instructors in the Solutions Manual, which can be located on the Instructor Resource Center website at www.pearsonhighered.com

Organization of the Book

The book is divided into sixteen chapters and four appendices. Chapter 1 is an introductory chapter that presents the basic concepts of reliability, reliability analysis, and reliability-based design. The patterns of failure of mechanical/structural and electrical/electronic systems, as well as the relationship between the factor of safety and reliability, are discussed. The failures of some prominent systems are summarized and a brief outline of the historical development of the subject of reliability engineering is provided. Chapters 2 to 5 present the fundamentals of probability and random variables. These chapters cover the necessary background material on probability and statistics. Chapter 2 deals with the basic concepts of probability theory, including the laws of probability and Bayes' rule. Chapter 3 presents the principles of random variables and probability distributions. The concepts of discrete and continuous random variables and the corresponding probability distributions as well as moments of random variables are presented. Chapter 4 introduces extremal distributions outlining their importance in engineering applications. Chapter 5 outlines the concepts of functions of random variables.

The time-dependent reliability of components and systems is presented in Chapter 6. The construction of the failure rate curve from empirical data and the analysis of series, parallel, and complex multicomponent systems are also discussed. The computation of hazard and reliability functions corresponding to different distributions of failure time is also presented. In Chapter 7, the statistical treatment of dimensions, including tolerances and clearances, is considered in the modeling of component geometry. Brittle, plastic, and fiber-reinforced materials are considered in the modeling of material strength. Strength-based reliability and the computation of reliability using interference theory are discussed in Chapter 8. The use of experimentally determined distributions for the reliability analysis of components is also outlined. The methods of analyzing multicomponent structures, including weakest-link and fail-safe systems, are also considered. Truss and frame examples are included for illustration. The reliability analysis and the methods of design of mechanical components and systems are considered in Chapter 9. The analysis and design of different types of machine components, including connecting rod, pressure vessel, helical spring, and fatigue design of shafts, are considered. The reliability analysis and design of multicomponent systems are illustrated by considering a gear train, cam-follower system, and four-bar mechanism.

Chapter 10 presents the Monte Carlo simulation method for the reliability analysis of engineering systems. The analyses of a contact stress problem, plate-clutch mechanism, and a function generating mechanism are presented as illustrative examples. The application of optimization techniques such as graphical procedures, calculus principles, and nonlinear and dynamic programming approaches are outlined along with illustrative examples in Chapter 11. The reliability allocation problem involving the minimization of total cost or maximization of system reliability for series–parallel systems is considered. The optimum designs of a two-bar truss, shoe brake, and gear train are presented as illustrative examples.

Chapter 12 discusses several types of design and safety review techniques, including the failure modes, event-tree, and fault-tree analyses. These methods aid in understanding how accidents occur, how their probabilities can be estimated, and how to reduce the probability of their occurrence. The concept of minimal cut-sets and its use in the reliability analysis of systems, with the aid of fault trees, is also introduced. The design of a machine tool gear train is presented to illustrate the various concepts. The analysis and plotting of failure time data and the aspects of accelerated life testing are discussed in Chapter 13. The concepts of magnified load, sudden-death testing, and sequential life testing are introduced. The roles of statistical inference and parameter estimation, confidence intervals, and plotting of reliability data are also discussed in detail.

The relationship between quality and reliability is discussed in Chapter 14. The concepts of six sigma approach, including its implementation, is presented. The basics of acceptance sampling, sampling plans, and process capability are discussed. The methods of statistical quality control, including the p-chart, $\overline{X}$-chart,

R-chart, and c-chart, are outlined. The idea of operating characteristic curve and the method of its construction are presented. The off-line quality control method, based on Taguchi's robust design approach, is presented for the design of devices and systems. The reliability analysis of maintainable and repairable systems is outlined in Chapter 15. The influence of imperfect maintenance is also examined. The optimal replacement strategy for minimum cost and the aspect of spare parts requirement are discussed. The availability analysis, based on Markovian approach, is presented for single components as well as series and parallel systems with one or more repair persons. Chapter 16 presents the methods of determining different warranty policies for a product including the length of warranty and the corresponding reserve fund needed to cover the warranty.

The four appendices cover additional topics relevant to the book. Specifically, Appendices A to C include the distribution function tables and an introduction to product liability is given in Appendix D.

Typical Syllabi

The material of the book provides flexible options for different types of reliability engineering courses at the senior or graduate level. Different emphases/orientations can be given to the course by covering specific set of chapters as indicated below (in addition to the basic material covered in Chapters 1 to 6):

- Chapters 7 to 11: For emphasis on structural and mechanical systems and design
- Chapters 12 to 16: For emphasis on manufacturing and quality
- Chapters 10 to 13: For emphasis on numerical and experimental aspects.

Acknowledgments

I would like to express my appreciation to the many students who took my courses on Reliability Engineering and Reliability-Based Design at the Indian Institute of Technology, Kanpur, Purdue University, West Lafayette, and University of Miami, Coral Gables, and whose queries and comments helped me in improving the presentation of the material. I would also like to thank the reviewers: Harry Millwater, University of Texas at San Antonio; Azim Houshyar, Western Michigan University; Tunc Aldemir, University of Illinois; and Paul Wlodkowsky, University of Maine–Orono. Finally, I would like to thank my wife, Kamala, for her patience, support, and encouragement in completing the book.

SINGIRESU S. RAO

CHAPTER 1

Introduction

What You Will Learn

In this chapter, you will learn

- The definition of reliability
- The importance of reliability in engineering and consumer products, components, and systems
- The pattern of time-dependent failures of most electronic and electromechanical components
- The strength-based failure modes of mechanical and structural components
- The relation between the factor of safety and reliability
- The importance of reliability management, including the design, production, testing, maintenance, and repair of components and systems
- Some examples of recent system failures from different areas of engineering
- A brief history of reliability, along with a comparison of the risks associated with different human activities
- An outline of reliability literature

Jerome Cardan (or Gerolamo Cardano), born in Pavia, Italy, on September 24, 1501, is considered to be the pioneer in the field of probability. Encouraged by his father, Fazio Cardano, who was a lawyer and professor of medicine in Milan as well as a friend of Leonardo da Vinci, Cardan began his university studies in 1520 at Pavia and completed them in 1526 at Padua with a doctorate in medicine. Cardan had an interesting personality. In 1534, while practicing medicine, he joined as a teacher of mathematics. He wrote on many subjects such as mathematics, astronomy, physics, music, and games of chance. He was put in prison in 1570 for publishing the horoscope of Christ. The book he wrote on gambling, *Liber de Ludo Aleae (Book on Games of Chance)*, is considered to be the first book on probability. The book describes numerous problems relating to the games of dice and cards, and includes tips on how to cheat and avoid being cheated. The book also introduces the concepts of probability, mathematical expectation, and the law of large numbers. Cardan discovered the method of solving a cubic equation, which has become known as the "Cardan's method." His multifaceted personality ended dramatically with his prediction of the day of his own death. When the day arrived he was still alive, so he committed suicide to preserve his reputation [1.248].

1.1 Uncertainty in Engineering

Uncertainty exists in most phenomena observed in the world. All engineering problems involve measurement or observation of multiple physical phenomena. For example, the design of an airplane involves the measurement of phenomena such as strength of materials (e.g., aluminum), dimensions of manufactured components, and loads associated with wind gusts.

For an uncertain quantity such as the gust load, multiple measurements (at different times) lead to multiple outcomes. Because of the inherent uncertainties associated with the parameters of an engineering system, the performance or output of most systems will also be uncertain. This leads to the problem of reliability of the system in the presence of uncertain inputs.

In practice, different types of uncertainty may be present in different systems. Broadly speaking, the various forms of uncertainty can be classified into two categories or types. The type of uncertainty which results from the fact that a system can behave in random ways is called stochastic uncertainty, also termed aleatory or irreducible uncertainty. The type of uncertainty which results from the lack of knowledge about a system is known as subjective or reducible or epistemic uncertainty.

Stochastic uncertainty can be of two types. The first one is the inherent randomness that exists in all physical parameters. When a random parameter is measured or observed a number of times, we are likely to find a different value each time. In practice, we can take care of this type of uncertainty by using a large number of measurements or observations so that the information used will have a high degree of confidence. However, the collection of large quantity of data on the random parameter may not be always feasible due to the limitations of high cost and impracticality. In some cases, we need to conduct the analysis of a system in the presence of only a limited amount of data. In such a case, the degree of confidence associated with the results predicted using only a small sample of data will be low.

The second type of uncertainty is associated with the modeling of the system. Any analysis used for predicting the behavior of the system is only approximate. In general, an accurate analysis requires a very complex computational model. In order to be able to use a computationally simple model, which implies some error in predicting the system behavior, the modeling uncertainty needs to be included as an additional uncertain factor.

Subjective or qualitative uncertainty describes the inherent vagueness of the system and its parameters. Some of the factors that contribute to subjective uncertainty include the skill and experience of construction workers and machine operators, the condition of manufacturing equipment, quality of maintenance and repair, and the influence of environment on the performance of the system. Although traditionally probability (and reliability) theory has been used to consider subjective uncertainty, recent research indicates that other theories such as fuzzy theory, interval analysis, and evidence theory can be used to model subjective uncertainty more realistically and accurately.

1.2 Definition of Reliability

Due to the increasing complexity of modern engineering systems, the concept of reliability has become a very important factor in the overall system design. Reliability is important because a designer must design machinery and equipment that will work both in theory and in practice. In order to express the reliability of a system in quantitative terms, it is necessary to develop a mathematical model of the overall system and analyze its performance under realistic operating conditions. The generally accepted definition of reliability is as follows:

Reliability is the probability of a device performing its function over a specified period of time and under specified operating conditions.

Thus, reliability can be viewed as a measure of successful performance of a system. The foregoing definition includes such diverse cases as that of a television that is switched on and off several times and is expected to function for years without failure, and that of a missile that must be in standby condition for long periods of time and then must respond just once, but it must do so without failure. In all the cases, the success in meeting reliability requirements is critically dependent on the extent to which reliability is designed and built into the system.

Reliability is an analytical problem that involves both statistical and engineering aspects. It must be given critical attention throughout the life of a system—including its development, design, production, quality control, shipping, installation, operation, and maintenance. It requires the integrated application of many disciplines: statistics, probability, materials engineering, circuit analysis, mechanical design, structural analysis, production engineering, and several other engineering disciplines. Reliability also requires sound organization for the discovery and correction of the causes of failure.

1.3 Importance of Reliability

Throughout the history of modern engineering, failures of systems have been observed in every field of engineering. For example, several failures and accidents related to bridges, dams, tankers, airplanes, space shuttles, nuclear power plants, and chemical plants have been observed in recent history. The importance of reliability of components and systems is also recognized at every stage of daily life, ranging from consumer products, such as TV sets, clothes washers and dryers, lawn mowers, and automobiles, to larger systems, such as trains and airlines.

In reliability studies, the distinction between a component and a system should be clearly understood because the same physical item can be treated as either a component or a system depending on the viewpoint. For example, an automobile engineer may view the engine, gearbox, rear axle, and so on, as components and the complete vehicle as a drivetrain system. However, the engine manufacturer treats the engine as a system having pistons, piston rings, cylinders, crankshaft, and so on, as its components. Thus, the same item, in this case the engine, can be regarded as a component or a system depending on the context. For the purpose of reliability analysis, we shall use the word *component* to denote an integral item which is nonmaintained, and the word *system* to indicate an assembly of components, which may be maintained or nonmaintained. By a *nonmaintained* component, we mean that if it fails, it is removed and discarded from the aggregation of components. According to this terminology, if a failed component is repaired to original standards and returned to service, it will be called a *new component*. Similarly if a system fails due to the failure of some of its components, it is removed and discarded from the population if it is a nonmaintained system. The failed components are replaced, and the system is put into service if it is a *maintained system*.

In the case of consumer products such as electric fans, clothes washers and dryers, TV sets, and automobiles, the simplest concept of reliability (in terms of length of service) is assessed against specified values at the product inspection stage. The products passing the inspection stage are then delivered to the customers. Usually the reliability of the product is related to the price of the product. The customer buys the product with the understanding that it might fail at a future time. Although the manufacturer gives a warranty to cover the failures of the product during its early stages of life, too many failures during the warranty period causes inconvenience to the customer and high cost of repair to the manufacturer. In addition, too many product failures, either during or after the warranty period, will mean a loss of reputation which might adversely affect the future business for the manufacturer.

In the case of large mission-oriented systems, such as missiles and space shuttles, reliability (program) plays a crucial role. These systems are composed of several subsystems and components. All the individual components must be designed properly to achieve the specified overall reliability of the system. Often, it is not the complex elements that cause troubles but the simple ones; a 10-cent resistor can and has aborted the flight of a 300,000-part, million-dollar missile. In fact, much of the common loss of reliability is not mainly due to the failure of complex parts, but due to the malfunction of simple parts such as faulty joints—electrical, hydraulic, mechanical. It is well-known that it is the failure of O-rings that caused the destruction of the space shuttle *Challenger*, which cost the lives of seven astronauts, including the life of a school teacher.

1.4 Pattern of Failures

Before studying the methods of improving the reliability of a component or system, we need to understand the patterns and modes of failure associated with the particular component or system. In practice, a major difference can be noted between the patterns of failure of electronic/mechanical systems and structural systems. Most electronic and mechanical components and systems deteriorate during use as a result of elevated temperatures, chemical changes, mechanical wear, fatigue, and a number of other reasons. Thus, the reliability of an electronic/mechanical system is closely associated with the life of the system. The time at which failure occurs will be the primary random variable in this case. On the other hand, many structural systems tend not to deteriorate with time, except by the mechanisms of fatigue and corrosion. In some cases, the structure may even get stronger with time; for example, the strength of concrete increases with time. Usually, a structural component fails when extreme loads are encountered. Hence, the pattern of failures of structural components depends on the pattern of loads exceeding the strengths of the components.

1.4.1 Component Failures

For electronic, mechanical, and electromechanical components, there are three basic failure types in which the rate of failure varies with time. The failure rate may be decreasing, constant, or increasing. In fact, the variation over time of the failure rate for most mechanical, electronic, and electromechanical components can be represented as shown in Fig. 1.1. This failure-rate curve is commonly known as the bathtub curve because of its shape. In most cases, we can tell much about the cause of failure and the reliability of the product by knowing the shape of its failure-rate curve. In general, a decreasing failure rate is observed in the early stages of the product life. However, this indicates the presence of substandard components due to manufacturing defects and poor quality control procedures. As these

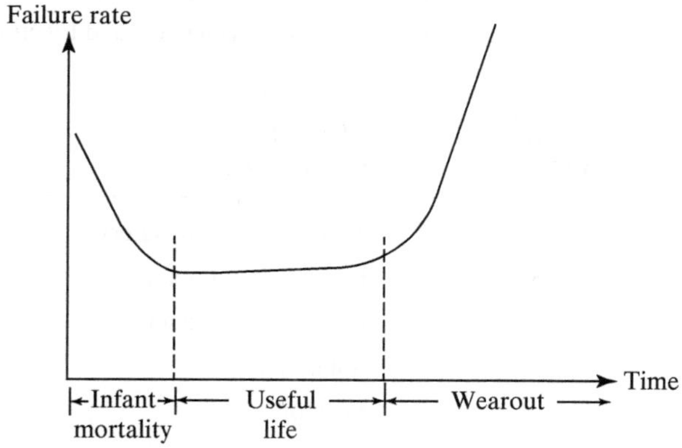

FIGURE 1.1 The bathtub curve.

defective products fail and are repaired or replaced, the failure rate decreases as time progresses. The constant failure rate indicates failures due to random causes. For example, mechanical failures caused by accidental overloads and electrical failures caused by transient circuit overloads occur randomly, usually, at a constant rate. The constant failure rates observed during the useful life of some mechanical and electrical components are given in Table 1.1. An increasing rate of failures is observed after the useful life of the product and the increase represents, in mechanical components, fatigue brought about by strength

TABLE 1.1 Representative Failure Rates of Some Mechanical and Electrical Components [1.185]

Mechanical Components	Failures Per Million Hours	Electrical Components	Failures Per Million Hours
Accelerometer	35.1	AC generator	0.81
Actuator	50.5	Ammeter/Voltmeter	26.0
Air compressor	6.0	Circuit breaker	1.2
Air pressure gauge	2.6	Coaxial connector	0.19
Ball bearing	1.1	DC generator	36.8
Boiler feed pump	0.42	Electric heater	2.3
Brake	4.3	Emergency light	2.0
Clutch	0.6	Fractional horse power motor	1.5
Differential	15.0	Incandescent lamp	18.6
Fan	2.8	Indicator	3.9
Gasket and seal	1.3	Large electric motor	0.9
Gear	0.17	Lead-acid battery	0.44
Gear shaft	6.7	Neon lamp	0.49

(Continued)

TABLE 1.1 Representative Failure Rates of Some Mechanical and Electrical Components [1.185] *(Continued)*

Mechanical Components	Failures Per Million Hours	Electrical Components	Failures Per Million Hours
Gyroscope	513.9	Nickel-cadmium battery	0.25
Heat exchanger	1.1	Printed circuit board	0.24
Hydraulic valve	9.3	Rechargeable battery	1.5
O-ring	2.4	Solder joint	0.001
Roller bearing	0.28	Solenoid	2.4
Shock absorber	0.81	Switch	107.3
Spring	5.0	Tachometer	10.7
Storage tank	1.6	Turbine/Generator	626.2
Thermostat	17.4	Voltage regulator	3.0

deterioration due to cyclic loading. In engineering product design, it is important to know the underlying failure patterns so that their causes can be understood and corrective actions can be taken to improve the product reliability. The experimental data on the failure of a component are usually represented by a frequency distribution curve, as illustrated by the following example.

EXAMPLE 1.1

The data collected on the lives of brakes of a particular model of an automobile (measured in miles of successful operation) are shown in Table 1.2.

TABLE 1.2 Lives of Brakes in Miles of Successful Operation

75,900	55,100	64,900	60,100	51,350	74,050	54,700
70,600	50,200	34,900	58,150	45,900	67,200	83,250
68,300	42,200	41,300	51,000	53,800	57,950	46,250
56,500	61,000	84,750	79,250	37,200	59,300	48,500
50,050	50,350	50,500	66,950	51,450	45,750	
26,100	33,000	56,850	47,400	47,650	59,950	
61,200	86,300	66,450	35,100	57,550	64,050	
55,450	71,450	77,450	55,850	67,850	53,350	
39,950	74,850	55,700	50,850	80,000	57,850	
64,550	66,000	45,450	62,250	60,750	67,500	

(Continued)

TABLE 1.2 Lives of Brakes in Miles of Successful Operation (*Continued*)

40,050	72,500	46,950	57,150	49,600	60,950
52,800	63,800	69,600	73,300	62,800	59,600
69,100	45,250	35,050	74,200	44,700	55,900
31,050	47,150	56,050	69,950	53,050	38,800
56,350	61,650	52,200	57,200	46,700	45,900
52,650	43,500	64,750	44,200	51,700	63,450

Show the distribution of the life of brakes on a *relative frequency diagram*. Also find the mean and standard deviation of the life of brakes.

Solution: Let X denote the life of brakes. To show the distribution of the life of brakes on a relative frequency diagram, we proceed as follows:

1. The lowest and the highest values of life (X) observed are identified as 26,100 and 86,300 miles, respectively. The nearest rounded-off numbers are chosen for plotting purpose as 25,000 and 90,000 miles. This gives the range of life of brakes as $90{,}000 - 25{,}000 = 65{,}000$ miles.
2. The range of life of brakes is divided into thirteen equally spaced intervals of 5000 miles each.
3. The number of lives observed in each interval is counted and the frequency table shown in Table 1.3 is set up.

TABLE 1.3 Frequency Table for the Lives of Brakes

Interval of Life (in Miles of Operation)	Number of Failures Observed in the Interval	Interval of Life (in Miles of Operation)	Number of Failures Observed in the Interval
25,001–30,000	1	60,001–65,000	14
30,001–35,000	3	65,001–70,000	10
35,001–40,000	5	70,001–75,000	7
40,001–45,000	8	75,001–80,000	4
45,001–50,000	11	80,001–85,000	2
50,001–55,000	16	85,001–90,000	1
55,001–60,000	18		

4. The number of lives observed in each interval is represented as a bar as shown in Fig. 1.2(a). This diagram is known as a *histogram,* or *bar chart*.
5. If the ordinate in Fig. 1.2(a) is nondimensionalized by dividing the number of lives observed in each interval by the total number of lives (100 in this case), we get Fig. 1.2(b). This diagram is known as the relative frequency diagram and shows the desired distribution of the life of brakes.

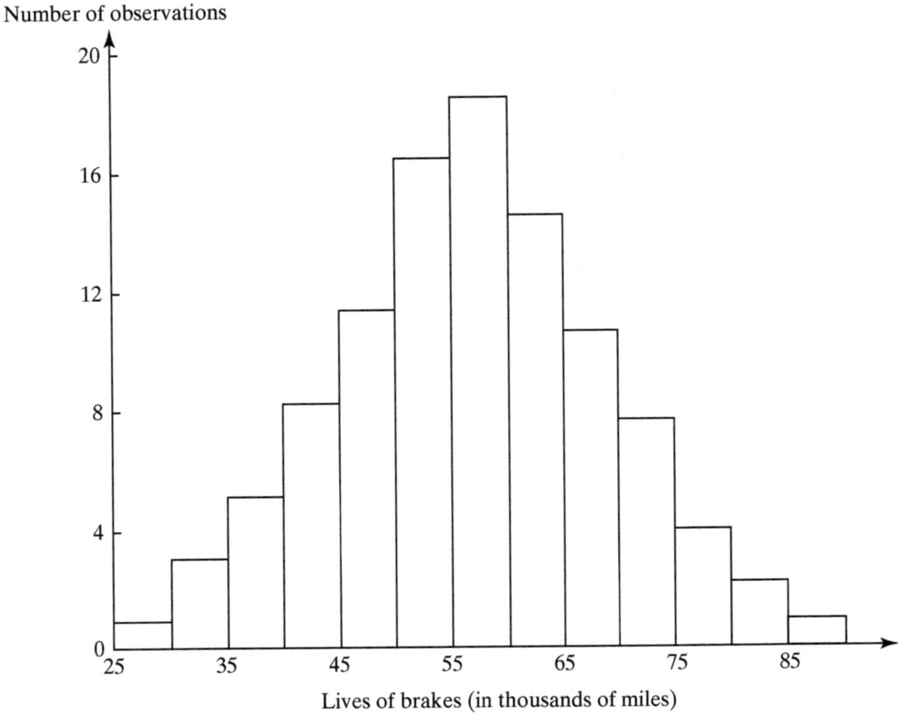

FIGURE 1.2 *(a)* Lives of brakes: Frequency of observations.

6. Often, the relative frequency diagram is approximated by a continuous curve as shown by the dotted line in Fig. 1.2(*c*). This diagram can be used to obtain what is known as the probability density function of the life of brakes by rescaling the ordinate such that the area under the curve is unity.

To determine the mean value of brake life, we compute the arithmetic average of the observed lives, X_i, $i = 1, 2, \ldots, 100$.

$$\begin{aligned}\overline{X} &= \text{mean life of brakes} = \frac{1}{100}\sum_{i=1}^{100} X_i \\ &= (75{,}900 + 70{,}600 + \cdots + 48{,}500)/100 = 56{,}669.5 \text{ miles}\end{aligned}$$

The standard deviation of the life of brakes, σ_X, can be determined as

$$\begin{aligned}\sigma_X &= \left\{\frac{1}{100}\sum_{i=1}^{100}(X_i - \overline{X})^2\right\}^{1/2} \\ &= \left\{\frac{1}{100}\left[(75{,}900 - 56{,}669.5)^2 + (70{,}600 - 56{,}669.5)^2 + \cdots + (48{,}500 - 56{,}669.5)^2\right]\right\}^{1/2} \\ &= 12{,}393.64 \text{ miles}\end{aligned}$$

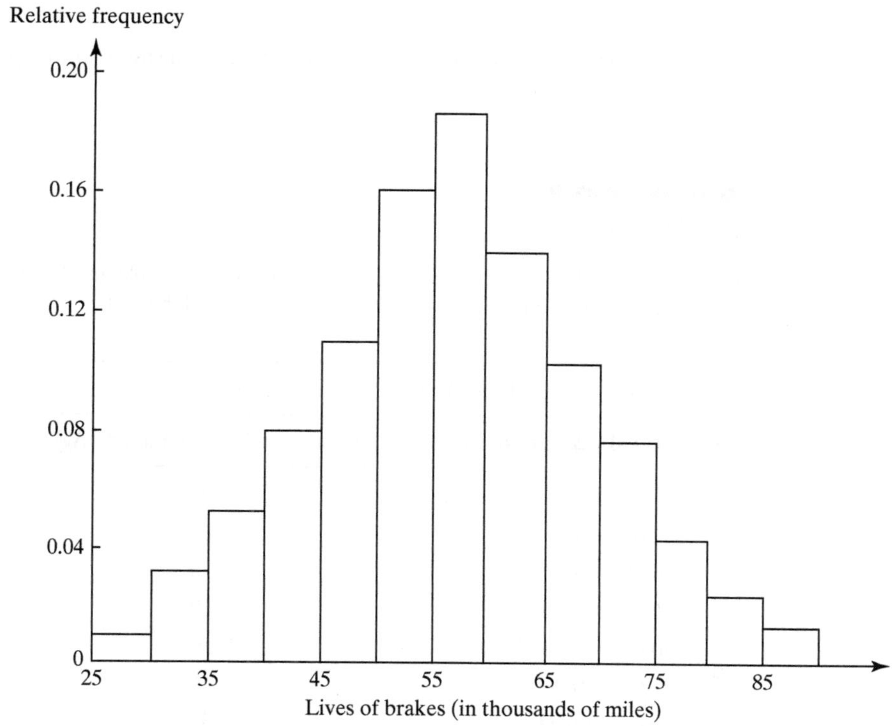

FIGURE 1.2 (*b*) Lives of brakes: Relative frequency diagram.

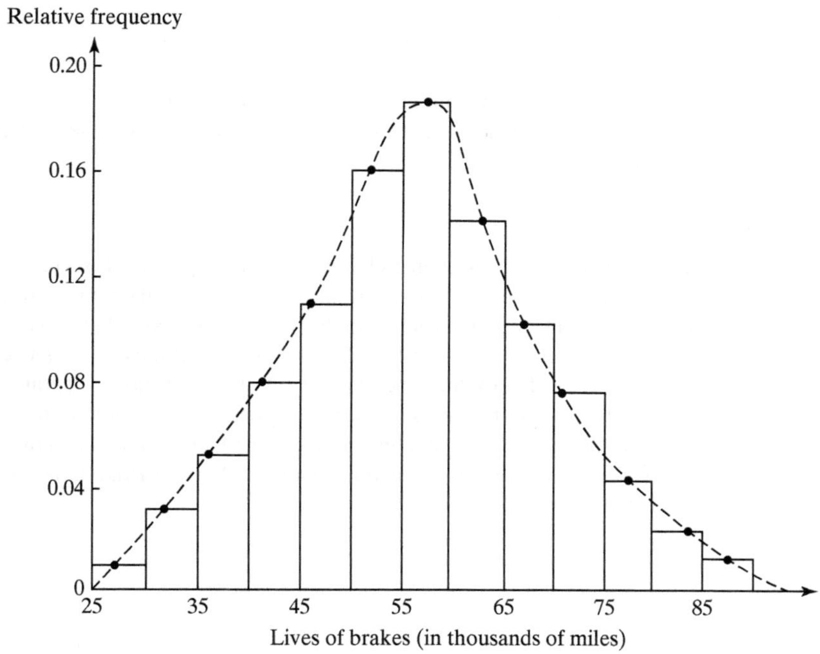

FIGURE 1.2 (*c*) Lives of brakes: Continuous frequency diagram.

As will be shown later, the standard deviation indicates a measure of the *spread out* of the life of brakes.

■

EXAMPLE 1.2

A manufacturer of electronic timers conducted a reliability test placing $N_0 = 500$ timers under identical operating conditions and observed the data shown in Table 1.4.

TABLE 1.4 Failure Time Data for Electronic Timers

Time Interval $(\Delta\tau)$ (Hours)	Number of Timers Failed (N_f)
0–5,000	105
5,001–10,000	85
10,001–15,000	70
15,001–20,000	60
20,001–25,000	50
25,001–30,000	40
30,001–35,000	35
35,001–40,000	30
40,001–45,000	15
45,001–50,000	10

1. Plot the histogram of the failure time of the electronic timers.
2. Plot the relative frequency and the cumulative distributions of the failure time of the electronic timers.

Solution:

1. The number of timers failed in each time interval is represented as a bar as shown in Fig. 1.3(*a*).
2. If the ordinate in Fig. 1.3(*a*) is normalized by dividing the number of failed timers observed in each time interval by the total number of timers tested (500 in this case) we get the relative frequency diagram shown in Fig. 1.3(*b*). When each relative frequency value (in each time interval) is divided by the time interval $(\Delta\tau)$, the resulting diagram, shown in Fig. 1.3(*c*), denotes the probability mass (or density) function of the failure time of the timers. The cumulative distribution of the failure time of the timers can be found by adding all the previous relative frequency values at any time (shown in Fig. 1.3(*d*)). All the calculations involved are indicated in Table 1.5.

■

1.4.2 Mechanical and Structural Failures

All mechanical and structural components have certain useful lives within which they perform satisfactorily, but beyond which they cannot be used. A mechanical or structural component is considered to have failed when it ceases to function properly for its intended use. One reason for the failure of a component is the magnitude and type of load that it is required to withstand. There are three basic types of loads: static, dynamic, and cyclic. Buckling, creep, fracture, crushing, tearing, spalling, abrading, and wear are some types of failures that result from different types of loads.

Another cause of failure is corrosion or chemical attack by the environment. General thinning, pitting, cavitation, hydrogen embrittlement, and intergranular corrosion are some types of damage caused by corrosion. Often, several modes of failure occur simultaneously.

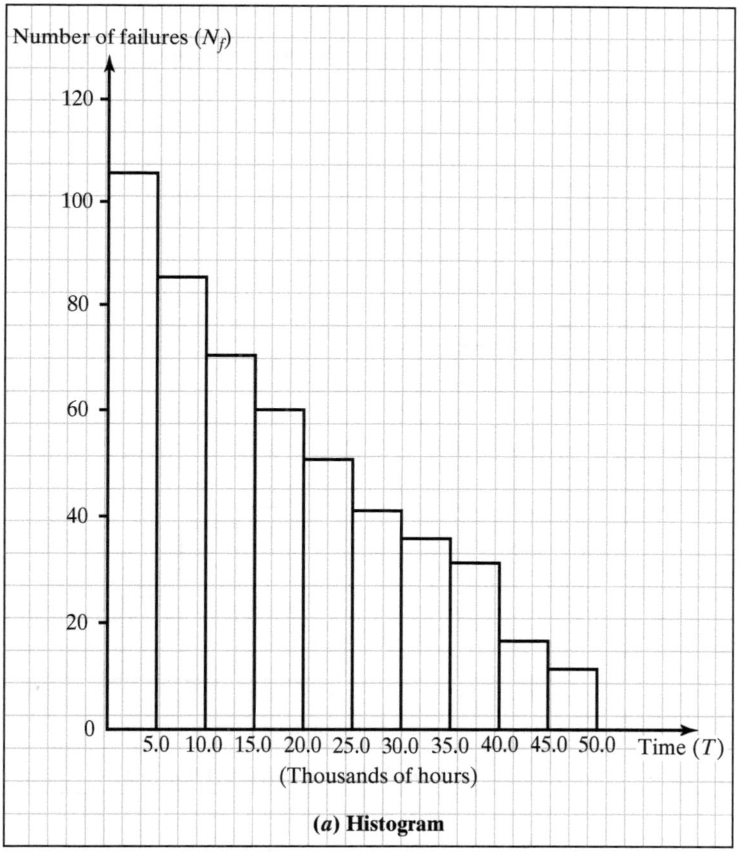

FIGURE 1.3 (*a*) Histogram

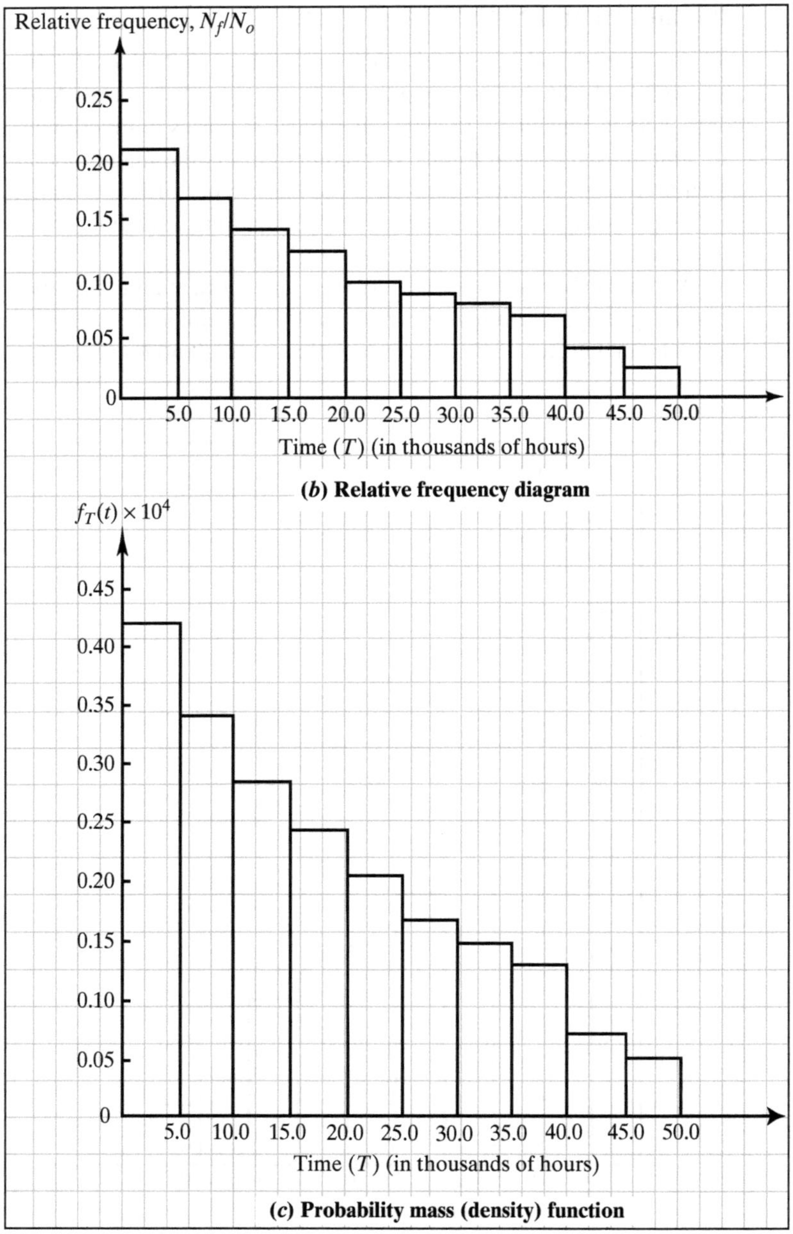

FIGURE 1.3 (*b*) Relative frequency diagram; (*c*) Probability mass (density) function.

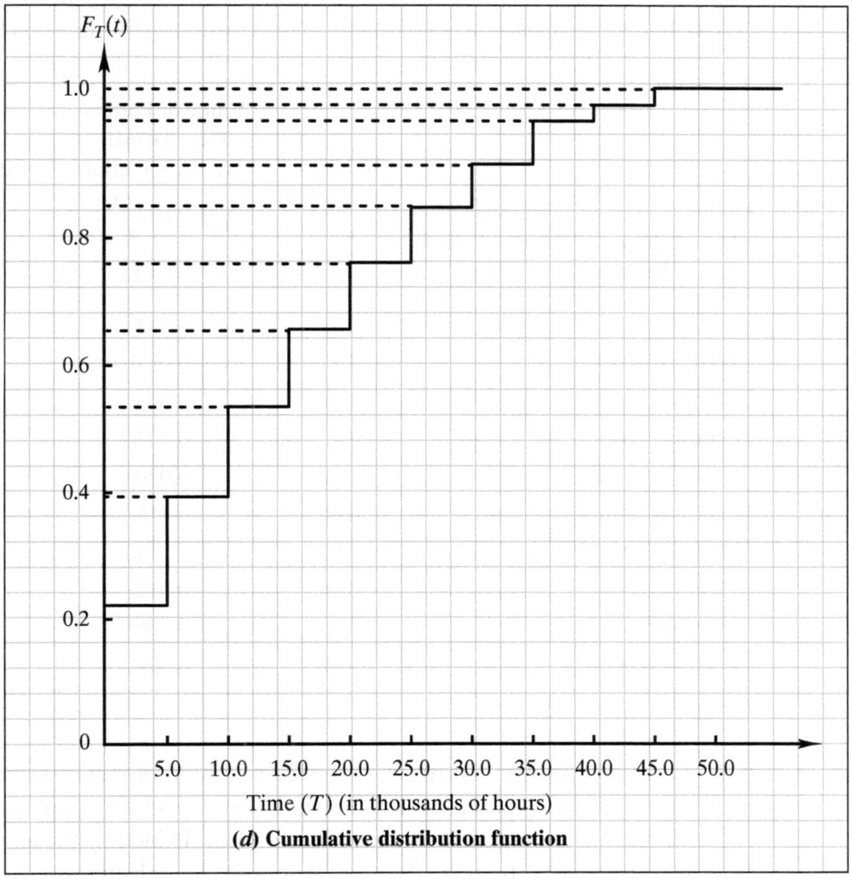

FIGURE 1.3 (*d*) Cumulative distribution function.

Some of the important failure modes and their characteristics can be summarized as follows [1.187].

1. *Static failure.* When excessive static loads beyond the ultimate strength are applied, most materials fail by fracture.
2. *Fatigue failure.* When cyclic loads are applied, materials can fail by fracture even when the magnitude of the maximum cyclic stress is far below the yield strength. Initially, failures usually begin with a crack at a point of stress concentration and end with the propagation of the crack, which causes the final fracture.
3. *Creep failure.* Creep refers to the steady flow of metal under a sustained load. The continuing creep deformation causes failure when either a limiting tolerable distortion is exceeded or rupture occurs.

TABLE 1.5 Computation of Probability Mass (Density) and Distribution Functions of the Failure Time of the Timers

Time Interval ($\Delta\tau$) (Hours)	Number of Timers Failed (N_f)	Relative Frequency $\left(\frac{N_f}{N_0}\right)$	Value of Failure Time Density (Mass) Function $f_T(t) = \frac{N_f}{N_0\,\Delta\tau}$ (Unit: 10^{-3})	Value of Cumulative Distribution Function $F_T(t) = \sum_{\tau=0}^{t} f_T(\tau)\cdot\Delta\tau$
0–5,000	105	0.21	0.042	0.21
5,001–10,000	85	0.17	0.034	0.38
10,001–15,000	70	0.14	0.028	0.52
15,001–20,000	60	0.12	0.024	0.64
20,001–25,000	50	0.10	0.020	0.74
25,001–30,000	40	0.08	0.016	0.82
30,001–35,000	35	0.07	0.014	0.89
35,001–40,000	30	0.06	0.012	0.95
40,001–45,000	15	0.03	0.006	0.98
45,001–50,000	10	0.02	0.004	1.00

4. *Corrosion failure.* Corrosion is the deterioration of metal surfaces under service or storage conditions due to direct chemical or electrochemical reaction with its environment. The corrosion damage is usually accelerated by the application of stress. In hydrogen embrittlement, the ductility of the metal decreases due to the absorption of hydrogen. This results in either brittle failure under static loads at low-strain rates or fracture failure under impact loads at high-strain rates.
5. *Wear failure.* Wear is the gradual mechanical destruction of a metal surface by contact with another metal or nonmetal surface. Wear occurs in all types of contact: sliding, rolling, or impact.
6. *Instability.* When a small change in work done by external forces exceeds the strain energy stored in the member, the system becomes unstable as in the case of buckling of columns and plates.

In practice, machine components experience several other types of failure as well. For example, manufacturing tolerances can produce higher stresses than the ideal structure. When two machine parts rub together fretting fatigue will occur, which can reduce the strength by 50 percent or more. Casting and welding defects, poor surface finishes, and overtightening or faulty bolting or riveting cause larger stresses. Improper grinding may embrittle a surface. Hammering a part to straighten it can introduce residual stresses. Each type of failure requires a different analysis, inspection, and test procedure.

1.5 Factor of Safety and Reliability

The quantity that is traditionally used to maintain a proper degree of safety in structural and mechanical design is the *factor of safety*. Generally the factor of safety is understood to be the ratio of the expected strength to the expected load. The strength of the component and the load acting are assumed to be unique in conventional design. However, in practice both the strength and load are variables, the values of which are scattered about their respective expected (or mean) values. This results in an overlap in the distributed values of strength and load that might lead to the failure of the system. In fact, the interference area between the strength and load distributions can be used to compute the probability of failure of the component. The conventional design methods based on the factor of safety are not rational in the sense that the same factor of safety might imply different values of reliability in different situations.

For illustration, consider a bolted joint in a machine. If experiments are conducted to find the tensile strength S and the load (stress) applied on the bolt L in several seemingly identical machines, the results would be as indicated in Figs. 1.4(*a*) and (*b*). These

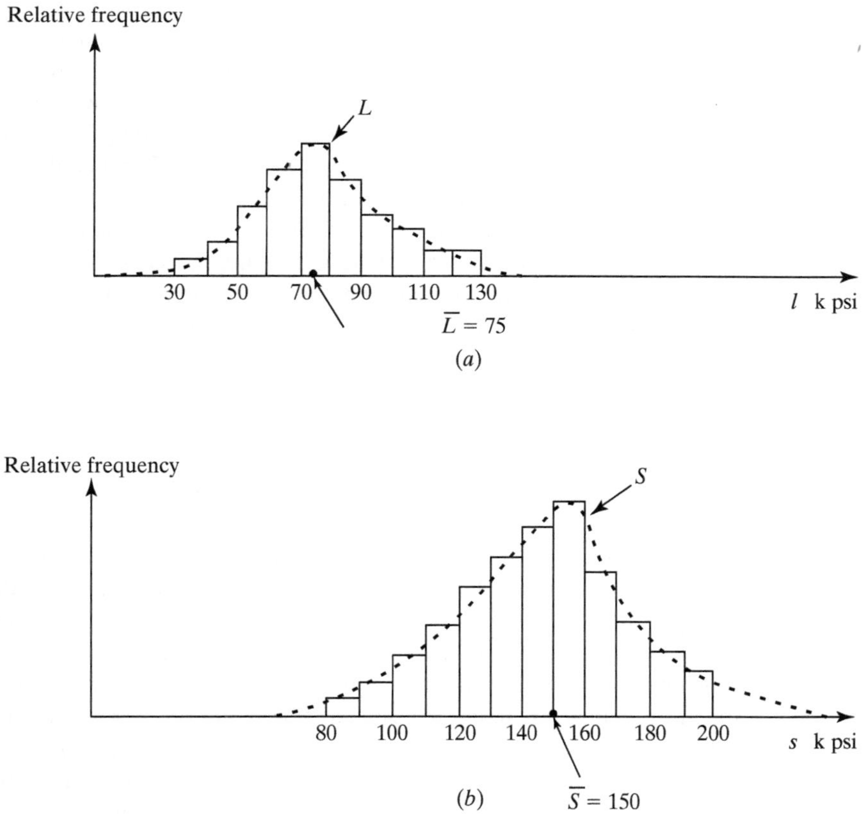

FIGURE 1.4 Distributions of load and strength shown separately.

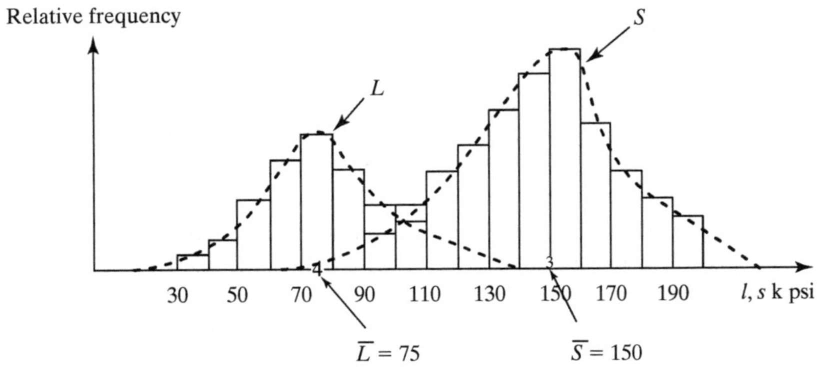

FIGURE 1.5 Distributions of load and strength shown in the same graph.

distributions (histograms) can be approximated by the curves shown by dotted lines. The average or expected values of the strength and load are indicated as $\overline{S} = 150\text{k}(\text{lb/in}^2)$ and $\overline{L} = 75\text{k}(\text{lb/in}^2)$. According to traditional design procedures, the computed factor of safety is greater than one $(n = \overline{S}/\overline{L} = 150/75 = 2)$ and hence the bolt will not fail (in theory). However, when the distributions of bolt strength and load are shown on the same graph, the result will be indicated as in Fig. 1.5. Since there is an overlap of the distributions, there is a finite probability of the load exceeding any specified value of the strength, that is, there is a finite probability of failure of the bolt.

For example, let a different material, having the same distribution but with the mean strength equal to a constant multiplied by that of the previous material $(c\,\overline{S})$, be used for the bolt. If the mean value of the applied load is taken as $c\,\overline{L}$ (changed from the previous value by the factor c) without any change in the distribution, the new factor of safety is given by

$$n = \frac{c\,\overline{S}}{c\,\overline{L}} = \frac{\overline{S}}{\overline{L}}$$

The distributions of strength and load, for the case of $c > 1$, are shown in Fig. 1.6 [1.190]. Since the probability of failure increases with the overlaping (or shaded) area between the strength and load distribution curves, the probability of failure of the component can be seen to be different in the two cases for the same factor of safety. Next, consider the case in which the strength of the bolt and the applied load follow different distributions with the mean values remaining the same as shown in Fig. 1.7 [1.190]. Here also the factor of safety remains the same, but the probability of failure changes because the overlapping area between the strength and load distributions reduces when both the standard deviations are reduced. Finally, it is possible for the mean values of bolt strength and applied load to be changed by the same factor with different values of standard deviations so that the factor of safety may remain unchanged but the probability of failure will be different [1.13, 1.190].

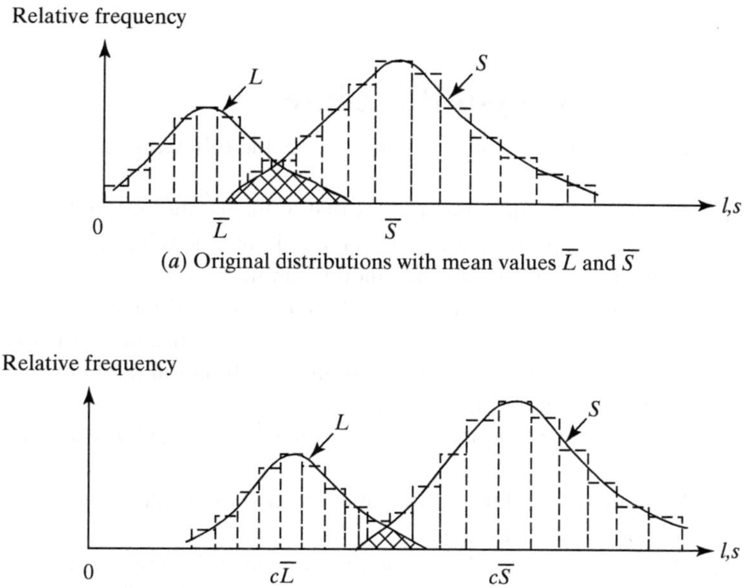

(a) Original distributions with mean values $\overline{L}$ and $\overline{S}$

(b) New distributions with increased mean values of $c\bar{L}$ and $c\bar{S}$ ($c > 1$)

FIGURE 1.6 Change in probability of failure due to changes in the mean values of load (L) and strength (S).

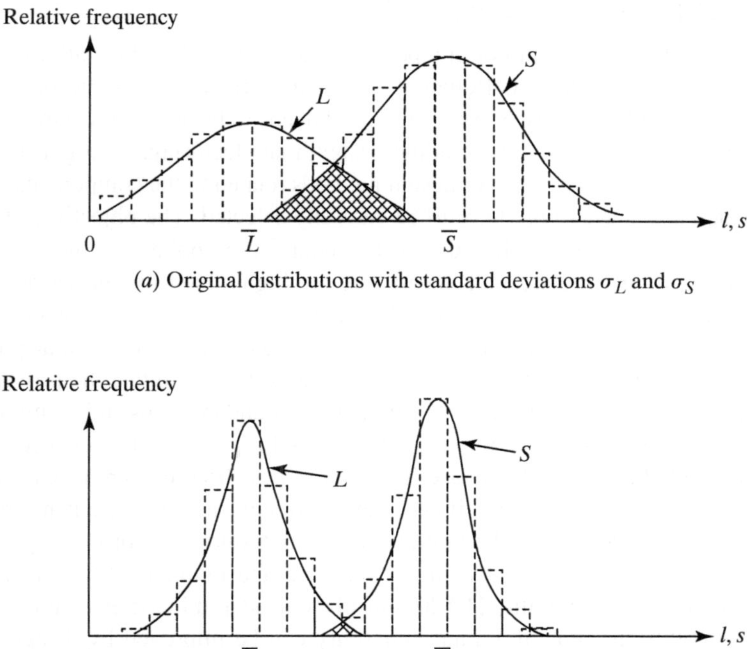

(a) Original distributions with standard deviations σ_L and σ_S

(b) New distributions with reduced standard deviations $k\,\sigma_L$ and $k\,\sigma_S$ ($k<1$)

FIGURE 1.7 Change in probability of failure due to changes in the standard deviations of load and strength.

1.6 Reliability Analysis Procedure

The reliability analysis of a system involves the following steps:

1. Identify the random parameters that influence the system behavior.
2. Identify the performance criteria that define the reliability of the system.
3. Collect data on each of the random parameters.
4. Construct a histogram of each of the random parameters.
5. Determine a suitable probability density or distribution function to describe each of the random parameters. Use parameter estimation methods to identify the values of the parameters of the distributions.
6. Using a suitable model for the probabilistic behavior, estimate the reliability of the system.
7. Find the risk and the consequence or cost associated with the estimated value of the reliability of the system.
8. If the predicted reliability and the associated risk are not acceptable, improve the reliability and risk by changing the design variables of the system, and repeat Steps 1 through 8.

1.7 Reliability Management

Reliability of a system or component is not achieved accidentally; it is to be built into the system or component. Reliability is an inherent characteristic of the system, similar to the system's capacity or power rating. It needs to be addressed at every stage of the product or system development including design, manufacturing, testing, and maintenance phases. In the design phase, proper design methods related to the components, materials, processes, tolerances, and so on, have to be carefully selected. The objectives at this stage are to ensure that well-established design procedures are applied, known materials and processes are used, and the areas of uncertainty are highlighted for further action. When the initial hardware manufacturing is completed, tests are to be carefully planned, executed, and data collected to generate confidence in the design. During the regular production stage, suitable quality control techniques are to be used to ensure that the proven design is repeated. Further testing may be done to eliminate substandard items and to improve confidence. Data collection, analysis, and both preventive and corrective maintenance actions are continued through the product-life phase. The reliabilities observed in practical use are compared with the initial predictions and, if necessary, corrective actions are taken. The data is also used to provide guidelines for the improvement and development of future products.

In many cases, a system is controlled and maintained by human operators. In such cases, human reliability [1.120] must be considered in the evaluation of the overall system reliability. In general, human reliability can be improved by designing such that the assembly, operation, and maintenance of the system are as straightforward as possible. Other factors that contribute to the improvement of human reliability include proper training of operators, low stress levels for operators, and proper positioning of gauge displays and controls. The development, design, and manufacture of a product should also take into account the product liability laws that hold the manufacturer responsible for the injuries and damages caused by a product that is judged to be defective and dangerous.[1]

Reliability can be estimated only by testing the product until it fails. Obviously, the more tests are conducted, the more confidence one can have in the estimated reliability value. Since testing involves money and time, one must consider a trade-off between the confidence level in the estimated reliability and the cost of additional testing. Thus the implementation and management of a total reliability program for the development of a new system requires a reliability group that possesses not only an understanding of the basic mathematics of reliability, but also a knowledge of design methods, economics, system interface problems, quality control and testing techniques, and human factors.

1.8 History of Reliability Engineering

Reliability engineering emerged as a separate discipline in the United States during the early 1950s. The complexity of problems associated with the electronic systems designed during the early 1940s for use in the war effort led to the growth of the theory of reliability. It was reported that during World War II nearly 60 percent of the airborne equipment shipped to the Far East was damaged on arrival, and 50 percent of the spare parts and equipment in storage became unserviceable before they were ever used. In 1949, about 70 percent of the electronic equipment possessed by the Navy was not operating properly.

In view of these difficulties with electronic components, the Air Force formed the ad hoc Group on Reliability of Electronic Equipment in December 1950 to study the situation and recommend measures that would increase the reliability of equipment and reduce maintenance. The Navy began a lengthy study on radio tubes in April 1951. The Army started a similar investigation in June 1951. In late 1952, the Department of Defense (DOD), with a view of coordinating the efforts of the Air Force, Navy, and Army, established the Advisory Group on Reliability of Electronic Equipment (AGREE). In June 1957, AGREE published its first report on reliability [1.189]. The report concluded that reliability testing must be made an integral part in the development of new systems. The new equipment was required to be tested for several hours in a high-stress environment that involves on and off switching, low and high temperatures, and steady and vibratory conditions. The idea was to detect most of the design defects at a sufficiently early stage so as to be able to correct them before the start of production. The report also recommended that the procuring agencies should accept the equipment only after getting the reliability demonstrated by the manufacturer. The reliability demonstration was to prove that a desired statistical confidence level has been achieved for the specified MTBF (mean time before failure) of the equipment. The report also provided detailed test plans for demonstrating various levels of statistical confidence. The AGREE report was accepted by the DOD and soon the AGREE recommendations became an accepted procedure [1.111]. Military equipment contractors and the related manufacturing companies were forced to invest in expensive environmental test equipment for

[1]An introduction to product liability, including the methods of preventing product liability, is given in Appendix D.

demonstrating the reliability of their products. They very soon realized that levels of reliability could be achieved that far exceed the levels attained earlier by traditional methods. The DOD reissued the AGREE report on testing as the military standard, MIL-STD-781: Reliability Demonstration—Exponential Distribution, which was revised as MIL-STD-781B in 1967 [1.169].

The AGREE report and its recommendations on reliability were later adopted by NASA and many other major contractors, suppliers, and purchasers of high technology equipment [1.169–1.175]. In 1965, the DOD issued MIL-STD-785B: Reliability Programs for Systems and Equipment, which was revised in 2008 [1.174]. This standard made compulsory the integration of reliability activity with the traditional engineering activities of design, development, and production. This was intended to detect and eliminate potential reliability problems at an early stage in the development of a new system.

Meanwhile, mechanical structures were becoming much more complex and started posing more difficult problems, especially in aerospace and military applications. Unlike electronic components and equipment for which large amount of failure data was available, reliability data was relatively scarce on mechanical structures. The safety aspects of structures were studied as early as 1929. Similarly, the study of fatigue life in materials and the related subject of extreme value theory applicable to material strengths and loads was started in the mid-1930s. Contributions to the reliability approach to static structural design were made by Freudenthal and Pugsley. Largely motivated by vibration problems encountered in the commercial jet aircraft, Birnbaum and Saunders presented an ingenious statistical model for life lengths of structures under dynamic loading [1.55].

While dealing with reliability-based design of machines and structures, we can study the relative importance of mechanical and structural failures from the point of view of loss of human lives (see Table 1.6). It can be seen that the risk to life from mechanical and structural failures is not substantial. In spite of this good record, designers endeavor to achieve even higher reliabilities for machines and structures.

TABLE 1.6 Representative Risks of Death [1.186]

Activity/Cause	Number of Deaths Per Year Per 10^6 Persons
Air travel	9
Rail travel	4
Water transport	9
Motor vehicles	300
Poison	20
Lightning	0.5
Fires	40
Machinery	10
Structural failures	0.2

1.9 Some Examples of System Failures

Engineering activity involves improvement/development of new products and systems. Although most designs are evolutionary in nature with continuous improvement of existing products/systems, some designs are truly revolutionary in nature. Often revolutionary designs involve higher risks. For example, the development of the space shuttle involved estimating, among other things, the temperature to which the shuttle will be exposed during its re-entry (with no real experimental data available). Some prominent historical failures of engineering systems are described in this section to indicate the following aspects related to reliability:

- A system can fail in a failure mode that was not considered in its design.
- Because most systems are comprised of thousands and, often, millions of components, the failure of even a simple component can lead to a catastrophic failure of the system when the operating conditions are inappropriate for that specific component.
- A system can fail due to the damage caused by extraneous conditions (or loads) that were not considered in standard design procedures.
- The use of materials, whose behavior was not understood completely, can lead to a system failure.
- Sometimes a system can fail because of human error (or poor management decision made with inadequate information).
- A system can fail due to poor or inadequate design of even simple components.
- A system can fail due to poor maintenance.
- A system can fail due to unexpected natural load or disaster conditions.

1.9.1 Collapse of Tacoma Narrows Bridge in 1940

The Tacoma Narrows Bridge constructed to connect Seattle and Tacoma at a cost of $6.4 million was opened on July 1, 1940, and collapsed on November 7, 1940. It was the third largest suspension bridge in the world with a 2800 ft long center span, two 1100 ft long side spans, one on each side of the center span, and two suspension cables from the towers anchored 1000 ft back towards the river banks. The width of the bridge was only 39 ft with two lanes. Thus its width-to-span ratio was nearly $1/350$ with flexibility nearly three times that of the Golden Gate Bridge. The bridge was nicknamed Galloping Gertie even before it was opened after observing its violent vibrations. After its opening, it became a challenge for motorists to cross the vibrating bridge many of whom experienced seasickness. Several attempts were made to dampen or stabilize the vibrations—including tying girders to 50-ton concrete blocks on riverbanks by heavy cables, installing dynamic vibration absorbers in the form of a piston-and-cylinder type damper, and connecting the main suspension cables to the midpoint of the deck through stay cables. All these methods proved futile. Several measurements and movie camera films of the vibrating bridge were made to study and understand its behavior. Even wind tunnel tests of the scale model of the bridge were conducted to find ways of controlling the vibrations of the bridge. Unfortunately, before any successful method of controlling vibrations was implemented, the bridge collapsed [1.242]. At 10:00 am on November 7, 1940, the bridge deck was observed to first rise and fall three feet (distance between crest and valley or amplitude) 38 times a minute and then started to twist in a wind with velocity of about 40 mph. By 10:30 am, the vibration amplitude increased to 25 ft (distance between crest and valley) and the suspension cables started to break. By 11:00 am, several sections of the deck and stiffening girder fell through 190 ft into the Puget Sound river. Subsequent study of the failure of the bridge established the cause as the aerodynamically induced (or self-excited) vibrations. A new bridge was constructed, with a 60 ft wide deck and four lanes and stiffening trusses in the form of a box to resist torsional forces, at a cost of $18 million and opened in 1950.

1.9.2 Crash of El Al *Boeing 747-200* in 1992

The freighter plane 747-200 operated by El Al Israel Airlines, bound for Tel Aviv from Amsterdam, crashed on October 4, 1992, within thirteen minutes after takeoff. The cause was first the inboard engine of the right wing became loose. Then the engine veered forcefully to the side and knocked off the outboard engine of the right wing. The plane lost altitude, made two wide clockwise circles over the city of Amsterdam, and crashed on a crowded 10-story apartment building killing all the 6 crew members and 50 people on the ground [1.243]. It has been found that the bolts used to connect the engine to the wing, also known as fuse pins, failed during the flight due to corrosion and cracking. Each of the four fuse pins that connected the engine strut to the wing was made of cadmium-plated steel and was in the form of a hollow cylinder of diameter 2.25 inches and length 4 inches. The design of the fuse pins was a delicate process. They were supposed to be sufficiently strong to endure the stresses induced during the various flight conditions, such as takeoff, climb, steady flight, flight under gusts and landing. At the same time, they were expected to break or snap easily upon striking the runway and force the engine to go away from the wings if the airplane was forced to belly land. This would prevent the hot engines coming into contact with any spilled fuel that might trigger a fire. Although *Boeing* improved the design of fuse pins later, the fuse pins used in their 747s produced during the period 1969 through 1980 were susceptible to corrosion and cracks due to the tiny nicks caused by the machining process used to produce them. The fuse pins were required to be periodically inspected for corrosion and cracks. They were supposed to be replaced by new, improved, fuse pins if there was any evidence of corrosion and cracking. The EI Al Airlines maintenance record indicated that the fuse pins of this particular airplane were inspected four months before the crash as per FAA regulations and revealed no problems. Subsequent to this accident, FAA implemented a more aggressive inspection process for the entire fleet of more than 900 Boeing 747 jets.

1.9.3 Disaster of Space Shuttle *Challenger* in 1986

The launch of *Challenger* on January 28, 1986, from Cape Canaveral was NASA's twenty fifth mission of the space shuttle program. It received special publicity because it carried, among its crew of seven persons, 37-year-old Christa McAuliffe from Concord High School as the first teacher-in-space. Within 73 seconds after its launch, *Challenger* exploded. The module containing the crew members got detached during explosion and freely fell from a height of nine miles into the Atlantic ocean. Apparently, although the crew survived the explosion, they died at the impact of the module with water. The launch was originally scheduled for January 22, 1986, postponed to January 24, then to January 25, and again to January 27. A problem with a hatch bolt and subsequent development of dangerous crosswinds at 30 knots on January 27 postponed the launch to January 28. When the temperature at the launch station (Cape Canaveral) dropped below freezing during the night of January 27, NASA managers were concerned as no previous shuttle was launched at temperatures less than 53° F. They conducted a late-night review of the prospects of the launch on January 28 with their contractors. At that time, some engineers from Morton Thiokol, a NASA contractor, expressed concerns about the ability of the O-rings in the shuttle's solid rocket boosters to act as seals. Since no experiments were conducted on O-rings below freezing temperature previously, and with NASA managers pushing for a launch without further delays, Thiokol managers overruled their own engineers and recommended the launch on January 28 by signing a statement assuring the safety of the O-rings at the colder temperatures. Due to the potential damage to the thermal tiles of the shuttle during reentry because of the ice formed on the shuttle, and the low temperatures that ranged from 19° F to 25° F, the launch was delayed from 9:38 am to 10:38 am and again to 11:38 am. After finding no abnormalities due to the ice buildup by the inspection team, the launch took place at 10:38 am (Fig. 1.8). After the explosion of *Challenger* within 73 seconds after the launch, NASA began its rescue operations. The crew module was found after six weeks on the ocean floor of the Atlantic Ocean with the bodies of all the crew members. President Ronald Reagon set up a Presidential Commission to investigate the disaster with the former secretary of state, William Rogers, as the chair of the commission. The commission released its 256-page report on June 6, 1986, pointing the reason for the accident as the failure of the joint between the two lower segments of the right solid rocket motor due to the failure of O-rings.

FIGURE 1.8 Space shuttle *Challenger* at lift-off (exploded 73 seconds after the launch).

1.9.4 Chernobyl Nuclear Power Plant Accident in 1986

Chernobyl nuclear power plant was located in Pripyat, Ukraine, part of the former USSR. The power plant, consisting of four reactors, was constructed during the period 1977–1983 with generating capacity of 1 GW (gigawatts) of electric power per reactor. The reactors No. 1, 2, 3, and 4 were commissioned in 1977, 1978, 1981, and 1983, respectively. Two more reactors (Nos. 5 and 6) were under construction at the time of the accident. All the reactors were graphite-moderated, water-cooled boiling water reactors known as RBMKs (acronym for the Russian *reaktor bolshoy moshchnosti kanalniy*). The No. 4 reactor suffered a catastrophic steam explosion that resulted in a fire and series of additional explosions and a final nuclear meltdown at 1:23:58 am on April 26, 1986 [1.244].

The accident is considered as the worst case in the history of nuclear power. Since there was no containment structure, the plume of radioactive fallout spread over parts of western Soviet Union, Eastern Europe, Scandinavia, and UK. Although nearly 200,000 people were evacuated due to the accident, the number of deaths caused by the accident could not be estimated accurately as most of the deaths are expected to occur over a long period of time. A 2005 report by the International Atomic Energy Agency attributed the deaths of 47 workers and 9 children (with thyroid cancer) and as many as 4,000 deaths in future to the accident.

Two conflicting official theories were given for the accident. The first one, published in 1986, blamed the negligence of the power plant operators. The second one, published in 1991, blamed the flaws in the RBMK reactor design. It is to be noted that the operator faults include switching off many of the reactor's safety systems and operation of the reactor with fewer than the minimum of 15 control rods inside the core zone, both in violation of the published guidelines. At the same time, the negligence and violation of the guidelines by the operators was attributed to the fact that the operators were mostly non-RBMK-qualified personnel who were not familiar with the special problems of the RBMK reactor, including its inherent design flaws. After the accident, the damaged reactor No. 4 was sealed off using a 200 meter-thick concrete. Because of the energy shortage, the Ukrainian government continued the operation of the other three reactors while halting the construction of reactor No. 5 and 6. Due to damage caused, beyond repair, to reactor No. 2 due to a fire, it was closed in 1991. Reactor No. 1 was decommissioned in 1996 because of a deal between the Ukrainian government and international organizations. Finally, reactor No. 3 was shut off in 2000 effectively closing the entire Chernobyl plant.

1.9.5 Mississippi River Bridge 9340 Collapse in 2007

Mississippi river bridge 9340 (also called I-35W highway) in Minneapolis, Minnesota, was an eight-lane steel truss arch bridge, which was opened in November 1967. It had fourteen spans extending over 1907 feet, carried 140,000 vehicles daily, and failed during the evening rush hour on August 1, 2007; it collapsed into the river and onto the riverbanks beneath. It involved nearly 100 vehicles with 13 people killed and 145 injured (Fig. 1.9) [1.245]. Although the bridge was inspected annually by the Minnesota Department of Transportation, the inspection was not completed in 2007 due to construction work. In 1990, the federal government gave the bridge a rating of "structurally deficient" citing significant corrosion in its bearings. It is to be noted that approximately 75,000 other bridges in United States were also given the same rating in 1990. The National Transportation

FIGURE 1.9 Mississippi bridge I-35W failure (with permission from the Minnesota Department of Transportation).

Safety Board, in its investigation report released on November 13, 2008, determined the undersized gusset plates (at a thickness of 0.5 inch) as the cause of the collapse of the bridge. The additional 2-inch concrete layer added to the road surface over the years (which increased the dead load by 20 percent) and the extraordinary weight of the construction equipment and material placed on the bridge (about 578,000 lb) just above its weakest point contributed to the failure. It was determined that the corrosion present in the bearings was not a significant factor in the failure of the bridge. A new bridge constructed to replace the collapsed one was opened on September 18, 2008.

1.9.6 Fukushima Nuclear Accident in 2011

The Fukushima nuclear reactor accident occurred on March 11, 2011, and it involved equipment failures and release of radioactive materials due to an earthquake of magnitude 9.0 on Richter scale followed by a tsunami with waves of height 14 meters. This accident is considered to be the second-largest nuclear accident after the Chernobyl accident but more complex due to the involvement of multiple reactors. The Fukushima nuclear plant was comprised of six separate boiling water reactors [1.246]. Reactors 5 and 6 were in a state of cold shutdown for planned maintenance while reactor 4 was defueled at the time of the earthquake. The remaining reactors, reactors 1, 2, and 3, were shut down automatically after the earthquake. Under normal operation, the uranium fuel rods undergo nuclear fission that causes a rise in temperature of the rods and the hot rods, in turn, convert water into steam to generate electricity. Water is used to cool the hot rods to 270° C. If cooling fails, the temperature of the rods could increase up to 1200° C. At such a temperature, the rods will meltdown. When the earthquake hit, the first safety system to prevent a meltdown got activated. A set of control rods were made to rise to the level of fuel rods to stop nuclear fission of the fuel rods. Because the fuel rods are still hot, water should be circulating around the fuel rods to cool them. However, water circulation did not take place because the power outage caused by the earthquake resulted in a shutdown of the generator that supplies power to the water pumps. Then a second safety system got activated which started an emergency diesel generator. This helped in spraying water on the hot fuel rods. But an hour later, without any warning, the emergency diesel generator stopped functioning. At about the same time, a tsunami with waves of 10–14 m height hit. It is to be noted that the nuclear power plant was protected by a seawall designed to withstand tsunamis with a wave height of 5.7 m only. Thus the entire plant, including the low-lying generators and electrical switchgear in the reactor basement as well as external pumps that supply cooling seawater, was flooded. It is believed that the emergency diesel generator stopped working due to the tsunami. Then a third safety system started operating. This system converts steam into water which, in turn, is used to cool the fuel rods. However, because the water level was too low, the temperature of the upper parts of the fuel rods continued to increase that resulted in a partial meltdown and a subsequent explosion that blew the top part of the containment building. Thus all the three safety systems of the reactors failed.

It was found that reactors 1, 2, and 3 suffered from partial meltdown; the subsequent hydrogen explosions destroyed the upper cladding of the buildings in which reactors 1, 3 and 4 were housed; and the explosion inside reactor 2 damaged its containment. In addition, multiple fires broke out at reactor 4. Although reactors 5 and 6 were shut down by submerging their fuel rods in pools of seawater, the fuel rods began to overheat as water levels in the pools dropped. Workers in the plant suffered radiation exposure. The generator at reactor 6 was restarted on March 17, 2011, which provided some cooling to reactors 5 and 6 which suffered the least damage. On March 20, 2011, grid power was restored to parts of reactors 1 to 4; but the machinery damaged by floods, fires, and explosions remained inoperable. The basements of reactors 1 to 4, flooded by radioactive water, continued to prevent access to carry out repairs.

Due to the fear of radiation leaks, people within a radius of 20 km around the plant were evacuated. Food grown in the area was banned for consumption. It was decided, in Tokyo, that tap water should not be used to prepare food for infants. It was found that the amounts of iodine-131 and caesium-137 isotopes released from the Fukushima plant were of the same orders of magnitude as those released from the Chernobyl disaster in 1986. On March 20, 2011, the Japanese government decided to decommission the power plant once the crisis was over.

1.9.7 Explosion of the First Jet Airplane *Comet*

The Comet, the first jet airplane, revolutionized air tavel around the world. The 36-seat *Comet 1*, built by De Havilland, was placed in service by the British Overseas Aircraft Corporation (BOAC), the British national carrier. The first flight took place on May 2, 1952, from London to Johannesburg (South Africa). At the time, the top cruising speed of a piston engine aircraft, such as DC-3, was about 180 mph. With the *Comet*, air travel made a revolutionary leap at 480 mph. In addition, the *Comet* airplane, equipped with jet engines, could climb faster and fly higher more quietly with less vibration. On May 2, 1953, the first anniversary of the jet aircraft service, a *Comet* exploded in midair while taking off from Calcutta, India. Another *Comet* exploded at an altitude of 27,000 ft while flying over the Mediterranean Sea on January 10, 1954. Another *Comet* exploded in midair on April 8, 1954, while taking off from Rome. In all these accidents, it was determined that no explosives were involved. Because of these tragic accidents, BOAC suspended all flights within two years. The subsequent thorough investigation revealed the primary cause of the accidents. During simulated flight conditions, the corner of a cabin window initially developed a crack, grew rapidly during continued flight simulation that ultimately caused the crack to propagate through the plane's metal skin. This resulted in the explosion of the pressurized airplane. All these failures were attributed to the use of square windows whose corners caused stress concentration and subsequent fatigue failure of the metal. This design fault was rectified by using round windows and reinforced window panels. *Comet 4*, improved from the previous failures, was introduced for trans-Atlantic jet passenger service in 1958. Although it experienced a modest success, the *Boeing 707*, which was introduced at about the same time (on October 26, 1958) enjoyed a greater success and ultimately dominated the market.

1.9.8 Breaking of the Tanker S. S. *Schenectady*

Fracture mechanics and brittle fracture played an important role in developing successful engineering systems. Its is well known that materials such as steel, which are considered to be ductile, can fracture in a brittle manner at sufficiently low temperatures. The brittle fracture is caused by stress concentrations around sharp corners and imperfections such as microscopic cracks and surface roughness due to machining. A spectacular example of failure due to brittle fracture is the breaking of S. S. *Schenectady* T-2 tanker (built by Kaiser Ship Building Co.) in two in the fitting-out dock on January 16, 1943 (Fig. 1.10). The Kaiser Shipbuilding Company built many ships, all known as Liberty ships built in the United States based on British design concept, during World War II. These ships were cheap and quick to build and 2751 were produced based on a single design during 1941 to 1945 at eighteen American shipyards. As many as 1500 of these ships suffered hull and deck cracks, including three that broke in half without warning. Subsequent investigations revealed the causes of failure of the ships: (1) most of the cracks started at square corners with substantial stress concentrations, (2) most of the ships were overloaded, (3) welds were produced by semiskilled workers, and (4) the steel used in the construction of the ships had low toughness. Particularly, when ships in the North Atlantic were exposed to temperatures that could fall below a critical point, the mechanism of failure changed from ductile to brittle resulting in the fracture of the hull. The building of Liberty ships was stopped after the delivery of the last one, S.S. *Albert M. Boe*, on October 30, 1945.

1.9.9 Crash of the Supersonic Aircraft *Concorde*

Concorde was the first supersonic transport (SST) that was produced through a collaboration between the United Kingdom and France. It was flown for the first time in 1969 and brought into commercial service in 1976. It was used for flights between London Heathrow and New York JFK and Washington Dulles by British Airways and for flights between Paris Charles de Gaulle and New York JFK and Washington Dulles by Air France. In terms of fuel efficiency, *Concorde* had 15.8 passenger miles per gallon of fuel while

FIGURE 1.10 S.S. *Schenectady* T-2 tanker broken into two (Used with permission from the U.S. Coast Guard.)

Boeing 747 had a value of 46.4 and DC-10 had a value of 53.6. Thus, *Concorde* was very expensive to fly. *Concorde* was made of mainly aluminum, had double-delta shaped wings with four Olympus engines, droopnose section for better visibility, a cruising speed of Mach 2.04 (1350 mph) for optimum fuel consumption, and fuel-regime autopilot and autothrottle that permitted hands-off control from climbout to landing. It took just under 3.5 hours to fly from New York to Paris whereas conventional (subsonic) aircrafts take eight hours. Its maximum cruise altitude was 18,300 m, or 60,039 ft.

The only fatal accident involving *Concorde* occurred to the Air France flight 4590, which crashed on July 25, 2000, in Gonesse (France) in which 100 passengers and 9 crew members on board and 4 people on the ground were killed. As per the official investigation of the accident, the crash was caused by a titanium strip that fell from a *DC-10* aircraft (Continental flight) during its takeoff a few minutes earlier. The metal strip caused a tire of the left main wheel bogie of *Concorde* to puncture during its takeoff. As a result, the tire exploded with a piece of rubber hitting the fuel tank. This impact resulted in a shock wave that made one of the fuel valves in the wing burst open. This caused a major fuel leak which triggered a fire due to the sparking electrical landing gear wiring. Although the crew shut down the affected engine, the other engine could not produce enough power to gain height or speed. As a result, the aircraft experienced violent descent and crashed into a hotel in Gonesse [1.247]. The Continental Airlines and one of its mechanics, who serviced the DC-10 before its flight, were found guilty of involuntary manslaughter by a French court on December 6, 2010. *Concorde* was made to retire at the end of its last flight on November 26, 2003, due to rising maintenance costs and a slump in the air travel caused by the earlier *Concorde* accident, as well as the 9-11 attacks on the World Trade Center, Pentagon, and Shanksville, Pennsylvania in 2001.

1.10 Numerical Solutions Using MATLAB and Excel

The solution of statistical, probability, and reliability problems can be found using the statistical toolbox of MATLAB. The statistical toolbox contains a library of programs or m-files which can be used for plotting histograms, computation of statistical parameters such as mean, mode, median, standard deviation, and skewness coefficient, plotting of probability mass (or density) functions, cumulative distribution functions of discrete and continuous distributions, computation of probabilities for any specific distribution or inverse distribution, extremal probabilities, maximum likelihood estimation, parameter estimation, and plotting of quality control charts.

Microsoft Excel has several functions that can be used for making statistical, probability, and reliability calculations, as well as for constructing a variety of charts and graphs, including histograms, to display data. The computation of probabilities and inverse probabilities of distributions such as normal, lognormal, t-distribution, F-distribution, chi-square, binomial, hypergeometric, Poisson, exponential, beta, and gamma can be performed using Excel.

The following examples are presented to illustrate the plotting of the histogram and the computation of the basic characteristics of any specified data.

EXAMPLE 1.3

Find the mean, median, mode, and standard deviation and plot the histogram of the data shown in Table 1.2 (Example 1.1) using MATLAB.

Solution: The MATLAB commands used and the resulting outputs are shown below.

```
>> x = [75900, 55100, 64900, 60100, 51350, 74050, 54700, 70600, 50200,34900,58150, 45900,↙
67200, 83250, 68300, 42200, 41300, 51000, 53800, 57950, 46250, 56500, 61000, 84750, 79250,↙
37200, 59300, 48500, 50050, 50350, 50500, 66950, 51450, 45750, 26100, 33000, 56850, 47400,↙
47650, 59950, 61200, 86300, 66450, 35100, 57550, 64050, 55450, 71450,  77450, 55850,↙
67850, 53350, 39950, 74850, 55700, 50850, 80000, 57850, 64550, 66000, 45450, 62250, 60750,↙
67500, 40050, 72500, 46950, 57150, 49600, 60950, 52800, 63800, 69600, 73300, 62800, 59600,↙
69100, 45250, 35050, 74200, 44700, 55900, 31050, 47150, 56050, 69950, 53050, 38800, 56350,↙
61650, 52200, 57200, 46700, 45900, 52650, 43500, 64750, 44200, 51700, 63450]

>> xmean = mean (x)

xmean =

  5.6670e+004
```

```
>> xmedian = median (x)

xmedian =

        56200

>> xmode = mode (x)

xmode =

        45900

>> xsd = std (x)

xsd =

   1.2456e+004

>> hist (x)
>> title ('Histogram of x')
```

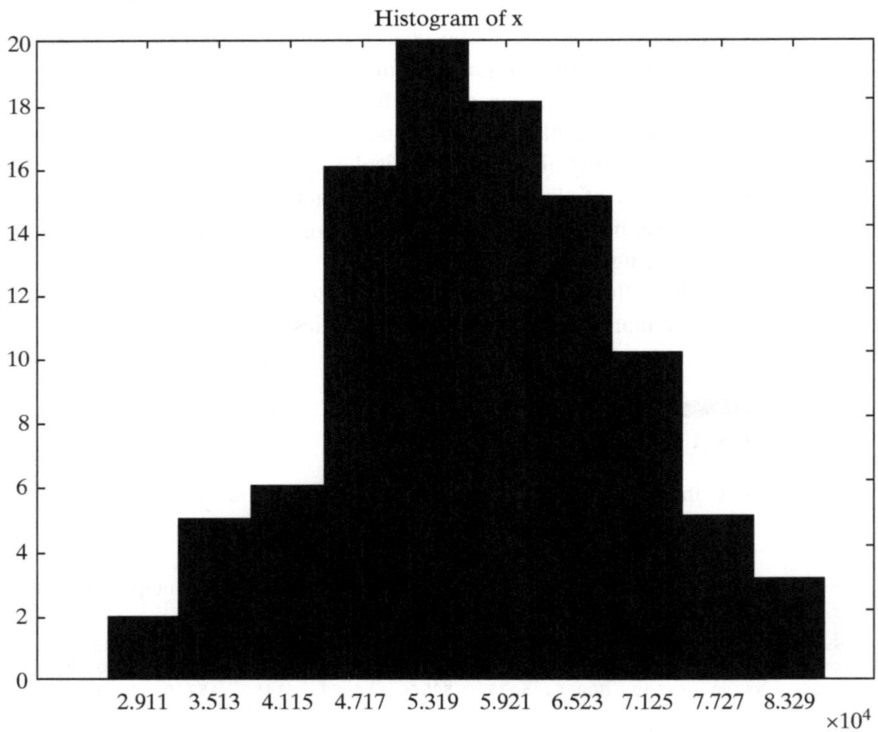

■

EXAMPLE 1.4

Plot the histogram and find the mean and standard deviation of the data shown in Table 1.2 (Example 1.1) using Excel.

Solution: First the data are entered (typed), one number at a time, in the first column (Column A) of the Excel spreadsheet. Then, go to the *Data* tab in the top line and select *Data Analysis*. This displays the *Analysis Tools* in a box. Click on *Histogram* in the analysis tools displayed. Press *OK* to see the dialog box, go to *Input Range* and specify the location of the data by typing "$A$1:$A$100". Next, select the output range by typing "C3" (Column C, line 3) to print the results to the right of Column C and below line 3. Select the *Chart Output* and click *OK* to obtain the histogram displayed. In the same display, type "=AVERAGE(A1:A100)" and "=STDEV(A1:A100)", one at a time, to find the mean or average value and the standard deviation as indicated in Fig. 1.11.

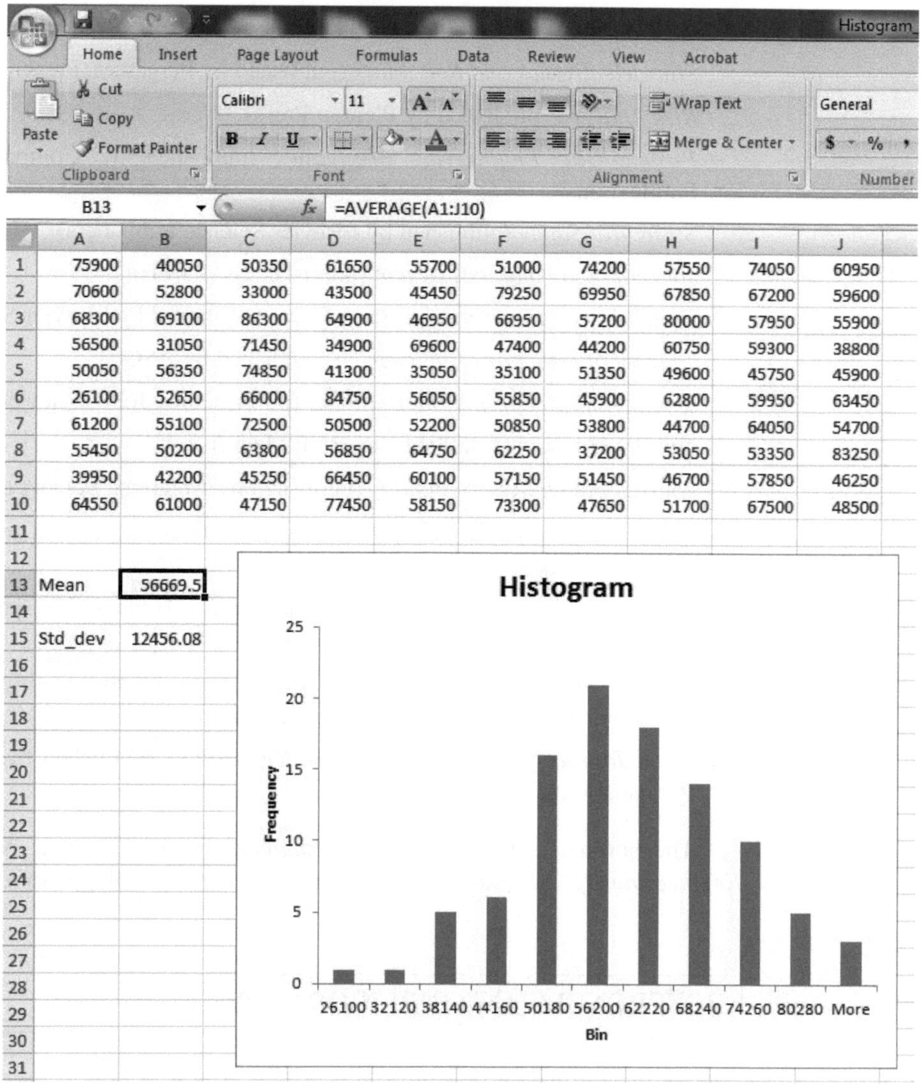

FIGURE 1.11 Excel screen shot of the histogram with values of mean and standard deviation.

■

1.11 Reliability Literature

The literature on reliability engineering is large and diverse. Several textbooks are available [1.1–1.126, 1.181–1.236] and numerous conference proceedings were published on the subject [1.127–1.1.165]. Also, several technical periodicals regularly or occasionally publish papers relating to reliability and reliability-based design. This is primarily because reliability affects so many disciplines, from electronic devices to machinery to airplane structures. The journals that regularly publish papers relating to reliability are:

IEEE Transactions on Reliability (published by IEEE Reliability Society)

Microelectronics and Reliability (published by Pergamon Press)

Risk Analysis (published by Plenum Press)

Probabilistic Engineering Mechanics (published by Computational Mechanics Publications)

Structural Safety (published by Elsevier Science Publishers)

Journal of Mechanical Design (published by ASME)

Reliability Engineering & System Safety (published by Elsevier Science Publishers)

Journal of Structural Engineering (published by ASCE)

Quality and Reliability International (published by John Wiley & Sons)

Quality Magazine (published by BNP Media)

Journal of Quality Technology (published by American Society of Quality)

Technometrics (published by American Statistical Association and American Society for Quality)

International Journal of Quality and Reliability Management (published by Emerald Group Publishing Ltd.)

Journal of Risk and Reliability, Proceedings of the Institution of Mechanical Engineers, Part O (published by Sage Publications)

Open Journal of Safety Science and Technology (published by Scientific Research Publishing, Inc.)

The journals such as *Nuclear Engineering and Design, Nuclear Safety*, and *Engineering Optimization* occasionally publish papers dealing with reliability.

REFERENCES AND BIBLIOGRAPHY

System Reliability and Design

1.1 I. Bazovsky, *Reliability Theory and Practice*, Prentice-Hall, Englewood Cliffs, NJ, 1961.

1.2 R. Billinton and R.N. Allan, *Reliability Evaluation of Engineering Systems: Concepts and Techniques,* Plenum Press, New York, 1983.

1.3 H.P. Bloch and F. K. Geitner, *An Introduction to Machinery Reliability Assessment*, Van Nostrand Reinhold, New York, 1990.

1.4 J. H. Bompas-Smith, *Mechanical Survival: The Use of Reliability Data*, McGraw-Hill, London, 1973.

1.5 A. D. S. Carter, *Mechanical Reliability*, Wiley, New York, 1972.

1.6 B. S. Dhillon, *Reliability Engineering in Systems Design and Operation*, Van Nostrand Reinhold, New York, 1983.

1.7 L. A. Doty, *Reliability for the Technologies*, Industrial Press, New York, 1985.

1.8 G. W. Dummer and R. C. Winton, *An Elementary Guide to Reliability*, 2nd ed., Pergamon Press, Oxford, 1974.

1.9 E. G. Frankel, *Systems Reliability and Risk Analysis*, Martinus Nijhoff Publishers, The Hague, The Netherlands, 1984.

1.10 H. Goldberg, *Extending the Limits of Reliability Theory*, John Wiley, New York, 1981.

1.11 D.L. Grosh, *A Primer of Reliability Theory*, John Wiley, New York, 1989.

1.12 R. P. Haviland, *Engineering Reliability and Long Life Design*, D. Van Nostrand Co., Princeton, NJ, 1964.

1.13 E. B. Haugen, *Probabilistic Approaches to Design*, John Wiley, New York, 1968.

1.14 E. B. Haugen, *Probabilistic Mechanical Design*, John Wiley, New York, 1980.

1.15 E. J. Henley and H. Kumamoto, *Designing for Reliability and Safety Control*, Prentice-Hall, Englewood Cliffs, NJ, 1985.

1.16 K. C. Kapur and L. R. Lamberson, *Reliability in Engineering Design*, John Wiley, New York, 1977.

1.17 G. Kivenson, *Durability and Reliability in Engineering Design*, Hayden Book Co., New York, 1971.

1.18 R. R. Landers, *Reliability and Product Assurance*, Prentice-Hall, Englewood Cliffs, NJ, 1963.

1.19 E. E. Lewis, *Introduction to Reliability Engineering*, John Wiley, New York, 1987.

1.20 C. Lipson, J. Kerawalla, and L. Mitchell, *Engineering Applications of Reliability*, The University of Michigan, Ann Arbor, MI, 1963.

1.21 W. Nelson, *How to Analyze Reliability Data*, American Society for Quality Control, Milwaukee, WI, 1983.

1.22 P. D. T. O'Connor, *Practical Reliability Engineering*, 4th ed., Wiley, New York, NY, 2002.

1.23 S. Osaki, *Stochastic System Reliability Modeling*, World Scientific Publishing Co., Singapore, 1985.

1.24 A. Pages and M. Gondran, *System Reliability: Evaluation and Prediction in Engineering*, Springer-Verlag, Berlin, 1986.

1.25 E. Pieruschka, *Principles of Reliability*, Prentice-Hall, Englewood Cliffs, NJ, 1963.

1.26 A. M. Polovko, *Fundamentals of Reliability Theory* (English translation), Academic Press, New York, 1968.

1.27 A. S. Pronikov, *Dependability and Durability of Engineering Products* (English translation), John Wiley, New York, 1973.

1.28 G. H. Sandler, *System Reliability Engineering*, Prentice-Hall, Englewood Cliffs, NJ, 1963.

1.29 M. L. Shooman, *Probabilistic Reliability: An Engineering Approach*, McGraw-Hill, New York, 1968.

1.30 J. N. Siddall, *Probabilistic Engineering Design: Principles and Applications*, Marcel Dekker, Inc., New York, 1983.

1.31 C. O. Smith, *Introduction to Reliability*, McGraw-Hill, New York, 1976.

1.32 E. J. Tangerman (ed.), "A Manual of Reliability," *Product Engineering*, May 16, 1960, pp: 65–96.

1.33 F. A. Tillman, C. L. Hwang, and W. Kuo, *Optimization of Systems Reliability*, Marcel Dekker Inc., New York, 1980.

1.34 P. A. Tobias and D. C. Trindade, *Applied Reliability*, Van Nostrand Reinhold, New York, 1986.

1.35 W. H. Von Alven (ed.), *Reliability Engineering*, Prentice-Hall, Englewood Cliffs, NJ, 1964.

1.36 L. W. Condra, *Reliability Improvement with Design of Experiments*, Marcel Dekker, Inc., New York, 1993.

1.37 A. Birolini, *Reliability Engineering: Theory and Practice*, 4th ed., Springer, Berlin, 2004.

1.38 R. Billinton and R. N. Allan, *Reliability Evaluation of Engineering Systems: Concepts and Techniques*, 2nd ed., Plenum Press, New York, 1992.

1.39 K. Kolowrocki, *Reliability of Large Systems*, Elsevier, Amsterdam, 2004.

1.40 R. D. Leitch, *Reliability Analysis for Engineers: An Introduction*, Oxford University Press, Oxford, 1995.

1.41 N. D. Singpurwalla, *Reliability and Risk: A Bayesian Perspective*, Wiley, Hoboken, NJ, 2006.

1.42 K. Neubeck, *Practical Reliability Analysis*, Prentice-Hall, NJ, 2004.

1.43 M. Rausand and A. Hoyland, *System Reliability Theory*, 2nd ed., Wiley, Hoboken, NJ, 2004.

1.44 B. S. Dhillon, *Mechanical Reliability Theory, Models and Applications*, American Institute of Aeronautics and Astronautics, Washington, DC, 1988.

1.45 H. Hecht, *Systems Reliability and Failure Prevention*, Artech House, Boston, MA, 2004.

1.46 J. W. Priest, *Engineering Design for Producibility and Reliability*, Marcel Dekker, New York, 1988.

Structural Reliability

1.47 G. Augusti, A. Baratta, and F. Casciati, *Probabilistic Methods in Structural Engineering*, Chapman and Hall, London, 1984.

1.48 V. V. Bolotin, *Statistical Methods in Structural Mechanics*, English translation, Holden Day, Inc., Oakland, CA, 1969.

1.49 P. T. Christensen and M. J. Baker, *Structural Reliability Theory and Its Applications*, Springer-Verlag, Berlin, 1982.

1.50 P. T. Christensen and Y. Murotsu, *Application of Structural Systems Reliability Theory*, Springer-Verlag, Berlin, 1986.

1.51 I. Elishakoff, *Probabilistic Methods in the Theory of Structures*, Wiley, New York, 1983.

1.52 G. C. Hart, *Uncertainty Analysis, Loads, and Safety in Structural Engineering*, Prentice-Hall, Englewood Cliffs, NJ, 1982.

1.53 E. Leporati, *The Assessment of Structural Safety*, English translation, Research Studies Press, Forest Grove, OR, 1979.

1.54 Y. K. Lin, *Probabilistic Theory of Structural Dynamics*, McGraw-Hill, New York, 1967.

1.55 H. O. Madsen, S. Krenk, and N. C. Lind, *Methods of Structural Safety*, Prentice-Hall, Englewood Cliffs, NJ, 1986.

1.56 R. E. Melchers, *Structural Reliability, Analysis and Prediction*, Ellis Horwood, Chichester, England, 1987.

1.57 T. Nakagawa, H. Ishikawa, and A. Tsurui (eds.), *Recent Studies on Structural Safety and Reliability*, Elsevier Applied Science, London, 1989.

1.58 J. W. Provan, *Probabilistic Fracture Mechanics and Reliability*, Martinus Nijhoff Publishers, Dordrecht, The Netherlands, 1987.

1.59 A. Halder and S. Mahadevan, *Reliability Assessment using Stochastic Finite Element Analysis*, Wiley, New York, 2000.

Reliability and Mechanical Design

1.60 O. Vinogradov, *Introduction to Mechanical Reliability: A Designer's Approach*, Hemisphere Publishing Corporation, New York, 1991.

1.61 T. A. Cruse (ed.), *Reliability-Based Mechanical Design*, Marcel Dekker, Inc., New York, 1997.

1.62 B. S. Dhillon, *Design Reliability: Fundamentals and Applications*, CRC Press, Boca Raton, FL, 1999.

1.63 A. C. Brombacher, *Reliability by Design*, Wiley, New York, 1992.

Reliability and Geotechnical Engineering

1.64 G. B. Baecher, *Reliability and Statistics in Geotechnical Engineering*, Wiley, Hoboken, NJ, 2003.

Reliability Handbooks

1.65 "Automotive Electronics Reliability Handbook," AE-9, Society of Automotive Engineers, Inc., Warrendale, PA., 1987.

1.66 W. G. Ireson, *Reliability Handbook*, McGraw-Hill, New York, 1966.

1.67 B. A. Kozlov and I. A. Ushakov, *Reliability Handbook*, Holt, Rinehart & Winston Inc., New York, 1970.

1.68 E. Nikolaidis, D. M. Ghiocel, and S. Singhal (eds.), *Engineering Design Reliability Handbook*, CRC Press, Boca Raton, FL, 2005.

1.69 W. E. Vesely, et al., *Fault Tree Handbook*, Microfishe, Systems and Reliability Research, Office of Nuclear Regulatory Research, U.S. Nuclear Regulatory Commission, Washington, DC, 1981.

1.70 D. I. Gertman and H. S. Blackman, *Human Reliability and Safety Analysis Data Handbook*, Wiley, New York, 1994.

1.71 Electronics Reliability Subcommittee, *Automotive Electronics Reliability Handbook*, Society of Automotive Engineers, Warrandale, PA, 1987.

1.72 D. Kececioglu, *Reliability Engineering Handbook*, Prentice-Hall, Englewood Cliffs, NJ, 1991.

1.73 D. J. Klinger, Y. Nakada, and M. A. Menendez (eds.), *AT&T Reliability Manual*, Van Nostrand Reinhold, New York, 1990.

1.74 A. G. Cannon and A. Bendell (eds.), *Reliability Data Banks*, Elsevier Applied Science, London, 1991.

1.75 I. A. Ushakov and R. A. Harrison, (I. A. Ushakov – Ed.), *Handbook of Reliability Engineering*, Wiley, New York, 1994.

Reliability and Maintainability

1.76 H. Ascher and H. Feingold, *Repairable Systems Reliability*, Marcel Dekker Inc., New York, 1984.

1.77 H. J. Bajaria, *Integration of Reliability, Maintainability and Quality Parameters in Design*, SP-533, Soc. of Automotive Eng., Warrendale, PA., 1983.

1.78 B. S. Dhillon, *Systems Reliability, Maintainability and Management*, Petrocelli Books Inc., New York, 1983.

1.79 B. S. Dhillon and H. Reiche, *Reliability and Maintainability Management*, Van Nostrand Reinhold, New York, 1985.

1.80 J. W. Foster, D. T. Phillips, and T. R. Rogers, *Reliability, Availability and Maintainability*, M/A Press, Beaverton, OR., 1981.

1.81 M. O. Locks, *Reliability, Maintainability, and Availability Assessment*, Spartan Books, dist. by Haydon Books Co., Rochelle Park, NJ, 1973.

1.82 D. J. Smith, *Reliability, Maintainability, and Risk: Practical Methods for Engineers*, 4th ed., Butterworth-Heinemann, Oxford, 1993.

1.83 H. Wang and H. Pham, *Reliability and Optimal Maintenance*, Springer-Verlag, London, 2006.

1.84 A. K. S. Jardine and A. H. C. Tsang, *Maintenance, Replacement, and Reliability: Theory and Applications*, CRC Press, Boca Raton, FL, 2005.

1.85 J. A. Nachlas, *Reliability Engineering: Probabilistic Models and Maintenance Methods*, CRC Press, Boca Raton, FL, 2005.

1.86 C. E. Ebeling, *An Introduction to Reliability and Maintainability Engineering*, McGraw-Hill, Boston, MA, 1997.

1.87 U. Dinesh Kumar et al., *Reliability Maintenance and Logistic Support: A Life Cycle Approach*, Kluwer Academic, Boston, MA, 2000.

1.88 A. K. S. Jardine, *Maintenance, Replacement and Reliability*, Wiley, New York, 1973.

1.89 R. T. Anderson and L. Neri, *Repairable Systems Reliability: Reliability-Centered Maintenance*, Elsevier Applied Science, London, 1990.

1.90 S. J. Thomas, *Improving Maintenance and Reliability Through Cultural Change*, Industrial Press, New York, 2005.

1.91 T. Nakagawa, *Advanced Reliability Models and Maintenance Policies*, Springer-Verlag, London, 2008.

1.92 K. Holmberg and A. Folkeson (eds.), *Operational Reliability and Systematic Maintenance*, Elsevier Applied Science, London, 1991.

1.93 D. J. Smith, *Reliability and Maintainability in Perspective*, 2nd ed., Wiley, New York, 1985.

Mathematical Methods of Reliability

1.94 L. J. Bain, *Statistical Analysis of Reliability and Life-Testing Models: Theory and Methods*, Marcel Dekker Inc., New York, 1978.

1.95 R. E. Barlow, F. Proschan, and L. C. Hunter, *Mathematical Theory of Reliability*, Wiley, New York, 1965.

1.96 B. V. Gnedenko, Y. K. Belyayev, and A. D. Solovyev, *Mathematical Methods of Reliability Theory*, English translation, Academic Press, New York, 1969.

1.97 D. K. Lloyd and M. Lipow, *Reliability: Management, Methods, and Mathematics*, Prentice-Hall, Englewood Cliffs, NJ, 1962.

1.98 N. H. Roberts, *Mathematical Methods in Reliability Engineering*, McGraw-Hill, New York, 1964.

1.99 L. Leemis, *Reliability: Probabilistic Models and Statistical Methods*, Prentice-Hall, Upper Saddle River, NJ, 1995.

1.100 B. L. Amstadter, *Reliability Mathematics: Fundamentals, Practices, Procedures*, McGraw-Hill, New York, 1971.

Reliability and Mechanisms

1.101 B. Z. Sandler, *Probabilistic Approach to Mechanisms*, Elsevier Science Publishing Co. Inc., Amsterdam, The Netherlands, 1984.

1.102 Y. Armijo, *Modern Statistical and Mathematical Methods in Reliability*, World Scientific, Singapore, 2005.

Reliability and Power Systems

1.103 G. J. Anders, *Probability Concepts in Electric Power Systems*, John Wiley, New York, 1990.

1.104 P. W. Beeker and F. Jensen, *Design of Systems and Circuits for Maximum Reliability or Maximum Production Yield*, McGraw-Hill, New York, 1977.

1.105 B. S. Dhillon, *Power System Reliability, Safety and Management*, Ann Arbor Science, Ann Arbor, MI., 1983.

1.106 R. Billinton and R. N. Allan, *Reliability Evaluation of Power Systems*, 2nd ed., Plenum Press, New York, 1996.

1.107 J. Endrenyi, *Reliability Modeling in Electric Power Systems*, Wiley, New York, 1978.

Reliability and Quality Control

1.108 R. H. W. Brook, *Reliability Concepts in Engineering Manufacture*, John Wiley, New York, 1972.

1.109 N. L. Enrick, *Quality Control and Reliability*, 7th ed., Industrial Press, New York, 1977.

1.110 B. S. Dhillon, *Quality Control, Reliability, and Engineering Design*, Marcel Dekker Inc., New York, 1985.

1.111 S. Halpern, *The Assurance Sciences: An Introduction to Quality Control and Reliability*, Prentice-Hall, Englewood Cliffs, NJ, 1978.

1.112 G. Taguchi, *Introduction to Quality Engineering*, American Supplier Institute, Dearborn, MI, 1986.

1.113 P. J. Ross, *Taguchi Techniques for Quality Engineering*, McGraw-Hill, New York, 1988.

1.114 M. S. Phadke, *Quality Engineering Using Robust Design*, Prentice-Hall, Englewood Cliffs, NJ, 1989.

1.115 B. S. Dhillon, *Reliability, Quality, and Safety for Engineers*, CRC Press, Boca Raton, FL, 2004.

1.116 K. R. Bhote and A. K. Bhote, *World Class Reliability*, American Management Association, New York, 2004.

Reliability and Electronics

1.117 N. B. Fuqua, *Reliability Engineering for Electronic Design*, Marcel Dekker, Inc., New York, 1987.

1.118 T. I. Bajunescu and M. I. Bazu, *Reliability of Electronic Components: A Practical Guide to Electronic Systems Manufacturing*, Springer, New York, 1999.

1.119 A. Christou, *Integrating Reliability into Microelectronics Manufacturing*, Wiley, New York, 1994.

Reliability, Human Factors, and Safety Systems

1.120 B. S. Dhillon, *Human Reliability: With Human Factors*, Pergamon Press, Oxford, 1986.

1.121 A. E. Green, *Safety Systems Reliability*, Wiley, New York, 1983.

1.122 M. Walter and W. Schneeweiss, *The Modeling World of Reliability/Safety Engineering*, LiLoLe-Verlag, Hagen, Germany, 2005.

Reliability and Warranties

1.123 M. U. Thomas, *Reliability and Warranties: Methods for Product Development and Quality Improvement*, CRC Press, Boca Raton, FL, 2006.

1.124 W. R. Blischke and D. N. P. Murthy, *Warranty Cost Analysis*, Marcel Dekker, New York, 1994.

1.125 W. R. Blischke and D. N. P. Murthy, *Product Warranty Handbook*, Marcel Dekker, New York, 1996.

1.126 R. B. James, *Warranties Planning, Analysis and Implementation*, McGraw-Hill, New York, 1994.

Conference Proceedings

1.127 *Advanced Techniques in Failure Analysis Symposium—1976*, Institute of Electrical and Electronic Engineers, New York, 1976.

1.128 A. H. S. Ang and M. Shinozuka (eds.), *Probabilistic Mechanics and Structural Reliability*, American Society of Civil Engineers, New York, 1979.

1.129 *Annals of Assurance Sciences*, Proceeding of the Annual Symposium on Reliability, published by Inst. of Electrical and Electronic Engineers, Inc., New York, 1969–1974.

1.130 G. Apostolakis, S. Garribba, and G. Volta (eds.), *Synthesis and Analysis Methods for Safety and Reliability Studies*, Plenum Press, New York, 1980.

1.131 *Avionics, Reliability, Its Techniques and Related Disciplines*, Adv. Group for Aerospace Research & Dev., AGARD-CP-261, 1979.

1.132 A. P. Basu (ed.), *Reliability and Quality Control*, North-Holland, Amsterdam, 1986.

1.133 S. B. Bennett, A. L. Ross, and P. Z. Zemanick (eds.), *Failure Prevention and Reliability*, Papers presented at the Design Eng. Tech. Conf., Chicago, IL, Sept. 26–28, 1977, American Soc. of Mech. Eng., New York, 1977.

1.134 K. L. Carper (ed.), *Forensic Engineering: Learning from Failures*, Amer. Soc. of Civil Eng., New York, 1986.

1.135 P. T. Christensen (ed.), *Reliability Theory and Its Application in Structural and Soil Mechanics*, Martinus Nijhoff Publishers, The Hague, The Netherlands, 1983.

1.136 *Composite Reliability*, ASTM STP-580, Amer. Soc. for Testing and Materials, Philadelphia, PA, 1975.

1.137 P. E. Doepker (ed.), *Failure Prevention and Reliability—1985*, Amer. Soc. of Mech. Eng., New York, 1985.

1.138 A. C. Gangadharan and S. J. Brown, Jr. (eds.), *Failure Data and Failure Analysis in Power and Processing Industries*, Amer. Soc. of Mech. Eng., New York, 1977.

1.139 K. Gibble (ed.), *Management Lessons from Engineering Failures*, Amer. Soc. of Civil Eng., New York, 1986.

1.140 K. Hennig (ed.), *Random Vibrations and Reliability*, Academic-Verlag, Berlin, 1983.

1.141 M. Grigoriu (ed.), *Risk, Structural Engineering and Human Error*, University of Waterloo Press, Waterloo, Canada, 1984.

1.142 D. Grouchko (ed.), *Operations Research and Reliability*, Gordon and Breach, New York, 1971.

1.143 E. J. Henley and J. W. Lynn (eds.), *Generic Techniques in Systems Reliability Assessment*, Noordhoff, Neyden, The Netherlands, 1976.

1.144 *International Seminar on Probabilistic and Extreme Load Design of Nuclear Plant Facilities*, Amer. Soc. of Civil Eng., New York, 1979.

1.145 G. M. Kurajian (ed.), *Failure Prevention and Reliability—1983*, Amer. Soc. of Mech. Eng., New York, 1983.

1.146 F. T. C. Loo (ed.), *Failure Prevention and Reliability—1981,* Amer. Soc. of Mech. Eng., New York, 1981.

1.147 S. M. Ma (ed.), *Effects of Deterioration on Safety and Reliability of Structures,* Amer. Soc. of Civil Eng., New York, 1986.

1.148 *Mechanical Reliability Concepts,* Amer. Soc. of Mech. Eng., New York, 1965.

1.149 *Methods for Assessing the Structural Reliability of Brittle Materials*, ASTM STP-844, Amer. Soc. for Testing and Materials, Philadelphia, PA, 1984.

1.150 W. D. Milestone (ed.), *Reliability, Stress Analysis and Failure Prevention Methods in Mechanical Design*, Amer. Soc. of Mech. Eng., New York, 1980.

1.151 T. Moan and M. Shinozuka (eds.), *Structural Safety and Reliability*, Elsevier Science Publishing Co. Inc., Amsterdam, The Netherlands, 1981.

1.152 A. S. Nowak (ed.), *Modeling Human Errors in Structural Design and Construction*, Amer. Soc. of Civil Eng., New York, 1986.

1.153 *"Power Plant Reliability"*, Adv. Group for Aerospace Research and Dev., AGARD-CP-215, 1977.

1.154 *Prevention of Structural Failures*, American Society for Metals, Metals Park, Ohio, 1978.

1.155 *Reliability Problems of Reactor Pressure Components*, vols. I and II, International Atomic Energy Agency, Vienna, 1978.

1.156 M. Shinozuka and J. T. P. Yao (eds.), *Proceedings of the Symposium on Probabilistic. Methods in Structural Engineering*, Amer, Soc. of Civil Eng., New York, 1981.

1.157 C. Sundararajan (ed.), *Reliability and Safety of Pressure Components*, Amer. Soc. of Mech. Eng., New York, PVP-Vol. 62, 1982.

1.158 J. L. Swedlow, T. A. Cruse, and J. C. Halpin (eds.), *Proceedings of the Colloquium on Structural Reliability: The Impact of Advanced Materials on Engineering Design*, Carnegie-Mellon University, Pittsburgh, PA., 1972.

1.159 C. P. Tsokos and I. N. Shimi (eds.), *The Theory and Applications of Reliability: With Emphasis on Bayesian and Nonparametric Methods*, vols. I and II, Academic Press, New York, 1977.

1.160 H. J. Wingerder (ed.), *Reliability Data Collection and Use in Risk and Availability Assessment*, Proceedings of the 5th EuRe Data Conference, Heidelberg, Springer-Verlag, Berlin 1986.

1.161 H. Pham (ed.), *Recent Advances in Reliability and Quality in Design*, Springer-Verlag, London, 2008.

1.162 P. R. Krishnaiah and C. R. Rao (eds.), *Quality Control and Reliability*, North-Holland (Elsevier Science), Amsterdam, 1988.

1.163 J.W. Evans and J. Y. Evans (eds.), *Product Integrity and Reliability in Design*, Springer-Verlag, London, 2001.

1.164 H. Pham (ed.), *Reliability Modeling, Analysis and Optimization*, World Scientific, Singapore, 2006.

1.165 N. Limnios and M. Nikulin (eds.), *Recent Advances in Reliability Theory*, Birkhauser, Boston, 2000.

Reliability Standards

1.166 U.S. Department of Defense, *Reliability Prediction of Electronic Equipment*, MIL-HDBK-217F, February 1995.

1.167 U.S. Department of Defense, *Electronic Reliability Design Handbook*, MIL-HDBK-338B, October 1998.

1.168 U.S. Department of Defense, *Failure Rate Sampling Plans and Procedures*, MIL-STD-690D, June 2005.

1.169 U.S. Department of Defense, *Reliability Tests Exponential Distribution*, MIL-STD-781B, November 1967.

1.170 U.S. Department of Defense, *Reliability Test Methods, Plans, and Environments for Engineering Development, Qualification, and Production*, MIL-HDBK-781A, April 1996.

1.171 U.S. Department of Defense, *Reliability Program Requirements for Space and Launch Vehicles*, MIL-STD-1543B, October 1988.

1.172 U.S. Department of Defense, *Procedures for Performing a Failure Mode Effects and Criticality Analysis*, MIL-STD-1629A, November 1980.

1.173 U.S. Department of Defense, *Reliability-Centered Maintenance (RCM) Requirements for Naval Aircraft, Weapon Systems, and Support Equipment*, January 1998; Superseded by: NAVAIR 00-25-403.

1.174 U.S. Department of Defense, *Reliability Program for Systems and Equipment Development and Production*, MIL-STD-785B, September 1980; Superseded by: *Reliability Program Standard for Systems Design, Development, and Manufacturing*, ANSI/GEIA-STD-0009-2008, November 2008.

1.175 U.S. Department of Defense, *Reliability Design Qualification and Production Acceptance Tests: Exponential Distribution*, October 1977.

1.176 U.S. Department of Defense, *Test and Evaluation of System Reliability, Availability, and Maintainability (A Primer)*, DoD 3235.1H, 3rd ed., March 1982.

1.177 The Aerospace Corporation, *Reliability Program Requirements for Space Systems*, Aerospace Report Number—TOR-2007(8583)-6889, July 2007.

1.178 National Aeronautics and Space Administration Goddard Space Flight Center, *Flight Assurance Procedure: Performing a Failure Mode and Effects Analysis*, NASA GSFC 431-REF-000370, August 1996.

1.179 National Aeronautics and Space Administration Jet Propulsion Laboratory, *Reliability Analysis Handbook*, JPL D-5703, July 1990.

1.180 Institute of Electrical and Electronics Engineers, *IEEE Standard Reliability Program for the Development and Production of Electronic Systems and Equipment*, IEEE 1332-1998, 1998.

Reliability Applications

1.181 "Rationalization of Safety and Serviceability Factors in Structural Codes," CIRIA (Construction Industry Research and Information Association), Report No. 63, London, 1977.

1.182 "Digital Evaluation and Generic Failure Analysis Data," MDR-10, Reliability Analysis Center, Rome Air Development Center, Griffiss Air Force Base, New York, 1978.

1.183 Y. Fukumoto and Y. Itoh, "Statistical Study of Experiments on Welded Beams," ASCE J. of the Structural Division, vol. 107, no. ST1, pp. 89–103, Jan. 1981.

1.184 W. H. Middendorf, *Design of Devices and Systems*, 2nd ed., Marcel Dekker, New York, 1990.

1.185 "Non-Electronic Parts Reliability Data," NPRD-1, Reliability Analysis Center, Rome Development Center, Griffiss Air Force Base, New York, 1978.

1.186 N. Rasmussen, et al., "Reactor Safety Study," WASH-1400-D, U.S. Atomic Energy Commission, Washington, DC, 1974.

1.187 N. A. Tiner, *Failure Analysis with the Electron Microscope*, Fox-Mathis Publications, Los Angeles, CA, 1973.

1.188 *Selected Papers by Alfred M. Freudenthal. Civil Engineering Classics*, American Society of Civil Engineers, New York, 1981.

1.189 "Reliability of Military Electronic Equipment," *Report of the Advisory Group on Reliability of Electronic Equipment (AGREE)*, Office of the Assistant Secretary of Defense, Washington, DC, June 4, 1957.

1.190 D. Kececioglu and D. Cormier, "Designing a Specified Reliability Directly into a Component," *Proc. of the Third Annual Aerospace Reliability and Maintainability Conf.*, Soc. of Automotive Engineers, Inc., New York, June 1964, pp. 546-565.

1.191 K. L. Wong, "The Bathtub Does Not Hold Water Any More", *Quality and Reliability Engineering International*, Vol. 4, pp. 279–282, 1988.

1.192 N. Schlager (ed.), *When Technology Fails*, Gale Research Inc., Detroit, MI, 1994.

1.193 J. R. Cogdell, *Modeling Random Systems*, Pearson Prentice Hall, Upper Saddle River, NJ, 2004.

1.194 L. W. Condra, *Reliability Improvement with Design of Experiments*, Marcel Dekker, New York, 1993.

1.195 H. Pham (ed.), *Reliability Modeling, Analysis, and Optimization*, World Scientific Publishing Co., Singapore, 2006.

1.196 E. Hollnagel (ed.), *The Reliability of Expert Systems*, Ellis Horwood, Chichester, UK, 1989.

1.197 J. Kogan, *Crane Design: Theory and Calculations of Reliability*, Wiley, New York, 1976.

1.198 H. Ascher and H. Feingold, *Repairable Systems Reliability*, Marcel Dekker, New York, 1984.

1.199 R. G. Miller, Jr., *Survival Analysis*, Wiley, New York, 1981.

1.200 J. D. Kalbfleisch and R. L. Prentice, *The Statistical Analysis of Failure Time Data*, Wiley, New York, 1980.

1.201 L. J. Bain and M. Engelhardt, *Statistical Analysis of Reliability and Life Testing Models: Theory*, Marcel Dekker, New York, 1991.

1.202 L. W. Mays, *Reliability Analysis of Water Distribution Systems*, American Society of Civil Engineers, New York, 1989.

1.203 W. R. Blischke and D. N. P. Murthy, *Reliability: Modeling, Prediction and Optimization*, Wiley, New York, 2000.

1.204 D. N. P. Murthy, *Weibull Models*, Wiley, Hoboken, NJ, 2004.

1.205 A. Haldar (Ed.), *Recent Developments in Reliability-Based Civil Engineering*, World Scientific Publishing, Singapore, 2006.

1.206 H. Pham, *Recent Advances in Reliability and Quality Engineering*, World Scientific Publishing, Singapore, 2001.

1.207 P. MacDiarmid et al., *Reliability Toolkit: Commercial Practices Edition*, Reliability Analysis Center and Rome Laboratory, Rome, NY, 1995.

1.208 M. Modarres, M. Kaminskiy and V. Krivtsov, *Reliability Engineering and Risk Analysis: A Practical Guide*, CRC Press, Boca Raton, FL, 1999.

1.209 J. C. Cluley, *Reliability in Instrumentation and Control*, Butterworth-Heinemann, Boston, MA, 1992.

1.210 E. Bauer, *Design for Reliability: Information and Computer-Based Systems*, IEEE Press and Wiley, Hoboken, NJ, 2010.

1.211 A. Birolini, *Reliability Engineering: Theory and Practice*, 4th ed., Springer-Verlag, Heidelberg, Germany, 2004.

1.212 R. D. Leitch, *Reliability Analysis for Engineers, An Introduction*, Oxford University Press, Oxford, 1995.

1.213 M. J. LuValle, B. G. Lefevre and S. R. Kannan, *Design and Analysis of Accelerated Tests for Mission Critical Reliability*, Chapman and Hall/CRC Press, Boca Raton, FL, 2004.

1.214 L. M. Klyatis and E. L. Klyatis, *Accelerated Quality and Reliability Solutions*, Elsevier, San Diego, CA, 2006.

1.215 A. A. Hattangadi, *Plant and Machinery Failure Prevention*, McGraw-Hill, New York, 2005.

1.216 A. Lisnianski and G. Levitin, *Multi-State System Reliability*, World Scientific, Singapore, 2003.

1.217 T. Nakagawa, *Shock and Damage Models in Reliability Theory*, Springer-Verlag, London, 2007.

1.218 F. J. Samaniego, *System Signatures and Their Applications in Engineering Reliability*, Springer Science+Business Media, New York, 2007.

Software Reliability

1.219 J. Musa, *Software Reliability Engineering: More Reliable Software Faster and Cheaper*, 2nd ed., AuthorHouse, 2005.

1.220 A. M. Neufelder, *Ensuring Software Reliability*, Marcel Dekker, New York, 1993.

1.221 M. Shooman, *Software Engineering: Design, Reliability, and Management*, McGraw Hill, New York, NY, 1987.

Fuzzy Reliability

1.222 K–Y. Cai, *Introduction to Fuzzy Reliability*, Kluwer Academic, Boston, 1996.

1.223 T. Onisawa and J. Kacprzyk (eds.), *Reliability and Safety Analyses under Fuzziness*, Physica-Verlag, Heidelberg, Germany, 1995.

Stochastic Models in Reliability

1.224 C–D. Lai and M. Xie, *Stochastic Ageing and Dependence for Reliability*, Springer Science + Business Media, New York, 2006.

1.225 T. Aven and U. Jensen, *Stochastic Models in Reliability*, Springer-Verlag, New York, 1999.

1.226 N. Ravichandran, *Stochastic Methods in Reliability Theory*, Wiley Eastern, New Delhi, India, 1990.

1.227 S. Osaki (Ed.), *Stochastic Models in Reliability and Maintenance*, Springer-Verlag, Berlin, 2002.

1.228 S. Nakamura and T. Nakagawa (eds.), *Stochastic Reliability Modeling, Optimization and Applications*, World Scientific, Singapore, 2010.

1.229 B. Gnedenko and I. Ushakov, *Probabilistic Reliability Engineering*, Wiley, New York, 1995.

1.230 I. B. Gertsbakh, *Statistical Reliability Theory*, Marcel Dekker, New York, 1989.

1.231 S. Osaki, *Stochastic System Reliability Modeling*, World Scientific, Singapore, 1985.

Reliability and Optimization

1.232 W. Kuo and M. J. Zuo, *Optimal Reliability Modeling: Principles and Applications*, Wiley, New York, 2003.

Reliability and Risk

1.233 M. Todinov, *Reliability and Risk Models: Setting Reliability Requirements*, Wiley, Chichester, UK, 2005.

1.234 M. T. Todinov, *Risk-Based Reliability Analysis and Genetic Principles for Risk Reduction*, Elsevier, San Diego, CA, 2007.

1.235 J. X. Wang and M. L. Roush, *What Every Engineer Should Know about Risk Engineering and Management*, Marcel Dekker, New York, 2000.

1.236 N. D. Singpurwalla, *Reliability and Risk: A Bayesian Perspective*, Wiley, Chichester, UK, 2006.

Engineering System Failures and Disasters

1.237 N. Schlaser (ed.), *When Technology Fails*, Gale Research, Detroit, MI, 1994.

1.238 D. Lawson, *Engineering Disasters—Lessons to be Learned*, American Society of Mechanical Engineers, New York, 2005.

1.239 J. Fortune and G. Peters, *Learning from Failure—The Systems Approach*, Wiley, Chichester, UK, 1995.

1.240 D. R. H. Jones (ed.), *Failure Analysis Case Studies III*, Elsevier, Amsterdam, The Netherlands, 2004.

1.241 V. Bignell, G. Peters, and C. Pym, *Catastrophic Failures*, Open University Press, McGraw-Hill Education, 1977.

1.242 Find on Internet: *Pictures of Tacoma Narrows Bridge disaster.*

1.243 Find on Internet: *El al Boeing 747 disaster photos.*

1.244 Find on Internet: *Chernobyl nuclear disaster pictures.*

1.245 National Transportation Safety Board, Accident Report, *Collapse of I-35W Highway Bridge Minneapolis, Minnesota August 1, 2007*, NTSB/HAR-08/03, PB 2008-916203, Washington, DC, November 14, 2008.

1.246 Find on Internet: *Fukushima nuclear disaster pictures.*

1.247 Find on Internet: *Concorde disaster pictures.*

1.248 C.C. Gillispie (ed.-in-chief), *Dictionary of Scientific Biography,* Vol. 3, Charles Seribner's Sons, New York, 1980.

REVIEW QUESTIONS

1.1 Give brief answers to the following:

1. Define reliability.
2. What is the difference between a component and a system?
3. What is a bathtub curve?
4. Why does the failure rate decrease in the early stages of a component's life?
5. Why are the failure rates constant during the useful lives of most components and systems?
6. Give reasons for the increasing failure rate of a component or a system after its useful life.
7. Give the failure rates of three mechanical components.
8. Indicate the failure rates of three electrical components.
9. What are the various modes of failure of a mechanical or a structural component?
10. Define the term "factor of safety." Why is it considered to be inadequate to indicate the safety of a system?
11. What is the AGREE report?
12. State two common activities of a human being and the associated risks of death.
13. Name two journals that publish papers on reliability.
14. Identify a specific engineering system and state possible causes of its failure.

1.2 Indicate whether each of the following statements is true or false:

1. The overlapping area between the probability density functions of strength and load represents the exact probability of failure of the system.
2. If standard deviations of both strength and load increase, the probability of failure increases.
3. Air travel is more dangerous to human life than road travel by automobile.
4. Fatigue failure involves smaller magnitudes of stress compared to those associated with static failure.
5. The failures of large complex systems are caused mainly by the failure of larger components or subsystems.

1.3 Fill in the blank with the proper word(s):

1. The initial period of the bath tub curve is known as the _____ mortality period.
2. The constant failure rate during the useful life of a system represents failures due to _____ causes.
3. During the wear out period, the failure rate _____ with an increase in time.
4. The factor of safety denotes the ratio of expected _____ and expected _____.

1.4 Match the following:

1. Static failure	**a.** Chemical reaction of material with environment
2. Fatigue failure	**b.** Flow of material with time under load
3. Creep failure	**c.** Load magnitude exceeds ultimate strength
4. Corrosion failure	**d.** Change in work done by external forces exceeds strain energy stored
5. Wear failure	**e.** Cyclic stress of smaller magnitude
6. Instability failure	**f.** Failure due to contact of materials

PROBLEMS

1.1 The following sample data represent the Young's modulus of steel in 10^6 lb/in^2:

25.1	26.5	27.4	28.4	29.1
29.9	31.2	32.3	34.7	27.2
28.1	29.2	29.8	30.1	31.3
32.5	26.9	30.3	25.9	27.6
28.5	29.3	30.4	31.4	32.7
29.4	30.5	31.6	32.8	28.3
25.4	28.6	29.5	30.6	31.8
33.4	28.3	28.7	29.6	30.8
31.9	33.8	30.9	29.6	27.7
26.6	26.8	27.8	28.9	29.7

a. Construct the histogram using intervals of 10^6 lb/in^2 with 25×10^6 lb/in^2 as the lower limit of the first interval.
b. Plot the relative frequency diagram.
c. Find the mean value and standard deviations of the Young's modulus.

1.2 The eccentricity of the applied load (in inches) in a sample of nominally identical columns is shown as follows:

0.410	0.050	0.090	0.195	0.345
0.155	0.320	0.120	0.290	0.065
0.275	0.230	0.140	0.265	0.215
0.070	0.115	0.305	0.435	0.130
0.535	0.110	0.205	0.085	0.135
0.125	0.185	0.480	0.175	0.145
0.380	0.165	0.255	0.180	0.240
0.220	0.105			

Plot the histogram and the corresponding frequency distribution curve.

1.3 The maximum load carried by 29 nominally identical welded beams (in kilo-newtons) before failure [1.183] are given in the following table:

176.0	157.0	131.0	143.1	154.1
164.8	124.6	150.0	161.9	129.3
168.0	140.9	123.1	165.1	139.1
170.9	151.3	141.3	152.8	128.2
186.9	129.3	153.6	166.9	139.2
147.4	156.9	136.0	147.3	

Construct the histogram and the corresponding relative frequency diagram.

1.4 The compressive strengths of a sample of concrete test cylinders are given in the following table [in $k(lb/in^2)$]:

5.9	6.2	5.8	7.8	6.5
6.3	8.9	5.3	3.7	1.4
2.1	6.8	9.1	4.3	3.2
7.2	6.1	5.7	4.9	2.6
3.4	6.8	8.3	5.1	7.3
8.2	7.7	5.4	3.7	4.5
4.1	5.6	6.4	6.7	7.9
6.9	7.5	5.2	4.3	6.6
5.4	6.4			

Plot the histogram and the corresponding probability density function. Also determine the mean and standard deviations of the compressive strength.

1.5 Bars made of two different grades of steel are used for reinforcing concrete in a construction project. Experiments conducted on a sample of the reinforcing bars gave the following results for the yield strength in $k(lb/in^2)$:

35.7	40.9	43.4	30.2	35.5
31.1	43.3	40.8	38.1	33.7
33.2	38.8	39.6	41.5	32.5
42.5	40.4	33.8	31.2	40.3
41.2	42.9	36.6	31.7	41.8
42.8	38.4	39.9	34.6	32.2
37.5	41.7	32.3	41.1	40.6
40.7	42.7	32.6	37.2	33.4
42.3	40.1	32.9	39.5	41.6
42.2	41.4	34.5	39.3	41.3
34.1	39.2			

a. Plot the histogram.
b. Plot the relative frequency diagram.
c. How do you interpret the shape of the relative frequency diagram?
d. What are the mean value and standard deviations of the yield strength?

1.6 The data on the compressive strength of 75 aluminum–lithium specimens (in MPa) are given below:

1.0335	0.9302	0.7923	0.7234	0.6752	1.6880	1.1231	1.1024
0.8337	0.9990	1.1506	1.1782	1.1989	1.2815	0.9784	1.3711
1.0886	0.9853	1.1506	0.8130	1.0197	1.6329	1.2471	1.0542
1.3711	0.9233	1.3849	1.2126	1.0335	1.1713	1.0886	0.7579
0.9164	0.8475	1.0886	0.9164	1.1920	1.1024	1.3367	0.5305
1.1851	1.4331	1.2402	1.0266	0.9370	1.1782	1.1368	
1.0404	1.0611	1.5227	1.4262	1.5020	1.3504	1.1024	
1.1575	1.2471	1.2678	1.1231	1.0266	1.3780	1.2609	
1.2126	1.0886	1.0748	1.5778	1.1644	1.0059	0.9715	
0.8337	1.3711	1.3298	0.6063	1.0611	1.0680	1.3091	

a. Plot the histogram.
b. Plot the relative frequency diagram.
c. Find the mean value and standard deviations of the compressive strength.
d. Find the percentage of specimens that fail below 1.0 MPa.

1.7 The annual flow rates, in cubic feet per second (cfs), in a river are observed to be as follows:

2015	1176	143	889	2016	4098	1811	697	521	834
1202	1341	1423	2583	3185	3825	6617	1564	3662	3707
1655	3219	1447	384	277	147	352	1421	2144	2317
3800	3051	166	229	1259	2091	235	318	503	591
1392	2185	2006	2084	263	1768	2305	3143	4120	398

a. Plot the histogram.
b. Plot the probability density function.
c. Find the mean value and standard deviations of the annual flow rate.
d. Determine the percentage of flow rates exceeding a value of 4000 cfs.

1.8 The joint temperatures (in °F) of the O-rings used in the rocket motor of the space shuttle measured during the test firings or launches are given below:

67	45	66	70	69	80	31	84	49	61
40	83	58	76	75	76	79	81	61	70
67	70	58	68	60	67	72	73	57	63

a. Plot the histogram.
b. Plot the relative frequency diagram of the joint temperatures.
c. Determine the mean value and standard deviations of the joint temperature.
d. Determine the percentage of joint temperatures falling below the freezing point of water.

1.9 The numbers of defective leaf springs identified in random samples of size 50 by the quality control department of a manufacturing process are given below:

8	11	7	10	6	13	13	5	6	7
9	6	8	6	8	11	4	7	6	6
10	3	4	8	8	6	7	5	4	7
18	20	17	11	12	18	16	17	14	13

a. Plot the histogram.
b. Plot the probability density function of the defective leaf springs.
c. Find the mean value and the standard deviations.
d. Determine the percentage of defective springs that fall outside the band defined by mean value ± 3 standard deviations.

1.10 The observed number of automobiles crossing an intersection per minute are given below:

8	12	12	12	9	8	9	6	6	4
11	13	5	12	5	17	9	9	9	7
13	15	7	11	9	9	7	12	9	11
8	13	7	5	15	7	8	9	8	6
4	14	12	6	6	9	10	12	15	7
12	3	8	7	6	8	16	7	8	9
4	7	3	6	14	7	13	11	12	10
8	14	11	8	11	5	15	10	9	10
6	5	10	5	6	8	5	12	15	9
13	8	9	11	13	15	19	6	7	10

a. Plot the histogram.
b. Find the probability density function of the number of automobiles.
c. Determine the sample mean and sample standard deviation.
d. Compute the percentage of number of automobiles exceeding a value of 15.

1.11 The failure data of disk drives observed in a laboratory test during the development stage are given below:

Hours of Operation (Unit: 100 hours)	Number of Failures (Initial Number of Units: 1000)
0–10.0	0
10.1–20.0	8
20.1–30.0	15
30.1–40.0	19
40.1–50.0	21
50.1–60.0	17
60.1–70.0	23
70.1–80.0	27
80.1–90.0	25
90.1–100.0	24

a. Plot the histogram.
b. Plot the relative frequency diagram.
c. Find the sample mean and sample standard deviation of lives of disk drives.
d. Find the expected reliability of disk drives at 8,000 hours of operation.

1.12 The failure data of 25 turbine blades, in terms of number of cycles to failure, are given below:

250	319	472	533	675
680	868	994	1165	1389
1760	2176	2860	3127	3990
4350	5280	6329	7756	8531
8846	9210	9564	10,320	11,685

a. Plot the histogram.
b. Plot the probability density function of the number of cycles to failure.
c. Find the sample mean and sample standard deviation.
d. Determine the percentage of turbine blades that have a life greater than 7500 cycles.

Note: *Use Internet for answering Problems 1.13–1.20.*

1.13 List five famous bridge failures and describe the reasons for each failure.

1.14 Indicate five airplane accidents along with the possible reasons.

1.15 Describe the top five worst engineering disasters in the last 100 years.

1.16 Describe the following disasters and indicate the possible reasons:

a. Sinking of the *Titanic*
b. *Estonia* ferry disaster
c. Mohave Generating Station accident in Nevada in 1971
d. Denver Airport baggage handling system failure in 1995

1.17 Give the details of the space shuttle *Columbia* disaster.

1.18 Describe the leaning tower of Pisa along with the attempts made to restore it.

1.19 Describe two most prominent power outages in the last 100 years in North America.

1.20 Describe the failure of a sports arena.

Solutions using MATLAB and Excel

1.21 Plot the histogram and find the mean and standard deviation of the data given in Problem 1.5 using MATLAB.

1.22 Plot the histogram and find the mean and standard deviation of the data given in Problem 1.10 using MATLAB.

1.23 Plot the histogram and find the mean and standard deviation of the data given in Problem 1.5 using Excel.

1.24 Plot the histogram and find the mean and standard deviation of the data given in Problem 1.10 using Excel.

CHAPTER 2

Basic Probability Theory

What You Will Learn

In this chapter, you will learn

- The fact that most phenomena in real life have a certain element of uncertainty
- The basic concepts of experiment, event, outcome, and mutually exclusive events
- The set theoretic concepts such union and intersection of sets and complement of a set, as well as sample points and sample space of an experiment
- The definitions of probability based on relative frequency of occurrence of an event and the axiomatic approach
- The laws of probability, the concept of statistically independent events, the total probability theorem, and Bayes' rule

Abraham De Moivre was born on May 26, 1667, in France and died on November 27, 1754, in London. He was the son of a surgeon and the family left France in 1685. He came across Newton's *Principia Mathematica* (1687) while he was tutoring for a living and became interested in mathematics. He used to tear a few pages at a time from *Principia*, stuff them in his pocket, and study them between his tutoring hours. Later, he became an intimate friend of Newton and it was said that when Newton was asked questions about *Principia*, he would direct them to De Moivre for the answers. De Moivre published his first book on probability, entitled *Doctrine of Chances*, in 1718 and dedicated it to Newton. On November 12, 1733, De Moivre presented a paper on binomial theorem that led him to the idea of what is now called the "normal probability curve." As a result, this date is usually taken as the date of origin of normal distribution theory. De Moivre was chosen to decide the controversy over the invention of differential calculus between Newton and Leibniz. His contributions to trigonometry reflect in two well-known theorems, which bear his name, on the expansion of trigonometric functions. According to an interesting story, which may not be true, shortly before his death De Moivre wanted to sleep 15 minutes more every day. He was getting 6 hours of sleep every day when he announced this and naturally he would reach the 24-hour limit in 72 days. Surprisingly he died on the 73rd day, in his sleep [2.6–2.9].

2.1 Introduction

Every phenomenon in real life has a certain element of uncertainty. For example, the strength of a beam, the wind velocity at a particular locality, the number of vehicles crossing a bridge, and the life of a machine cannot be predicted exactly. These phenomena are chance dependent and one has to resort to *probability theory* to describe the characteristics of such phenomena [2.1–2.3]. This chapter provides the basic concepts of the probability theory that are necessary for the continuity of presentation in this book.

Before introducing the concept of probability, it is necessary to define the terms *experiment* and *event*. An experiment denotes the act of performing something the outcome of which is subject to uncertainty and not known exactly. For example, tossing a coin, rolling a die, and measuring the yield strength of steel can be called experiments. The number of possible outcomes in an experiment may be finite or infinite depending on the nature of the experiment. The outcome is a head or a tail in the case of tossing a coin, and any one of the numbers 1, 2, 3, 4, 5, and 6 in the case of rolling a die. On the other hand, the outcome may be any real number, say, between 28,000 and 32,000 lb/in^2 in the case of measuring the yield strength of steel.

An event represents the outcome of a single experiment. For example, realizing a head on tossing a coin, getting an even number (2 or 4 or 6) on rolling a die, and observing the yield strength of steel to be greater than 30,000 lb/in^2 in a measurement can be called events. Several possible outcomes (events) may be realized for a given experiment. For example, in the case of rolling a die, some of the events that may be realized are as follows:

1. getting number 1,
2. getting number 4,
3. getting an odd number (1 or 3 or 5),
4. getting a number greater than 3,
5. getting numbers 2 and 3 simultaneously (impossible event), and
6. getting any of the numbers 1, 2, 3, 4, 5, and 6 (certain event).

2.2 Mutually Exclusive Events

If the occurrence of an event precludes the occurrence of other events in a given experiment, the events are called *mutually exclusive*. This means that two events are mutually exclusive if both events could not occur simultaneously. The event of tail and the event of head occurring simultaneously in a toss of a coin is an example of mutually exclusive events.

2.3 Set Theory

The outcome of an experiment can be represented conveniently using the relationships of set theory. A set is a well-defined collection of objects, so that given any object, we can say whether it belongs to the set or not. Each object of the set is called an element of the set. The set of all elements under consideration is called a universal set and the set containing no element is called an empty or null set.

Union of two sets. The union of two sets A and B is defined as the set of all elements that belong to A or B or both, and is denoted by the symbol $A \cup B$, which reads as "A-union-B." For example, if A denotes the set of numbers (1, 2, 3) and B denotes the numbers (1, 3, 4), the union of A and B represents the set of numbers (1, 2, 3, 4).

Intersection of two sets. The intersection of two sets A and B is defined as the set of elements that belong to both A and B, and is denoted by the symbol $A \cap B$, which reads as "A-intersection-B." For example, if A denotes the set of numbers (1, 2, 3) and B denotes the numbers (1, 3, 4), the intersection of A and B represents the set of numbers (1, 3).

Complement of a set. The complement of a set A, denoted as $\overline{A}$, is the set of all those elements of the universal set which do not belong to A. For example, if the universal set contains the hundred numbers $1, 2, \ldots, 100$, and the set A consists of the numbers $1, 2, \ldots, 10$, the complement of A consists of the remaining ninety numbers so that $\overline{A} = (11, 12, \ldots, 100)$.

A set A is said to be a subset of a set B if whenever an element x belongs to A, it also belongs to B, and we denote it as $A \subset B$ which reads as "A belongs to B or A is contained in B."

Commutative rule. The union and intersection of sets are commutative, that is, if A and B are two sets, then $A \cup B = B \cup A$ and $A \cap B = B \cap A$

Associative rule. The union and intersection of sets are associative, that is, if A, B, and C are three sets, then $(A \cup B) \cup C = A \cup (B \cup C)$ and $(A \cap B) \cap C = A \cap (B \cap C)$

Distributive rule. The union and intersection of sets are distributive, that is, if A, B, and C are three sets, then $(A \cup B) \cap C = (A \cap C) \cup (B \cap C)$ and $(A \cap B) \cup C = (A \cup C) \cap (B \cup C)$

de Morgan's rule. This rule states that the complement of unions (intersections) is equal to the intersections (unions) of the respective complements. Thus, if A and B are two events and $\overline{A}$ and $\overline{B}$ are their complements, then de Morgan's rule states that $\overline{A \cup B} = \overline{A} \cap \overline{B}$ and $\overline{A \cap B} = \overline{A} \cup \overline{B}$

2.4 Sample Points and Sample Space

The outcome of an experiment or trial can be represented conveniently with the help of set theory. A sample space defines the set which includes all the possible outcomes of an experiment. For example, in the die-rolling experiment, the sample space consists of the numbers $\{1, 2, 3, 4, 5, 6\}$. Similarly, if S represents the yield strength of steel, then its sample space is given by $\{0 < s < \infty\}$. Thus, the sample space may be finite or infinite depending on whether the number of outcomes is finite or infinite, respectively. Each element of the sample space is called a sample point. Thus, a sample point represents a possible outcome of a trial. An event can be defined as a subset of the sample space. It is often convenient to associate a geometric figure called a Venn diagram with the ideas of sample space (set) and event (subset).

EXAMPLE 2.1

A thermal power plant buys four boilers. If there is a 50 percent chance that each boiler works for a year without any breakdown, identify the possible events at the end of the first year.

Solution: Since each boiler can be either in good or operating condition (G) or in failed condition (B) after one year, the possible states at the end of one year can be represented as follows:

Boiler Condition after One Year of Operation

Condition of First Boiler	Condition of Second Boiler	Condition of Third Boiler	Condition of Fourth Boiler	Total Number of Boilers in Good Condition
G	G	G	G	4
G	G	G	B	3
G	G	B	G	3
G	B	G	G	3
B	G	G	G	3
G	G	B	B	2
G	B	B	G	2
B	B	G	G	2
B	G	G	B	2
G	B	G	B	2
B	G	B	G	2
G	B	B	B	1
B	G	B	B	1
B	B	G	B	1
B	B	B	G	1
B	B	B	B	0

Each of these outcomes, namely, GGGG, GGGB, . . . , BBBB can be called a sample point. Thus, the sample space of the experiment (of observing the condition of four boilers at the end of one year) consists of sixteen sample points, and it can be represented in a Venn diagram as shown in Fig. 2.1. Some of the possible events and sample points corresponding to them can be identified as follows:

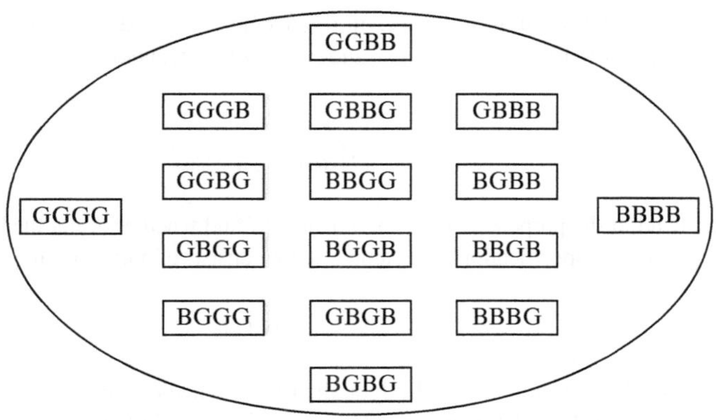

FIGURE 2.1 Sample space corresponding to the state of boilers at the end of the year.

Event	Physical Meaning of the Event	Sample Points Corresponding to the Event				
E1	All four boilers in operating condition	GGGG	____	____	____	____
E2	Two boilers in operating condition	GGBB	GBBG	BBGG	BGGB	GBGB
		BGBG				
E3	One boiler in operating condition	GBBB	BGBB	BBGB	BBBG	____
E4	None in operating condition	BBBB	____	____	____	____
E5	At least three in operating condition	GGGG	GGGB	GGBG	GBGG	BGGG
E6	At most two in operating condition	GGBB	GBBG	BBGG	BGGB	GBGB
		BGBG	GBBB	BGBB	BBGB	BBBG
		BBBB	____	____	____	____

■

EXAMPLE 2.2

The yield strength of steel in tension is being measured in a laboratory experiment. Identify the possible sample spaces corresponding to the measured strength of the steel.

Solution: Different sample spaces can be identified based on the resolution of the measuring instrument used, the purpose of the experiment, and the range of the yield strength of steel. If the measuring instrument is assumed to be very accurate and capable of finding any value of the yield strength (x), the sample space, S, consists of all positive values so that

$$S = \{x|_{x>0}\}$$

because negative values are not possible for the yield strength of steel. If the measuring instrument can only give the yield strength values in multiples of thousands of pounds per square inch, the sample space is given by

$$S = \{x|_{x=0,\ 1000,\ 2000,\ 3000,\ 4000,\ \ldots}\}$$

If the lower and upper bounds specified on the acceptable yield strength of steel in the production process are known to be 26,000 psi and 30,000 psi, respectively, the sample space can be stated as

$$S = \{x|_{26000 \le x \le 30000}\}$$

On the other hand, if the purpose of the experiment is to find whether or not the batch of steel produced satisfies the production specifications, the sample space consists of only two sample points (or outcomes):

$$S = \{yes, no\}$$

If the purpose of the experiment is to find, qualitatively, whether the yield strength can be classified as low (values smaller than 26,000 psi), medium (values that lie in the range 27,000 psi to 29,000 psi), or high (values larger than 29,000 psi), the sample space will be composed of three sample points (outcomes) as

$$S = \{low, medium, high\}$$

■

2.5 Definition of Probability

Probability can be defined in several ways. We shall consider two definitions, namely, the relative frequency and the axiomatic definitions, in this section.

2.5.1 Relative Frequency (Statistical) Definition

The probability of occurrence of an event E is defined as the ratio of the number of occurrences of the event E to the total number of trials. Thus $P(E)$ is defined as

$$P(E) = \lim_{N \to \infty} \left(\frac{n}{N} \right) \tag{2.1}$$

where n denotes the number of trials in which the event E has occurred and N the total number of trials. Since the minimum and maximum possible values of n are 0 and N, respectively, we have

$$0 \le P(E) \le 1 \tag{2.2}$$

where $P(E) = 0$ indicates that it is impossible to realize the event E while $P(E) = 1$ represents that the occurrence of the event E is certain. For example, the probability associated with the event of realizing both the head and the tail on tossing a coin is zero (impossible event), while the probability of the event that a rolled die will show up any number between 1 and 6 is one (certain event).

2.5.2 Axiomatic Definition

According to axiomatic approach, the probability of occurrence of an event E is defined as a number $P(E)$ such that $P(E)$ obeys the following three axioms or postulates

1. $P(E) \ge 0$
2. $P(E) = 1$ if E is a certain event
3. If E_1 and E_2 are mutually exclusive events, then

$$P(E_1 \cup E_2) = P(E_1) + P(E_2) \tag{2.3}$$

Since the foregoing three axioms are essentially assumptions, they need no proof. However, these axioms and the resulting theory must be useful and consistent with the practical problems.

Although any of the two definitions of probability can be used for the development of probability theory, the axiomatic approach has been used widely in the literature.

2.6 Laws of Probability

2.6.1 Union and Intersection of Two Events

If E_1 and E_2 denote two events, the union of E_1 and E_2 (denoted as $E_1 \cup E_2$ or less commonly as $E_1 + E_2$) means the occurrence of E_1 or E_2 or both. The intersection of E_1 and E_2 (denoted as E_1E_2 or $E_1 \cap E_2$) means the joint occurrence of E_1 and E_2. The union and the intersection operations are most easily explained in terms of the Venn diagram as shown in Fig. 2.2. Here, the set E_1 is composed of the disjoint subsets A_1 and A_2 and set E_2 is composed of the disjoint subsets A_2 and A_3. The subset A_2 denotes the points common to E_1 and E_2, while the subset A_1 represents points in E_1 but not in E_2, and A_3 represents points that belong to E_2 but not E_1. The union of the sets E_1 and E_2 denotes the points belonging to the subsets A_1, A_2, and A_3, and the intersection of the sets E_1 and E_2 denotes the points belonging to the subset A_2. For example, if E_1 denotes the shortage of materials and E_2 denotes the shortage of skilled workers in an automobile plant, $E_1 \cup E_2$ is the shortage of materials or skilled workers, or both, and $E_1 \cap E_2$ is the shortage of materials and skilled workers. Mathematically, we can write that

$$P(E_1 \cup E_2) = P(E_1) + P(E_2) - P(E_1E_2) = \text{total crossed area in Fig. 2.2} \tag{2.4}$$

and

$$P(E_1 \cap E_2) = P(E_1E_2) = \text{double crossed area in Fig. 2.2} \tag{2.5}$$

2.6.2 Mutually Exclusive Events

When the two sets E_1 and E_2 have no common elements, the areas do not overlap in the Venn diagram as shown in Fig. 2.3, and they are called disjoint or mutually exclusive sets. In this case, the two sets contain no common elements, and hence their intersection is a *null set*.[1] In other words, the two events E_1 and E_2 cannot occur simultaneously if they are mutually exclusive. For example, if E_1 denotes rain in a locality and E_2 denotes drought in

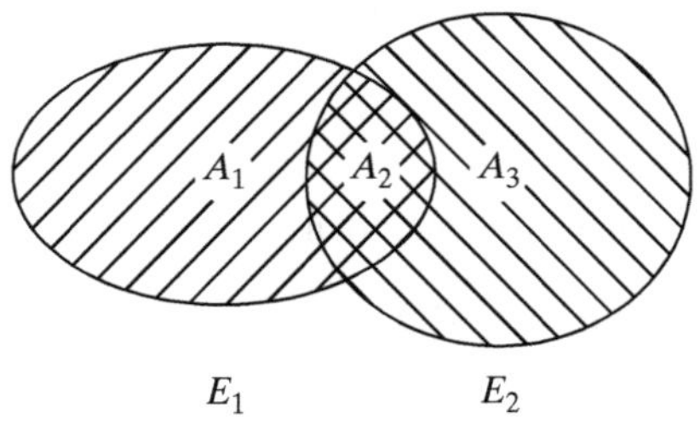

FIGURE 2.2 Union of the events E_1 and E_2.

[1]A set with no elements is called a *null set*.

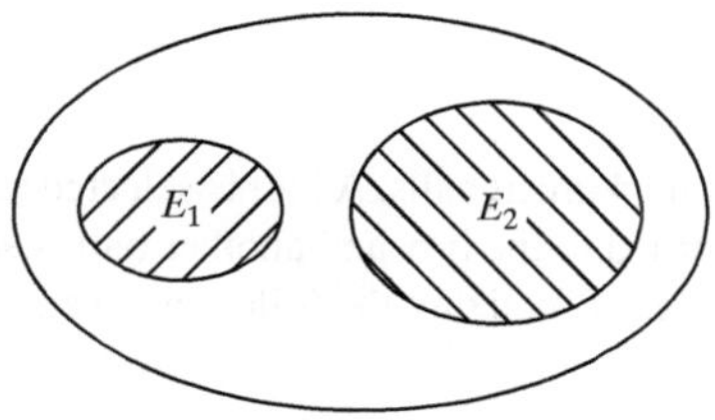

FIGURE 2.3 Venn diagram of two mutually exclusive events.

the same locality, then the Venn diagram has no overlapping region. In this case, E_1 and E_2 are disjoint or mutually exclusive. Hence the intersection $E_1 \cap E_2$ will be a null set, that is, the probability of simultaneous occurrence of E_1 and E_2 will be zero, that is

$$P(E_1 \cap E_2) = 0. \tag{2.6}$$

and, from Eq. (2.4), we get

$$P(E_1 \cup E_2) = P(E_1) + P(E_2) \tag{2.7}$$

EXAMPLE 2.3

When two dice are rolled, find the probability of realizing the sum of two numbers less than seven.

Solution: A sum of less than seven implies the sum of 6, 5, 4, 3, or 2. There are five ways of getting the sum of numbers as 6. These are given by

$$6: (1, 5), (2, 4), (3, 3), (4, 2), (5, 1)$$

Similarly, the sum of numbers equal to 5, 4, 3, and 2 can be obtained in different ways as indicated below:

$$5: (1, 4), (2, 3), (3, 2), (4, 1)$$
$$4: (1, 3), (2, 2), (3, 1)$$
$$3: (1, 2), (2, 1)$$
$$2: (1, 1)$$

Thus, the total number of ways of realizing a sum less than seven is $5 + 4 + 3 + 2 + 1 = 15$. Noting that the sample space of the outcome when two dice are rolled is $6^2 = 36$, the probability of realizing the sum less than seven is given by:

$$P(\text{sum less than } 7) = \frac{15}{36} = \frac{5}{12}.$$

■

EXAMPLE 2.4

A box contains 3 white balls, 2 red balls, and 4 green balls. A ball is selected randomly from the box and its color is noted. The ball is then returned to the box and another ball is selected randomly and its color is noted. Determine the probabilities of realizing the following events:

1. Two white balls
2. First ball red and the second ball green

Solution: Since the first ball is returned before the second ball is selected, the events will be independent.

1. The number of white balls is 3 out of a total of 9 balls in the box. Thus, the probability of selecting two white balls (with replacement) is

$$\begin{aligned} P(\text{Ball 1 white and ball 2 white}) &= P(\text{White ball}) \cdot P(\text{White ball}) \\ &= \frac{3}{9} \cdot \frac{3}{9} = \frac{1}{9} \end{aligned}$$

2. The number of red balls is 2 out of 9 and the number of green balls is 4 out of 9. Thus, the probability of selecting the first ball red and the second ball green is

$$\begin{aligned} P(\text{Ball 1 red and ball 2 green}) &= P(\text{Red ball}) \cdot P(\text{Green ball}) \\ &= \frac{2}{9} \cdot \frac{4}{9} = \frac{8}{81}. \end{aligned}$$

■

EXAMPLE 2.5

Find (1) the probability of having one boiler left operative at the end of the first year and (2) the probability of having at least three boilers operative at the end of the first year in Example 2.1.

Solution: Among the 16 possible outcomes listed in Example 2.1, only one can be realized at the end of the first year. If one possibility (event) is realized, none of the other 15 possibilities (events) can be realized. Hence, the 16 events are mutually exclusive. Because $P(G) = P(B) = 1/2$, all the 16 events are equally likely to occur and hence the probability of realizing any one of the 16 possibilities is $1/16$.

1. Now, if we define the event E as "exactly one boiler operating at the end of the first year," it consists of the four sample points GBBB, BGBB, BBGB, and BBBG.

$$\therefore P(E_1) = \frac{4}{16} = \frac{1}{4}$$

Similarly, we can obtain that

$$P(E_2) = P(\text{Exactly two boilers operating}) = \frac{6}{16} = \frac{3}{8}$$

$$P(E_3) = P(\text{Exactly three boilers operating}) = \frac{4}{16} = \frac{1}{4}$$

$$P(E_4) = P(\text{Exactly four boilers operating}) = \frac{1}{16}$$

$$P(E_0) = P(\text{No boilers operating}) = \frac{1}{16}$$

2. If E denotes the event that "at least three boilers in operating condition," it implies that either all four are in good condition or any three are in good condition. Thus, the event E is equivalent to the union of E_3 and E_4.

$$\therefore P(E) = P(E_3 \cup E_4) = P(E_3) + P(E_4) - P(E_3 \cap E_4) = \frac{1}{4} + \frac{1}{16} = \frac{5}{16}$$

since E_3 and E_4 are mutually exclusive events.

■

2.6.3 Complementary Events

Associated with any event E there is another event known as the complementary event (designated as $\overline{E}$). If the sample space corresponding to the event E is represented as S in the Venn diagram of Fig. 2.4, the remainder of the space $\overline{S}$ denotes the complement or negation of E, $\overline{E}$. It indicates that E did not occur. Hence the whole space represents the sample spaces of the two events E and $\overline{E}$. If $P(E)$ and $P(\overline{E})$ are their respective probabilities, then

$$P(E) + P(\overline{E}) = 1 \tag{2.8}$$

For example, if E denotes the event "appearance of number 6 upon rolling of a die," then $\overline{E}$ represents the event "appearance of any other number." Hence, for a fair die,

$$P(E) = \frac{1}{6} \qquad \text{and} \qquad P(\overline{E}) = 1 - P(E) = \frac{5}{6}$$

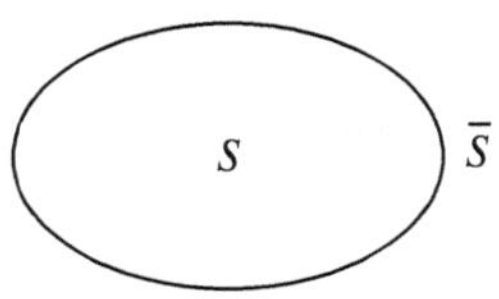

S = Sample space of E
$\overline{S}$ = Sample space of $\overline{E}$

FIGURE 2.4 Sample space of an event and its complement.

EXAMPLE 2.6

A small town has two banks A and B. It is estimated that 45 percent of the potential customers do business only with bank A, 30 percent only with bank B, and 15 percent with both banks A and B. The remaining 10 percent of the customers do business with none of the banks. If E_1 (E_2) denotes the event of a randomly selected customer doing business with bank A (B), find the following probabilities: (i) $P(\overline{E}_1)$, (ii) $P(E_2)$, (iii) $P(E_1 \cap E_2)$, (iv) $P(E_1 \cup E_2)$, (v) $P(\overline{E}_1\overline{E}_2)$, (vi) $P(\overline{E}_1 \cup E_2)$

Solution:

1. The customers doing business with bank A include those who do business exclusively with bank A and those who do business with both banks A and B. Thus, $P(E_1) = 0.45 + 0.15 = 0.6$ and $P(\overline{E}_1) = 1 - P(E_1) = 1 - 0.6 = 0.4$.
2. The customers doing business with bank B include those who do business exclusively with bank B and those who do business with both banks A and B. Hence $P(E_2) = 0.30 + 0.15 = 0.45$.
3. $P(E_1 \cap E_2) = P(E_1E_2) = 0.15$
4. $P(E_1 \cup E_2) = P(E_1) + P(E_2) - P(E_1E_2) = 0.60 + 0.45 - 0.15 = 0.90$
5. $P(\overline{E}_1\overline{E}_2) = P(\overline{E_1 \cup E_2}) = 1 - P(E_1 \cup E_2) = 1 - 0.9 = 0.1$
6. $P(\overline{E}_1 \cup E_2) = P(\overline{E}_1) + P(E_2) - P(\overline{E}_1E_2) = [1 - P(E_1)] + P(E_2) - P(\overline{E}_1 \cap E_2)$ where the event $\overline{E}_1 \cap E_2$ denotes the customers who do not do business with bank A, but do business with bank B (i.e., 30 percent).
   Hence, $P(\overline{E}_1 \cup E_2) = (1 - 0.6) + 0.45 - 0.30 = 0.55$.

■

EXAMPLE 2.7

The inspection of a batch of 730 castings produced in an engine production plant for the presence of defects yielded the following data:

Type of Defect	Location of Defect: Inside	Location of Defect: on Surface	Total
Blowholes	470	130	600
Cracks	55	75	130
Total	525	205	730

For a casting selected randomly from the batch, determine the following:

1. Probability that the casting has a blowhole.
2. Probability that the casting has a defect inside.
3. Probability that the casting has a blowhole inside.
4. Probability that the casting has a defect inside and a blowhole.
5. Probability that the casting has a defect inside or a blowhole or both.

Solution: Let the event B denote the presence of a blowhole anywhere in the casting, B_i the presence of a blowhole inside the casting, D_i the presence of a defect (blowhole or crack) inside the casting, and C the presence of a crack anywhere in the casting.

1. Since the total number of blowholes inside and on the surface of the casting is $470 + 130 = 600$, we have

$$P(B) = \frac{600}{730}$$

2. Because 470 castings have blowholes inside and 55 castings have cracks inside, we have

$$P(D_i) = \frac{470 + 55}{730} = \frac{525}{730}$$

3. As there are 470 castings with blowholes inside,

$$P(B_i) = \frac{470}{730}$$

4. Because there are a total of 600 castings with blowholes (with 470 inside blowholes plus 130 with on-surface blowholes), and there are a total of 525 castings with inside defects (470 with blowholes plus 55 with cracks inside), as shown in Venn diagram of Fig. 2.5, we can find

$$P(D_i \cap B) = \frac{470}{730}$$

5. The probability that a casting has a defect inside or a blowhole or both is given by

$$\begin{aligned} P(D_i \cup B) &= P(D_i) + P(B) - P(D_i \cap B) \\ &= \frac{525}{730} + \frac{600}{730} - \frac{470}{730} = \frac{655}{730} \end{aligned}$$

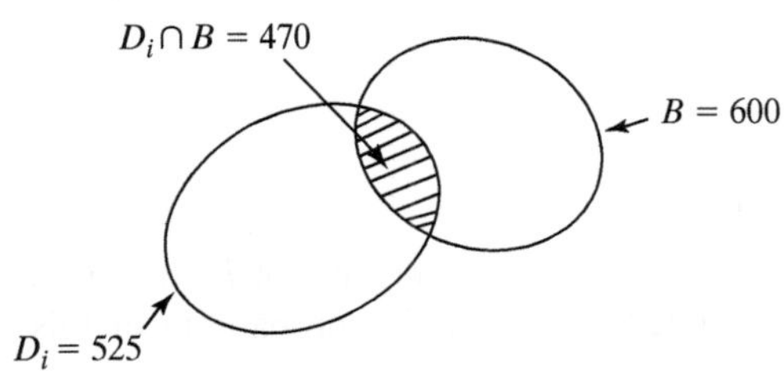

FIGURE 2.5 Venn diagram representation of defective castings.

■

EXAMPLE 2.8

The inspection of a batch of laminated composite beams produced in a construction company for defects yielded the following data:

Number of Defects	Proportion of Beams with Defects Inside	Proportion of Beams with Defects on Surface	Total
0	0.40	0.15	0.55
1	0.10	0.05	0.15
2	0.07	0.03	0.10
3	0.06	0.02	0.08
4	0.02	0.03	0.05
5 or more	0.03	0.04	0.07
Total	0.68	0.32	1.00

Determine the probability that the beam has a defect on the surface or it has 4 or more defects.

Solution: Let S_1 denote an event that the beam has a defect on the surface and D_1 the event that the beam has 4 or more defects. From the data given,

$$\begin{aligned} P(S_1) &= 0.32 \\ P(D_1) &= P(4 \text{ defects}) + P(5 \text{ or more defects}) = 0.05 + 0.07 = 0.12 \\ P(S_1 \cap D_1) &= P(\text{beam has a defect on the surface and the number of defects is 4 or more}) \\ &= 0.03 + 0.04 = 0.07 \end{aligned}$$

Thus, the probability that the beam has a defect on the surface or it has 4 or more defects can be found as

$$P(S_1 \cup D_1) = P(S_1) + P(D_1) - P(S_1 \cap D_1) = 0.32 + 0.12 - 0.07 = 0.37$$

■

2.6.4 Conditional Probability

The occurrence of the event E_2 when it is known that E_1 has occurred is known as the conditional event and is denoted as $E_2|E_1$. The probability of this conditional event is defined as

$$P\left(E_2 \middle| E_1\right) = \frac{P(E_1 E_2)}{P(E_1)}; \qquad P(E_1) > 0 \tag{2.9}$$

For example, consider an applied mathematics class consisting of 35 male and 15 female students. Among the male students, 15 are science majors and 20 are engineering majors. Among the female students, 5 major in science and 10 major in engineering. Let us consider the following events: E_1 = student is a female and E_2 = student is a science major so that $P(E_1) = 15/(15 + 35) = 3/10$ and $P(E_2) = (5 + 15)/(15 + 35) = 2/5$. The situation is shown in Fig. 2.6(*a*). If a randomly selected student is known to be a female (i.e., given the event E_1), the probability of that student majoring in science is given by $5/15 = 1/3$. This is shown graphically in Fig. 2.6(*b*). As soon as the student is known to be a female, the area corresponding to male students will be zero and the proportion of science majors among female students will be $5/15 = 1/3$. The probability of finding a randomly selected student to be a female majoring in science can be obtained as $P(E_1E_2) = P(E_2|E_1)P(E_1) = (1/3)(3/10) = 1/10$. This is a general rule and the probability of occurrence of both E_1 and E_2 is given by the product of $P(E_1)$ and $P(E_2)$:

$$\therefore P(E_1E_2) = P(E_1) \cdot P\left(E_2 \middle| E_1\right) = P(E_2) \cdot P\left(E_1 \middle| E_2\right) \tag{2.10}$$

From Eq. (2.10), we find that if $P(E_1) = 0$ or $P(E_2) = 0$, then $P(E_1E_2) = 0$. Further, if E_1 and E_2 are mutually exclusive, $P(E_1E_2) = 0$ and hence $P(E_1|E_2) = 0$ from Eq. (2.9).

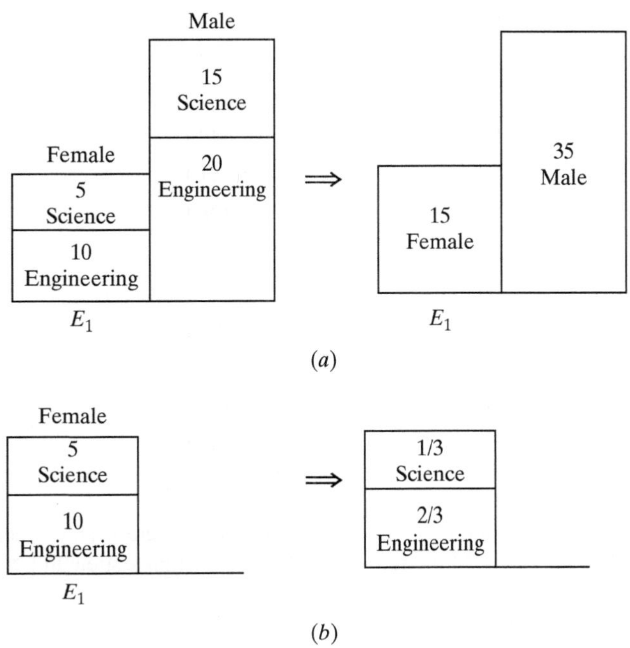

FIGURE 2.6 Representation of different groups of students.

EXAMPLE 2.9

Experiments conducted to find the yield and ultimate strengths of bolted timber joints yielded the following results:

Specimen	Yield Strength (lb)	Ultimate Strength (lb)
1	3410	4120
2	3520	4240
3	2800	4010
4	2950	4050
5	3100	4230
6	3080	3970
7	3250	4290
8	2900	3980
9	2880	3770
10	3140	4060
11	2580	3730
12	2990	4020
13	3210	4310
14	3160	4190
15	2770	3790
16	2890	4040
17	3380	4420
18	3470	4410
19	3090	4080
20	2630	3680
21	2860	3990
22	2740	3810
23	3130	4270
24	2880	3780
25	3210	4310

Find the conditional probabilities $P(U_a | Y_a)$ and $P(Y_a | U_a)$ where Y_a and U_a denote the events of realizing acceptable values of yield strength $(\geq 3000\,\text{lb})$ and ultimate strength $(\geq 4000\,\text{lb})$, respectively.

Solution: From the data, we find that

$P(Y_a U_a)$ = Probability of realizing yield strength greater than or equal to 3000 pounds and ultimate strength greater than or equal to 4000 pounds $= 12/25$.

Similarly, we find that

$$P(Y_a\overline{U}_a) = \frac{1}{25}$$

$$P(\overline{Y}_a U_a) = \frac{4}{25}$$

$$P(\overline{Y}_a\overline{U}_a) = \frac{8}{25}$$

$P(Y_a)$ = Probability of realizing the yield strength greater than or equal to 3000 lb $= 13/25$.

$P(U_a)$ = Probability of realizing the ultimate strength greater than or equal to 4000 lb $= 16/25$.

Thus, we can compute the required probabilities as

$$P\left(U_a \middle| Y_a\right) = \frac{P(U_a Y_a)}{P(Y_a)} = \frac{\left(\frac{12}{25}\right)}{\left(\frac{13}{25}\right)} = \frac{12}{13}$$

and

$$P\left(Y_a \middle| U_a\right) = \frac{P(Y_a U_a)}{P(U_a)} = \frac{\left(\frac{12}{25}\right)}{\left(\frac{16}{25}\right)} = \frac{3}{4}$$

■

2.6.5 Statistically Independent Events

If the occurrence of the event E_1 in no way affects the probability of occurrence of the event E_2, then the events E_1 and E_2 are said to be statistically independent. In this case, we have,

$$P\left(E_1 \middle| E_2\right) = \frac{P(E_1 E_2)}{P(E_2)} = P(E_1) \tag{2.11}$$

and

$$P(E_1 E_2) = P(E_1) \cdot P(E_2) \tag{2.12}$$

For example, if $P(E_1) = P(\text{shortage of steel in a project}) = 0.4$, and $P(E_2) = (\text{realizing the head on tossing a coin}) = 0.7$, then obviously E_1 and E_2 are statistically independent and $P(E_1 E_2) = P(E_1) \cdot P(E_2) = 0.28$.

2.6.6 General Laws

The union, intersection, and conditional events defined in terms of two events can be generalized for more than two events. For example, the union of three events E_1, E_2, and E_3 can be obtained as follows:

$$\begin{aligned} P(E_1 \cup E_2 \cup E_3) &= P(E_1 \cup W), \qquad \text{where} \qquad W = E_2 \cup E_3 \\ &= P(E_1) + P(W) - P(E_1 W) \\ &= P(E_1) + P(E_2) + P(E_3) - P(E_2E_3) - P(E_1)\cdot P(W) \\ &= P(E_1) + P(E_2) + P(E_3) - P(E_1E_2) - P(E_2E_3) \\ &\qquad\qquad - P(E_3E_1) + P(E_1E_2E_3) \end{aligned} \tag{2.13}$$

In terms of the Venn diagram, the union of E_1, E_2, and E_3 is given by the area bounded by the thick line and their intersection is given by the shaded region shown in Fig. 2.7. If the sample space of E_1 is denoted by $A_1 + A_4 + A_6 + A_7$, that of E_2 by $A_2 + A_4 + A_5 + A_7$, and that of E_3 by $A_3 + A_5 + A_6 + A_7$, the area bounded by the thick line can be obtained as follows:

Area bounded by the thick line

$$\begin{aligned} &= A_1 + A_2 + A_3 + A_4 + A_5 + A_6 + A_7 \\ &= (A_1 + A_4 + A_6 + A_7) + (A_2 + A_4 + A_5 + A_7) \\ &\quad + (A_3 + A_5 + A_6 + A_7) - (A_4 + A_5 + A_6 + 2A_7) \\ &= P(E_1) + P(E_2) + P(E_3) \\ &\quad - [\underbrace{(A_4 + A_7)}_{P(E_1E_2)} + \underbrace{(A_5 + A_7)}_{P(E_2E_3)} + \underbrace{(A_6 + A_7)}_{P(E_1E_3)} - \underbrace{A_7}_{P(E_1E_2E_3)}] \end{aligned} \tag{2.14}$$

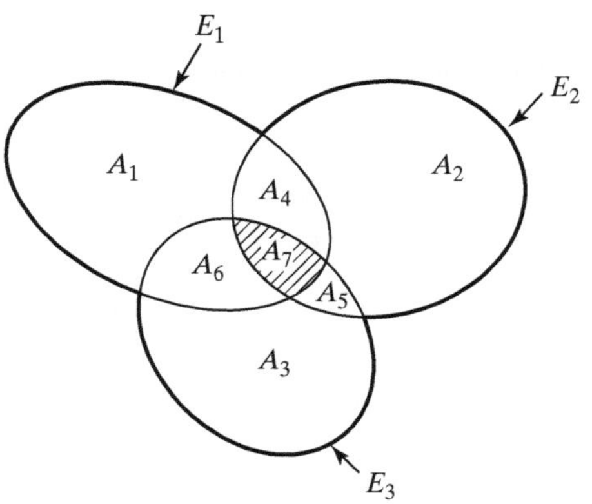

FIGURE 2.7 Union of the three events E_1, E_2, and E_3 in a Venn diagram.

Thus, Eq. (2.13) can also be derived geometrically. The probability of a union of n events $E_1, E_2, \ldots, E_n$ can be derived in a similar manner to be[2]

$$\begin{aligned}
P(E_1 \cup E_2 \cup E_3 \cup \cdots \cup E_n) = {} & \left[P(E_1) + P(E_2) + P(E_3) + \cdots + P(E_n)\right] \\
& \qquad \binom{n}{1} = n \text{ terms} \\
& - \left[P(E_1E_2) + P(E_1E_3) + \cdots + P(E_iE_j)\right] \\
& \qquad \binom{n}{2} \text{terms, } i \neq j \\
& + \left[P(E_1E_2E_3) + P(E_1E_2E_4) + \cdots + P(E_iE_jE_k)\right] \\
& \qquad \binom{n}{3} \text{terms, } i \neq j \neq k \\
& + \cdots + (-1)^{n-1}\left[P(E_1E_2E_3\cdots E_n)\right] \\
& \qquad \binom{n}{n} = 1 \text{ term}
\end{aligned} \tag{2.15}$$

Thus, there are

$$\binom{n}{1} + \binom{n}{2} + \binom{n}{3} + \cdots + \binom{n}{n-1} + \binom{n}{n} = 2^n - 1 \tag{2.16}$$

terms in the expansion of $P(E_1 \cup E_2 \cup \cdots \cup E_n)$.

In the case of mutually exclusive or disjoint events, all possible intersections are zero, and Eq. (2.15) simplifies to

$$P(E_1 \cup E_2 \cup E_3 \cup \cdots \cup E_n) = P(E_1) + P(E_2) + P(E_3) + \cdots + P(E_n) \tag{2.17}$$

We can define the intersection of three events E_1, E_2, and E_3 as

$$P(E_1E_2E_3) = P(W) \cdot P\left(E_3 \middle| W\right) \tag{2.18}$$

where W represents the simultaneous occurrence of E_1 and E_2. Thus

$$P(E_1E_2E_3) = P(E_1) \cdot P\left(E_2 \middle| E_1\right) \cdot P\left(E_3 \middle| E_1E_2\right) \tag{2.19}$$

[2]The notation $P(E_iE_j)i \neq j$ is used to denote the probability of all combinations of events taken two at a time. For example, if there are four events E_1, E_2, E_3, and E_4, the second term of Eq. (2.15) is $-\left[P(E_1E_2) + P(E_1E_3) + P(E_1E_4) + P(E_2E_3) + P(E_2E_4) + P(E_3E_4)\right]$.

Successive application of this technique leads to the general result:

$$P(E_1E_2E_3\ldots E_n) = P(E_1)\cdot P\left(E_2\middle|E_1\right)\cdot P\left(E_3\middle|E_1E_2\right)\cdots P\left(E_n\middle|E_1E_2E_3\cdots E_{n-1}\right) \tag{2.20}$$

Thus, the probability of intersection of n events is expressed as the product of one independent probability and $(n-1)$ conditional probabilities.

2.7 Total Probability Theorem

If $B_1, B_2, \ldots, B_n$ are a set of mutually exclusive, collectively exhaustive events, the probability of another event A can always be expanded as

$$\begin{aligned} P(A) &= P(A\cap B_1) + P(A\cap B_2) + \cdots + P(A\cap B_n) \\ &= \sum_{i=1}^{n} P(A\cap B_i) \\ &= \sum_{i=1}^{n} P\left(A\middle|B_i\right)\cdot P(B_i) \end{aligned} \tag{2.21}$$

This is called the theorem of total probability and $P(A)$ is given by the crossed area in the Venn diagram shown in Fig. 2.8.

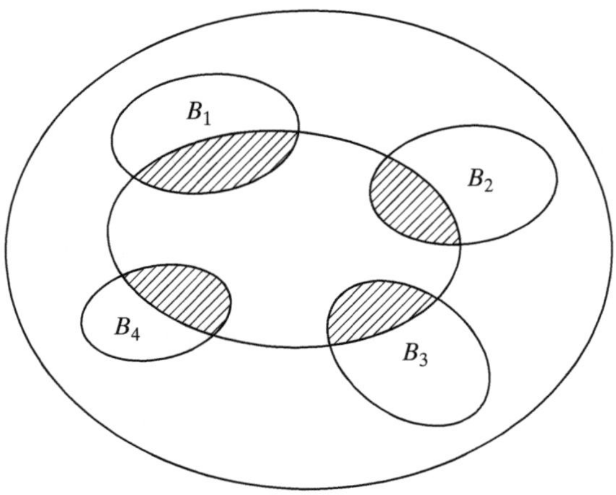

FIGURE 2.8 Venn diagram representation of the total probability theorem.

EXAMPLE 2.10

In Example 2.1, find the probability of realizing at least three boilers in operating condition after one year.

Solution: Let E_1 = GGGG, E_2 = GGGB, E_3 = GGBG, E_4 = GBGG, E_5 = BGGG, E_6 = GGBB, E_7 = GBBG, E_8 = BBGG, E_9 = BGBG, E_{10} = BGGB, E_{11} = GBGB, E_{12} = GBBB, E_{13} = BGBB, E_{14} = BBGB, E_{15} = BBBG, E_{16} = BBBB, and $P(E_i) = 1/16$ for $i = 1$ to 16.

Now, we define the event A as realizing at least three boilers in operating condition after one year. Applying the total probability theorem, we obtain

$$P(A) = \sum_{i=1}^{16} P\left(A \middle| E_i\right) P(E_i)$$

$$= \left(1 \cdot \frac{1}{16}\right) + \left(1 \cdot \frac{1}{16}\right) + \left(1 \cdot \frac{1}{16}\right) + \left(1 \cdot \frac{1}{16}\right) + \left(1 \cdot \frac{1}{16}\right)$$

$$+ \left(0 \cdot \frac{1}{16}\right) + \left(0 \cdot \frac{1}{16}\right) + \left(0 \cdot \frac{1}{16}\right) + \left(0 \cdot \frac{1}{16}\right) + \left(0 \cdot \frac{1}{16}\right)$$

$$+ \left(0 \cdot \frac{1}{16}\right) + \left(0 \cdot \frac{1}{16}\right) + \left(0 \cdot \frac{1}{16}\right) + \left(0 \cdot \frac{1}{16}\right) + \left(0 \cdot \frac{1}{16}\right)$$

$$+ \left(0 \cdot \frac{1}{16}\right) = \frac{5}{16}$$

■

EXAMPLE 2.11

A batch of 1000 piston rings manufactured in an engine manufacturing facility contains 40 defective ones (e.g., that do not meet the specified size tolerances). Two piston rings are randomly selected from the batch, one at a time, without replacement. If E_i denotes the event that the ith piston ring selected is defective $(i = 1, 2)$, determine the values of $P(E_1)$ and $P(E_2)$.

Solution: The probability of the event E_1 is given by

$$P(E_1) = \frac{40}{1000} = 0.040$$

To find the probability of the event E_2, first we note that the two events E_1 and E_2 do not appear to be independent. The reason is that if the first piston ring is found to be defective, then the probability of finding the second piston ring to be defective will be lower because

$$P\left(E_2 \middle| E_1\right) = \frac{39}{999} = 0.039$$

Since the first piston ring selected can be a good one, we have

$$P(\overline{E}_1) = \frac{960}{1000} = 0.960$$

Then the computation of the probability of the event E_2 requires the consideration of the occurrences of both E_1 and $\overline{E}_1$:

$$P(E_2) = P\left(E_2\Big|_{E_1}\right)P(E_1) + P\left(E_2\Big|_{\overline{E}_1}\right)P(\overline{E}_1) = \frac{39}{999}\frac{40}{1000} + \frac{40}{999}\frac{960}{1000} = \frac{40}{1000} = 0.040$$

Notes:

1. The result shows that the probability of finding the second piston ring to be defective is the same as the probability of finding the first piston ring to be defective.
2. Because $P(E_2|_{E_1}) \neq P(E_2)$, the two events E_1 and E_2 are not independent.

■

EXAMPLE 2.12

In a thermal power plant, four components—the feed water pump, boiler, turbine, and generator—are connected one after another (in series) as shown in Fig. 2.9. The probabilities that the different components function successfully are also indicated in Fig. 2.9. The thermal power plant generates electric power only when each of the four components is functional. Determine the probability that the four components are functional and the electric power is generated in the power plant. Assume that the failure of a component is independent of the failures of other components.

Solution: Let E_i denote the event that the ith component is functional $(i = 1, 2, 3, 4)$. Since the events E_1, E_2, E_3, and E_4 are independent, the probability that all components are functional and the electric power is generated by the power plant is given by

$$P(E_1 \cap E_2 \cap E_3 \cap E_4) = P(E_1)P(E_2)P(E_3)P(E_4) = (0.9)(0.95)(0.85)(0.99) = 0.7195$$

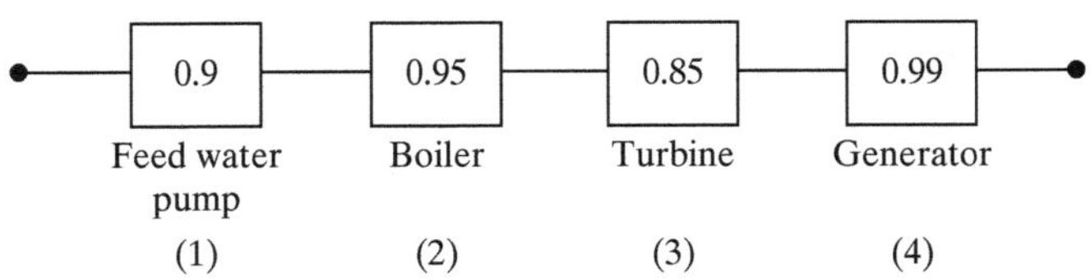

FIGURE 2.9 Series connection of components.

■

EXAMPLE 2.13

An airplane has two engines so that in case one of them fails to function, the airplane can safely land at the nearest airport with the other engine functioning. For the computation of the probabilities, the engines can be modeled to be in parallel as shown in Fig. 2.10. If the probability that the engine 1

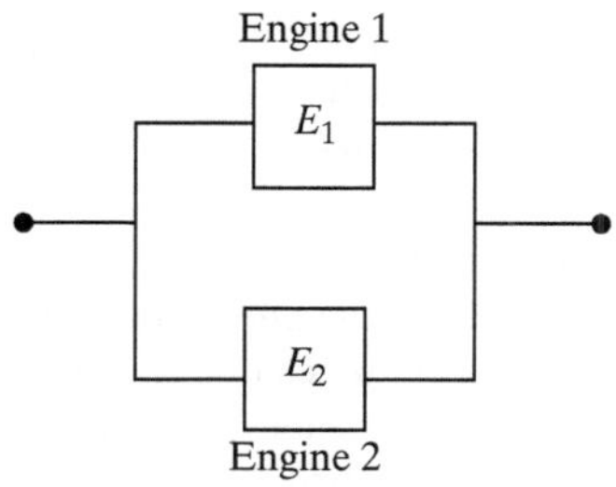

FIGURE 2.10 Parallel model of engines.

functions, $P(E_1)$, is 0.98 and the probability that the engine 2 functions, $P(E_2)$, is 0.99, determine the probability that the airplane functions. Assume that the engines failure are independent.

Solution: Let A denote the event that the airplane functions. Because the engine fails only if both engines fail, the probability that the airplane functions (event A) can be computed as

$$P(A) = 1 - P(\text{both engines fail}) = 1 - P(\overline{E}_1 \cap \overline{E}_2) \qquad (E_1)$$

Since the engine failures are independent, Eq. (E_1) can be expressed as

$$P(A) = 1 - P(\overline{E}_1)P(\overline{E}_2) = 1 - \{1 - P(E_1)\}\{1 - P(E_2)\} \qquad (E_2)$$

Using the known values of $P(E_1)$ and $P(E_2)$, Eq. (E_2) gives

$$P(A) = 1 - (1 - 0.98)(1 - 0.99) = 0.9998$$

■

EXAMPLE 2.14

In the manufacture of a fiber-reinforced laminated composite material (f.l.c.m.), the following probabilities can be associated with the failure of f.l.c.m. structural components that are subject to different levels of delamination:

Probability of Failure of the Structural Component	Level of Delamination in the f.l.c.m.
0.20	High
0.05	Medium
0.01	Low

In a batch of composite material manufactured, 10 percent of the material is found to have high levels of delamination, 30 percent to medium levels of delamination, and 60 percent to low levels of delamination. For a f.l.c.m. structural component using this batch of composite material, indicate the various events associated with the failure of the structural component as a tree diagram. Also, determine the probability that the structural component fails.

Solution: Let E_1, E_2, and E_3 denote the events that the composite material is subject to high, medium, and low levels of delamination, respectively. The probabilities associated with the three possible levels of delamination of the composite material can be shown as branches of a tree as in Fig. 2.11. For a f.l.c.m. structural component that is subject to a high level of delamination, the probability that it will fail (F) is 0.20; hence the probability that it will not fail (NF) is 0.80. These probabilities are shown as branches emanating from the event E_1 in Fig. 2.11. Similarly, the probabilities associated with failure (F) and no failure (NF) associated with the occurrence of events E_2 and E_3 are also shown in Fig. 2.11. Finally, the probability of failure of the f.l.c.m. structural component can be found as

$$P(F) = \sum_{i=1}^{3} P\left(F \Big|_{E_i}\right) P(E_i) = 0.1(0.2) + 0.3(0.05) + 0.6(0.01) = 0.041$$

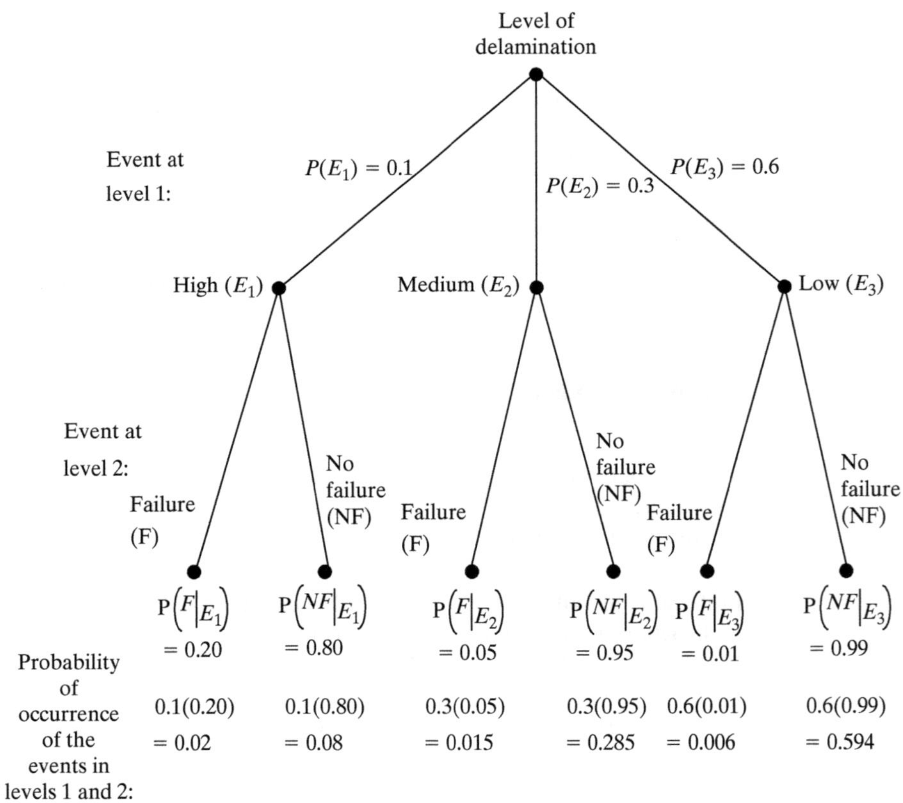

FIGURE 2.11 Levels of delamination of composite material.

■

2.8 Bayes' Rule

One of the most useful formulas in the theory of applied probability can be derived by using the concept of conditional probability as follows [2.4, 2.5]. If A and B are two events, since $P(A \cap B) = P(B \cap A)$ we obtain

$$P\left(A \middle| B\right) P(B) = P\left(B \middle| A\right) P(A) \tag{2.22}$$

that is,

$$P\left(B \middle| A\right) = \frac{P\left(A \middle| B\right) \cdot P(B)}{P(A)} \tag{2.23}$$

Equation (2.23) is called the Bayes' rule. This can be generalized by using the total probability theorem given by Eq. (2.21). Thus,

$$P\left(B_j \middle| A\right) = \frac{P\left(A \middle| B_j\right) \cdot P(B_j)}{P(A)} \tag{2.24}$$

and, this can be rewritten, using Eq. (2.21), as

$$P\left(B_j \middle| A\right) = \frac{P\left(A \middle| B_j\right) \cdot P(B_j)}{\sum_{i=1}^{n} P\left(A \middle| B_i\right) \cdot P(B_i)} \tag{2.25}$$

Here the quantity $P(B_j | A)$ is called the posterior probability and $P(A | B_i)$ is called prior probability. The utility of the generalized Bayes' rule, Eq. (2.25), can be explained through the following example.

Let A denote the event of "collapse of a building." The building might collapse due to an earthquake (event B_1), wind loading (event B_2), or any other cause (event B_3). Let the probabilities of occurrence of these events at a locality, in a particular year, be known as $P(B_1)$, $P(B_2)$, and $P(B_3)$. From the details of design, we will be knowing the probabilities $P(A | B_i)$, $i = 1, 2, 3$, that is, the probability of the collapse of the building given the occurrence of an earthquake, wind load, or any other cause. Now, let us assume that the event A has already occurred and we would like to know the probability of earthquake being the cause of collapse of the building. In such a case, we can find $P(B_1 | A)$ by applying the generalized Bayes' rule.

EXAMPLE 2.15

In an engine manufacturing plant, 70 percent of the crankshafts are ground by machinist R and 30 percent by machinist S. It is known from past experience that the crankshafts ground by machinists R and S contain 2 percent and 3 percent defective units, respectively. If a randomly selected crankshaft is found to be defective, find the probability that it was ground by the machinist R.

Solution: We define the following events: A = finding the crankshaft to be defective, B_1 = crankshaft is ground by the machinist R, and B_2 = crankshaft is ground by the machinist S. From the given data, we have $P(B_1) = 0.7$, $P(B_2) = 0.3$, $P(A|_{B_1}) = 0.02$, and $P(A|_{B_2}) = 0.03$. The application of the generalized Bayes' rule gives

$$P\left(B_1\middle|_A\right) = \frac{P\left(A\middle|_{B_1}\right)P(B_1)}{\sum_{i=1}^{2} P\left(A\middle|_{B_i}\right)P(B_i)} = \frac{(0.02)(0.7)}{(0.02)(0.7) + (0.03)(0.3)} = \frac{14}{23}$$

■

EXAMPLE 2.16

An automobile transmission can fail due to three types of problems—gear failure, bearing failure, or shaft failure—with probabilities 0.3, 0.5, and 0.2, respectively. The probabilities of a transmission failure given a gear failure is 0.5, given a bearing failure is 0.8, and given a shaft failure is 0.6. If a transmission fails, what is the most likely cause?

Solution: Denote the events of gear failure, bearing failure, and shaft failure as G, B, and S, respectively, and the failure of the automobile transmission as F. The most likely cause of the failure of the automobile transmission is the one that corresponds to the largest value among $P(G|_F)$, $P(B|_F)$, and $P(S|_F)$. The probability of failure of an automobile transmission, $P(F)$, can be found as

$$\begin{aligned} P(F) &= P(F|_G)P(G) + P(F|_B)P(B) + P(F|_S)P(S) \\ &= 0.5(0.3) + 0.8(0.5) + 0.6(0.2) = 0.67 \end{aligned}$$

Then the desired conditional probabilities are given by Eq. (2.23):

$$\begin{aligned} P(G|_F) &= \frac{P(F|_G)P(G)}{P(F)} = \frac{(0.5)(0.3)}{0.67} = 0.2239 \\ P(B|_F) &= \frac{P(F|_B)P(B)}{P(F)} = \frac{(0.8)(0.5)}{0.67} = 0.5970 \\ P(S|_F) &= \frac{P(F|_S)P(S)}{P(F)} = \frac{(0.6)(0.2)}{0.67} = 0.1791 \end{aligned}$$

It can be noted that $P(G|_F) + P(B|_F) + P(S|_F) = 0.2239 + 0.5970 + 0.1791 = 1.0$ because one of the three types of problems is responsible for the failure of the transmission. Because $P(B|_F)$ has the largest value, the most likely cause of failure of the automobile transmission is the bearing failure.

■

REFERENCES AND BIBLIOGRAPHY

2.1 T. T. Soong, *Probabilistic Modeling and Analysis in Science and Engineering*, John Wiley, New York, 1981.

2.2 R. L. Scheaffer and J. T. McClave, *Probability and Statistics for Engineers*, 2nd ed., Duxbury Press, Boston, MA, 1986.

2.3 A. H. S. Ang and W. H. Tang, *Probability Concepts in Engineering Planning and Design*: Vol. 1: *Basic Principles*, John Wiley, New York, 1975.

2.4 L. L. Lapin, *Probability and Statistics for Modern Engineering*, Brooks/Cole Publishing Co.–Engineering Div., Monterey, CA, 1983.

2.5 W. W. Hines and D. C. Montgomery, *Probability and Statistics in Engineering and Management Science*, 2nd ed., John Wiley, New York, 1980.

2.6 I. Todhunter, *A History of the Mathematical Theory of Probability*, G. E. Stechert & Co., New York, 1931.

2.7 I. Asimov, *Asimov's Biographical Encyclopedia of Science and Technology*, Doubleday & Co., Garden City, NY, 1972.

2.8 A. C. King, and C. B. Read, *Pathways to Probability*, Holt, Rinehart & Winston, New York, 1963.

2.9 C. C. Gillispie (ed.-in-chief), *Dictionary of Scientific Biography*, vol. 3, Charles Scribner's Sons, New York, 1980.

2.10 A. F. Karr, *Probability*, Springer-Verlag, New York, 1993.

2.11 S. Lipschutz, *Schaum's Outline of Theory and Problems of Probability* (Electronic), McGraw-Hill, New York, 2000.

2.12 F. Mosteller, *Probability: A First Course*, Addison-Wesley, Reading, MA, 1970.

2.13 J. J. Kinney, *Probability: An Introduction with Statistical Applications*, Wiley, New York, 1997.

2.14 T. C. Fry, *Probability and Its Engineering Uses*, 2nd ed., Van Nostrand, Princeton, NJ, 1965.

2.15 J. D. Enderle, D. C. Farden, and D. J. Krau, *Basic Probability Theory for Biomedical Engineers*, Morgan & Claypool Publishers, San Rafael, CA, 2006.

2.16 S. M. Ross, *Introduction to Probability Models*, Academic Press, San Diego, CA, 2003.

2.17 D. P. Bertsekas and J. N. Tsitsiklis, *Introduction to Probability*, Athena Scientific, Belmont, MA, 2002.

2.18 P. Z. Peebles, Jr., *Probability, Random Variables and Random Signal Principles*, McGraw-Hill, New York, 1987.

2.19 G. P. Beaumont, *Probability and Random Variables*, Ellis Horwood, Chichester, England, 1986.

2.20 A. Papoulis, *Probability and Random Variables and Stochastic Processes*, McGraw-Hill, New York, 1984.

2.21 A. M. Breipohl, *Probabilistic Systems Analysis*, Wiley, New York, 1970.

2.22 B. Fristedt and L. Gray, *A Modern Approach to Probability Theory*, Birkhauser, Boston, MA, 1997.

2.23 J. Hajek and V. Dupac, *Probability in Science and Engineering*, Academic Press, New York, 1967.

2.24 J. Galambos, *Introductory Probability Theory*, Marcel Dekker, New York, 1984.

2.25 P. G. Hoel, S. C. Port and C. J. Stone, *Introduction to Probability Theory*, Houghton Miffin Co., Boston, MA, 1971.

2.26 L. B. Page, *Probability for Engineering with Applications to Reliability*, Computer Science Press, New York, 1989.

2.27 H. Gordon, *Discrete Probability*, Springer-Verlag, New York, 1997.

2.28 W. Feller, *An Introduction to Probability Theory and Its Applications*, Vol. I, 3rd ed., Revised Printing, Wiley, New York, 1968.

2.29 E. Beltrami, *What is Random?*, Copernicus, Springer-Verlag, New York, 1999.

2.30 A. G. Bluman, *Probability Demystified*, McGraw-Hill, New York, 2005.

2.31 K. L. Chung and F. AitSahlia, *Elementary Probability Theory*, 4th ed., Springer-Verlag, New York, 2003.

2.32 K. Burdzy, *The Search for Certainty: On the Clash of Science and Philosophy of Probability*, World Scientific Publishing Co., Singapore, 2009.

REVIEW QUESTIONS

2.1 Give brief answers to the following:

1. Give five engineering examples involving uncertainty.
2. What is the difference between an experiment and an event?
3. What are mutually exclusive events? Give two examples of mutually exclusive events.
4. What are universal and null sets?
5. Define the union and intersection of two sets.
6. What is the complement of a set?
7. Explain the commutative, associative, and distributive rules of set theory?
8. What is de Morgan's rule?
9. What is a sample point?
10. Define sample space.
11. What is a Venn diagram?
12. What is the statistical definition of probability?
13. State the axiomatic definition of probability.
14. What is conditional probability?
15. What is meant by statistically independent events?
16. What is the total probability theorem?
17. What is Bayes' rule?
18. Give two engineering applications of Bayes' theorem.

2.2 Indicate whether each of the following statements is true or false:

1. Swimming in a river and seeing a crocodile represent a pair of independent events.
2. The number of sample points when three dice are rolled is 216.
3. The outcome of any experiment can be represented using set theory.

4. The probability of realizing the sum of the two numbers as 10 upon rolling two dice is $\frac{1}{6}$.
5. The probability of realizing doubles upon rolling two dice is $\frac{1}{6}$.

2.3 Fill in the blank with the proper word(s):

1. Tossing a coin and then tossing another coin represent _____ events.
2. When two coins are tossed, the sample space consists of the events _____.
3. The set of all possible outcomes of an experiment is called the _____ space.
4. When an event cannot occur, its probability is _____.
5. It is known that 15 percent of pressure vessels have defect of type 1 and 60 percent have defect of type 2. The probability that a randomly chosen pressure vessel has defect of type 1 given that it has defect of type 2 is _____.
6. If a particular brand of automobile comes in eight different colors for the body, four different colors for the interior, and two different colors for the sidewall tires, the total number of color combinations available to a customer is _____.
7. If an automobile license plate consists of three letters and two digits and if repetitions of the letters and digits are permitted, the total number of different license plates that can be made are _____.
8. If an automobile license plate consists of three letters and two digits and if repetitions of the letters and digits are not permitted, the total number of different license plates that can be made are _____.
9. If the occurrence of an event implies the nonoccurrence of another event, the events are called mutually _____.
10. The union of sets A and B indicates elements that belong to A or B or _____.
11. The intersection of sets A and B indicates elements that belong to _____ A and B.
12. The compliment of a set A is the universal set _____ the events of set A.
13. A box contains a black ball, a red ball, and a yellow ball. The probability that one ball is red when two balls are selected randomly from the box without replacement is _____.
14. The probability of realizing a queen when a single card is drawn at random from a standard deck of cards is _____.
15. Two dice are rolled. The probability of realizing the sum of the two numbers as 4 if it is known that the sum of the two numbers is less than 6 is _____.
16. When a die is rolled, the probability of realizing the number 6 if it is given that the number realized is an odd number is _____.
17. Two dice are rolled. The probability of realizing the sum of the two numbers as 8 if it is known that each of the numbers is an even number is _____.
18. A thermal power plant has five boilers, each with a probability of failure of 0.1. Assuming the failures of the boilers to be independent, the probability of exactly three boilers failing is given by _____.

2.4 Select the most appropriate answer from the multiple choices given:

1. The following pair of events are dependent:
   a. Driving a car and involving in an accident
   b. Driving a car and experiencing rain on the way
   c. Driving a car and noticing an increase in the price of gold

2. The following is a discrete random variable:
a. Distance travelled in an automobile
b. Quantity of fuel used during a trip by an automobile
c. Number of cars crossing a toll booth in an hour

3. When a die is rolled, the probability of realizing the number greater than or equal to 3 is:

a. $\frac{1}{2}$ **b.** $\frac{2}{3}$ **c.** $\frac{1}{3}$

4. The probability of getting at least one head on tossing two coins is:

a. $\frac{3}{4}$ **b.** $\frac{1}{4}$ **c.** $\frac{1}{2}$

5. A box contains a black ball, a red ball, a green ball, and a blue ball. When a ball is selected from the box randomly, the probability that the color of the ball is either black or blue is:

a. $\frac{3}{4}$ **b.** $\frac{1}{4}$ **c.** $\frac{1}{2}$

6. A letter in the word RELIABILITY is selected randomly. The probability that the letter is a *i* is:

a. $\frac{1}{11}$ **b.** $\frac{3}{11}$ **c.** $\frac{8}{11}$

7. A box contains six coins—a penny, a nickel, a dime, a quarter, a half-dollar, and a dollar. If a coin is selected at random, the probability that the value of the coin is more than 10 cents is:

a. $\frac{1}{2}$ **b.** $\frac{2}{3}$ **c.** $\frac{1}{3}$

8. The number of different ways in which five different flower plants can be planted in a row is:
a. 5! **b.** 5^5 **c.** 1

9. The number of different ways in which five different flower plants can be selected from a group of 7 different flower plants for planting in a row is:
a. 630 **b.** 42 **c.** 21

10. A machine shop has 7 different lathes and 4 different milling machines. If a job requires operations on 3 different lathes and 2 different milling machines, the number of different possibilities for completing the job is:
a. 210 **b.** 105 **c.** 2520

2.5 Four cranes, each with a probability of failure of 0.1, are used in a construction project. Match the following events related to the failure of the cranes during a construction project:

1. All four cranes fail	**a.** 0.0486
2. Three cranes fail	**b.** 0.2916
3. Two cranes fail	**c.** 0.0001
4. One crane fails	**d.** 0.6561
5. No crane fails	**e.** 0.0036

2.6 Match the following:

1. Complement	**a.** $A \cap B = B \cap A$
2. de Morgan rule	**b.** $\overline{A}$
3. Commutative	**c.** $(A \cup B) \cap C = (A \cap C) \cup (B \cap C)$
4. Distributive	**d.** $\overline{A \cup B} = \overline{A} \cap \overline{B}$

2.7 Match the following:

1. $P(E_1 \cup E_2) = P(E_1) + P(E_2)$	**a.** Total probability theorem
2. $P(E_1 \cup E_2) = P(E_1) + P(E_2) - P(E_1 \cap E_2)$	**b.** For statistically independent events
3. $P\left(E_1 \middle\vert_{E_2}\right) = P(E_1)$	**c.** For mutually exclusive events
4. $P(E_2)P\left(E_1 \middle\vert_{E_2}\right) = P(E_1)P\left(E_2 \middle\vert_{E_1}\right)$	**d.** Bayes' rule
5. $P(A) = \sum_{i=1}^{n} P\left(A \middle\vert_{B_i}\right)P(B_i)$	**e.** For any two events
6. $P(B_j \vert A) = \dfrac{P\left(A \middle\vert_{B_j}\right)P(B_j)}{\sum_{j=1}^{n} P\left(A \middle\vert_{B_j}\right)P(B_j)}$	**f.** Probability of occurrence of both events

2.8 Match the following:

A box contains 5 white, 6 black, and 7 red balls. Two balls are randomly selected from the box, the second one selected after replacing the first ball. Match the probabilities of the following events:

1. Both balls are black	**a.** $\left(\frac{5}{18}\right)\left(\frac{7}{17}\right)$
2. Both balls are white	**b.** $\left(\frac{5}{18}\right)\left(\frac{4}{17}\right)$
3. Both balls are red	**c.** $\left(\frac{7}{18}\right)\left(\frac{6}{17}\right)$
4. First ball is white and the second one is red	**d.** $\left(\frac{6}{18}\right)\left(\frac{7}{17}\right)$
5. First ball is black and the second one is red	**e.** $\left(\frac{6}{18}\right)\left(\frac{5}{17}\right)$

PROBLEMS

2.1 Explain the validity of the commutative, associative, distributive, and de Morgan's rules stated in Section 2.3 using Venn diagrams.

2.2 The following events are defined after observing the scores of five students in a mathematics examination: A = at most four students failed to secure a score of 90 percent, B = exactly three students secured a score of 90 percent, and C = all five students failed to secure a score of 90 percent. Interpret the following events: (i) $A \cup B$, (ii) $A \cup C$, (iii) $B \cup C$, (iv) $A \cap B$, (v) $A \cap C$, (vi) $B \cap C$, (vii) $\overline{A} \cap \overline{C}$, (viii) $\overline{A} \cup \overline{B}$, (ix) $\overline{A} \cup \overline{C}$, (x) $\overline{B} \cap \overline{C}$.

2.3 A batch of 40 castings of an engine contains three defective units. If a sample of five castings is randomly selected from the batch, determine:

a. The number of different sample of size five that can be drawn from the batch of 40 castings.
b. The number of different samples of size five that contain two defective castings.

Assume that a casting, once selected, is not returned to the batch during the process of selecting a sample of five castings.

2.4 A batch of 100 spot welds in a construction project are inspected by selecting a sample of five welds randomly. If 8 welds are defective in the batch, determine the following:

a. How many different samples are possible?
b. How many samples of five welds contain exactly one defective weld?
c. How many samples of five welds contain at least one defective weld?

2.5 The production of a machine part involves five different operations on a lathe and four different operations on a milling machine. If the operations can be performed in any order, how many different possibilities exist to produce the machine part?

2.6 The production of a machine part involves the operations of machining on a lathe, polishing on grinding machine, and painting in sequence. If there are four lathes, three grinding machines, and two painting stations, determine the number of different routings (the lathe number, followed by the grinding machine number, followed by the paint station number) possible for producing the machine part.

2.7 The tensile strength of laminated composite plates (x) is measured in pounds per square inch. If two events A and B are defined as

$$A = \{x|_{x>10.5}\}, \quad B = \{x|_{x<15.0}\}$$

describe the following events:

a. $\overline{A}$ **b.** $\overline{B}$ **c.** $A \cup B$ **d.** $A \cap B$

2.8 Figure 2.12 shows the events A, B, and C in a Venn diagram. Show the following events, each by shading the appropriate region in the Venn diagram separately:

a. $\overline{B}$ **b.** $A \cap B$ **c.** $A \cap B \cap C$
d. $(A \cap B) \cup C$ **e.** $(\overline{A \cap B}) \cup C$

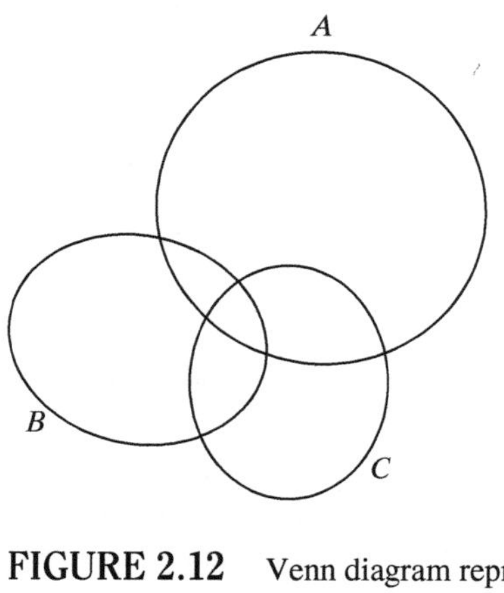

FIGURE 2.12 Venn diagram representation of events A, B, and C.

2.9 The laminated composite beams produced in a manufacturing plant are known to contain defects such as cracks, debonding of fibers, incorrect ply orientation, low volume fraction and delamination. An inspection of a batch of laminated composite beams for different number of defects resulted in the following data:

Type of Defect	Proportion of Composite Beams
No defects	0.30
One defect	0.25
Two defects	0.15
Three defects	0.10
Four defects	0.15
Five defects	0.05

For a composite beam selected randomly from this batch, determine the following:

a. Probability that it contains no defects.
b. Probability that it contains at most one defect.
c. Probability that it contains at least two types of defects.
d. Probability that it contains three or more types of defects.

2.10 If $P(E_1|_{E_2}) = 0.4$, $P(E_1|_{\overline{E}_2}) = 0.5$, and $P(E_2) = 0.6$, find the value of $P(E_1)$.

2.11 A lot of 50 forged connecting rods contains 5 that have surface cracks. If three connecting rods are selected at random, without replacement, find the probability that the third connecting rod selected is the only one with a surface crack.

2.12 If $P(E_1) = 0.4$, $P(E_2) = 0.3$, and $P(E_1 \cap E_2) = 0.1$, determine the following probabilities:
(a) $P(E_1 \cup E_2)$, (b) $P(\overline{E}_1)$, (c) $P(\overline{E}_2)$, (d) $P(\overline{E}_1 \cap \overline{E}_2)$, and (e) $P(E_1 \cap \overline{E}_2)$.

2.13 If the events E_1, E_2, and E_3 are mutually exclusive with $P(E_1) = 0.1$, $P(E_2) = 0.3$, and $P(E_3) = 0.5$, determine the values of: (a) $P(E_1 \cap E_3)$, (b) $P(E_1 \cup E_2 \cup E_3)$, (c) $P(E_1 \cap E_2 \cap E_3)$, and (d) $P(\overline{E}_1 \cap \overline{E}_2 \cap \overline{E}_3)$.

2.14 A batch of 200 specimens of fiber-reinforced composite material, when tested for strength and density, yielded the following data:

	High Density	Low Density
High strength	151	11
Low strength	18	20

a. For a specimen randomly selected from the batch, determine the following:
  i. Probability that both its strength and density are low.
  ii. Probability that its strength is high and density is low.

b. Consider the event E_1 that a specimen has a high strength and the event E_2 that a specimen has a high density. Are the events E_1 and E_2 mutually exclusive?

2.15 Draw the Venn diagram to represent events E_1, E_2, and E_3 that satisfy the following probabilities: (a) $P(E_1|_{E_2}) = 1$, and (b) $P(E_1|_{E_3}) = 1$ and $P(E_2|_{E_3}) = 0$.

2.16 An automobile assembly plant collected the following data on the failure of welded joints:

	Failure Due to Inadequate Size	
	Yes	**No**
Failure due to incorrect location		
Yes	147	41
No	72	5

A welded joint is considered to find whether it is at the correct location and has the adequate size. From the data given, determine the following:

a. Probability that the location of the welded joint is incorrect.
b. Probability that the size of the welded joint is inadequate.
c. Probability that the welded joint is at an incorrect location given that its size is inadequate.
d. Probability that the welded joint has inadequate size given that its location is incorrect.

2.17 A machinist produced 100 shafts according to the specification 1" $\pm$ 0.001". During inspection, the diameters of 85 shafts were found to be within the tolerance limits and 15 were found to be outside the tolerance band. If 6 shafts are randomly selected, find the probability of finding the diameter of at least one shaft falling outside the tolerance limits.

2.18 One hundred castings were inspected for defects and 85 were found to be free of defects. Among the 15 defective castings, 9 were found to have blowholes, and 12 were found to have sand inclusions. (a) How many castings have exactly one type of defect? (b) How many castings have at least one type of defect? (c) How many castings have both types of defects?

2.19 The computer laboratory in a university has 40 microcomputers. Of these, 25 are connected to a dot matrix printer, 10 are connected to a laser printer, and 8 are not connected to any printer. (a) How many microcomputers are connected to at least one type of printer? (b) How many are connected to both types of printers?

2.20 Three girls—Kathy, Cindy, and Nancy—compete in a swimming race. If each girl is equally likely to win the race, find (a) the probability of Kathy or Cindy winning the race, (b) the probability of Nancy winning the race, and (c) the probability of Kathy not winning the race.

2.21 Six gearboxes are assembled in a machine shop. It is known that two have defects. If two gearboxes are selected at random, what is the probability that (a) both will be defective, (b) both will not be defective, (c) one will be defective and the other will not be defective, (d) the second one will be defective given the first one is not defective, and (e) the second one will not be defective given that the first one is not defective?

2.22 In a high school senior class, 15 percent of the students secured an SAT score of 1200 or higher. It is known from past experience that 60 percent of the students whose SAT score is at least 1200 select engineering as major while 10 percent of those students whose SAT score is less than 1200 select engineering as major in college. If a randomly selected student is known to choose engineering, what is the probability that his/her SAT score is at least 1200?

2.23 The castings for the engine block in a diesel engine manufacturing facility are supplied by two different companies. Company A supplies 200 castings each day out of which 2 percent do not meet the specifications. Company B supplies 300 castings each day out of which 3 percent do not satisfy the specifications. (a) What is the probability that a randomly selected casting was supplied by company B? (b) What is the probability that a randomly selected casting will not meet the specifications? (c) If a casting does not meet the specifications, what is the probability that it was supplied by company A? (d) If a casting meets the specifications, what is the probability that it was supplied by company B?

2.24 Determine whether the events A and B are independent for the following situations:

a. $P(A|_B) = 0.6, P(B) = 0.5$, and $P(A) = 0.6$
b. $P(A|_B) = 0.4, P(B) = 0.5$, and $P(A) = 0.6$
c. $P(B|_A) = 0.7, P(B) = 0.5$, and $P(A) = 0.6$

2.25 If $P(A|_B) = 0.4$ and $P(B|_A) = 0.6$, determine the values of $P(A)$ and $P(B)$ if the events A and B are independent.

2.26 If the events A and B are mutually exclusive with $P(A) = 0.4$ and $P(B) = 0.6$, determine whether the events A and B are independent.

2.27 Two boilers are used to supply steam for the operation of a steam turbine. Each boiler can completely supply the steam to the turbine 80 percent of the time. The probability of failure of each boiler is 0.05 and that of both boilers failing simultaneously is 0.03. Find the probability of the steam turbine getting the complete supply of steam and one of the boilers failing simultaneously.

2.28 Suppose that 15 percent of all people have a high-cholesterol level. It is found that 5 percent of people with a high-cholesterol level get a heart attack while only 1 percent of those without a high-cholesterol level get heart attack. If a randomly selected person gets a heart attack, find the probability of that person having a high-cholesterol level.

2.29 Among the 30 students enrolled in an Engineering Design course, 23 belong to mechanical engineering branch and 7 belong to other branches of engineering. If 2 students are randomly chosen from this class, what is the probability of both the students belonging to a branch other than mechanical engineering?

2.30 Suppose that 15 percent of all people have a high-cholesterol level. A particular type of cholesterol test, when used on a person having a high-cholesterol level, detects it correctly 95 percent of the time and fails to detect it 5 percent of the time. The same test, when used on a person without a high-cholesterol level, wrongly indicates a high-cholesterol level 4 percent of the time. (a) If the test shows a high-cholesterol level for a randomly selected person, what is the probability of that person actually not having a high-cholesterol level? (b) What is the probability of a randomly selected person having a high-cholesterol level and being indicated as such by the test? (c) What is the probability of a randomly selected person not having a high-cholesterol level but the test indicating the person as having a high-cholesterol level? (d) Do you consider the test to be reliable?

2.31 A boiler has three independent safety valves which open and let steam off to the atmosphere in case of a high-pressure buildup. The boiler is considered to be safe if at least two valves open in the case of a high-pressure buildup. If the probability of each valve opening in case of high-pressure buildup is 0.9, find the probability of the boiler to be in the safe condition in case of a high-pressure buildup.

2.32 A gear can fail due to either excessive bending stress or excessive surface wear. Let B and W denote the failures in these modes. If it is known that in a particular application, $P(B) = 0.0005$, $P(W) = 0.001$ and $P(B|_W) = 0.1$, find (a) the probability of failure of the gear, and (b) the probability of a gear having only excessive bending stress but not excessive surface wear.

2.33 A small river is found to be polluted due to the chemical waste from two major industries, A and B. Due to the strict standards enforced by the EPA, the probability of controlling the pollutions from A and B are 0.7 and 0.85, respectively, in the next two years. It is estimated that, if only one of the industries complies with the strict standards of EPA, there is a 75 percent probability of controlling the pollution in the river to an acceptable level. (a) Find the probability of controlling the pollution in the river in the next two years. (b) If the pollution in the river is not controlled in the next two years, what is the probability that it is caused entirely by the industry A?

2.34 An automatic spot-welding machine is used in an assembly plant. It is known that 5 percent of the welds made by the machine are defective. If 10 welds made by the machine are inspected at random, what is the probability of finding (a) no defective welds, (b) exactly one defective weld, (c) at least one defective weld, and (d) no more than one defective weld?

2.35 Consider a thermal power plant composed of four subsystems, namely, feed water pumps, boilers, turbines, and generators, and connected as shown in Fig. 2.13. Each subsystem consists of two identical units connected in parallel to increase its reliability. The power plant functions and generates power only when at least one unit of each subsystem between the points 1 and 2 functions. The probability that each unit functions is given in Fig. 2.13. Assuming that the probability of the functioning of a unit does not depend on whether or not the other units (and other subsystems as well) are functional, determine the probability that the power plant generates power.

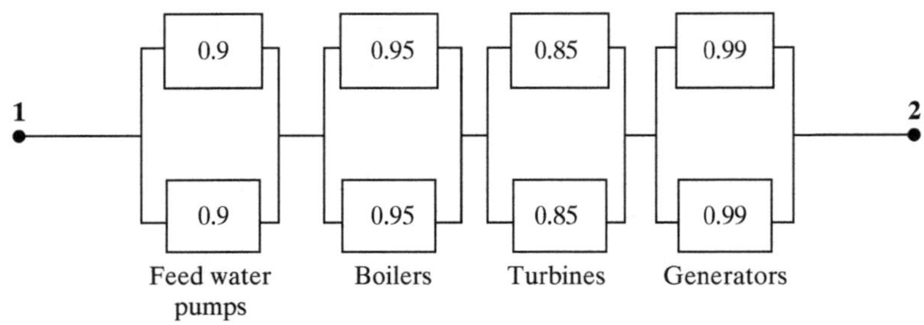

FIGURE 2.13 Parallel subsystems connected in series.

2.36 Consider a thermal power plant composed of four types of components or units, namely, feed water pump, boiler, turbine, and generator, and connected as shown in Fig. 2.14. The power plant functions and generates power only when each component or unit

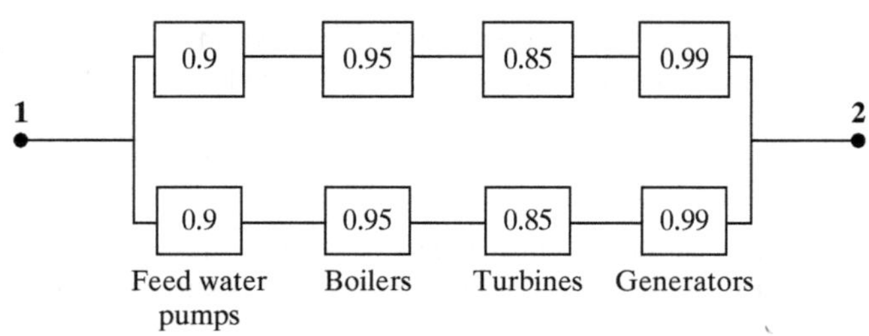

FIGURE 2.14 Series subsystems connected in parallel.

between the points 1 and 2 functions. Two subsystems, each composed of all the four components, are connected in parallel to increase its reliability. The probability that each unit functions is given in Fig. 2.14. Assuming that the probability of the functioning of a unit does not depend on whether or not the other units (and other subsystems as well) are functional, determine the probability that the power plant generates power.

2.37 Among the graduate students of a university, 26 percent receive loans, 55 percent receive research assistantships, and 8 percent receive both. What percent of students receive either a loan or an assistantship? What percent receive neither a loan nor an assistantship?

2.38 The percentage of mechanical-, civil-, aeronautical-, industrial-, and chemical-engineering students in the course, Engineering Mathematics, is, respectively, 25, 20, 15, 22, and 18. The probability of a mechanical-, civil-, aeronautical-, industrial-, and chemical-engineering students getting the grade A is 0.7, 0.6, 0.8, 0.5, and 0.65, respectively. If a student chosen at random at the end of the semester earned an A in the course, find the probability of the student belonging to mechanical engineering.

2.39 The machine used to detect blowholes in castings detects not only 84 percent of all blowholes, but also incorrectly indicates a blowhole in 2 percent of good castings. It is known, from past experience, that 5 percent of all castings will have blowholes. Find the probability that a casting which the machine indicates to have a blowhole is in fact a good one.

2.40 A crane has two cables C_1 and C_2 to lift a load as shown in Fig. 2.15. The cable C_2 is designed to be slightly longer than cable C_1 so that C_1 carries the load under normal circumstances while cable C_2 is intended to carry the load only when C_1 fails. From past experience, it is known that the probability of failure of cable C_1 is 0.01 and the probability of failure of cable C_2 when it is required to carry the load all by itself is 0.25. (a) What is the probability of the crane lifting the load? (b) What is the probability of the crane failing to lift the load? (c) If the load is being lifted by the crane, what is the probability that none of the cables have failed?

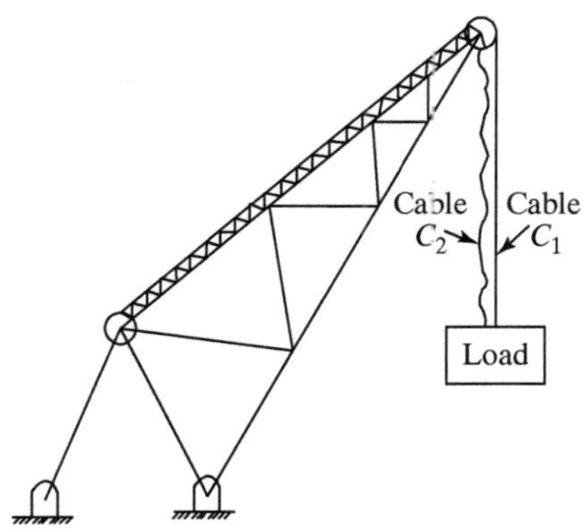

FIGURE 2.15 Crane lifting a load using two cables.

2.41 In a batch of 800 connecting rods produced by forging in a production plant, 6 are known to be defective. Two connecting rods are selected randomly from the batch. If E_i denotes the event that the ith connecting rod selected is defective, $i = 1, 2$, determine whether the events E_1 and E_2 are independent for the following situations:

a. The first connecting rod selected is not returned to the batch before selecting the second one.
b. The first connecting rod selected is returned to the batch before selecting the second one.

2.42 A 6-bit number used in genetic algorithm (used for the minimization of a function) is composed of six digits with each digit being a 0 or a 1. For a randomly generated number, each bit is equally likely to be 0 or 1. Assuming that the bits are independent, determine the following:

a. The probability that the number is 000000.
b. The probability that the number is 111111.
c. The probability that the number is 110000.
d. The probability that two bits are 1 and four bits are 0.

2.43 A computer numerically controlled (CNC) machine tool can fail due to three types of problems—hardware, software, or controls. The probabilities of hardware, software, and controls failures are given by 0.5, 0.3 and 0.2, respectively. The probabilities of the failure of the CNC machine tool given a hardware problem is 0.7, given a software problem is 0.5 and given a controls problem is 0.4. If a CNC machine tool fails, determine the most likely cause.

2.44 The inspection of castings using an X-ray machine in a production plant is known to have a 98 percent chance of correctly identifying the defective ones and 1 percent chance of incorrectly identifying a good casting as a defective one. The production plant has evidence that only 0.5 percent of the castings are defective. Determine the following:

a. Probability of a randomly selected casting is identified to be a defective one.
b. Probability of a randomly selected casting to be really nondefective if it is identified to be nondefective by the inspection process.

CHAPTER 3

Random Variables and Probability Distributions

What You Will Learn

In this chapter, you will learn

- The concept of a random variable
- The difference between discrete and continuous random variables
- The probability distribution function of a random variable as an integral of the corresponding probability mass or density function
- The definitions of mean, mode, and median, the standard deviation and coefficient of variation, and the skewness coefficient
- The idea of moments of a random variable and the importance of moment functions in practical applications
- The concepts of joint density and distribution functions for multiple random variables, the method of computing marginal density function from the joint density function, and the moments of jointly distributed random variables
- Some commonly used probability distributions, namely, binomial and Poisson distributions for discrete random variables and normal and lognormal distributions for continuous random variables
- The central limit theorem that provides justification for using normal distribution in several practical applications

Pafnutii Lvovitch Chebyshev (or Tchebysheff or Tshebysheff) was born on May 16, 1821, and died on December 8, 1894, in Russia. He graduated from Moscow University with a degree in mathematics in 1841. He was one of the most famous Russian mathematicians, and he made numerous important contributions to the theory of numbers, algebra, theory of probability, analysis, and applied mathematics. Chebyshev won a silver medal in 1841 for deriving an error estimate in the Newton-Raphson iterative method. He was very curious about mechanical inventions during his childhood and it was stated that during his very first lesson in geometry he saw its application to mechanics. His technological inventions include a calculating machine built in the late 1870s. When his father became very poor during the famine of 1840, Chebyshev helped support his family. He devoted his life to scientific teaching and never married. He became interested in the theory of numbers and stated the Chebyshev problem relating probability to the theory of numbers. The Chebyshev inequality, which gives the probability of an event occurring not more than a specified distance from the average, is one of the popular theorems in the theory of probability [3.6, 3.9].

3.1 Introduction

An event has been defined as the possible outcome of an experiment. Let us assume that a random event is the measurement of a quantity X, which takes on various values x within the range $-\infty < x < \infty$. Such a quantity (like X) is called a *random variable*. We denote a random variable by a capital letter and the particular value taken by it by a lowercase letter. Random variables are of two types: (i) discrete and (ii) continuous. If the random variable is allowed to take only discrete values $x_1, x_2, \ldots, x_n$, it is called a *discrete random variable.* On the other hand, if the random variable is permitted to take any real value in a specified range, it is called a *continuous random variable.* For example, the number of people who report for work on a particular day at a factory is a discrete random variable, whereas the weight of a machined metal part can be treated as a continuous random variable.

3.2 Probability Mass Function for Discrete Random Variables

Corresponding to each value x_i that a discrete random variable X can take, we can associate a probability of occurrence $P(x_i)$. We can describe the probabilities associated with the random variable X by a table of values, but it will be easier to write a general formula that permits one to calculate $P(x_i)$ by substituting the appropriate value of x_i. Such a formula is called the *probability mass function* of the random variable X, and is usually denoted as $p_X(x_i)$ or simply as $p(x_i)$. Thus, the function which gives the probability of realizing the random variable $X = x_i$ is called the probability mass function $p_X(x_i)$

$$p(x_i) = p_X(x_i) = P(X = x_i) \tag{3.1}$$

3.3 Cumulative Distribution Function for Discrete Random Variables

Although a random variable X is completely described by the probability mass function, it is often convenient to deal with another related function known as the probability distribution function. The probability that the value of the random variable X is less than or equal to some number x is defined as the cumulative distribution function $F_X(x)$. Thus,

$$F_X(x) = P(X \le x) = \sum_i p_X(x_i) \tag{3.2}$$

where the summation extends over those values of i for which $x_i \le x$. Since the distribution function is a *cumulative probability,* it is also termed as the *cumulative distribution function.*

EXAMPLE 3.1

Find the probability mass and distribution functions for the number realized when a fair die is thrown.

Solution: Since each face is equally likely to show up, the probability of realizing any number between 1 and 6 is $1/6$.

$$P(X = 1) = P(X = 2) = \ldots = P(X = 6) = \frac{1}{6}$$

and hence,

$$p_X(1) = p_X(2) = \ldots = p_X(6) = \frac{1}{6}$$

The analytical form of $F_X(x)$ is

$$F_X(x) = x/6 \quad \text{for} \quad 1 \leq x \leq 6$$

Correspondingly, the probability mass and distribution functions are shown in Fig. 3.1 where X is the random variable denoting the number realized after throwing the die.

It can be seen that for any discrete random variable, the distribution function will be a *step function.* If the least possible value of the variable X is L and the greatest possible value is U, then

$$F_X(x) = 0 \quad \text{for all} \quad x < L$$

and

$$F_X(x) = 1 \quad \text{for all} \quad x > U$$

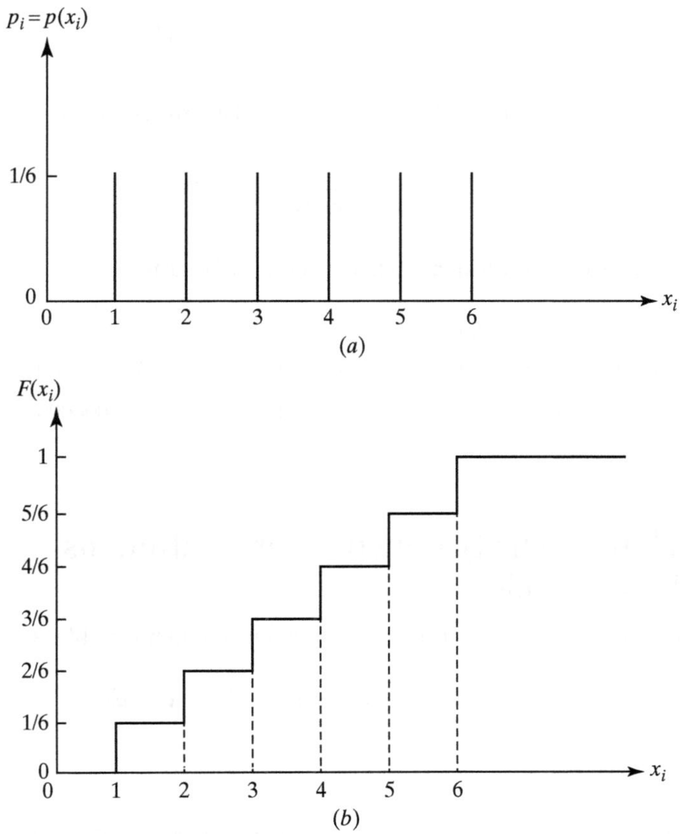

FIGURE 3.1 (*a*) Probability mass function, (*b*) Cumulative distribution function.

■

EXAMPLE 3.2

A batch of 250 castings of an engine, containing 22 units with blowholes, are inspected for defects. A casting is considered defective if it contains a blowhole. A sample of two castings is selected from the batch randomly, without replacement, for inspection. If the number of defective castings in the sample is denoted by the random variable X, find the probability mass and cumulative distribution functions of X.

Solution: The probability mass function of X can be found as follows:

$p_X(0) = P(X = 0) =$ Probability of finding zero defective castings among the two units inspected

$$= \frac{228}{250}\frac{227}{249} = 0.831422$$

$p_X(1) = P(X = 1) =$ Probability of finding exactly one defective casting among the two units inspected

$$= 2\left(\frac{228}{250}\frac{22}{249}\right) = 0.161157$$

$p_X(2) = P(X = 2) =$ Probability of finding both the castings inspected to be defective

$$= \frac{22}{250}\frac{21}{249} = 0.007422$$

Thus, the cumulative distribution function of X can be found as:

$$F_X(0) = P(X \le 0) = p_X(0) = 0.831422$$
$$F_X(1) = P(X \le 1) = p_X(0) + p_X(1) = 0.831422 + 0.161157 = 0.992579$$
$$F_X(2) = P(X \le 2) = p_X(0) + p_X(1) + p_X(2) = 1.000000$$

■

3.4 Probability Density Function for Continuous Random Variables

The probability density function of a continuous random variable is defined by[1]

$$f_X(x)\,dx = P(x \le X \le x + dx) \tag{3.3}$$

[1]The probability density function can also be defined as $f_X(x)dx = P(x - dx/2 \le X \le x + dx/2) =$ probability of realizing X in the infinitesimal interval $(x - dx/2, x + dx/2)$.

which is equal to the probability of detecting X in the infinitesimal interval $(x, x + dx)$. The distribution function of X is defined as the probability of detecting X less than or equal to x, that is,

$$F_X(x) = P(X \leq x) = \int_{-\infty}^{x} f(x')\, dx' \tag{3.4}$$

where the condition $F(-\infty) = 0$ has been used. As the upper limit of the integral goes to infinity, we have

$$\int_{-\infty}^{\infty} f_X(x)\, dx = F_X(\infty) = 1 \tag{3.5}$$

This is called the normalization condition. $f_X(x)$ is called the probability density function to distinguish it from the probability mass function defined in the case of a discrete random variable.

From the definition of probability distribution function, we get

$$F_X(x + \Delta x) - F_X(x) = P(x \leq X \leq x + \Delta x) \tag{3.6}$$

and the probability of detecting the variable X in the finite interval $[x_1, x_2]$ may be written as

$$\begin{aligned} P(x_1 \leq X \leq x_2) &= \int_{x_1}^{x_2} f_X(x)\, dx = P(X \leq x_2) - P(X \leq x_1) \\ &= F_X(x_2) - F_X(x_1) \end{aligned} \tag{3.7}$$

It can be seen that if the random variable X is continuous and if the first derivative of the distribution function exists, the probability density function $f_X(x)$ is given by the first derivative of $F_X(x)$.

$$f_X(x) = \frac{dF_X(x)}{dx} = \lim_{\Delta x \to 0} \frac{F_X(x + \Delta x) - F_X(x)}{\Delta x} \tag{3.8}$$

A typical probability density function and the corresponding distribution function are shown in Fig. 3.2.

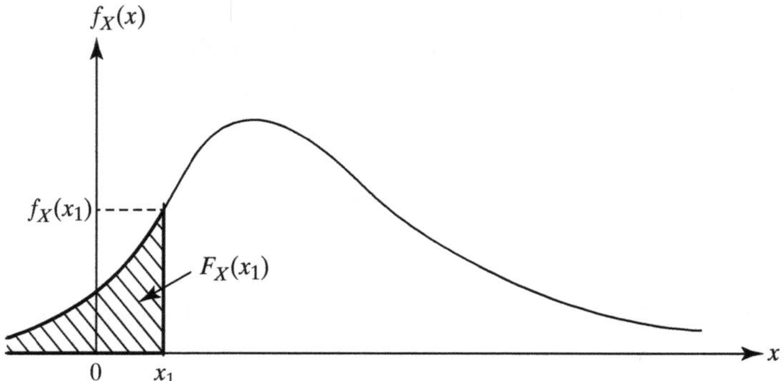

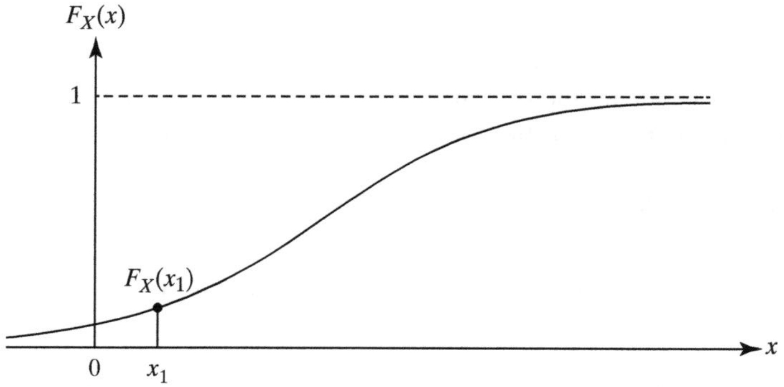

FIGURE 3.2 (*a*) Probability density function, (*b*) Probability distribution function.

EXAMPLE 3.3

The probability density function of the wind load acting on a water tank (X, in kips) is given by

$$f_X(x) = \begin{cases} \dfrac{3x}{50 \times 10^4}(100 - x); & 0 \le x \le 100 \text{ kips} \\ 0; & \text{otherwise} \end{cases}$$

Find the probability of realizing the wind load to be less than 20 kips or greater than 80 kips.

Solution: If $E_1(E_2)$ denotes the event of realizing the wind load to be less than 20 kips (greater than 80 kips), the required probability is given by

$$P(E_1 \cup E_2) = P(E_1) + P(E_2) - P(E_1 E_2) = P(E_1) + P(E_2)$$

since the probability of simultaneous occurrence of E_1 and E_2 is zero. Here

$$\begin{aligned}
P(E_1) &= P(X \leq 20) = \int_{x=0}^{20} f_X(x)\,dx \\
&= \frac{3}{50 \times 10^4} \int_0^{20} x(100 - x)\,dx = \frac{3}{50 \times 10^4} \left(50x^2 - \frac{1}{3}x^3 \right) \bigg|_0^{20} \\
&= \frac{3}{50 \times 10^4} \left[50(400) - \frac{1}{3}(8000) \right] \\
&= 0.104
\end{aligned}$$

$$\begin{aligned}
P(E_2) &= P(X \geq 80) = \int_{80}^{100} f_X(x)\,dx \\
&= \frac{3}{50 \times 10^4} \int_{80}^{100} x(100 - x)\,dx = \frac{3}{50 \times 10^4} \left(50x^2 - \frac{1}{3}x^3 \right) \bigg|_{80}^{100} \\
&= \frac{3}{50 \times 10^4} \left[50(10^4) - \frac{1}{3}(10^6) - 50(80)^2 + \frac{1}{3}(80)^3 \right] \\
&= 0.104
\end{aligned}$$

Hence the probability of realizing X less than 20 kips or greater than 80 kips is given by $0.104 + 0.104 = 0.208$.

■

EXAMPLE 3.4

Consider the probability density function of the wind load given in Example 3.3. If the wind load is known to be greater than 20 kips, find the probability of its value being larger than 80 kips.

Solution: The desired conditional probability can be expressed as

$$P\left(X \geq 80 \Big|_{X \geq 20} \right) = \frac{P(X \geq 80)}{P(X \geq 20)}$$

Expressing $P(X \geq 20)$ as $1 - P(X < 20)$, we find that

$$P\left(X \geq 80 \Big|_{X \geq 20} \right) = \frac{P(X \geq 80)}{1 - P(X < 20)} = \frac{0.104}{1 - 0.104} = \frac{0.104}{0.896} = 0.1161$$

■

EXAMPLE 3.5

The diameter of a rod produced on a lathe (X) during mass production, with the specification of the nominal diameter as 20 mm, is known to follow the following distribution:

$$f_X(x) = \begin{cases} 0 & ; x < 20 \text{ mm} \\ 15e^{-15(x-20)} & ; x \geq 20 \text{ mm} \end{cases}$$

Determine the following:

1. The probability distribution function of the diameter of the rod.
2. If rods with diameters larger than 20.2 mm are not acceptable, determine the proportion of rods that are accepted and that are rejected.

Solution:

1. The probability distribution function of the diameter of the rod (X) is given by

$$F_X(x) = 0 \quad \text{for} \quad x < 20 \text{ mm}$$

and

$$F_X(x) = \int_0^x f_X(x')dx' = \int_{20}^x 15e^{-15(x'-20)}\,dx' = -e^{-15(x'-20)}\Big|_{20}^{x} = 1 - e^{-15(x-20)}$$

for $x \geq 20$ mm

2. The proportion of rods acceptable is given by

$$P(20 < x < 20.2) = \int_{20}^{20.2} f_X(x)\,dx = -e^{-15(x-20)}\Big|_{20}^{20.2} = -e^{-15(20.2-20)} + e^{-15(20-20)}$$

$$= 1 - e^{-3} = 0.950213$$

$$= 95.0213\%$$

The proportion of rods rejected is given by

$$P(X > 20.2) = \int_{20.2}^{\infty} f_X(x)\,dx = -e^{-15(x-20)}\Big|_{20.2}^{\infty} = e^{-15(20.2-20)} = e^{-3} = 0.049787$$

$$= 4.9787\%$$

■

3.5 Mean, Mode, and Median

The probability density or distribution function of a random variable contains all the information about the variable. However, in many cases we require only the gross properties but not the entire information about the random variable. In such cases we can specify the range of the variable, a measure of its central value, or an indicator of its spread.

3.5.1 Mean

The *mean value*, also termed as the *expected value, mathematical expectation*, or *average*, is used to describe the central tendency of a random variable. The mode and median are also used for the same purpose.

Discrete Case. Let us assume that there are n trials of an experiment in which the random variable X is observed to take on the value x_1 (n_1 times), x_2 (n_2 times), ..., x_m (n_m times) so that $n_1 + n_2 + \ldots + n_m = n$. Then the mean value or the arithmetic mean of X, denoted as $\overline{X}$, is given by

$$\overline{X} = \frac{\sum_{k=1}^{m} x_k n_k}{n} = \sum_{k=1}^{m} x_k \left(\frac{n_k}{n}\right) = \sum_{k=1}^{m} x_k \cdot p_X(x_k) \tag{3.9}$$

where (n_k/n) is the relative frequency of occurrence of x_k and is same as the probability mass function $p_X(x_k)$. Hence, in general, the mean value, which is also denoted as $E(X)$, the expected value, of the discrete random variable can be expressed as

$$\overline{X} = E(X) = \sum_{\text{all } i} x_i p_X(x_i) \tag{3.10}$$

Continuous Case. If $f_X(x)$ is the density function of a continuous random variable X, the mean is given by[2]

$$\mu = E(X) = \int_{-\infty}^{\infty} x\, f_X(x)\, dx \tag{3.11}$$

According to the definition of a random variable, any function of a random variable is itself a random variable. Thus if $h(x)$ is an arbitrary function of x, the expected value of $h(x)$ is defined as

$$E[h(x)] = \sum_i h(x_i)\, p_X(x_i), \quad \text{if } X \text{ is discrete}$$
$$= \int_{-\infty}^{\infty} h(x) f_X(x)\, dx, \quad \text{if } X \text{ is continuous} \tag{3.12}$$

[2]The symbol μ is generally used to denote the mean value of a continuous random variable and $\overline{X}$ that of a discrete random variable.

EXAMPLE 3.6

What is the expected value of the upturned face of a single die?

Solution:

$$\begin{aligned}\overline{X} = E(X) &= \sum_{\text{all } i} x_i p_X(x_i) \\ &= 1(1/6) + 2(1/6) + 3(1/6) + 4(1/6) + 5(1/6) + 6(1/6) = 3.5\end{aligned}$$

■

EXAMPLE 3.7

Find the mean value of the Rockwell hardness number X of a particular type of steel for which the density function is given by

$$f_X(x) = \begin{cases} 0, & \text{for } x < 50 \\ 1/20, & \text{for } 50 \leq x \leq 70 \\ 0, & \text{for } x > 70 \end{cases}$$

Solution:

$$\begin{aligned}\mu = E[X] &= \int_{-\infty}^{\infty} x f_X(x)\, dx \\ &= \int_{50}^{70} x(1/20)\, dx = (1/20)\left(\frac{x^2}{2}\right)_{50}^{70} = 60\end{aligned}$$

■

3.5.2 Mode

Discrete Case. Mode is the value of X at which the probability mass function $p_X(x_i)$ has the highest value. In essence, mode is the most probable value of the random variable.

Continuous Case. Mode is the value of X at which $df_X(x)/dx = 0$, that is, the value of x corresponding to the peak value of the probability density function.

3.5.3 Median

Discrete Case. Median is the number in the middle.

Continuous Case. Median (x_{50}) is the value of X at which the cumulative distribution function has a value of 0.5:

$$0.5 = \int_{-\infty}^{x_{50}} f_X(x)\, dx$$

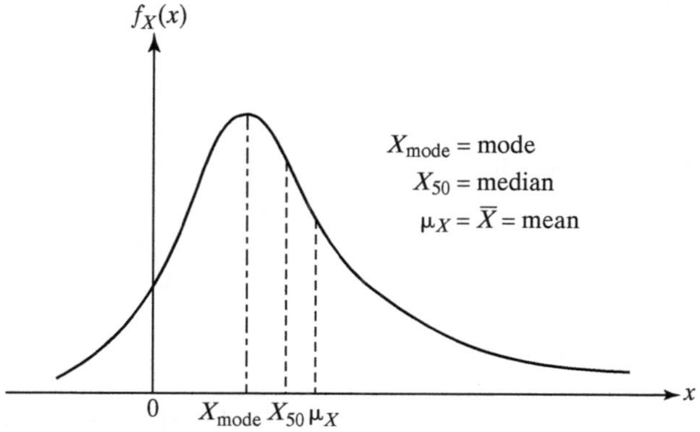

FIGURE 3.3 Mean, mode and median of X.

The significance of mean, mode, and median of a random variable is illustrated in Fig. 3.3.

EXAMPLE 3.8

Find the mean, mode, and median of the random variable whose distribution function is given by[3]

$$F_X(x) = 1 - \exp\left[-\frac{1}{2}kx^2\right], \qquad 0 \leq x < \infty$$

where k is the parameter (constant) of the distribution.

Solution: The probability density function can be obtained by differentiating the distribution function as

$$f_X(x) = \frac{dF_X(x)}{dx} = kx\exp\left[-\frac{1}{2}kx^2\right]$$

We take the integral of this density function from 0 to ∞ as

$$\int_0^\infty f_X(x)\,dx = F_X(x)\Big|_0^\infty = 1$$

and find that it satisfies the normality condition. The mean of X is given by

$$\mu_X = \int_0^\infty x f_X(x)\,dx = 2\int_0^\infty \frac{1}{2}kx^2\exp\left[-\frac{1}{2}kx^2\right]dx$$

[3]This is a one-parameter distribution called the Rayleigh distribution.

By making a change of variable as

$$\frac{1}{2}kx^2 = y$$

the expression for μ_X can be written as[4]

$$\mu_X = \sqrt{\frac{2}{k}}\int_0^\infty \sqrt{y}e^{-y}dy = \sqrt{\frac{\pi}{2k}}$$

To obtain the mode of X, we set

$$\frac{d\,f_X(x)}{dx} = \frac{d}{dx}\left[kx\exp\left\{-\frac{1}{2}kx^2\right\}\right] = k\exp\left\{-\frac{1}{2}kx^2\right\} - k^2x^2\exp\left\{-\frac{1}{2}kx^2\right\} = 0$$

and solve for x to obtain

$$X_{\text{mode}} = \sqrt{\frac{1}{k}}$$

The median of X can be obtained from the relation

$$0.5 = \int_0^{x_{50}} f_X(x)\,dx = F_X(x)\Big|_0^{x_{50}} = \left\{1 - \exp\left[-\frac{1}{2}kx^2\right]\right\}_0^{x_{50}}$$

that is,

$$0.5 = \exp\left(-\frac{1}{2}kx_{50}^2\right)$$

Thus

$$x_{50} = \left\{\frac{\log_e 4}{k}\right\}^{1/2}$$

■

3.6 Standard Deviation and Skewness Coefficient

The expected value or mean is a measure of the central tendency, indicating the location of the distribution on the coordinate axis representing the random variable. A measure of the variability of the random variable is usually given by a quantity known as the standard deviation. Another quantity, which not only gives a measure of the variability but also a measure of the symmetry of the density function is called the *skewness coefficient.*

[4]From calculus, the value of the definite integral can be obtained as $\int_0^\infty \sqrt{x}e^{-nx}\,dx = \frac{1}{2n}\sqrt{\pi/n}$.

3.6.1 Standard Deviation

The mean square deviation or variance of a random variable X is defined as

$$\begin{aligned}\sigma_X^2 &= \text{var}(X) = E[(X - \mu_X)^2] \\ &= E[X^2 - 2X\mu_X + \mu_X^2] \\ &= E[X^2] - 2\mu_X E[X] + E[\mu_X^2] \\ &= E[X^2] - \mu_X^2 \end{aligned} \tag{3.13}$$

and the standard deviation as

$$\sigma_X = +\sqrt{\text{var}(X)} = \sqrt{E(X^2) - \mu_X^2} \tag{3.14}$$

The coefficient of variation is a measure of dispersion in nondimensional form and is defined as

$$\text{Coefficient of variation of } X = \gamma_X = \frac{\text{standard deviation}}{\text{mean}} = \frac{\sigma_X}{\mu_X} \tag{3.15}$$

The variance of a random variable is analogous to the moment of inertia of a weight about its centroid. Figure 3.4 shows two density functions with the same mean μ_X, but with different variances. As can be seen, the variance or standard deviation is a measure of the variability of a random variable (or breadth of the density function).

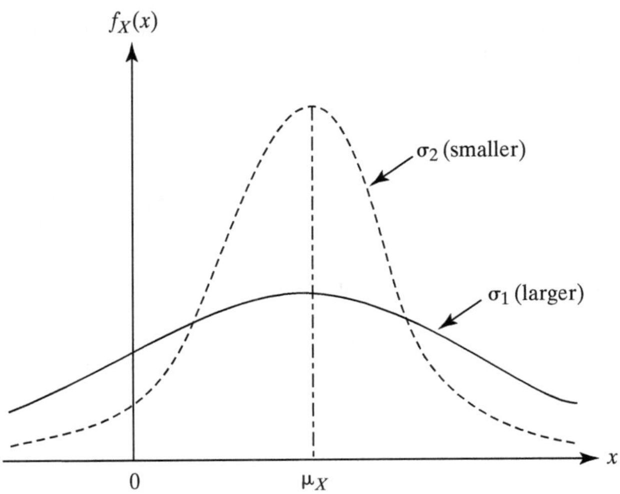

FIGURE 3.4 Probability density functions with different variances.

EXAMPLE 3.9

The number of automobiles arriving at a tollbooth per minute has the following distribution:

Number of automobiles, x	2	3	4	5	6	7	8	9
Probability mass function, $p_X(x)$	0.06	0.09	0.15	0.22	0.18	0.16	0.10	0.04

Find the mean and standard deviation of the number of automobiles arriving per minute.

Solution: Denoting the number of automobiles arriving at the tollbooth per minute as X, the mean value of X can be determined as

$$E(X) = 2(0.06) + 3(0.09) + 4(0.15) + 5(0.22) + 6(0.18) + 7(0.16) + 8(0.10) + 9(0.04) = 5.45$$

The variance of X can be computed as

$$\begin{aligned} V(X) &= \sigma_X^2 = E(X^2) - \{E(X)\}^2 \\ &= 2^2(0.06) + 3^2(0.09) + 4^2(0.15) + 5^2(0.22) + 6^2(0.18) + 7^2(0.16) \\ &\quad + 8^2(0.10) + 9^2(0.04) - (5.45)^2 = 32.91 - 29.7025 = 3.2075 \end{aligned}$$

Thus, the standard deviation of X is given by $\sigma_X = 1.7909$.

■

EXAMPLE 3.10

For the probability density given in Example 3.5 for the diameter of a rod produced on a lathe (X), find the mean and standard deviation of X.

Solution:

$$E(X) = \int_{20}^{\infty} x f_X(x)\,dx = \int_{20}^{\infty} x\,15 e^{-15(x-20)}\,dx \tag{E1}$$

Equation (E1) can be integrated by parts to obtain

$$E(X) = -xe^{-15(x-20)} - \frac{e^{-15(x-20)}}{15}\Big|_{20}^{\infty} = 20 + 0.066667 = 20.066667 \text{ mm}$$

The variance of X is given by

$$V(X) = \int_{20}^{\infty} (x - 20.066667)^2 f_X(x)\,dx = 0.004444 \text{ (mm)}^2$$

Thus, the standard deviation of X will be

$$\sigma_X = \sqrt{V(X)} = 0.066667 \text{ mm}$$

■

EXAMPLE 3.11

An experiment is conducted to find the tensile strength of steel by testing 1000 specimens. The range of the tensile strength observed in the tests is divided into sixteen equal intervals and the resulting histogram is shown in Fig. 3.5. The mid-range values of the intervals and the corresponding number of specimens that gave the tensile strength within the intervals are given below:

Interval number, i	1	2	3	4	5	6	7	8	9	10	11	12	13	14	15	16
Midpoint of the interval, x_i (kpsi)	61	62	63	64	65	66	67	68	69	70	71	72	73	74	75	76
Frequency (Number of tests in the interval i), f_i	4	19	25	32	82	106	135	153	140	130	77	47	29	13	5	3

Find the sample mean and sample standard deviation of the tensile strength of steel.

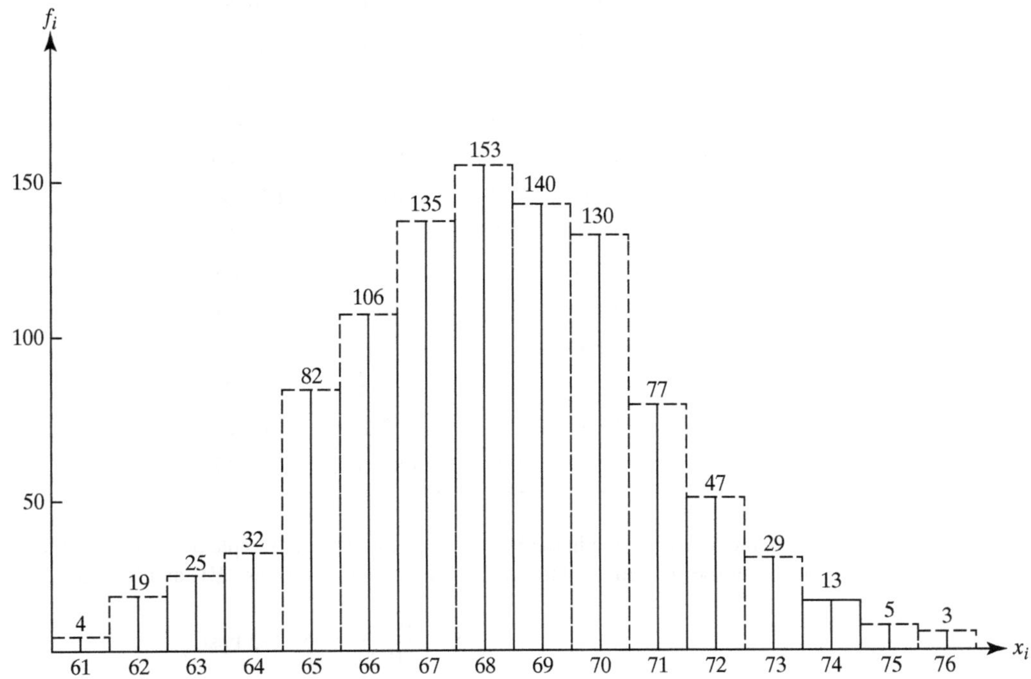

FIGURE 3.5 Histogram of the tensile strength of steel (kpsi).

Solution: For the grouped data, the sample mean and sample standard deviation can be determined as follows:

$$\overline{X} = \frac{1}{N}\sum_{i=1}^{1000} x_i = \frac{1}{N}\sum_{i=1}^{16} f_i x_i \tag{E1}$$

$$s_X = \sqrt{\frac{1}{N-1}\left\{(x_i - \overline{X})^2\right\}} = \sqrt{\frac{1}{N-1}\left\{\sum_{i=1}^{N} x_i^2 - \frac{1}{N}\left(\sum_{i=1}^{N} x_i\right)^2\right\}}$$

$$= \sqrt{\frac{1}{N-1}\left\{\sum_{i=1}^{16} f_i\, x_i^2 - \left[\frac{1}{N}\left(\sum_{i=1}^{16} f_i x_i\right)^2\right]\right\}} \tag{E2}$$

The computations involved in Eqs. (E1) and (E2) are shown in Table 3.1.

TABLE 3.1 Computations Involved in Eqs. (E1) and (E2)

i	x_i	x_i^2	f_i	$x_i f_i$	$x_i^2 f_i$
1	61	3721	4	244	14884
2	62	3844	19	1178	73036
3	63	3969	25	1575	99225
4	64	4096	32	2048	131072
5	65	4225	82	5330	346450
6	66	4356	106	6996	461736
7	67	4489	135	9045	606015
8	68	4624	153	10404	707472
9	69	4761	140	9660	666540
10	70	4900	130	9100	637000
11	71	5041	77	5467	388157
12	72	5184	47	3384	243648
13	73	5329	29	2117	154541
14	74	5476	13	962	71188
15	75	5625	5	375	28125
16	76	5776	3	228	17328
Sum			1000	68113	4646417

From the results shown in the last row of Table 3.1, the sample mean and sample standard deviation of the tensile strength of steel can be determined, using Eqs. (E1) and (E2), as

$$\overline{X} = \frac{68113}{1000} = 68.113 \text{ kpsi}$$

$$s_X = \sqrt{\frac{1}{999}\left\{4{,}646{,}417 - \frac{1}{1000}(68{,}113)^2\right\}} = \sqrt{\frac{1}{999}\{4{,}646{,}417 - 4{,}639{,}380.769\}}$$

$$= 2.6539 \text{ kpsi}$$

■

3.6.2 Skewness Coefficient

The expected value of the cube of the deviation of the random variable from its expected value (also known as the third moment of the distribution about the mean) is taken as a measure of the skewness of the distribution

$$E[(X - \mu_X)^3] = \int_{-\infty}^{\infty} (x - \mu_X)^3 f_X(x)dx, \quad \text{if } X \text{ is continuous}$$

$$= \sum_{\text{all } i} (x_i - \overline{X})^3 p_X(x_i), \quad \text{if } X \text{ is discrete} \tag{3.16}$$

The value of $E[(X - \mu_X)^3]$ can be positive or negative as illustrated in Fig. 3.6. The skewness coefficient is defined as

$$\text{Skewness coefficient} = \frac{E[(X - \mu_X)^3]}{\sigma_X^3} \tag{3.17}$$

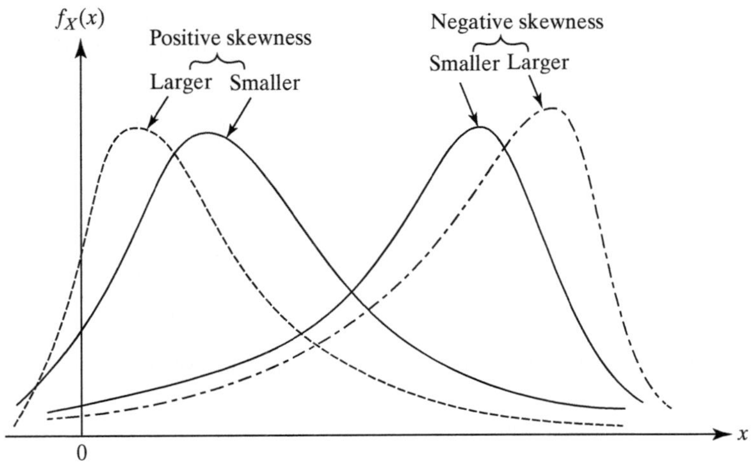

FIGURE 3.6 Probability density functions with positive and negative skewness.

EXAMPLE 3.12

Find the coefficient of variation and the skewness coefficient of the wind-induced load (X) whose density function is given by

$$f_X(x) = \frac{3}{50 \times 10^4} x(100 - x); \qquad 0 \le x \le 100 \text{ kips}$$

Solution: The mean value of X is given by

$$\mu_X = \int_0^{100} x f_X(x)dx = \frac{3}{50 \times 10^4} \int_0^{100} (100x^2 - x^3)dx$$

$$= \frac{3}{50 \times 10^4} \left(\frac{100}{3} x^3 - \frac{1}{4} x^4 \right) \Big|_0^{100}$$

$$= \frac{3}{50 \times 10^4} \left[\frac{100}{3} (10^6) - \frac{1}{4} (10^8) \right] = 50$$

The expected value of X^2 is

$$E[X^2] = \int_0^{100} x^2 f_X(x)dx = \frac{3}{50 \times 10^4} \int_0^{100} (100x^3 - x^4)dx$$

$$= \frac{3}{50 \times 10^4} \left(\frac{100}{4} x^4 - \frac{1}{5} x^5 \right) \Big|_0^{100}$$

$$= \frac{3}{50 \times 10^4} \left[25(10^8) - \frac{1}{5}(10^{10}) \right] = 3000$$

$$\therefore \sigma_X = \sqrt{E[X^2] - \mu_X^2} = \sqrt{3000 - 2500} = 22.3607$$

Thus the coefficient of variation of X will be

$$\gamma_X = \frac{\sigma_X}{\mu_X} = \frac{22.3607}{50} = 0.4472$$

The third moment of the distribution about the mean is given by

$$E[(X - \mu_X)^3] = \int_0^{100} (x - 50)^3 f_X(x)dx$$

$$= \frac{3}{50 \times 10^4} \int_0^{100} (x - 50)^3 (100x - x^2)\, dx$$

$$= \frac{3}{50 \times 10^4} \int_0^{100} (-x^5 + 250x^4 - 22{,}500x^3 + 875{,}000x^2 - 12{,}500{,}000x)\, dx$$

$$= \frac{3}{50 \times 10^4} \left[-\frac{x^6}{6} + \frac{250}{5}x^5 - \frac{22{,}500}{4}x^4 + \frac{875{,}000}{3}x^3 - \frac{12{,}500{,}000}{2}x^2 \right] \Bigg|_0^{100}$$

$$= 0$$

The skewness coefficient of X is given by Eq. (3.17)

$$\text{Skewness coefficient} = \frac{0}{\sigma_X^3} = 0$$

■

3.7 Moments of Random Variables

The moments of a distribution are the expected values of the power of the random variable that have the given distribution. The rth moment about the origin, denoted as μ_r', is given by

$$\mu_r' = \int_{-\infty}^{\infty} x^r f_X(x)\, dx \tag{3.18}$$

which can be termed as the expected value of x^r, $E(X^r)$. The zeroth order moment μ_o' is the area under the density function which is, of course, unity. The first order moment can be seen to be same as the mean. It can be seen that if the density function $f_X(x)$ is an even function about the origin, the moments of order, 1, 3, 5, . . . , are zero.

The moments about the origin for a discrete random variable that takes on the values $x_1, x_2, x_3, \ldots, x_n$ are given by

$$\mu_r' = \sum_{i=1}^{n} x_i^r p(x_i) \tag{3.19}$$

The moments about an arbitrary point k are defined as

$$E[(x-k)^r] = \int_{-\infty}^{\infty} (x-k)^r f_X(x)\, dx, \qquad \text{for a continuous random variable}$$

$$= \sum_{i=1}^{n} (x_i - k)^r p(x_i), \qquad \text{for a discrete random variable} \tag{3.20}$$

If $k = \mu_1' =$ mean, the moments are called central moments and are given by

$$\mu_r = E[(x - \mu_1')^r] = \int_{-\infty}^{\infty} (x - \mu_1')^r f_X(x)\, dx \tag{3.21}$$

Accordingly,

$$\begin{aligned}\mu_1 &= \text{first moment about the mean} = \int_{-\infty}^{\infty} (x - \mu_1')^2 f_X(x)\, dx \\ &= \mu_1' - \mu_1' = 0 \qquad (3.22)\end{aligned}$$

and

$$\begin{aligned}\mu_2 &= \text{second moment about the mean} = \int_{-\infty}^{\infty} (x - \mu_1')^2 f_X(x)\, dx \\ &= \int_{-\infty}^{\infty} (x^2 - 2x\mu_1' + \mu_1'^2) f_X(x)\, dx \\ &= \mu_2' - (\mu_1')^2 = \text{var}(X) \qquad (3.23)\end{aligned}$$

It can be seen that for a density function which is symmetrical about the mean, the moments $\mu_1, \mu_3, \mu_5, \ldots,$ are zero. Although the higher-order moments give additional information about the random variable, they are generally not required since μ_X and σ_X reveal enough information about the variable for most engineering purposes.

3.8 Importance of Moment Functions—Chebyshev Inequality

The moment function defined by Eqs. (3.18) to (3.21) represent the gross properties of the probabilistic nature of a random variable. In most applications, only the first few moments are generally used. To illustrate the application of the moment functions, we consider an inequality known as the *Chebyshev inequality.* According to the Chebyshev inequality, the following relation is valid for any random variable X regardless of its distribution

$$P\left[\,|x - \mu_X| \geq t\sigma_X\right] \leq \frac{1}{t^2} \qquad (3.24)$$

where t is any positive constant. In other words, the Chebyshev inequality states that the probability of the absolute value of the difference between the random variable and its mean exceeding a constant t times the standard deviation is not greater than t^{-2}. For $t = 2$, Eq. (3.24) gives

$$P\left[\,|x - \mu_X| \geq 2\sigma_X\right] \leq \frac{1}{2^2} = 0.2500$$

and for $t = 4$, we get

$$P\left[\,|x - \mu_X| \geq 4\sigma_X\right] \leq \frac{1}{4^2} = 0.0625$$

Thus, the probability of exceeding a stated magnitude of deviation from the mean tends to decrease as σ_X decreases. The inequality (3.24) can be applied to simple design problems as indicated in the following example.

EXAMPLE 3.13

Find the area of cross section of the uniform bar shown in Fig. 3.7 under a probabilistic load P with $\mu_P = 0$ and $E(P^2) = \sigma_P^2$. Assume the cross-sectional area of the bar, A, and the permissible stress, x^*, to be deterministic.

Solution: If X denotes the uniform stress at any section in the bar, we have

$$X = \frac{P}{A} \tag{E1}$$

Since A is deterministic $(\mu_A = A \text{ and } \sigma_A = 0)$, we obtain

$$\mu_X = 0, \qquad \sigma_X = \frac{\sigma_P}{A} \tag{E2}$$

The bar will fail whenever $|X| \geq x^*$. Thus, the probability of failure of the bar can be estimated using the inequality (3.24) by setting

$$x^* = t\sigma_X \tag{E3}$$

Equations (E3) and (E2) give

$$t = \frac{x^*}{\sigma_X} = \frac{x^* A}{\sigma_P} \tag{E4}$$

The application of inequality (3.24) yields

$$P\left[|X| \geq x^*\right] \leq \frac{\sigma_P^2}{x^{*2} A^2}$$

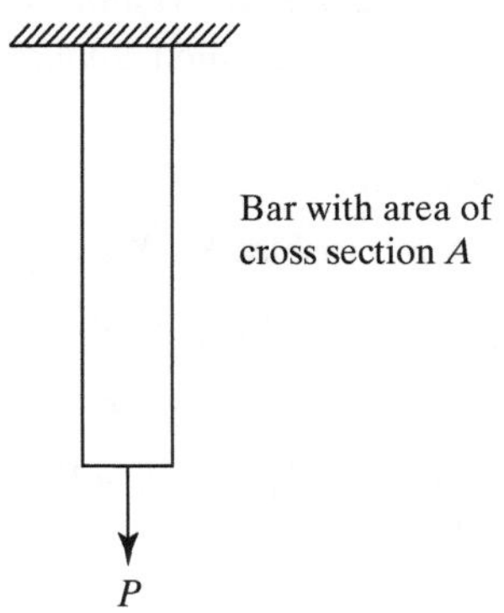

FIGURE 3.7 Bar subjected to an axial force P.

This shows that in order to design the bar for a specified upper bound on the probability of failure, say $1/t^2$, the area of cross section of the bar has to be equal to

$$A = \frac{t \cdot \sigma_P}{x^*}$$

For example, a probability of failure of 1 in 100 implies that $t = 10$ and the corresponding area is given by

$$A = \frac{10 \cdot \sigma_P}{x^*}$$

■

The Chebyshev inequality (3.24) applies to any distribution. It can be shown that if the probability density function is unimodal,[5] the following inequality, known as *Gauss inequality,* holds [3.1]

$$P\left[\,|x - \mu_X| \geq t\sigma_X\right] \leq \frac{4(1 + \alpha^2)}{9(t - |\alpha|)^2} \tag{3.25}$$

for $t > |\alpha|$ where $\alpha = (\mu_X - X_{\text{mode}})/\sigma_X$. If the probability density function is symmetric and unimodal, the mean and the mode coincide $(\alpha = 0)$ and the Gauss inequality reduces to the *Camp-Meidell inequality* [3.1]:

$$P\left[\,|x - \mu_X| \geq t\sigma_X\right] \leq \frac{4}{9}\frac{1}{t^2} \text{ for } t > 0 \tag{3.26}$$

It can be seen that this inequality is sharper than Chebyshev inequality by a factor of $4/9$.

3.9 Jointly Distributed Random Variables

When two or more random variables are being considered simultaneously, their joint behavior is determined by a joint probability distribution function. The probability distributions of single random variables are called *univariate* distributions and the distributions that involve two random variables are called *bivariate* distributions. In general, if a distribution involves more than one random variable, it is called a *multivariate* distribution.

3.9.1 Joint Density and Distribution Functions

The joint density function of n continuous random variables $X_1, X_2, \ldots, X_n$ is defined [3.8] as

$$\begin{aligned} f_{X_1,\ldots,X_n}(x_1, x_2, \ldots, x_n) \cdot dx_1 \ldots dx_n &= P(x_1 \leq X_1 \leq x_1 + dx_1, \\ &\quad x_2 \leq X_2 \leq x_2 + dx_2, \ldots, x_n \leq X_n \leq x_n + dx_n) \\ &= \text{Probability that a random point } x_1', x_2', \ldots, x_n' \text{ will fall in the} \\ &\quad \text{infinitesimal region defined by} \\ &\quad\quad x_1 \leq x_1' \leq x_1 + dx_1, \ldots, x_n \leq x_n' \leq x_n + dx_n \end{aligned} \tag{3.27}$$

[5]A *unimodal distribution* is one that has a single peak in its range of definition.

For independent random variables, the joint density function is given by the product of individual or marginal density functions as

$$f_{X_1,\ldots,X_n}(x_1,\ldots,x_n) = f_{X_1}(x_1)\ldots f_{X_n}(x_n) \tag{3.28}$$

The joint distribution function

$$F_{X_1,X_2,\ldots,X_n}(x_1,x_2,\ldots,x_n)$$

associated with the density function of Eq. (3.27) is given by

$$\begin{aligned} F_{X_1,\ldots,X_n}(x_1,\ldots,x_n) &= P[X_1 \le x_1,\ldots,X_n \le x_n] \\ &= \int_{-\infty}^{x_1}\cdots\int_{-\infty}^{x_n} f_{X_1,\ldots,X_n}(x_1',x_2',\ldots,x_n')\cdot dx_1'\cdot dx_2'\ldots dx_n' \end{aligned} \tag{3.29}$$

If $X_1, X_2, \ldots, X_n$ are independent random variables, we have

$$F_{X_1,\ldots,X_n}(x_1,\ldots,x_n) = F_{X_1}(x_1)\cdot F_{X_2}(x_2)\ldots F_{X_n}(x_n) \tag{3.30}$$

It can be seen that the joint density function can be obtained by differentiating the joint distribution function as

$$f_{X_1,\ldots,X_n}(x_1,\ldots,x_n) = \frac{\partial^n}{\partial x_1\,\partial x_2\ldots\partial x_n}[F_{X_1,\ldots,X_n}(x_1,\ldots,x_n)] \tag{3.31}$$

3.9.2 Obtaining the Marginal or Individual Density Function from the Joint Density Function

Let the joint density function of two random variables X and Y be denoted by $f(x,y)$ and the marginal or individual density functions of X and Y by $f_X(x)$ and $f_Y(y)$, respectively. Take the infinitesimal rectangle with corners located at the points $(x,y),(x+dx,y),(x,y+dy)$, and $(x+dx,y+dy)$. The probability of a random point (x',y') falling in this rectangle is $f_{X,Y}(x,y)\cdot dx\cdot dy$. The integral of such probability elements with respect to y (for a fixed value of x) is the sum of the probabilities of all the mutually exclusive ways of obtaining the points lying between x and $x+dx$. Let the lower- and upper- limits of y be $a_1(x)$ and $b_1(x)$: Then

$$P[x \le x' \le x+dx] = \left[\int_{a_1(x)}^{b_1(x)} f_{X,Y}(x,y)\cdot dy\right]\cdot dx = f_X(x)\cdot dx$$

$$\therefore f_X(x) = \int_{y_1=a_1(x)}^{y_2=b_1(x)} f_{X,Y}(x,y)\cdot dy \tag{3.32}$$

Similarly, we can show that

$$f_Y(y) = \int_{a_2(y)}^{b_2(y)} f_{X,Y}(x,y) \cdot dx \tag{3.33}$$

where $a_2(y)$ and $b_2(y)$ denote the lower- and upper- limits of x for the specified value of y.

EXAMPLE 3.14

The joint density function of two random variables X and Y is given by

$$f(x, y) = \begin{cases} 12 \exp(-4x - 2y); & 0 \le x \le y,\ 0 \le y < \infty \\ 0; & \text{otherwise} \end{cases}$$

Determine whether X and Y are independent.

Solution: The marginal density functions of X and Y can be computed as (see Fig. 3.8 for the ranges of x and y):

$$f(x) = 12\int_{y=x}^{\infty} e^{-4x-2y} dy = 12e^{-4x}\int_{x}^{\infty} e^{-2y} dy$$

$$= 12e^{-4x}\left(\frac{e^{-2y}}{-2}\right)\Bigg|_{x}^{\infty} = 6e^{-6x}$$

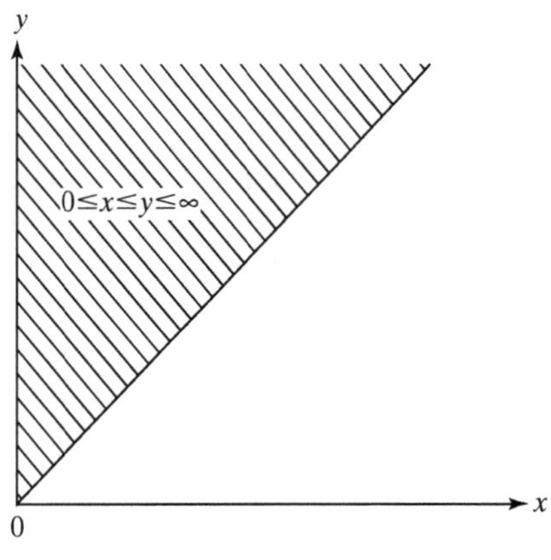

FIGURE 3.8 Ranges of x and y.

and

$$f(y) = 12\int_{x=0}^{y} e^{-4x-2y}dx = 12e^{-2y}\int_{0}^{y} e^{-4x}dx$$

$$= 12e^{-2y}\left(\frac{e^{-4x}}{-4}\right)\Bigg|_0^y = 3e^{-2y}(1 - e^{-4y})$$

Since $f(x, y) \neq f(x)f(y)$, the variables X and Y are not independent.

■

3.10 Moments of Jointly Distributed Random Variables

If $Z = g(X, Y)$ is a function of two jointly distributed random variables, its expected value is given by

$$E[Z] = E[g(X,Y)] = \sum_{\text{all}\, x_i} \sum_{\text{all}\, y_j} g(x_i, y_j)\, p_{X,Y}(x_i, y_j)$$

if X and Y are discrete, and

$$E[Z] = \int_{-\infty}^{\infty}\int_{-\infty}^{\infty} g(x,y)\cdot f_{X,Y}(x,y)\cdot dx\cdot dy \tag{3.34}$$

if X and Y are continuous.

In particular, if we take

$$g(x, y) = x^l y^m \tag{3.35}$$

the moment about the origin of order $l + m$ of the random variables X and Y is given by

$$E[X^l Y^m] = \int_{-\infty}^{\infty}\int_{-\infty}^{\infty} x^l y^m f_{X,Y}(x, y)\cdot dx\cdot dy \tag{3.36}$$

Similarly, by taking

$$g(x, y) = (x - \mu_X)^l (y - \mu_Y)^m \tag{3.37}$$

the central moment of order $l + m$ of the variables X and Y can be obtained as

$$E[g(x, y)] = E[(X - \mu_X)^l (Y - \mu_Y)^m]$$

$$= \int_{-\infty}^{\infty}\int_{-\infty}^{\infty} (x - \mu_X)^l (y - \mu_Y)^m f_{X,Y}(x, y)\, dx\, dy \tag{3.38}$$

The most valuable central moments are the second-order moments, namely, the moments obtained by taking $(l = 2, m = 0)$, $(l = 0, m = 2)$, and $(l = 1, m = 1)$. For $l = 2, m = 0$, we obtain the variance of X as

$$E[(X - \mu_X)^2] = \text{var}[X] = \int_{-\infty}^{\infty} (x - \mu_X)^2 \cdot f_X(x) \cdot dx \tag{3.39}$$

For $l = 0, m = 2$, we obtain the variance of Y as

$$E[(X - \mu_Y)^2] = \text{var}[Y] = \int_{-\infty}^{\infty} (y - \mu_Y)^2 \cdot f_Y(y) \cdot dy \tag{3.40}$$

For $l = 1, m = 1$, we obtain the covariance of X and Y as

$$E[(X - \mu_X)(Y - \mu_Y)] = \text{cov}(X, Y) \tag{3.41}$$

$$= \int_{-\infty}^{\infty} \int_{-\infty}^{\infty} (x - \mu_X)(y - \mu_Y) \cdot f_{X,Y}(x, y) \cdot dx \cdot dy = \sigma_{X,Y}$$

The correlation coefficient $\rho_{X,Y}$ for the random variables is defined as

$$\rho_{X,Y} = \frac{\text{cov}(X, Y)}{\sigma_X \cdot \sigma_Y} \tag{3.42}$$

and it can be proven that $-1 \leq \rho_{X,Y} \leq 1$. The physical meaning of the correlation coefficient is that its value will be nearly unity provided the two random variables are linearly related, and will be nearly zero if the two random variables are not linearly related. The significance of the correlation coefficient is illustrated in Fig. 3.9.[6]

3.11 Probability Distributions

There are several types of probability distributions for describing various types of discrete and continuous random variables. Some of the common distributions and their characteristics are summarized in Table 3.2. In any physical problem, one chooses a particular type of probability distribution depending on the:

1. nature of the problem;
2. underlying assumptions associated with the distribution;
3. shape of the graph between $f(x)$ or $F(x)$ and x obtained after plotting the available data, and;
4. convenience and simplicity afforded by the distribution in subsequent computations.

[6] If experiments are conducted to find the values of two related random variables X and Y (such as the yield strength and ultimate strength of steel), we realize the values x_i and y_i in the ith experiment. If the experimental values realized in all the experiments, (x_i, y_i), are plotted in the (X, Y) space, the resulting diagram is known as the *scattergram*.

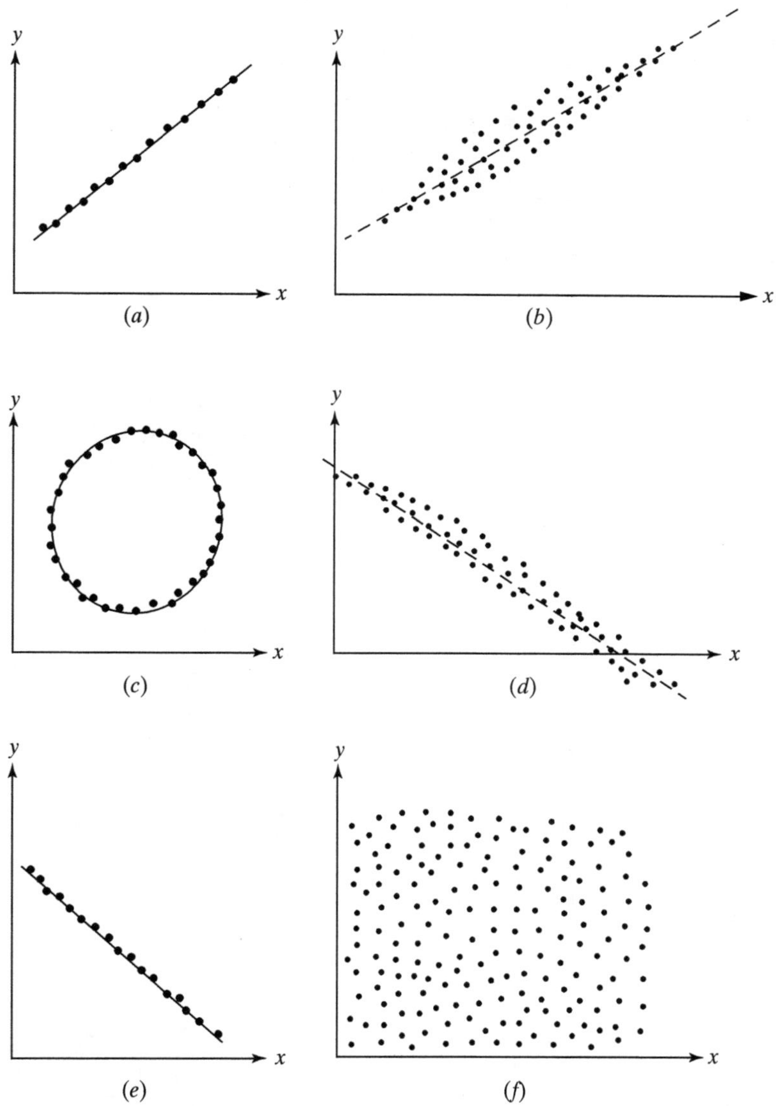

FIGURE 3.9 (*a*) $\rho_{X,Y} = 1$ (*X* and *Y* are linearly related with positive slope), (*b*) $\rho_{X,Y} =$ positive (*X* and *Y* are slightly linearly related with positive slope), (*c*) $\rho_{X,Y} = 0$ (*X* and *Y* are not linearly related), (*d*) $\rho_{X,Y} =$ negative (*X* and *Y* are slightly linearly related with negative slope), (*e*) $\rho_{X,Y} = -1$ (*X* and *Y* are linearly related with negative slope), (*f*) $\rho_{X,Y} = 0$ (*X* and *Y* are not linearly related).

The properties of some of the more commonly used distributions are presented in the following sections and the characteristics of some additional distributions are summarized in Table 3.3.

TABLE 3.2 Characteristics of Common Probability Distributions

Distribution	Probability Mass or Density Function, $p(x)$ or $f(x)$	Range of x	Expected Value, $E(X)$	Variance, σ^2	Comments
Binomial	$\frac{n!}{x!(n-x)!}p^x(1-p)^{n-x}$	$x = 1, 2, \ldots, n$; $0 \leq p \leq 1$	np	$np(1-p)$	n = number of trials; p = probability of success in each trial
Poisson	$\frac{e^{-\lambda}\lambda^x}{x!}$	$x = 0, 1, 2, \ldots$; $\lambda > 0$	λ	λ	λ = average number of successes
Hypergeometric	$\frac{\binom{k}{x}\binom{N-k}{n-x}}{\binom{N}{n}}$	$x = 0, 1, 2, \ldots, n$	np	$np(1-p)\left(\frac{N-n}{N-1}\right)$	N = lot size (total number of items) k = number of good items in the lot $N - k$ = number of bad items in the lot n = sample size (number of items selected randomly from the lot)
Uniform (rectangular)	$\frac{1}{b-a}$	$a \leq x \leq b$	$\frac{a+b}{2}$	$\frac{(b-a)^2}{12}$	$-\infty < a,b < \infty$
Triangular	$\frac{2}{b-a}\left(\frac{x-a}{c-a}\right)$ for $a \leq x \leq c$ and $\frac{2}{b-a}\left(\frac{b-x}{b-c}\right)$ for $c \leq x \leq b$	$a \leq x \leq b$	$\frac{1}{3}(a+b+c)$	$\frac{1}{18}(a^2+b^2+c^2$ $-ab-bc-ca)$	$a \leq c \leq b$
Exponential	$\lambda e^{-\lambda x}$	$o \leq x < \infty$	$\frac{1}{\lambda}$	$\frac{1}{\lambda^2}$	$\lambda > 0$
Gaussian (normal)	$\frac{1}{\sqrt{2\pi}\sigma}\exp\left\{-\frac{(x-\mu)^2}{2\sigma^2}\right\}$	$-\infty < x < \infty$	μ	σ^2	$-\infty < \mu < \infty; \sigma > 0$

TABLE 3.3 Characteristics of Additional Probability Distributions

Distribution	Probability Density Function, $f(x)$	Range of x	Expected Value, $E(X)$	Variance, σ^2	Comments
Rayleigh	$kx \exp\left\{-\frac{k}{2}x^2\right\}$	$0 \le x < \infty$	$\sqrt{\frac{\pi}{2k}}$	$\frac{2}{k}\left(1 - \frac{\pi}{4}\right)$	$k > 0$
Lognormal	$\frac{1}{\sqrt{2\pi}\sigma x}\exp\left\{-\frac{(\ln x - \mu)^2}{2\sigma^2}\right\}$	$0 < x < \infty$	$\exp\left(\mu + \frac{\sigma^2}{2}\right)$	$\exp(2\mu + 2\sigma^2) - \exp(2\mu + \sigma^2)$	$-\infty < \mu < \infty$
Weibull	$\frac{\beta}{\eta - \varepsilon}\left(\frac{x - \varepsilon}{\eta - \varepsilon}\right)^{\beta - 1} \exp\left\{-\left(\frac{x - \varepsilon}{\eta - \varepsilon}\right)^{\beta}\right\}$	$\varepsilon \le x < \infty$	$\varepsilon + (\eta - \varepsilon)\Gamma\left(1 + \frac{1}{\beta}\right)$	$(\mu - \varepsilon)^2\left\{\Gamma\left(1 + \frac{2}{\beta}\right) - \Gamma^2\left(1 + \frac{1}{\beta}\right)\right\}$	$\Gamma()$ = gamma function ε = lower bound on x β = shape parameter η = characteristic value
Gamma	$\frac{\lambda^k}{\Gamma(k)}x^{k-1}e^{-\lambda x}$	$0 \le x < \infty$	$\frac{k}{\lambda}$	$\frac{k}{\lambda^2}$	$k > 0; \lambda > 0$; $\Gamma()$ = gamma function
Beta	$\frac{1}{\beta(p, q)}\frac{(x - a)^{p-1}(b - x)^{q-1}}{(b - a)^{p+q-1}}$	$a \le x \le b$	$a + \frac{p}{p + q}(b - a)$	$\frac{pq(b - a)^2}{(p + q)^2(p + q - 1)}$	p, q = positive $\beta(p, q)$ = beta function $= \frac{\Gamma(p) \cdot \Gamma(q)}{\Gamma(p + q)}$

3.11.1 Binomial Distribution

Repeated independent trials of a random experiment are called Bernoulli trials if there are only two possible outcomes for each trial and their probabilities remain the same throughout the trials. Let S (success) and F (failure) denote the two outcomes and p and q the corresponding probabilities. Clearly

$$p + q = 1 \tag{3.43}$$

The sample space of each individual trial is formed by the two sample points S and F. The sample space of n-Bernoulli trials contains 2^n sample points. Since the trials are independent, the probabilities multiply. For example,

$$P(SSFS) = ppqp = p^3q$$

Let S_n denote the number of successes in n trials. Then it can be shown that

$$P(S_n = k) = \binom{n}{k} p^k q^{n-k} \tag{3.44}$$

Since Eq. (3.44) represents the kth term in the binomial expansion of $(p + q)^n$, it is called the binomial distribution. It can also be shown that

$$E[S_n] = \overline{S}_n = np \tag{3.45}$$

$$\text{var}[S_n] = E[(S_n - \overline{S}_n)^2] = npq \tag{3.46}$$

EXAMPLE 3.15

A thermal power plant buys four boilers. If the probability of a boiler functioning without failure for a year is 0.7, find the probability mass and distribution functions of the status of the boilers at the end of the year.

Solution: Here $p = 0.7$, $q = 0.3$, and $n = 4$.

$$\begin{aligned}
P(S_n = 0) &= 1(0.7)^0(0.3)^4 = 0.0081 \\
P(S_n = 1) &= 4(0.7)^1(0.3)^3 = 0.0756 \\
P(S_n = 2) &= 6(0.7)^2(0.3)^2 = 0.2646 \\
P(S_n = 3) &= 4(0.7)^3(0.3)^1 = 0.4116 \\
P(S_n = 4) &= 1(0.7)^4(0.3)^0 = 0.2401
\end{aligned}$$

The probability mass and distribution functions are shown in Figs. 3.10(*a*) and 3.10(*b*), respectively.

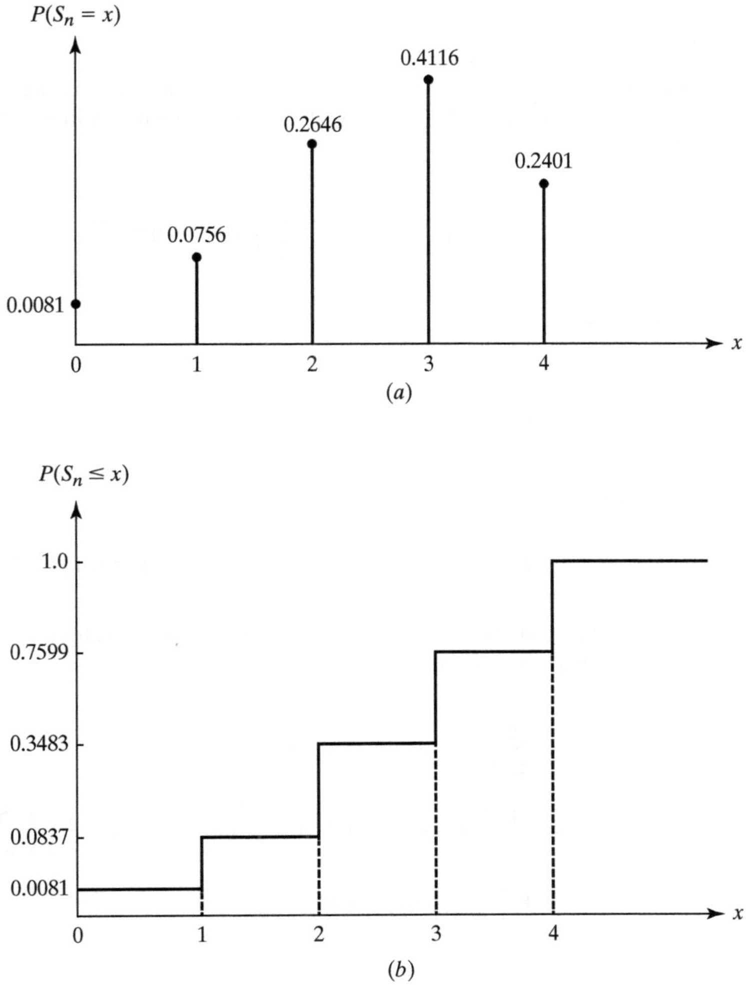

FIGURE 3.10 Probability mass and distribution functions of X in Example 3.15.

■

EXAMPLE 3.16

There is a 5 percent chance of dimensional distortion in each of the forged connecting rods produced in an engine manufacturing plant.

1. Determine the probability of finding exactly two connecting rods with dimensional distortion when 15 connecting rods are inspected. Assume that the connecting rods are independent with respect to dimensional distortions.

2. Determine the probability that the number of connecting rods with dimensional distortion lies between 2 and 5 when 15 connecting rods are inspected.

Solution:

1. If the number of connecting rods with dimensional distortion in the 15 connecting rods inspected is denoted as X, then X follows binomial distribution with $p = 0.05$ and $n = 15$. Then

$$P(X = 2) = \binom{15}{2}(0.05)^2(0.95)^{13}$$

Noting that

$$\binom{15}{2} = \frac{15!}{2!\,13!} = 105$$

we obtain

$$P(X = 2) = 105\,(0.0025)\,(0.513342) = 0.134752$$

2. The probability of finding the number of connecting rods with dimensional distortion to lie between 2 and 5 is given by

$$\begin{aligned} P(2 \le X \le 5) &= \sum_{x=2}^{5} \binom{15}{x}(0.05)^x\,(0.95)^{15-x} \\ &= \binom{15}{2}(0.05)^2(0.95)^{13} + \binom{15}{3}(0.05)^3(0.95)^{12} + \binom{15}{4}(0.05)^4(0.95)^{11} \\ &\quad + \binom{15}{5}(0.05)^5(0.95)^{10} \end{aligned}$$

Noting that

$$\binom{15}{3} = \frac{15!}{3!\,12!} = 455, \binom{15}{4} = \frac{15!}{4!\,11!} = 1365, \binom{15}{5} = \frac{15!}{5!\,10!} = 3003,$$

we find

$$\begin{aligned} P(2 \le X \le 5) &= 0.134752 + 455\,(125 \times 10^{-6})\,(0.540360) \\ &\quad + 1365\,(625 \times 10^{-8})\,(0.568800) + 3003\,(3125 \times 10^{-10})\,(0.598737) \\ &= 0.134752 + 0.030733 + 0.004852 + 0.000562 = 0.170899 \end{aligned}$$

■

3.11.2 Poisson Distribution

In many engineering problems, certain isolated events occur at any point in time or space. The occurrences of these events are affected only by chance. The following examples indicate such problems:

1. Fatigue cracks may occur along a long metal component such as an airplane wing.
2. Telephone calls may be received at any time over a specified period of time.
3. Earthquakes may occur at any time and at any point within a seismic zone.
4. Automobiles may arrive at a tollbooth at any time.
5. Automobile accidents can occur at any time along a given highway.

An important characteristic of Poisson distribution is that only the occurrences of the event are counted but not its non-occurrences. Note that in binomial distribution, both the occurrences and non-occurrences of an event, are counted. Thus, in situations when only the occurrences of an event are counted, Poisson distribution can be used. We can try to apply binomial distribution to events occurring randomly in time or space by dividing the time or space into small intervals and assuming that the event will either occur or not occur in each interval. Each event, in this case is considered to be a Bernoulli trial (with only two possibilities). However, since a possibility exists for the occurrence of the event more than once in any small time or space interval, the binomial distribution is not very convenient to use. Hence, the application of Poisson distribution is more appropriate to such cases.

In Poisson distribution, the occurrence of events is assumed to follow a Poisson process that is based on the following assumptions:

1. An event can occur randomly at any time or at any point in space.
2. The occurrence(s) of an event in any time interval is independent of those occurring in other time intervals.
3. The probability of occurrence of an event in a subinterval is the same for all subintervals and proportional to the length of the subinterval.

The random variable X that denotes the number of events in an interval is a Poisson random variable with parameter $\lambda > 0$ with λ representing the mean rate of occurrence of the event, assumed to be a constant. The probability mass function of X is given by

$$f_X(x) = \frac{e^{-\lambda}\lambda^x}{x!}; x = 0, 1, 2, \ldots \tag{3.47}$$

Notes:

1. The sum of the probabilities given by Eq. (3.47) is 1. This can be seen by considering

$$\sum_{x=0}^{\infty} \frac{e^{-\lambda}\lambda^x}{x!} = e^{-\lambda}\sum_{x=0}^{\infty} \frac{\lambda^x}{x!} \tag{3.48}$$

Since the summation on the right-hand side of Eq. (3.48) denotes the Taylor's series expansion of e^{λ} evaluated at λ, the right-hand side of Eq. (3.48) will be equal to 1.

2. Proper units are to be used for λ (the mean rate of occurrence of the event in the given interval or the mean value of X) in calculating the probabilities, means, and variances of a Poisson random variable. For example, if the average number of fatigue cracks in a meter length of a continuous weld is 1.8, the average number of fatigue cracks in a four-meter length of continuous weld will be $1.8 \times 4 = 7.2$. The following examples illustrate this point.

EXAMPLE 3.17

The average number of defects in a long underground cable per 1000 meter length of cable per year is 0.07. Find the probability mass functions of the following variables:

1. Number of defects in a cable of length 500 meters in a period of 5 years.
2. Number of defects in a cable of length 2000 meters in a period of 8 years.

Solution:

1. The average number of defects for a cable of length 500 meters for a period of 5 years is given by

$$\lambda_1 = \frac{0.07}{1000}(500)(5) = 0.175 \text{ defects/500 meters/5 years}$$

The probability mass function of the random variable X, the number of defects in 500 meters long cable in 5 years, can be expressed as

$$f_X(x) = \frac{e^{-\lambda_1}(\lambda_1)^x}{x!} = \frac{e^{-0.175}(0.175)^x}{x!}; x = 0, 1, 2, \ldots \quad \text{(E1)}$$

2. The average number of defects for a cable of length 2000 meters for a period of 8 years is given by

$$\lambda_2 = \frac{0.07}{1000}(2000)(8) = 1.12 \text{ defects/2000 meters/8 years}$$

The probability mass function of the random variable X, the number of defects in 2000 meters long cable in 8 years, can be expressed as

$$f_X(x) = \frac{e^{-\lambda_2}(\lambda_2)^x}{x!} = \frac{e^{-1.12}(1.12)^x}{x!}; x = 0, 1, 2, \ldots \quad \text{(E2)}$$

The probability mass functions given by Eqs. (E1) and (E2) are shown in Figs. 3.11(a) and 3.11(b), respectively.

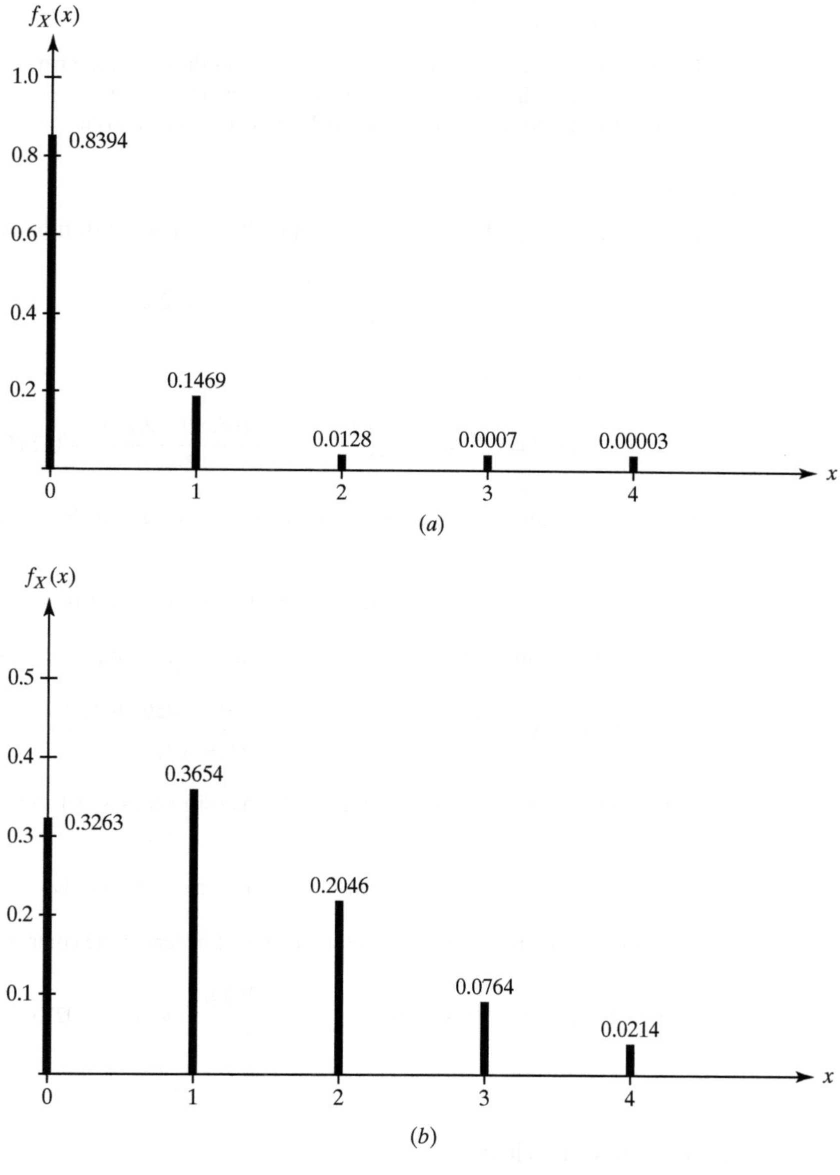

FIGURE 3.11 Probability mass functions given by Eqs. (E1) and (E2) in Example 3.17.

■

EXAMPLE 3.18

Fiber-reinforced composite bars are subject to flaws such as voids, debonding, and breakage of fibers. Assuming that the number of flaws follows a Poisson distribution with a mean of 1.8 flaws per meter, determine the following:

1. Probability of finding exactly 2 flaws in 1 meter of the composite bar.
2. Probability of finding 5 flaws in 3 meters of the composite bar.
3. Probability of finding at least 1 flaw in 2 meters of the composite bar.

Solution:

1. Since the number of flaws in a meter length follows a Poisson distribution, we have

$$f_X(x) = \frac{e^{-\lambda}\lambda^x}{x!};\ x = 0, 1, 2, \ldots$$

where $\lambda = E(X) = 1.8$. Thus,

$$P(X = 2) = \frac{e^{-1.8}(1.8^2)}{2!} = \frac{(0.165299)(3.24)}{2} = 0.267784$$

2. Denoting the number of flaws in a 3 meter length of composite bar as X, X follows a Poisson distribution with

$$E(X) = 3 \text{ meters} \times 1.8 \text{flaws/meter} = 5.4 \text{ flaws}$$

Then the probability of finding 5 flaws in 3 meters of the composite bar can be found as

$$P(X = 5) = \frac{e^{-5.4}(5.4^5)}{5!} = \frac{(0.00451658)(4591.65024)}{(5)(4)(3)(2)} = 0.172821$$

3. Denoting the number of flaws in 2 meters of composite bar as X, X follows a Poisson distribution with

$$E(X) = 2 \text{ meters} \times 1.8 \text{ flaws/meter} = 3.6 \text{ flaws}$$

Then the probability of finding at least 1 flaw in 2 meters of the composite bar is given by

$$P(X \geq 1) = 1 - P(X = 0) = 1 - \frac{e^{-3.6}(3.6^0)}{0!} = 1 - 0.0273237 = 0.972676$$

■

3.11.3 Normal Distribution

The density function of a normally distributed random variable X (also known as Gaussian distribution) is given by

$$f_X(x) = \frac{1}{\sqrt{2\pi}\sigma_X} \exp\left[-\frac{1}{2}\left(\frac{x - \mu_X}{\sigma_X}\right)^2\right] \quad (3.49)$$

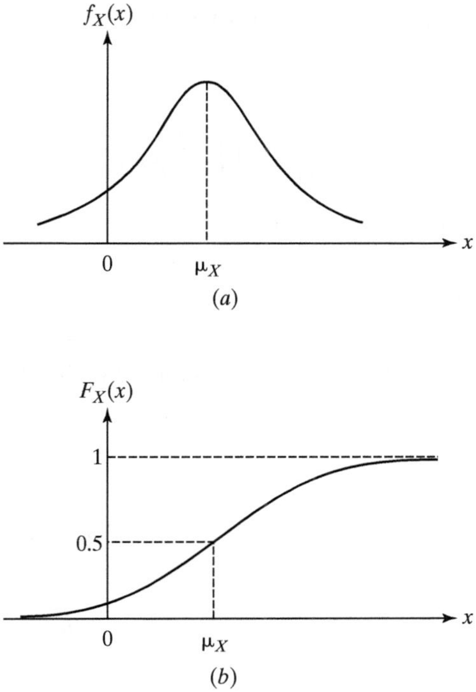

FIGURE 3.12 Normal distribution. (*a*) Density function, (*b*) Distribution function.

and X is identified as $N(\mu_X, \sigma_X)$. The parameters of the distribution μ_X and σ_X denote, respectively, the mean value and standard deviation of the variable X. The density function and the corresponding distribution function are shown in Fig. 3.12. The normal distribution has the following properties:

1. Any linear function of a normally distributed random variable is also normally distributed.
2. If

$$E[X] = 0, \qquad \text{then}$$
$$E[X^{2n+1}] = 0, \qquad \text{and}$$
$$E[X^{2n}] = 1 \cdot 3 \cdot 5 \ldots \cdot (2n - 1) \cdot \sigma_X^{2n} \tag{3.50}$$

Hence all the moments can be deduced from the second moment of a normal variate.[7]

[7]A normally distributed random variable is also known as a *normal variate*.

Joint normal law for n independent random variables.

$$f_{X_1, X_2, \ldots, X_n}(x_1, x_2, \ldots, x_n) = \frac{1}{\sqrt{(2\pi)^n} \cdot \sigma_1 \cdot \sigma_2 \cdot \sigma_n} \exp\left[-\frac{1}{2} \sum_{k=1}^{n} \left(\frac{x_k - \mu_k}{\sigma_k} \right)^2 \right] = f_{X_1}(x_1) \cdot f_{X_2}(x_2) \ldots f_{X_n}(x_n) \tag{3.51}$$

where $\sigma_i = \sigma_{X_i}$ and $\mu_i = \mu_{X_i}$.

If correlation between X_k and X_j is not zero. In this case, the joint density function is given by

$$f_{X_1, X_2, \ldots, X_n}(x_1, x_2, \ldots, x_n) = \frac{1}{\sqrt{(2\pi)^n |[K]|}} \exp\left[-\frac{1}{2} \sum_{j=1}^{n} \sum_{k=1}^{n} \left\{ [K]^{-1} \right\}_{jk} (x_j - \mu_{X_j}) \cdot (x_k - \mu_{X_k}) \right] \tag{3.52}$$

where

$$K_{X_j X_k} = E[\,(x_j - \mu_{X_j}) \cdot (x_k - \mu_{X_k})\,] = \int_{-\infty}^{\infty} \int_{-\infty}^{\infty} (x_j - \mu_{X_j}) \cdot (x_k - \mu_{X_k}) \cdot f_{X_j, X_k}(x_j, x_k) \cdot dx_j \cdot dx_k$$

$= \text{covariance between } X_j \text{ and } X_k,$

$$[K] = \text{correlation matrix} = \begin{bmatrix} K_{11} & K_{12} & \ldots & K_{1n} \\ K_{21} & K_{22} & \ldots & K_{2n} \\ \vdots & \vdots & & \vdots \\ K_{n1} & K_{n2} & \ldots & K_{nn} \end{bmatrix}$$

and

$$\left\{ [K]^{-1} \right\}_{jk} = jk\text{th element of } [K]^{-1}.$$

It is to be noted that

$$K_{X_j, X_k} = \begin{cases} 0, & \text{for } j \neq k \\ \sigma_{X_j}^2, & \text{for } j = k \end{cases}$$

in case there is no correlation between X_j and X_k.

Standard Normal Distribution. A Gaussian distribution with parameters $\mu = 0$ and $\sigma = 1$ is called the *standard normal distribution* and is identified as $N(0, 1)$. The density function of a standard normal variate Z is given by

$$f_Z(z) = \frac{1}{\sqrt{2\pi}} e^{-1/2z^2}; \qquad -\infty < z < \infty \tag{3.53}$$

and is symmetric about its mean $\mu = 0$, as shown in Fig. 3.13. The distribution function of the standard normal variate Z is commonly denoted as $\Phi(z)$ and is given by

$$\Phi(z) = F_Z(z) = \int_{-\infty}^{z} \frac{1}{\sqrt{2\pi}} e^{-1/2z^2} dz \tag{3.54}$$

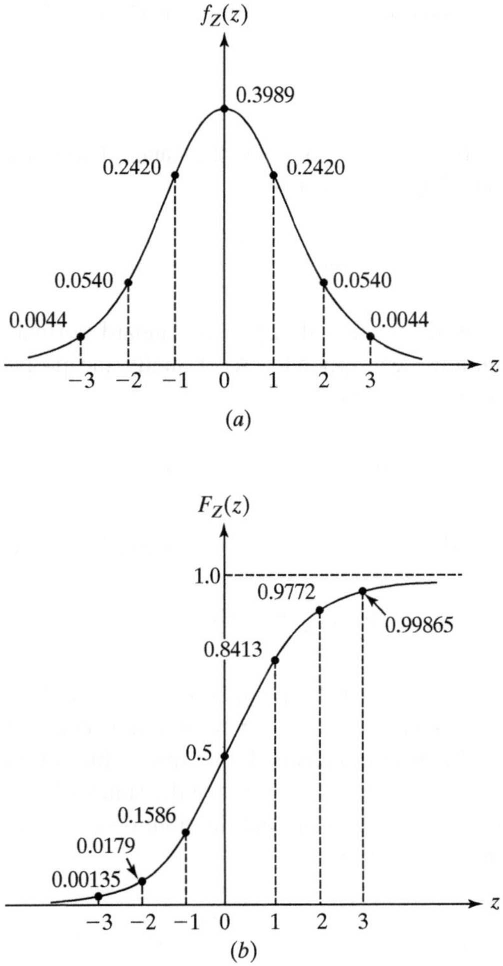

FIGURE 3.13 (*a*) Standard normal density function, (*b*) Standard normal distribution function.

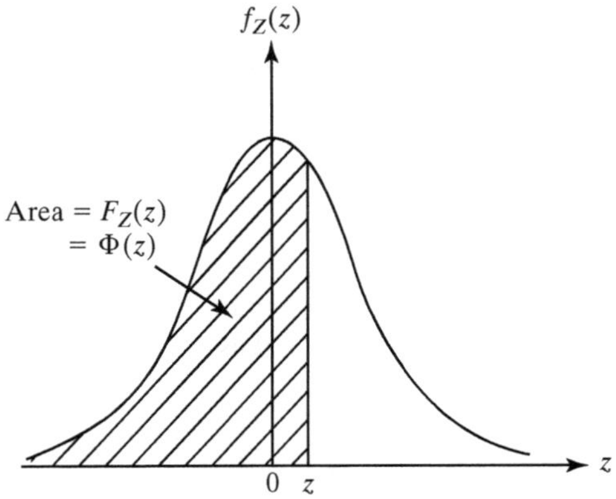

FIGURE 3.14 Distribution function of the standard normal variate Z.

This is shown in Fig. 3.14. If $\Phi(z_p) = p$ is given, the standard normal variate z_p corresponding to the cumulative probability (p) is denoted as

$$z_p = \Phi^{-1}(p) \tag{3.55}$$

The values of the distribution function $\Phi(z)$ of a standard normal variate are given as tables of normal distribution (see Appendix A). Usually the probabilities are given in tables only for positive values of z and

$$\Phi(-z) = 1 - \Phi(z) \tag{3.56}$$

due to the symmetry of the density function about zero. Similarly, we can find that

$$z_p = \Phi^{-1}(p) = -\Phi^{-1}(1 - p) \tag{3.57}$$

In many practical applications, it will be of interest to find probabilities either up to a particular value of standard deviation from $-\infty$ or within a specified number of standard deviations from the mean. Such values can be found from tables of normal distribution [or from Fig. 3.13(b)] and are shown in Fig. 3.15. Once the standard normal table of $\Phi(z)$ is available, the probabilities of any other normal distribution can be determined readily as follows. For a nonstandard X with $N(\mu, \sigma)$,

$$P(l \leq X \leq u) = \frac{1}{\sqrt{2\pi} \cdot \sigma} \int_l^u e^{-1/2\{(x-\mu)/\sigma\}^2} \cdot dx \tag{3.58}$$

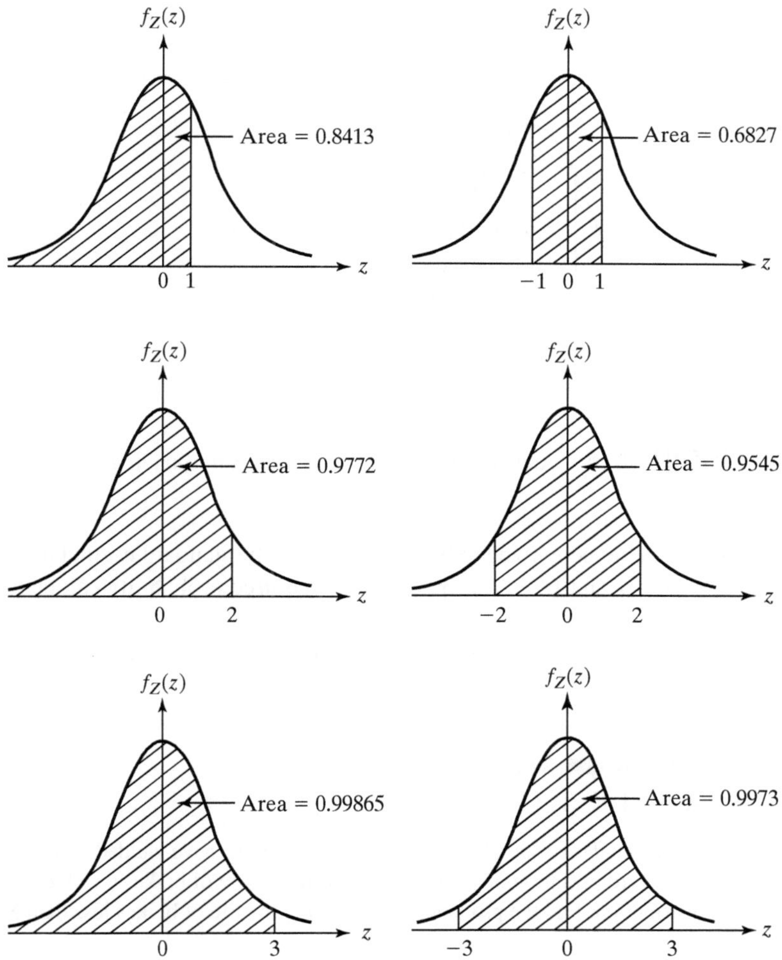

FIGURE 3.15 Representative probabilities associated with standard normal distribution.

This represents the area under the density function between l and u as shown in Fig. 3.16. By defining a new variable (standard normal variate) z as

$$z = \frac{x - \mu}{\sigma}, \quad \text{we find} \quad dz = \frac{dx}{\sigma},$$

and Eq. (3.58) can be rewritten as

$$P(l \leq X \leq u) = \frac{1}{\sigma\sqrt{2\pi}} \int_{(l-\mu)/\sigma}^{(u-\mu)/\sigma} e^{-1/2\, z^2} \cdot \sigma\, dz = \frac{1}{\sqrt{2\pi}} \int_{(l-\mu)/\sigma}^{(u-\mu)/\sigma} e^{-1/2\, z^2} \cdot dz \quad (3.59)$$

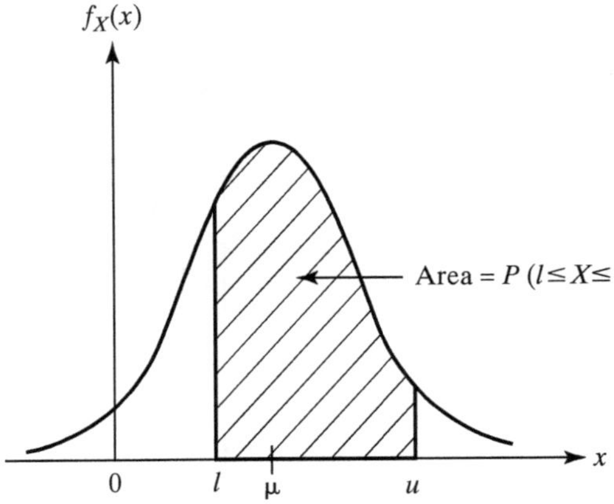

FIGURE 3.16 Area under the density function corresponding to $P(l \leq X \leq u)$.

which can be recognized to be the area under the standard normal density function between $(u - \mu)/\sigma$ and $(l - \mu)/\sigma$. Thus, the required probability can be found as

$$P(l \leq X \leq u) = \Phi\left(\frac{u - \mu}{\sigma}\right) - \Phi\left(\frac{l - \mu}{\sigma}\right) \tag{3.60}$$

EXAMPLE 3.19

Tests conducted on a sample of 100 automobile brakes have yielded a mean value of 56,669.5 miles and a standard deviation of 12,393.64 miles for the life of brakes (see Example 1.1). Assuming normal distribution, find the probability of realizing the life of brakes less than 50,000 miles.

Solution: Denoting the life of brakes as X, we have $\mu_X = 56{,}669.5$ miles and $\sigma_X = 12{,}393.64$ miles. The corresponding standard normal variate (z) is defined as

$$z = \frac{x - \mu_X}{\sigma_X} = \frac{x - 56{,}669.5}{12{,}393.64}$$

and the required probability is given by

$$P(X < 50{,}000) = P\left(z < \frac{50{,}000 - 56{,}669.5}{12{,}393.64}\right) = P(z < -0.5381)$$

Since the probability density function is symmetric about $z = 0$, the probability of realizing z less than -0.5381 is same as the probability of realizing z greater than $+0.5381$. This gives

$$P(z < -0.5381) = P(z > 0.5381) = 1 - P(z < 0.5381) = 1 - 0.7054 = 0.2946$$

■

EXAMPLE 3.20

If the tensile strength of steel described in Example 3.11 follows normal distribution, find the values of the probability density function at different discrete values of strength x_i and compare them with the corresponding relative frequency values observed (f_i/N).

Solution: Using the sample mean $(\overline{X} = 68.113 \text{ kpsi})$ and sample standard deviation $(s_X = 2.6539 \text{ kpsi})$ as population mean (μ_X) and population standard deviation (σ_X), respectively, the normal probability density function, given by Eq. (3.49), becomes:

$$f_X(x) = \frac{1}{2.6539\sqrt{2\pi}} \exp\left[-\frac{1}{2}\left(\frac{x - 68.113}{2.6539}\right)\right] \tag{E1}$$

The values of $f_X(x_i)$, $i = 1,2,\ldots,16$, given by Eq. (E1) are compared with the observed relative frequency values $f_i/N = f_i/1000$ in Table 3.4.

TABLE 3.4 Comparison of Theoretical (Normal) and Experimental Values of Density Function

i	x_i	Eq. (E1), $f_X(x_i)$	Experimental, $f_i/1000$
1	61	0.00	0.004
2	62	0.01	0.019
3	63	0.02	0.025
4	64	0.05	0.032
5	65	0.08	0.082
6	66	0.11	0.106
7	67	0.14	0.135
8	68	0.15	0.153
9	69	0.14	0.140
10	70	0.12	0.130
11	71	0.08	0.077
12	72	0.05	0.047
13	73	0.03	0.029
14	74	0.01	0.013
15	75	0.01	0.005
16	76	0.00	0.003

The graph of the probability density function, Eq. (E1), is shown Fig. 3.17.

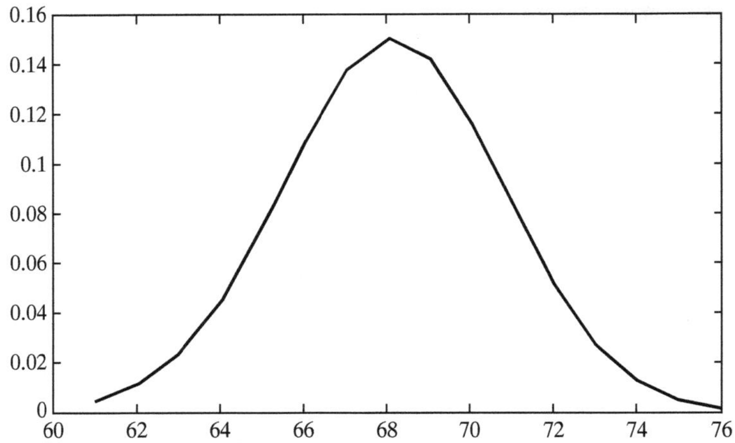

FIGURE 3.17 Probability density function given by Eq. (E1).

■

3.11.3 Lognormal Distribution

A random variable X is said to follow lognormal distribution if $Y = \ln X$ follows normal distribution. Thus

$$f(y) = \frac{1}{\sqrt{2\pi}\,\sigma_Y} \exp\left[-\frac{1}{2}\left(\frac{y-\mu_Y}{\sigma_Y}\right)^2\right]; \quad -\infty < y < \infty \tag{3.61}$$

Since $Y = \ln X$, Eq. (3.61) can be rewritten, in terms of X, as [3.2]

$$f(x) = \frac{1}{\sqrt{2\pi}\,x\sigma_Y} \exp\left[-\frac{1}{2}\left(\frac{\ln x-\mu_Y}{\sigma_Y}\right)^2\right]; \quad x \geq 0 \tag{3.62}$$

where

$$\sigma_Y^2 = \ln\left[\left(\frac{\sigma_X}{\mu_X}\right)^2 + 1\right] \tag{3.63}$$

and

$$\mu_Y = \ln \mu_X - \frac{1}{2}\sigma_Y^2 \tag{3.64}$$

The following example illustrates the application of lognormal distribution.

EXAMPLE 3.21

If the life of brakes described in Example 3.19 is assumed to follow lognormal distribution with mean and standard deviations of 56,669.5 and 12,393.64 miles, respectively, find the probability of realizing the life of brakes less than 50,000 miles.

Solution: Let X denote the life of brakes with mean and standard deviations of $\mu_X = 56{,}669.5$ miles and $\sigma_X = 12{,}393.64$ miles. Since $Y = \ln X$ follows normal distribution, the mean and standard deviations of the variable Y are given by

$$\mu_Y = \ln \mu_X - \frac{\sigma_Y^2}{2}$$

where

$$\sigma_Y^2 = \ln\left[\left(\frac{\sigma_X}{\mu_X}\right)^2 + 1\right]$$

In the present case,

$$\sigma_Y^2 = \ln\left[\left(\frac{12{,}393.64}{56{,}669.5}\right)^2 + 1\right] = 0.0467$$

and hence $\sigma_Y = 0.2162$ and

$$\mu_Y = \ln 56{,}669.5 - \frac{0.0467}{2} = 10.9450 - 0.0234 = 10.9216$$

The standard normal variate corresponding to a brake life of 50,000 miles can be determined as

$$z = \frac{\ln x - \mu_Y}{\sigma_Y} = \frac{\ln 50{,}000 - 10.9216}{0.2162} = -0.4710$$

Thus the required probability is given by

$$\begin{aligned} P(X < 50{,}000) &= P(z < -0.4710) = P(z > 0.4710) \\ &= 1 - P(z \le 0.4710) = 1 - 0.6808 = 0.3192 \end{aligned}$$

■

EXAMPLE 3.22

If the tensile strength of steel described in Example 3.11 follows lognormal distribution, find the values of the probability density function at different discrete values x_i and compare them with the corresponding relative frequency values observed (f_i/N).

Solution: Using the sample mean $(\overline{X} = 68.113 \text{ kpsi})$ and sample standard deviation $(s_X = 2.6539 \text{ kpsi})$ as population mean (μ_X) and population standard deviation (σ_X) of the corresponding normal distribution, respectively, the lognormal probability density function is given by Eq. (3.62). The parameters of the lognormal distribution can be computed, using Eqs. (3.63) and (3.64), as

$$\sigma_Y{}^2 = \ln\left\{\left(\frac{\sigma_X}{\mu_X}\right)^2 + 1\right\} = \ln\left\{\left(\frac{2.6539}{68.113}\right)^2 + 1\right\} = 0.00151321$$

or

$$\sigma_X = 0.0389$$

$$\mu_Y = \ln \mu_X - \frac{1}{2}\sigma_Y{}^2 = \ln 65.113 - 0.5\,(0.00151321) = 4.2204$$

Thus, the probability density function of the lognormal distribution, Eq. (3.62), becomes

$$f_X(x) = \frac{1}{(0.0389)\, x\sqrt{2\pi}}\exp\left[-\frac{1}{2}\left(\frac{\ln x - 4.2204}{0.0389}\right)\right]; \quad x \geq 0 \qquad \text{(E1)}$$

The values of $f_X(x_i), i = 1, 2, \ldots, 16$, given by Eq. (E1) are compared with the observed relative frequency values $f_i/N = f_i/1000$ in Table 3.5.

TABLE 3.5 Comparison of Theoretical (Lognormal) and Experimental Values of Density Function

i	x_i	**Eq. (E1), $f_X(x_i)$**	**Experimental, $f_i/1000$**
1	61	0.00	0.004
2	62	0.01	0.019
3	63	0.02	0.025
4	64	0.05	0.032
5	65	0.08	0.082
6	66	0.11	0.106
7	67	0.14	0.135
8	68	0.15	0.153
9	69	0.14	0.140
10	70	0.11	0.130
11	71	0.08	0.077
12	72	0.05	0.047
13	73	0.03	0.029
14	74	0.01	0.013
15	75	0.01	0.005
16	76	0.00	0.003

The graph of the probability density function, Eq. (E1), is shown in Fig. 3.18. It can be seen that the values of the probability density functions using normal and lognormal distribution models for the tensile strength of steel (results in column 3 of Tables 3.4 and 3.5) are very close to one another. The two density functions are shown on the same graph in Fig. 3.19. The results indicate that the tensile strength of steel, given by the experimental data in Example 3.11, can be modeled as either a normal or a lognormal random variable.

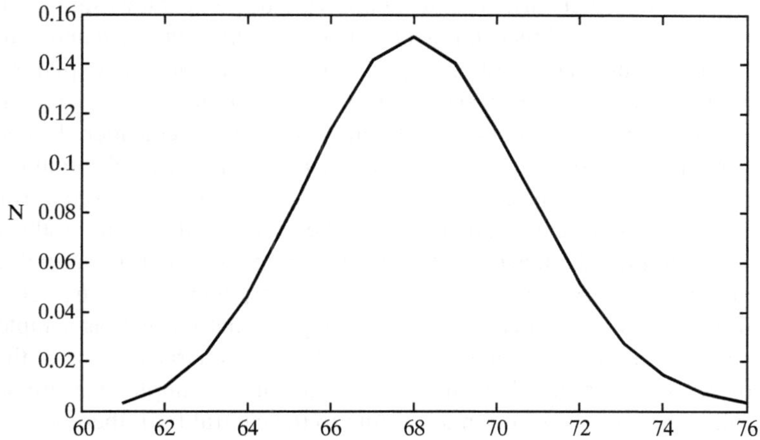

FIGURE 3.18 Probability density function given by Eq. (E1).

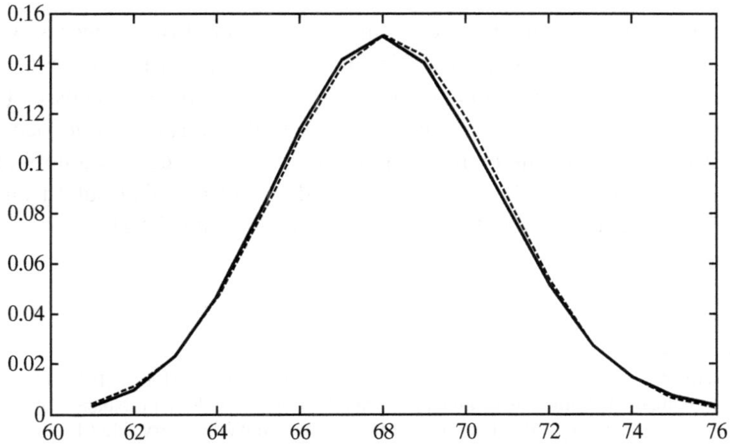

FIGURE 3.19 Comparison of normal and lognormal density functions.

■

3.12 Central Limit Theorem

If $X_1, X_2, \ldots, X_n$ are n mutually independent random variables with finite mean and variance, then the sum

$$S_n = \sum_{i=1}^{n} X_i \tag{3.65}$$

tends to a normal variable if no single variable contributes significantly to the sum as n tends to infinity. This is known as the *central limit theorem* and is useful in several engineering applications [3.2, 3.3]. The central limit theorem explains why many physical phenomena can be described, approximately, by normal distribution. For example, the tensile strength of an alloy steel component can be considered to be influenced by the percentages of the alloying elements such as manganese, chromium, nickel, and silicon, the heat treatment it received, and the machining process used during its production. If each of these effects tends to add to the others in determining the value of the tensile strength, then the tensile strength can be approximated by normal distribution according to the central limit theorem. Another important use of the central limit theorem is in the area of statistical inference. Many estimators such as sample mean and sample variance are used to make inferences about the characteristics of the population. Since these estimators involve sums or averages of the sample measurements, the estimators follow normal distribution according to the central limit theorem when the sample size is sufficiently large. This result is useful when repeated samplings are used to make inferences.

3.13 Normal Approximation to Binomial Distribution

In binomial experiments, each trial yields either 0 or 1 success with a probability of $q = 1 - p$ and p, respectively. If there are a total of n trials, we can consider each trial as an independent observation and hence the total number of successes (r) in n trials can be considered as a sum of n independent observations. If n is large, according to the central limit theorem, we can approximate the binomial variate r as a normal variate with mean np and variance $np(1 - p)$. Thus, standard normal distribution tables can be used to approximate the binomial probabilities[8] (see Problem 3.54).

---

[8]This result was presented on November 12, 1733, by Abraham De Moivre (1667–1754, English) in a paper titled "Approximatio ad Summam Terminorum Binomii $(a + b)^n$ in Seriem Expansi." As such, November 12, 1733, is usually taken as the date of the origin of the "normal probability curve" [3.6].

3.14 Numerical Solutions Using MATLAB and Excel

3.14.1 MATLAB Functions for Discrete and Continuous Probability Distributions

In MATLAB, the values of the probability density function y of one-parameter family of distributions such as exponential distribution, Poisson distribution, and Rayleigh distribution can be found at any specified value of x using the following MATLAB function:

```
y=pdf(name,x,a)
```

where *name* identifies the probability distribution, and a the value of the parameter of the distribution. The name to be used is exp or Exponential for exponential distribution (with a denoting the mean), poiss or Poisson for Poisson distribution (with a representing the mean), and rayl or Rayleigh for Rayleigh distribution (with a indicating the scale parameter). For a two-parameter probability distribution with parameters a and b, the value of the probability density function at any specified value of x can be found using the statement:

```
y=pdf(name,x,a,b)
```

where *name* is used to identify the distribution as bino or Binomial for binomial distribution (with a denoting the number of trials and b indicating the probability of success for each trial), norm or Normal for normal distribution (with a and b denoting the mean and standard deviation, respectively) and logn or Lognormal for lognormal distribution (with parameters a and b).

In addition, for a specified set of values x, the values of a specified probability density or distribution function can be obtained and plotted as illustrated in the following examples.

EXAMPLE 3.23

Generate and plot the values of binomial density (mass) and distribution functions for $x = 1, 2, \ldots, 10$ using MATLAB. Assume $n = 10$ and $p = 0.7$.

Solution: The program statements (m-file) and the results given by MATLAB program are shown below:

```
clear all;clc; close all;

x=0:10;
n=10;
p=0.7;
y=binopdf(x,10,0.7)

y =
  Columns 1 through 8

    0.0000  0.0001  0.0014  0.0090  0.0368  0.1029  0.2001  0.2668
```

```
  Columns 9 through 11

    0.2335  0.1211  0.0282

plot(x,y,'*') ;
```

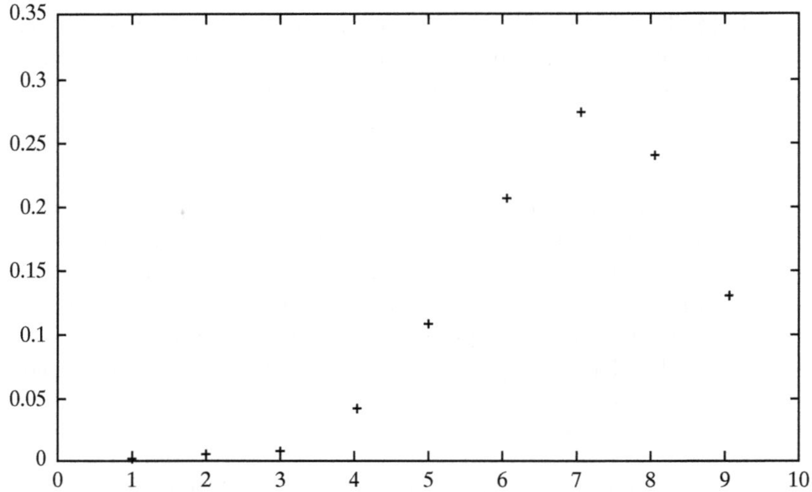


```
z=binocdf(x,n,p)

z =

  Columns 1 through 8

    0.0000  0.0001  0.0016  0.0106  0.0473  0.1503  0.3504  0.6172

  Columns 9 through 11

    0.8507  0.9718  1.0000

figure ;
plot(x,z,'ro');
```

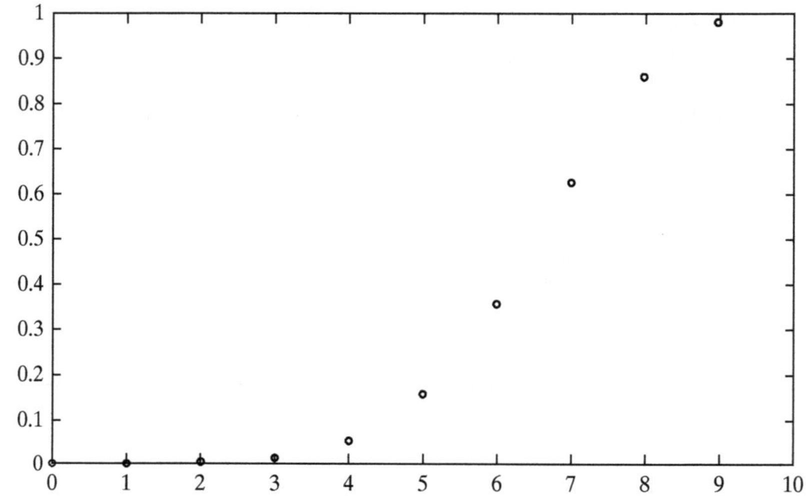

```
figure ;
stairs(X,Z)
```

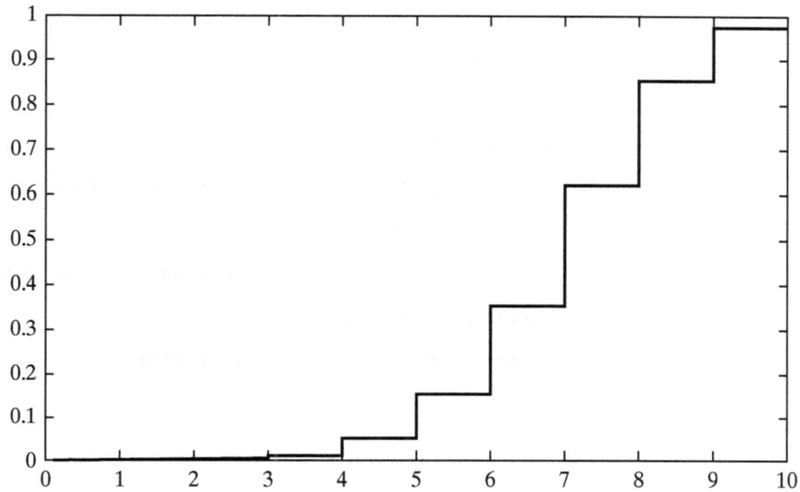

■

EXAMPLE 3.24

Find the values of the inverse binomial density (mass) function, with parameters $n = 10$ and $p = 0.7$, for the following values of density function: (i) 0.8 and (ii) 0.05 and 0.95.

Solution: The MATLAB code and the results given below:

```
clear all; clc
 n=10    ;
p=0.7    ;
y=0.8    ;
x=binoinv(y,n,p)

 x=      8

y=binoinv([0.05,0.95],10,0.7)

 y=      5   9
```

■

EXAMPLE 3.25

Find the values of the Poisson density function having a mean value of 4 for $x = 1, 2, 3, \ldots, 20$ and also plot the corresponding density function using MATLAB.

Solution: The MATLAB code and the results are shown below:

```
clear all;clc;

mu=4;
x=0:20;
y=poisspdf(x,mu)

y =

  Columns 1 through 8

   0.0183  0.0733  0.1465  0.1954  0.1954  0.1563  0.1042  0.0595

  Columns 9 through 16

   0.0298  0.0132  0.0053  0.0019  0.0006  0.0002  0.0001  0.0000

  Columns 17 through 21

   0.0000  0.0000  0.0000  0.0000  0.0000

plot(x,y,'*');
```

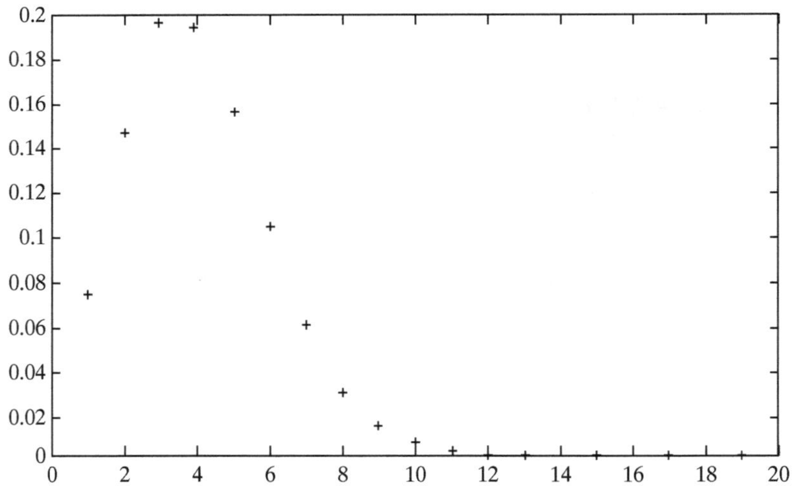

■

EXAMPLE 3.26

Plot the density function of normal distribution, having a mean 10 and a standard deviation 2, between the values of 2 and 18, and also find the area under the resulting density function using MATLAB.

Solution: The MATLAB code and the results are indicated below:

```
clear all;clc;
format long

specs=[2 18];
mu=10;
sigma=2;
y=normspec(specs,mu,sigma)

 y= 0.999936657516334
```

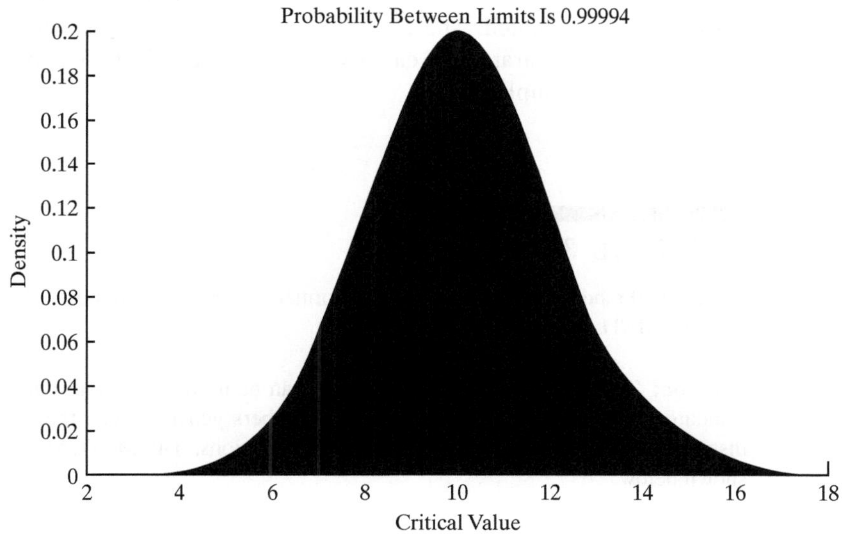

■

EXAMPLE 3.27

Find the values of a random variable X that follows normal distribution with mean 0 and standard deviation 0.25 corresponding to the cumulative distribution values of 0.001 and 0.999 using MATLAB.

Solution: The MATLAB code and the results are given below:

```
clear all;clc;

p=[0.001 0.999] ;
mu=0 ;
sigma=0.25 ;
x=norminv(p,mu,sigma)

 x=[ -0.7726  0.7726]
```

■

3.14.2 Random Numbers, Fitting Data to Distributions and Confidence Intervals[9]

MATLAB can be used to generate random numbers following different types of probability distributions so that the performance of an engineering system can be numerically simulated. In addition, data obtained for any probabilistic quantity can be plotted on the probability paper corresponding to any specific distribution. If all the data points fall on a straight line, the probabilistic quantity is confirmed to follow the distribution of the probability paper. Otherwise, the quantity does not follow the distribution of the probability paper. If probabilistic data are fitted into a specific distribution, the parameters of the distribution (that define the distribution mathematically) can be estimated. The confidence intervals, at any specified confidence level, for the estimated parameters can also be obtained. All these aspects are illustrated in the following examples.

EXAMPLE 3.28

Generate 20 random numbers that follow normal distribution with mean 100 and standard deviation 10 using MATLAB.

Solution: The MATLAB function *normrnd* can be used to generate random numbers following standard normal distribution. The random numbers generated can then be converted to normal distribution of specified mean and standard deviations. The MATLAB program and results are shown below:

```
clear all; clc;
mu = 100
sd = 10
x = normrnd (0,1,20,1)
for i = 1:20
    y(i) = x(i) * sd + mu;
end
y

mu =

   100

sd =

    10

y =
```

[9]The theory behind the generation of random numbers following different probability distributions can be found in Chapter 10. The theory behind the construction of probability papers for different distributions, the estimation of parameters of an assumed distribution for a given set of data, and the method of finding confidence intervals for the estimated parameters can be found in Chapter 13.

```
Columns 1 through 15

 95.6744  83.3442  101.2533  102.8768  88.5353  111.9092  111.8916  99.6237  103.2729↙
101.7464  98.1329  107.2579  94.1168  121.8319  98.6360

 Columns 16 through 20

 101.1393  110.6677  100.5928  99.0435  91.6765
```

■

EXAMPLE 3.29

Plot the data of Young's modulus of steel given in Problem 1.1 on normal probability paper.

Solution: The command used and the resulting plot are shown below:

```
clear all; clc;
x = [25.1; 26.5; 27.4; 28.4; 29.1; 29.9; 31.2; 32.3; 34.7; 27.2; 28.1;...
    29.2; 29.8; 30.1; 31.3; 32.5; 26.9; 30.3; 25.9; 27.6; 28.5; 29.3;...
    30.4; 31.4; 32.7; 29.4; 30.5; 31.6; 32.8; 28.3; 25.4; 28.6; 29.5;...
    30.6; 31.8; 33.4; 28.3; 28.7; 29.6; 30.3; 31.9; 33.8; 30.9; 29.6;...
    27.7; 26.6; 26.8; 27.8; 28.9; 29.7]
y = normplot (x)
```

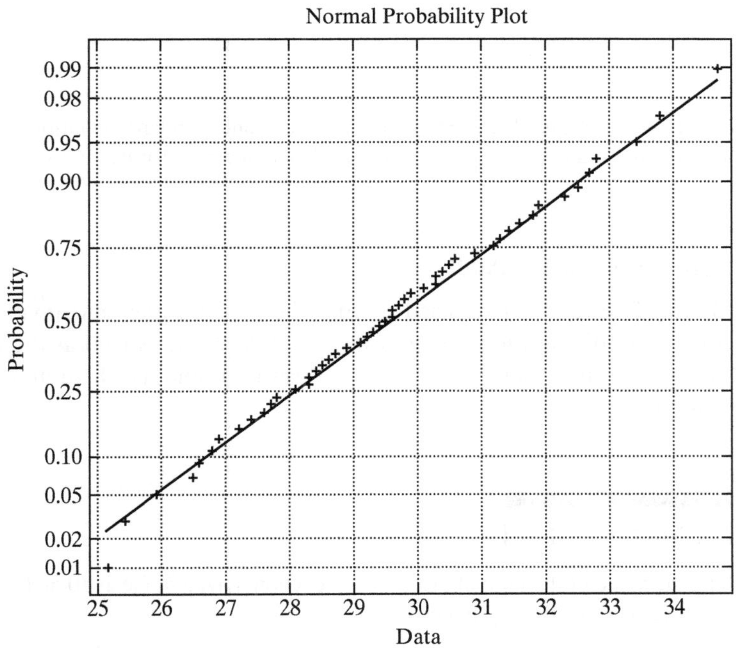

Note: Since all the data points fall on a straight line, the data follows normal distribution.

■

EXAMPLE 3.30

Assuming that the data of Young's modulus of steel given in Problem 1.1 follows normal distribution, find the parameters of the distribution, mean μ and standard deviation σ, and the associated 95 percent confidence interval for each of the parameters using MATLAB.

Solution: The MATLAB command used and the results given by the program are shown below:

```
clear all; clc;
x = [25.1; 26.5; 27.4; 28.4; 29.1; 29.9; 31.2; 32.3; 34.7; 27.2; 28.1;...
    29.2; 29.8; 30.1; 31.3; 32.5; 26.9; 30.3; 25.9; 27.6; 28.5; 29.3;...
    30.4; 31.4; 32.7; 29.4; 30.5; 31.6; 32.8; 28.3; 25.4; 28.6; 29.5;...
    30.6; 31.8; 33.4; 28.3; 28.7; 29.6; 30.3; 31.9; 33.8; 30.9; 29.6;...
    27.7; 26.6; 26.8; 27.8; 28.9; 29.7]
[mu, sigma, muci, sigmaci] = normfit(x)

mu =

   29.5660

sigma =

   2.2123

muci =

   28.9373
   30.1947

sigmaci =

   1.8480
   2.7568
```

Note: *mu* and *sigma* denote the estimated mean and standard deviation, respectively, while *muci* and *sigmaci* indicate the 95 percent confidence intervals for *mu* and *sigma,* respectively.

■

3.14.3 Solutions using Excel

Microsoft Excel can be used to find the probabilities and inverse probabilities. The cumulative distribution function values, $P[X \leq x]$, as well as the inverse probabilities, values of x such that $P[X \leq x] = p$, can be found as illustrated in the following two examples.

EXAMPLE 3.31

Find the value of the normal distribution function, having a mean 40 and standard deviation 2.5, at the value of $x = 50$ using Excel.

Solution: The command to be used in the Excel spreadsheet, along with the result, is shown below:

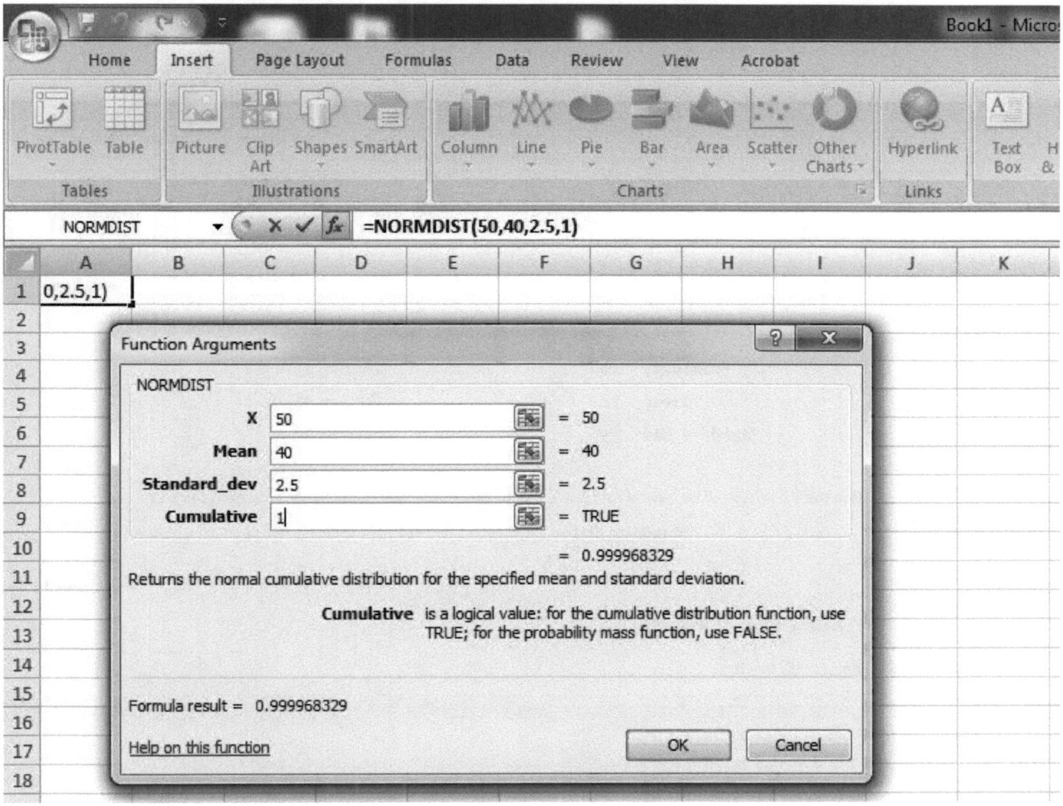

■

EXAMPLE 3.32

Find the value of a normally distributed variable x corresponding to a cumulative distribution function value of 0.9 when the mean and standard deviation of the variate are 40 and 2.5, respectively.

Solution: The command to be used in the Excel spreadsheet and the result are shown below:

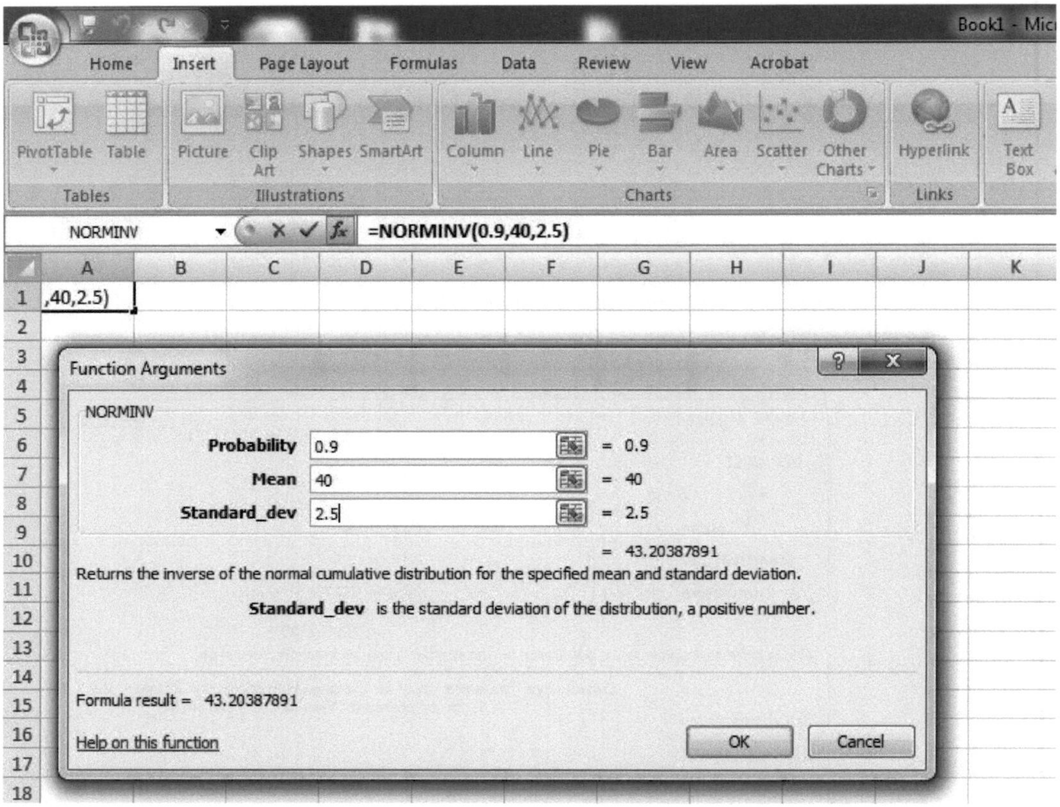

■

REFERENCES AND BIBLIOGRAPHY

3.1 G. P. Wadsworth and J. G. Bryan, *Introduction to Probability and Random Variables,* McGraw-Hill, New York, 1960.

3.2 G. C. Hart, *Uncertainty Analysis, Loads, and Safety in Structural Engineering,* Prentice-Hall, Englewood Cliffs, NJ, 1982.

3.3 W. Mendenhall, *Introduction to Probability and Statistics,* 3rd ed., Duxbury Press, Belmont, CA., 1971.

3.4 I. Olkin and J. W. Pratt, *A Multivariate Chebyshev Inequality,* Annals Math. Stat., Vol. 29, 1958, pp. 201–211.

3.5 E. Parzen, *Modern Probability Theory and Its Applications,* John Wiley, New York, 1960.

3.6 A. C. King and C. B. Read, *Pathways to Probability,* Holt, Rinehart & Winston, New York, 1963.

3.7 S. S. Rao and C. P. Reddy, "Mechanism Design by Chance Constrained Programming Techniques," *Mech. Mach. Theor.*, Vol. 14, 1979, pp. 413–424.

3.8 A. H. S. Ang and W. H. Tang, *Probability Concepts in Engineering Planning and Design,* Vol. 1: *Basic Principles,* John Wiley, New York, 1975.

3.9 C. C. Gillispie (ed.-in-chief), *Dictionary of Scientific Biography,* Vol. 3, Charles Scribner's Sons, New York, 1980.

REVIEW QUESTIONS

3.1 Give brief answers to the following:

1. Define the various measures of central tendency of a random variable.
2. Define the terms "standard deviation" and "variance."
3. What is the measure used to indicate the degree of nonsymmetry of a distribution?
4. How is the skewness coefficient defined for (a) a discrete distribution and (b) a continuous distribution?
5. What is a random variable?
6. Explain the difference between a discrete and a continuous random variable. Give two examples of each.
7. What is the difference between a histogram and a density function?
8. Define the terms: mean, mode, and median.
9. What is the difference between unimodal and multimodal distributions?
10. What is a probability mass function?
11. How are the probability density and probability distribution functions related?
12. What is the significance of probability distribution function?
13. Define the *r*th central moment of a distribution function.
14. Why is the area under a density function equal to one?
15. Can the probability density function have a value greater than one for any distribution?
16. Under what conditions will the mean, mode, and median of a random variable be identical?
17. Explain the notation: $N(\mu, \sigma)$.
18. What is a standard normal variate?
19. How is the root mean square value of a random variable related to its standard deviation?
20. What are the characteristics of Gaussian distribution?
21. What is central limit theorem and what is its use?
22. What is a Bernoulli trial?
23. Define binomial distribution.
24. What is the difference between variance and covariance?
25. Define correlation coefficient.
26. What is Chebyshev inequality? Give two practical uses of this inequality.
27. What are univariate, bivariate, and multivariate distributions?
28. Define a marginal density function.
29. How do you find the marginal density function from a joint density function?

3.2 Indicate whether each of the following statements is true or false:

1. The probability of a certain event is one.
2. The probability of occurrence of an event can be larger than one.
3. The mathematical expectation of a random variable is same as the mean value.
4. The skewness coefficient can also give a measure of the variability of the random variable.

5. The Chebyshev inequality represents the application of the first and second moments of a random variable.
6. Bivariate distribution involves two independent random variables X and Y.
7. The number of sample points in n Bernoulli trials is n^2.
8. Moment of any order of a normally distributed random variable can be found once the mean and standard deviation are known.
9. The outcome of any experiment is subject to uncertainty.
10. The number of possible outcomes of an experiment is always finite.
11. The events of an experiment need not be independent.

3.3 Fill in the blank with the proper word(s):

1. If a coin is tossed 100 times, the mean or expected value of the number of heads that occur is ______.
2. If a coin is tossed 100 times, the standard deviation of the number of heads that occur is ______.
3. If 5 percent of the welds are known to be defective in a welding process, the average number of defective welds in a system composed of 1000 welds is ______.
4. If 5 percent of the welds are known to be defective in a welding process, the standard deviation of defective welds in a system composed of 1000 welds is ______.
5. If 5 percent of the welds are known to be defective in a welding process, the approximate range of the number of defective welds (corresponding to $\pm 3\sigma$ values) in a system composed of 1000 welds is ______ to ______.
6. Mode is the value of X at which the probability ______ function has the largest value.
7. The standard deviation of a random variable is a measure of its ______.
8. A measure of the symmetry of the probability density function is called ______.
9. Mean, mode, and median coincide in ______ distribution.
10. $f_{X,Y}(x, y)\, dx\, dy$ gives the probability of realizing ______.
11. Repeated independent trials of a random experiment, each with only two possible outcomes, are called ______ trials.
12. The probability distribution of a linear function of several normally distributed random variables is ______.
13. The mathematical expectation and variance of a standard normal variate are given by ______ and ______, respectively.
14. The central limit theorem states that the sum of several random variables following different distributions follows ______ distribution.
15. The outcome of an experiment is called a(n)______.

3.4 Select the most appropriate answer from the multiple choices given:

1. Four bulldozers, each with a probability of failure of 0.5, are used in a construction project. The probability that three bulldozers fail during a construction project is:
   a. 0.25 **b.** 0.125 **c.** 0.0625
2. Four bulldozers, each with a probability of failure of 0.5, are used in a construction project. The probability that two bulldozers fail during the construction project is:
   a. 0.375 **b.** 0.25 **c.** 0.75
3. Four bulldozers, each with a probability of failure of 0.5, are used in a construction project. The probability that only one bulldozer fails during the construction project is:
   a. 0.25 **b.** 0.125 **c.** 0.5

4. Four bulldozers, each with a probability of failure of 0.5, are used in a construction project. The probability that none of the bulldozers fail during the construction project is:
a. 0.0625 **b.** 0.5 **c.** 1

5. Four bulldozers, each with a probability of failure of 0.5, are used in a construction project. The probability that all four bulldozers fail during the construction project is:
a. 0.0625 **b.** 0.5 **c.** 1

6. The median of a random variable is:
a. Average value **b.** Middle value **c.** Largest value

7. The second order central moment of a random variable is given by:
a. $\int_{-\infty}^{\infty} x^2 f_X(x)\,dx$ **b.** $\int_{-\infty}^{\infty} (x-\mu_X)^2 f_X(x)\,dx$ **c.** $\int_{-\infty}^{0} (x-\mu_X)^2 f_X(x)\,dx$

8. Gauss inequality is applicable to:
a. Uniform distribution **b.** Normal distribution **c.** Unimodal function

9. The correlation coefficient is a number in the range:
a. 0 to 1 **b.** -1 to 1 **c.** 0 to ∞

10. For a lognormally distributed random variable X:
a. ln X follows normal distribution
b. X follows normal distribution
c. $\frac{1}{X}$ follows normal distribution

3.5 Match the following probabilities of realizing the events upon rolling a die:

1. Probability of getting a number less than 2 **a.** $\frac{1}{2}$

2. Probability of getting a number greater than 4 **b.** $\frac{5}{6}$

3. Probability of getting an odd number **c.** $\frac{1}{6}$

4. Probability of getting a number in the range 3 to 6 **d.** $\frac{1}{3}$

5. Probability of getting a number less than 6 **e.** $\frac{2}{3}$

3.6 Match the following relations in the context of normal distribution:

1. $\frac{1}{\sqrt{2\pi}} e^{-\frac{1}{2}z^2}$ **a.** Normally distributed variable X with mean 10 and standard deviation 2

2. $\Phi(-z)$ **b.** 0.5

3. $\Phi^{-1}(p)$ **c.** Standard normal variate z

4. $\frac{x-\mu_X}{\sigma_X}$ **d.** $-\Phi^{-1}(1-p)$

5. N(10, 2) **e.** Probability density function of a standard normal variate

6. $\int_{-\infty}^{0} f_Z(z)\,dz$ **f.** $1-\Phi(z)$

3.7 Match the following:

1. $p_X(x_i)$	**a.** Expected value of the function $h(X)$ where X is a continuous random variable
2. $f_X(x)$	**b.** Expected value of the continuous random variable X
3. $\sum_{\text{all } x_i < x} p_X(x_i)$	**c.** Expected value of the function $h(X)$ where X is a discrete random variable
4. $\int_{-\infty}^{x} f_X(x')\,dx'$	**d.** Expected value of the discrete random variable X
5. $\sum_{\text{all } i} x_i p_X(x_i)$	**e.** Probability distribution function of the continuous random variable X
6. $\sum_{\text{all } i} h(x_i)\, p_X(x_i)$	**f.** Probability distribution function of the discrete random variable X
7. $\int_{-\infty}^{\infty} x f_X(x)\,dx$	**g.** Probability density function of the continuous random variable X
8. $\int_{-\infty}^{\infty} h(x)\, f_X(x)\,dx$	**h.** Probability mass function of the discrete random variable X

PROBLEMS

3.1 An experiment consists of tossing a fair coin 5 times. (a) Find the probability mass and distribution functions for the number of heads realized. (b) Find the probability of realizing heads at least 3 times out of the 5 trials.

3.2 The number of airplanes landing at an airport in any minute X is given by the following distribution:

x	$p(x)$
0	0.06
1	0.14
2	0.20
3	0.25
4	0.18
5	0.12
6	0.05

a. Find the mean, standard deviation, and skewness coefficient of X.
b. What is the probability of having more than 3 landings in any minute?
c. What is the probability of having the number of landings between 2 and 5 in any minute?

3.3 Depending on the surface finish requirements, a machinist uses different cutting speeds on a lathe. From past experience, the various cutting speeds used during machining operations and the probabilities of using them are given as follows:

Cutting Speed, X (m/min)	Probability
20	0.04
30	0.10
40	0.16
50	0.15
60	0.14
70	0.12
80	0.09
90	0.08
100	0.06
110	0.03
120	0.02
130	0.01

a. Plot the probability mass and distribution functions.
b. Find the mean, mode, and median of the random variable.

3.4 For the probability mass function given in Problem 3.3, determine $E(X^2)$, $E(X^3)$, standard deviation, and skewness coefficient.

3.5 In a manufacturing plant, a batch of 750 pistons is known to contains 30 pistons that do not meet the tolerance specification on the diameter. Two pistons are selected randomly from the batch, without replacement, for inspection of the diameters. If X is a random variable denoting the number of pistons that do not conform to the tolerance specification, determine the probability mass and cumulative distribution functions of X.

3.6 Consider the following cumulative distribution function of a random variable X:

$$F_X(x) = \begin{cases} 0; x < -5 \\ 0.1; -5 \leq x < 1 \\ 0.8; 1 \leq x < 4 \\ 1; x \geq 4 \end{cases}$$

Find the probability mass function of X.

3.7 The strength of fiber-reinforced composite beams is a random variable with the following cumulative distribution function:

$$F_X(x) = \begin{cases} 0\ ; x < 10{,}000 \text{ psi} \\ 0.5\ ; 10{,}000 \text{ psi} \leq x < 12{,}000 \text{ psi} \\ 0.7\ ; 12{,}000 \text{ psi} \leq x < 15{,}000 \text{ psi} \\ 1\ ; x \geq 15{,}000 \text{ psi} \end{cases}$$

Find the following probabilities:

a. $P(X \leq 5{,}000 \text{ psi})$
b. $P(X \leq 11{,}500 \text{ psi})$
c. $P(X \leq 15{,}000 \text{ psi})$

3.8 The number of emails received per hour by a college admissions office on a typical working day has the following distribution:

Number of emails, X	12	13	14	15	16	17	18	19
Probability mass function, $p_X(x)$	0.03	0.11	0.23	0.21	0.17	0.14	0.09	0.02

Determine the mean and standard deviation of the number of emails received per hour.

3.9 The possible values of a random variable X are given by $\{0, 1, 3, 5, 7, c\}$ where c is unknown. If each possible value of X is equally likely and the mean of X is 4.5, determine the value of c.

3.10 During the morning commute to work, a person has to cross 20 traffic lights. If the probability that each traffic light is green as the person approaches that traffic light is 0.2, determine the following:

a. The probability of finding exactly five traffic lights green during the commute.
b. The probability of finding the number of traffic lights green during the commute lies between 3 and 6.

3.11 The number of flaws in a continuous welded joint in a construction project is 0.05 per meter length. If the number of flaws in the weld follows a Poisson distribution, find the following:

a. Probability of finding exactly one flaw in a 5 meter length of the weld.
b. Probability of finding at most two flaws in a 5 meter length of the weld.
c. Probability of finding at least one flaw in a two meter length of the weld.

3.12 The number of flaws in the castings used in internal combustion (I.C.) engines is known to follow Poisson distribution with a mean of 0.01 flaws per cubic foot of the casting. If a casting with a volume of 50 cubic feet is used in an I.C. engine, determine the following:

a. Probability that there are no flaws in the 50 cubic feet casting.
b. If 5 of the 50 cubic feet castings are bought by a power generating company, the probability that none of the castings has a flaw.
c. If 5 of the 50 cubic feet castings are bought by a power generating company, the probability that at most one of the castings has a flaw.

3.13 The number of cars arriving at a tollbooth in a minute and their probabilities are given below:

Number of Cars Arriving in a Minute (X)	**Probability**
1	0.025
2	0.075
3	0.125
4	0.150
5	0.200
6	0.275
7	0.100
8	0.050

a. Find the probability distribution function of X.
b. Find the mean, mode, and median of the number of cars arriving in a minute.

3.14 For the probability mass function given in Problem 3.13, determine $E(X^2)$, $E(X^3)$, standard deviation, and skewness coefficient.

3.15 The diameter of a forged crankshaft is specified as 1.2 ± 0.02 in. Past experience with the forging process gives a mean value of 1.2 in and a standard deviation of 0.005 in. Find the percent of discarded crankshafts. Assume normal distribution for the diameter of the crankshafts.

3.16 The life of a heat exchanger (in thousands of hours) is represented by the probability density function

$$f(x) = ae^{-ax}; \quad x \geq 0$$

Determine (a) mean, (b) mode, (c) median, (d) standard deviation, and (e) skewness coefficient of the life of the heat exchanger.

3.17 An O-ring used to seal the leakage of fluid in a pressure vessel is subjected to a random pressure X [in k(lb/in^2)] which can be approximated by the distribution

$$f(x) = \begin{cases} \dfrac{x}{10}; & 0 \leq x \leq 2 \\ \dfrac{10 - x}{40}; & 2 \leq x \leq 10 \end{cases}$$

Find (a) mean, (b) mode, (c) median, (d) standard deviation, and (e) skewness coefficient of the pressure induced on the O-ring.

3.18 The force applied on an engine brake, X (in kips), can be modeled as a random variable following the density function

$$f(x) = \begin{cases} \dfrac{x}{40}; & 0 \leq x \leq 8 \\ \dfrac{10 - x}{10}; & 8 \leq x \leq 10 \end{cases}$$

Find: (a) mean, (b) mode, (c) median, (d) standard deviation, and (e) skewness coefficient of the force applied on the brake.

3.19 The probability density function of the failure time of an ammeter (X) is given by

$$f_X(x) = \frac{1}{500} \exp\left(-\frac{x}{500}\right); x > 0; x \text{ in hours}$$

Determine the following:

a. The probability that the life of the ammeter is more than 2000 hours.
b. The probability that the life of the ammeter lies between 500 and 1500 hours.
c. The probability that the life of the ammeter is less than 500 hours.
d. The number of hours of operation at which 10% of the ammeters will fail.

3.20 Find the probability density function of a random variable X for which the cumulative distribution function is given by

$$F_X(x) = 1 - \exp(-5x); \ x > 0$$

3.21 The probability density function of the diameter of a shaft machined on a lathe is given by

$$f_X(x) = 5 \exp\{-5(x - 2)\}; \ x > 2 \text{ inches}$$

Although the nominal diameter of the shaft is 2 inches, due to the presence of several uncertain factors during machining, shafts are produced with diameters larger than 2 inches. Determine the following:

a. Mean and standard deviation of the diameter of the shaft.
b. Probability that the shaft diameter exceeds 2.2 inches.

3.22 The probability density function of the volume of motor oil added to an engine (X) is given by

$$f_X(x) = \frac{100}{99x^2}; 1 \text{ cm}^3 \le x \le 100 \text{ cm}^3$$

Determine the following:

a. Mean and standard deviation of the volume of the motor oil added to the engine.
b. If the cost of the motor oil is \$0.45 per cm^3, the mean cost of the motor oil added to the engine.
c. The probability that the volume of the motor oil added to the engine exceeds 80 cm^3.

3.23 The number of emails received at a personal computer can be modeled as a Poisson process with a mean value of 5 per hour. Determine the following:

a. Probability that no email is received in an interval of 30 minutes.
b. Probability that the time elapsed between the arrival of two consecutive emails lies between 15 and 30 minutes.
c. The duration of time (ΔT) for which the probability that no email arrives during the time ΔT is 0.80.

3.24 The coefficient of kurtosis (flatness) of a random variable X is defined as

$$\delta = \frac{E[(X - \mu)^4]}{\sigma_X^4}$$

Find the coefficient of kurtosis for the life of the heat exchanger stated in Problem 3.16.

3.25 Find the coefficient of kurtosis for the cutting speed described in Problem 3.3.

3.26 Find the probability mass and distribution functions for the sum of numbers realized when two fair dice are thrown. Also find the mean and standard deviations of the sum of numbers realized.

3.27 Experiments conducted on the fatigue life of helical springs indicated a mean value of 6×10^4 cycles and a standard deviation of 2×10^3 cycles. Assuming that the probability density function for fatigue life is not known, find the ranges within which the fatigue life will lie with probabilities of at least 0.5, 0.6, 0.7, 0.8, 0.9, and 0.9999. Use the Chebyshev inequality.

3.28 Plot a graph between

$$P[\mu - k\sigma \le X \le \mu + k\sigma]$$

and k for the following cases:

1. Normal distribution for X with $0 \le k \le 10$;
2. Exponential distribution for X with $0 \le k \le 10$;
3. Uniform distribution for X with $0 \le k \le 10$; and
4. Chebyshev inequality with $1 \le k \le 10$. What do you observe from these graphs?

3.29 The probabilistic characteristics of the bending strength (X_1) and the surface wear strength (X_2) of a gear tooth are given by $\mu_1 = 20{,}000$ psi, $\sigma_1 = 2{,}000$ psi, $\mu_2 = 60{,}000$ psi, and $\sigma_2 = 3{,}000$ psi. If the bending and surface wear strengths are required to lie in the ranges 16,000 to 24,000 psi and 54,000 to 66,000 psi, respectively, with a minimum probability of 0.6, determine the necessary correlation coefficient between the bending and surface wear strengths of the gear tooth. Use the generalized Chebyshev inequality which can be stated in the case of two random variables X_1 and X_2 as [3.4]:

$$P\left[\,|X_1 - \mu_1| \geq t\sigma_1 \quad \text{or} \quad |X_2 - \mu_2| \geq t\sigma_2\right] \leq \frac{1}{t^2}\left(+\sqrt{1 - \rho_{1,2}^2}\right)$$

where μ_i and σ_i are, respectively, the mean value and the standard deviation of $X_i\,(i = 1, 2)$ and $\rho_{1,2}$ is the correlation coefficient of X_1 and X_2.

3.30 The bathroom scales manufactured by a company exhibit a standard deviation of 0.5 lb when the mean weight is 100 lb. Determine the probability of a 100 lb weight being in error less than or equal to ± 2.0 lb using the following: (a) Chebyshev inequality, (b) Camp-Meidell inequality, and (c) Gauss inequality assuming that the mean and mode are separated by one standard deviation.

3.31 The times to failure, in hours of operation, of the differential and the rear axle of a truck (T_1 and T_2) can be approximated by the joint density function

$$f(t_1, t_2) = 5 \times 10^{-10} \exp\left\{-(10^{-4}t_1 + 5 \times 10^{-6}t_2)\right\}; \qquad t_1 \geq 0; \qquad t_2 \geq 0$$

a. Find the marginal density functions of T_1 and T_2.
b. Find the mean values of T_1 and T_2.

3.32 The joint density function of two random variables is given by

$$f(x, y) = \frac{1}{102}(x + 3x^2y)$$

for $1 \leq x \leq 4$ and $1 \leq y \leq 2$. (a) Find the marginal density functions of X and Y. (b) Find the mean values of X and Y. (c) Find the correlation coefficient of X and Y.

3.33 Experiments are conducted on fiber-reinforced plastic specimens to find the load at which the first crack appeared and the load at which the specimen broke and the following values are observed:

Specimen Number	Load at Which First Crack Appeared, X_1 (Newtons)	Load at Which Specimen Broke, X_2 (Newtons)
1	5025	6075
2	5050	6225
3	5075	6125
4	5100	6175
5	5125	6150
6	5150	6225
7	5175	6175

(*Continued*)

(*Continued*)

Specimen Number	Load at Which First Crack Appeared, X_1 (Newtons)	Load at Which Specimen Broke, X_2 (Newtons)
8	5200	6200
9	5225	6175
10	5250	6100
11	5275	6125
12	5300	6175
13	5325	6200
14	5350	6175
15	5375	6125
16	5400	6150
17	5425	6225
18	5450	6300
19	5475	6275
20	5500	6300

a. Plot the scattergram,

b. Find μ_{X_1}, μ_{X_2}, σ_{X_1}, σ_{X_2}, and ρ_{X_1,X_2}.

3.34 The following numbers indicate the grades received by the students in two mid-semester examinations of the course "Reliability-Based Design":

Student	Examination 1 (X_1)	Examination 2 (X_2)
1	30	50
2	35	40
3	40	45
4	45	55
5	50	45
6	50	70
7	55	60
8	60	50
9	65	70
10	70	45
11	70	60
12	75	60
13	75	70
14	80	60
15	80	70
16	85	60
17	85	75
18	90	85
19	90	40
20	95	60

a. Plot the data in a scattergram.

b. Find μ_{X_1}, μ_{X_2}, σ_{X_1}, and σ_{X_2}.

c. Find the correlation coefficient between X_1 and X_2.

3.35 The surface finish achieved in a machining operation at different cutting speeds of the machine tool is indicated in the following table:

Cutting Speed (rpm)	Surface Finish Achieved (the Higher the Number, the Better the Finish)
200	4
220	11
240	5
260	3
280	6
300	2
320	12
340	12
360	15
380	12
400	17
420	20
440	18
460	21
480	18
500	24

a. Plot the data in a scattergram.
b. Find the correlation between the surface finish and the cutting speed.

3.36 The presence of clearances in joints of a linkage makes it necessary to consider the link lengths and the clearances as random variables [3.7]. Figure 3.20(*a*) shows two adjacent links in their nominal (ideal) positions while Fig. 3.20(*b*) shows their positions with clearances. If r denotes the radius of the clearance zone at the joint, the pin of link 2 (P_2) can lie anywhere in the circular zone. Then the equivalent length of link 1 can be expressed as

$$l_{1e} = \sqrt{(l_1 + x)^2 + y^2} = l_1 + x$$

since $y \ll l_1 + x$ where x and y denote the coordinates of the pin P_2 with respect to the origin O_2 shown in Figs. 3.20(*b*) and 3.20(*c*). Assuming that the pin P_2 can lie anywhere in the clearance zone with equal probability, the joint density function of x and y can be expressed as

$$f(x, y) = \begin{cases} \dfrac{1}{\pi r^2}; & x^2 + y^2 \le r^2 \\ 0; & \text{otherwise} \end{cases} \tag{E1}$$

Since the computation of the mean and standard deviation of l_{1e} requires the mean and standard deviation of x, determine these quantities $(\mu_x$ and $\sigma_x)$ using the probability density function of Eq. (E1).

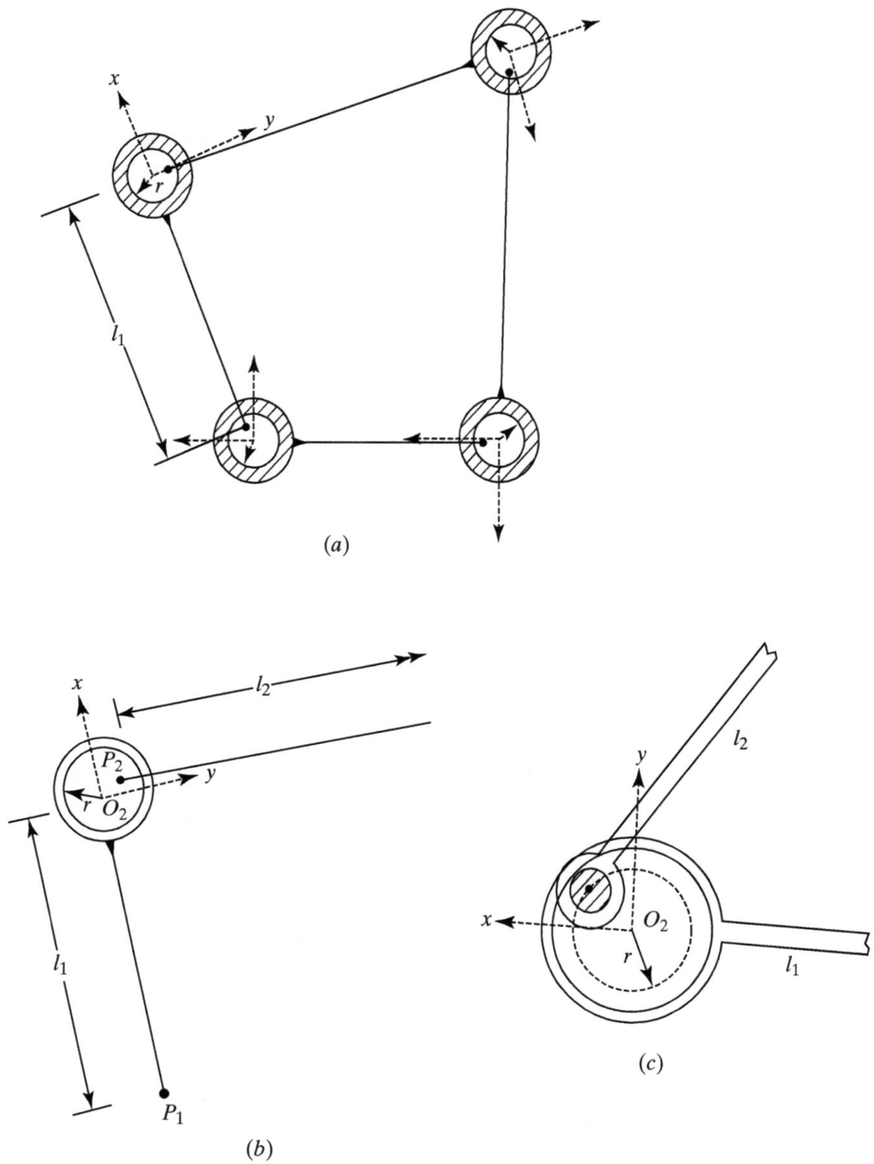

FIGURE 3.20 Clearances in the joints of a linkage.

3.37 Find the probability distribution function corresponding to the following density functions:

a. $f(x) = \begin{cases} \dfrac{1}{b-a}; & a \le x \le b \\ 0; & \text{otherwise} \end{cases}$

b. $f(x) = \begin{cases} \lambda e^{-\lambda x}; & x \geq 0 \\ 0; & \text{otherwise} \end{cases}$

Also sketch the density and distribution functions in each case.

3.38 The electric motors manufactured by a company are found to have an average life of 10^4 hours. Assuming an exponential distribution, find the fraction of motors that would fail before (a) the average life and (b) half the average life.

3.39 The welds produced by an automatic welding machine are found to be defective 1 percent of the time. If six welds are randomly selected, find the probability of finding the number of defective welds to be (a) equal to four, (b) at least one, (c) no more than two, and (d) equal to three.

3.40 The quality control department rejects the whole batch of resistors whenever it finds two or more defective units in a sample of 15 resistors tested. If a batch of resistors contain 1 percent defective units, determine the following: (a) the mean and standard deviation of the defective resistors in a sample of 15 resistors chosen from the batch, (b) the probability of rejecting the entire batch of resistors, and (c) the probability of finding no defective resistor when a sample of 15 resistors are tested.

3.41 The probability density function of a random variable is given by

$$f_X(x) = ax^2 + bx + c; \qquad 0 \leq x \leq 1$$

where a, b, and c are constants. (a) If the mean and standard deviation of X are specified as 0.6 and 0.1, respectively, determine the constants a, b, and c. (b) Find the skewness coefficient of X.

3.42 The number of cycles of stress variation (X) a particular plastic material can withstand before failure is given by:

$$f(x) = \frac{1}{4}e^{-\frac{x}{4}}; \qquad x \geq 0$$

where x is measured in thousands of cycles. Find the following: (a) The probability of the material withstanding more than 40,000 cycles without failure given that it has already survived 30,000 cycles. (b) The probability of the plastic material failing between 30,000 and 40,000 cycles.

3.43 The load raised by a screw jack (X) can be approximated by a uniform distribution:

$$f(x) = \frac{1}{b - a}; \qquad a \leq x \leq b$$

If the mean value and standard deviation of the load are known as μ and σ, determine the smallest and the largest values of the load raised by the screw jack.

3.44 The strength of a leather belt (X) follows a triangular distribution given by

$$f(x) = \begin{cases} \dfrac{4}{(b-a)^2}(x - a); & a \leq x \leq \dfrac{a+b}{2} \\ \dfrac{4}{(b-a)^2}(b - x); & \dfrac{a+b}{2} \leq x \leq b \end{cases}$$

If the mean value and standard deviation of the strength are known as μ and σ, determine the lower- and upper- bounds on the strength of belt.

3.45 The magnitude of earthquakes in a region (measured in the Richter scale) can be modeled by the density function

$$f(x) = \frac{4}{\pi(4 + x^2)}, \qquad 0 \leq x < \infty$$

Find (a) the average magnitude of the earthquakes, (b) the probability of the magnitude of the next earthquake exceeding a value of 4.0, and (c) the mode and the median of X.

3.46 The stress induced in a machine during its operation has a mean value of 25,000 lb/in^2 with a standard deviation of 1,000 lb/in^2. The material of the machine has a mean strength of 35,000 lb/in^2 with a standard deviation of 2,000 lb/in^2. Find the probability of failure of the machine assuming that both the induced stress and the material strength follow normal distribution.

3.47 An elevator is supported by a wire rope that can withstand a mean stress of 25,000 lb/in^2 with a standard deviation of 500 lb/in^2. The load on the elevator is a random variable that causes a stress with a mean value of 20,000 lb/in^2 and a standard deviation of 2,000 lb/in^2. If the load on the elevator and the strength of the rope are normally distributed, find the probability of failure of the rope.

3.48 The life of an automobile battery is known to follow normal distribution with a mean of 2000 days and a standard deviation of 200 days. (a) Find the percentage of batteries which fail before 1500 days. (b) What should be the guarantee period to be specified by the manufacturer so that the manufacturer will encounter only 10 percent failures of the battery?

3.49 The tolerances on the diameter of a shaft are specified as 2 ± 0.05 in. The shafts produced by a machinist are found to have a mean value of 2.01 in and a standard deviation of 0.02 in. Find the fraction of the machined shafts that need to be discarded.

3.50 The point reached by the end effector of a robot manipulator along a line is given by 10 ± 0.5 in. What is the probability of the end effector reaching a point beyond 10.3 in? What is the probability that the point reached is beyond 10.3 in given that it is beyond 10.1 in?

3.51 The lives of tires produced by a tire manufacturer A are approximately normally distributed with $\mu = 60{,}000$ miles and $\sigma = 5000$ miles. An automobile manufacturer B requires that at least 95 percent of the tires should have a life greater than 50,000 miles. Can the automobile manufacturer B use the tires produced by the tire manufacturer A?

3.52 The error in the altitude of a helicopter predicted by an altimeter can be approximated as a normal variate with mean 200 m and standard deviation 100 m. Find the following probabilities for the error: (a) less than or equal to -200 m, (b) greater than or equal to 300 m, (c) between -200 m and 150 m, (d) less than -100 m or greater than 250 m.

3.53 The diameters of a batch of ball bearings are known to follow normal distribution with a mean 4.0 in and a standard deviation 0.1 in. If a ball bearing is chosen randomly, find the probability of realizing the following event: (a) diameter between 3.8 in and 4.3 in, (b) diameter smaller than 3.9 in, (c) diameter larger than 4.2 in.

3.54 The performance of a certain batch of electric resistance strain gauges is being tested using a random sample of 1000 strain gauges. If past manufacturing experience indicates the reliability of the strain gauges as 0.97, find the probability of observing 25 or more defective strain gauges in the sample of 1000 tested using a normal approximation to the binomial distribution [3.5].

3.55 Write a computer program, in the form of a subroutine, to find the value of a standard normal variate z given the corresponding cumulative distribution function value, $\Phi(z)$. Using this program, generate the values of z when $\Phi(z)$ varies between 0.1 and 0.9 in increments of 0.1.

3.56 A missile is fired to hit a target point O. Due to uncertainties associated with the firing, the missile can hit any point within radius a of the target point with equal probability (see Fig. 3.21) so that

$$f(x, y) = \begin{cases} \dfrac{1}{\pi a^2}; & \text{for } x^2 + y^2 \leq a^2 \\ 0; & \text{otherwise} \end{cases}$$

a. Find the marginal density functions $f(x)$ and $f(y)$.
b. What is the probability of hitting any point within a radius of $a/4$ around the target point O?
c. What is the probability of hitting any point, within a distance of $a/4$ from the y-axis?

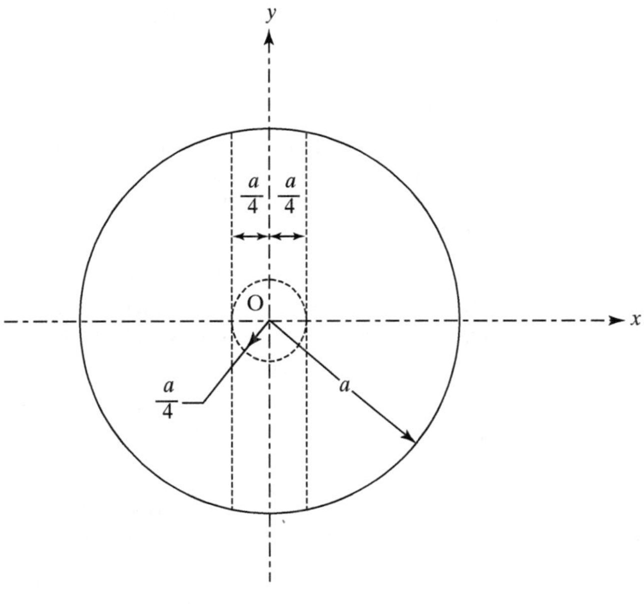

FIGURE 3.21 Area surrounding the target of a missile.

3.57 From the sample data of the Young's modulus given in Problem 1.1; determine the probability of realizing the Young's modulus greater than 30×10^6 lb/in^2 assuming (a) normal distribution and (b) lognormal distribution.

3.58 From the compressive strength data of concrete cylinders given in Problem 1.4, find the probability of realizing the compressive strength between 3 and 6 k(lb/in^2) assuming (a) normal distribution and (b) lognormal distribution.

3.59 Find the probability of realizing the first crack below a load of 5200 N and the breakage of the specimen above a load of 6200 N for the fiber-reinforced plastic specimens in Problem 3.33.

3.60 The beta density function is given by

$$f(x) = \frac{\Gamma(p+q)}{\Gamma(p)\Gamma(q)} \frac{(x-a)^{p-1}(b-x)^{q-1}}{(b-a)^{p+q-1}}; \qquad a \le x \le b; p > 0, q > 0$$

Plot this function for the following cases: (a) $a = 0, b = 1, p = 1, q = 2$; (b) $a = 0, b = 1, p = 2, q = 2$; (c) $a = 0, b = 1, p = 0.5, q = 2$; (d) $a = 0, b = 1, p = 2, q = 4$; (e) $a = 0, b = 1, p = 3, q = 3$; (f) $a = 0, b = 1, p = 4, q = 2$.

3.61 The gamma density function is given by

$$f(x) = \frac{\lambda^k}{\Gamma(k)} x^{k-1} e^{-\lambda x}; \qquad 0 \le x < \infty; \qquad \lambda > 0$$

Plot this function for the following cases: (a) $k = 3, \lambda = 1$; (b) $k = 3, \lambda = 2$; (c) $k = 3, \lambda = 4$; (d) $k = 2, \lambda = 1$; (e) $k = 1, \lambda = 1$; (f) $k = 0.5, \lambda = 1$.

3.62 Find the parameters of beta distribution for the automobile brake life data given in Example 1.1.

3.63 Find the parameters of gamma distribution for the automobile brake life data given in Example 1.1.

MATLAB and Excel Problems

3.64 Using the values of the lognormal probability density function corresponding to the values 0.00, 0.01, 0.02, . . . , 6.00 of the random variable $y = \ln x$, plot the probability density function in the range $0 \le y \le 6$ using MATLAB. Assume the parameters of the lognormal distribution (of y) as 4.00 and 0.25.

3.65 Using the values of the lognormal probability distribution function corresponding to the values 0.00, 0.01, 0.02, . . . , 6.00 of the random variate $y = \ln x$, plot the probability distribution function in the range $0 \le y \le 6$ using MATLAB. Assume the parameters of the lognormal distribution (of y) as 4.00 and 0.25.

3.66 Find the values of the lognormally distributed variable corresponding to the probability values of $p = 0.0, 0.1, 0.2, \ldots, 1.0$ using MATLAB. The parameters of the lognormal distribution (of y) are 4.00 and 0.25.

3.67 Plot the exponential probability density and distributions in the range of the random variable $0 \le x \le 40$ using MATLAB. The mean value of the random variable is 4.0.

3.68 Find the mean value, standard deviation, and skewness of a lognormally distributed variable (y) with parameters 4.00 and 0.25 using Excel.

3.69 Find the mean value and standard deviation of a uniformly distributed variable in the range $0 \le x \le 10$ using Excel.

3.70 Find the values of the probability density and distribution functions of a lognormally distributed variable y, with parameters 4.00 and 0.25, corresponding to $y = 4.0$ using Excel.

CHAPTER 4

Extremal Distributions

What You Will Learn

In this chapter, you will learn

- The need for the consideration of extreme values of random variables in several practical engineering problems
- The determination of the probability distributions of minimum and maximum values of a random variable
- The definitions of order statistics and asymptotic distributions
- The three types of asymptotic distributions for maximum value as well as minimum value
- The concepts of the return period, exceedance, and characteristic value
- The fitting of extremal distributions to experimental data using the least-squares approach
- The generalized extreme value distribution and its statistical parameters

Waloddi Weibull was born on June 18, 1887 in Sweden. He joined the Royal Swedish Coast Guard in 1904 and left the military service in 1917. He studied at the Royal Institute of Technology and at Stockholm University. Later he became a consulting engineer and invented ball bearings and electric hammers. He published his first scientific paper on the propagation of explosive waves in 1914 and became a full professor at the Royal Institute of Technology in 1924. He received his doctorate from the University of Uppsala in 1932. He published two important papers on the statistical distribution of material strength in late 1930s. The probability distribution he proposed for describing material strength has become known as Weibull distribution, and it finds application in reliability, statistics, fatigue, and many other fields. After his retirement from the Royal Institute of Technology in 1953, he became a consultant to the U.S. Air Force Materials Laboratory, published a book on Fatigue Testing and Analysis of Results in 1961, and accepted a position as a visiting professor at Columbia University's Institute for the Study of Fatigue and Reliability. In 1972, the American Society of Mechanical Engineers awarded a gold medal to Weibull with the citation, "a pioneer in the study of fracture, fatigue, and reliability who has contributed to the literature for over thirty years". Weibull received many other awards, honors, and medals before he died in France on October 12, 1979 [4.14].

4.1 Introduction

In many engineering applications, we will be concerned with the largest or smallest value of a random variable. Success or failure of a system may rest solely on its ability to function under the *maximum demand* (load) or *minimum capacity* (strength), and not simply on the typical values. Floods, wind, temperature, solar radiation, and floor loadings are variables for which the largest value in a sequence may be critical to several engineering systems. Similarly, the smallest value of strength of materials will also be critical in some situations. The following examples indicate situations where a knowledge of the minimum or maximum value of a random variable is required.

1. In structural and mechanical design, we need to know the minimum strength of the material and the maximum load acting on the structure for a safe design.
2. In flood control programs, such as the design of dams, we need to know the intensity of the maximum flood of a river, whereas in irrigation and water supply programs a knowledge of the minimum flow will be a matter of concern.
3. The largest amount of traffic experienced in a day is important in the design of highways.
4. The corrosion of a metal is associated with small pits caused by chemical agents. Thus, the corrosion resistance of a component is governed by the size of the largest pit present.
5. The safe design of offshore platforms involves a knowledge of the maximum wave heights experienced in the ocean.
6. The development of a new suburb is encouraged when the maximum pollution level caused by the industrial and automobile exhausts in the region is limited to small values.
7. The design of high-rise buildings is influenced by the maximum wind or earthquake load expected in the region.
8. The *size* plays a major role in the strength of any material. According to a widely accepted theory, the body of a material (such as a machine component) is considered to be composed of several small pieces or elemental volumes. The material (component) is considered to have failed whenever any one of the small pieces (elemental volumes) fails. This is called the *weakest link model* since the strength of the total volume is governed by that of the elemental volume that has the smallest strength.
9. The maximum vertical gust velocity encountered by an aircraft is important for a safe design.

A systematic study of the maximum or minimum values of the random variables influencing the design may lead to a better design. Further, the concept of finite-life design can also be incorporated and thus a better use of materials can be made and technological obsolescence can be cared for. Gumbel [4.1] has found that the use of extreme values is much more reliable than the median whose value increases with the sample size. Wirsching and Jones [4.7] have used extreme values in reliability analysis and design.

To illustrate the concept of extreme value distribution, assume that we have the wind velocity record for the past 40 years on an hourly basis. This means that we have a total of $40 \times 365 \times 24 = 350{,}400$ data points (assuming 365 days per year). Let the mean value corresponding to this data be 14 mph and the distribution of the wind velocity be $f_X(x)$ as shown in Fig. 4.1. Suppose we want to know about the distribution followed by the largest wind velocity measured on a yearly basis. This time, we have one

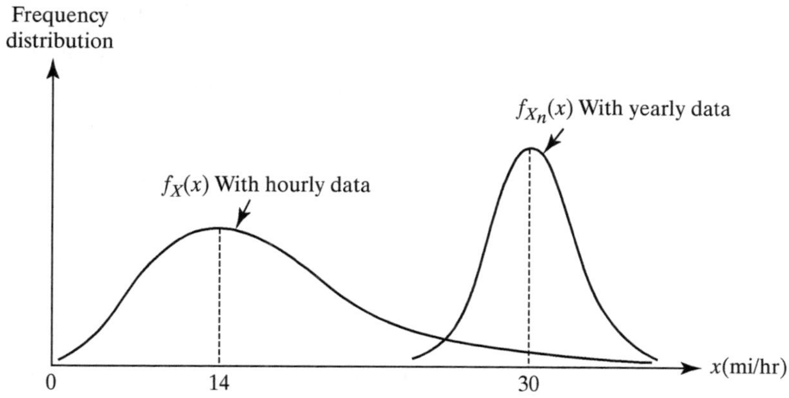

FIGURE 4.1 Probability density functions with hourly and yearly data.

maximum wind velocity for each year and a total of 40 data points only. Naturally, the distribution of this extreme wind velocity will be shifted toward the right with a higher mean (30 mph) as indicated in Fig. 4.1. Our interest is to find the distribution of the largest value, X_n (and sometimes, the smallest value X_1) in terms of the original (parent) distribution, $f_X(x)$.

4.2 Extreme Value Distributions in Terms of Parent Distribution

To express the density functions of the smallest and largest values of a random variable in terms of the parent density function, consider a random variable X denoting the yield strength of steel. Let n tests be conducted to find the yield strength of steel as X_i', $i = 1, 2, \ldots, n$. Arrange the results in an increasing order so that the smallest value is designated as X_1' and the largest value as X_n'. If the n tests are repeated, we obtain, after rearranging the values in an increasing order, a different set of values as $X_1'', X_2'', \ldots, X_n''$. Similarly, the repetition of the n tests a number of times yields the results $(X_1''', X_2''', \ldots, X_n''')$, and so on. Let the results of all the tests, that is, $X_1', X_1'', X_1''', \ldots, X_2', X_2'', X_2''', \ldots, X_n', X_n'', X_n''', \ldots$, fit into a distribution, $F_X(x)$, which is known as the (parent) distribution function of the random variable X. If we fit separate distributions for the smallest (i.e., for $X_1', X_1'', X_1''', \ldots$) and for the largest values only (i.e., for $X_n', X_n'', X_n''', \ldots$), the resulting distributions, $F_{X_1}(x)$ and $F_{X_n}(x)$, are called the extreme value distributions for the smallest and the largest values of X, respectively. It is possible to express the extreme value distributions $F_{X_1}(x)$ and $F_{X_n}(x)$ in terms of the parent distribution $F_X(x)$.

To find the distribution function of the smallest value X_1, we use the relation

$$1 - F_{X_1}(x) = P(X_1 > x) = P(X_1 > x)P(X_2 > x) \cdots P(X_n > x) \tag{4.1}$$

if all X_i are independent. This equation indicates that the probability of realizing X_1 to be greater than x is same as the probability of realizing all $X_i(i = 1, 2, \ldots, n)$ to be greater than x. If all the X_i are identically distributed with a common distribution function $F_X(x)$, then

$$1 - F_{X_1}(x) = (P(X > x))^n = (1 - F_X(x))^n$$

or

$$F_{X_1}(x) = 1 - (1 - F_X(x))^n \tag{4.2}$$

Similarly, the distribution function of the largest value X_n is given by

$$\begin{aligned} F_{X_n}(x) &= P(X_n \leq x) = P(\text{all} X_i \leq x) \\ &= P(X_1 \leq x)P(X_2 \leq x) \cdots P(X_n \leq x) \end{aligned}$$

if all X_i are independent. This means that the probability of realizing X_n less than or equal to x is same as the probability of realizing all $X_i(i = 1, 2, \ldots, n)$ to be less than or equal to x. If all the X_i are identically distributed with the distribution function $F_X(x)$,

$$F_{X_n}(x) = (F_X(x))^n \tag{4.3}$$

Order Statistics. When the sample values $(X_1, X_2, \ldots, X_n)$ of a random variable are arranged in an increasing order as $X_1 \leq X_2 \leq \cdots \leq X_i \leq \cdots \leq X_n$, the ith element X_i is called the ith order statistic of the sample. Thus, the distributions of the smallest and the largest values can be called the distributions of the first and the nth order statistics, respectively.

EXAMPLE 4.1

The time elapsed between the arrival of customers at an automobile repair shop can be assumed to follow an exponential distribution with a mean of 10 minutes. A knowledge of the maximum time elapsed between the arrival of customers will be of interest in deciding the number of mechanics to be made available in the shop. Assuming that the time elapsed between the arrival of customers is independent from one customer to the next, compute the probability that the maximum time elapsed between customers exceeds 12 minutes over a period covering the arrival of 15 customers.

Solution: The parent distribution for the time elapsed between the arrival of customers X is given by

$$F_X(x) = 1 - e^{-\lambda x}$$

where $\lambda = 1/\mu_X = 1/10 = 0.1$. Since 15 customers imply 14 intercustomer periods $(n = 14)$, the probability of realizing the maximum intercustomer time $(Y \equiv X_{14})$ to be less than 12 minutes is given by Eq. (4.3):

$$F_Y(12) = (F_X(12))^{14} = (1 - e^{-0.1(12)})^{14} = 0.006622$$

Hence, the probability that the maximum time elapsed between the arrival of customers exceeding 12 minutes is $1.0 - 0.006622 = 0.993378$.

■

EXAMPLE 4.2

The strength of a welded joint, in $k(\text{lb/in}^2)$, is distributed normally as

$$f_X(x) = \frac{1}{\sqrt{2\pi}\,7.5} \exp\left[-\frac{1}{2}\left(\frac{x - 75}{7.5}\right)^2\right]$$

Find the probability distribution and density functions for the largest strength realized in a sample of n welded joints. Plot these functions for $n = 1$, 10, and 20.

Solution: The probability distribution function corresponding to the largest strength in a sample of n welded joints is given by Eq. (4.3)

$$F_{X_n}(x_n) = [F_X(x_n)]^n \tag{E1}$$

where

$$F_X(x_n) = \int_{-\infty}^{x_n} f(x)\,dx = \int_{-\infty}^{x_n} \frac{1}{7.5\sqrt{2\pi}} \exp\left[-\frac{1}{2}\left(\frac{x - 75}{7.5}\right)^2\right] dx \tag{E2}$$

This gives

$$F_{X_n}(x_n) = \left[\int_{-\infty}^{x_n} \frac{1}{7.5\sqrt{2\pi}} \exp\left[-\frac{1}{2}\left(\frac{x - 75}{7.5}\right)^2\right] dx\right]^n \tag{E3}$$

Defining a standard normal variate z_n as

$$z_n = \frac{x_n - 75}{7.5}$$

Equation (E3) can be expressed as

$$F_{X_n}(x_n) = F_{Z_n}(z_n) = [\Phi(z_n)]^n \tag{E4}$$

The probability density function, $f_{X_n}(x_n)$, can be obtained from Eq. (E4) as

$$f_{X_n}(x_n) = n[\Phi(z_n)]^{n-1} f_{Z_n}(z_n) \tag{E5}$$

With the help of standard normal tables, Eqs. (E4) and (E5) can be plotted for any specific value of n. The result is shown in Fig. 4.2.

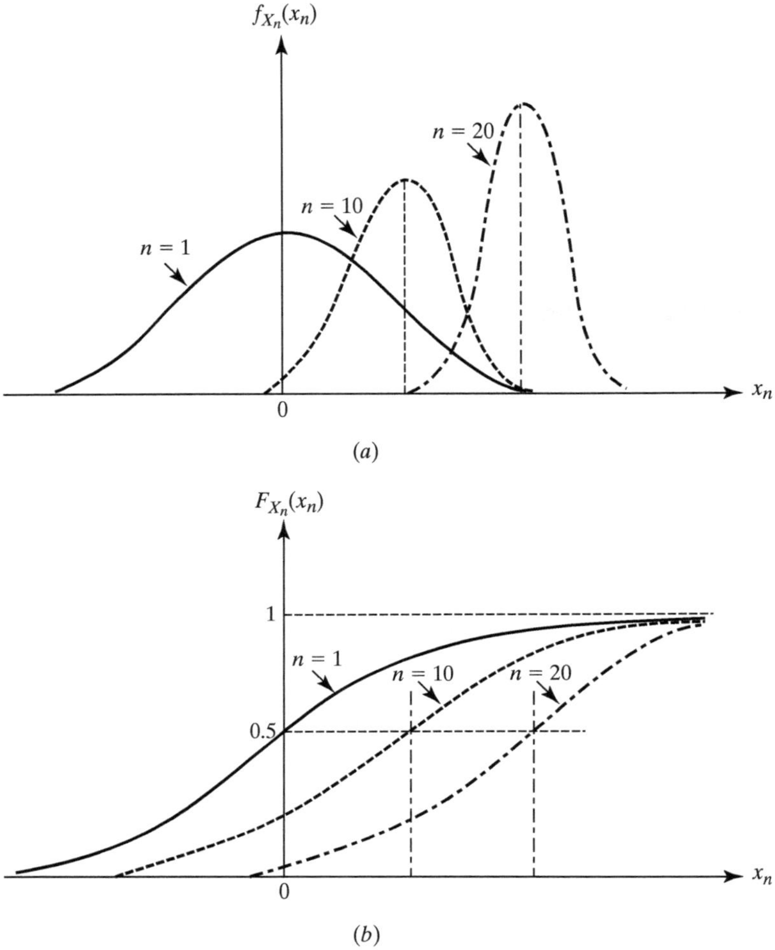

FIGURE 4.2 Probability density and distribution functions for largest strength.

■

4.3 Asymptotic Distributions

If the random variables X_i, $i = 1, 2, \ldots, n$, are independent and follow a common distribution $F_X(x)$, the shape of the distribution function of the extreme values X_1 or X_n becomes increasingly insensitive to the exact shape of the common distribution function $F_X(x)$, as n tends to infinity. The limiting forms of the distribution functions of the extreme values X_1 or X_n as $n \to \infty$ are known as asymptotic distributions. The asymptotic distributions often describe the behavior of the random variable X_1 or X_n reasonably well even when the exact shape of the parent distribution function $F_X(x)$ is not known precisely. The asymptotic forms of $F_{X_1}(x)$ and $F_{X_n}(n)$ are classified into three types based on the general features of the tail part of the distribution of the random variable X [4.1]. Different types of asymptotic distributions have been used for different engineering applications.

4.4 Type-I Asymptotic Distributions

4.4.1 Maximum Value

The type-I asymptotic distribution for the maximum values is useful whenever the right tail of the parent density function $f_X(x)$ is unbounded $(x \rightarrow \infty)$ and is of an exponential type. In such a case, we can express $F_X(x)$ as

$$F_X(x) = 1 - e^{-h(x)} \tag{4.4}$$

where $h(x)$ increases with x monotonically. The distributions such as normal, lognormal, and gamma belong to this category. If $F_X(x)$ is of the form given by Eq. (4.4), the asymptotic distribution function for the largest value $Y \equiv X_n$ is given by

$$F_Y(y) = \exp\left[-e^{-a(y-w)}\right]; \qquad -\infty < y < \infty \tag{4.5}$$

where the parameters of the distribution, a and w, can be determined from the observed data. The probability density function of Y can be obtained as

$$f_Y(y) = a \exp\left[-a(y - w) - e^{-a(y-w)}\right]; \qquad -\infty < y < \infty \tag{4.6}$$

It can be shown that the mode (Y_{mode}), mean (μ_Y), and the variance (σ_Y^2) of Y are given by

$$Y_{\text{mode}} = w \tag{4.7}$$

$$\mu_Y = w + \frac{\gamma}{a} \tag{4.8}$$

$$\sigma_Y^2 = \frac{\pi^2}{6a^2} \tag{4.9}$$

where $\gamma \approx 0.577$ is the Euler's constant. This distribution has a positive skewness coefficient (γ_1) with $\gamma_1 \approx 1.1396$ as shown in Fig. 4.3.

4.4.2 Smallest Value

The type-I asymptotic distribution for the minimum values is useful whenever the left tail of the parent density function $f_X(x)$ is unbounded $(-\infty < x)$ and decreases to zero toward the left in an exponential form. For example, the normal distribution belongs to this category. In this case, the asymptotic distribution function for the smallest value, $Z \equiv X_1$, is given by

$$F_Z(z) = 1 - \exp\left[-e^{a(z-w)}\right]; \qquad -\infty < z < \infty \tag{4.10}$$

where a and w are the parameters of the distribution. The density function can be obtained by differentiating Eq. (4.10) as

$$f_Z(z) = a \exp\left[a(z - w) - e^{a(z-w)}\right]; \qquad -\infty < z < \infty \tag{4.11}$$

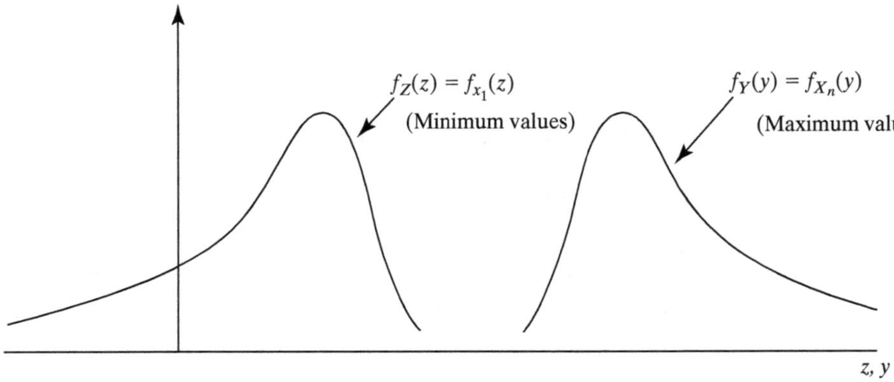

FIGURE 4.3 Typical shapes of type-I asymptotic distributions.

The mode (Z_{mode}), mean (μ_Z), and the variance (σ_Z^2) of Z can be found as

$$Z_{\text{mode}} = w \tag{4.12}$$

$$\mu_Z = w - \frac{\gamma}{a} \tag{4.13}$$

$$\sigma_Z^2 = \frac{\pi^2}{6a^2} \tag{4.14}$$

This distribution has a negative skewness coefficient (γ_1) with $\gamma_1 \approx -1.1396$, as shown in Fig. 4.3.

Notes:

1. It can be seen that the type-I asymptotic distributions for largest and smallest values are mirror images of each other. Typical shapes of $f_Y(y)$ and $f_Z(z)$ are shown in Fig. 4.3.
2. The type-I distributions are also known as Gumbel distributions.
3. The type-I maximum value distribution is useful, for example, to describe the maximum flood in a river.
4. The type-I minimum value distribution is useful, for example, to describe the strength of brittle materials where the specimen is assumed to consist of several microscopic volumes of material. In this case, the specimen fails whenever the weakest microscopic volume fails.

4.5 Type-II Asymptotic Distributions

4.5.1 Maximum Value

The type-II asymptotic distribution for the maximum values is useful whenever the parent distribution $F_X(x)$ is defined over the range $0 < x < \infty$ and approaches one as $x \to \infty$ according to the relation

$$F_X(x) = 1 - \alpha\left(\frac{1}{x}\right)^m; \qquad x > 0 \tag{4.15}$$

where $\alpha > 0$ and $m > 0$ are the parameters of the distribution. In this case, $F_Y(y)$ can be expressed as

$$F_Y(y) = \exp\left[-\left(\frac{w}{y}\right)^m\right]; \qquad m > 0, w > 0, y > 0 \tag{4.16}$$

so that

$$f_Y(y) = \frac{m}{w}\left(\frac{w}{y}\right)^{m+1} e^{-(w/y)^m}; \qquad m > 0, w > 0, y > 0 \tag{4.17}$$

The mode, mean, and variance of Y are given by

$$Y_{\text{mode}} = \frac{wm}{m + 1} \tag{4.18}$$

$$\mu_Y = w\Gamma\left(1 - \frac{1}{m}\right); \qquad m > 1 \tag{4.19}$$

and

$$\sigma_Y^2 = w^2\left[\Gamma\left(1 - \frac{2}{m}\right) - \Gamma^2\left(1 - \frac{1}{m}\right)\right]; \qquad m > 2 \tag{4.20}$$

where $\Gamma(x)$ is the gamma function defined as

$$\Gamma(x) = \int_0^\infty y^{x-1}e^{-y}dy \tag{4.21}$$

If x is a positive integer, the gamma function is given by

$$\Gamma(x) = (x - 1)! \tag{4.22}$$

4.5.2 Smallest Value

The type-II asymptotic distribution for the minimum values is useful whenever the parent distribution $F_X(x)$ is defined over the range $-\infty < x \leq 0$ and has a form similar to that of Eq. (4.15). In this case, $F_Z(z)$ can be expressed as

$$F_Z(z) = 1 - \exp\left[-\left|\frac{w}{z}\right|^m\right]; \qquad z \leq 0 \tag{4.23}$$

where $w > 0$ and $m > 0$ are the parameters of the distribution.

Notes:

1. The type-II maximum value distribution is useful for representing annual maximum winds and other meteorological and hydrological phenomena.
2. The type-II minimum value distribution is not commonly used since the required parent distribution shape is not commonly observed in practical applications.

4.6 Type-III Asymptotic Distributions

4.6.1 Maximum Value

The type-III asymptotic distribution for the maximum values is useful whenever the parent distribution $F_X(x)$ is defined over the range $-\infty < x \le \omega$ and has the form

$$F_X(x) = 1 - a(\omega - x)^m; \qquad x \le \omega \tag{4.24}$$

where $a > 0$ and $m > 0$ are the parameters of the distribution. Notice that Eq. (4.24) denotes a uniform distribution when $m = 1$ and a triangular distribution when $m = 2$. In this case, the type-III maximum value distribution can be derived to be of the form [4.1, 4.10]

$$F_Y(y) = \exp\left[-\left(\frac{\omega - y}{\omega - v}\right)^m\right]; \qquad y \le \omega \tag{4.25}$$

with

$$f_Y(y) = \frac{m}{\omega - v}\left(\frac{\omega - y}{\omega - v}\right)^{m-1} \exp\left[-\left(\frac{\omega - y}{\omega - v}\right)^m\right]; \qquad y \le \omega \tag{4.26}$$

where $v < \omega$.

4.6.2 Smallest Value

The type-III minimum value distribution can be used whenever the parent distribution is defined over the range $\varepsilon \le x < \infty$ and has the form

$$F_X(x) = a(x - \varepsilon)^{\beta}; \qquad x \ge \varepsilon \tag{4.27}$$

where $a > 0$ and $\beta > 0$ are the parameters of the distribution. In this case, the type-III asymptotic distribution for the minimum value can be obtained as [4.1, 4.10]

$$F_Z(z) = 1 - \exp\left\{-\left(\frac{z - \varepsilon}{\eta - \varepsilon}\right)^{\beta}\right\}; \qquad z \ge \varepsilon, \beta > 0, \eta > \varepsilon \tag{4.28}$$

where β is called the shape parameter $(\beta > 0)$, η the characteristic value (or scale factor) with $\eta \ge \varepsilon$, and ε the minimum permissible value of Z. Larger values of the shape parameter β correspond to distributions that are skewed to the right and smaller values correspond to distributions that are skewed to the left. The values of β in the range $3.3 < \beta < 3.5$ represent distributions that are approximately symmetric and hence can be used as a good approximation to the normal distribution. The value of $\beta = 1$ corresponds to the exponential distribution. The characteristic value η serves a role similar to the mean value and denotes a value of Z below which lie 63.2 percent of the observations.

By differentiating Eq. (4.28), we obtain the density function as

$$f_Z(z) = \frac{\beta}{\eta - \varepsilon}\left(\frac{z - \varepsilon}{\eta - \varepsilon}\right)^{\beta - 1} \exp\left\{-\left(\frac{z - \varepsilon}{\eta - \varepsilon}\right)^{\beta}\right\}; \qquad z \ge \varepsilon \tag{4.29}$$

In this case, the median, mode, mean, and variance of Z are given by

$$Z_{50} = \varepsilon + (\eta - \varepsilon)(\ln 2)^{1>\beta} \tag{4.30}$$

$$Z_{\text{mode}} = \varepsilon + (\eta - \varepsilon)\left(\frac{\beta - 1}{\beta}\right)^{1/\beta} \tag{4.31}$$

$$\mu_Z = \varepsilon + (\eta - \varepsilon)\Gamma\left(1 + \frac{1}{\beta}\right) \tag{4.32}$$

and

$$\sigma_Z^2 = (\eta - \varepsilon)^2\left\{\Gamma\left(1 + \frac{2}{\beta}\right) - \Gamma^2\left(1 + \frac{1}{\beta}\right)\right\} \tag{4.33}$$

where $\Gamma(\ )$ is the gamma function defined in Eqs. (4.21) and (4.22).

Note:
The distribution indicated by Eqs. (4.28) and (4.29) is known as the three-parameter Weibull distribution. When $\varepsilon = 0$, Eq. (4.27) denotes gamma distribution. For this case $(\varepsilon = 0)$, the type-III minimum value distribution denotes the two-parameter Weibull distribution, which is useful in the study of material strength in tension and fatigue. In this case, the parameters $\beta > 0$ and $\eta > 0$, are called the slope and scale parameters, respectively.

4.7 Return Period

The *return period* is defined as the average elapsed time between occurrences of an event with a specified magnitude or greater. For example, a 20-year earthquake intensity is an intensity that is equaled or exceeded, on the average, once every 20 years over a long period of time. It does not mean that an exceedance occurs every 20 years, but that the average time between exceedances is 20 years. An *exceedance* is an event with a magnitude equal to or greater than a certain value. The concept of return period can also be applied to minimum values such as droughts, low flows, and shortages. In this case, the return period would be the average time between events with a certain magnitude or less. Such an event can still be called an exceedance in the sense that the severity of a drought (or an earthquake) exceeds some specified level. Regardless of whether the return period refers to an event greater than a specified value or to an event less than a specified value, it can be related to the probability of exceedance. If an exceedance occurs every 20 years, on the average, then the probability that the event occurs in any given year is $1/20 = 0.05$ or 5 percent. Thus, the return period T and the probability P are related as

$$T = \frac{1}{P} \tag{4.34}$$

EXAMPLE 4.3

The probability distribution function of the yearly maximum wind velocity, in mi/h, at a particular locality is represented as

$$F(x) = \exp\left[-\exp\left(-\frac{x-60}{20}\right)\right]$$

Find the return period of a yearly maximum wind velocity of 100 mi/h.

Solution: The probability of realizing the yearly maximum wind velocity greater than 100 mi/h is given by

$$P(x > 100) = 1 - P(x \le 100) = 1 - F(100)$$
$$= 1 - \left[\exp\left(-\exp\left(-\frac{100-60}{20}\right)\right)\right] = 1 - \exp\left[-\exp\left(-2\right)\right] = 0.1266$$

Thus, the return period is given by

$$T = \frac{1}{P(x > 100)} = \frac{1}{0.1266} = 7.8989 \text{ years}$$

■

4.8 Characteristic Value

If the random variable T denotes the time between consecutive occurrences of an event E with a specified magnitude x or greater, then the mean value of T, $\overline{T}(x)$, gives the return period:

$$\overline{T}(x) = \frac{1}{1 - F(x)} \tag{4.35}$$

where $F(x)$ is the cumulative distribution function indicating the probability of occurrence of the event E with a magnitude smaller than x. Thus, the expected number of occurrences of event E with magnitude x or greater in one time period will be $1/\overline{T}(x) = 1 - F(x)$, and in k time periods it will be

$$k\left(\frac{1}{\overline{T}(x)}\right) = k[1 - F(x)] \tag{4.36}$$

The particular value of the random variable X, denoted as x_{cl}, is called the *characteristic largest value* over a period of k time periods if the mean value of the number of exceedances of x_{cl} in k time periods is unity. The characteristic largest value gives an idea of the central location of the possible largest values. By setting Eq. (4.36) to unity, we obtain

$$k[1 - F(x_{cl})] = 1$$

or

$$F(x_{cl}) = 1 - \frac{1}{k} \tag{4.37}$$

If we consider k time periods, the distribution of the largest value of x_{cl} is given by Eq. (4.3):

$$F_{X_k}(x_{cl}) = F^k(x_{cl}) = \left(1 - \frac{1}{k}\right)^k \tag{4.38}$$

Thus the probability of exceeding the characteristic largest value in k time periods will be

$$1 - F_{X_k}(x_{cl}) = 1 - \left(1 - \frac{1}{k}\right)^k \tag{4.39}$$

As k tends to infinity, Eq. (4.39) converges to the value $1 - e^{-1} \approx 0.6321$. In a similar manner, the *characteristic smallest value* x_{cs} can be obtained from

$$kF(x_{cs}) = 1$$

or

$$F(x_{cs}) = \frac{1}{k} \tag{4.40}$$

If we consider k time periods, the probability of realizing a value smaller than the characteristic smallest value in k time periods is given by Eq. (4.2):

$$F_{X_1}(x_{cs}) = 1 - [1 - F(x_{cs})]^k = 1 - \left(1 - \frac{1}{k}\right)^k \tag{4.41}$$

Again, it can be seen that for large values of k (k tending to infinity), Eq. (4.41) converges to the value, $1 - e^{-1} \approx 0.6321$.

4.9 Fitting Extremal Distributions to Experimental Data

In practice, the observed data is plotted on a graph paper called the probability paper. This paper provides scaled ordinates such that the cumulative distribution function plots as a straight line. Thus, the probability paper reduces the comparison between the data and the probability model to a comparison between the data points and a straight line. The probability paper can also be used to find the parameters of the distribution from the observed data. Chapter 15 describes the construction and use of probability papers in a greater detail.

To derive the necessary relationships, we first consider the extremal distributions corresponding to the maximum values. Taking natural logarithms of Eq. (4.5) for type-I largest values, we obtain

$$\ln F_Y(y) = -\exp[-a(y - w)] \tag{4.42}$$

By taking natural logarithms once again, Eq. (4.42) gives

$$y = -\frac{1}{a}\ln\ln\frac{1}{F_Y(y)} + w \tag{4.43}$$

This equation represents a straight line of the form

$$v = ay + b \tag{4.44}$$

where $v = -\ln \ln [1/F_Y(y)]$ and $b = -aw$. Hence, a plot of v versus y will be a straight line for type-I distribution for the largest values. Similarly, the type-I distribution for minimum values, Eq. (4.10), gives

$$\ln [1 - F_Z(z)] = -e^{a(z-w)}$$

or

$$\ln \frac{1}{1 - F_Z(z)} = e^{a(z-w)} \tag{4.45}$$

By taking natural logarithms, Eq. (4.45) leads to

$$z = \frac{1}{a} \ln \ln \frac{1}{1 - F_Z(z)} + w$$

or

$$u = az + b \tag{4.46}$$

where $u = \ln \ln \{1/[1 - F_Z(z)]\}$ and $b = -aw$. Thus a plot of u versus z will be a straight line for type-I distribution for the smallest values.

To plot the experimental data, let there be N (observed) values for the random variable y. Arrange the data in an increasing order such that $y_1 \le y_2 \le \cdots \le y_i \le \cdots \le y_N$ where y_1 and y_N denote, respectively, the smallest and the largest observed values. Then the fractions $\{i/(N + 1)\}$, $i = 1, 2, \ldots, N$ are computed and the points $\{y_i, i/(N + 1)\}$, $i = 1, 2, \ldots, N$ are plotted on the extremal probability paper after suitable transformation. The fraction, $\{i/(N + 1)\}$, is considered the cumulative probability, $F(y_i)$ corresponding to the observed value y_i for the purpose of plotting. Once the different points are plotted, a straight line can be fitted through these points using the least-squares method. The error between the actual, experimental point and the corresponding point on the straight line can also be computed. The magnitude of this error can be used as a measure to determine which of the distributions fits the observed data best. The procedure is illustrated in Example 4.4.

4.9.1 Least-squares Fit

Let the equation approximating the data points (x_i, y_i), $i = 1, 2, \ldots, N$ be represented as

$$y = ax + b \tag{4.47}$$

The error between the value predicted by Eq. (4.47) and the experimental value observed at x_i, namely y_i, is given by

$$y_i - ax_i - b$$

and the function $f(a, b)$ to be minimized in the least-squares method can be constructed as

$$f(a, b) = \sum_{i=1}^{N} (y_i - ax_i - b)^2 \tag{4.48}$$

The necessary conditions for minimizing f yield

$$\frac{\partial f}{\partial a} = \sum_{i=1}^{N} 2(y_i - ax_i - b)(-x_i) = 0$$

or,

$$a\left(\sum_{i=1}^{N} x_i^2\right) + b\left(\sum_{i=1}^{N} x_i\right) = \left(\sum_{i=1}^{N} x_i y_i\right) \tag{4.49}$$

and

$$\frac{\partial f}{\partial b} = \sum_{i=1}^{N} 2(y_i - ax_i - b)(-1) = 0$$

or,

$$a\left(\sum_{i=1}^{N} x_i\right) + b(N) = \left(\sum_{i=1}^{N} y_i\right) \tag{4.50}$$

The solution of Eqs. (4.49) and (4.50) gives

$$a = \frac{\left(\sum_{i=1}^{N} x_i\right)\left(\sum_{i=1}^{N} y_i\right) - N\left(\sum_{i=1}^{N} x_i y_i\right)}{\left(\sum_{i=1}^{N} x_i\right)^2 - N\left(\sum_{i=1}^{N} x_i^2\right)} \tag{4.51}$$

$$b = \frac{\left(\sum_{i=1}^{N} x_i\right)\left(\sum_{i=1}^{N} x_i y_i\right) - \left(\sum_{i=1}^{N} x_i^2\right)\left(\sum_{i=1}^{N} y_i\right)}{\left(\sum_{i=1}^{N} x_i\right)^2 - N\left(\sum_{i=1}^{N} x_i^2\right)} \tag{4.52}$$

EXAMPLE 4.4

The following load data (in newtons) has been obtained experimentally: 13,800; 18,875; 10,450; 12,550; 15,175; 14,025; 11,875; 13,150; 16,850; 12,300; 13,425; 14,700. Find the extremal distribution corresponding to the type-I largest value for this data:

Solution:

1. Rearrange the experimental data in an increasing order, $y_1 < y_2 < \cdots < y_N$, as shown in the second column of Table 4.1.
2. Calculate the probability distribution function $F_Y(y)$:

$$F_Y(y) = \frac{i}{N+1}$$

as shown in the third column of Table 4.1.

TABLE 4.1 Computations for Type-I Largest Values

Serial Number i	Ordered Value of Load, Newtons (y_i)	Cumulative Probability $F(y_i) = \frac{i}{N+1}$	$\frac{1}{F(y_i)}$	$-\ln\ln\frac{1}{F(y_i)}$
1	10,450	1/13	13.00	−0.9419
2	11,875	2/13	6.50	−0.6269
3	12,300	3/13	4.33	−0.3828
4	12,550	4/13	3.25	−0.1644
5	13,150	5/13	2.60	0.0455
6	13,425	6/13	2.17	0.2572
7	13,800	7/13	1.86	0.4796
8	14,025	8/13	1.63	0.7226
9	14,700	9/13	1.44	1.0004
10	15,175	10/13	1.30	1.3380
11	16,850	11/13	1.18	1.7894
12	18,875	12/13	1.08	2.5252

3. Compute the ordinates v_i corresponding to the data (abscissa) y_i as

$$v_i = -\ln\ln\frac{1}{F_Y(y_i)}$$

and are shown in Table 4.1.

4. The least-squares fit is given by

$$v = ay + b$$

where the constants a and b are given by equations similar to Eqs. (4.51) and (4.52). This results in

$$v = 0.000446322y - 5.714323$$

5. The parameters of the distribution are given by

$$a = 0.000446322$$

and

$$w = -\frac{b}{a} = \frac{5.714323}{0.000446322} = 12803.139$$

Thus, the type-I extremal distribution corresponding to the largest values for the given load data can be expressed as

$$F_Y(y) = \exp\left(-\exp\left[-0.000446322\,(y - 12803.139)\right]\right)$$

■

4.10 Generalized Extreme Value Distribution

The generalized extreme value distribution is developed to represent all the three types (types I, II, and III) of extreme value distributions. It can be used to model the smallest or largest values among a large set of independent, identically distributed random variables.

The generalized extreme value distribution is defined in terms of three parameters by the cumulative distribution function

$$F(x) = \exp\left\{-\left[1 + k\frac{(x-\mu)}{\sigma}\right]^{-\frac{1}{k}}\right\} \tag{4.53}$$

with

$$1 + k\frac{(x-\mu)}{\sigma} > 0 \tag{4.54}$$

The probability density function corresponding to Eq. (4.53) is given by

$$f(x) = \frac{1}{\sigma}\exp\left\{-\left[1 + k\frac{(x-\mu)}{\sigma}\right]^{-\frac{1}{k}}\right\}\left[1 + k\frac{(x-\mu)}{\sigma}\right]^{-\frac{1}{k}-1} \tag{4.55}$$

where μ is called the location parameter, $\sigma > 0$ is called the scale parameter, and k is called the shape parameter. The value of the shape parameter $k > 0$ corresponds to the type-II extremal distribution, $k < 0$ corresponds to the type-III extremal distribution, and $k = 0$ corresponds to type-I extremal distribution. Types I, II, and III extreme value distributions are also known as Gumbel, Frechet, and Weibull distributions, respectively. Some of the important statistical parameters of the generalized extreme value distribution are given below:

$$E\{X\} = \mu - \frac{\sigma}{k} + \frac{\sigma}{k}g_1 \tag{4.56}$$

$$Var[X] = \frac{\sigma^2}{k^2}(g_2 - g_1^2) \tag{4.57}$$

$$\text{Mode}\,[X] = \mu + \frac{\sigma}{k}\{(1+k)^{-k} - 1\} \tag{4.58}$$

$$\text{Skewness}\,[X] = \begin{cases} \dfrac{g_3 - 3g_1g_2 + 2g_1^3}{(g_2 - g_1^2)^{\frac{3}{2}}}; k > 0 \\ \dfrac{12\sqrt{6}\,\zeta(3)}{\pi^3}; k = 0 \\ \dfrac{-g_3 + 3g_1g_2 - 2g_1^3}{(g_2 - g_1^2)^{\frac{3}{2}}}; k < 0 \end{cases} \tag{4.59}$$

where

$$g_i = \Gamma(1 - ik); \quad i = 1, 2, 3 \tag{4.60}$$

and $\zeta(x)$ is Riemann zeta function. The generalized extreme value density function is shown graphically in Example 4.7 for $k < 0$, $k = 0$, and $k > 0$.

Notes:

1. The generalized extreme value distribution given by Eq. (4.53) corresponds to the maximum values. The generalized extreme value distribution for minimum values can be obtained by substituting $-x$ in place of x in Eq. (4.53) and subtracting the resulting expression from 1.

2. The range of Gumbel distribution is unlimited, the range of Frechet distribution has a lower limit, and the reversed Weibull distribution has an upper limit. The cumulative distribution functions corresponding to Gumbel (or type I), Frechet (or type II), and reversed Weibull (or type III) distributions can be expressed as

Gumbel:
$$F(x) = \exp\left\{-\exp\left[-\left(\frac{x-\mu}{\sigma}\right)\right]\right\}; x = \text{real} \tag{4.61}$$

Frechet:
$$\begin{aligned} F(x) &= -\exp\left[-\left(\frac{x-\mu}{\sigma}\right)^{-\alpha}\right]; x > \mu \\ &= 0; \qquad x \le \mu \end{aligned} \tag{4.62}$$
where $k = \alpha^{-1}$

Reversed Weibull:
$$F(x) = \begin{cases} \exp\left[-\left\{-\frac{x-\mu}{\sigma}\right\}^{\alpha}\right]; x < \mu \\ 1 \qquad ; x \ge \mu \end{cases} \tag{4.63}$$
where $\sigma > 0$ and $k = -\alpha^{-1}$; $\alpha > 0$

4.11 Numerical Solutions Using MATLAB and Excel

MATLAB has programs (m-files) for finding the values of the probability density and/or distribution functions of several types of extreme value distributions for any specified value(s) of the random variable. By finding these values, plots of density and/or distribution functions of specific extremal distributions can be generated. The following examples illustrate the procedure.

EXAMPLE 4.5

Find the values of the density and distribution functions of Weibull probability distribution with shape parameter $\beta = 1$ and scale parameter $\eta = 1$ using MATLAB.

Solution: The MATLAB program and the results are shown below:

```
>>x=0.6;

>>y=weibpdf(x,1,1)

y=

 0.5488

>>z=weibcdf(x,1,1)

z=

 0.4512
```

■

EXAMPLE 4.6

By generating the values of the probability density and distributions of Weibull distribution at 100 values in the interval $x = (0.001, 2.5)$, plot the corresponding density and distribution functions. Plot four sets of graphs using the following values for the parameters of the distribution: (i) $k = \beta = 2$ and $\lambda = \eta = 0.5$, (ii) $k = \beta = 2$ and $\lambda = \eta = 1$, (iii) $k = \beta = 2$ and $\lambda = \eta = 2$, and (iv) $k = \beta = 2$ and $\lambda = \eta = 5$.

Solution: The m-file and the plots are shown below:

```
clear all;close all; clc

x = linspace(0.001,2.5,100)  ;

y1 = weibpdf(x,2,0.5)  ;
y2 = weibpdf(x,2,1)      ;
y3 = weibpdf(x,2,2)   ;
y4 = weibpdf(x,2,5)      ;
plot(x,y1,'k-',x,y2,'k:',x,y3,'k*-',x,y4,'k--') ;
legend({'\lambda=2, k=0.5' '\lambda=2, k=1' '\lambda=2, k=2' '\lambda=2,
k=5'})    ;
axis([0 2.5 0 3])

figure
y1 = weibcdf(x,2,0.5)  ;
y2 = weibcdf(x,2,1)      ;
y3 = weibcdf(x,2,2)   ;
y4 = weibcdf(x,2,5)      ;
plot (x,y1,'k-',x,y2,'k:',x,y3,'k*-',x,y4,'k--') ;
legend({'\lambda=2, k=0.5' '\lambda=2, k=1' '\lambda=2, k=2' '\lambda=2,
k=5'})     ;
axis([0 2 0 1.5])
```

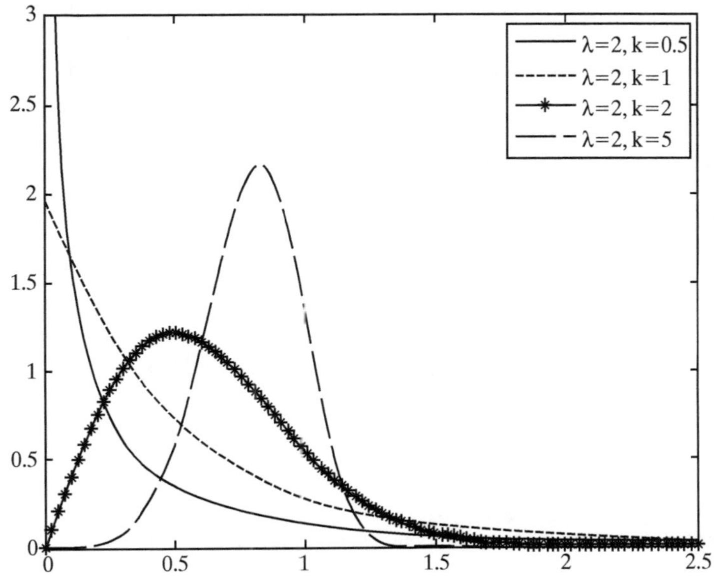

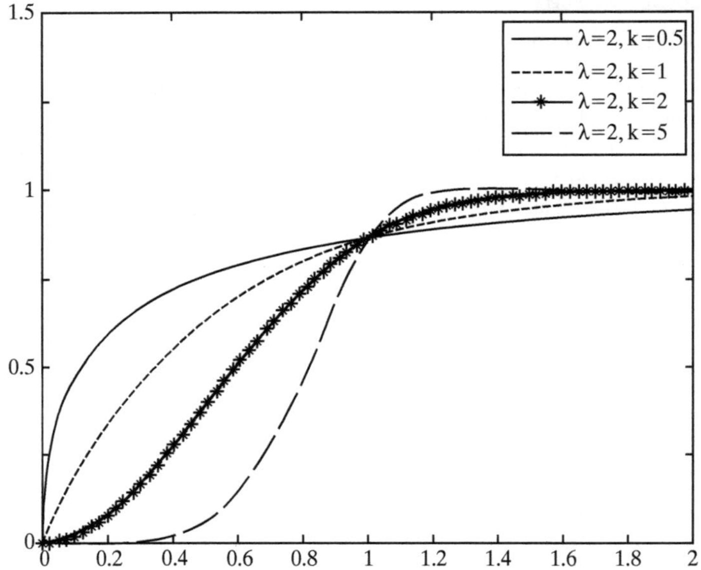

■

EXAMPLE 4.7

By generating the values of the probability density and distributions of generalized extreme value distribution at 1000 values in the interval $x = (-2, 5)$, plot the corresponding density and distribution functions. Plot three sets of graphs using the following values for the parameters of the distribution: (i) $k = -0.5, \sigma = 1, \mu = 0$, (ii) $k = 0, \sigma = 1, \mu = 0$, and (iii) $k = 1, \sigma = 1, \mu = 0$. Note that the values of $k < 0$, $k = 0$, and $k > 0$ correspond to type I, II, and III extreme value distributions, respectively.

Solution: The m-file and the plots are shown below:

```
clear all;close all; clc

x = linspace(-2,5,1000)  ;
y1 = gevpdf(x,-0.5,1,0)  ;
y2 = gevpdf(x,0,1,0)     ;
y3 = gevpdf(x,1,1,0)     ;
plot(x,y1,'k-',x,y2,'k:',x,y3,'k--') ;
legend({'k<0,Type III' 'k=0 Type I' 'k>0, Type II'}) ;

figure ;
x = linspace(-2,5,1000)  ;
y1 = gevcdf(x,-0.5,1,0)  ;
y2 = gevcdf(x,0,1,0)     ;
y3 = gevcdf(x,1,1,0)     ;
plot(x,y1,'k-',x,y2,'k:',x,y3,'k--') ;
legend({'k<0,Type III' 'k=0 Type I' 'k>0, Type II'}) ;
```

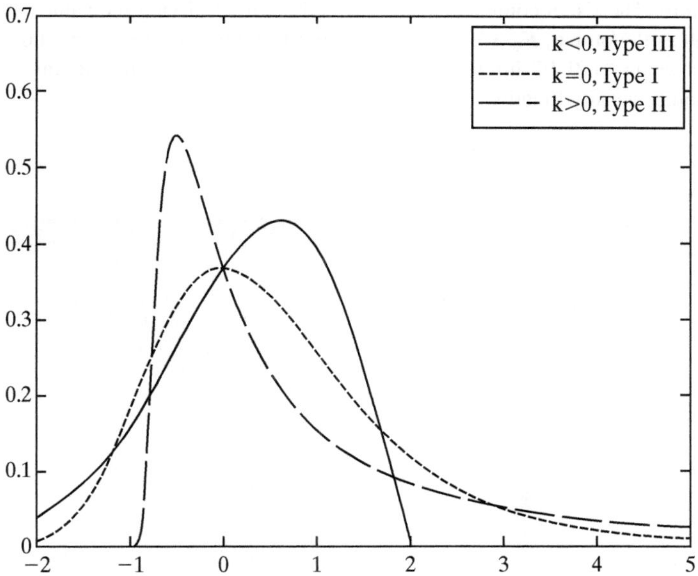

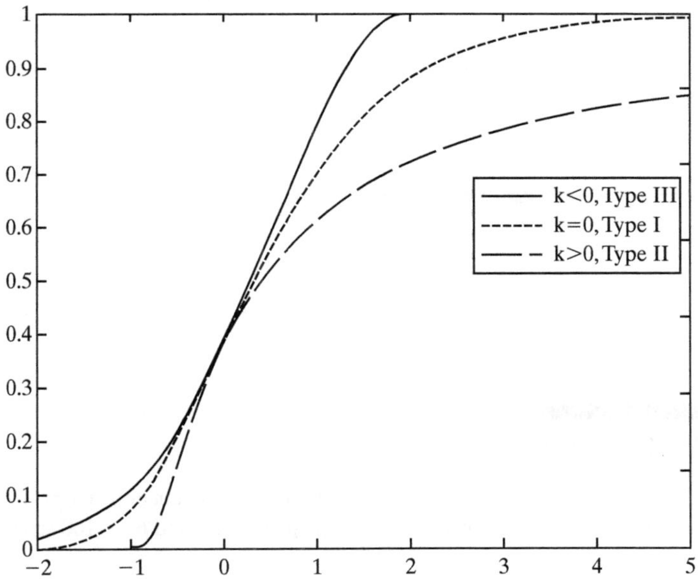

■

EXAMPLE 4.8

Using Excel, find the value of Weibull cumulative distribution function at $x = 105$ when the parameters of the distribution are given by $\alpha = 20$ (shape parameter) and $\beta = 100$ (characteristic value).

Solution: The Excel command used, the method of specifying the values of x, and the parameters as well as the output, $F_X(x = 105) = 0.92958139$, given by Excel are shown below. Note that the logical value "TRUE" implies the cumulative distribution while the value "FALSE" implies the probability density function.

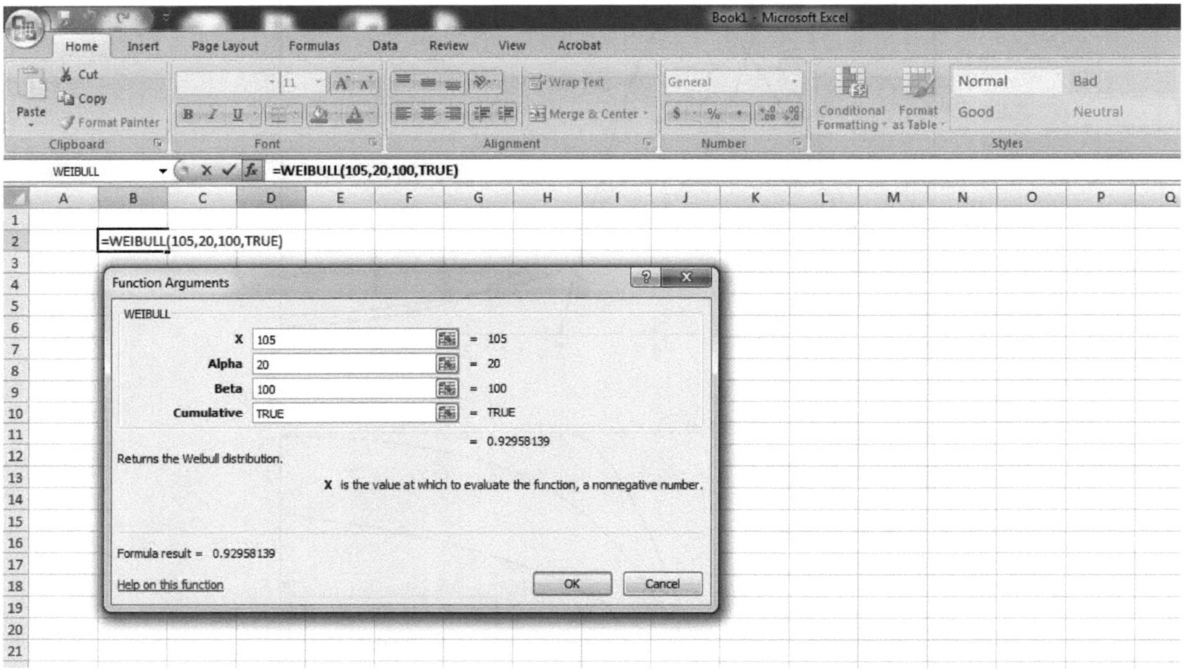

■

EXAMPLE 4.9

Using Excel, find the value of Weibull probability density function at $x = 105$ when the parameters of the distribution are given by $\alpha = 20$ (shape parameter) and $\beta = 100$ (characteristic value).

Solution: The Excel command used, the method of specifying the values of x, and the parameters as well as the output, $f_X(x = 105) = 0.035588864$, given by Excel are shown below. Note that the logical value "TRUE" implies the cumulative distribution while the value "FALSE" implies the probability density function.

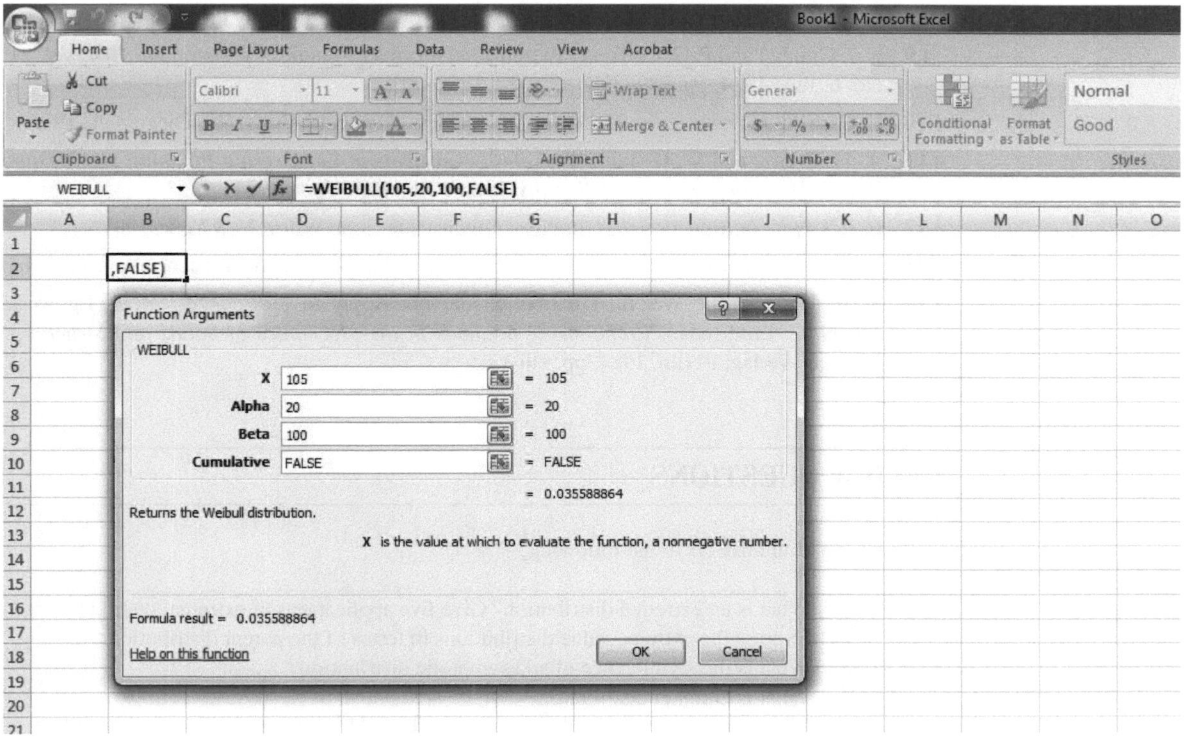

■

REFERENCES AND BIBLIOGRAPHY

4.1 E. J. Gumbel, *Statistics of Extremes,* Columbia University Press, New York, 1958.

4.2 W. Weibull, "A Statistical Distribution Function of Wide Applicability," *ASME J. Appl. Mech.,* Vol. 18, 1951, pp. 293–297.

4.3 T. T. Soong, *Probabilistic Modeling and Analysis in Science and Engineering,* John Wiley, New York, 1981.

4.4 J. R. Benjamin and C. A. Cornell, *Probability, Statistics and Decision for Civil Engineers,* McGraw-Hill, New York, 1970.

4.5 H. C. S. Thom, "Distributions of Extreme Winds in the United States," *J. Structural Div., Proc. ASCE,* Vol. 86, No. ST4, April 1960, pp. 11–24.

4.6 S. S. Rao and B. D. Gupta, "Application of Extremal Distributions in the Design of Thermal Systems," *J. Mech. Design, Trans. ASME,* Vol. 102, 1980, pp. 481–489.

4.7 P. H. Wirsching and L. H. Jones, "On the Use of the Extreme Value Distribution in Reliability Analysis and Design," *J. Eng. Industry, Trans. ASME,* Vol. 98, August 1976, pp. 1080–1085.

4.8 C. T. Haan, *Statistical Methods in Hydrology,* The Iowa State University Press, Ames, IA, 1977.

4.9 E. J. Gumbel and P. G. Carlson, "Extreme Values in Aeronautics," *J. Aeronauti. Sci.,* Vol. 21, No. 6, June 1954, pp. 389–398.

4.10 E. Castillo, *Extreme Value Theory in Engineering,* Academic Press, San Diego, CA, 1988.

4.11 A. H. S. Ang and W. H. Tang, *Probability Concepts in Engineering Planning and Design,* Vol. II: *Decision, Risk, and Reliability,* John Wiley, New York, 1984.

4.12 E. E. Lewis, *Introduction to Reliability Engineering,* John Wiley, New York, 1987.

4.13 A. C. King and C. B. Read, *Pathways to Probability,* Holt, Rinehart & Winston, New York, 1963.

4.14 R. A. Heller, "The Weibull Distribution did not Apply to its Founder," in S. Eggwertz and N. C. Lind (eds.), *Probabilistic Methods in the Mechanics of Solids and Structures,* Springer-Verlag, Berlin, 1984, pp. xiii-xvi.

REVIEW QUESTIONS

4.1 Give brief answers to the following:

1. What is an extremal distribution? Give five applications of extremal distributions.
2. Express the extreme value distributions in terms of the parent distribution.
3. What is the significance of an asymptotic distribution?
4. What is Gumbel distribution?
5. Where are extremal distributions of type I used?
6. Give a practical application of type-II maximum value distribution.
7. What are the practical applications of type-III distribution for the smallest value?
8. What is the significance of the return period?
9. Express the return period in terms of the probability distribution function.
10. How do you construct an extremal probability paper?
11. What is a characteristic largest value?
12. Under what conditions Weibull distribution can be used to represent the smallest value of a random variable?
13. Which parameter value determines the type of distribution described by the generalized extreme value distribution?

4.2 Indicate whether each of the following statements is true or false:

1. The ith order statistic corresponding to the sample values $(X_1, X_2, \ldots, X_n)$ of the random variable X is the ith largest value.
2. The distribution of the largest value among a sample of size n of a random variable X is called the distribution of the nth order statistic.
3. The return period is given by the reciprocal of the probability of occurrence of the event.
4. The generalized extreme value distribution can be used to model the smallest as well as the largest values.

4.3 Fill in the blank with the proper word(s):

1. The distribution of the smallest value among a sample of size n of a random variable X is called the distribution of the ______ order statistic.

2. Asymptotic distribution is the distribution function of the extreme values X_1 or X_n as n goes to ______.
3. There are ______ types of asymptotic distributions.
4. Type I distributions are called ______ distributions.
5. ______ period denotes the average elapsed time between occurrences of an event with a specified magnitude or greater.
6. The characteristic largest value gives an idea of the central location of the possible ______ values.
7. The least-squares fit minimizes the sum of the squares of the error between ______ and ______ values.
8. The generalized extreme value distribution is defined in terms of ______ parameters.

4.4 Select the most appropriate answer from the multiple choices given:

1. The minimum value of the following random variable is required in practical designs:
   a. Wind velocity **b.** Intensity of flood **c.** Strength of material
2. The maximum value of the following random variable is required in practical designs:
   a. Earthquake ground acceleration
   b. Ambient temperature for computing heating load
   c. Strength of material
3. The probability distribution of the largest value, $F_{X_n}(x)$, among n values of X is given by
   a. $F_{X_n}(x) = F_X(x)$ **b.** $F_{X_n}(x) = \{F_X(x)\}^n$ **c.** $F_{X_n}(x) = \{F_X(x)\}^{-n}$
4. The return period T is given by:
   a. $\dfrac{1}{1 - F_X(x)}$ **b.** $\dfrac{1}{F_X(x)}$ **c.** $1 - F_X(x)$
5. From the generalized extreme value distribution given, the extreme value distribution for minimum values can be obtained by:
   a. using no modification
   b. using $-x$ in place of x and subtracting the result from 1
   c. using $-x$ in place of x

4.5 Match the given distributions with their uses:

1. Type I minimum value	**a.** Strength of brittle materials
2. Type I maximum value	**b.** Material strength in fatigue
3. Type II minimum value	**c.** Annual maximum wind
4. Type II maximum value	**d.** Not commonly used
5. Type III minimum value	**e.** Maximum flood in a river

4.6 Match the given type of extremal distribution with its corresponding name:

1. Type I extremal distribution	**a.** Weibull distribution
2. Type II extremal distribution	**b.** Gumbel distribution
3. Type III extremal distribution	**c.** Frechet distribution

PROBLEMS

4.1 The load required to be lifted by a screw jack is uniformly distributed between 2000 lb and 3000 lb. Determine the probability distribution and density functions, $F(x_n)$ and $f(x_n)$, for the largest load lifted in a sample of n screw jacks. Plot the functions $F(x_n)$ and $f(x_n)$ for $n = 1, 5, 10, 15$, and 20.

4.2 For the screw jack stated in Problem 4.1, determine the probability distribution and density functions, $F(x_1)$ and $f(x_1)$, for the smallest load lifted in a sample of n screw jacks. Plot the functions $F(x_1)$ and $f(x_1)$ for $n = 1, 5, 10, 15$, and 20.

4.3 The fatigue life of a bolted joint, in thousands of cycles, is given by

$$f(x) = \begin{cases} \dfrac{x - 600}{10000} & \text{for } 600 \le x \le 700 \\ \dfrac{800 - x}{10000} & \text{for } 700 \le x \le 800 \end{cases}$$

Find the probability distribution and density functions, $F(x_n)$ and $f(x_n)$, corresponding to the largest fatigue in a sample of n bolted joints. Plot the functions $F(x_n)$ and $f(x_n)$ for $n = 1, 10$, and 20.

4.4 For the bolted joint described in Problem 4.3, find the probability distribution and density functions, $F(x_1)$ and $f(x_1)$, corresponding to the smallest fatigue life in a sample of n bolted joints and plot them for $n = 1, 10$, and 20.

4.5 The probability density function of the time to failure, in hours, of a disk brake is given by

$$f(x) = 2 \times 10^{-5} e^{-2\times 10^{-5} x}$$

Find the probability distribution and density functions, $F(x_n)$ and $f(x_n)$, for the largest failure time realized in a sample of n disk brakes. Plot the functions $F(x_n)$ and $f(x_n)$ for $n = 1, 10$, and 20.

4.6 For the disk brakes considered in Problem 4.5, find the probability distribution and density functions, $F(x_1)$ and $f(x_1)$, for the smallest failure time realized in a sample of n disk brakes. Plot the functions $F(x_1)$ and $f(x_1)$ for $n = 1, 10$, and 20.

4.7 The strength of structural steel, z in k(lb/in^2), is described by Weibull distribution with parameters $\varepsilon = 100$, $\beta = 5$, and $\eta = 150$ so that

$$f(z) = \frac{1}{10}\left(\frac{z - 100}{50}\right)^4 \exp\left[-\left(\frac{z - 100}{50}\right)^5\right]$$

Determine the mean and standard deviations of the strength of steel.

4.8 A mainframe computer has a constant breakdown rate of one breakdown every 96 hours of continuous operation. Find the probability of solving a problem that requires 120 hours of continuous operation of the computer over a period of 30 days.

4.9 The probability distribution function of the yearly maximum flood, in m^3/s, in a river is given by

$$F(x) = \exp\left[-\exp\left(-\frac{x - 25}{5}\right)\right]$$

Find the intensity of the flood that corresponds to a return period of 50 years.

4.10 The maximum daily demand of electric power from a thermal power station is known to follow extremal distribution for the largest value with a mean of 5 MW and a standard deviation of 1 MW. What is the probability of the daily demand of electric power exceeding a value of 8 MW if the extremal distribution is of (a) type I, (b) type II, and (c) type III?

4.11 For the thermal power station described in Problem 4.10, determine the particular value of the daily demand of electric power which will be met 99 percent of the time.

4.12 The minimum monthly flow rate of water in a river follows extremal distribution for the smallest value with a mean of 3 million gallons per hour and a standard deviation of 1 million gallons per hour. Find the probability of having a flow rate of less than 1 million gallons per hour in any month assuming the extremal distribution to be of (a) type I, (b) type II, and (c) type III.

4.13 For the river considered in Problem 4.12, determine the value of the monthly minimum flow rate which will be guaranteed to occur 95 percent of the time.

4.14 It is required to design a landing gear of an aircraft to withstand an impact load which occurs no more than 1 percent of the time during its 20-year design life. Find the probability of occurrence of one or more impact loads of this magnitude or larger in any one year. Also find the probability of occurrence of more than one impact load of this or larger magnitude during the design life of the landing gear.

4.15 Derive linear relationships for plotting type I, type II, and type III asymptotic distributions corresponding to the largest value.

4.16 Derive linear relationships for plotting type I, type II, and type III asymptotic distributions corresponding to the smallest value.

4.17 The AISI 1060 annealed steel has a mean tensile strength of 620 MPa with a standard deviation of 40 MPa. Assuming a minimum tensile strength of 500 MPa, determine the Weibull parameters.

4.18 The following data represent the yearly maximum value of wind-induced load on a building frame in kilonewtons:

43.30	54.49	58.71	48.18
53.02	47.38	42.96	47.92
63.52	40.78	55.77	55.61
45.93	45.05	41.31	53.95
48.26	50.37	58.83	46.08
50.51	54.91	48.21	49.76
49.57	51.28	44.67	57.31
43.93	39.91	67.72	37.59
46.77	53.29	43.11	48.84
59.12	67.59	34.93	

Fit this data into the following distributions and compare the results:

1. Normal distribution
2. Type I, type II, and type III distributions for largest values.

4.19 The following data indicate the minimum values of strength, in k(lb/in^2), found for various batches of steel:

48.71	79.87	56.66	51.17	60.17
52.92	59.37	61.78	75.47	64.32
65.32	53.71	48.31	57.48	44.99
57.37	52.26	56.34	71.44	68.89
69.31	58.62	66.91	56.78	82.14
40.27	54.23	54.82	56.11	60.37
59.98	44.31	47.32	67.12	67.80
55.18	50.89	63.77	53.34	52.18
43.12	73.18	70.22	58.71	59.41
62.33	57.93	55.83	49.24	

Fit this data into the following distributions and compare the results:

1. Normal distribution
2. Type I, type II, and type III distributions for smallest values.

4.20 The hourly data of outside dry bulb temperature in degrees (°C) in New Delhi, India, are given as follows:

45.1	33.8	20.1
43.2	32.7	18.9
42.4	31.9	17.9
41.8	30.4	16.8
41.3	29.3	41.0
28.2	40.1	27.0
39.2	26.1	38.0
25.2	36.9	23.9
35.8	23.1	34.7
22.2	34.1	21.6

Find the extremal distributions corresponding to type I, type II, and type III largest values for this data.

4.21 The hourly data of outside solar radiation in New Delhi, India, are shown as follows:

86.2	66.8	29.9
85.0	63.2	26.1
84.5	59.9	22.6
83.1	56.3	19.5
82.4	52.0	15.4
81.5	48.2	11.3
79.0	44.8	9.6
76.6	40.7	7.2
72.7	37.4	70.3
32.5		

Determine the extremal distributions corresponding to type I, type II, and type III largest values for the data.

4.22 The yearly maximum values of dry bulb temperature in degrees (°C) in Poona, India, over 18 years are given as follows:

40.25	40.48	40.61
40.79	41.02	41.20
41.43	41.64	41.80
42.00	42.24	42.41
42.63	42.82	43.05
43.24	43.39	43.67

Fit the data into the following distributions and compare the results:

1. Normal distribution
2. Type I, type II, and type III distributions for the largest values.

4.23 The maximum gust accelerations, measured in g units, experienced by airplanes during flight periods of 250 hours each in a particular region are given as follows [4.9]:

0.73	0.61	0.50	0.65	0.66	0.80
1.80	1.35	1.10	1.00	1.05	1.00
1.30	1.00	1.65	1.30	0.98	0.91
0.80	1.20	0.91	0.82	1.05	1.40
0.70	1.25				

If a precision instrument, which can withstand a maximum acceleration of 2.00 g's, is transported by an airplane in the region, find the probability that the instrument will be damaged using the following distributions:

1. Normal distribution
2. Type I distribution for the largest value
3. Type II distribution for the largest value
4. Type III distribution for the largest value.

4.24 Write a computer program to find the parameters of type I, type II, and type III distributions corresponding to the largest value using experimental data. Use this program to determine the extremal distributions for the data of Problem 4.18.

4.25 Write a computer program to find the parameters of type I, type II, and type III distributions corresponding to the smallest value using experimental data. Use this program to determine the extremal distributions for the data of Problem 4.19.

4.26 Find the expected value, mode, variance, and skewness of a generalized extreme value distribution with parameters $k = -0.5$, $\mu = 1$, and $\sigma = 0$.

4.27 Find the expected value, mode, variance, and skewness of a generalized extreme value distribution with parameters $k = 0$, $\mu = 1$, and $\sigma = 0$.

4.28 Find the expected value, mode, variance, and skewness of a generalized extreme value distribution with parameters $k = 0.5$, $\mu = 1$, and $\sigma = 0$.

MATLAB and Excel Problems

4.29 The strength of a welded joint is known to follow normal distribution with mean value and standard deviation given by 75 kpsi and 7.5 kpsi, respectively. Find the probability density and distribution functions for the largest strength realized in a sample of n welded joints using MATLAB. Plot the resulting functions for n equal to 1, 5, and 10.

4.30 The strength of a welded joint is known to follow normal distribution with mean value and standard deviation given by 75 kpsi and 7.5 kpsi, respectively. Find the probability density and distribution functions for the smallest strength realized in a sample of n welded joints using MATLAB. Plot the resulting functions for n equal to 1, 5, and 10.

4.31 By evaluating the values of the Weibull probability density function at 100 values in the interval $x = (0.001, 2.5)$, plot the corresponding density function using Excel. Plot four graphs using the following values of the parameters of the distribution:

i. $k = \beta = 2$ and $\lambda = \eta = 0.5$
ii. $k = \beta = 2$ and $\lambda = \eta = 1$
iii. $k = \beta = 2$ and $\lambda = \eta = 2$
iv. $k = \beta = 2$ and $\lambda = \eta = 5$

4.32 By evaluating the values of the Weibull probability distribution function at 100 values in the interval $x = (0.001, 2.5)$, plot the corresponding distribution function using Excel. Plot four graphs using the following values of the parameters of the distribution:

i. $k = \beta = 2$ and $\lambda = \eta = 0.5$
ii. $k = \beta = 2$ and $\lambda = \eta = 1$
iii. $k = \beta = 2$ and $\lambda = \eta = 2$
iv. $k = \beta = 2$ and $\lambda = \eta = 5$

CHAPTER 5

Functions of Random Variables

What You Will Learn

In this chapter, you will learn

- The determination of probability characteristics of functions of random variables whose probability distributions are known
- The method of finding probability distributions of monotonically increasing and monotonically decreasing functions in one random variable with numerical examples
- The probability characteristics of functions such as sum, product, and quotient of two random variables
- The moments of a function of several random variables and the partial derivative rule based on Taylor's series expansion of a nonlinear function
- The use of moment generating function in finding the mean and standard deviation of a function of several random variables
- The probability description of several nonlinear functions, each expressed in terms of multiple random variables with known probability distributions

Reverend Thomas Bayes was born in London in 1702 and died in Tunbridge Wells, England, in 1761. He was the eldest son of Ann Bayes and Joshua Bayes and was educated privately, as was usual with Non-conformists at that time. Bayes' father was a respected theologian of dissent and Bayes became first his father's assistant and later spent most of his life as minister at the chapel in Tunbridge Wells. His mathematical work, although small in quantity, is of very high quality. He was elected as a Fellow of the Royal Society of London in 1742. He devoted much time to probability while preaching in the church, and is well known for introducing the concept of inverse probability, more commonly known as the "Bayes' rule." This was presented in his brief *An Essay Towards Solving a Problem in the Doctrine of Chances*, published in 1763 after Bayes's death. This concept helps in estimating the probabilities of the causes by which an observed event may have been produced. Bayes' rule can be used in several applications including inspection processes [5.10, 5.11].

5.1 Introduction

In engineering, most phenomena involve functional relationships in which a dependent variable is expressed in terms of one or more independent variables. For example, the stress in a bar is a function of the load and the cross-sectional area, or the work done is a function of the force and the distance through which the force is moved. If any of the independent variables are random, the dependent variable will also be random. The probability distribution of the dependent variable as well as its moments can be derived from those of the independent random variables. The techniques presented in this chapter will permit us to determine the probability characteristics of random variables that are functionally dependent on some others with known probability characteristics.

5.2 Functions of a Single Random Variable

If X is a random variable, any other variable Y defined as a function of X will also be a random variable. If $f_X(x)$ and $F_X(x)$ denote, respectively, the probability density and distribution functions of X, the problem is to find the density function $f_Y(y)$ and the distribution function $F_Y(y)$ of the random variable Y. Let the functional relation be expressed as

$$Y = g(X) \tag{5.1}$$

By definition, the distribution function of Y is the probability of realizing Y less than or equal to y:

$$F_Y(y) = P(Y \le y) = P(g(x) \le y) = \int_{g(x) \le y} f_X(x)\, dx \tag{5.2}$$

where the integration is to be carried over all values of x for which $g(x) \le y$. For example, if the functional relation between x and y is as shown in Fig. 5.1, the range of integration will be $\Delta x_1 + \Delta x_2 + \Delta x_3 + \cdots$. The probability density function of Y is given by

$$f_Y(y) = \frac{d}{dy}\left[F_Y(y)\right] \tag{5.3}$$

Equations (5.2) and (5.3) represent general relationships and are applicable to any function $g(x)$. Specifically, if $g(x)$ is a monotonically increasing or decreasing function as shown

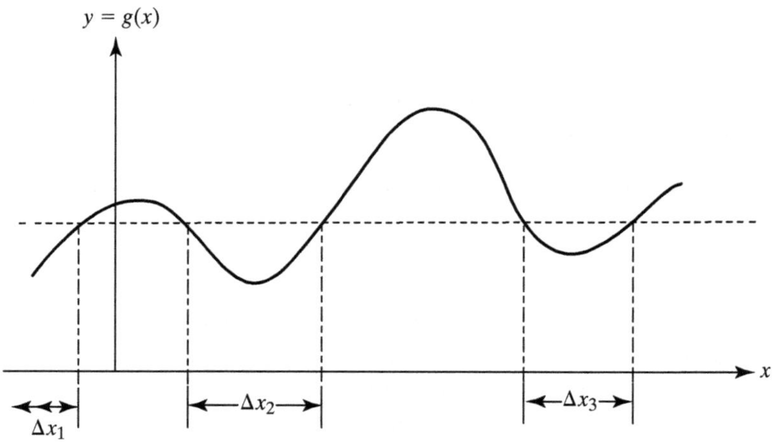

FIGURE 5.1 Range of integration in Eq. (5.2).

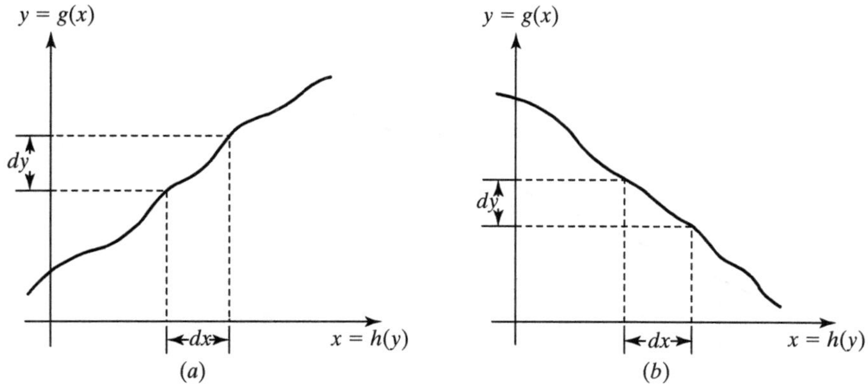

FIGURE 5.2 (*a*) Monotonically increasing, (*b*) monotonically decreasing.

in Fig. 5.2, the density function $f_Y(y)$ can be expressed in a more convenient way. In this case, the inverse relation of Eq. (5.1) can be denoted as

$$x = g^{-1}(y) = h(y) \tag{5.4}$$

where x can be seen to be a single valued function of y. Differentiation of Eq. (5.4) gives

$$dx = \left|\frac{dh}{dy}\right| dy = |h'(y)|dy \tag{5.5}$$

where the absolute sign is used to take care of both the increasing and the decreasing type of functions (to avoid negative values for the density and distribution functions). Equations (5.2) and (5.5) lead to

$$F_Y(y) = \int_{-\infty}^{y} f_X[h(y)]|h'(y)|dy \tag{5.6}$$

and the differentiation of $F_Y(y)$ gives

$$f_Y(y) = f_X[h(y)]|h'(y)| \tag{5.7}$$

The application of Eqs. (5.2) and (5.7) is demonstrated through the following examples.

EXAMPLE 5.1

Derive the probability density function of Y in terms of $f_X(x)$ where the relation between X and Y is given by $Y = a + bX, a > 0, b > 0$.

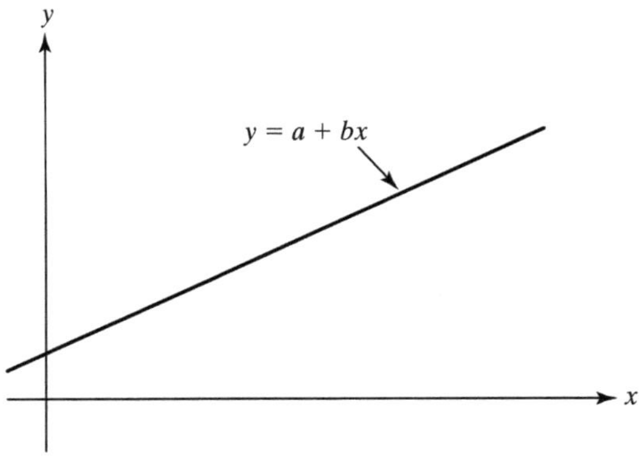

FIGURE 5.3 A monotonically increasing linear function.

Solution: The function $y = a + bx$ is monotonically increasing as shown in Fig. 5.3. The inverse relation can be expressed as

$$x = \frac{y - a}{b} = h(y) \quad \text{and} \quad \frac{dh}{dy} = \frac{1}{b}$$

Equation (5.7) gives

$$f_Y(y) = f_X[h(y)]|h'(y)| = \frac{1}{b} f_X\left(\frac{y - a}{b}\right)$$

■

EXAMPLE 5.2

Derive the probability density function $f_Y(y)$ in terms of $f_X(x)$ when X and Y are related as

$$Y = (X + 1)^2$$

Solution: The function $y = (x + 1)^2$, represented graphically in Fig. 5.4, can be seen to be a non-monotonic function. Using Eqs. (5.2) and (5.3), we obtain

$$F_Y(y) = P[(X + 1)^2 \leq y] = \int_{\text{all } x \text{ where } (x+1)^2 \leq y} f_X(x)\, dx$$

that is,

$$F_Y(y) = \int_{-\sqrt{y}-1}^{+\sqrt{y}-1} f_X(x)\, dx \tag{E1}$$

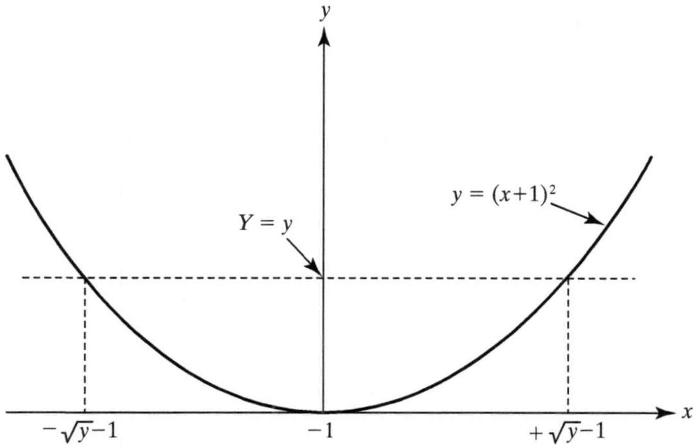

FIGURE 5.4 Graphical representation of the function $y = (x + 1)^2$.

and

$$f_Y(y) = \frac{dF_Y(y)}{dy} = \frac{d}{dy}\left\{\int_{-\sqrt{y}-1}^{+\sqrt{y}-1} f_X(x)\,dx\right\} \tag{E2}$$

Using the formula of differentiation under integral sign

$$\frac{d}{da}\int_p^q f_X(x, a)\,dx = \int_p^q \frac{\partial}{\partial a}[f_X(x, a)]\,dx + f_X(q, a)\frac{dq}{da} - f_X(p, a)\frac{dp}{da}$$

Equation (E2) gives

$$f_Y(y) = \begin{cases} \dfrac{f_X(\sqrt{y} - 1) + f_X(-\sqrt{y} - 1)}{2\sqrt{y}}; & y > 0 \\ 0; & y \le 0 \end{cases} \tag{E3}$$

■

EXAMPLE 5.3

The energy (E) stored in a flywheel of mass moment of inertia I rotating at an angular velocity Ω is given by

$$E = \frac{1}{2} I\,\Omega^2$$

If I is deterministic and Ω is a random variable following normal distribution $N(0, 1)$ determine the probability density function of the energy E.

Solution: The energy stored in the flywheel can be expressed as $E = a\,\Omega^2$ where $a = I/2$. This gives

$$\omega = \pm\sqrt{\frac{e}{a}} \quad \text{and} \quad \left|\frac{d\omega}{de}\right| = \frac{1}{2\sqrt{ae}} \tag{E1}$$

The probability density function of the energy E can be obtained from Eq. (5.7):

$$f_E(e) = \left\{ f_\Omega\left(\sqrt{\frac{e}{a}}\right) + f_\Omega\left(-\sqrt{\frac{e}{a}}\right) \right\} \frac{1}{2\sqrt{ae}} \tag{E2}$$

Since

$$f_\Omega(\omega) = \frac{1}{\sqrt{2\pi}} \exp\left\{-\frac{1}{2}\omega^2\right\} \tag{E3}$$

Equation (E2) gives

$$f_E(e) = \frac{1}{\sqrt{2\pi a e}} \exp\left\{-\frac{e}{2a}\right\}; \qquad e \geq 0 \tag{E4}$$

This density function is called the chi-square distribution with one degree of freedom and its general shape is shown in Fig. 5.5.

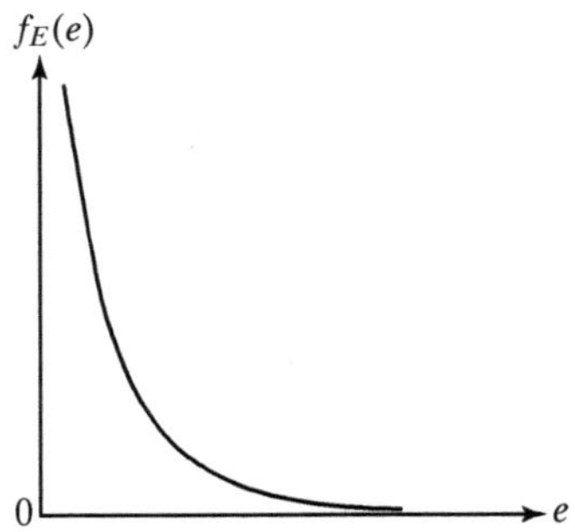

FIGURE 5.5 Chi-square distribution with one degree of freedom.

■

If x is a discrete random variable with probability mass function $p_X(x)$, then the probability mass and distribution functions of $Y = g(X)$ are given by

$$p_Y(y) = \sum_{\text{all } x \text{ such that } g(x_i)=y} p_X(x_i) \tag{5.8}$$

and

$$F_Y(y) = \sum_{\text{all } x \text{ such that } g(x_i)\leq y} p_X(x_i) \tag{5.9}$$

EXAMPLE 5.4

Find the probability mass function of y when $y = x^2$ and

$$p_X(x) = \begin{cases} \dfrac{27}{1093}\left(\dfrac{1}{3}\right)^x; & x = -3, -2, -1, 0, 1, 2, 3 \\ 0; & \text{otherwise} \end{cases}$$

Solution: Since $x = -3, -2, -1, 0, 1, 2, 3$, and $y = x^2$, the possible values of y are 0, 1, 4, and 9. Using Eq. (5.8), we obtain

$$p_Y(0) = \sum_{x^2=0} p_X(x) = p_X(0) = \frac{27}{1093}$$

$$p_Y(1) = \sum_{x^2=1} p_X(x) = p_X(-1) + p_X(1) = \frac{27}{1093}(3) + \frac{27}{1093}\left(\frac{1}{3}\right) = \frac{90}{1093}$$

$$p_Y(4) = \sum_{x^2=4} p_X(x) = p_X(-2) + p_X(2) = \frac{27}{1093}(9) + \frac{27}{1093}\left(\frac{1}{9}\right) = \frac{246}{1093}$$

$$p_Y(9) = \sum_{x^2=9} p_X(x) = p_X(-3) + p_X(3) = \frac{27}{1093}(27) + \frac{27}{1093}\left(\frac{1}{27}\right) = \frac{730}{1093}$$

and $p_Y(y) = 0$ for all other values of y.

■

5.3 Functions of Two Random Variables

Consider a function of two random variables X_1 and X_2 as

$$Y = g(X_1, X_2) \tag{5.10}$$

If X_1 and X_2 are discrete random variables with joint probability mass function $p_{X_1,X_2}(x_{1i}, x_{2j})$, the probability mass function of Y can be found as

$$p_Y(y) = P(Y = y) = P[g(X_1, X_2) = y] = \sum_{i}\sum_{j}\limits_{g(x_{1i}, x_{2j})=y} p_{X_1,X_2}(x_{1i}, x_{2j}) \tag{5.11}$$

and the cumulative distribution function as

$$F_Y(y) = \sum_{i}\sum_{j}\limits_{g(x_{1i}, x_{2j})\le y} p_{X_1,X_2}(x_{1i}, x_{2j}) \tag{5.12}$$

If X_1 and X_2 are continuous random variables with the joint density function $f_{X_1, X_2}(x_1, x_2)$, the probability distribution function of Y is given by

$$F_Y(y) = \iint\limits_{g(x_1, x_2) \leq y} f_{X_1, X_2}(x_1, x_2)\, dx_1\, dx_2 \tag{5.13}$$

If Eq. (5.10) can be solved for X_1 as

$$X_1 = g^{-1}(Y, X_2) \tag{5.14}$$

Equation (5.13) can be expressed as

$$F_Y(y) = \int_{x_2=-\infty}^{\infty} \int_{y=-\infty}^{y} f_{X_1, X_2}(g^{-1}, x_2) \left| \frac{\partial g^{-1}}{\partial y} \right| dy\, dx_2 \tag{5.15}$$

The probability density function of Y can be obtained from Eq. (5.15) as

$$f_Y(y) = \int_{-\infty}^{\infty} f_{X_1, X_2}(g^{-1}, x_2) \left| \frac{\partial g^{-1}}{\partial y} \right| dx_2 \tag{5.16}$$

5.3.1 Sum of Two Random Variables

If

$$Y = X_1 + X_2 \tag{5.17}$$

the probability distribution function of Y is given by

$$F_Y(y) = P[Y \leq y] = \iint\limits_{x_1 + x_2 \leq y} f_{X_1, X_2}(x_1, x_2)\, dx_1\, dx_2$$

$$= \int_{-\infty}^{\infty} \int_{-\infty}^{(y - x_2)} f_{X_1, X_2}(x_1, x_2)\, dx_1\, dx_2 \tag{5.18}$$

where $f_{X_1, X_2}(x_1, x_2)$ is the joint density function of X_1 and X_2 and the integration is over the crossed region shown in Fig. 5.6. Differentiation of Eq. (5.18) gives

$$f_Y(y) = \frac{dF_Y(y)}{dy} = \int_{-\infty}^{\infty} f_{X_1, X_2}(y - x_2, x_2)\, dx_2 \tag{5.19}$$

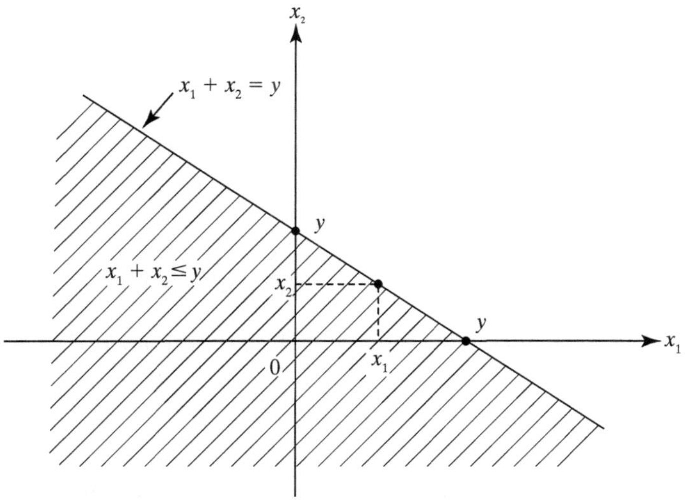

FIGURE 5.6 Region defined by $x_1 + x_2 \leq y$.

If X_1 and X_2 are independent, $f_{X_1, X_2}(x_1, x_2) = f_{X_1}(x_1) f_{X_2}(x_2)$ and Eq. (5.19) becomes

$$f_Y(y) = \int_{-\infty}^{\infty} f_{X_1}(y - x_2) f_{X_2}(x_2)\, dx_2 \tag{5.20}$$

EXAMPLE 5.5

The number of automobiles produced per day in each of the two plants by an automobile manufacturer is a random variable following Poisson distribution. Find the probability distribution of the total number of automobiles produced per day.

Solution: Let X_1 and X_2 denote the number of automobiles produced per day in the two plants with

$$p_{X_1}(x_1) = \frac{(\alpha)^{x_1}}{x_1!} e^{-\alpha} \tag{E1}$$

and

$$p_{X_2}(x_2) = \frac{(\beta)^{x_2}}{x_2!} e^{-\beta} \tag{E2}$$

where α and β are the parameters of the distribution indicating the mean values (and also the variances) of X_1, X_2, respectively. If $Y = X_1 + X_2$, the probability mass function of Y can be determined as

$$p_Y(y) = \sum_{\substack{i \ j \\ x_{1i} + x_{2j} = y}} \sum p_{X_1, X_2}(x_{1i}, x_{2j}) = \sum_{\text{all}\, x_{1i}} p_{X_1, X_2}(x_{1i}, y - x_{1i}) \tag{E3}$$

If X_1 and X_2 are assumed to be independent random variables,

$$p_Y(y) = \sum_{\text{all } x_1} p_{X_1}(x_1)\, p_{X_2}(y - x_1)$$

$$= \sum_{\text{all } x_1} \left[\frac{(\alpha)^{x_1}}{x_1!} e^{-\alpha} \right] \left[\frac{(\beta)^{y-x_1}}{(y - x_1)!} e^{-\beta} \right]$$

$$= e^{-(\alpha+\beta)} \left\{ \sum_{\text{all } x_1} \frac{(\alpha)^{x_1} (\beta)^{y-x_1}}{x_1!(y - x_1)!} \right\} \tag{E4}$$

Since the quantity in brackets in Eq. (E4) is the binomial expansion of $(\alpha + \beta)^y / y!$, we obtain

$$p_Y(y) = \frac{(\alpha + \beta)^y}{y!} e^{-(\alpha+\beta)} \tag{E5}$$

This shows that the total number of automobiles produced per day also follows Poisson distribution with the parameter $(\alpha + \beta)$.

■

EXAMPLE 5.6

Find the probability density function of sum of two independent normally distributed random variables.

Solution: Let X_1 and X_2 be the normal variates with

$$f_{X_1}(x_1) = \frac{1}{\sqrt{2\pi}\, \sigma_{X_1}} e^{-\frac{1}{2}\{(x_1 - \mu_{X_1})/\sigma_{X_1}\}^2} \tag{E1}$$

and

$$f_{X_2}(x_2) = \frac{1}{\sqrt{2\pi}\, \sigma_{X_2}} e^{-\frac{1}{2}\{(x_2 - \mu_{X_2})/\sigma_{X_2}\}^2} \tag{E2}$$

The substitution of Eqs. (E1) and (E2) into Eq. (5.20) gives

$$f_Y(y) = \frac{1}{2\pi \sigma_{X_1} \sigma_{X_2}} \int_{-\infty}^{\infty} \exp\left[-\frac{1}{2} \left\{ \left(\frac{y - x_2 - \mu_{X_1}}{\sigma_{X_1}} \right)^2 + \left(\frac{x_2 - \mu_{X_2}}{\sigma_{X_2}} \right)^2 \right\} \right] dx_2 \tag{E3}$$

which can be rewritten as

$$f_Y(y) = \frac{1}{2\pi \sigma_{X_1} \sigma_{X_2}} \exp\left[-\frac{1}{2} \left\{ \left(\frac{y - \mu_{X_1}}{\sigma_{X_1}} \right)^2 + \left(\frac{\mu_{X_2}}{\sigma_{X_2}} \right)^2 \right\} \right]$$

$$\times \int_{-\infty}^{\infty} \exp\left[-\frac{1}{2} \left\{ \left(\frac{x_2}{\sigma_{X_1}} \right)^2 - 2x_2 \left(\frac{y - \mu_{X_1}}{\sigma_{X_1}^2} \right) + \left(\frac{x_2}{\sigma_{X_2}} \right)^2 - 2x_2 \left(\frac{\mu_{X_2}}{\sigma_{X_2}^2} \right) \right\} \right] dx_2 \tag{E4}$$

The integral in Eq. (E4) can be expressed as

$$\int_{-\infty}^{\infty} \exp\left[-\frac{1}{2}\left\{c_1 x_2^2 - 2\, x_2 c_2\right\}\right] dx_2$$

$$= \int_{-\infty}^{\infty} \exp\left[-\frac{1}{2} c_1 \left\{x_2^2 - 2\frac{c_2}{c_1} x_2 + \left(\frac{c_2}{c_1}\right)^2 - \left(\frac{c_2}{c_1}\right)^2\right\}\right] dx_2 \tag{E5}$$

where

$$c_1 = \left(\frac{1}{\sigma_{X_1}^2} + \frac{1}{\sigma_{X_2}^2}\right) \tag{E6}$$

and

$$c_2 = \left(\frac{\mu_{X_2}}{\sigma_{X_2}^2} + \frac{y - \mu_{X_1}}{\sigma_{X_1}^2}\right) \tag{E7}$$

By defining a new random variable w as

$$w = x_2 - \frac{c_2}{c_1} \tag{E8}$$

the integral of Eq. (E5) can be written as

$$e^{(c_2^2/2c_1)} \int_{-\infty}^{\infty} e^{-1/2 c_1 w^2} dw = \sqrt{2\frac{\pi}{c_1}} e^{(c_2^2/2c_1)} \tag{E9}$$

Thus, Eqs. (E4) and (E9) lead to

$$f_Y(y) = \frac{1}{2\pi\sigma_{X_1}\sigma_{X_2}} \exp\left[-\frac{1}{2}\left\{\left(\frac{y - \mu_{X_1}}{\sigma_{X_1}}\right)^2 + \left(\frac{\mu_{X_2}}{\sigma_{X_2}}\right)^2\right\}\right]$$

$$\times \sqrt{\frac{2\pi\sigma_{X_1}^2\sigma_{X_2}^2}{(\sigma_{X_1}^2 + \sigma_{X_2}^2)}} \exp\left\{\frac{1}{2}\left[\frac{\mu_{X_2}\sigma_{X_1}^2 + (y - \mu_{X_1})\sigma_{X_2}^2}{\sigma_{X_1}^2\sigma_{X_2}^2}\right]^2 \left(\frac{\sigma_{X_1}^2\sigma_{X_2}^2}{\sigma_{X_1}^2 + \sigma_{X_2}^2}\right)\right\} \tag{E10}$$

which upon simplification yields

$$f_Y(y) = \frac{1}{\sqrt{2\pi(\sigma_X^{2_1} + \sigma_X^{2_2})}} \exp\left[-\frac{1}{2}\left\{\frac{y - (\mu_{X_1} + \mu_{X_2})}{\sqrt{(\sigma_X^{2_1} + \sigma_X^{2_2})}}\right\}^2\right] \tag{E11}$$

This equation shows that Y is also a normal variate with

$$\mu_Y = \mu_{X_1} + \mu_{X_2} \tag{E12}$$

and

$$\sigma_Y^2 = \sigma_{X_1}^2 + \sigma_{X_2}^2 \tag{E13}$$

■

Note: This result can be generalized to the case of a linear sum of n random variables. If

$$Y = \sum_{i=1}^{n} a_i X_i \tag{5.21}$$

where a_i are constants and X_i are independent normal variates $N(\mu_{X_i}, \sigma_{X_i})$, then Y also is a normal variate $N(\mu_Y, \sigma_Y)$ with

$$\mu_Y = \sum_{i=1}^{n} a_i \mu_{X_i} \tag{5.22}$$

and

$$\sigma_Y^2 = \sum_{i=1}^{n} a_i^2 \sigma_{X_i}^2 \tag{5.23}$$

5.3.2 Product of Two Random Variables

Let

$$Y = X_1 X_2 \tag{5.24}$$

From Eq. (5.13), the probability distribution function of Y can be expressed as

$$F_Y(y) = \iint_{x_1 x_2 \leq y} f_{X_1, X_2}(x_1, x_2)\, dx_1\, dx_2 \tag{5.25}$$

The region of integration, in which $x_1 x_2 \leq y$, is shown in Fig. 5.7. Thus, Eq. (5.25) can be rewritten as

$$F_Y(y) = \int_0^{\infty} \int_{-\infty}^{(y/x_2)} f_{X_1, X_2}(x_1, x_2)\, dx_1\, dx_2 + \int_{-\infty}^{0} \int_{(y/x_2)}^{\infty} f_{(X_1, X_2)}(x_1, x_2)\, dx_1\, dx_2 \tag{5.26}$$

If X_1 and X_2 are independent, Eq. (5.26) reduces to

$$F_Y(y) = \int_0^{\infty} F_{X_1}\left(\frac{y}{x_2}\right) f_{X_2}(x_2)\, dx_2 + \int_{-\infty}^{0} \left[1 - F_{X_1}\left(\frac{y}{x_2}\right)\right] f_{X_2}(x_2)\, dx_2 \tag{5.27}$$

Differentiation of Eq. (5.27) gives the probability density function of Y as

$$f_Y(y) = \frac{dF_Y(y)}{dy} = \int_{-\infty}^{\infty} f_{X_1}\left(\frac{y}{x_2}\right) f_{X_2}(x_2) \left|\frac{1}{x_2}\right| dx_2 \tag{5.28}$$

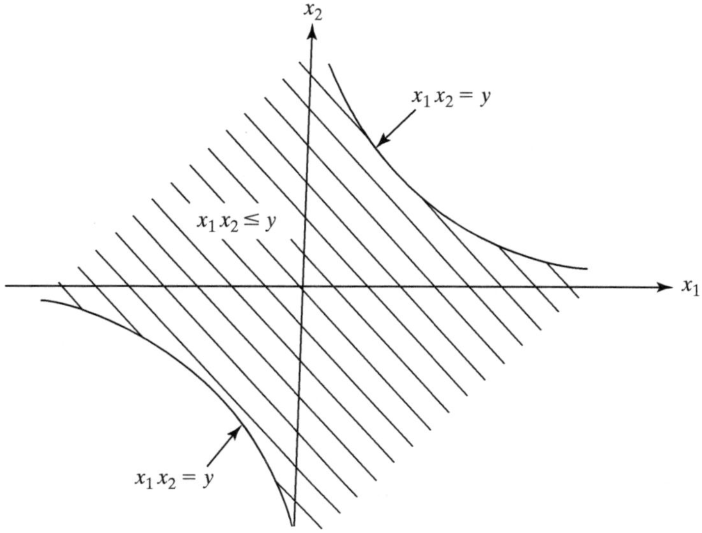

FIGURE 5.7 Region in which $x_1 x_2 \leq y$.

5.3.3 Quotient of Two Random Variables

Let

$$Y = \frac{X_1}{X_2} \tag{5.29}$$

Then Eq. (5.16) gives

$$f_Y(y) = \int_{-\infty}^{\infty} |x_2| \, f_{X_1, X_2}(y\, x_2, x_2) \, dx_2 \tag{5.30}$$

The use of Eq. (5.30) is illustrated through the following example.

EXAMPLE 5.7

The tensile load acting on a uniform bar (L) and the cross-sectional area of the bar (A) are assumed to be independent random variables with the distributions

$$f_L(l) = \begin{cases} \alpha e^{-\alpha l}; & l > 0 \\ 0; & \text{elsewhere} \end{cases} \tag{E1}$$

and

$$f_A(a) = \begin{cases} \beta e^{-\beta a}; & a > 0 \\ 0; & \text{elsewhere} \end{cases} \tag{E2}$$

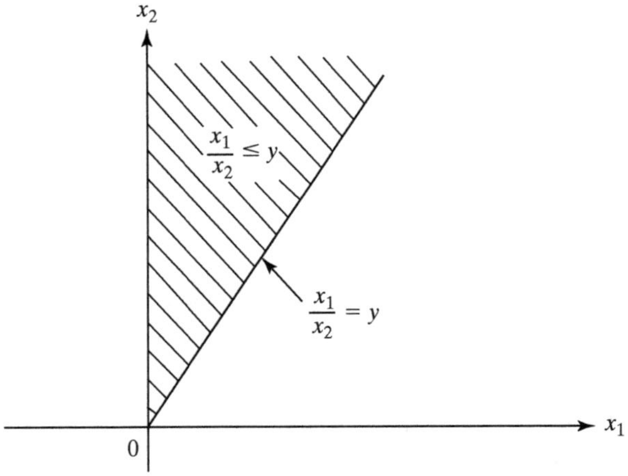

FIGURE 5.8 Region of integration in Eq. (E3).

Find the probability density function of the stress induced in the bar.

Solution: Since the stress induced in the bar (Y) is given by $Y = X_1/X_2$ where $X_1 = L$ and $X_2 = A$, the distribution function of Y is given by

$$F_Y(y) = \iint_{x_1/x_2 \le y} f_{X_1,X_2}(x_1, x_2)\, dx_1\, dx_2 \tag{E3}$$

where the region of integration, for possible values of x_1 and x_2, is shown in Fig. 5.8. Equation (E3) can be rewritten as

$$F_Y(y) = \begin{cases} \int_0^\infty \int_0^{(x_2 y)} f_{X_1,X_2}(x_1, x_2)\, dx_1\, dx_2; & y > 0 \\ 0; & \text{elsewhere} \end{cases} \tag{E4}$$

Since X_1 and X_2 are independent, Eq. (E4) becomes

$$\begin{aligned} F_Y(y) &= \int_0^\infty \int_0^{(x_2 y)} f_{X_1}(x_1) f_{X_2}(x_2)\, dx_1\, dx_2 \\ &= \begin{cases} \int_0^\infty F_{X_1}(x_2 y) f_{X_2}(x_2)\, dx_2; & y > 0 \\ 0; & \text{elsewhere} \end{cases} \end{aligned} \tag{E5}$$

The probability density function can be found by differentiating Eq. (E5) as

$$f_Y(y) = \begin{cases} \int_0^\infty x_2 f_{X_1}(x_2 y) f_{X_2}(x_2)\, dx_2; & y > 0 \\ 0; & \text{elsewhere} \end{cases} \tag{E6}$$

Substitution of Eqs. (E1) and (E2) into Eq. (E6) gives, for $y > 0$,

$$f_Y(y) = \int_0^\infty a\alpha e^{-\alpha a y} \beta e^{-\beta a} da = \alpha\beta \int_0^\infty a e^{-(\alpha y + \beta) a} \, da$$

$$= \alpha\beta \frac{1}{(\alpha y + \beta)^2} \tag{E7}$$

■

5.4 Function of Several Random Variables

Consider the function

$$Y = g(X_1, X_2, \ldots, X_n) \tag{5.31}$$

The distribution function of Y can be obtained by generalizing Eq. (5.13) as

$$F_Y(y) = \int \cdots \int_{g(x_1, x_2, \ldots, x_n) \le y} f_{X_1, X_2, \ldots, X_n}(x_1, x_2, \ldots, x_n) dx_1 \, dx_2 \ldots dx_n \tag{5.32}$$

By solving Eq. (5.31) as

$$x_1 = g^{-1} = g^{-1}(y, x_2, x_3, \ldots, x_n) \tag{5.33}$$

Equation (5.32) can be expressed as

$$F_Y(y) = \int_{-\infty}^{\infty} \cdots \int_{-\infty}^{\infty} \int_{-\infty}^{g^{-1}} f_{X_1, \ldots, X_n}(x_1, \ldots, x_n) \, dx_1 \ldots dx_n$$

$$= \int_{-\infty}^{\infty} \cdots \int_{-\infty}^{\infty} \int_{-\infty}^{y} f_{X_1, X_2, \ldots, X_n}(g^{-1}, x_2, \ldots, x_n) \left| \frac{\partial g^{-1}}{\partial y} \right| dy \, dx_2 \ldots dx_n \tag{5.34}$$

Differentiation of Eq. (5.34) gives the density function of Y as

$$f_Y(y) = \int_{-\infty}^{\infty} \cdots \int_{-\infty}^{\infty} f_{X_1, X_2, \ldots, X_n}(g^{-1}, x_2, \ldots, x_n) \left| \frac{\partial g^{-1}}{\partial y} \right| dx_2 \ldots dx_n \tag{5.35}$$

5.5 Moments of a Function of Several Random Variables

As seen in Sections 5.3 and 5.4, the probability distribution of a function of several random variables can be derived from the known probability distributions of the original random variables. However, the derivations are quite involved, particularly when the functions are nonlinear. In such cases, the moments of the function of the random variables can be determined in terms of the moments of the basic random variables rather easily. In most cases, the task will be simple because only the first two moments, namely the mean and the variance, are of primary importance.

5.5.1 Mean and Variance of a Linear Function

Let

$$Y = a_1 X_1 + a_2 X_2 + \cdots + a_n X_n \tag{5.36}$$

where a_i are constants and X_i are random variables with means $\mu_i = \mu_{X_i}$ and standard deviations $\sigma_i = \sigma_{X_i}$. The mean value of Y is, by definition,

$$\begin{aligned}
\mu_Y = E[Y] &= \int_{-\infty}^{\infty} y\, f_Y(y)\, dy \\
&= \int_{-\infty}^{\infty} \cdots \int_{-\infty}^{\infty} (a_1 x_1 + \cdots + a_n x_n) f_{X_1,\ldots,X_n}(x_1,\ldots,x_n)\, dx_1 \ldots dx_n \\
&= a_1 \int_{-\infty}^{\infty} x_1\, f_{X_1}(x_1) dx_1 + \cdots + a_n \int_{-\infty}^{\infty} x_n\, f_{X_n}(x_n)\, dx_n \\
&= a_1 \mu_1 + \cdots + a_n \mu_n
\end{aligned} \tag{5.37}$$

where the marginal density functions, $f_{X_i}(x_i)$, were obtained from the joint density function $f_{X_1, X_2, \ldots, X_i, \ldots, X_n}(x_1, x_2, \ldots, x_i, \ldots, x_n)$ as

$$\begin{aligned}
f_{X_i}(x_i) = \int_{x_1=-\infty}^{\infty} \cdots \int_{x_{i-1}=-\infty}^{\infty} \int_{x_{i+1}=-\infty}^{\infty} \\
\cdots \int_{x_n=-\infty}^{\infty} f_{X_1, \ldots, X_{i-1}, X_i, X_{i+1}, \ldots, X_n}(x_1, \ldots, x_{i-1}, x_i, x_{i+1}, \ldots, x_n) \\
dx_1 \ldots dx_{i-1}\, dx_{i+1} \ldots dx_n
\end{aligned} \tag{5.38}$$

Equation (5.37) shows that the mean value of a linear sum of several random variables is the linear sum of the mean values. The variance of Y is given by

$$\begin{aligned}
\text{var}[Y] = \sigma_Y^2 &= E[(Y - \mu_Y)^2] \\
&= E[\{(a_1 X_1 + \cdots + a_n X_n) - (a_1 \mu_1 + \cdots + a_n \mu_n)\}^2] \\
&= E[\{a_1(X_1 - \mu_1) + \cdots + a_n(X_n - \mu_n)\}^2] \\
&= E[a_1^2(X_1 - \mu_1)^2 + \cdots + a_n^2(X_n - \mu_n)^2 + 2a_1 a_2(X_1 - \mu_1)(X_2 - \mu_2) \\
&\quad + 2a_1 a_3(X_1 - \mu_1)(X_3 - \mu_3) + \cdots + 2a_1 a_n(X_1 - \mu_1)(X_n - \mu_n) \\
&\quad + 2a_2 a_3(X_2 - \mu_2)(X_3 - \mu_3) + \cdots]
\end{aligned} \tag{5.39}$$

It can be seen that the first set of n terms on the right-hand side of Eq. (5.39) denotes the variances and the remaining terms represent the covariances. This gives

$$\mathrm{var}[Y] = \sum_{i=1}^{n} a_i^2 \,\mathrm{var}[X_i] + \sum_{i=1}^{n} \sum_{\substack{j=1 \\ i \neq j}}^{n} a_i a_j \,\mathrm{cov}[x_i, x_j]$$

or,

$$\sigma_Y^2 = \sum_{i=1}^{n} a_i^2 \sigma_i^2 + \sum_{i=1}^{n} \sum_{\substack{j=1 \\ i \neq j}}^{n} a_i a_j \rho_{ij} \sigma_i \sigma_j \tag{5.40}$$

where σ_i^2 is the variance of X_i and ρ_{ij} is the correlation coefficient between X_i and X_j.

5.5.2 Mean and Variance of Sum of Two Random Variables

If $Y = X_1 \pm X_2$, Eqs. (5.37) and (5.40) can be specialized to obtain

$$\mu_Y = \mu_{X_1} \pm \mu_{X_2} \tag{5.41}$$

and

$$\sigma_Y = \begin{cases} (\sigma_{X_1}^2 + \sigma_{X_2}^2)^{1/2}; & X_1 \text{ and } X_2 \text{ uncorrelated} \\ (\sigma_{X_1}^2 + \sigma_{X_2}^2 \pm 2\rho\sigma_{X_1}\sigma_{X_2})^{1/2}; & X_1 \text{ and } X_2 \text{ correlated} \end{cases} \tag{5.42}$$

where ρ is the correlation coefficient between X_1 and X_2.

5.5.3 Mean and Variance of Product of Two Random Variables

If $Y = X_1 X_2$, then the mean value of Y is given by

$$\mu_Y = E[X_1 X_2] = \int_{-\infty}^{\infty} \int_{-\infty}^{\infty} x_1 x_2 \, f_{X_1, X_2}(x_1, x_2) \, dx_1 \, dx_2 \tag{5.43}$$

If X_1 and X_2 are independent,

$$f_{X_1, X_2} = f_{X_1}(x_1) f_{X_2}(x_2) \tag{5.44}$$

and Eq. (5.43) can be rewritten as

$$\mu_Y = \int_{-\infty}^{\infty} x_1 f_{X_1}(x_1) dx_1 \int_{-\infty}^{\infty} x_2 f_{X_2}(x_2) \, dx_2 = \mu_{X_1} \mu_{X_2} \tag{5.45}$$

The variance of Y can be determined as

$$\begin{aligned} \sigma_Y^2 &= E[(Y - \mu_Y)^2] = E[X_1^2 X_2^2 + \mu_{X_1}^2 \mu_{X_2}^2 - 2\mu_{X_1}\mu_{X_2} X_1 X_2] \\ &= E[X_1^2 X_2^2] + \mu_{X_1}^2 \mu_{X_2}^2 - 2\mu_{X_1}\mu_{X_2} E[X_1 X_2] \end{aligned} \tag{5.46}$$

where, for independent variables,

$$E\left[X_1^2 X_2^2\right] = \int_{-\infty}^{\infty} \int_{-\infty}^{\infty} x_1^2 x_2^2 f_{X_1,X_2}(x_1, x_2)\, dx_1\, dx_2 = \mu_{X_1^2}\, \mu_{X_2^2} \tag{5.47}$$

For normally distributed variables, we have (see Example 5.9)

$$E\left[X^2\right] = \mu_X^2 + \sigma_X^2 \tag{5.48}$$

and hence Eq. (5.46) yields

$$\sigma_Y = \left[\mu_{X_1}^2 \sigma_{X_2}^2 + \mu_{X_2}^2 \sigma_{X_1}^2 + \sigma_{X_1}^2 \sigma_{X_2}^2\right]^{1/2} \tag{5.49}$$

If X_1 and X_2 are correlated with a correlation coefficient of ρ, the standard deviation of Y [5.2] is given by

$$\sigma_Y = \mu_{X_1} \mu_{X_2} \left\{ \left(\frac{\sigma_{X_1}^2}{\mu_{X_1}^2} + \frac{\sigma_{X_2}^2}{\mu_{X_2}^2} + \frac{\sigma_{X_1}^2 \sigma_{X_2}^2}{\mu_{X_1}^2 \mu_{X_2}^2} \right) \left(1 + \rho^2\right) \right\}^{1/2} \tag{5.50}$$

5.5.4 Mean and Variance of Quotient of Two Random Variables

If $Y = X_1/X_2$, the mean and variance of Y can be obtained, assuming normal distribution for X_1 and X_2 as [5.2]

$$\mu_Y = \frac{\mu_{X_1}}{\mu_{X_2}} \tag{5.51}$$

and

$$\sigma_Y = \begin{cases} \dfrac{1}{\mu_{X_2}} \left[\dfrac{\mu_{X_1}^2 \sigma_{X_2}^2 + \mu_{X_2}^2 \sigma_{X_1}^2}{\mu_{X_2}^2 + \sigma_{X_2}^2} \right]^{1/2}; & X_1 \text{ and } X_2 \text{ uncorrelated} \\[2ex] \dfrac{\mu_{X_1}}{\mu_{X_2}} \left[\dfrac{\sigma_{X_1}^2}{\mu_{X_1}^2} + \dfrac{\sigma_{X_2}^2}{\mu_{X_2}^2} - 2\rho \dfrac{\sigma_{X_1}\sigma_{X_2}}{\mu_{X_1}\mu_{X_2}} \right]^{1/2}; & X_1 \text{ and } X_2 \text{ correlated} \end{cases} \tag{5.52}$$

5.5.5 Mean and Variance of a General Nonlinear Function of Several Random Variables

Consider a nonlinear function of several random variables

$$Y = g(X_1, X_2, \ldots, X_n) \tag{5.53}$$

The mean and variance of Y are, by definition,

$$\mu_Y = E[Y] = \int_{x_1=-\infty}^{\infty} \cdots \int_{x_n=-\infty}^{\infty} g(x_1, x_2, \ldots, x_n) \times f_{X_1, X_2, \ldots, X_n}(x_1, x_2, \ldots, x_n)\, dx_1\, dx_2 \ldots dx_n \tag{5.54}$$

and

$$\sigma_Y^2 = E[(Y - \mu_Y)^2]$$
$$= \int_{x_1=-\infty}^{\infty} \cdots \int_{x_n=-\infty}^{\infty} [g(x_1, x_2, \ldots, x_n) - \mu_Y]^2$$
$$\times f_{X_1, X_2, \ldots, X_n}(x_1, x_2, \ldots, x_n)dx_1\, dx_2 \ldots dx_n \tag{5.55}$$

In many engineering applications, the joint density function of $X_1, X_2, \ldots, X_n$ may not be known, but the mean and variance of the random variables X_i may be known. Furthermore, even when the joint density function is known, the multiple integrations involved in Eqs. (5.54) and (5.55) may be difficult to perform. In these cases, we would like to find the approximate mean and variance of the function g. For this, we expand the function $g(X_1, X_2, \ldots, X_n)$ in a Taylor's series about the mean values $\mu_1 = \mu_{X_1}$, $\mu_2 = \mu_{X_2}, \ldots, \mu_n = \mu_{X_n}$ as

$$Y = g(\mu_1, \mu_2, \ldots, \mu_n) + \sum_{i=1}^{n} \frac{\partial g}{\partial X_i}\bigg|_{\mu_1, \mu_2, \ldots, \mu_n} (X_i - \mu_i)$$
$$+ \frac{1}{2} \sum_{i=1}^{n} \sum_{j=1}^{n} \frac{\partial^2 g}{\partial X_i\, \partial X_j}\bigg|_{\mu_1, \mu_2, \ldots, \mu_n} (X_i - \mu_i)(X_j - \mu_j) + \cdots \tag{5.56}$$

where the partial derivatives are evaluated at the mean values of X_i as

$$\frac{\partial g}{\partial X_i}\bigg|_{\mu_1, \mu_2, \ldots, \mu_n} = \frac{\partial g}{\partial X_i}(X_1 = \mu_1, \ldots, X_n = \mu_n) \tag{5.57}$$

By neglecting terms involving second- and higher-order derivatives of g in Eq. (5.56), we obtain

$$Y \approx g(\mu_1, \mu_2, \ldots, \mu_n) + \sum_{i=1}^{n} \frac{\partial g}{\partial X_i}\bigg|_{\mu_1, \ldots, \mu_n} (X_i - \mu_i) \tag{5.58}$$

Since $g(\mu_1, \ldots, \mu_n)$ and $\frac{\partial g}{\partial X_i}\Big|_{\mu_1, \ldots, \mu_n}$ are constants, Eq. (5.58) gives the mean value of Y as

$$\mu_Y = E[Y] \approx g(\mu_1, \ldots, \mu_n) \tag{5.59}$$

Similarly, the variance of Y can be obtained as

$$\sigma_Y^2 \approx \sum_{i=1}^{n} a_i^2 \operatorname{var}[X_i] + \sum_{\substack{i=1 \\ i \neq j}}^{n} \sum_{j=1}^{n} a_i a_j \operatorname{cov}[X_i, X_j] \tag{5.60}$$

where

$$a_i = \frac{\partial g}{\partial X_i}\bigg|_{\mu_1, \ldots, \mu_n} \tag{5.61}$$

and

$$a_j = \left. \frac{\partial g}{\partial X_j} \right|_{\mu_1, \ldots, \mu_n} \tag{5.62}$$

If the variables X_i are statistically independent, then

$$\sigma_Y^2 \approx \sum_{i=1}^{n} a_i^2 \, \sigma_{X_i}^2 \tag{5.63}$$

The procedure adopted in deriving Eqs. (5.59), (5.60), and (5.63) is known as the *partial derivative rule*.

Note: The accuracy of expressions for the mean and variance of Y can be improved by including higher-order derivative terms in Eq. (5.56). For example, if the second derivative terms are also included, the mean value of Y can be determined as

$$\mu_Y = E[Y] \approx g(\mu_1, \ldots, \mu_n) + \frac{1}{2} \sum_{i=1}^{n} \sum_{j=1}^{n} \left. \frac{\partial^2 g}{\partial X_i \partial X_j} \right|_{\mu_1, \ldots, \mu_n} \text{cov}[X_i, X_j] \tag{5.64}$$

If X_i are independent, then Eq. (5.64) becomes

$$\mu_Y = E[Y] \approx g(\mu_1, \ldots, \mu_n) + \frac{1}{2} \sum_{i=1}^{n} \left. \frac{\partial^2 g}{\partial X_i^2} \right|_{\mu_1, \ldots, \mu_n} \text{var}[X_i] \tag{5.65}$$

EXAMPLE 5.8

The inertia force developed in a reciprocating engine (F) is given by (see Fig. 5.9):

$$F = \underbrace{mr\omega^2 \cos \omega t}_{\text{primary force}} + \underbrace{m \frac{r^2\omega^2}{l} \cos 2\omega t}_{\text{secondary force}} \tag{E1}$$

where m = mass of the piston, r = length of the crank, l = length of the connecting rod, ω = angular velocity of the crank, and t = time. If the mean values and coefficients of variation of the parameters

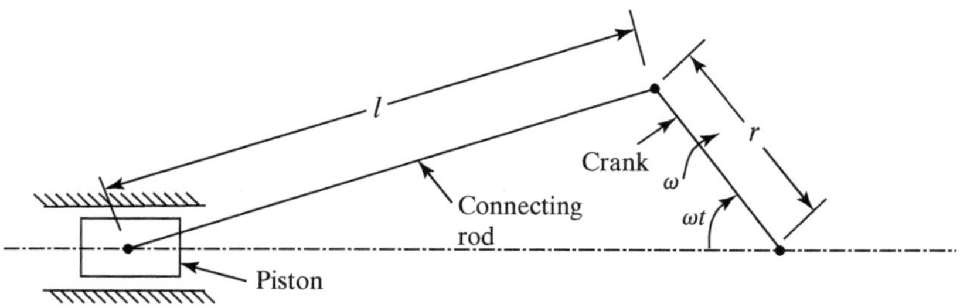

FIGURE 5.9 A reciprocating engine.

are given by $\overline{m} = 1$ kg, $\gamma_m = 0.10$, $\overline{r} = 7.5$ cm, $\gamma_r = 0.05$, $\overline{l} = 30$ cm, $\gamma_l = 0.05$, $\overline{\omega} = 1500$ rpm, and $\gamma_\omega = 0.10$, find the first-order mean and standard deviation of F when $t = 0.003333$ s. Assume that all the random variables are statistically independent.

Solution: The first-order approximate mean value of the inertia force at $t = 0.003333$ s can be obtained as

$$\overline{F} = \overline{m}\,\overline{r}\,\overline{\omega}^2 \cos \overline{\omega} t + \overline{m} \frac{\overline{r}^2 \overline{\omega}^2}{\overline{l}} \cos 2\,\overline{\omega} t$$

$$= (1)(0.075)(157.08)^2 \cos(30°) + \left(\frac{1}{0.3}\right)(0.075)^2 (157.08)^2 \cos(60°)$$

$$= 1833.9515 \text{ N}$$

where the mean value of the angular velocity of the crank is

$$\overline{\omega} = \frac{1500(2\pi)}{60} = 50\,\pi = 157.08 \,\frac{\text{rad}}{\text{s}}$$

The partial derivatives of F with respect to the random variables, along with their values at the mean values of random variables, are given by

$$\frac{\partial F}{\partial m} = r\omega^2 \cos \omega t + \frac{r^2\omega^2}{l} \cos 2\omega t = 1833.9514$$

$$\frac{\partial F}{\partial r} = m\omega^2 \cos \omega t + \frac{2mr\omega^2}{l} \cos 2\omega t = 27536.9522$$

$$\frac{\partial F}{\partial \omega} = 2mr\omega \cos \omega t - mr\omega^2 t \sin \omega t + \frac{2mr^2\omega}{l} \cos 2\omega t - \frac{2mr^2\omega^2 t}{l} \sin 2\omega t = 17.5951$$

$$\frac{\partial F}{\partial l} = -\frac{mr^2\omega^2}{l^2} \cos 2\omega t = -771.0664$$

Noting that the standard deviations of the random variables are $\sigma_m = \overline{m}\gamma_m = (1)(0.1) = 0.1$ kg, $\sigma_r = \overline{r}\gamma_r = (0.075)(0.05) = 0.00375$ m, $\sigma_\omega = \overline{\omega}\gamma_\omega = (157.08)(0.1) = 15.708$ rad/s, and $\sigma_l = \overline{l}\gamma_l = (0.3)(0.05) = 0.015$ m, the first-order approximate variance of F can be obtained as

$$\sigma_F^2 = \left(\frac{\partial F}{\partial m}\right)^2 \sigma_m^2 + \left(\frac{\partial F}{\partial r}\right)^2 \sigma_r^2 + \left(\frac{\partial F}{\partial \omega}\right)^2 \sigma_\omega^2 + \left(\frac{\partial F}{\partial l}\right)^2 \sigma_l^2$$

$$= (1833.9514)^2 (0.1)^2 + (27536.9522)^2 (0.00375)^2 + (17.5951)^2 (15.708)^2 + (-771.0664)^2 (0.015)^2 = 120818.9366$$

Thus

$$\sigma_F = 347.5902 \text{ N}$$

Note: The second-order approximate mean value of the inertia force can be determined as follows:

$$\frac{\partial^2 F}{\partial m^2} = 0, \frac{\partial^2 F}{\partial r^2} = \frac{2m\omega^2}{l}\cos 2\omega t = 82247.088,$$

$$\begin{aligned}\frac{\partial^2 F}{\partial \omega^2} &= 2mr\cos\omega t - 4mr\omega t\sin\omega t - mr\omega^2 t^2\cos\omega t \\ &\quad + \frac{2mr^2}{l}\cos 2\omega t - \frac{8mr^2\omega t}{l}\sin 2\omega t - \frac{4mr^2\omega^2 t^2}{l}\cos 2\omega t \\ &= 0.4656\end{aligned}$$

$$\frac{\partial^2 F}{\partial l^2} = \frac{2mr^2\omega^2}{l^3}\cos 2\omega t = 5140.443$$

$$\begin{aligned}\overline{F} &= \overline{m}\,\overline{r}\,\overline{\omega}^2\cos\overline{\omega}t + \frac{\overline{m}\,\overline{r}^2\overline{\omega}^2}{\overline{l}}\cos 2\overline{\omega}t \\ &\quad + \frac{1}{2}\left(\frac{\partial^2 F}{\partial m^2}\right)\sigma_m^2 + \frac{1}{2}\left(\frac{\partial^2 F}{\partial r^2}\right)\sigma_r^2 + \frac{1}{2}\left(\frac{\partial^2 F}{\partial \omega^2}\right)\sigma_\omega^2 + \frac{1}{2}\left(\frac{\partial^2 F}{\partial l^2}\right)\sigma_l^2 \\ &= 1602.6315 + 231.3200 + 0 + \frac{1}{2}(82247.088)(0.00375)^2 \\ &\quad + \frac{1}{2}(0.4656)(15.708)^2 + \frac{1}{2}(5140.443)(0.015)^2 = 1949.9904 \text{ N}\end{aligned}$$

■

5.6 Moment-Generating Function

The evaluation of the mean values and standard deviations in Sections 5.5.1 to 5.5.4 involved lengthy expressions, especially when the random variables are correlated. It will be convenient to introduce a function, known as the moment-generating function, to derive expressions for the means and standard deviations of functions of random variables.[1] If X is a random variable, then the quantity $e^{X\theta}$ is a function of X for each fixed value of θ. The moment-generating function of X, denoted as $M(\theta)$, is defined as the expected value of the function $e^{X\theta}$:

$$M(\theta) = E\left[e^{X\theta}\right] = \int_{-\infty}^{\infty} e^{x\theta} f_X(x)\,dx \tag{5.66}$$

[1]In fact, the moment-generating function of a random variable can also be used to generate all the moments of the random variable. Thus, the moment-generating function can be considered as an alternate way of describing the probability law of a random variable.

Differentiating Eq. (5.66) r times with respect to θ gives

$$\frac{d^r M(\theta)}{d\theta^r} = \int_{-\infty}^{\infty} x^r e^{x\theta} f_X(x)\, dx \tag{5.67}$$

and setting $\theta = 0$,

$$\begin{aligned} \frac{d^r M(0)}{d\theta^r} &= \int_{-\infty}^{\infty} x^r f_X(x)\, dx = E[X^r] = \mu_r' \\ &= r\text{th moment of } X \text{ about the origin} \end{aligned} \tag{5.68}$$

Equation (5.68) can be used to obtain moments of any order of the distribution. The moments of a function of a random variable $g(X)$ can also be evaluated using the moment-generating function of X. In this case, the moment-generating function of $g(X)$ is defined as

$$M(\theta) = E\left[e^{\theta g(x)}\right] = \int_{-\infty}^{\infty} e^{\theta g(x)} f_X(x)\, dx \tag{5.69}$$

The rth moment of $g(X)$ about the origin can be obtained as

$$E[\{g(X)\}^r] = \frac{d^r M(0)}{d\theta^r} = \int_{-\infty}^{\infty} \{g(x)\}^r f_X(x)\, dx \tag{5.70}$$

The moment-generating function can also be defined for multiple random variables. Let X_1 and X_2 be two random variables with joint density function $f_{X_1, X_2}(x_1, x_2)$. Then

$$\begin{aligned} M(\theta_1, \theta_2) &= E\left[e^{\theta_1 X_1} e^{\theta_2 X_2}\right] \\ &= \int_{-\infty}^{\infty} \int_{-\infty}^{\infty} e^{\theta_1 x_1 + \theta_2 x_2} f_{X_1, X_2}(x_1, x_2)\, dx_1\, dx_2 \end{aligned} \tag{5.71}$$

The rth moment of X_1, for example, can be found by differentiating the moment-generating function r times with respect to θ_1 and then setting $\theta_1 = \theta_2 = 0$. Similarly, the joint moment $E\left[X_1^p X_2^q\right]$ can be obtained by differentiating $M(\theta_1, \theta_2)$ p times with respect to θ_1 and q times with respect to θ_2 and then setting $\theta_1 = \theta_2 = 0$.

5.6.1 Moments of Normally Distributed Variables

Since normal distribution is commonly used in many engineering applications, we shall now consider the moments of normally distributed variables using moment-generating functions. For a normal variate X with mean μ and standard deviation σ, the moment-generating function is given by

$$M(\theta) = E\left[e^{\theta X}\right] = \frac{1}{\sqrt{2\pi}\,\sigma} \int_{-\infty}^{\infty} e^{\theta x - 1/2\{(x-\mu)/\sigma\}^2}\, dx \tag{5.72}$$

The term in the exponent can be rewritten as

$$\theta x - \frac{1}{2}\left(\frac{x-\mu}{\sigma}\right)^2 = -\frac{1}{2\sigma^2}\left[x - (\mu + \theta\,\sigma^2)\right]^2 + \mu\theta + \frac{1}{2}\sigma^2\theta^2 \tag{5.73}$$

Thus, Eq. (5.72) becomes

$$M(\theta) = \exp\left(\mu\theta + \frac{1}{2}\sigma^2\theta^2\right)$$

$$\times \int_{-\infty}^{\infty} \frac{1}{\sqrt{2\pi}\,\sigma} \exp\left\{-\frac{1}{2\sigma^2}\left[x - (\mu + \sigma^2\theta)\right]^2\right\} dx \tag{5.74}$$

The expression under the integral sign can be identified as the probability density function of a normal variate with mean $(\mu + \sigma^2\theta)$ and standard deviation σ, and thus integrates to 1. Thus the moment-generating function of the normal variate X is given by

$$M(\theta) = e^{(\mu\theta + 1/2\sigma^2\theta^2)} \tag{5.75}$$

EXAMPLE 5.9

Derive an expression for the mean value of $Y = X^2$ when X is normally distributed with mean μ and standard deviation σ.

Solution: The mean value of the function Y is given by

$$E[X^2] = \mu_{X^2} = \frac{\partial^2 M(0)}{\partial\theta^2}$$

The moment-generating function of X is given by Eq. (5.75):

$$M(\theta) = \exp(\theta\,\mu + 1/2\sigma^2\theta^2) = e^w$$

where

$$w = \theta\mu + 1/2\,\sigma^2\theta^2$$

Differentiation of $M(\theta)$ gives

$$\frac{\partial M(\theta)}{\partial\theta} = (\mu + \sigma^2\theta)\,e^w$$

$$\frac{\partial^2 M(\theta)}{\partial\theta^2} = \sigma^2 e^w + (\mu + \sigma^2\theta)^2 e^w$$

Thus

$$E[X^2] = \mu_{X^2} = \frac{\partial^2 M(0)}{\partial\theta^2} = \sigma^2 + \mu^2$$

■

5.7 Functions of Several Random Variables

In Sections 5.2 and 5.3, we considered the derivation of the probability density function of a random variable defined as a function of other random variable(s). In many cases, we encounter several functions involving multiple random variables. The joint probability density and distribution functions of variables defined as functions of several other random variables can be derived as follows. Let

$$Y_k = g_k(X_1, X_2, \ldots, X_n), \; k = 1, 2, \ldots, m, \; m \le n \tag{5.76}$$

Then the joint distribution function $F_{Y_1, Y_2, \ldots, Y_m}(y_1, y_2, \ldots, y_m)$, by definition, is given by

$$\begin{aligned} F_{Y_1, Y_2, \ldots, Y_m}(y_1, y_2, \ldots, y_m) &= P(Y_1 \le y_1, Y_2 \le y_2, \ldots, Y_m \le y_m) \\ &= \int_{x_1} \int_{x_2} \cdots \int_{x_n} f_{X_1, X_2, \ldots, X_n}(x_1, x_2, \ldots, x_n) dx_1\, dx_2 \ldots dx_n \end{aligned} \tag{5.77}$$

where the integrals have to be evaluated over the domain of the n-dimensional $(X_1, X_2, \ldots, X_n)$–space in which the inequalities

$$g_k(x_1, x_2, \ldots, x_n) \le y_k, \qquad k = 1, 2, \ldots, m$$

are satisfied. The joint density function is given by

$$f_{Y_1, Y_2, \ldots, Y_m}(y_1, y_2, \ldots, y_m) = \frac{\partial^m F_{Y_1, Y_2, \ldots, Y_m}(y_1, y_2, \ldots, y_m)}{\partial y_1\, \partial y_2 \ldots \partial y_m} \tag{5.78}$$

Equations (5.77) and (5.78) are valid for any kind of functions g_k, $k = 1, 2, \ldots, m$. If the set of Eqs. (5.76) has a unique solution for some m-variables as

$$X_k = h_k(Y_1, Y_2, \ldots, Y_m; X_{m+1}, X_{m+2}, \ldots, X_n), \quad k = 1, 2, \ldots, m \tag{5.79}$$

then by changing the variables on the right-hand side of Eq. (5.77), we obtain

$$\begin{aligned} F_{Y_1, Y_2, \ldots, Y_m}(y_1, y_2, \ldots, y_m) &= \int_{-\infty}^{y_1} dy_1 \int_{-\infty}^{y_2} dy_2 \cdots \int_{-\infty}^{y_m} dy_m \int_{x_{m+1}=-\infty}^{\infty} \cdots \int_{x_n=-\infty}^{\infty} \\ & f_{X_1, X_2, \ldots, X_n}(h_1, h_2, \ldots, h_m; x_{m+1}, \ldots, x_n) \left|J\right| dy_1 \ldots dy_m\, dx_{m+1} \ldots dx_n \end{aligned} \tag{5.80}$$

where J denotes the Jacobian given by

$$J = \text{determinant of} \begin{bmatrix} \frac{\partial h_1}{\partial y_1} & \frac{\partial h_2}{\partial y_1} & \cdots & \frac{\partial h_m}{\partial y_1} \\ \frac{\partial h_1}{\partial y_2} & \frac{\partial h_2}{\partial y_2} & \cdots & \frac{\partial h_m}{\partial y_2} \\ \cdot & \cdot & \cdots & \cdot \\ \cdot & \cdot & \cdots & \cdot \\ \cdot & \cdot & \cdots & \cdot \\ \frac{\partial h_1}{\partial y_m} & \frac{\partial h_2}{\partial y_m} & \cdots & \frac{\partial h_m}{\partial y_m} \end{bmatrix} \tag{5.81}$$

By differentiating Eq. (5.80), we can find the joint probability density function as

$$\begin{aligned} &f_{Y_1, Y_2, \ldots, Y_m}(y_1, y_2, \ldots, y_m) \\ &\quad = f_{X_1, X_2, \ldots, X_n}(h_1, h_2, \ldots, h_m; x_{m+1}, x_{m+2}, \ldots, x_n)\,|J| \end{aligned} \tag{5.82}$$

EXAMPLE 5.10

Find the joint density function of Y_1 and Y_2 where $Y_1 = X_1 X_2$, $Y_2 = X_1/X_2$ and the joint density function of X_1 and X_2 is given by

$$f_{X_1, X_2}(x_1, x_2) = \frac{1}{x_1^2 x_2^2}; \quad 1 \le x_1 < \infty; \quad 1 \le x_2 < \infty$$

Solution: Since $Y_1 = g_1(X_1, X_2) = X_1 X_2$ and $Y_2 = g_2(X_1, X_2) = X_1/X_2$, we obtain

$$X_1 = h_1(Y_1, Y_2) = \sqrt{Y_1 Y_2} \quad \text{and} \quad X_2 = h_2(Y_1, Y_2) = \sqrt{\frac{Y_1}{Y_2}}$$

The Jacobian matrix is given by

$$[J] = \begin{bmatrix} \frac{\partial h_1}{\partial y_1} & \frac{\partial h_2}{\partial y_1} \\ \frac{\partial h_1}{\partial y_2} & \frac{\partial h_2}{\partial y_2} \end{bmatrix} = \begin{bmatrix} \frac{1}{2}\sqrt{\frac{y_2}{y_1}} & \frac{1}{2}\frac{1}{\sqrt{y_1 y_2}} \\ \frac{1}{2}\sqrt{\frac{y_1}{y_2}} & -\frac{1}{2}\frac{\sqrt{y_1}}{y_2^{3/2}} \end{bmatrix}$$

so that

$$|[J]| = \left| -\frac{1}{4}\frac{1}{y_2} - \frac{1}{4}\frac{1}{y_2} \right| = \frac{1}{2y_2}.$$

The application of Eq. (5.82) gives

$$f_{Y_1,Y_2}(y_1, y_2) = f_{X_1,X_2}(h_1, h_2)\,|[J]| = \frac{1}{h_1^2 h_2^2}\,\frac{1}{2y_2} = \frac{1}{(y_1 y_2)\left(\dfrac{y_1}{y_2}\right)}\,\frac{1}{2y_2} = \frac{1}{2y_1^2 y_2}$$

■

5.8 Numerical Solutions Using MATLAB

EXAMPLE 5.11

The mean values and variances of two jointly distributed normal random variables X and Y are given by $\overline{X} = \mu_X = 1, \overline{Y} = \mu_Y = 2$, $\sigma_X = 0.8, \sigma_Y = 0.5$, and $\sigma_{X,Y} = 0.4$. Determine the following using MATLAB:

1. Plot of the joint probability density function, $f_{X,Y}(x, y)$
2. Plot of the joint distribution function, $F_{X,Y}(x, y)$

Solution: The MATLAB program and the results obtained are shown below:

```
clear all ;clc  ;

mu = [1  2]  ;
Sigma = [0.8  0.4;  0.4  0.5]  ;
x1 = -2 : 0.2 : 4   ;
x2 = -1 : 0.2 : 5   ;

[X1,X2] = meshgrid(x1,x2)  ;
F = mvnpdf([X1(:) X2(:)],mu,Sigma)  ;
F = reshape(F,length(x2),length(x1))    ;
surf(x1,x2,F)  ;
caxis([min(F(:))-0.5*range(F(:)), max(F(:))])  ;
axis([-2 4 -1 5 0 0.4])  ;

xlabel('x1')  ; ylabel('x2')  ; zlabel('Probability Density')  ;

F = mvncdf([X1(:) X2(:)], mu, Sigma)  ;
F = reshape(F,length(x2),length(x1))  ;
figure
surf(x1,x2,F)  ;
caxis([min(F(:))-0.5*range(F(:)), max(F(:))])  ;
axis([-2 4 -1 5 0 1])  ;
xlabel('x1')  ;ylabel('x2') ;zlabel('Cumulative Probability')  ;
```

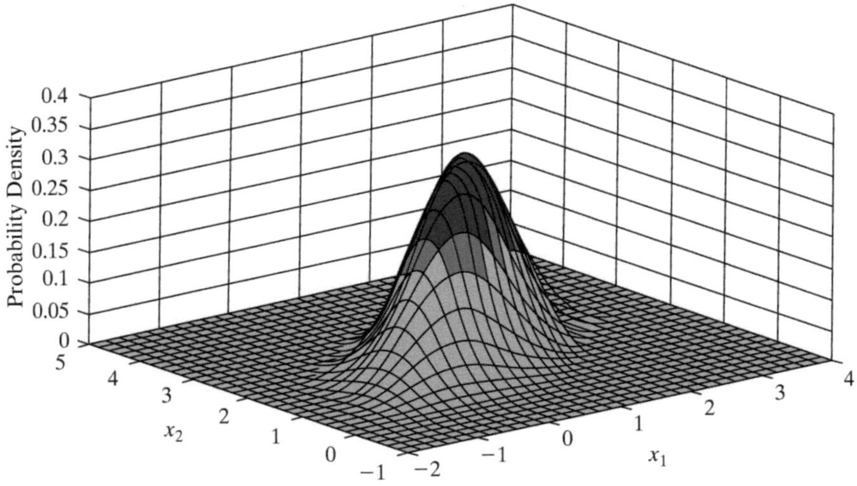

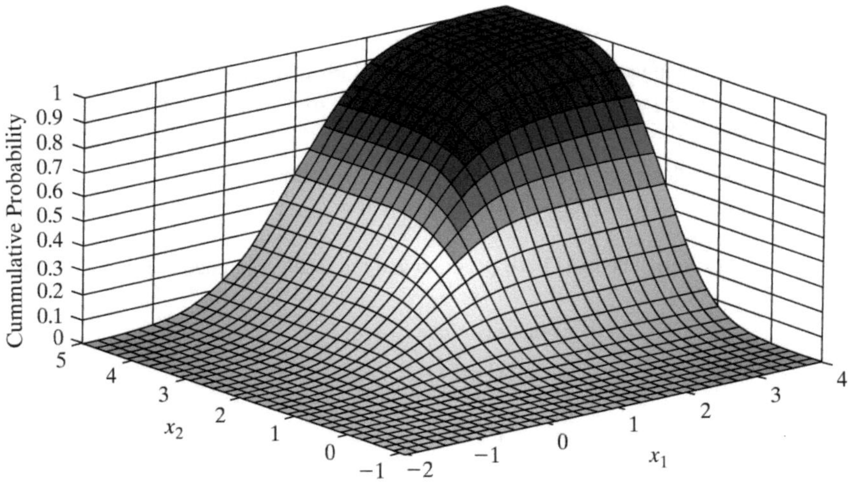

■

EXAMPLE 5.12

Experiments are conducted on eight different batches of a particular grade of steel to find the modulus of elasticity (E) and modulus of rigidity (G). The experimental results obtained on the eight batches of steel are shown below:

Variable \ Batch of Steel	1	2	3	4	5	6	7	8
E (GPa)	208.1	214.3	213.8	193.7	198.8	210.2	205.5	210.4
G (GPa)	78.9	82.5	72.2	83.4	71.6	81.9	79.8	82.2

Using MATLAB, find the following quantities:

1. Covariance and correlation matrices between the two random variables.
2. Assuming that E and G are jointly distributed normal random variables with the mean values and covariances given by the experimental results, determine the probability of simultaneously realizing the value of E less than 210 GPa and G less than 80 GPa.

Solution: The MATLAB program developed and the results obtained are shown below:

```
clear all ; clc ; format short

x1 = [208.1    78.9;
      214.3    82.5;
      213.8    72.2;
      193.7    83.4;
      198.8    71.6;
      210.2    81.9;
      205.5    79.8;
      210.4    82.2]

C = cov(x1)

R = corrcoef(x1)

X = [210  80]

mu = [206.85  79.0625]

sigma = [7.2579  0.23749;  0.23749  4.6580]

mvncdf(X,mu,sigma)

x1 =

  208.1000   78.9000
  214.3000   82.5000
  213.8000   72.2000
  193.7000   83.4000
  198.8000   71.6000
  210.2000   81.9000
  205.5000   79.8000
  210.4000   82.2000

C =

   52.6771    0.0564
    0.0564   21.6970

R =

    1.0000    0.0017
    0.0017    1.0000

X =

  210   80

mu =

  206.8500   79.0625

sigma =

    7.2579    0.2375
    0.2375    4.6580

ans =

    0.5901

>>
```

■

REFERENCES AND BIBLIOGRAPHY

5.1 W. W. Hines and D. C. Montgomery, *Probability and Statistics in Engineering and Management Science*, 2nd ed., John Wiley, New York, 1980.

5.2 E. B. Haugen, *Probabilistic Approaches to Design*, John Wiley, New York, 1968.

5.3 G. P. Wadsworth and J. G. Bryan, *Applications of Probability and Random Variables*, 2nd ed., McGraw-Hill, New York, 1974.

5.4 M. Kaufman and A. H. Seidman (eds.), *Handbook of Electronics Calculations for Engineers and Technicians*, 2nd ed., McGraw-Hill, New York, 1988.

5.5 E. C. Guyer and D. L. Brownell (eds.), *Handbook of Applied Thermal Design*, McGraw-Hill, New York, 1989.

5.6 J. E. Shigley and C. R. Mischke, *Mechanical Engineering Design*, 5th ed., McGraw-Hill, New York, 1989.

5.7 C. E. Wilson, *Noise Control: Measurement, Analysis, and Control of Sound and Vibration*, Harper & Row, New York, 1989.

5.8 A. H. S. Ang and W. H. Tang, *Probability Concepts in Engineering Planning and Design*, Vol. 1, *Basic Principles*, John Wiley, New York, 1975.

5.9 T. T. Soong, *Probabilistic Modeling and Analysis in Science and Engineering*, John Wiley, New York, 1981.

5.10 A. C. King and C. B. Read, *Pathways to Probability*, Holt, Rinehart & Winston, New York, 1963.

5.11 C. C. Gillispie (ed.-in-chief), *Dictionary of Scientific Biography*, Vol. 1, Charles Scribner's Sons, New York, 1980.

REVIEW QUESTIONS

5.1 Give brief answers to the following:

1. How is the partial derivative rule derived?
2. Find the mean and variance of the function $Y = g(X_1, X_2, \ldots, X_n)$ using the partial derivative rule.
3. Express the probability distribution function of Y in terms of the probability density function of X where $Y = g(X)$.
4. Find the probability density function of $Y = 4X - 3$ when the density function of X is given by $f_X(x) = 6e^{-6x}$.
5. If the probability distributions of two independent normal variates X_1 and X_2 are given by $N(100, 10)$ and $N(80, 50)$, respectively, find the probability density function of $Y = X_1 + X_2$.
6. If the probability distributions of two independent normal variates X_1 and X_2 are given by $N(100, 10)$ and $N(80, 50)$, respectively, find the probability density function of $Y = X_1 - X_2$.
7. What is a Jacobian?
8. What is the use of a moment-generating function?
9. Define the moment-generating function of a random variable X.

10. What is a partial derivative rule?
11. Express the standard deviation of the function $g = g(X_1, X_2, \ldots, X_n)$ in terms of the joint density function of $X_1, X_2, \ldots, X_n$.

5.2 Indicate whether each of the following statements is true or false:

1. If x is a random variable, the function $y = a + bx + cx^2$, where a, b, and c are constants, will also be a random variable.
2. A function $f(x)$ is considered to be monotonically increasing if $f(x + dx) \geq f(x)$ for any positive dx.
3. The partial derivative rule is an exact method for finding the mean value and standard deviation of any function of multiple random variables.
4. The moment-generating function can be considered as an alternate way of describing the probability characteristics of a random variable.
5. The function $\int_{-\infty}^{\infty} |x_2| f_{X_1, X_2}(yx_2, x_2)\, dx_2$ denotes the probability density function of $\frac{X_1}{X_2}$.

5.3 Fill in the blank with the proper word(s):

1. If $Y = a_1X_1 + a_2X_2 + \ldots + a_nX_n$ where $a_1, a_2, \ldots, a_n$ are constants, μ_i = mean value of X_i and σ_i = standard deviation of X_i, the mean value of Y, μ_Y, is given by ______.
2. If $Y = a_1X_1 + a_2X_2 + \ldots + a_nX_n$ where $a_1, a_2, \ldots, a_n$ are constants, μ_i = mean value of X_i and σ_i = standard deviation of X_i, the variance of Y, $\text{Var}[Y]$, is given by ______.
3. The partial derivative rule is based on the expansion of the function $Y = Y(X_1, X_2, \ldots, X_n)$ in Taylor's series about the ______ values of the random variables $X_1, X_2, \ldots, X_n$.
4. The accuracy of the partial derivative rule can be ______ by including higher-order derivative terms in the Taylor's series expansion of the function $Y = Y(X_1, X_2, \ldots, X_n)$.
5. The moment-generating function can be used to generate all ______ of the random variable.

5.4 Select the most appropriate answer from the multiple choices given:

1. If $f_X(x)$ is the probability density function of the random variable X and $Y = g(X)$, then the integral $\int_{g(x) \leq y} f_X(x)\, dx$ denotes:

   a. Probability density function of Y
   b. Probability distribution function of Y
   c. 1 – Probability distribution function of Y

2. The relation $\frac{d}{dy}[F_Y(y)]$ indicates:

   a. Probability density function of Y
   b. Probability distribution function of Y
   c. 1 – Probability distribution function of Y

3. The mean value, μ_Y, of $Y = g(X)$ is given by:

a. $\int_0^{\infty} y f_Y(y) dy$ **b.** $\int_{-\infty}^{0} y f_Y(y) dy$ **c.** $\int_{-\infty}^{\infty} y f_Y(y) dy$

4. The moment-generating function of the normal variate X is given by:

a. $e^{\left(\mu\theta + \frac{1}{2}\sigma^2\theta^2\right)}$ **b.** $e^{\mu\theta}$ **c.** $e^{\frac{1}{2}\sigma^2\theta^2}$

5. The expression $\dfrac{\partial^m F_{Y_1, Y_2, \ldots, Y_m}(y_1, y_2, \ldots, y_m)}{\partial y_1\, \partial y_2 \ldots \partial y_m}$ denotes:

a. Probability distribution function of $Y_1, Y_2, \ldots, Y_m$
b. Probability density function of $Y_1, Y_2, \ldots, Y_m$
c. Probability mass function of $Y_1, Y_2, \ldots, Y_m$

6. The determinant of the matrix

$$\begin{bmatrix} \dfrac{\partial h_1}{\partial y_1} & \dfrac{\partial h_2}{\partial y_1} & \cdots & \dfrac{\partial h_m}{\partial y_1} \\ \vdots & & & \\ \dfrac{\partial h_1}{\partial y_m} & \dfrac{\partial h_2}{\partial y_m} & \cdots & \dfrac{\partial h_m}{\partial y_m} \end{bmatrix}$$

is called:

a. Jacobian matrix **b.** Hessian matrix **c.** Correlation matrix

PROBLEMS

5.1 The ultimate strength of cast iron (S_u) is related to its Brinell hardness number (H_b) as

$$S_u = 0.23\, H_b - 12.5\,\text{k}(\text{lb/in}^2)$$

If H_b is uniformly distributed in the range 140 to 150, determine the probability density function of S_u.

5.2 The ratio of tensions in a belt drive is given by $r = e^{f\theta}$ where f is the coefficient of friction and θ is the angle of contact between the belt and the pulley. Find the probability density function of r when f follows uniform distribution in the range 0.2 to 0.3 and $\theta = 1$.

5.3 Find the probability density function of $x = \sin\theta$ when θ is uniformly distributed in the range $-\dfrac{\pi}{2}$ to $\dfrac{\pi}{2}$.

5.4 Find the probability density function of $y = \sqrt{x}$ when the probability density function of x is given by

$$f_X(x) = e^{-x}; \qquad x \geq 0$$

5.5 The probability distributions of two independent random variables are given by

$$f_{X_i}(x_i) = \begin{cases} e^{-x_i}; & x_i > 0 \\ 0; & \text{otherwise} \end{cases}$$

for $i = 1, 2$. Find the probability density function of $Y = X_1 + X_2$.

5.6 The joint probability mass function of three discrete random variables is given in the following table:

(x_1, x_2, x_3)	$p_{X_1, X_2, X_3}(x_1, x_2, x_3)$
(0, 0, 0)	1/8
(1, 0, 0)	3/32
(0, 1, 0)	3/16
(0, 0, 1)	1/32
(1, 1, 0)	5/32
(1, 0, 1)	1/16
(0, 1, 1)	7/32
(1, 1, 1)	1/8

Find the probability mass function of $Y = X_1 + X_2 + X_3$.

5.7 A function Y is defined as

$$Y = \sum_{i=1}^{n} c_i X_i$$

where c_i are constants and X_i are a set of independent random variables following exponential distribution:

$$f_{X_i}(x_i) = \begin{cases} \lambda_i e^{-\lambda_i x_i}; & x_i \geq 0 \\ 0; & \text{otherwise} \end{cases}$$

for $\lambda_i > 0$ and $i = 1, 2, \ldots, n$.

1. Find the probability density function of Y.
2. Find $f_Y(y)$ when $\lambda_i = \lambda$ for $i = 1, 2, \ldots, n$.
3. Find $f_Y(y)$ when $\lambda_i = \lambda$ and $c_i = 1$ for $i = 1, 2, \ldots, n$.
4. Show that the random variable Y follows gamma distribution in part 3.

5.8 If the diameter of a shaft (X) is normally distributed with mean μ and standard deviation σ, determine the probability density function of the area of cross section of the shaft, $f_Y(y)$, where $Y = \frac{\pi}{4} X^2$.

5.9 A set of two mutually independent random variables X_i follows uniform distribution in the range 0 to 1 so that

$$f_{X_i}(x_i) = \begin{cases} 1; & 0 \le x_i \le 1 \\ 0; & \text{otherwise} \end{cases}$$

for $i = 1, 2$. Find the probability density function of $Y = X_1 + X_2$.

5.10 The efficiency (η) of a V-threaded screw is given by

$$\eta = \frac{\cos\beta - \mu\tan\alpha}{\cos\beta + \mu\cot\alpha}$$

where α = helix angle of the screw, β = half of the thread angle, and μ = coefficient of friction. If α, β, and μ are random with mean values and coefficients of variation given by

$$\overline{\alpha} = 10°, \overline{\beta} = 30°, \overline{\mu} = 0.1, \gamma_\alpha = 0.05, \gamma_\beta = 0.05, \text{ and } \gamma_\mu = 0.1,$$

determine the first-order mean and standard deviation of η.

5.11 The pressure, p, in kgf/cm^2, at which the oil film breaks down so that metal to metal contact occurs in a journal bearing is given by (see Fig. 5.10)

$$p = \frac{ZN}{475 \times 10^6}\left(\frac{d}{c}\right)^2 \frac{l}{d + l}$$

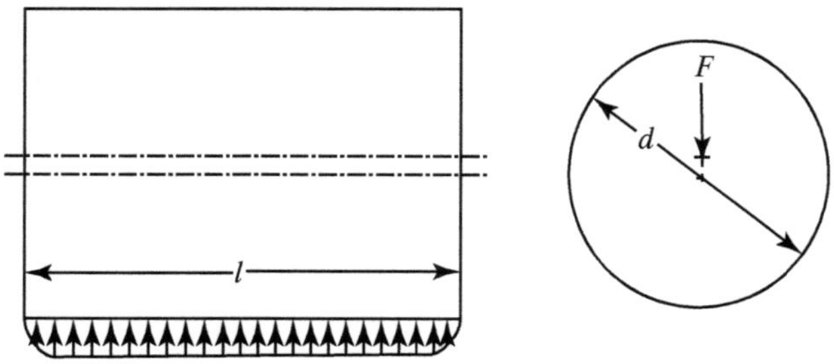

FIGURE 5.10 A journal bearing.

where Z = absolute viscosity of the lubricant in centipoise, N = speed of the journal in rev/min, d = diameter of the journal in cm, c = difference between the diameters of bushing and journal in cm, and l = length of the bearing in cm. The bearing parameters are known to be random variables for a bearing used in a reciprocating compressor with

$$\overline{l} = 15 \text{ cm}, \overline{d} = 7.5 \text{ cm}, \overline{c} = 0.02 \text{ cm}, \overline{Z} = 60, \overline{N} = 300, \gamma_l = \gamma_d = \gamma_c = \gamma_Z = \gamma_N = 0.1.$$

Compute the first-order mean and standard deviation of the pressure, p.

5.12 The length of the belt needed (l) in a belt drive (as shown in Fig. 5.11) is given by

$$l = \left[4\,c^2 - (d_2 - d_1)^2\right]^{1/2} + \frac{1}{2}\,(d_2\,\theta_2 + d_1\,\theta_1)$$

where

$$\theta_1 = \pi - 2\,\sin^{-1}\left(\frac{d_2 - d_1}{2\,c}\right)$$

and

$$\theta_2 = \pi + 2\,\sin^{-1}\left(\frac{d_2 - d_1}{2\,c}\right)$$

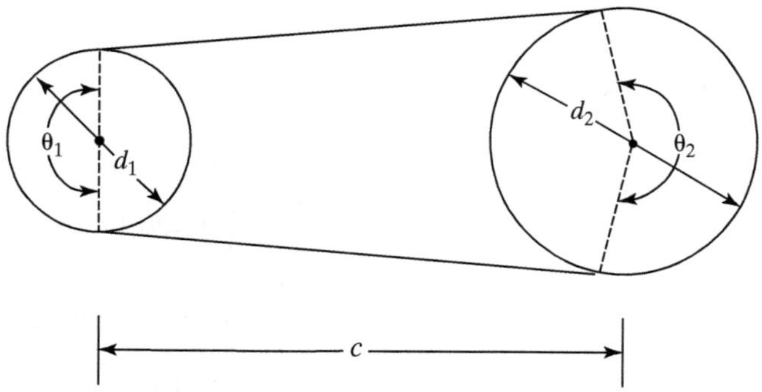

FIGURE 5.11 A belt drive.

where d_1 = diameter of the smaller pulley, d_2 = diameter of the larger pulley, and c = center distance. If d_1, d_2, and c are random with $\overline{d}_1 = 10$ cm, $\sigma_{d_1} = 0.1$ cm, $\overline{d}_2 = 20$ cm, $\sigma_{d_2} = 0.25$ cm, $\overline{c} = 100$ cm, and $\sigma_c = 0.5$ cm, determine the mean and standard deviation of the length of belt.

5.13 The maximum contact pressure developed (p) between two cylinders under a load F can be expressed as

$$p = \sqrt{\frac{FE\,(d_1 + d_2)}{\pi l\,(1 - v^2)\,d_1\,d_2}}$$

where d_1, d_2 = diameters of cylinders, v = Poisson's ratio, E = Young's modulus, and l = axial length of contact between the cylinders. If $(\overline{F}, \sigma_F) = (10{,}000, 2000)$ N, $(\overline{l}, \sigma_l) = (0.2, 0.02)$ m, $(\overline{E}, \sigma_E) = (2.06, 0.103) \times 10^6 \text{N/m}^2$, $(v, \sigma_v) = (0.3, 0.015)$, $(\overline{d}_1, \sigma_{d_1}) = (0.1, 0.01)$ m and $(\overline{d}_2, \sigma_{d_2}) = (0.25, 0.025)$ m, determine the first-order mean and standard deviations of the maximum contact pressure, p.

5.14 The actuating force (F) necessary to apply the brake shown in Fig. 5.12 is given by

$$F = \frac{M_n - M_f}{a}$$

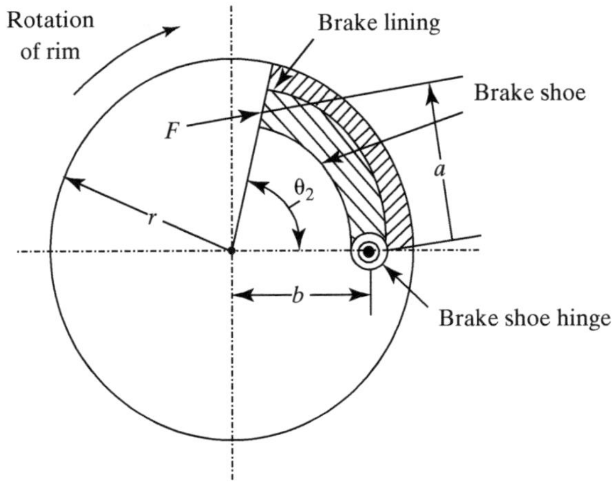

FIGURE 5.12 A brake.

where

$$M_n = p_a trb\left(\frac{\theta_2}{2} - \frac{1}{4}\sin 2\,\theta_2\right)$$

$$M_f = p_a trf\left(r - r\cos\,\theta_2 - \frac{b}{2}\sin^2\theta_2\right)$$

with p_a = maximum pressure exerted on the brake lining, t = face width of the brake shoes, r = inner radius of the brake drum, b = radius of the brake shoe, f = coefficient of friction, a = linear distance between the brake shoe hinge and the line of application of the actuating force, and θ_2 = angle between the brake shoe hinge and the point of application of the actuating force (see Fig. 5.12). Find the mean value and standard deviation of F when $(\bar{p}_a, \sigma_{p_a}) = (10^5, 10^4)$ Pa, $(\bar{t}, \sigma_t) = (30, 1)$ mm, $(\bar{r}, \sigma_r) = (150, 10)$ mm, $(\bar{b}, \sigma_b) = (120, 4)$ mm, $(\bar{f}, \sigma_f) = (0.35, 0.035)$, $(\bar{a}, \sigma_a) = (200, 10)$mm, and $(\bar{\theta}_2, \sigma_{\theta_2}) = (120°, 3°)$.

5.15 An electrical network consists of three resistors in series as shown in Fig. 5.13. If the resistances are independent normal variates with $R_1 = N(50, 2)$ ohms, $R_2 = N(100, 5)$ ohms, and $R_3 = N(150, 10)$ ohms, find the probability density function of the total resistance of the network.

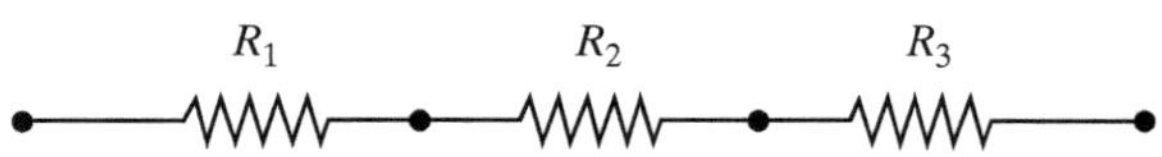

FIGURE 5.13 Three resistors in series.

5.16 A gear is to be mounted on a shaft using an interference fit as shown in Fig. 5.14. If the diameters of the shaft and the hole are given by $D = N(2, 0.05)$ inch and $d = N(1.95, 0.05)$ inch, determine the probability density function of the interference between the two parts.

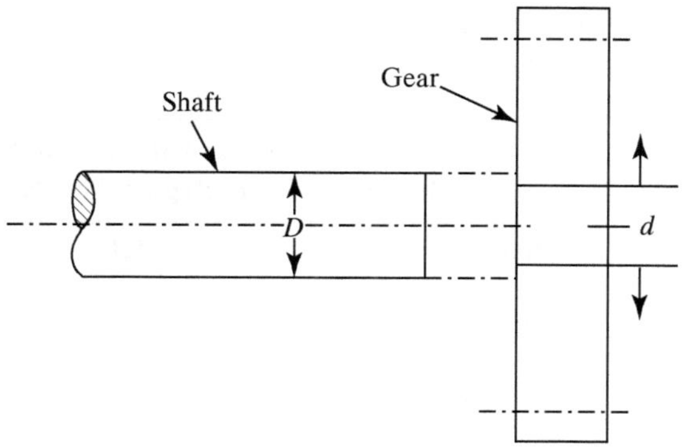

FIGURE 5.14 A gear mounted on a shaft.

5.17 A method of determining the probability density function of the function $Y = g(X_1, X_2)$ in terms of the joint density function, $f_{X_1, X_2}(x_1, x_2)$, can be summarized [5.1] as follows:

1. Introduce a new random variable Z as $Z = h(X_1, X_2)$ where the function $h(X_1, X_2)$ is suitably selected so that X_1 and X_2 can be expressed in terms of Y and Z as $X_1 = p(Y, Z)$ and $X_2 = q(Y, Z)$.
2. Find the joint density function of Y and Z as

$$f_{Y,Z}(y, z) = f_{X_1, X_2}\left[p(y, z), q(y, z)\right] \left|J(y, z)\right|$$

where $\left|J(y, z)\right|$ is the Jacobian of the transformation defined as

$$J(y, z) = \begin{vmatrix} \dfrac{\partial p}{\partial y} & \dfrac{\partial p}{\partial z} \\ \dfrac{\partial q}{\partial y} & \dfrac{\partial q}{\partial z} \end{vmatrix}$$

3. Determine the probability density function of Y as

$$f_Y(y) = \int_{-\infty}^{\infty} f_{Y,Z}(y, z)\, dz$$

Using this method, find the probability density function of $Y = X_1/X_2$ when the joint density function of X_1 and X_2 is given by

$$f_{X_1, X_2}(x_1, x_2) = \begin{cases} 25\, e^{-5(x_1 + x_2)}; & x_1 \geq 0, x_2 \geq 0 \\ 0; & \text{otherwise} \end{cases}$$

Hint: Define Z as $Z = X_1 + X_2$.

5.18 An automobile repair shop performs different types of repair. The arrival rates of customers with different types of repair requests follow Poisson distribution with the following data:

Type of Repair	Mean Value of Number of Repairs Per Day
Brakes	5.1
Radiator	3.2
Air conditioning	2.3
Transmission	1.6
Fuel pump	4.5

Find the probability distribution function of the total number of repairs per day.

5.19 Two links, of lengths X_1 and X_2, are connected as shown in Fig. 5.15. If X_1 and X_2 follow uniform distributions over the ranges 0 to 1 and 0 to 2, respectively, determine the probability density function of the total length $X_1 + X_2$. Also find the probability of realizing $X_1 + X_2$ greater than or equal to 2.5.

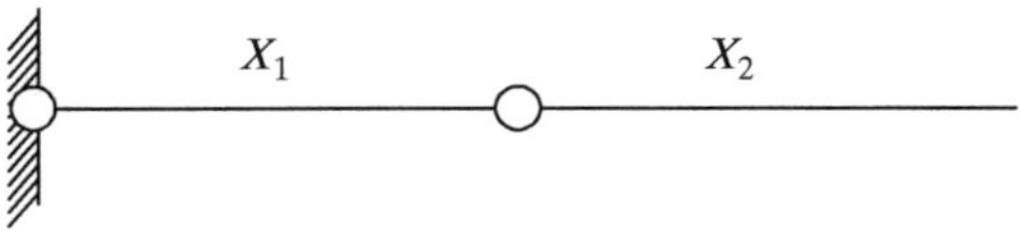

FIGURE 5.15 Two links connected in series.

5.20 The heat transfer rate q, in BTU/h, through a uniform one-dimensional fin with an insulated end (Fig. 5.16) is given by

$$q = \sqrt{hPkA}\,(T_s - T_\infty)\tanh mL$$

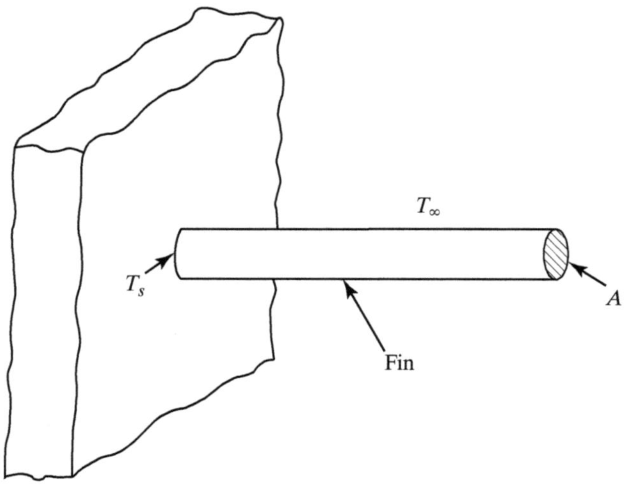

FIGURE 5.16 A uniform fin.

where h = heat transfer coefficient (BTU/h-ft^2-deg F), P = perimeter of the fin cross section (ft), k = thermal conductivity of the fin (BTU/h-ft-deg F), A = area of cross section of the fin (ft^2), T_s = temperature at the root (deg F), T_∞ = atmospheric temperature (deg F), $m = \sqrt{hP/kA}$ ft^{-1}, and L = length of the fin (ft). For a steel circular fin with diameter $(\overline{D}, \sigma_D) = (1, 0.1)$ in; $(\overline{h}, \sigma_h) = (20, 2)$ BTU/h-ft^2-deg F, $(\overline{k}, \sigma_k) = (40, 2)$ BTU/h-ft-degF, $(\overline{T}_S, \sigma_{T_S}) = (200, 5)$ deg F, $(\overline{T}_\infty, \sigma_{T_\infty}) = (70, 3)$ deg F, $(\overline{L}, \sigma_L) = (40, 2)$ in, determine the mean value and standard deviation of q.

5.21 The pressure drop in the fluid flowing through a circular pipe (Δp) is given by

$$\Delta p = f \frac{L}{D} \frac{V^2}{2} \rho$$

where f = friction factor, L = length of pipe, D = diameter of pipe, V = velocity of fluid, and ρ = density of fluid. If $(\overline{f}, \sigma_f) = (0.02, 0.004)$, $(\overline{L}, \sigma_L) = (2000, 20)$ m, $(\bar{\rho}, \sigma_\rho) = (1000, 10)$ kg/m^3, $(\overline{D}, \sigma_D) = (0.1, 0.001)$ m, and $(\overline{V}, \sigma_V) = (10, 1)$ m/s, find the mean value and standard deviation of Δp.

5.22 The random motion of electrons in a resistor generates small noise terminal voltage, V_n, known as Johnson or white noise, and is given by [5.4]

$$V_n = \sqrt{4kTR\,\Delta f} \text{ volts}$$

where k = Boltzmann constant = 1.38×10^{-23} J/K, T = absolute temperature (K), R = resistance (ohms), and Δf = bandwidth (Hz). If $(\overline{R}, \sigma_R) = (1, 0.01)$ MΩ, $(\overline{T}, \sigma_T) = (27, 1)$°C, and $(\overline{\Delta f}, \sigma_{\Delta f}) = (10, 0.1)$ kHz, determine the mean value and standard deviation of the noise voltage generated.

5.23 The sound pressure level, L_p, in decibels, due to an air jet nozzle is given by [5.7]

$$L_p = 168.6 + 20 \log\left(\frac{d}{r}\right) + 40 \log\left(1 - P^{-0.2857}\right)$$

where d = nozzle diameter, r = distance from the nozzle where sound pressure is measured, P = pressure ratio = $p_i/p_a = 1 + p_i'/p_a$, p_i' = inlet gauge pressure, and p_a = atmospheric pressure. If $(\overline{d}, \sigma_d) = (0.25, 0.025)$ in, $(\bar{r}, \sigma_r) = (36, 4)$ in, $(\overline{p}_i', \sigma_{p_i'}) = (60, 1)$ lb/in^2, and $(\overline{p}_a, \sigma_{p_a}) = (14.696, 0.1)$ lb/in^2, find the mean and standard deviation of the sound pressure level.

5.24 Find the mean and variance of $Z = X_1 X_2$ using the bivariate normal moment-generating function of X_1 and X_2 (similar to Eq. (5.75)):

$$m(t_1, t_2) = \exp\left\{t_1\, \mu_{X_1} + t_2\, \mu_{X_2} + \frac{1}{2}\left(t_1^2 \sigma_{X_1}^2 + 2\,\rho\, t_1 t_2 \sigma_{X_1} \sigma_{X_2} + t_2^2 \sigma_{X_2}^2\right)\right\}$$

Neglect correlation between X_1 and X_2.

5.25 Derive an expression for the probability density function of $Y = X_1 + X_2 + \ldots + X_n$ where $X_i, i = 1, 2, \ldots, n$ are independent normally distributed random variables. Also find the mean and standard deviation of Y.

5.26 Given $f_X(x) = \dfrac{1}{\beta} \exp\left(-\dfrac{1}{\beta} x\right)$, find the expression of $f_Y(y)$ where $Y = X^2$.

5.27 Given, $f_X(x) = e^{-x}$, find the expression of $f_Y(y)$ where $Y = X^{\frac{1}{4}}$.

5.28 If X follows uniform distribution in the interval 0 to 1, find the probability density function of $Y = X^2$.

5.29 Using the moment-generating function given in Problem 5.24, find the mean and standard deviation of $Z = X_1 X_2$ by including the correlation between X_1 and X_2.

5.30 Derive the moment-generating function of a random variable X which follows the distribution

$$f_X(x) = \begin{cases} \lambda e^{-\lambda x}; & x \geq 0 \\ 0; & \text{otherwise} \end{cases}$$

Also find the mean and variance of X using the moment-generating function.

5.31 The moment-generating function for a discrete random variable X is defined as

$$M(\theta) = \sum_{\text{summation on all } x \text{ for which } p_X(x) > 0} e^{\theta x} p_X(x)$$

where $p_X(x)$ is the probability mass function of X. Using this, find the moment-generating function of X following binomial distribution. Also find the mean and standard deviation of X using the moment-generating function.

5.32 Show that the product and quotient of two statistically independent random variables following lognormal distributions also follow lognormal distributions.

MATLAB Problems

5.33 The data on the number of occurrences of hurricanes (X) and the number of days on which the maximum temperature exceeds 100°F (Y) in each year have been collected over a period of ten years at a particular location on the East Coast of the United States. The data are shown below:

Year	**1**	**2**	**3**	**4**	**5**	**6**	**7**	**8**	**9**	**10**
Number of hurricanes	2	1	0	2	0	1	0	0	2	1
Number of days with temp. > 100°F	24	19	21	18	14	15	17	20	21	23

Find the covariance and correlation coefficient matrices between the random variables X and Y using MATLAB.

5.34 Experiments are conducted on eight different batches of a particular grade of steel to find the yield strength (S_y) and the ultimate tensile strength (S_{ut}). The experimental results obtained on the eight batches of steel are shown below:

Variable \ **Batch of Steel**	**1**	**2**	**3**	**4**	**5**	**6**	**7**	**8**
S_y (MPa)	181	173	167	171	174	187	191	199
S_{ut} (MPa)	290	323	228	323	314	312	324	337

Using MATLAB, find the following quantities:

a. Covariance and correlation matrices between the two random variables.

b. Assuming that S_y and S_{ut} are jointly distributed normal random variables with the mean values and covariances given by the experimental results, determine the probability of simultaneously realizing the value of S_y less than 175 MPa and S_{ut} less than 300 MPa.

5.35 An experiment is conducted at a tollbooth on a North–South highway for nearly 4 hours to find the frequencies of different number of Northbound and Southbound automobiles crossing the tollbooth in different 16-second intervals. The results of the experiment are shown in the following table. In this table, for example, the number in the third row and fourth column (22) denotes the number of 16-second intervals in which 2 Southbound and 3 Northbound automobiles were observed to cross the tollbooth.

Number of Northbound Automobiles	**0**	**1**	**2**	**3**	**4**	**5**
Number of Southbound Automobiles						
0	1	6	13	43	51	32
1	2	8	16	38	59	41
2	12	41	42	22	18	7
3	4	32	19	20	12	14
4	51	63	30	15	6	8
5	17	21	41	55	15	10

Find the covariance and correlation matrices between the random variables using MATLAB.

CHAPTER 6

Time-Dependent Reliability of Components and Systems

What You Will Learn

In this chapter, you will learn

- That a definite pattern exists for the failure rate of most mechanical and electrical components and systems
- The concepts of infant mortality, operating period and wear out periods in the failure-rate or bath tub curve
- The estimation of failure rate from empirical data
- The concept of hazard function and its relation to reliability and probability of failure
- The definition of mean time to failure (MTTF) and its relationship to reliability
- The reliability and hazard functions when the failure rate follows exponential, normal, lognormal, Weibull, gamma, Rayleigh, and uniform distributions
- The computation of the expected residual (remaining) life of components and systems
- The computation of the reliability of series, parallel, and series–parallel models of multicomponent systems
- The computation of reliability of complex multicomponent systems using enumeration, conditional probability, and cut-set methods
- The reliability enhancement of series and parallel systems

Karl Friedrich Gauss was born in Brunswick, Germany, on 30 April, 1777. He is considered to be one of the three greatest mathematicians that history has produced, the other two being Archimedes and Newton. He was born to a poor family; his father was a bricklayer and a gardener. It is said that Gauss corrected an error in calculating his father's wages at the age of three. His doctoral thesis, in mathematics, dealt with the proof of the "Fundamental Theorem of Algebra," which states that every polynomial equation of degree one or higher has a root. He made many contributions to mathematics, astronomy, and physics. The numbers of the form $x + iy$, where x and y are integers and $i = \sqrt{-1}$, are known as Gaussian integers. The unit of magnetic flux density is named gauss in honor of his contributions to physics. His work on the theory of errors led him to discover the "method of least squares" and later the named or Gaussian distribution curve. Although the normal curve was originally studied by De Moivre, Gauss was the first one to justify it by means of the theory of probability. Gauss was made the director of Göttingen Observatory in 1807. He used to get involved with his work so deeply that even when a servant rushed to tell him that his wife was dying, he replied, "ask her to wait until I finish the work." Gauss died in Göttingen on 23 February, 1855.

6.1 Introduction

As stated earlier, reliability is the probability of successful performance of a system at any time. The system may be an electric fan which is expected to start and stop, and run perhaps for years without failure, or a missile which is expected to be in standby condition for long periods of time and then respond just once without failure. The successful performance of any system depends on the extent to which reliability is designed and built into it. In practice, it is observed that even seemingly identical systems, operating under similar conditions, fail at different times. A very simple example is a lightbulb. Different lightbulbs, even taken from the same batch of production and put into operation under identical conditions, will have different lives. Thus, the failure of such systems can only be described probabilistically. Since most engineering systems are composed of several components, the relationship between component reliability and system reliability must be understood. This chapter deals with the reliability analysis of components and systems, including the problem of reliability allocation in multicomponent systems.

6.2 Failure Rate versus Time Curve

Experience with a large number of mechanical and electronic systems has shown that their failure characteristics follow definite patterns. A plot of failure rate versus time of a typical component such as an electric motor or a transistor is shown in Fig. 6.1. The curve shown in Fig. 6.1 is also called the "bathtub curve" due to its shape. Surprisingly, the life of human beings also was found to follow a similar curve. In Fig. 6.1, the initial failures during the time 0 to t_1 are caused due to reasons such as defective materials, poor design, bad workmanship, poor quality control, or poor transportation. This period is called the *infant mortality* or *burn-in* period and is characterized by a rapidly declining failure rate and is generally covered by the warranty or guarantee period given by the manufacturer. As the defective components are replaced by newer ones, the reliability improves. The next phase of failures occur in the *normal* (*operating*) period t_1 to t_2. These failures, also

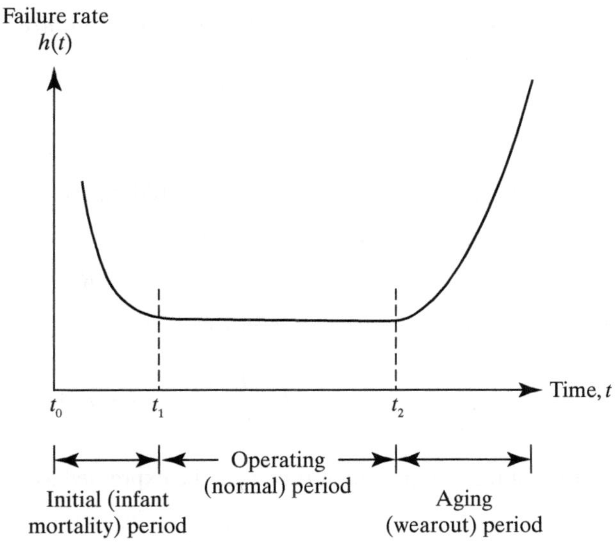

FIGURE 6.1 Bathtub curve.

known as *random failures*, occur due to random overloads and other extreme operating conditions. The operating period is characterized by a relatively constant failure rate. After time t_2, *aging* or *wearout failures* occur due to excessive wear after the expected useful design life has been exceeded. The wearout period is characterized by a rapidly increasing rate of failure. From a design point of view, the operating period will be of interest.

6.3 Reliability and Hazard Functions

Consider a fixed number of identical components tested simultaneously under similar operating conditions. Let N = number of identical components tested, $N_s(t)$ = number of components surviving at time t, and $N_f(t)$ = number of components failed in time t. As the test proceeds, the number of surviving components gets smaller and the number of failed components gets larger. Then the reliability of the component at time t can be defined as

$$R(t) = \frac{N_s(t)}{N} = \frac{N - N_f(t)}{N} = 1 - \frac{N_f(t)}{N} \tag{6.1}$$

Since N is fixed, Eq. (6.1) gives

$$\frac{dR(t)}{dt} = -\frac{1}{N}\frac{dN_f(t)}{dt} \tag{6.2}$$

The rate at which components fail can be found as

$$\frac{dN_f(t)}{dt} = -N\frac{dR(t)}{dt} \tag{6.3}$$

The *failure rate*, also known as the *instantaneous rate of failure, hazard function*, or *hazard rate*, $h(t)$, is defined as

$$h(t) = \frac{1}{N_s(t)}\frac{dN_f(t)}{dt} \tag{6.4}$$

Dividing both sides of Eq. (6.3) by $N_s(t)$, we find the failure rate as

$$h(t) = \frac{1}{N_s(t)}\frac{dN_f(t)}{dt} = -\frac{N}{N_s}\frac{dR(t)}{dt} = -\frac{1}{R(t)}\frac{dR(t)}{dt} \tag{6.5}$$

Integration of Eq. (6.5) gives, using the relation $R(0) = 1$,

$$\int_0^t h(x)\,dx = -\int_0^t \frac{dR(x)}{R(x)} = -\ln R(t) \tag{6.6}$$

Thus, the reliability of the component at time t can be expressed as

$$R(t) = \exp\left[-\int_0^t h(x)\,dx\right] \tag{6.7}$$

The failure rate can also be expressed in terms of the probability distribution function of life of the component as follows. The distribution function of life or time to failure T of the component is given by

$$F_T(t) = \int_0^t f_T(x)\, dx \tag{6.8}$$

and the reliability of the component by

$$R(t) = P(T > t) = 1 - P(T \leq t) = 1 - F_T(t) = \exp\left[-\int_0^t h(x)\, dx\right] \tag{6.9}$$

This gives

$$\ln\left[1 - F_T(t)\right] = -\int_0^t h(x)\, dx \tag{6.10}$$

which, upon differentiation, yields

$$-\frac{f_T(t)}{1 - F_T(t)} = -h(t)$$

that is,

$$h(t) = \frac{f_T(t)}{R(t)} = \frac{f_T(t)}{1 - F_T(t)} \tag{6.11}$$

EXAMPLE 6.1

Derive expressions for the failure rate and reliability function starting from the definition of the failure time density in the case of the experimental observation of N identical components placed into service at time $t = 0$.

Solution: The failure time density function, $f_T(t)$, is defined by the relation:

$$f_T(t)\, dt = \text{fractional number of failures in the time interval } (t, t + dt) = \frac{N_f(t)}{N\, dt}$$

The failure or hazard rate, $h(t)$, is the instantaneous rate of failure:

$$h(t) = \lim_{\Delta t \to 0} \frac{R(t) - R(t + \Delta t)}{R(t)\Delta t} = \frac{N_f(t)}{N_s(t)\Delta t}$$

The reliability function, $R(t)$, is given by

$$R(t) = \frac{N_s(t)}{N} = \frac{f_T(t)}{h(t)}$$

Since

$$R(t) = 1 - F(t), \quad \frac{dR(t)}{dt} = -\frac{dF(t)}{dt} = -f_T(t)$$

the probability of failure of a component or system in the time interval $[t_1, t_2]$ can be expressed in terms of the reliability function as

$$\int_{t_1}^{t_2} f_T(t)\, dt = -\int_{t_1}^{t_2} \frac{dR(t)}{dt}\, dt = -\int_{t_1}^{t_2} dR(t) = R(t_1) - R(t_2)$$

The failure or hazard rate of a component in the time interval $[t_1, t_2]$ is given by the probability of failure per unit time at time t_1 given that the component has not failed until the beginning of the interval, t_1:

$$\text{Failure rate at } t_1 = h(t_1) = \frac{R(t_1) - R(t_2)}{R(t_1)\,(t_2 - t_1)}$$

■

6.4 Modeling of Failure Rates

First, we consider the portion of the failure rate curve over the component life between times t_1 and t_2. During this time, chance or random failures occur and failure cannot be attributed to deterioration of the strength of the components over time. The failure rate during this period can be assumed as a constant (λ)

$$h(t) = \lambda \text{ failures per unit time} \tag{6.12}$$

From Eq. (6.11), we have[1]

$$f(t) = h(t)[1 - F(t)] = h(t) \exp\left[-\int_0^t h(\tau)d\tau\right] = \lambda e^{-\lambda} \tag{6.13}$$

and

$$R(t) = \exp\left[-\int_0^t h(\tau)d\tau\right] = e^{-\lambda t} \tag{6.14}$$

Equation (6.13) denotes an exponential density function.

Next, we consider the wearout portion of the failure rate curve (after time t_2). In general, this is a nonlinear curve. If a linear variation is assumed, for simplicity, we have

$$h(t) = c_1 t + c_2 \tag{6.15}$$

[1]The notations $f(t)$ and $F(t)$ are used in place of $f_T(t)$ and $F_T(t)$ for simplicity, to denote the probability density and distribution functions of T.

If λ and λ_{max} denote the failure rates at $t = t_2$ and $t = t_3$, respectively, we have

$$c_1 = \frac{\lambda_{max} - \lambda}{t_3 - t_2} \tag{6.16}$$

and

$$c_2 = \frac{\lambda\, t_3 - \lambda_{max}\, t_2}{t_3 - t_2} \tag{6.17}$$

The probability density function of the failure time can be determined, using Eqs. (6.11) and (6.9), as

$$f(t) = h(t)\exp\left[-\int_0^t h(\tau)d\tau\right] = (c_1 t + c_2)\exp\left[-\frac{1}{2}c_1 t^2 - c_2 t\right] \tag{6.18}$$

and

$$R(t) = \frac{f(t)}{h(t)} = \exp\left[-\frac{1}{2}c_1 t^2 - c_2 t\right] \tag{6.19}$$

The density function given by Eq. (6.18) degenerates to Rayleigh density function if $c_2 = 0$.

Similarly, the early failure portion of the failure rate function, corresponding to failures due to manufacturing defects, can also be approximated, for simplicity, by a linear relation as

$$h(t) = c_1 t + c_2 \tag{6.20}$$

where c_1 and c_2 are constants. If λ_0 and λ denote the failure rates at times t_0 and t_1, respectively, the constants c_1 and c_2 can be evaluated as

$$c_1 = \frac{\lambda_0 - \lambda}{t_0 - t_1} \tag{6.21}$$

and

$$c_2 = \frac{\lambda\, t_0 - \lambda_0 t_1}{t_0 - t_1} \tag{6.22}$$

The probability density and reliability functions of the failure time can be expressed as in Eqs. (6.18) and (6.19) with c_1 and c_2 determined from Eqs. (6.21) and (6.22). The representation of failure time using different probability distributions and the resulting reliability and hazard functions are considered in Section 6.7.

6.5 Estimation of Failure Rate from Empirical Data

When the failure time data of a component or a system are available, we can find the reliability and failure rate functions numerically. For this, first the time interval or the duration of the test is divided into segments of length Δt each; then the statistical definition is used to estimate the reliability function at time t as

$$R(t) = \frac{N_s(t)}{N} \tag{6.23}$$

where the initial value of time t is taken as zero. From the known number of components that failed in the interval $(t, t + \Delta t)$, the failure rate can be computed as

$$h(t) = \frac{N_s(t) - N_s(t + \Delta t)}{N_s(t)\,\Delta t} \tag{6.24}$$

The following example is considered to illustrate the procedure.

EXAMPLE 6.2

The experimental data obtained with a set of 10,000 identical components is shown in columns 1 and 2 of Table 6.1. Plot the corresponding failure rate versus time curve.

TABLE 6.1 Failure data of Example 6.2

Time of Observation t (Hours)	Number of Components Operating at Time t	Reliability $R(t) = \frac{N_s(t)}{N}$	Rate of Change of Reliability $-\frac{dR(t)}{dt} = -\frac{\Delta R}{\Delta t}$	Failure Rate $h(t) = -\frac{dR(t)}{dt}\frac{1}{R(t)}$
0	N = 10,000	1.0	—	—
50	8,880	0.8880	0.002240	0.002240
100	8,300	0.8300	0.001160	0.001306
150	7,918	0.7918	0.000764	0.000920
200	7,585	0.7585	0.000666	0.000842
250	7,274	0.7274	0.000622	0.000820
300	6,968	0.6968	0.000612	0.000842
350	6,668	0.6668	0.000600	0.000862
400	6,375	0.6375	0.000586	0.000878
450	6,088	0.6088	0.000574	0.000900
500	5,808	0.5808	0.000560	0.000920
550	5,535	0.5535	0.000546	0.000940
600	5,269	0.5269	0.000532	0.000962
650	5,011	0.5011	0.000516	0.000980
700	4,760	0.4760	0.000502	0.001002
750	4,517	0.4517	0.000486	0.001022
800	4,237	0.4237	0.000560	0.001240
850	3,864	0.3864	0.000746	0.001760
900	3,396	0.3396	0.000936	0.002422
950	2,819	0.2819	0.001154	0.003398
1000	2,219	0.2219	0.001200	0.004256

Solution: The computational details are shown in columns 3, 4, and 5 of Table 6.1. The failure rate (entries in the last column of Table 6.1) versus time (entries in the first column of Table 6.1) curve is shown in Fig. 6.2.

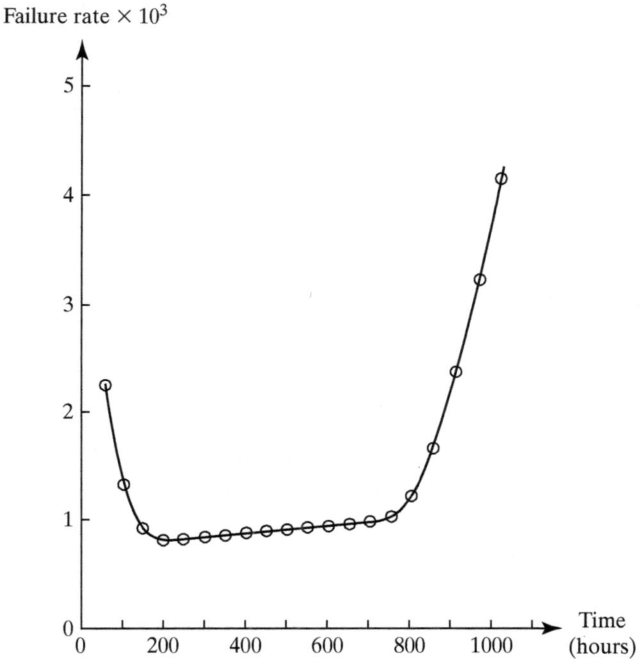

FIGURE 6.2 Failure rate–time curve.

■

6.6 Mean Time to Failure (MTTF)

The mean time to failure (MTTF) is used to denote the expected time to failure of a nonrepairable system. Sometimes, the mean time to failure is also called mean time before failure (MTBF). However, this term is not commonly used because the abbreviation MTBF is frequently used for the mean time between failures.

When the times to failure of n identical nonrepairable components or systems are observed to be $t_1, t_2, \ldots, t_n$, the mean time to failure (MTTF) is given by

$$\text{MTTF} = \frac{1}{n} \sum_{i=1}^{n} t_i \tag{6.25}$$

As the sample size n is large in practice, the time to failure (or equivalently, the life), T, is considered a random variable with a probability density function, $f_T(t)$. The expected or mean value of time to failure or the expected life of the component or system is given by

$$\text{MTTF} = \mu_T = \int_0^\infty t f_T(t)\, dt \tag{6.26}$$

which can be rewritten, using the relation $f_T(t) = -\dfrac{dR(t)}{dt}$, as

$$\text{MTTF} = -\int_{t=0}^{\infty} t\frac{dR}{dt}\,dt = -t\,R(t)\,|_0^{\infty} + \int_{t=0}^{\infty} R(t)\,dt \tag{6.27}$$

The term $t\,R(t)$, on the right-hand side of Eq. (6.27), will be zero at $t = 0$. Since all components/systems fail after a finite time, we have $R(t) \to 0$ as $t \to \infty$. Thus, Eq. (6.27) gives

$$\text{MTTF} = \mu_T = \int_{t=0}^{\infty} R(t)\,dt \tag{6.28}$$

EXAMPLE 6.3

The reliability of a forging press is given by

$$R(t) = \begin{cases} \left(1 - \dfrac{t}{T_{\max}}\right)^3; & 0 \le t \le T_{\max} \\ 0; & t > T_{\max} \end{cases} \tag{E1}$$

Determine the following:

1. Probability density function of the failure time (T).
2. Probability distribution function of the failure time (T).
3. Failure rate function, $h(t)$.
4. MTTF of the forging press.

Solution:

1. The probability density function of the failure time or life of the forging press (T) is given by

$$f_T(t) = -\frac{dR(t)}{dt} = -\left[3\left(1 - \frac{t}{T_{\max}}\right)^2\left(-\frac{1}{T_{\max}}\right)\right] = \frac{3}{T_{\max}}\left(1 - \frac{t}{T_{\max}}\right)^2 \tag{E2}$$

2. The probability distribution function of T can be found as

$$F_T(t) = 1 - R(t) = \begin{cases} 1 - \left(1 - \dfrac{t}{T_{\max}}\right)^3; & 0 \le t \le T_{\max} \\ 0; & t > T_{\max} \end{cases} \tag{E3}$$

3. The hazard rate function is given by

$$h(t) = \frac{f_T(t)}{R(t)} = \frac{\dfrac{3}{T_{\max}}\left(1 - \dfrac{t}{T_{\max}}\right)^2}{\left(1 - \dfrac{t}{T_{\max}}\right)^3} = \frac{3}{T_{\max} - t} \tag{E4}$$

4. The mean time to failure of the forging press can be computed as

$$\text{MTTF} = \int_0^{\infty} R(t)\,dt = \int_0^{T_{\max}} \left(1 - \frac{t}{T_{\max}}\right)^3 dt \tag{E5}$$

By denoting $y = 1 - \frac{t}{T_{max}}$ so that $dy = -\frac{dt}{T_{max}}$, Eq. (E5) can be written as

$$\text{MTTF} = -T_{max} \int_1^0 y^3\, dy = \frac{T_{max}}{4} \tag{E6}$$

■

6.7 Reliability and Hazard Functions for Different Distributions

The variations of the reliability and hazard functions with time depend on the probability distributions assumed for the failure time. For example, the hazard function may remain constant, decrease or increase with time depending on the probability distribution of the failure time as shown in the following subsections.

6.7.1 Exponential Distribution

If the failure rate of a component is constant with respect to time, the probability distribution function of the failure time (or life) of the component, T, follows exponential distribution. In exponential distribution, the probability distribution function of failure time, T, is given by

$$F_T(t) = 1 - e^{-\lambda t} \tag{6.29}$$

where λ is a constant denoting the failure rate. Exponential distribution can also be interpreted to represent a Poisson process in which the time T before the first occurrence of the event (failure) can be represented as

$$P(T > t) = \frac{e^{-\lambda t}(\lambda t)^0}{0!} = e^{-\lambda t} \tag{6.30}$$

Note that in Eq. (6.30), λt denotes the average number of occurrences of the event (failure) in time t with λ denoting the failure rate or the number of failures per unit time. Equation (6.30) gives the probability distribution function of T as

$$F_T(t) = P(T \le t) = 1 - P(T > t) = 1 - e^{-\lambda t} \tag{6.31}$$

Note that Eqs. (6.29) and (6.31) are identical. In this case, the probability density function of failure time is given by

$$f_T(t) = \frac{dF_T(t)}{dt} = \lambda e^{-\lambda t} \tag{6.32}$$

The reliability function can be written as

$$R(t) = P(T > t) = 1 - F_T(t) = e^{-\lambda t} \tag{6.33}$$

The failure rate or hazard rate function can be expressed as

$$h(t) = \frac{f_T(t)}{R(t)} = \frac{\lambda e^{-\lambda t}}{e^{-\lambda t}} = \lambda \tag{6.34}$$

The variations of the functions $f_T(t)$, $F_T(t)$, $R(t)$, and $h(t)$ with time t are shown in Figs. 6.3 (*a*)–(*d*), respectively.

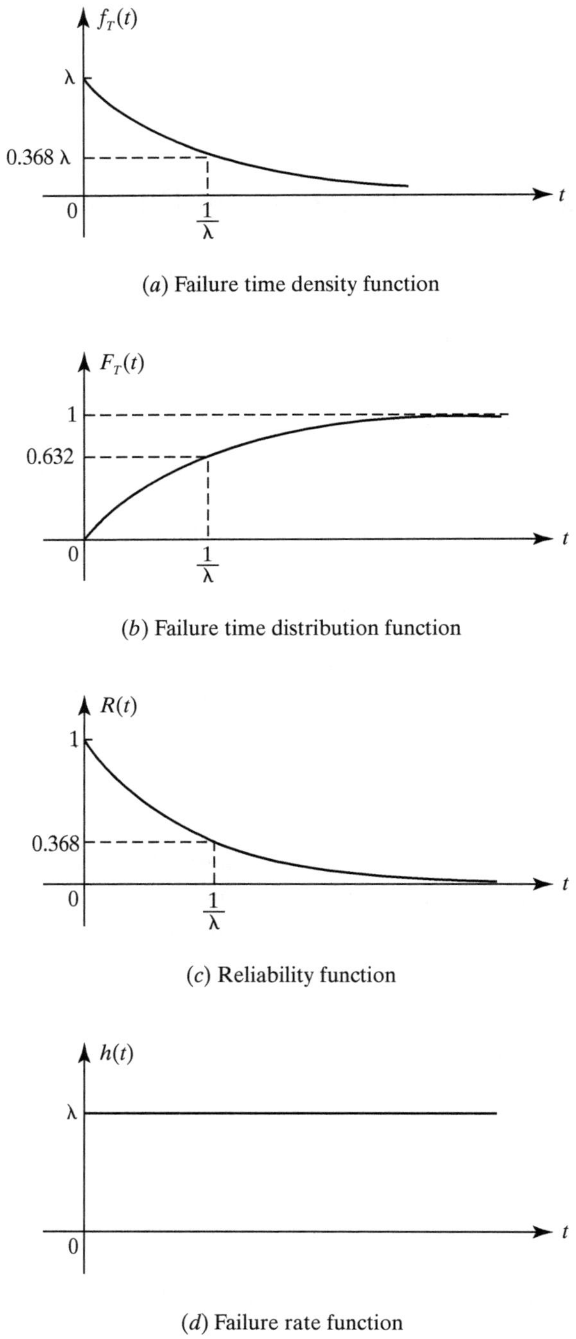

FIGURE 6.3 Exponential distribution.

EXAMPLE 6.4

Find the mean and standard deviation of a system whose failure time (T) follows the exponential distribution:

$$f_T(t) = \lambda e^{-\lambda t};\ t \geq 0, \lambda > 0$$

Solution: The reliability of the system is given by

$$R(t) = 1 - F_T(t) = e^{-\lambda t}$$

Thus, the mean time to failure of the system can be found from Eq. (6.28) as

$$\text{MTTF} = \mu_T = \int_0^\infty R(t)\, dt = \int_0^\infty e^{-\lambda t}\, dt = -\frac{1}{\lambda}\left(e^{-\lambda t}\right)_0^\infty = -\frac{1}{\lambda}\left(e^{-\infty} - e^0\right) = \frac{1}{\lambda}$$

The standard deviation of the failure time, T, can be determined as

$$V(T) = \sigma_T^2 = \int_{t=0}^\infty (t - \mu)^2 f_T(t)\, dt = \frac{1}{\lambda^2}$$

or

$$\sigma_T = \frac{1}{\lambda}$$

■

EXAMPLE 6.5

An electromechanical system has a constant failure rate of $\lambda = 0.01$/hour. Determine the following:

1. Probability that the system will fail within the first 25 hours of operation.
2. If the system has performed successfully for 200 hours, the probability that it will fail during the next 25 hours of operation.

Solution:

1. Since the failure rate is constant, the failure time, T, follows exponential distribution. The probability that the system will fail within the first 25 hours of operation is given by

$$P(T \leq 25) = F_T(25) = \int_0^{25} f_T(t)\, dt = 1 - e^{-\lambda t} = 1 - e^{-(0.01)(25)} = 1 - e^{-0.25} = 0.221199$$

2. If the system has survived for 200 hours, the probability that it will fail in the next 25 hours can be determined as

$$P(T \leq 225|_{T>200}) = \frac{P[(T \leq 225) \cap (T > 200)]}{P(T > 200)}$$

$$= \frac{P(200 \leq T \leq 225)}{1 - P(T \leq 200)} = \frac{\int_{200}^{225} f_T(t)\, dt}{1 - F_T(200)} = \frac{F_T(225) - F_T(200)}{1 - F_T(200)}$$

$$= \frac{(1 - e^{-0.01(225)}) - (1 - e^{-0.01(200)})}{1 - (1 - e^{-0.01(200)})} = \frac{e^{-2.00} - e^{-2.25}}{e^{-2.00}}$$

$$= 1 - \frac{e^{-2.25}}{e^{-2.00}} = 1 - e^{-0.25} = 0.221199$$

Note: This example shows the property that for any system following exponential failure time distribution, the probability that it will fail during some interval of time in the future is independent of its age.

■

6.7.2 Normal Distribution

For many components, the failure time can be assumed to follow normal distribution. If the failure time follows normal distribution, the probability distribution, reliability and hazard functions cannot be expressed in closed form. The probability density function of the failure time, T, is given by

$$f_T(t) = \frac{1}{\sqrt{2\,\pi}\,\sigma} \exp\left\{-\frac{(t-\mu)^2}{2\sigma^2}\right\} \tag{6.35}$$

where μ = MTTF is the mean value and σ is the standard deviation of the failure time (T). The probability distribution function of T can be expressed as

$$F_T(t) = \int_{\tau=-\infty}^{t} \frac{1}{\sqrt{2\,\pi}\,\sigma} \exp\left\{-\frac{(\tau-\mu)^2}{2\,\sigma^2}\right\} d\tau \tag{6.36}$$

By introducing the standard normal variate as $z = \dfrac{t-\mu}{\sigma}$, the probability distribution function can be expressed as

$$F_T(t) = F_T(z) = \Phi(z) \equiv \Phi\left(\frac{t-\mu}{\sigma}\right) \tag{6.37}$$

The reliability can be expressed as

$$R(t) = 1 - F_T(t) = 1 - \Phi\left(\frac{t-\mu}{\sigma}\right) \tag{6.38}$$

The failure rate function is given by

$$h(t) = \frac{f_T(t)}{R(t)} = \frac{\dfrac{1}{\sqrt{2\,\pi}\,\sigma} \exp\left\{-\dfrac{1}{2}\left(\dfrac{t-\mu}{\sigma}\right)^2\right\}}{1 - \Phi\left(\dfrac{t-\mu}{\sigma}\right)} \tag{6.39}$$

The variations of the probability density, probability distribution, reliability, and hazard functions with time corresponding to normally distributed failure time are shown in Figs. 6.4(*a*)–(*d*), respectively.

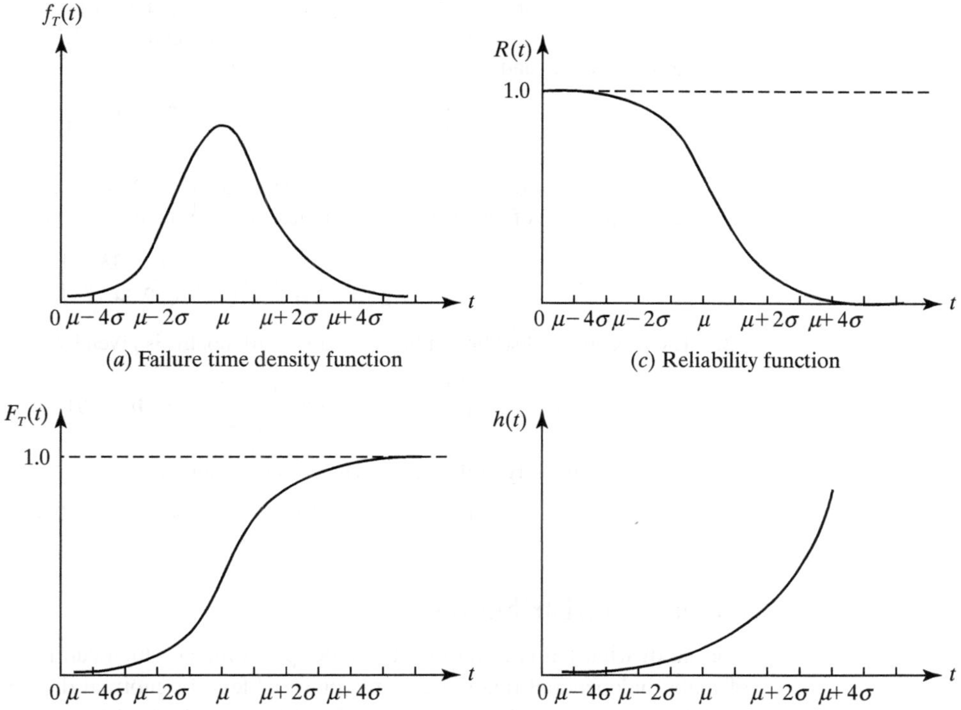

(*a*) Failure time density function (*c*) Reliability function (*b*) Failure time distribution function (*d*) Failure rate function

FIGURE 6.4 Normal distribution.

EXAMPLE 6.6

It is observed that 80 percent of the automobile batteries produced by a manufacturer fail between 30 and 36 months of operation. If the lives of the batteries are assumed to follow normal distribution, determine the following:

1. Probability density function of the lives of the batteries.
2. Reliability of batteries at the end of 40 months of operation.

Solution:

1. Let μ and σ denote, respectively, the mean and standard deviation of the lives of the batteries. Assuming that 10 percent of the batteries fail before 30 months and 10 percent after 36 months of operation, we have

$$\begin{aligned} z_1 &= \frac{30 - \mu}{\sigma}, \quad \Phi(z_1) = 0.1 \\ z_2 &= \frac{36 - \mu}{\sigma}; \quad \Phi(z_2) = 0.9 \end{aligned} \tag{E1}$$

From standard normal tables, we find the values of the standard normal variate corresponding to $\Phi(z_1) = 0.1$ and $\Phi(z_2) = 0.9$ as $z_1 = -1.282$ and $z_2 = 1.282$, respectively. Using the definitions of z_1 and z_2 from Eq. (E1), we obtain

$$z_1 = \frac{30 - \mu}{\sigma} = -1.282, \quad z_2 = \frac{36 - \mu}{\sigma} = 1.282 \tag{E2}$$

The solution of Eq. (E2) gives $\mu = 33$ months and $\sigma = 2.34$ months. Thus, the probability density function of the lives of the batteries (T) is given by

$$f_T(t) = \frac{1}{2.34\sqrt{2\,\pi}} \exp\left\{-\frac{1}{2}\left(\frac{t - 33}{2.34}\right)^2\right\} \tag{E3}$$

2. The probability distribution function at $t = 40$ months is given by

$$F_T(40) = \Phi\left(\frac{40 - \mu}{\sigma}\right) = \Phi\left(\frac{40 - 33}{2.34}\right) = \Phi(2.9914) = 0.9986 \tag{E4}$$

Thus, the reliability of the batteries at $t = 40$ months is

$$R(40) = 1 - F_T(40) = 1 - 0.9986 = 0.0014 \tag{E5}$$

■

6.7.3 Lognormal Distribution

Normal distribution is developed to model phenomena which can be considered to be sums of many independent random actions (or variables). Lognormal distribution is developed to model phenomena, which can be considered to be products of many independent random factors (or variables). An example of a multiplicative phenomenon is the fatigue damage in which the damage at any cycle of loading can be considered to be randomly proportional to the damage caused by the previous cycle of loading. Consider a random variable X given by a product of n independent random variables $Y_1, Y_2, \ldots, Y_n$, which can take only positive values as

$$X = Y_1\, Y_2 \ldots Y_n \tag{6.40}$$

By taking natural logarithms of both sides of Eq. (6.40), we obtain

$$\ln X = \ln Y_1 + \ln Y_2 + \cdots + \ln Y_n \tag{6.41}$$

Since Y_i, $i = 1, 2, \ldots, n$ are random variables, $\ln Y_i$, $i = 1, 2, \ldots, n$ are also random variables. Thus, $\ln X$ is given by the sum of n independent random variables. According to central limit theorem, $\ln X$ tends to follow normal distribution as n becomes large. By defining

$$Y = \ln X \tag{6.42}$$

the problem is to find the probability distribution of

$$X = e^Y \tag{6.43}$$

when Y follows normal distribution. The random variable X whose logarithm follows normal distribution is said to follow lognormal distribution. Expressing the normally distributed random variable by its density function

$$f_Y(y) = \frac{1}{\sqrt{2\,\pi}\,\sigma} \exp\left\{-\frac{1}{2}\left(\frac{y - \mu}{\sigma}\right)^2\right\}; \quad -\infty \leq y \leq \infty \tag{6.44}$$

the probability density function of the lognormally distributed random variable X can be written as

$$f_X(x) = \frac{1}{\sqrt{2}\,\pi\,\sigma\,x}\exp\left\{-\frac{1}{2}\left(\frac{\ln x - \mu}{\sigma}\right)^2\right\}; \quad x \geq 0 \tag{6.45}$$

where

$$\mu \equiv \mu_Y = \mu_{\ln X} = E[\ln X], \quad \sigma \equiv \sigma_Y = \sigma_{\ln X} = \sqrt{Var[\ln Y]} \tag{6.46}$$

The expected value and standard deviation of X can be determined as follows:

$$E(X) = E(e^Y) = \int_{-\infty}^{\infty} e^y f_Y(y)\, dy = \int_{-\infty}^{\infty} \frac{1}{\sqrt{2}\,\pi\,\sigma}\exp\left\{y - \frac{1}{2}\left(\frac{y-\mu}{\sigma}\right)^2\right\} dy \tag{6.47}$$

The exponent on the right-hand side can be rearranged to obtain

$$E(X) = \exp\left(\mu + \frac{\sigma^2}{2}\right)\int_{-\infty}^{\infty} \frac{1}{\sqrt{2}\,\pi\,\sigma}\exp\left[-\frac{1}{2\,\sigma^2}\{y - (\mu + \sigma^2)\}\right] dy \tag{6.48}$$

Thus, the mean of the lognormal distribution can be expressed as

$$E(X) = \exp\left(\mu + \frac{\sigma^2}{2}\right) \tag{6.49}$$

Using a similar procedure, we can find the expected value of X^2 as

$$E(X^2) = E(e^{2Y}) = \exp\{2(\mu + \sigma^2)\} \tag{6.50}$$

The standard deviation of X will be

$$\sigma_X^2 = Var(X) = E(X^2) - [E(X)]^2 = \left(e^{2\mu+\sigma^2}\right)\left(e^{\sigma^2} - 1\right) \tag{6.51}$$

When the failure time (T) follows lognormal distribution:

The probability density function of the failure time is given by

$$f_T(t) = \frac{1}{\sqrt{2}\,\pi\,\sigma\,t}\exp\left\{-\frac{1}{2}\left(\frac{\ln t - \mu}{\sigma}\right)^2\right\}; \quad t \geq 0 \tag{6.52}$$

where $\mu \equiv \mu_{\ln T}$ and $\sigma \equiv \sigma_{\ln T}$. The probability distribution function of T can be determined as

$$F_T(t) = \int_{\tau=0}^{t} \frac{1}{\sqrt{2}\,\pi\,\sigma\,\tau}\exp\left\{-\frac{1}{2}\left(\frac{\ln \tau - \mu}{\sigma}\right)^2\right\} d\tau \tag{6.53}$$

which can be expressed in terms of standard normal variate z as

$$F_T(t) = P(T \leq t) = P\left[z \leq \frac{\ln t - \mu}{\sigma}\right] \tag{6.54}$$

The reliability function is given by

$$R(t) = P(T > t) = 1 - F_T(t) = P\left[z > \frac{\ln t - \mu}{\sigma}\right] \tag{6.55}$$

The hazard function can be determined as

$$h(t) = \frac{f_T(t)}{R(t)} = \frac{\phi\left(\dfrac{\ln t - \mu}{\sigma}\right)}{\sigma\, t\, R(t)} \tag{6.56}$$

where ϕ is the standard normal probability density function given by

$$\phi(z) = \frac{1}{\sqrt{2\pi}} \exp\left(-\frac{1}{2} z^2\right); \quad -\infty < z < \infty \tag{6.57}$$

The failure time density function, $f_T(t)$, failure time distribution function, $F_T(t)$, reliability function, $R(t)$, and the failure rate function, $h(t)$, given by Eqs. (6.52), (6.54), (6.55), and (6.56) are shown in Figs. 6.5(*a*)–(*d*), respectively.

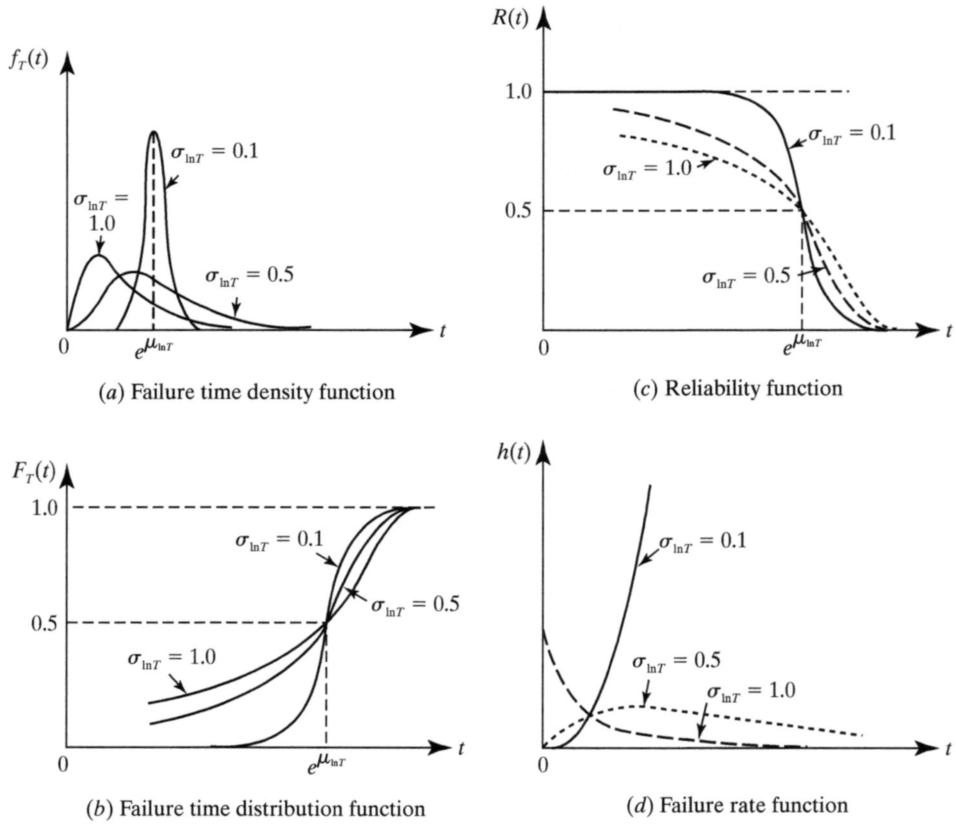

FIGURE 6.5 Lognormal distribution.

Notes:

1. The range of X is 0 to ∞ whereas that of Y is $-\infty$ to ∞. When $X = 1$, $Y = 0$ and when $X > 1$, Y is positive. When X varies between 0 and 1, Y varies in the range $-\infty$ to 0.
2. The parameters used in Eq. (6.45) are the mean and standard deviation of Y or $\ln X$, not those of X.
3. The probability density function of X can be expressed in terms of the median of X, θ_X, as

$$f_X(x) = \frac{1}{\sqrt{2\pi}\,\sigma_{\ln X}\, x} \exp\left\{-\frac{1}{2\sigma_{\ln X}^2}\left(\ln\left(\frac{x}{\theta_X}\right)\right)^2\right\}; \quad x \geq 0 \tag{6.58}$$

where

$$\theta_X = \exp(\theta_Y) \equiv \exp(\mu_Y) \tag{6.59}$$

with θ_X and θ_Y denoting the medians of the random variables X and Y, respectively.

4. The mean and standard deviation of X can be found either by integrating its probability density function $f_X(x)$ directly or by using Eq. (6.43) along with the probability density function of Y. The median θ_X and the standard deviation σ_X of X can be expressed as

$$\mu_X = \theta_X \exp\left(\frac{\sigma_{\ln X}^2}{2}\right) \tag{6.60}$$

$$\sigma_X^2 = \mu_X^2\left\{\exp(\sigma_{\ln X}^2) - 1\right\} \tag{6.61}$$

5. Because lognormal distribution is related to normal distribution through Eq. (6.43), the probabilities associated with X can be computed using standard normal distribution tables. The probability density function $f_X(x)$ and the probability distribution function $F_X(x)$ of the lognormally distributed variable X can be expressed in terms of the probability density function $f_Z(z)$ and the probability distribution function $F_Z(z)$ of the standard normal variate Z as

$$f_X(x) = \frac{1}{x\,\sigma_{\ln X}} f_Z\left(\frac{\ln\left(\frac{x}{\theta_X}\right)}{\sigma_{\ln X}}\right) \equiv \frac{1}{x\,\sigma_{\ln X}} f_Z(z); \quad x \geq 0 \tag{6.62}$$

where

$$z = \frac{1}{\sigma_{\ln X}} \ln \frac{x}{\theta_X} \tag{6.63}$$

and

$$F_X(x) = P[X \leq x] = P[\ln X \leq \ln x] = P[Y \leq \ln x] = F_Y(\ln x) \tag{6.64}$$

Since Y is $N(\mu_Y, \sigma_Y)$ or $N(\ln \theta_X, \sigma_{\ln X})$, Eq. (6.64) can be expressed as

$$F_X(x) = F_Z\left(\frac{\ln x - \ln \theta_X}{\sigma_{\ln X}}\right) = F_Z\left(\frac{\ln\left(\frac{x}{\theta_X}\right)}{\sigma_{\ln X}}\right) = F_Z(z) \tag{6.65}$$

where z is given by Eq. (6.63).

6. In many practical problems, the mean and standard deviation of the variable X are available as μ_X and σ_X instead of the mean and standard deviations, $\mu = \mu_Y$ and σ_Y, of $Y = \ln X$. In such a case, Eqs. (6.60) and (6.61) can be solved to find

$$\sigma_Y^2 = \sigma_{\ln X}^2 = \ln\left\{1 + \frac{\sigma_X^2}{\mu_X^2}\right\} \tag{6.66}$$

and

$$\mu_Y = \mu_{\ln X} = \ln\left\{\mu_X e^{-\frac{1}{2}\sigma_Y^2}\right\} = \ln \mu_X - \frac{1}{2}\sigma_Y^2 \tag{6.67}$$

EXAMPLE 6.7

The ultimate strength of steel (X) is modeled by a lognormal distribution. A set of experimental results yielded the values of the mean and standard deviation of X as $\mu_X = 40{,}500$ psi and $\sigma_X = 5100$ psi. Determine the following: (1) Probability density function of X, and (2) $P(X > 45{,}000 \text{ psi})$.

Solution: Equations (6.60) and (6.61) can be used to find the values of θ_X and $\sigma_{\ln X}$:

$$40{,}500 = \theta_X \exp\left(\frac{\sigma_{\ln X}^2}{2}\right) \tag{E1}$$

$$\sigma_X^2 = \mu_X^2 \left\{\exp(\sigma_{\ln X}^2) - 1\right\} \tag{E2}$$

or

$$\exp\left(\sigma_{\ln X}^2\right) = \frac{\sigma_X^2}{\mu_X^2} + 1 = \frac{(5100)^2}{(40{,}500)^2} + 1 = 1.015857$$

or

$$\sigma_{\ln X}^2 = 2.761730$$

or

$$\sigma_{\ln X} = 1.661845 \tag{E3}$$

Equations (E1) and (E3) yield

$$\theta_X = \frac{40{,}500}{\exp\left(\frac{2.761730}{2}\right)} = 10{,}180.1218 \text{ psi}$$

1. The probability density function of X can be expressed, using Eq. (6.58), as

$$f_X(x) = \frac{1}{\sqrt{2\pi}\,(1.661845)\,x} \exp\left(-\frac{1}{2(2.761730)} \ln^2\left(\frac{x}{10180.1218}\right)\right)$$

$$= \frac{1}{4.165632\,x} \exp\left(-\frac{1}{5.523460} \ln^2\left(\frac{x}{10180.1218}\right)\right); \quad x \ge 0 \tag{E4}$$

2.

$$P(X > 45{,}000) = 1 - P(X \le 45{,}000) = 1 - F_X(45{,}000)$$

where

$$F_X(x = 45000) = F_Z\left(\frac{\ln\left(\frac{x}{\theta_X}\right)}{\sigma_{\ln X}}\right) = F_Z\left(\frac{\ln\left(\frac{45000}{10180.1218}\right)}{1.661845}\right) = F_Z\,(0.894322) = 0.81442$$

Thus $P(X > 45{,}000) = 1 - 0.81442 = 0.18558$

■

EXAMPLE 6.8

The failure time (X) of a machine component is known to follow lognormal distribution with parameters $\mu = \mu_{\ln X} = E[\ln X] = 4$ units of time and $\sigma = \sigma_{\ln X} = 2$ units of time. Determine the reliability and hazard rate of the machine component at $x = 100$ units of time.

Solution: The reliability of the machine component is given by Eq. (6.55):

$$R(100) = P\left[z > \frac{\ln x - \mu}{\sigma}\right] = P\left[z > \frac{\ln 100 - 4}{2}\right]$$
$$= P[z > 0.3026] = 0.3725$$

The hazard rate can be computed, using Eq. (6.56), as

$$h(100) = \frac{\phi\left(\frac{\ln x - \mu}{\sigma}\right)}{x\,\sigma\,R(100)} = \frac{\phi\left(\frac{\ln 100 - 4}{2}\right)}{(100)(2)(0.3725)} = \frac{0.378}{74.5} = 0.0051 \text{ failures/unit of time}$$

■

6.7.4 Weibull Distribution

The Weibull distribution is widely used in reliability calculations for modeling the probability distribution of the failure time of a variety of components and systems. It can be used for components and systems that exhibit monotonically decreasing, constant, or monotonically increasing failure rates. Weibull distribution can be defined either with two or three parameters.

Two-parameter Weibull distribution: The probability density function of the failure time, T, can be expressed as

$$f(t) = \frac{\beta}{\eta}\left(\frac{t}{\eta}\right)^{\beta-1} e^{-(\frac{t}{\eta})^{\beta}}; \quad t > 0 \tag{6.68}$$

where $\beta > 0$ and $\eta > 0$ are the parameters of the distribution, known as the shape and scale parameters, respectively. The scale parameter η is also known as characteristic life. The probability distribution function of the failure time is given by

$$F(t) = \int_0^t f(\tau)\,d\tau = \int_0^t \frac{\beta}{\eta}\left(\frac{\tau}{\eta}\right)^{\beta-1} e^{-(\frac{\tau}{\eta})^{\beta}}\,d\tau \tag{6.69}$$

Denoting

$$y = -\left(\frac{\tau}{\eta}\right)^{\beta} \tag{6.70}$$

so that

$$dy = -\frac{\beta}{\eta}\left(\frac{\tau}{\eta}\right)^{\beta-1} d\tau \tag{6.71}$$

Equation (6.69) yields

$$F(t) = -\int_0^t e^y\, dy = \left(-e^y\right)_0^t = 1 - e^{-\left(\frac{t}{\eta}\right)^{\beta}} \tag{6.72}$$

The corresponding reliability function is given by

$$R(t) = 1 - F(t) = e^{-\left(\frac{t}{\eta}\right)^{\beta}} \tag{6.73}$$

The failure rate or hazard function can be determined as

$$h(t) = \frac{f(t)}{R(t)} = \frac{\frac{\beta}{\eta}\left(\frac{t}{\eta}\right)^{\beta-1}\exp\left\{-\left(\frac{t}{\eta}\right)^{\beta}\right\}}{\exp\left\{-\left(\frac{t}{\eta}\right)^{\beta}\right\}} = \frac{\beta}{\eta}\left(\frac{t}{\eta}\right)^{\beta-1} \tag{6.74}$$

The variations of the failure time density function, failure time distribution function, reliability function, and failure rate function with time are shown in Figs. 6.6(*a*)–(*d*), respectively.

The distribution function of the failure time T, $F(t)$, reduces to exponential distribution when $\beta = 1$ and Rayleigh distribution when $\beta = 2$. The probability density function of the Weibull distribution for different values of β are shown in Fig. 6.6. It is possible that Weibull distribution can be made to denote a normal distribution approximately by selecting the value of the shape parameter β suitably.

Since Weibull distribution can be used to represent a variety of distributions such as exponential, Rayleigh, and normal, it is widely used in the reliability modeling (to denote the failure time distributions) of several types of components and systems in engineering. It can be seen that for $\beta < 1$, the hazard rate will be a decreasing function of time and denotes the initial or early failure rate region. When $\beta = 1$, the hazard rate will be a constant and denotes a constant failure rate region. When $\beta > 1$, the hazard rate will be a monotonically increasing function of t and hence it represents the wear-out period without any upper limit.

The mean and standard deviation of the Weibull distribution can be found as

$$\overline{T} = E[T(\text{time to failure})] = \eta\,\Gamma\left(1 + \frac{1}{\beta}\right) \tag{6.75}$$

and

$$\sigma_T^2 = Var[T] = \eta^2\left[\Gamma\left(1 + \frac{2}{\beta}\right) - \left\{\Gamma\left(1 + \frac{1}{\beta}\right)\right\}^2\right] \tag{6.76}$$

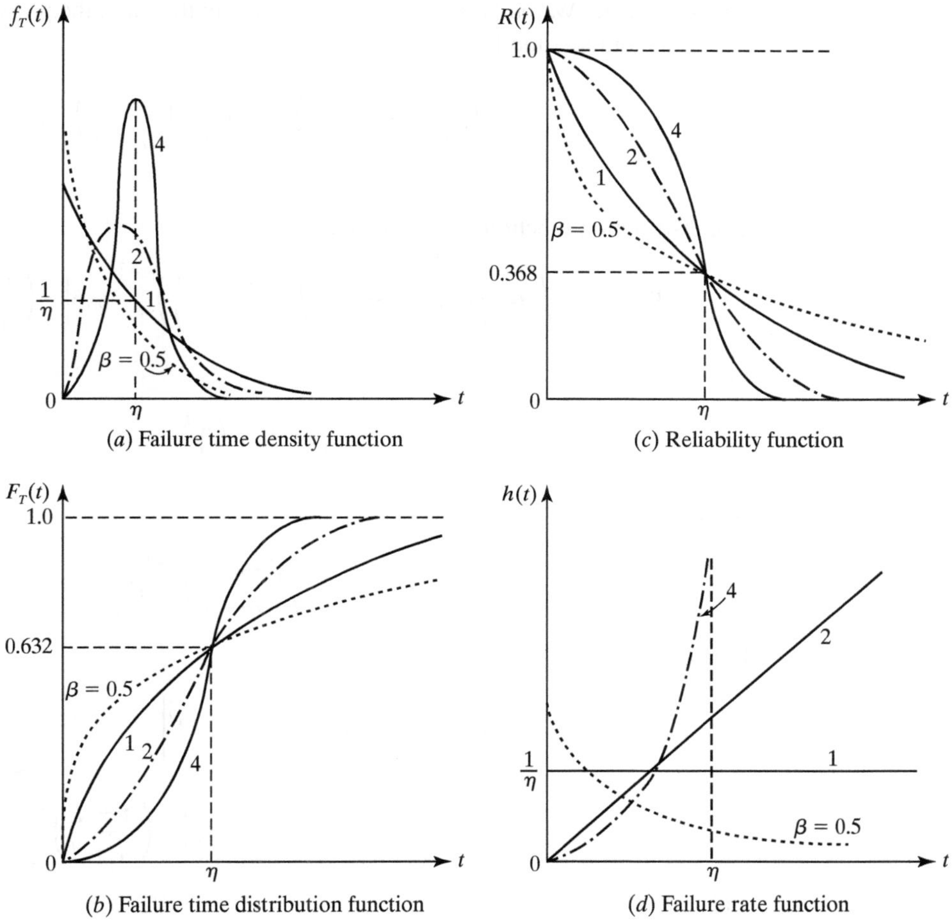

(*a*) Failure time density function

(*c*) Reliability function

(*b*) Failure time distribution function

(*d*) Failure rate function

FIGURE 6.6 Weibull distribution.

where

$$\Gamma(n) = \int_0^\infty x^{n-1} e^{-x}\, dx \tag{6.77}$$

and

$$\Gamma(n)\eta^n = \int_0^\infty x^{n-1} \eta^{-\frac{x}{n}}\, dx \tag{6.78}$$

Three-parameter Weibull distribution: The three-parameter Weibull distribution can be used to model the failure time distribution of components and systems for which certain time must elapse before failure can occur. For example, if it is necessary for certain amount (or threshold) of wear to take place before the component or system can fail, the three-parameter Weibull distribution can be used. In this case, the probability density function of the failure time (T) is defined as

$$f_T(t) = \begin{cases} \dfrac{\beta}{\eta - l}\left(\dfrac{t - l}{\eta - l}\right)^{\beta-1} \exp\left[-\left(\dfrac{t - l}{\eta - l}\right)^{\beta}\right]; & t \geq l \\ 0; & t < l \end{cases} \tag{6.79}$$

The distribution function of T is given by

$$F_T(t) = \int_l^t f_T(\tau)\, d\tau = \frac{\beta}{\eta - l}\int_{\tau=l}^{t}\left(\frac{\tau - l}{\eta - l}\right)^{\beta-1} \exp\left[-\left(\frac{\tau - l}{\eta - l}\right)^{\beta}\right] d\tau \tag{6.80}$$

Let

$$x = \left(\frac{\tau - l}{\eta - l}\right)^{\beta}$$

so that

$$dx = \frac{\beta}{\eta - l}\left(\frac{\tau - l}{\eta - l}\right)^{\beta-1} d\tau$$

Hence,

$$F_T(t) = \int_{x'=0}^{x} e^{-x'}\, dx' = -e^{-x'}\Big|_0^x$$

$$= 1 - e^{-x} = 1 - \exp\left[-\left(\frac{t - l}{\eta - l}\right)^{\beta}\right] \tag{6.81}$$

and

$$R(t) = 1 - F_T(t) = \exp\left[-\left(\frac{t - l}{\eta - l}\right)^{\beta}\right] \tag{6.82}$$

The failure rate can be found as

$$h(t) = \frac{f_T(t)}{R(t)} = \frac{\beta}{\eta - l}\left(\frac{t - l}{\eta - l}\right)^{\beta - 1} \tag{6.83}$$

EXAMPLE 6.9

The time to failure, in hours, of a bearing in a gearbox is modeled by Weibull distribution with parameters $\eta = 6000$ hours and $\beta = 0.5$.

1. Find the mean time to failure of the bearing.
2. Find the probability that the life of the bearing is at least 7000 hours.
3. If the reliability of the bearing is required to be at least 0.90, find the design life of the bearing.

Solution:

1. The expected value of the failure time or MTTF of the bearing is given by

$$E(T) = \text{MTTF} = \eta\Gamma\left(1 + \frac{1}{\beta}\right) = 6000\,\Gamma(1 + 2) = 6000(2!) = 12{,}000 \text{ hours}$$

2. The probability that the bearing life is at least 7000 hours can be determined as

$$P(T > 7000) = 1 - F_T(7000) = \exp\left\{-\left(\frac{t}{\eta}\right)^{\beta}\right\} = \exp\left\{-\left(\frac{7000}{6000}\right)^{0.5}\right\} = e^{-1.0801}$$

$$= 0.3395$$

3. The design life (t) that corresponds to a reliability of 0.9 can be found from

$$R(t) = \exp\left\{-\left(\frac{t}{\eta}\right)^{\beta}\right\} = \exp\left\{-\left(\frac{t}{6000}\right)^{0.5}\right\} = 0.9$$

Taking logarithms on both sides of the equation, we obtain

$$-\left(\frac{t}{6000}\right)^{0.5} = \ln 0.9 = -0.10536$$

which yields $t = 66.6044$ hours.

■

EXAMPLE 6.10

The time to failure, in hours, of a bearing in a gearbox is modeled by Weibull distribution with parameters $\eta = 6000$ hours and $\beta = 0.5$. If it is known that the bearing fails only after a wear-in period of 1000 hours, determine the following:

1. The probability that the life of the bearing is at least 7000 hours.
2. Design life of the bearing if the reliability of the bearing is required to be at least 0.90.

Solution:

1. The probability that the bearing life is at least 7000 hours can be determined as

$$P(T > 7000) = 1 - F_T(7000) = \exp\left\{-\left(\frac{t - l}{\eta - l}\right)^{\beta}\right\} = \exp\left\{-\left(\frac{7000 - 1000}{6000 - 1000}\right)^{0.5}\right\} = e^{-1.095445}$$

$$= 0.3344$$

2. The design life (t) that corresponds to a reliability of 0.9 can be found from

$$R(t) = \exp\left\{-\left(\frac{t - l}{\eta - l}\right)^{\beta}\right\} = \exp\left\{-\left(\frac{t - 1000}{6000 - 1000}\right)^{0.5}\right\} = 0.9$$

Taking logarithms on both sides of the equation, we obtain

$$-\left(\frac{t - 1000}{5000}\right)^{0.5} = \ln 0.9 = -0.10536$$

which yields $t = 1055.5$ hours.

■

EXAMPLE 6.11

The failure time of a mechanical component follows Weibull distribution with $\beta = 3$, $\eta = 2500$, and $l = 1000$. Find the reliability of the component and the failure rate for an operating time of 2000 hours.

Solution: By substituting the given data into Eqs. (6.82) and (6.83), we obtain

$$R(2000) = \exp\left[-\left(\frac{2000 - 1000}{2500 - 1000}\right)^3\right] = e^{-0.2963} = 0.7436$$

and

$$h(2000) = \frac{3(2000 - 1000)^{3-1}}{(2500 - 1000)^3} = \frac{3(1000)^2}{(1500)^3} = 0.00089 \text{ failures per hour}$$

■

6.7.5 Gamma Distribution

The gamma distribution, similar to the Weibull distribution, can also be used to model the decreasing, constant, or increasing hazard rates. The gamma distribution for the failure time, $f_T(t)$, is defined by

$$f(t) = \frac{t^{\beta-1}}{\eta^\beta \Gamma(\beta)} e^{-\frac{t}{\eta}} \tag{6.84}$$

where $\beta > 0$ is called the shape parameter, $\eta > 0$ is called the scale parameter, and $\Gamma(\alpha)$ is the gamma function defined by

$$\Gamma(\alpha) = \int_0^\infty t^{\alpha-1} e^{-t}\, dt \tag{6.85}$$

which, for integer values of α, reduces to

$$\Gamma(\alpha) = (\alpha - 1)! \tag{6.86}$$

The probability distribution function, $F_T(t)$, corresponding to Eq. (6.84) can be expressed as

$$F(t) = \int_0^t \frac{\xi^{\beta-1}}{\eta^\beta \Gamma(\beta)} e^{-\frac{\xi}{\eta}} d\xi \tag{6.87}$$

Using $z = \dfrac{\xi}{\eta}$ with $d\xi = \eta\, dz$, Eq. (6.87) can be expressed as

$$F(t) = \frac{1}{\Gamma(\beta)} \int_0^{\frac{t}{\eta}} z^{\beta-1} e^{-z}\, dz \tag{6.88}$$

Noting that the right-hand side is same as the incomplete gamma function, $I\left(\dfrac{t}{\eta}, \beta\right)$, the cumulative distribution function of the failure time is given by

$$F(t) = I\left(\frac{t}{\eta}, \beta\right) \tag{6.89}$$

The reliability function can be expressed as

$$R(t) = 1 - F(t) = 1 - \int_0^t \frac{\xi^{\beta-1}}{\eta^\beta \Gamma(\beta)} e^{-\frac{\xi}{\eta}}\, d\xi \tag{6.90}$$

which can be rewritten as

$$R(t) = \frac{1}{\Gamma(\beta)} \int_{\frac{t}{\eta}}^{\infty} z^{\beta-1} e^{-z}\, dz \tag{6.91}$$

In this case, the hazard rate becomes

$$h(t) = \frac{f(t)}{R(t)} \tag{6.92}$$

where $f(t)$ and $R(t)$ are given by Eqs. (6.84) and (6.91), respectively. Typical graphs of the functions $f(t)$, $F(t)$, $R(t)$, and $h(t)$ are shown in Figs. 6.7(*a*)–(*d*), respectively.

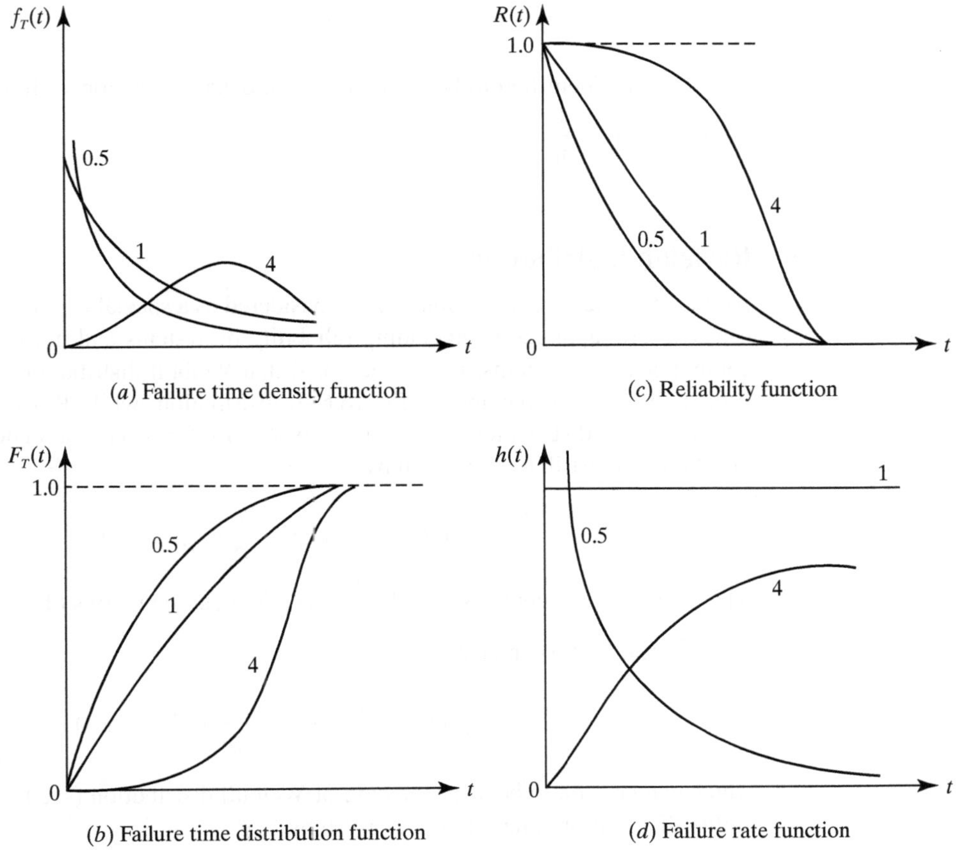

(*a*) Failure time density function

(*c*) Reliability function

(*b*) Failure time distribution function

(*d*) Failure rate function

FIGURE 6.7 Gamma distribution (for $\beta = 0.5$, 1 and 4).

The mean and standard deviation of gamma distribution are given by

$$\mu_T = \beta\eta \tag{6.93}$$

and

$$\sigma_T = \beta\eta^2 \tag{6.94}$$

Note: If the shape parameter β is an integer, the cumulative distribution and reliability functions, Eqs. (6.87) and (6.90), reduce to

$$F(t) = 1 - e^{-\frac{t}{\eta}} \sum_{k=0}^{n-1} \frac{1}{k!}\left(\frac{t}{\eta}\right)^k \tag{6.95}$$

$$R(t) = e^{-\frac{t}{\eta}} \sum_{k=0}^{n-1} \frac{1}{k!}\left(\frac{t}{\eta}\right)^k \tag{6.96}$$

In this case, the distribution, Eq. (6.95), is called the Erlangian distribution.

When $\beta = 1$, the density function given by Eq. (6.84) reduces to

$$f(t) = \frac{1}{\eta} e^{-\frac{t}{\eta}} \tag{6.97}$$

which can be identified to be the exponential density function with the failure rate given by $h(t) = \lambda = \dfrac{1}{\eta}$.

6.7.6 Rayleigh Distribution

Although Rayleigh distribution can be considered as a special case of Weibull distribution, it has several applications including reliability of systems and noise problems related to communications systems. It is to be noted that Weibull distribution is a three-parameter (Eq. (6.79)) or two-parameter (Eq. (6.68)) distribution, while Rayleigh distribution is a one-parameter distribution. The probability density function of a random variable following Rayleigh distribution is given by

$$f_T(t) = ct \exp\left(-\frac{c}{2}t^2\right); \quad t \geq 0 \tag{6.98}$$

where c is the parameter. By redefining the parameter c in terms of another parameter α as $c = \dfrac{2}{\alpha^2}$, Eq. (6.98) can be expressed as

$$f_T(t) = \frac{2}{\alpha^2} t \exp\left(-\frac{1}{\alpha^2}t^2\right); \quad t \geq 0 \tag{6.99}$$

which can be seen to be a special case of Weibull distribution (see Eq. (6.68)). The probability distribution function of T is given by

$$F_T(t) = \int_0^t f_T(\tau)\, d\tau = \int_0^t c\tau \exp\left(-\frac{c}{2}\tau^2\right) d\tau = 1 - \exp\left(-\frac{c}{2}t^2\right); \quad t \geq 0 \tag{6.100}$$

The reliability function, $R(t)$, is given by

$$R(t) = 1 - F_T(t) = \exp\left(-\frac{c}{2}t^2\right) \tag{6.101}$$

The hazard function, $h(t)$, of the random variable T following Rayleigh distribution can be determined as

$$h(t) = \frac{f_T(t)}{R(t)} = \frac{ct\exp\left(-\frac{c}{2}t^2\right)}{\exp\left(-\frac{c}{2}t^2\right)} = ct \tag{6.102}$$

Thus, the hazard function increases linearly with time. Typical probability density, distribution, reliability, and hazard functions corresponding to Rayleigh distribution are shown in Figs. 6.8(*a*)–(*d*), respectively. The mean and standard deviation of T are given by

$$\mu_T = \sqrt{\frac{\pi}{2c}} \tag{6.103}$$

and

$$\sigma_T = \left\{\frac{2}{c}\left(1 - \frac{\pi}{4}\right)\right\}^{\frac{1}{2}} \tag{6.104}$$

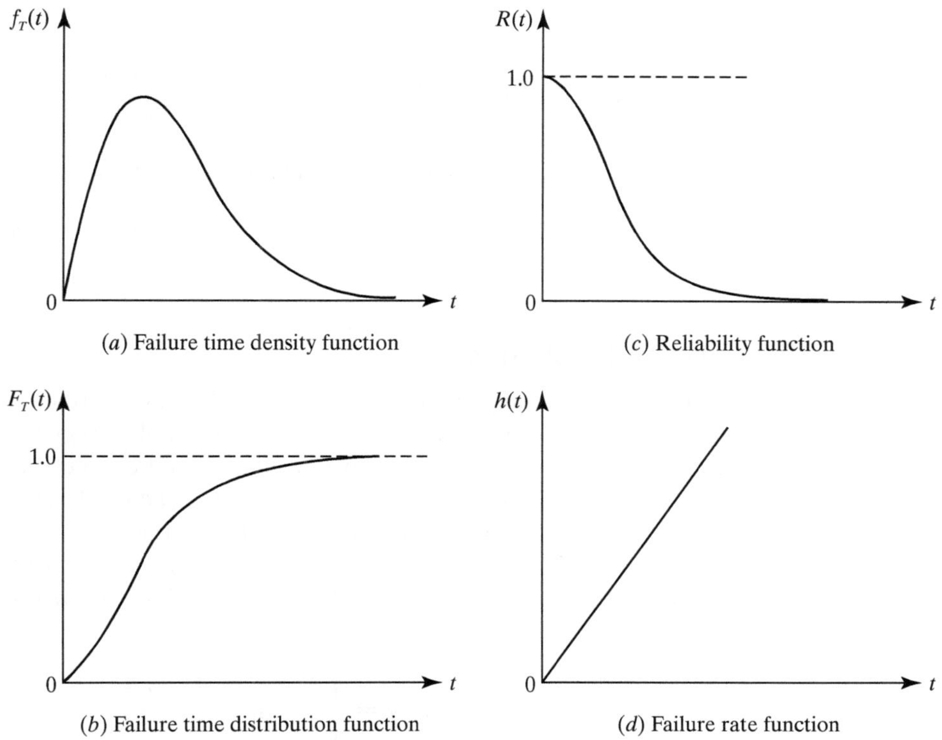

FIGURE 6.8 Rayleigh distribution.

EXAMPLE 6.12

Based on the laboratory tests, an automobile battery manufacturing firm determined that the failure time of its batteries follows Rayleigh distribution with the failure rate given by

$$h(t) = 2.5 \times 10^{-3} t; \quad t \text{ in months}$$

Determine the following:

1. Mean and standard deviation of the failure time of the batteries.
2. Reliability of the batteries at the end of two years.

Solution:

1. The known failure rate function

$$h(t) = 2.5 \times 10^{-3} t = ct$$

gives the value of $c = 2.5 \times 10^{-3}$. The mean and standard deviation of the failure time (T) can be found as

$$\mu_T = \sqrt{\frac{\pi}{2c}} = \sqrt{\frac{\pi}{2\,(2.5 \times 10^{-3})}} = 25.0663 \text{ months}$$

and

$$\sigma_T = \left\{\frac{2}{c}\left(1 - \frac{\pi}{4}\right)\right\}^{\frac{1}{2}} = \left\{\frac{2}{2.5 \times 10^{-3}}\left(1 - \frac{\pi}{4}\right)\right\}^{\frac{1}{2}} = 13.1027 \text{ months}$$

2. The reliability of the batteries at the end of two years, or 24 months, is given by

$$R(24) = e^{-\frac{1}{2}ct^2} = e^{-\frac{1}{2}(2.5\times 10^{-3})(24)^2} = e^{-0.72} = 0.4867$$

■

6.7.7 Uniform Distribution

The uniform (or rectangular) distribution is the simplest distribution and is used in situations where there is no evidence for the random variable to give unequal probabilities for assuming different possible values within a specified interval. Some examples include the live load acting on a bridge within a range and the duration of a flight between two cities within a range. A continuous random variable T is said to follow uniform distribution in the interval $a \le t \le b$ if it is equally likely to assume any value in the interval. The probability density function of T will be a constant in the interval a to b and is given by

$$f_T(t) = \begin{cases} \dfrac{1}{b-a}; & a \le t \le b \\ 0 \quad ; & \text{elsewhere} \end{cases} \tag{6.105}$$

Note that the area under the density function given by Eq. (6.105) is unity. The probability distribution function of T is given by the integral of the probability density function:

$$F_T(t) = \int_{-\infty}^{t} f_T(\tau)\, d\tau = \int_{a}^{t} \frac{1}{b-a}\, d\tau = \begin{cases} 0; & t < a \\ \dfrac{t-a}{b-a}; & a \le t \le b \\ 1; & t > b \end{cases} \tag{6.106}$$

Equations (6.105) and (6.106) are shown graphically in Figs. 6.9 (*a*) and (*b*), respectively.

The mean and variance of the random variable T can be determined as

$$\mu_T = \int_{-\infty}^{\infty} t f_T(t)\, dt = \int_a^b t \frac{1}{b-a}\, dt = \frac{1}{b-a}\left(\frac{t^2}{2}\right)\Bigg|_a^b = \frac{a+b}{2} \tag{6.107}$$

$$V(T) = \sigma_T^2 = \int_{-\infty}^{\infty} (t-\mu_T)^2 f_T(t)\, dt = \int_a^b \left(t - \frac{a+b}{2}\right)^2 \frac{1}{b-a}\, dt \tag{6.108}$$

$$= \left[\left(\frac{t^3}{3}\right)\Bigg|_a^b - \left(\frac{a+b}{2}\right)\left(\frac{t^2}{2}\right)\Bigg|_a^b + \left(\frac{a+b}{2}\right)^2 (t)\Bigg|_a^b\right] = \frac{(b-a)^2}{12}$$

Thus the standard deviation of T is given by

$$\sigma_T = \frac{b-a}{\sqrt{12}} \tag{6.109}$$

If t denotes time, the reliability or survivor function, $R(t)$, is given by

$$R(t) = 1 - F_T(t) = 1 - \frac{t-a}{b-a} = \frac{b-t}{b-a} \tag{6.110}$$

and the hazard rate, $\lambda(t)$, by

$$\lambda(t) = \frac{f_T(t)}{R(t)} = \frac{1}{b-a}\,\frac{b-a}{b-t} = \frac{1}{b-t} \tag{6.111}$$

The graphs of the reliability and failure rate functions are shown in Figs. 6.9 (*c*) and (*d*), respectively.

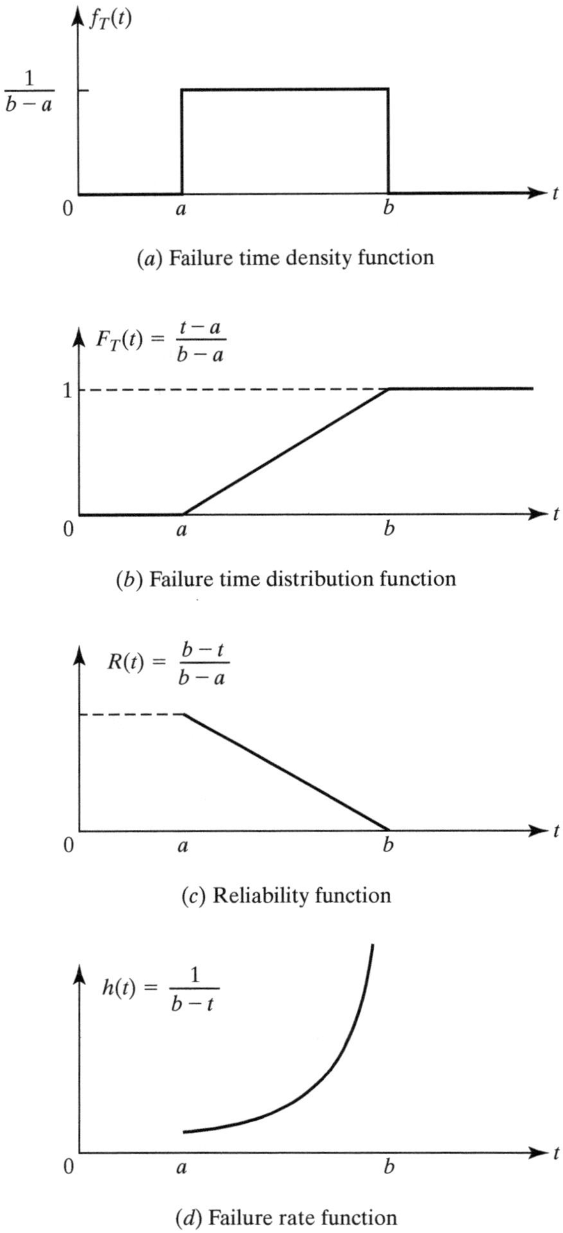

FIGURE 6.9 Uniform distribution.

6.8 Expected Residual Life

In practice, when a component or system is observed to be in operating condition at time t, we will be interested in knowing about its remaining life. The expected residual life function, $L(t)$, can be used for this purpose. If T is a random variable denoting the life of the component or system, its expected residual life function is defined as

$$L(t) = \text{Expected value of the life remaining beyond } t = E[T - t] \quad (6.112)$$

The function $L(t)$ can be expressed in terms of the conditional probability density function for any value of time $\tau \geq t$ as

$$f_{T|_{T\geq t}}(\tau) = \frac{f(\tau)}{R(t)}; \quad \tau \geq t \quad (6.113)$$

The conditional expectation of life greater than t or the conditional expected value of $T|_{T\geq t}$ can be computed as

$$E[T|_{T\geq t}] = \int_{\tau=t}^{\infty} \tau f_{T|_{T\geq t}}(\tau)\, d\tau = \int_{\tau=t}^{\infty} \tau \frac{f(\tau)}{R(t)}\, d\tau \quad (6.114)$$

Because the component, product, or system has already survived up to time t, the expected residual life (beyond the value of t) can be found from

$$L(t) = E[(T - t)|_{T\geq t}] = \int_{\tau=t}^{\infty} (\tau - t) \frac{f(\tau)}{R(t)}\, d\tau \quad (6.115)$$

Notes:

1. Equation (6.115) can be rewritten as follows. By expressing the right-hand side of Eq. (6.115) as two terms, we obtain

$$L(t) = \int_{\tau=t}^{\infty} \tau \frac{f(\tau)}{R(t)}\, d\tau - \int_{\tau=t}^{\infty} t \frac{f(\tau)}{R(t)}\, d\tau \quad (6.116)$$

Noting that

$$\int_{\tau=t}^{\infty} f(\tau)\, d\tau = R(t) \quad (6.117)$$

The expected residual life, $L(t)$, can be expressed as

$$L(t) = \frac{1}{R(t)} \int_{\tau=t}^{\infty} \tau f(\tau)\, d\tau - t \quad (6.118)$$

2. Another equivalent expression for Eq. (6.115) can be derived by integrating the right-hand side by parts as

$$L(t) = \frac{1}{R(t)} \left\{ -(\tau - t)\left(\int_{\tau}^{\infty} f(\tau)\, d\tau \right)\Bigg|_{\tau=t}^{\infty} + \int_{\tau=t}^{\infty} R(\tau)\, d\tau \right\} \quad (6.119)$$

Using Eq. (6.117), Eq. (6.119) can be written as

$$L(t) = \frac{1}{R(t)} \left\{ -(\tau - t)R(\tau)\big|_{\tau=t}^{\infty} + \int_{\tau=t}^{\infty} R(\tau)\, d\tau \right\} \quad (6.120)$$

For most probability distributions, particularly those arising in practical applications, $R(t)$ will approach zero as $t \to \infty$ in such a way that the product $t\,R(t) \to 0$. Thus, Eq. (6.120) can be simplified as

$$L(t) = \frac{1}{R(t)} \int_{\tau=t}^{\infty} R(\tau)\, d\tau \tag{6.121}$$

3. Equation (6.121) can be manipulated to express $R(t)$ in terms of the residual life function $L(t)$ as (see Problem 6.51):

$$R(t) = \frac{L(0)}{L(t)} \left\{ - \int_0^t \frac{d\tau}{L(\tau)} \right\} \tag{6.122}$$

4. Because $R(0) = 1$, we can observe that $L(0) = E[T]$, that is, the residual life of a new component or system is same as the mean of the (unconditional) random variable T.

EXAMPLE 6.13

Find the expected residual life of a system whose failure time distribution is given by

$$f_T(t) = \lambda\, e^{-\lambda t} \tag{E1}$$

Solution: The probability distribution function $F_T(t)$ corresponding to the probability density function given by Eq. (E1) can be found as

$$F_T(t) = \int_0^t f_T(\tau)\, d\tau = \int_0^t \lambda\, e^{-\lambda\tau}\, d\tau = 1 - e^{-\lambda t} \tag{E2}$$

and hence the reliability function is given by

$$R(t) = 1 - F_T(t) = e^{-\lambda t} \tag{E3}$$

The expected residual life of the component can be determined from Eq. (6.121):

$$L(t) = \frac{1}{R(t)} \int_{\tau=t}^{\infty} R(\tau)\, d\tau = \frac{1}{e^{-\lambda t}} \int_{\tau=t}^{\infty} e^{-\lambda\tau} d\tau = \frac{1}{e^{-\lambda t}} \left(- \frac{e^{-\lambda\tau}}{\lambda} \right)_{\tau=t}^{\infty} = \frac{1}{e^{-\lambda t}} \frac{e^{-\lambda t}}{\lambda} = \frac{1}{\lambda} \tag{E4}$$

■

EXAMPLE 6.14

The probability density function of the failure time of a feed water pump used in a thermal power plant is given by a beta distribution with the parameters $\alpha = 3$ and $\beta = 2$:

$$f_T(t) = \begin{cases} \dfrac{\Gamma(5)}{\Gamma(3)\Gamma(2)} t^2(1 - t); & 0 \le t \le 1 \\ 0; & \text{otherwise} \end{cases} \tag{E1}$$

where t is in years. If a pump was observed to be in operating condition at $t = 0.5$ year, determine its expected residual life.

Solution: Noting that

$$\frac{\Gamma(5)}{\Gamma(3)\Gamma(2)} = \frac{4!}{2!\,1!} = 12$$

Equation (E1) can be expressed as

$$f_T(t) = \begin{cases} 12t^2(1-t); & 0 \le t \le 1 \\ 0; & \text{otherwise} \end{cases} \tag{E2}$$

The probability distribution function corresponding to Eq. (E2) is given by

$$F_T(t) = \int_0^t f_T(\tau)\,d\tau = \int_0^t 12\,(\tau^2 - \tau^3)\,d\tau = 12\left(\frac{\tau^3}{3} - \frac{\tau^4}{4}\right)_0^t \tag{E3}$$

$$= t^3(4 - 3t); 0 \le t \le 1$$

and hence the reliability function becomes

$$R(t) = 1 - F_T(t) = 1 - 4t^3 + 3t^4$$

The expected residual life of the feed water pump can be determined from Eq. (6.121) as

$$L(t) = \frac{1}{R(t)} \int_{\tau=t}^{\infty} R(\tau)\,d\tau = \frac{1}{1 - 4t^3 + 3t^4} \int_t^1 \left\{1 - 4\tau^3 + 3\tau^4\right\} d\tau$$

$$= \frac{1}{1 - 4t^3 + 3t^4}\left\{\tau - \tau^4 + \frac{3}{5}\tau^5\right\}_t^1 = \frac{\frac{3}{5} - t + t^4 - \frac{3}{5}t^5}{1 - 4t^3 + 3t^4}$$

The residual life at $t = 0.5$ year can be found as

$$L(0.5) = \frac{\frac{3}{5} - 0.5 + (0.5)^4 - \frac{3}{5}(0.5)^5}{1 - 4(0.5)^3 + 3(0.5)^4} = \frac{0.13125}{0.6875} = 0.190909 \text{ year}$$

■

6.9 Series Systems

A series system is one that fails whenever any one of its components fails. A set of electrical switches connected in series as shown in Fig. 6.10 can be considered to be a series system, since the failure of any switch in making contact will prevent the flow of current from one end to the other (i.e., from A to B). The components need not be physically connected in series for the system to be called a series system. For example, the four tires of an automobile are in series for the purpose of reliability calculation because the failure

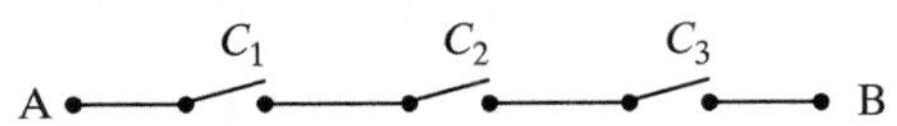

FIGURE 6.10 Series system.

TABLE 6.2 Possible states of switches connected in series

	Condition of Switch			Probability of Occurrence of the Event	Condition of the System
Event	C_1	C_2	C_3		
1	Closed	Closed	Closed	$R_1 R_2 R_3$	Operating
2	Closed	Closed	Open	$R_1 R_2 (1 - R_3)$	Failed
3	Closed	Open	Closed	$R_1 (1 - R_2)R_3$	Failed
4	Open	Closed	Closed	$(1 - R_1)R_2 R_3$	Failed
5	Closed	Open	Open	$R_1(1 - R_2)(1 - R_3)$	Failed
6	Open	Closed	Open	$(1 - R_1)R_2(1 - R_3)$	Failed
7	Open	Open	Closed	$(1 - R_1)(1 - R_2)R_3$	Failed
8	Open	Open	Open	$(1 - R_1)(1 - R_2)(1 - R_3)$	Failed

of any one makes the car inoperative. In Fig. 6.10, a switch is considered to be in operating (failed) condition if it is in a closed (open) position. If R_i denotes the reliability and $(1 - R_i)$ the probability of failure of switch C_i, then the various possibilities at any time t are shown in Table 6.2. The reliability of the system can be seen to be $R_s = R_1 R_2 R_3$. It is to be noted that the switches C_1, C_2, and C_3 were assumed to be independent in this analysis.

If we consider n components $C_1, C_2, \ldots,$ and C_n in series, the reliability of the series system (R_s), assuming the components to be independent, is given by

$$R_s = R_1 R_2 \cdots R_n = \prod_{i=1}^{n} R_i \tag{6.123}$$

where R_i denotes the reliability of component C_i. If the failure time of the ith component is given by t_i, $i = 1, 2, \ldots, n$, the failure time of the series system t_s is given by

$$t_s = \min_{1 \le i \le n} t_i \tag{6.124}$$

since the system fails as soon as one of its components fails. If $F_i(t)$ denotes the probability distribution function of the failure time of the ith component, the failure time distribution of the series system can be found as

$$F_s(t) = P(t_s \le t) = 1 - P(t_s > t) = 1 - \prod_{i=1}^{n} P(t_i > t) = 1 - \prod_{i=1}^{n} \left[1 - F_i(t) \right] \tag{6.125}$$

Thus the probability density function of the failure time of the series system is given by

$$f_s(t) = \frac{dF_s(t)}{dt} = \sum_{j=1}^{n} \frac{\partial F_s}{\partial F_j} \frac{\partial F_j}{\partial t} = \sum_{j=1}^{n} f_j(t) \prod_{i=1,\, i \ne j}^{n} \left[1 - F_i(t) \right] \tag{6.126}$$

6.9.1 Failure Rate of the System

If the failure time of the ith component follows exponential distribution, with a constant failure rate $\lambda_i (i = 1, 2, \ldots, n)$, then

$$R_i(t) = e^{-\lambda_i t} \tag{6.127}$$

and

$$R_s(t) = \prod_{i=1}^{n} R_i(t) = e^{-\left(\sum_{i=1}^{n} \lambda_i\right)t} = e^{-\lambda_s t} \tag{6.128}$$

where the failure rate of the system λ_s can be identified as

$$\lambda_s = \sum_{i=1}^{n} \lambda_i \tag{6.129}$$

Thus, the failure time of the system also follows exponential distribution with the failure rate of the system λ_s given by the sum of the component failure rates. In fact, Eq. (6.129) can be shown to be valid for any failure time distribution. If the failure rate of component i is given by $h_i(t)$, its reliability is given by Eq. (6.7) as

$$R_i(t) = \exp\left[-\int_0^t h_i(x)\,dx\right] \tag{6.130}$$

Equations (6.123) and (6.130) lead to

$$\begin{aligned} R_s(t) &= \prod_{i=1}^{n} R_i(t) = \prod_{i=1}^{n} e^{-\int_0^t h_i(x)\,dx} \\ &= e^{-\sum_{i=1}^{n} \int_0^t h_i(x)\,dx} = e^{-\int_0^t \left(\sum_{i=1}^{n} h_i(x)\right) dx} \end{aligned} \tag{6.131}$$

This shows that the system failure rate, $h_s(t)$, is given by

$$h_s(t) = \sum_{i=1}^{n} h_i(t) \tag{6.132}$$

If $h_i(t)$ is constant, as in the case of exponential failure time distribution, then

$$h_s(t) = \sum_{i=1}^{n} h_i \tag{6.133}$$

which shows that the system failure rate is also a constant.

6.9.2 MTBF of the System

The MTBF of a series system can be computed as

$$\text{MTBF} = \int_0^\infty R(t)\,dt = \int_0^\infty \left(\prod_{i=1}^{n} R_i(t)\right) dt \tag{6.134}$$

If the failure rates of the components follow exponential distribution, Eq. (6.134) gives

$$\text{MTBF} = \int_0^{\infty} e^{-\left(\sum_{i=1}^{n} \lambda_i\right) t}\, dt$$

$$= \left(-\frac{1}{\sum_{i=1}^{n} \lambda_i} \right) \left[e^{-\left(\sum_{i=1}^{n} \lambda_i\right) t} \right]_0^{\infty} = \frac{1}{\sum_{i=1}^{n} \lambda_i} \tag{6.135}$$

EXAMPLE 6.15

A system consists of five components in series, each having a reliability of 0.97. Find the reliability of the system.

Solution: The system reliability is given by

$$R_s = \prod_{i=1}^{5} R_i = (0.97)^5 = 0.8587$$

It can be seen that the greater the number of components, the less reliable the system becomes. Thus, a series system will have a reliability smaller than the reliability of any of its components.

■

6.10 Parallel Systems

A parallel system is one that fails only when all of its components fail. For example, a set of n switches connected in parallel will constitute a parallel system since all the switches must fail before the system is rendered useless for the flow of current from one end to the other. Consider a parallel system consisting of three independent switches as shown in Fig. 6.11. The eight possible outcomes of the system at time t are given in Table 6.3.

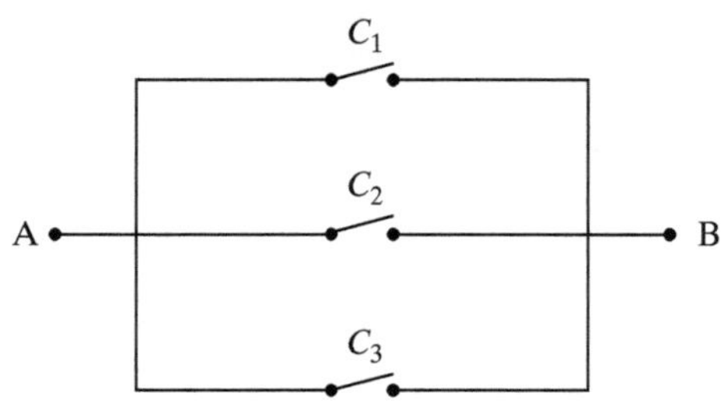

FIGURE 6.11 Parallel system.

TABLE 6.3 Possible states of switches connected in parallel

Event	Condition of Switch C_1	C_2	C_3	Probability of Occurrence of the Event	Condition of System
1	Closed	Closed	Closed	$R_1R_2R_3$	Operating
2	Closed	Closed	Open	$R_1R_2(1-R_3)$	Operating
3	Closed	Open	Closed	$R_1(1-R_2)R_3$	Operating
4	Open	Closed	Closed	$(1-R_1)R_2R_3$	Operating
5	Closed	Open	Open	$R_1(1-R_2)(1-R_3)$	Operating
6	Open	Closed	Open	$(1-R_1)R_2(1-R_3)$	Operating
7	Open	Open	Closed	$(1-R_1)(1-R_2)R_3$	Operating
8	Open	Open	Open	$(1-R_1)(1-R_2)(1-R_3)$	Failed

The reliability of the parallel system is given by the sum of the probabilities of realizing the first seven events, that is,

$$R_p = R_1R_2R_3 + R_1R_2(1-R_3) + R_1R_3(1-R_2) + R_2R_3(1-R_1)$$
$$+ R_1(1-R_2)(1-R_3) + R_2(1-R_1)(1-R_3) + R_3(1-R_1)(1-R_2) \quad (6.136)$$

In this case, the system reliability can be expressed more conveniently as

$$R_p = 1 - P_f = 1 - (1-R_1)(1-R_2)(1-R_3) = 1 - \prod_{i=1}^{3}(1-R_i) \quad (6.137)$$

This relation can be generalized as follows. If n components $C_1, C_2, \ldots,$ and C_n are connected in parallel, the reliability of the system is given by

$$R_p = 1 - P_f = 1 - P\,(\text{all components failing simultaneously})$$
$$= 1 - \prod_{i=1}^{n}(1-R_i) = 1 - \prod_{i=1}^{n}F_i \quad (6.138)$$

where $F_i = 1 - R_i$ denotes the probability of failure of the ith component at time t.

EXAMPLE 6.16

A parallel system consists of 3 components A, B, and C. If the reliabilities of the three components are $R_A = 0.9$, $R_B = 0.8$, and $R_C = 0.7$, find the system reliability.

Solution: Equation (6.138) gives

$$R_p = 1 - (1-R_A)(1-R_B)(1-R_C) = 1 - (0.1)(0.2)(0.3) = 1 - 0.006 = 0.994$$

This indicates that a larger number of components gives a higher reliability. It can also be seen that the reliability of a parallel system will always be larger than the reliability of any of its components.

■

If the failure time of the ith component is denoted as t_i, $i = 1, 2, \ldots, n$, the failure time of the parallel system t_p is given by

$$t_p = \max_{1 \le i \le n} t_i \tag{6.139}$$

since the system fails only when all the components fail. The probability distribution function of the failure time of the parallel system $F_p(t)$ can be expressed in terms of the probability distribution functions of the failure times of the individual components $F_i(t)$ as

$$F_p(t) = P(t_p \le t) = \prod_{i=1}^{n} F_i(t) \tag{6.140}$$

Thus, the probability density function of the failure time of the parallel system is given by

$$f_p(t) = \frac{dF_p(t)}{dt} = \sum_{j=1}^{n} f_j(t) \prod_{i=1, i \ne j}^{n} F_i(t) \tag{6.141}$$

6.10.1 Failure Rate of the System

The system failure rate, $h_p(t)$, is given by

$$h_p(t) = \frac{f(t)}{1 - F(t)} = \frac{\sum_{j=1}^{n} f_j(t) \prod_{i=1, i \ne j}^{n} F_i(t)}{1 - \prod_{i=1}^{n} F_i(t)} \tag{6.142}$$

It can be seen that the system failure rate, $h_p(t)$, is not equal to the sum of the component failure rates.

6.10.2 MTBF of the System

Consider a parallel system with n components. Let the probability density function of the ith component be exponential with a failure rate of λ_i, $i = 1, 2, \ldots, n$. Then the MTBF of the system can be computed as

$$\begin{aligned} \text{MTBF} &= \int_0^\infty R(t)\, dt = \int_0^\infty \left[1 - \prod_{i=1}^{n} \{1 - R_i(t)\} \right] dt \\ &= \int_0^\infty \left[1 - \prod_{i=1}^{n} \left\{ 1 - e^{-\lambda_i t} \right\} \right] dt \\ &= \int_0^\infty \left[1 - (1 - e^{-\lambda_1 t})(1 - e^{-\lambda_2 t}) \cdots (1 - e^{-\lambda_n t}) \right] dt \end{aligned} \tag{6.143}$$

Since Eq. (6.143) cannot be simplified any further, we consider the special case of $n = 2$. For this case, Eq. (6.143) gives

$$\text{MTBF} = \int_0^\infty \left[1 - \left(1 - e^{-\lambda_1 t} - e^{-\lambda_2 t} + e^{-(\lambda_1+\lambda_2)t}\right)\right] dt$$

$$= \frac{1}{\lambda_1} + \frac{1}{\lambda_2} - \frac{1}{\lambda_1 + \lambda_2} \tag{6.144}$$

where the relation

$$\int_0^\infty e^{-\lambda t}\,dt = \left[-\frac{1}{\lambda}e^{-\lambda t}\right]\Bigg|_0^\infty = \frac{1}{\lambda} \tag{6.145}$$

has been used in deriving Eq. (6.144). Similarly, for $n = 3$, Eq. (6.143) yields

$$\text{MTBF} = \frac{1}{\lambda_1} + \frac{1}{\lambda_2} + \frac{1}{\lambda_3} - \frac{1}{\lambda_1 + \lambda_2} - \frac{1}{\lambda_2 + \lambda_3} - \frac{1}{\lambda_1 + \lambda_3} + \frac{1}{\lambda_1 + \lambda_2 + \lambda_3} \tag{6.146}$$

6.11 (k, n) Systems

A system consisting of n components is called a (k, n) system if it can operate with any k components in operating condition. For example, in a two-engine aircraft, it may be possible to fly the aircraft (probably with less efficiency, but without any loss to property and people) if any one engine is in operating condition. In this case, the aircraft is considered to function as a $(1, 2)$ system. Similarly, in a six-cylinder automobile, it may be possible to drive the car with only five spark plugs firing. In this case, it can be called a $(5, 6)$ system.

If the failure times of all the components of a (k, n) system have the same probability distribution function $F(t)$, then the reliability of the system can be determined as

$$R_k(t) = \sum_{i=k}^{n} \binom{n}{i} [1 - F(t)]^i [F(t)]^{n-i} \tag{6.147}$$

and hence the probability distribution function of the system is given by

$$F_k(t) = 1 - R_k(t) = \sum_{i=0}^{k-1} \binom{n}{i} [1 - F(t)]^i [F(t)]^{n-i} \tag{6.148}$$

Equation (6.148) can be differentiated to find the probability density function of the failure time of the system as

$$f_k(t) = \frac{dF_k(t)}{dt} = \frac{n!}{(n-k)!(k-1)!} [F(t)]^{n-k} [1 - F(t)]^{k-1} f(t) \tag{6.149}$$

where $f(t) = \dfrac{dF(t)}{dt}$ denotes the probability density function of the failure time of the components.

6.11.1 MTBF of the System

If the failure times of individual components follow the same exponential distribution, the reliability of a (k, n) system is given by Eq. (6.147) so that the MTBF becomes

$$\text{MTBF} = \int_0^\infty R_k(t)\,dt = \int_0^\infty \left[\sum_{i=k}^{n} \binom{n}{i} \{e^{-\lambda t}\}^i \{1 - e^{-\lambda t}\}^{n-i}\right] dt \tag{6.150}$$

For example, for a $(1, 2)$ system, Eq. (6.150) gives

$$\begin{aligned} \text{MTBF} &= \int_0^\infty \left[\binom{2}{1} e^{-\lambda t}(1 - e^{-\lambda t}) + \binom{2}{2}(e^{-\lambda t})^2\right] dt \\ &= \int_0^\infty \left(2e^{-\lambda t} - e^{-2\lambda t}\right) dt = \frac{3}{2\lambda} \end{aligned} \tag{6.151}$$

6.12 Mixed Series and Parallel Systems

It can be noted from Eqs. (6.123) and (6.138) that the reliability of a series system is always smaller than the component reliability while that of a parallel system is always larger than that of the component reliability, that is,

$$R_s < R_i; \quad i = 1, 2, \ldots, n$$

and

$$R_p > R_i; \quad i = 1, 2, \ldots, n \tag{6.152}$$

This indicates that the reliability of a system can be improved by connecting the components in parallel. Other methods of improving the system reliability include placing the components in a combination of series and parallel manner as shown in Figs. 6.12 and 6.13. For the system shown in Fig. 6.12, called the parallel–series system, the reliability R_0 is given by

$$\begin{aligned} R_0 &= (R_{E_1} + R_{E_2} - R_{E_1} R_{E_2}) \\ &= (R_1 R_2 R_3 + R_4 R_5 R_6 - R_1 R_2 R_3 R_4 R_5 R_6) \end{aligned} \tag{6.153}$$

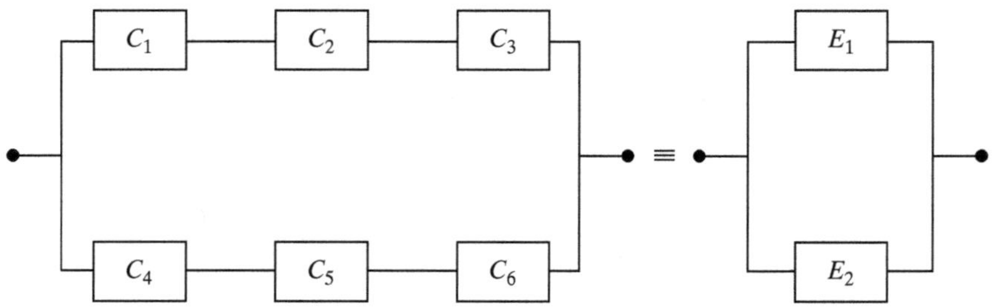

FIGURE 6.12 Components connected in series-parallel.

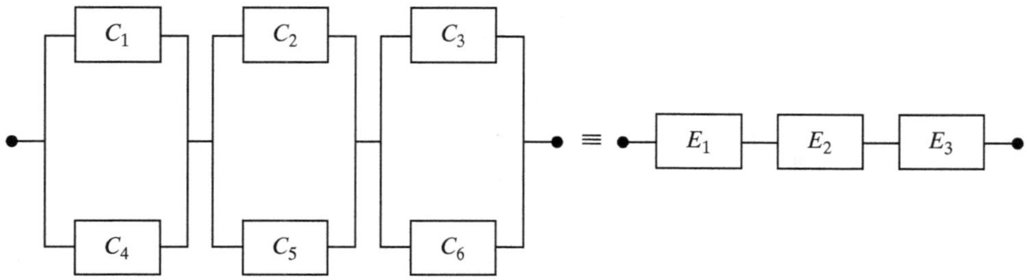

FIGURE 6.13 Components connected in series-parallel.

where

$$R_{E_1} = R_1 R_2 R_3$$
$$R_{E_2} = R_4 R_5 R_6$$

and $R_i = R_{C_i}$; $i = 1, 2, \ldots, 6$. If all the components have identical exponential failure time distributions as $F_i(t) = F(t) = 1 - e^{-\lambda t}$, Eq. (6.153) reduces to

$$R_0 = 2R^3 - R^6 = 2(1-F)^3 - (1-F)^6 = 2e^{-3\lambda t} - e^{-6\lambda t} \tag{6.154}$$

For this case, the MTBF of the system can be determined as

$$\text{MTBF} = \int_0^\infty R_0(t)\,dt = \int_0^\infty \left[2e^{-3\lambda t} - e^{-6\lambda t}\right] dt = \frac{2}{3\lambda} - \frac{1}{6\lambda} = \frac{1}{2\lambda} \tag{6.155}$$

The reliability R_0 of the system shown in Fig. 6.13, known as the series–parallel system, can be expressed as

$$\begin{aligned} R_0 &= R_{E_1} R_{E_2} R_{E_3} \qquad (6.156) \\ &= \left[(R_1 + R_4 - R_1 R_4)(R_2 + R_5 - R_2 R_5)(R_3 + R_6 - R_3 R_6)\right] \end{aligned}$$

where $R_i = R_{C_i}$, $i = 1, 2, \ldots, 6$. If all the components have identical exponential distributions as $F_i(t) = F(t) = 1 - e^{-\lambda t}$, Eq. (6.156) reduces to

$$\begin{aligned} R_0 &= (2R - R^2)^3 = (2e^{-\lambda t} - e^{-2\lambda t})^3 \\ &= 8e^{-3\lambda t} - 12e^{-4\lambda t} + 6e^{-5\lambda t} - e^{-6\lambda t} \qquad (6.157) \end{aligned}$$

The MTBF of the system can be found as

$$\begin{aligned} \text{MTBF} &= \int_0^\infty R_0(t)\,dt = \int_0^\infty \left[8e^{-3\lambda t} - 12e^{-4\lambda t} + 6e^{-5\lambda t} - e^{-6\lambda t}\right] dt \\ &= \frac{7}{10\lambda} \qquad (6.158) \end{aligned}$$

6.13 Complex Systems

If a system has a more complex configuration than the series–parallel structures discussed so far, additional evaluation techniques are needed to determine its reliability. A typical system that does not have a series–parallel configuration is shown

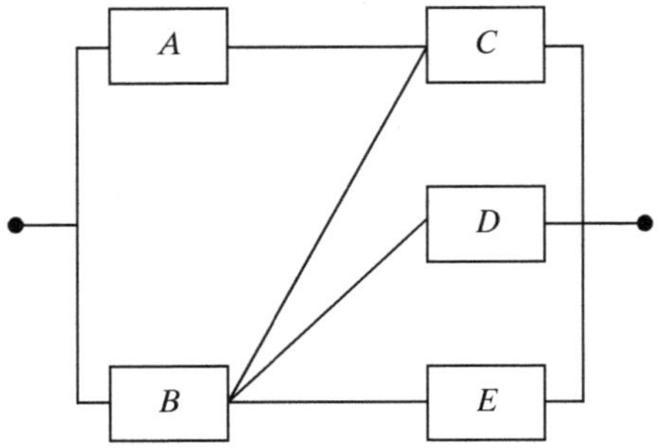

FIGURE 6.14 A typical complex system.

in Fig. 6.14. There are several methods available for finding the reliability of such systems [6.3, 6.5, 6.6]. We shall consider three methods, namely, the *enumeration, conditional probability*, and the *cut-set* approaches, in the following sections. It is to be noted that all of these methods make use of the same basic concept, but present the details in different ways. These methods are applicable for the reliability analysis of series–parallel systems as well.

6.13.1 Enumeration Method

The basis of the enumeration method is the observation that each component of the system can be either in operating condition or in failed condition. We enumerate the various possible events by considering all combinations of the conditions of different components of the system. The events which lead to the operation of the complete system are then identified by inspection and the reliability of the system is computed by finding the probability of realizing all the mutually exclusive events which lead to the operation of the complete system. Example 6.17 illustrates the procedure.

EXAMPLE 6.17

Find the reliability of the system shown in Fig. 6.14 using the enumeration method.

Solution: Denoting the operating condition of a component X as X and its failed condition as $\overline{X}$, the possible events of the system are identified as shown in Table 6.4. In this table, the events corresponding to the operation of the system are identified by the letter "O" and those corresponding to the failed condition are indicated by "F". If the reliability of component X is R_X and the probability of failure is $(1 - R_X)$, the reliability of the system can be found as

$$R_0 = \text{probability of realizing the events} \quad 1\text{–}6,\ 8\text{–}10,\ 12\text{–}16,\ 18, \text{ and } 21\text{–}23$$

TABLE 6.4 Possible events for the system of Fig. 6.14

Description of the System	Event and Condition of Various Components	Status of the System
No component fails	1. $ABCDE$	O
One component fails	2. $\bar{A}BCDE$	O
	3. $A\bar{B}CDE$	O
	4. $AB\bar{C}DE$	O
	5. $ABC\bar{D}E$	O
	6. $ABCD\bar{E}$	O
Two components fail	7. $\overline{AB}CDE$	F
	8. $\bar{A}B\bar{C}DE$	O
	9. $\bar{A}BC\bar{D}E$	O
	10. $\bar{A}BCD\bar{E}$	O
	11. $A\overline{BC}DE$	F
	12. $A\bar{B}C\bar{D}E$	O
	13. $A\bar{B}CD\bar{E}$	O
	14. $AB\overline{CD}E$	O
	15. $AB\bar{C}D\bar{E}$	O
	16. $ABC\overline{DE}$	O
Three components fail	17. $AB\overline{CDE}$	F
	18. $A\bar{B}C\overline{DE}$	O
	19. $A\overline{BC}D\bar{E}$	F
	20. $A\overline{BCD}E$	F
	21. $\bar{A}BC\overline{DE}$	O
	22. $\bar{A}B\bar{C}D\bar{E}$	O
	23. $\bar{A}B\overline{CD}E$	O
	24. $\overline{AB}CD\bar{E}$	F
	25. $\overline{AB}C\bar{D}E$	F
	26. $\overline{ABC}DE$	F
Four components fail	27. $A\overline{BCDE}$	F
	28. $\bar{A}B\overline{CDE}$	F
	29. $\overline{AB}C\overline{DE}$	F
	30. $\overline{ABC}D\bar{E}$	F
	31. $\overline{ABCD}E$	F
Five components fail	32. $\overline{ABCDE}$	F

This can be expressed as

$$\begin{aligned} R_0 = {} & R_A R_B R_C R_D R_E + (1 - R_A) R_B R_C R_D R_E \\ & + (1 - R_B) R_A R_C R_D R_E + (1 - R_C) R_A R_B R_D R_E \\ & + (1 - R_D) R_A R_B R_C R_E + (1 - R_E) R_A R_B R_C R_D \\ & + (1 - R_A)(1 - R_C) R_B R_D R_E + (1 - R_A)(1 - R_D) R_B R_C R_E \\ & + (1 - R_A)(1 - R_E) R_B R_C R_D + (1 - R_B)(1 - R_D) R_A R_C R_E \\ & + (1 - R_B)(1 - R_E) R_A R_C R_D + (1 - R_C)(1 - R_D) R_A R_B R_E \\ & + (1 - R_C)(1 - R_E) R_A R_B R_D + (1 - R_D)(1 - R_E) R_A R_B R_C \\ & + (1 - R_B)(1 - R_D)(1 - R_E) R_A R_C + (1 - R_A)(1 - R_D)(1 - R_E) R_B R_C \\ & + (1 - R_A)(1 - R_C)(1 - R_E) R_B R_D + (1 - R_A)(1 - R_C)(1 - R_D) R_B R_E \end{aligned} \tag{E1}$$

This equation can be simplified to obtain

$$\begin{aligned} R_0 = {} & R_B R_C R_D R_E - R_A R_B R_C \\ & - R_B R_C R_D - R_B R_C R_E - R_B R_D R_E + R_A R_C + R_B R_C \\ & + R_B R_D + R_B R_E \end{aligned} \tag{E2}$$

It can be noted that the method becomes quite tedious as the number of components increases. ■

6.13.2 Conditional Probability Method

In this method, a critical component of the system, C_{cri}, which possibly prevents the decomposition of the system into a series–parallel system is identified and the reliability of the system is computed as follows:

$$\begin{aligned} \begin{pmatrix}\text{Reliability of}\\ \text{the system}\end{pmatrix} = {} & \begin{pmatrix}\text{Reliability of the}\\ \text{system with component } C_{\text{cri}}\\ \text{in operating condition}\end{pmatrix}\begin{pmatrix}\text{Probability of}\\ \text{component } C_{\text{cri}} \text{ in}\\ \text{operating condition}\end{pmatrix} \\ & + \begin{pmatrix}\text{Reliability of the}\\ \text{system with component } C_{\text{cri}} \text{ in}\\ \text{failed condition}\end{pmatrix}\begin{pmatrix}\text{Probability of}\\ \text{component } C_{\text{cri}} \text{ in}\\ \text{failed condition}\end{pmatrix} \end{aligned} \tag{6.159}$$

The method is illustrated through the following example.

EXAMPLE 6.18

Find the reliability of the system shown in Fig. 6.14 using the conditional probability approach.

Solution: For this system, we choose the component C as the critical component. Then the system can be divided into two subsystems: one with component C in operating condition and the other with component C in failed condition. This subdivision is indicated in Figs. 6.15(*a*) and (*b*). When C is in

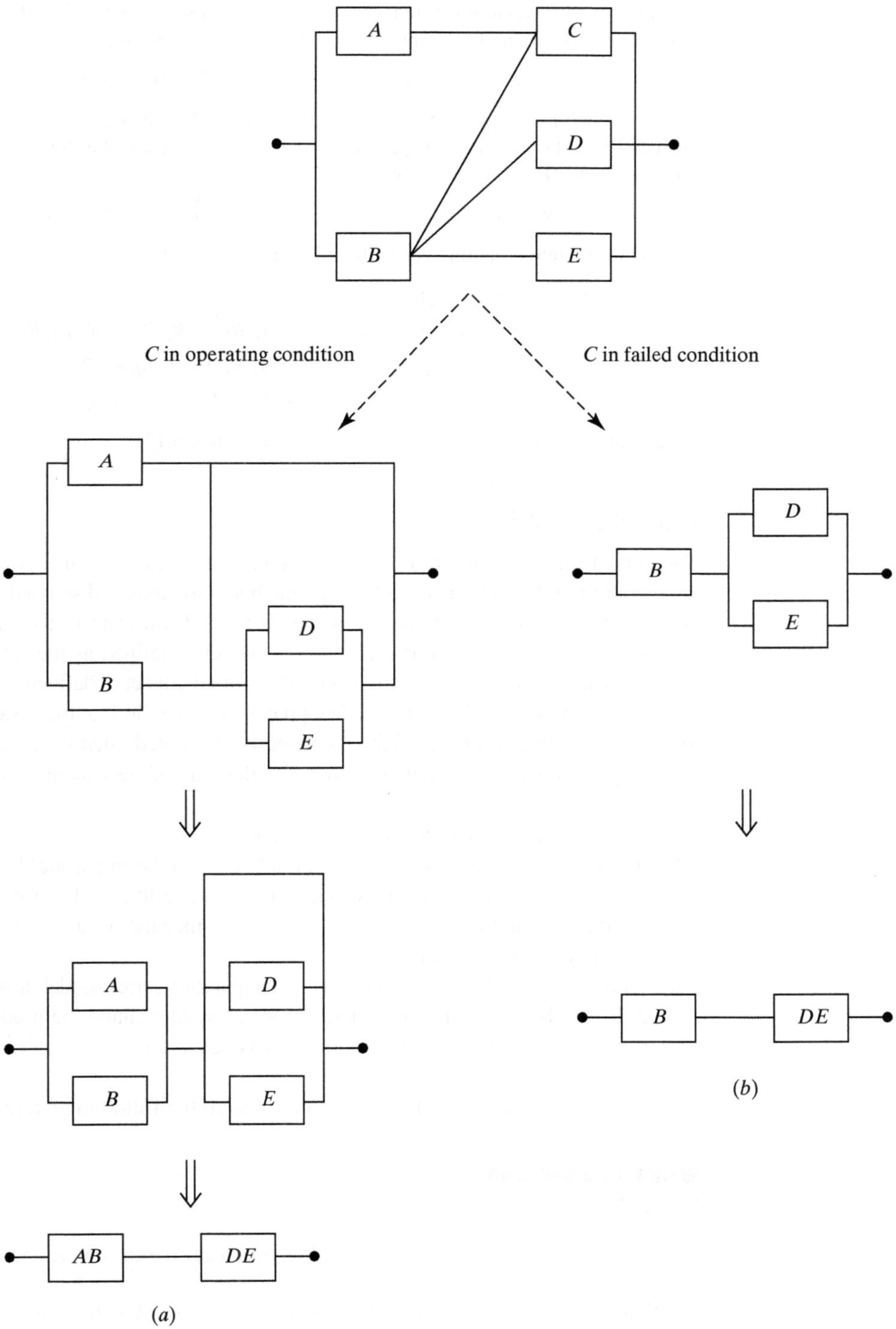

FIGURE 6.15 Subdivision of the complex system.

operating condition, the subsystem can be modeled as a series–parallel configuration as indicated in Fig. 6.15(*a*). The reliability of this system (R_1) can be evaluated as

$$R_1 = R_{AB}\,R_{DE} = [1 - (1 - R_A)(1 - R_B)](1) = R_A + R_B - R_A\,R_B \tag{E1}$$

since the reliability R_{DE} is always equal to unity. Similarly, when C is in failed condition, the subsystem can be modeled as a series–parallel configuration as indicated in Fig. 6.15(*b*). The reliability of this subsystem (R_2) can be computed as

$$R_2 = R_B\,R_{DE} = R_B[1 - (1 - R_D)(1 - R_E)] = R_B\,R_D + R_B\,R_E - R_B\,R_D\,R_E \tag{E2}$$

Finally, the system reliability can be found, using Eq. (6.159), as

$$\begin{aligned} R_0 &= R_1\,R_C + R_2(1 - R_C) \\ &= (R_A + R_B - R_A\,R_B)R_C + (R_B\,R_D + R_B\,R_E - R_B\,R_D\,R_E)(1 - R_C) \\ &= R_B\,R_C\,R_D\,R_E - R_A\,R_B\,R_C - R_B\,R_C\,R_D - R_B\,R_C\,R_E \\ &\quad - R_B\,R_D\,R_E + R_A\,R_C + R_B\,R_C + R_B\,R_D + R_B\,R_E \end{aligned} \tag{E3}$$

which can be seen to be identical to Eq. (E2) of Example 6.17.

■

6.13.3 Cut-set Method

The cut-set method is more useful in computing the reliability of a system compared to the enumeration and conditional probability methods previously discussed. The cut-set method is more convenient for computer implementation and can identify the various distinct failure modes of the system in a simple manner. A cut set is defined as the set of components that, when failed, will cause the system to fail. A minimal cut set is defined as the set of minimum number of components that, when failed, guarantees the failure of the system. This also implies that if any component of the minimal cut set has not failed, then the system will not fail. The following procedure is adopted in finding the reliability of the system using the cut-set method.

1. Identify the minimal cut sets of the system.
2. Model the components of each minimal cut set to be in parallel because, by definition, every component of the cut set must fail for the failure of the system.
3. Assume that the various cut sets are in series since the occurrence of any cut set implies the failure of the system.
4. Find the reliability of the system using the parallel–series model developed in steps 2 and 3. It is to be noted that the concept of series system cannot be used directly because the same component may appear in more than one cut set.

The procedure to be adopted is illustrated through the following example.

EXAMPLE 6.19

Find the reliability of the system shown in Fig. 6.14 using the cut-set method.

Solution: By inspection, the independent minimal cut sets of Fig. 6.14 can be identified as *AB, BC,* and *CDE*. These cut sets are denoted as C_1, C_2, and C_3 and are assumed to be connected in series as

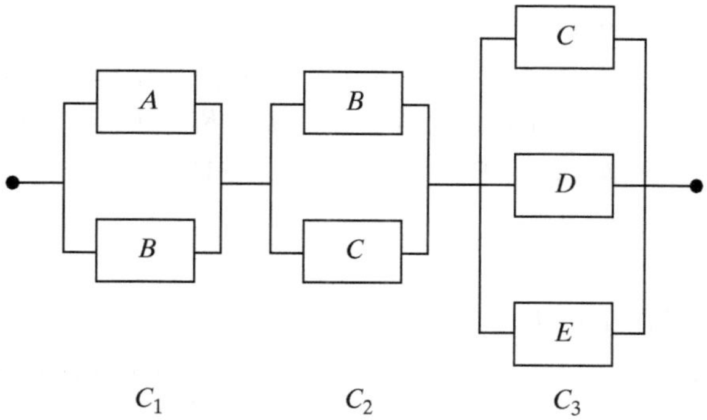

FIGURE 6.16 Modeling in the cut-set method.

indicated in Fig. 6.16, since the occurrence of any one cut set implies the failure of the system. It will be more convenient to find the probability of failure of the system P_{fs} as:

$$P_{fs} = P(C_1 \cup C_2 \cup C_3) = P(C_1) + P(C_2) + P(C_3) - P(C_1 \cap C_2) - P(C_1 \cap C_3) - P(C_2 \cap C_3) + P(C_1 \cap C_2 \cap C_3) \quad \text{(E1)}$$

where the probabilities of occurrence of the various events can be expressed as

$$P(C_1) = P_f(A)P_f(B) \quad \text{(E2)}$$

$$P(C_2) = P_f(B)P_f(C) \quad \text{(E3)}$$

$$P(C_3) = P_f(C)P_f(D)P_f(E) \quad \text{(E4)}$$

$$P(C_1 \cap C_2) = P(C_1)P(C_2) = P_f(A)P_f(B)P_f(C) \quad \text{(E5)}$$

$$P(C_1 \cap C_3) = P(C_1)P(C_3) = P_f(A)P_f(B)P_f(C)P_f(D)P_f(E) \quad \text{(E6)}$$

$$P(C_2 \cap C_3) = P(C_2)P(C_3) = P_f(B)P_f(C)P_f(D)P_f(E) \quad \text{(E7)}$$

$$P(C_1 \cap C_2 \cap C_3) = P(C_1)P(C_2)P(C_3) = P_f(A)P_f(B)P_f(C)P_f(D)P_f(E) \quad \text{(E8)}$$

and $P_f(X)$, for example, denotes the probability of failure of component X. Also note that if a component appears in more than one cut set, its probability of failure is represented only once. For example, in Eq. (E5), the component B appears in both the minimal cut sets C_1 and C_2; however, its probability of failure $P_f(B)$ is represented only once on the right-hand side of Eq. (E5). Thus, the probability of failure of the system P_{fs} can be expressed using Eqs. (E1) to (E8) as

$$P_{fs} = P_f(A)P_f(B) + P_f(B)P_f(C) + P_f(C)P_f(D)P_f(E) - P_f(A)P_f(B)P_f(C) - P_f(B)P_f(C)P_f(D)P_f(E) \quad \text{(E9)}$$

By using the relation

$$P_f(X) = 1 - R_X \quad \text{(E10)}$$

where R_X denotes the reliability of component X, Eq. (E9) can be expressed as

$$\begin{aligned} P_{fs} &= (1 - R_A)(1 - R_B) + (1 - R_B)(1 - R_C) + (1 - R_C)(1 - R_D)(1 - R_E) \\ &\quad - (1 - R_A)(1 - R_B)(1 - R_C) - (1 - R_B)(1 - R_C)(1 - R_D)(1 - R_E) \\ &= 1 - R_A R_C + R_A R_B R_C - R_B R_D - R_B R_E + R_B R_D R_E \\ &\quad - R_B R_C + R_B R_C R_D + R_B R_C R_E - R_B R_C R_D R_E \end{aligned} \quad \text{(E11)}$$

Finally, the reliability of the system can be computed as

$$R_0 = 1 - P_{fs} = R_A R_C + R_B R_C + R_B R_D + R_B R_E - R_A R_B R_C - R_B R_D R_E - R_B R_C R_D - R_B R_C R_E + R_B R_C R_D R_E \tag{E12}$$

which can be seen to be identical to Eq. (E2) of Example 6.17.

■

6.14 Reliability Enhancement

In general, the problem of "reliability allocation" deals with the determination of reliabilities of individual components such that the overall system reliability is equal to a specified value while satisfying any restrictions present on weight, volume, or cost [6.4]. This problem is usually considered at the design stage. The problem of "reliability enhancement" arises when the reliability of an existing system is found to be inadequate. The reliability of a system can be enhanced from a current value R_0 to a new value $R_0 + dR_0$ by improving the reliability of one or more of its components. In this section, we consider the improvement of system reliability by increasing the reliability of the most sensitive component. The weight, volume, or cost constraints also are to be considered in selecting a suitable component for reliability enhancement. The following procedure can be used for the enhancement of reliability of a system.

6.14.1 Series System

Assuming that the desired system reliability $R_0 + dR_0$ can be achieved by improving the reliability of any one component, we have

$$\begin{aligned} R_0 + dR_0 &= R_1 R_2 \cdots R_{i-1}(R_i + dR_i)R_{i+1} \cdots R_n \\ &= \prod_{j=1}^{n} R_j + \left(\prod_{j=1, j \neq i}^{n} R_j \right) dR_i; \quad i = 1, 2, \ldots, n \end{aligned} \tag{6.160}$$

where R_j is the reliability of jth component and dR_i is the improvement made in the reliability of ith component. Since

$$R_0 = \prod_{j=1}^{n} R_j \tag{6.161}$$

Equation (6.160) gives

$$dR_i = \frac{dR_0}{\left(\prod\limits_{j=1, j \neq i}^{n} R_j \right)}; \quad i = 1, 2, \ldots, n \tag{6.162}$$

With no constraint. If there are no constraints, we select the component that is most sensitive to system reliability for reliability enhancement. The rate of change of the system reliability R_0 with respect to the reliability of kth component R_k can be obtained, from Eq. (6.161), as

$$\frac{\partial R_0}{\partial R_k} = \prod_{j=1, j \neq k}^{n} R_j = \frac{R_0}{R_k}; \quad k = 1, 2, \ldots, n \tag{6.163}$$

Thus, the most sensitive component i which can be used for reliability enhancement is given by

$$\frac{R_0}{R_i} = \max_{k=1,2,\ldots,n} \left(\frac{R_0}{R_k} \right) \tag{6.164}$$

The reliability increment dR_i necessary to achieve the overall system reliability improvement of dR_0 is given by Eq. (6.162).

With constraint. Usually, the improvement in the reliability of a component involves extra weight, volume, or cost. Assuming, for example, the cost per unit reliability of ith component as c_i, the problem can be stated as follows:

> *Consider the improvement in the reliability of ith component dR_i necessary to increase the system reliability by dR_0 to minimize the cost, $c_i\, dR_i$.*

Assume that the reliability of the system can be improved from R_0 to $R_0 + dR_0$ by increasing the reliability either of component i from R_i to $R_i + dR_i$ or of component j from R_j to $R_j + dR_j$. Thus, Eq. (6.162) gives, in the absence of any cost constraint,

$$dR_0 = dR_i \left(\prod_{k=1,\, k \neq i}^{n} R_k \right) = dR_j \left(\prod_{k=1,\, k \neq j}^{n} R_k \right)$$

that is,

$$dR_i \frac{R_0}{R_i} = dR_j \frac{R_0}{R_j}$$

that is,

$$dR_i = \frac{R_i}{R_j} dR_j \tag{6.165}$$

The cost involved in achieving the new system reliability $R_0 + dR_0$ can be expressed as

$$c_i\, dR_i = \left(\frac{c_i\, R_i}{c_j\, R_j} \right) c_j\, dR_j \tag{6.166}$$

This shows that the cost involved in improving the reliability of ith component will be less than that of jth component if $c_i\, R_i$ is less than $c_j\, R_j$. Thus, the component i to be selected for system reliability enhancement, for minimum cost, is given by

$$c_i\, R_i = \min_{j=1,2,\ldots,n} c_j\, R_j \tag{6.167}$$

with dR_i given by Eq. (6.162).

6.14.2 Parallel System

The reliability of a parallel system, R_0, can be expressed in terms of the component reliabilities R_i as

$$R_0 = 1 - \prod_{i=1}^{n} (1 - R_i) \tag{6.168}$$

Assuming that the desired system reliability $R_0 + dR_0$ can be achieved by increasing the reliability of any one component, we obtain

$$\begin{aligned} R_0 + dR_0 &= 1 - (1 - R_1)(1 - R_2) \\ &\quad \cdots (1 - R_{i-1})(1 - R_i - dR_i)(1 - R_{i+1}) \cdots (1 - R_n) \\ &= 1 - \left[\prod_{k=1,\, k \neq i}^{n} (1 - R_k) \right] (1 - R_i - dR_i) \\ &= 1 - \prod_{k=1}^{n} (1 - R_k) + dR_i \left[\prod_{k=1,\, k \neq i}^{n} (1 - R_k) \right]; \quad i = 1, 2, \dots, n \end{aligned} \tag{6.169}$$

In view of Eq. (6.168), Eq. (6.169) gives

$$dR_i = \frac{dR_0}{\left[\prod_{k=1,\, k \neq i}^{n} (1 - R_k) \right]}; \quad i = 1, 2, \dots, n \tag{6.170}$$

With no constraint. If there are no constraints, we select the component that is most sensitive to system reliability for reliability enhancement. The sensitivity of the system reliability to changes in component reliability can be found as

$$\frac{\partial R_0}{\partial R_k} = \prod_{j=1,\, j \neq k}^{n} (1 - R_j) = \frac{1 - R_0}{1 - R_k}; \quad k = 1, 2, \dots, n \tag{6.171}$$

Thus, the most sensitive component i which can be used for reliability enhancement is given by

$$\frac{1 - R_0}{1 - R_i} = \max_{k=1,2,\dots,n} \left(\frac{1 - R_0}{1 - R_k} \right) \tag{6.172}$$

The incremental reliability dR_i necessary to achieve the overall system reliability improvement of dR_0 is given by Eq. (6.170).

With constraint. As before, let the cost per unit reliability of ith component be c_i and the problem be to find the value of dR_i necessary to achieve the incremental system reliability dR_0 so as to minimize the cost, $c_i\, dR_i$. If the reliability of the system can be improved from R_0 to $R_0 + dR_0$ by increasing the reliability either of component i from R_i to $R_i + dR_i$ or of component j from R_j to $R_j + dR_j$, Eq. (6.170) gives

$$dR_0 = dR_i \left[\prod_{k=1,\, k \neq i}^{n} (1 - R_k) \right] = dR_j \left[\prod_{k=1,\, k \neq j}^{n} (1 - R_k) \right]$$

that is,

$$\frac{dR_i}{(1 - R_i)} \prod_{k=1}^{n} (1 - R_k) = \frac{dR_j}{1 - R_j} \prod_{k=1}^{n} (1 - R_k)$$

that is,

$$dR_i = \left(\frac{1 - R_i}{1 - R_j} \right) dR_j \tag{6.173}$$

The cost involved in achieving the new system reliability $R_0 + dR_0$ can be expressed as

$$c_i \, dR_i = \frac{c_i(1 - R_i)}{c_j(1 - R_j)} c_j \, dR_j \tag{6.174}$$

This shows that the cost involved in improving the reliability of ith component will be less than that of jth component if $c_i \, (1 - R_i)$ is less than $c_j \, (1 - R_j)$. Thus, the component i to be selected for system reliability enhancement, for minimum cost, is given by

$$c_i \, (1 - R_i) = \min_{j=1,2,\ldots,n} c_j(1 - R_j) \tag{6.175}$$

with dR_i given by Eq. (6.170).

6.15 Reliability Allocation—AGREE Method

Several procedures can be adopted for allocating reliability to individual components in a complex system. The reliability allocation methods based on an optimization criterion are presented in Chapter 11. A simpler method, suggested by AGREE, is presented in this section. The AGREE (Advisory Group on Reliability of Electronic Equipment) method of allocating the component failure rates was given for aircraft electronic systems [6.1, 6.2]. It is also applicable to other systems that can be decomposed into a series of independent subsystems.

The reliability of the system $R_0(t)$, which consists of k subsystems (in series), is given by

$$R_0(t) = \prod_{i=1}^{k} R_i(t) \tag{6.176}$$

where $R_i(t)$ is the reliability of the ith subsystem. If the ith subsystem has a mean life of m_i (with a constant failure rate $\lambda_i = 1/m_i$) and is required to operate over a time period t_i $(t_i < m_i)$, then the reliability of the ith subsystem at time t_i is given by

$$R_i = e^{-t_i/m_i} \tag{6.177}$$

and its probability of failure by $(1 - R_i)$. If w_i denotes the probability of failure of the total system due to the failure of the ith subsystem, then the probability of simultaneous survival of the system and the failure of the ith subsystem is given by

$$1 - w_i \, [1 - R_i(t_i)]$$

Thus the reliability of the total system can be expressed as

$$R_0 = \prod_{i=1}^{k} \left\{ 1 - w_i [1 - R_i(t_i)] \right\} \tag{6.178}$$

Equations (6.177) and (6.178) yield

$$R_0 = \prod_{i=1}^{k} \left\{ 1 - w_i \left[1 - e^{-t_i/m_i} \right] \right\} \tag{6.179}$$

Since the expected operating time of a subsystem (t_i) is much shorter than its mean life (m_i), the term in square brackets in Eq. (6.179) can be approximated as

$$1 - e^{-t_i/m_i} \approx 1 - \left[1 - \frac{t_i}{m_i} \right] = \frac{t_i}{m_i}$$

and hence Eq. (6.179) can be replaced by

$$\begin{aligned} R_0 &\approx \prod_{i=1}^{k} \left\{ 1 - \frac{w_i\, t_i}{m_i} \right\} \approx \prod_{i=1}^{k} e^{-w_i t_i / m_i} \\ &= \exp\left[-\sum_{i=1}^{k} \frac{w_i\, t_i}{m_i} \right] \end{aligned} \tag{6.180}$$

where the equivalence

$$e^{-w_i t_i/m_i} \approx 1 - \frac{w_i\, t_i}{m_i}$$

has been used in Eq. (6.180). Assuming that the ith subsystem consists of n_i identical components in series, each with a mean life of T_i or failure rate of $1/T_i$, the failure rate of the ith subsystem (λ_i) can be expressed as

$$\lambda_i = \frac{1}{m_i} = \frac{n_i}{T_i}$$

or

$$T_i = m_i\, n_i \tag{6.181}$$

By noting that

$$\begin{aligned} \frac{w_i\, t_i}{m_i} &= \frac{w_i\, t_i\, n_i}{T_i} = n_i \left(\frac{w_i\, t_i}{T_i} \right) \\ &= \frac{w_i\, t_i}{T_i} + \frac{w_i\, t_i}{T_i} + \cdots \ (\text{total } n_i \text{ terms in the sum}) \\ &= \sum_{j=1}^{n_i} \frac{w_i\, t_i}{T_i} \end{aligned}$$

and $\sum_{i=1}^{k} n_i = N =$ total number of components in the system, Eq. (6.180) can be rewritten as

$$R_0 \approx \exp\left[-\sum_{i=1}^{k} \frac{w_i t_i}{m_i}\right]$$

$$\approx \exp\left[-\sum_{i=1}^{k}\sum_{j=1}^{n_i} \frac{w_i t_i}{T_i}\right] \qquad (6.182)$$

If each component contributes equally to the system reliability, we have

$$\frac{w_i t_i}{T_i} = c = \text{constant}$$

and Eq. (6.182) becomes

$$R_0 \approx e^{-cN} = \left[e^{-c}\right]^N = \left[e^{-w_i t_i / T_i}\right]^N$$

or

$$\ln R_0 = N\left(-\frac{w_i t_i}{T_i}\right) = -\frac{w_t t_i N}{n_i m_i}$$

or

$$m_i = -\frac{N w_i t_i}{n_i \ln R_0} \qquad (6.183)$$

EXAMPLE 6.20

A system is composed of five subsystems with details as indicated in the following table:

Subsystem (i)	Number of Components (n_i)	Operating Period in Hours (t_i)	Probability of System Failure Due to Failure of Subsystem i (w_i)
1	5	10	0.15
2	2	25	0.10
3	8	5	0.20
4	6	20	0.05
5	4	15	0.25

Determine the mean lives of the components of various subsystems (m_i) so as to have a system reliability of 0.99 using the AGREE method.

Solution: The total number of components is $N = \sum_{i=1}^{5} n_i = 5 + 2 + 8 + 6 + 4 = 25$. The application of Eq. (6.183) yields

$$m_1 = -\frac{N\,w_1\,t_1}{n_1 \ln R_0} = -\frac{25(0.15)(10)}{5 \ln 0.99} = 746.2687 \text{ hours}$$

$$m_2 = -\frac{N\,w_2\,t_2}{n_2 \ln R_0} = -\frac{25(0.10)(25)}{2 \ln 0.99} = 3109.4527 \text{ hours}$$

$$m_3 = -\frac{N\,w_3\,t_3}{n_3 \ln R_0} = -\frac{25(0.20)(5)}{8 \ln 0.99} = 310.9453 \text{ hours}$$

$$m_4 = -\frac{N\,w_4\,t_4}{n_4 \ln R_0} = -\frac{25(0.05)(20)}{6 \ln 0.99} = 414.5937 \text{ hours}$$

$$m_5 = -\frac{N\,w_5\,t_5}{n_5 \ln R_0} = -\frac{25(0.25)(15)}{4 \ln 0.99} = 2332.0896 \text{ hours}$$

■

6.16 Numerical Solutions Using MATLAB and Excel

EXAMPLE 6.21

The automobile batteries produced by a manufacturer are found to follow normal distribution with mean life 33 months and standard deviation 2.34 months. Find the reliability and failure rate of the batteries at the end of 40 months of operation using MATLAB.

Solution: The reliability and failure rates corresponding to normal distribution are given by Eqs. (6.38) and (6.39), respectively. The MATLAB program developed to find the solution and the results are given below.

```
clear all ; clc ;

format long
mu = 33

sd = 2.34

time = 40

z = (time - mu)/sd

R = 1 - normcdf(z)

h = normpdf(z,0,1)/R
```

```
mu =

    33

sd =

  2.340000000000000

time =

    40

z =

  2.991452991452992

R =

  0.001388266413960

h =

  3.275155867816572
```

■

EXAMPLE 6.22

The failure time of an electromechanical component is known to follow normal distribution with parameters $\mu = 5000$ hours (mean) and $\sigma = 400$ hours (standard deviation). Using MATLAB, find the expected residual life of the component at 4000 hours of operation.

Solution: The expected residual life can be found using Eq. (6.121). The trapezoidal rule is used for evaluating the integral in Eq. (6.121). The MATLAB program developed to find the residual life of the component and the result obtained are shown below.

```
clear all ; clc ;

mu = 5000
sd = 400
time = 4000
tmax = 8000
R0 = 1 - normcdf(time,mu,sd)

t(1) = 4000
np = 81
delta = (tmax-t(1))/(np-1)

for i = 2:np
    t(i) = t(1) +(i-1)*delta  ;
end
t

for i = 1:np
    R(i) = 1-normcdf(t(i),mu,sd)  ;
end
R

sum = 0 ;
for j=2 : np-1
    sum = sum + 2*R(j) ;
end
sum = sum + R(1) + R(np)

Int = (delta/2)*sum

LifeLeft = (1/R0)*Int

mu=

  5000

sd=

  400

time=

  4000

tmax=

  8000

R0=

  0.9938
```

```
sum=

  40.0324

Int=

   1.0008e+003

LifeLeft=

   1.0071e+003

>>
```

■

EXAMPLE 6.23

The failure time of an I.C. engine follows a two-parameter Weibull distribution with parameters $\beta = 4$ and $\eta = 2000$. Find the reliability and failure rate of the engine for an operating time of 2500 hours using Excel.

Solution: The reliability and failure rate corresponding to a two-parameter Weibull distribution are given by Eqs. (6.73) and (6.74), respectively. The solution found using Excel is shown below. Note that the parameters are labeled as α and β for β and η, respectively, in Excel.

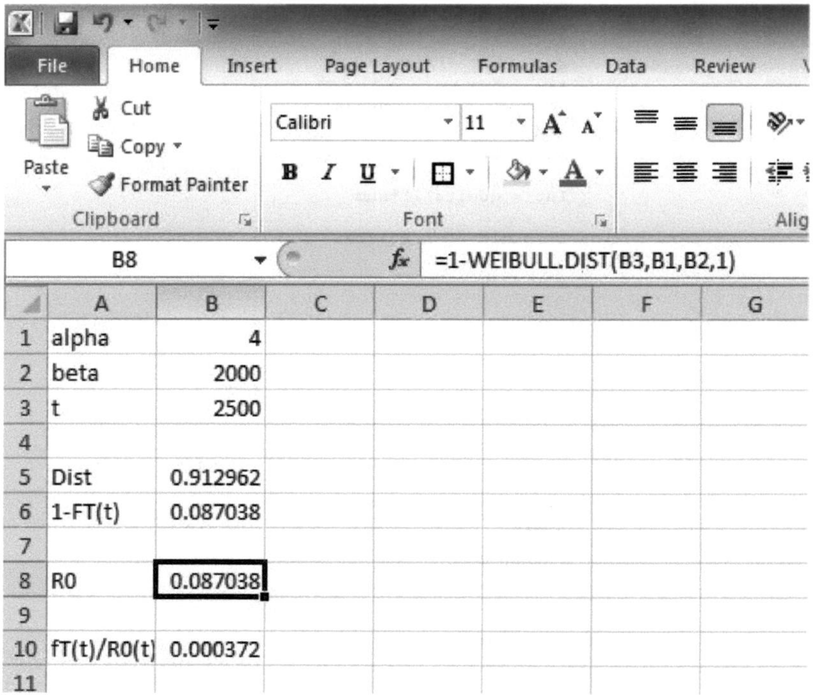

■

REFERENCES AND BIBLIOGRAPHY

6.1 D. L. Grosh, *A Primer of Reliability Theory*, John Wiley, New York, 1989.

6.2 K. C. Kapur and L. R. Lamberson, *Reliability in Engineering Design*, John Wiley, New York, 1977.

6.3 E. E. Lewis, *Introduction to Reliability Engineering*, John Wiley, New York, 1987.

6.4 J. G. Rau, *Optimization and Probability in Systems Engineering*, Van Nostrand Reinhold, New York, 1970.

6.5 L. A. Doty, *Reliability for the Technologies*, 2nd ed., Industrial Press, New York, 1989.

6.6 R. Billinton and R. N. Allan, *Reliability Evaluation of Engineering Systems: Concepts and Techniques*, 2nd ed., Plenum Press, New York, 1992.

6.7 A. C. King and C. B. Read, *Pathways to Probability*, Holt, Rinehart & Winston, Inc., New York, 1963.

6.8 I. Asimov, *Asimov's Biographical Encyclopedia of Science and Technology*, Doubleday & Co. Inc., Garden City, New York, 1972.

6.9 D. C. Montgomery and G. C. Runger, *Applied Statistics and Probability for Engineers*, 4th ed., John Wiley, Hoboken, NJ, 2007.

6.10 E. A. Elsayed, *Reliability Engineering*, Addison Wesley Longman, Inc., Reading, MA, 1996.

6.11 D. B. Bunday, *Statistical Methods in Reliability Theory and Practice*, Ellis Horwood, Chichester, England, 1991.

6.12 M. Modarres, M. Kaminskiy, and V. Krivtsov, *Reliability Engineering and Risk Analysis. A Practical Guide*, 2nd ed., CRC Press, Boca Raton, FL, 2010.

REVIEW QUESTIONS

6.1 Give brief answers to the following:

1. What is a bathtub curve?
2. What is infant mortality period?
3. What are the causes for failure of components during the wear-out period?
4. What are the causes of failure during the operating period of components?
5. Define the hazard function.
6. How is the reliability function related to the failure rate?
7. Which distribution function is commonly used to model failure time during the operating period of a component?
8. Suggest a method for estimating the failure rate of a component from experimental data.
9. Define MTBF.
10. What is the relationship between MTBF and the probability density function of failure time of a component?
11. Express the reliability of a series system in terms of its component reliabilities.
12. How is the system-failure rate related to the component-failure rates for a series system?
13. Find the reliability of a parallel system in terms of its component reliabilities.

14. What is a (*k*, *n*) system?
15. Give two practical examples each for a series, a parallel, and a (*k*, *n*) system.
16. What are series–parallel and parallel–series systems?
17. What is meant by a complex system? List three methods for finding the reliability of a complex system.
18. Define a cut set. What is the difference between a cut set and a minimal cut set?
19. What is the AGREE method?
20. State some methods of enhancing the reliability of a multicomponent system.
21. Define the residual life function.
22. Indicate the nature of reliability function when the failure time follows uniform distribution.

6.2 Indicate whether each of the following statements is true or false:

1. Even seemingly identical systems, operating under similar conditions, can fail at different times.
2. Human lives follow the bathtub curve.
3. The life of a system during the wear-out period has an increasing failure rate.
4. Mean time before failure is same as the expected life.
5. A series system is also called a fail-safe system.
6. The reliability of a series system is always greater than the reliability of any of its components.
7. The reliability of a parallel system is never smaller than the reliability of any of its components.
8. A parallel system is called a redundant system.
9. The probability distribution function of the failure time of a parallel system, in terms of the probability distribution functions of the failure times of its components $F_i(t)$, is given by $\prod_{i=1}^{n} F_i(t)$.
10. The enumeration method involves identification of the failure status of the system by inspection.

6.3 Fill in the blank with the proper word(s):

1. Reliability can be considered as the probability of ______ performance of the system at any time.
2. The failure rate versus time graph of most mechanical and electronic systems is called the ______ curve.
3. Failures due to defective materials or poor quality control usually take place during the ______ period.
4. Failures during the infant mortality period are generally covered by the manufacturer in the form of ______.
5. The reliability of a series system, in terms of the reliability of its components R_i, is given by ______.
6. If the failure rates of components 1 and 2 connected in series are 10^{-3}/hour and 10^{-4}/hour, the failure rate of the system is ______.
7. A (k, n) system is one which functions even when any ______ components fail.

8. For a $(1, 2)$ system with identical components, the expression

$$\sum_{i=1}^{2} \binom{2}{1} [1 - F(t)]^i [F(t)]^{2-i}$$

denotes the ______ of the system at time t.

9. If failure rates of components 1 and 2 connected in parallel are 10^{-3}/hour and 10^{-4}/hour, the failure rate of the system is ______.

10. The conditional probability method of finding the reliability of a complex system involves identifying a (N) ______ component.

11. The ______ method is more convenient for computer implementation in finding the reliability of a complex system.

6.4 Select the most appropriate answer from the multiple choices given:

1. The main reason for the failure of different lightbulbs at different times is due to:
a. Random life of bulbs **b.** Infant mortality of bulbs **c.** Poor quality of bulbs

2. Failures during the normal operating period are usually due to:
a. Poor quality control **b.** Wearing of the system **c.** Random life of the system

3. If the number of component that fail in time t is N_f out of N, the reliability of the component is:
a. $1 - \dfrac{N_f}{N}$ **b.** $\dfrac{N_f}{N}$ **c.** $\dfrac{N}{N_f}$

4. If the number of component that fail in time t is N_f out of N, the hazard rate is given by:
a. $\dfrac{dN_f}{dt}$ **b.** $\dfrac{d(N - N_f)}{dt}$ **c.** $\dfrac{1}{(N - N_f)} \dfrac{dN_f}{dt}$

5. The following relation represents the hazard rate:
a. $-\dfrac{dR}{dt}$ **b.** $\dfrac{f_T(t)}{R(t)}$ **c.** $\dfrac{f_T(t)}{F_T(t)}$

6. If t_i denotes the life of component i, the failure time of a series system with n components is given by:
a. $\min_{1 \le i \le n} (t_i)$ **b.** $\max_{1 \le i \le n} (t_i)$ **c.** $\prod_{i=1}^{n} t_i$

7. The failure rate of a series system, composed of components with constant failure rates λ_i, is given by:
a. $\prod_{i=1}^{n} \lambda_i$ **b.** $\sum_{i=1}^{n} \lambda_i$ **c.** $\max_{1 \le i \le n} (\lambda_i)$

8. If R_i is the reliability of component i, the reliability of a parallel system is given by:
a. $\prod_{i=1}^{n} (1 - R_i)$ **b.** $\sum_{i=1}^{n} R_i$ **c.** $1 - \prod_{i=1}^{n} (1 - R_i)$

9. If t_i denotes the life of component i, the failure time of a parallel system with n components is given by:

a. $\min_{1 \le i \le n} (t_i)$ **b.** $\max_{1 \le i \le n} (t_i)$ **c.** $\prod_{i=1}^{n} t_i$

10. The reliability of a complex system can be found using:
a. Cut-set approach
b. Product of component reliabilities
c. Sum of component reliabilities

11. The AGREE method is useful for:
a. Computing reliability **b.** Allocating reliability **c.** Enhancing reliability

6.5 For a system with constant failure rate λ, match the following:

1. Ratio of probability density and reliability functions	**a.** $\lambda e^{-\lambda t}$
2. Probability distribution function of failure time	**b.** $e^{-\lambda t}$
3. Probability density function of failure time	**c.** $1 - e^{-\lambda t}$
4. Reliability function	**d.** $\frac{1}{\lambda}$
5. MTBF	**e.** λ

6.6 If the life of a component follows exponential distribution with MTBF = 1000 hours, match the following:

1. Failure rate	**a.** e^{-2}
2. Reliability at t = 1000 hours	**b.** $10^{-3} e^{-1}$
3. Value of probability distribution function at t = 1000 hours	**c.** 10^{-3}
4. Value of probability density function at t = 1000 hours	**d.** $1 - e^{-1}$
5. Reliability at t = 2000 hours	**e.** e^{-1}

6.7 If the constant failure rates of two components are given by $\lambda_1 = 0.0005/\text{hour}$ and $\lambda_2 = 0.0010/\text{hour}$, match the following:

1. Failure rate of series system	**a.** 666.6667
2. MTBF of series system	**b.** 0.0015
3. Reliability of series system at t = 1000 hour	**c.** 2333.3333
4. MTBF of parallel system	**d.** 0.7513
5. Reliability of parallel system at t = 1000 hour	**e.** 0.2231

PROBLEMS

6.1 The hazard function of electric motors is given by $h(t) = 5 \times 10^{-8} t^{0.6}$ failures per hour. If 200 electric motors are tested, how many motors are expected to fail in the first 1000 hours?

6.2 Twenty automobile brakes are tested in the laboratory and the following failure times (in thousands of miles of driving) were observed: 18, 30, 48, 67, 80, 90, 100, 110, 120, 133,

142, 155, 159, 165, 175, 182, 190, 197, 205, 210. (a) Plot the probability density, probability distribution, reliability, and failure rate functions based on the data. (b) Which continuous distribution closely matches the data?

6.3 In order to determine the fatigue life of a composite material, 50 specimens are tested to a cyclic load and the following data are obtained:

Number of Cycles (Unit: 10^4 Cycles)	**Number of Specimens Failed**
50	14
100	25
150	36
200	45
250	42
300	47
350	50

a. Plot the graphs of $f_T(t)$, $F_T(t)$, $R(t)$, and $h(t)$.
b. Find an analytical equation for $h(t)$ that matches closely with the test results.
c. Find the MTTF of the composite material using the analytical equation of $h(t)$.

6.4 The data on the lives of 1000 identical components is given in the following table:

Time Interval (100 Hours)	**Number of Failures in the Interval**
1	223
2	68
3	57
4	53
5	50
6	46
7	44
8	46
9	77
10	142

Plot the failure rate versus time curve.

6.5 Find the MTTF of a component which has a reliability of 0.99 for an operating time of 1000 hours.

6.6 The reliability function of a system is given by $R(t) = \exp(-t^2)$. Find the corresponding

a. failure rate function.
b. probability density function of the failure time.

6.7 Find the reliability of an engine for an operating time of 500 hours if the failure rate is 4 per 10^6 hours.

6.8 If the failure rate of a component is 1 in 10^4 hours, determine the following:

a. reliability function of the component.
b. reliability function of a system with four components in series.
c. reliability function of a system with four components in parallel.

6.9 The probability density function of failure time of a component is shown in Fig. 6.17. Determine the corresponding failure rate and reliability functions.

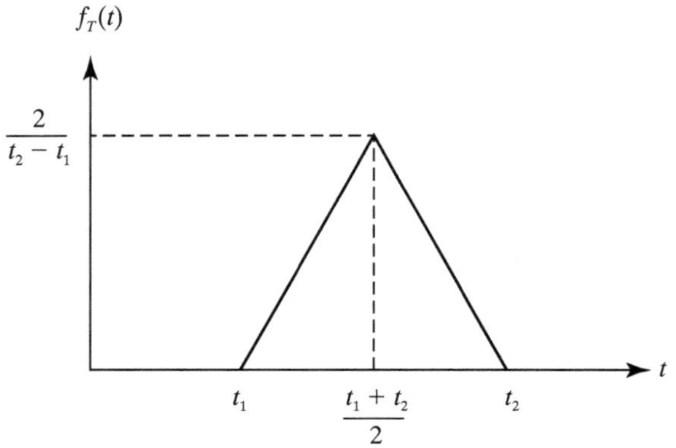

FIGURE 6.17 Triangular distribution.

6.10 The probability density function of failure time of a component is given by Rayleigh distribution

$$f_T(t) = at \exp\left[-\frac{a\,t^2}{2}\right]; \quad t \geq 0$$

Show that the failure rate and reliability functions of the component are given by

$$h(t) = at \qquad \text{and} \qquad R(t) = \exp\left[-\frac{a\,t^2}{2}\right]$$

6.11 The failure rate of a mechanical system is found to increase linearly with time as $h(t) = c\,t$ where c is a positive constant. Determine the following:

a. Probability density function of the failure time of the system.
b. Probability distribution function of the failure time.
c. Reliability of the system at time t.
d. MTTF of the system.

6.12 The life of an electronic component is found to follow exponential distribution. Determine the median life of the component.

Hint: Median life corresponds to the time at which 50 percent of the components will fail (or survive).

6.13 The probability density function of the failure time (T) of a component is given by $f_T(t) = \lambda\, e^{-\lambda t}$. Find the mean, standard deviation, and the coefficient of variation of the failure time T.

6.14 The life of windshield wipers of an automobile are found to follow exponential distribution with a failure rate of 0.0005 failures per hour.

a. Find the reliability of the wipers at 2000 hours.
b. If it is desired to have no more than 5 percent probability of failure of the wipers, how often the windshield wipers need to be replaced? Assume that the automobile is driven for 3 hours per day and there are 80 rainy days in a year.

6.15 The failure time of a mechanical component is approximated by the following distribution:

$$f_T(t) = \lambda\, e^{-\lambda(t-t_0)}; \quad t \geq 0 \tag{E1}$$

where $t_0 > 0$ denotes a constant. Equation (E1) indicates that the component will not fail until the operating period t exceeds a wear-in or threshold period t_0. Plot the graphs of $f_T(t), F_T(t), R(t)$, and $h(t)$ corresponding to Eq. (E1).

6.16 The failure time of the tires (in terms of the miles of operation) used in an automobile is found to follow normal distribution with a mean of 40,000 miles and standard deviation of 8000 miles.

a. Find the reliability of a tire at 50,000 miles.
b. Ignoring all other failure modes, find the reliability of the automobile considering all the four tires at 50,000 miles.
c. If it is desired to have no more than 5 percent probability of failure of the automobile (ignoring all failure modes other than the tire failures), how often the tires need to be replaced? Assume that the automobile is driven, on the average, for 250 miles a day.

6.17 The failure time of a machine component, in hours of operation, is known to follow the following distribution:

$$f_T(t) = t\, e^{-t^2}; \quad t \geq 0$$

a. Find the cumulative distribution function, $F_T(t)$.
b. Find the reliability function, $R(t)$.
c. Find the failure rate function, $h(t)$.

6.18 The life of a voltmeter, in hours, has exponential distribution with a value of $\lambda = 0.0005$. Determine the following:

a. Probability that the voltmeter's life is at least 8000 hours.
b. Probability that the voltmeter's life is at most 8000 hours.

6.19 The life of a roller bearing used in a rolling machine (in hours) is known to follow Weibull distribution with parameters $\beta = 2$ and $\eta = 800$ hours. Determine the following:

a. Mean and standard deviation of the roller bearing.
b. Probability that the roller bearing fails before 400 hours.

6.20 The life of a X-ray machine, in hours, is known to follow Weibull distribution with parameters $\beta = 3$ and $\eta = 1000$ hours.

a. Find the mean and standard deviation of the life of the machine.
b. Find the probability that the X-ray machine fails before 250 hours.

6.21 The life of a magnetic resonance imaging (MRI) machine (X) follows lognormal distribution with $\mu_Y = 10$ hours and $\sigma_Y = 2$ hours. Determine the following:

a. The probability density and cumulative distribution functions of the life of the MRI machine.
b. The mean and standard deviation of the life of the MRI machine.
c. The value of life that is exceeded by 99 percent of the MRI machines.

6.22 A lognormally distributed random variable X has the following parameters: $\mu_Y = -2$ and $\sigma_Y = 2$. Find the following:

a. $P(400 \leq X \leq 600)$
b. Value of x such that $P(X < x) = 0.05$
c. Mean and standard deviation of X.

6.23 The lognormal distribution is defined by

$$f_T(t) = \frac{1}{\sqrt{2\pi}\,\sigma\,t}\exp\left\{-\frac{1}{2}\left(\frac{\ln t - \mu}{\sigma}\right)^2\right\}; t \geq 0$$

where $\mu \equiv \mu_{\ln T}$ and $\sigma \equiv \sigma_{\ln T}$. Find the median of T.

6.24 If the probability distribution function of the failure time of disk drives is given by

$$F_T(t) = 1 - \frac{6}{5}e^{-2t} + \frac{1}{5}e^{-3t}$$

Determine the functions $f_T(t)$, $R(t)$, and $h(t)$.

6.25 If the expected residual life, $L(t)$, of a system is a constant $\alpha = 1/\lambda$ for all values of t, find the probability distribution of the life of the system.

6.26 A series system consists of five components with failure rates of 0.0001, 0.0003, 0.0002, 0.0005, and 0.0008 per hour. Find the reliability of the system at 100 hours. If it is desired to increase the 100-hour reliability of the system by 5 percent, select a single component and its new failure rate to accomplish the task under the following conditions:

a. When there are no cost constraints.
b. When a unit reliability improvement costs \$500, \$400, \$600, \$1000, and \$800 for components 1, 2, 3, 4, and 5, respectively.

6.27 A series system is composed of five components with failure rates of 0.0005, 0.0001, 0.0008, 0.004, and 0.0002 per hour. Find the reliability of the system at 200 hours.

6.28 A simplified model of a thermal power plant is shown in Fig. 6.18 where the failure rate of each component is also indicated. Determine the reliability of the power plant after (a) 1000 hours, and (b) 2000 hours.

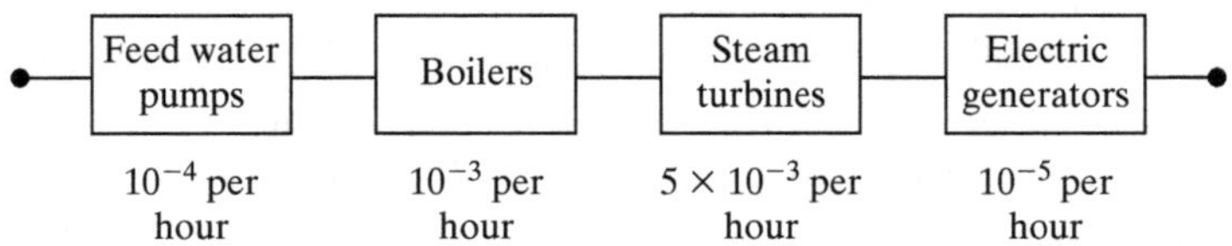

FIGURE 6.18 Model of a thermal power plant.

6.29 A series system is composed of n identical components, each having a failure rate of 10^{-3} per hour. Find the reliability of the system at time t. Also, plot the reliability function $R(t)$ for $n = 1, 2, 5$, and 10.

6.30 Two types of components, one with a failure rate of 10^{-4} per hour (type A) and the other with a failure rate of 10^{-5} per hour (type B), are arranged to form parallel systems. Five components of type A are used to form a parallel system X. Find the minimum number of components of type B needed to form a parallel system Y, which matches with the system X in terms of reliability.

6.31 Ten identical components are connected in parallel to achieve a system reliability of 0.95. Determine the reliability of each component.

6.32 Ten identical components are connected in parallel to achieve a system reliability of 0.9. Determine the additional number of components to be added in parallel to increase the reliability to 0.95.

6.33 A parallel system is composed of n identical components, each having a failure rate of 10^{-3} per hour. Determine the reliability of the system at time t. Also, plot the reliability function, $R(t)$, for $n = 1, 2, 5$, and 10.

6.34 Solve problem 6.27 by assuming that the components are connected in parallel.

6.35 A standby system is a parallel system where all the components except one (known as the primary component) are kept in a waiting condition. When the primary component fails, the standby component is switched into use in place of the failed component. Assuming the switch to be 100 percent reliable, the system life T is given by $T = T_1 + T_2$ where T_1 and T_2 denote, respectively, the lives of primary and standby components that follow the distributions:

$$f_{T_1}(t_1) = \lambda_1 \exp(-\lambda_1 t_1) \qquad \text{and} \qquad f_{T_2}(t_2) = \lambda_2 \exp(-\lambda_2 t_2)$$

Derive the following functions for the life of the standby system (T):

a. probability density function.
b. reliability function.
c. probability distribution function.
d. hazard function.

6.36 The reliability function of a standby system with three components can be expressed as

$$R(t) = \frac{\lambda_2\lambda_3 \exp(-\lambda_1 t)}{(\lambda_2 - \lambda_1)(\lambda_3 - \lambda_1)} + \frac{\lambda_1\lambda_3 \exp(-\lambda_2 t)}{(\lambda_1 - \lambda_2)(\lambda_3 - \lambda_2)} + \frac{\lambda_1\lambda_2 \exp(-\lambda_3 t)}{(\lambda_1 - \lambda_3)(\lambda_2 - \lambda_3)}$$

where the failure time density function of the components is given by

$$f_{T_i}(t_i) = \lambda_i \exp(-\lambda_i t_i); \quad i = 1, 2, 3$$

Show that the MTTF of the system is given by

$$\text{MTTF} = \frac{1}{\lambda_1} + \frac{1}{\lambda_2} + \frac{1}{\lambda_3}$$

Also, show that the MTTF of a standby system is larger than that of a simple parallel system.

6.37 A standby system consists of a total of three components with $\lambda_1 = 0.0005$, $\lambda_2 = 0.001$, and $\lambda_3 = 0.005$ per hour. Assuming a 100 percent reliability of the switch, determine the following:

a. the reliability of the system at 100 hours.
b. MTBF of the system.
c. MTBF of the system if the same components are connected as a simple parallel system.

6.38 A parallel system is composed of five components with failure rates of 0.009, 0.007, 0.006, 0.005 and 0.008 per hour. Find the reliability of the system at 100 hours. If it is desired to increase the 100-hour reliability of the system by 2 percent, select a single component and its new failure rate to accomplish the task under the following conditions:

1. When there are no cost constraints.
2. When a one-percent improvement in the 100-hour component reliability costs \$500, \$400, \$600, \$1000 and \$800 for components 1, 2, 3, 4 and 5, respectively.

6.39 Find the reliability of the system shown in Fig. 6.19 using the enumeration method.

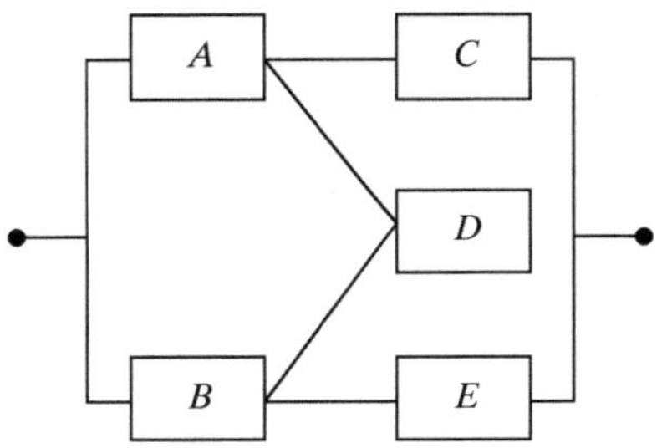

FIGURE 6.19 Complex system of Problems 6.39-6.41.

6.40 Find the reliability of the system shown in Fig. 6.19 using the conditional probability approach.

6.41 Find the reliability of the system shown in Fig. 6.19 using the cut-set method.

6.42 Find the reliability of the system shown in Fig. 6.20 using the enumeration method.

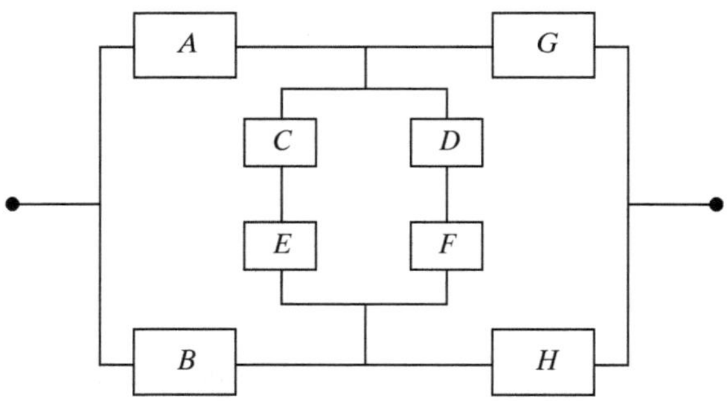

FIGURE 6.20 Complex system of Problems 6.42-6.44.

6.43 Determine the reliability of the system shown in Fig. 6.20 using the conditional probability approach.

6.44 Evaluate the reliability of the system shown in Fig. 6.20 using the cut-set method.

6.45 A system is composed of five identical components each having a failure rate of 0.0001 per hour. Find the reliability of the system at 100 hours and 200 hours if at least two components must operate for the survival of the system.

6.46 The series–parallel system shown in Fig. 6.21 functions if at least one of the components in each parallel unit functions. If the failure rates of the components A, B, C, and D are given by 0.5 λ, 1.0 λ, 2.0 λ, and 3.0 λ, respectively, find (a) the reliability of the system, and (b) MTTF of the system.

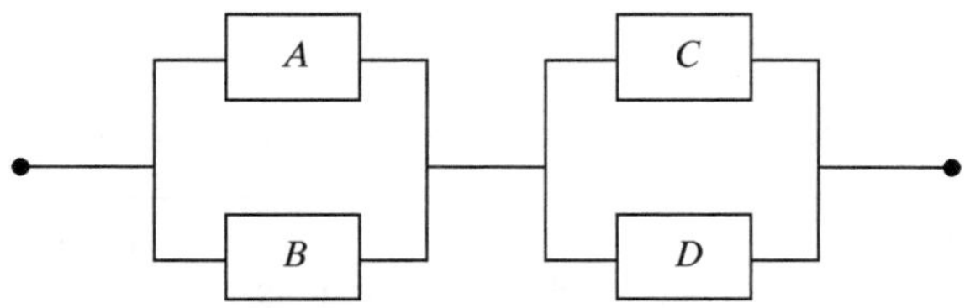

FIGURE 6.21 Series-parallel system of Problem 6.46.

6.47 A thermal power plant, consisting of three turbines, can supply the minimum required power to an industrial complex if at least two turbines operate. If the failures of the three turbines are independent and their failure rates are λ, 1.5λ, and 2.5λ, determine the probability that the power plant supplies the minimum required power at time t.

6.48 Find the reliability of the system shown in Fig. 6.22 when $R_A = 0.95$, $R_B = 0.98$, $R_C = 0.75$, $R_D = 0.85$, and $R_E = 0.90$.

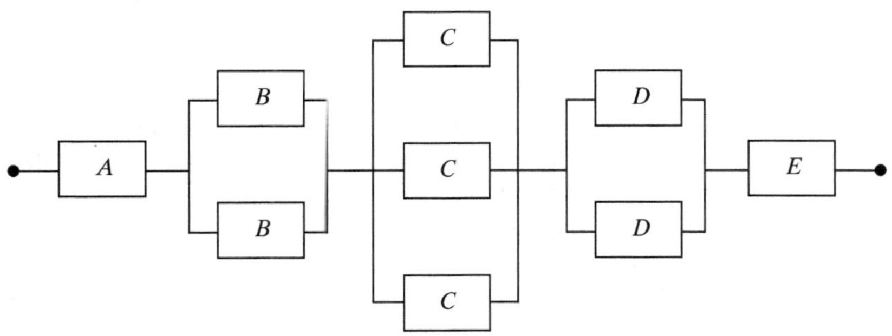

FIGURE 6.22 Series-parallel system of Problem 6.48.

6.49 Six identical components are connected in two different ways as shown in Figs. 6.23(a) and (b). Identify the configuration which yields a higher value of reliability.

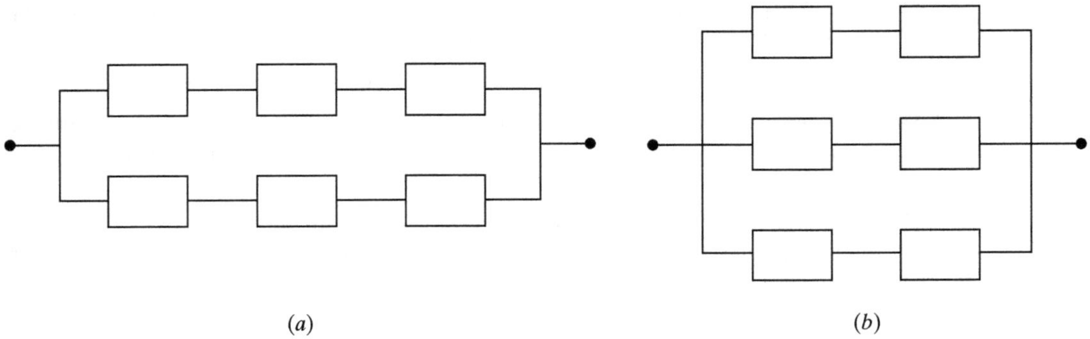

FIGURE 6.23 Components connected in two ways.

6.50 A complex system consists of six subsystems with details as indicated in the following table:

Subsystem (i)	Number of Components (n_i)	Operating Time (t_i, Hours)	Probability of System Failure Due to Failure of Subsystem i (w_i)
1	4	10	0.10
2	12	5	0.05
3	8	20	0.25
4	20	15	0.20
5	16	25	0.30
6	6	30	0.15

Find the required failure rates of the components of various subsystems in order to have an overall reliability of 0.995 using the AGREE method.

6.51 Derive Eq. (6.122) for the reliability function from the residual life given in Eq. (6.121).

MATLAB and Excel Problems

6.52 The failure time of an I.C. engine follows Weibull distribution with parameters $\beta = 4$, $\eta = 2000$, and $l = 1000$. Find the reliability and failure rate of the engine for an operating time of 2500 hours using MATLAB.

6.53 The failure time of an electronic component is known to follow normal distribution with parameters $\mu = 4000$ hours (mean) and $\sigma = 500$ hours (standard deviation). Using MATLAB, find the expected residual life of the component at 5000 hours of operation.

6.54 The automobile brakes produced by a manufacturer are found to follow normal distribution with a mean operating life of 50,000 miles and standard deviation of 5,000 miles. Find the reliability and failure rate of the brakes at the end of 50,000 miles of operation using MATLAB.

6.55 Solve Problem 6.52 using Excel.

6.56 The failure rate of an electromechanical component is given by $h(t) = 0.0025\,t$ (t in months). Determine the reliability of the component at the end of two years using Excel.

CHAPTER 7

Modeling of Geometry, Material Strength, and Loads

What You Will Learn

In this chapter, you will learn

- That the analysis and design of any mechanical or structural component is influenced by the geometry (dimensions), strength of the material used, and the loads acting on the component
- The tolerances obtainable on metal products produced by common manufacturing processes, rolled I-beams, and variations in mechanical and cross-sectional properties of steel parts
- The relative costs associated with different tolerances
- The assembly of parts when the tolerances on dimensions follow normal distribution
- The variability in the material strength parameters, including Young's modulus, Poisson's ratio, and yield strength, of structural steel
- The statistical models for brittle, plastic, and fiber-bundle models of material strength
- A probabilistic model, based on Palmgren-Miner hypothesis, for the fatigue strength of materials
- The modeling of dead loads, live loads, wind loads, and earthquake loads
- The statistical analysis of wind loads based on type II extreme value distribution for largest values
- The loads resulting from earthquake ground acceleration and the use of response spectrum in finding the influence of an earthquake on a structure

Ernst Abbe was a German physicist who was born on January 23, 1840, and died on January 14, 1905. He was the son of a master spinner who worked on his feet 16 hours every day with no break for meals. Abbe won scholarships and was also helped by his father's employer in getting education. He received his doctorate at the University of Jena at the age of 21 for a dissertation on thermodynamics. He became a professor at the University of Jena in 1870 and the director of astronomical and meteorological observatories in 1878. From 1866 onwards, Abbe devoted his time to theoretical and practical research in optics and astronomy. He was closely associated with the Carl Zeiss' company. He was an outstanding scientific genius combined with practical inventiveness. The paper he wrote in 1863, "On the Law of Distribution of Errors in Observation Series," formed a basis for his discovery of the χ^2-distribution, which can be used to represent a random variable that is the sum of squares of several independent normally distributed random variables. The χ^3-distribution also plays an important role in hypothesis testing (i.e., goodness-of-fit tests) [7.51].

7.1 Introduction

The analysis and design of any mechanical or structural component is influenced by the geometry or dimensions of the component, the strength of the material used, and the loads acting on the component. Although the nominal or design dimensions are known precisely, the actual dimensions will be random in nature. When dimensions are produced on a machine tool, the limitations of the machine and the operator lead to variations in the dimensions in the form of tolerances. Similarly, when structural members are rolled in a mill, the shaping rollers will gradually wear, resulting in variations in the cross-sectional dimensions of the rolled member. Since it is not possible to change the rollers frequently, some variations in the dimensions of the rolled members are unavoidable. The errors in fabrication and assembly also cause variations in the overall dimensions of multicomponent mechanical and structural systems.

Variations occur in the mechanical properties of materials due to uncertainties in the chemistry and geometric shapes, as well as the manufacturing and fabrication procedures used. The mechanical strength is also influenced by the loading and strain rates applied to the component. The load acting on a machine or structure is considered to be static if the magnitude of the load is constant, or dynamic if the load is variable, with respect to time. The static loads include dead loads due to self weights of components and live loads acting on the structure. These loads are subject to variability due to fluctuations in the specific weights of materials, as well as changes in the working loads acting on the structure. The dynamic loads include wind, snow, wave, and earthquake induced loads. These loads are to be treated as random processes for the reliability analysis of structures.

This chapter deals with the probabilistic description of geometric dimensions of components, strengths of materials, and loads acting on structures.

7.2 Modeling of Geometry

In mechanical design, the geometric parameters such as shaft diameters, lengths and thicknesses of parts, hole diameters, profiles of gear teeth and cams, and center distances between shafts are of prime importance. Similarly, in structural design the cross-sectional dimensions of beams and lengths, widths, and thicknesses of plates and shells play an important role. Although nominal values of the geometric variables are used in design, physical realization of these values during manufacturing is based on the specification of suitable tolerances on these values. This is due to the fact that all the machining processes such as turning, drilling, and milling and the production processes such as casting, forging, and rolling are subject to change over time. By collecting and statistically analyzing the data from the manufactured components, suitable probability density function can be determined for a particular class of dimensions like length of parts, angular measurements, center distances, diameters, and so on.

In the absence of directly measured data, approximate distributions, based on previous related studies, can be used for the dimension of interest. It has been observed that many manufacturing operations generate normally distributed dimensions if they are controlled. The tolerance limits on dimensions of components are set so as to conform to the capabilities of the manufacturing process except for a small fraction α of the items that will have to be rejected. If a dimension is approximately normally distributed, a good design will have the sample mean approximately equal to the nominal value and the sample standard deviation such that only a small allowable fraction α of the parts manufactured fall outside the tolerance specification limits. Thus the tolerance limits of a dimension, on the average, should permit acceptance of at least $(1 - \alpha)$ of the parts manufactured.

7.2.1 Tolerances on Finished Metal Products

Depending on the tolerances to be achieved on the dimensions of a manufactured product, a suitable machining or manufacturing process is to be used. The tolerances commonly obtainable on finished metal products are documented in the literature. The best tolerances obtainable in (in inches) different processes are given by $\begin{matrix}+0.002\\-0.000\end{matrix}$ in drilling, ± 0.001 in turning, shaping, and milling, ± 0.0002 in grinding, ± 0.00005 in lapping, ± 0.02 in flame cutting, ± 0.005 in sawing, ± 0.001 in stamping and cold rolling, ± 0.03 in/in in forging, ± 0.002 in wire drawing, ± 0.005 in hot extrusion, $\pm \frac{1}{32}$ in/ft in sand casting, ± 0.002 in/in in die casting, and ± 0.01 in spot welding, fusion welding, as well as heat treatment. The common tolerances achievable in the dimensions of steel structural components such as beams and columns with I, L, C, or Z section are given by the *AISC Manual of Steel Construction* [7.2]. For example, the tolerances obtainable on I-beams are shown in Fig. 7.1.

As reported by Alpsten [7.3], the variability of the area moments of inertia (I) of steel I-beams can be expressed as $I = \begin{matrix}+0.09\\-0.08\end{matrix} I_{\text{nominal}}$ for strong axis and $I = \begin{matrix}+0.06\\-0.10\end{matrix} I_{\text{nominal}}$ for weak axis, and the variability of the section modulus (Z) can be denoted as $Z = \begin{matrix}+0.09\\-0.04\end{matrix} Z_{\text{nominal}}$ for strong axis and $Z = \begin{matrix}+0.09\\-0.10\end{matrix} Z_{\text{nominal}}$ for weak axis.

The tolerances on diameters (d) of aluminum wire and rod are documented in several publications [7.4−7.7, 7.44]. For example, the tolerances on some aluminum wires and rods (in inches) are given by ± 0.0005 for $d \le 0.035$ inch and ± 0.0010 for $0.036 \le d \le 0.064$ inch in the case of drawn wire, ± 0.0015 for $0.375 \le d \le 0.500$ inch, ± 0.0020 for $0.501 \le d \le 1.000$ inch, and ± 0.0060 for $2.001 \le d \le 3.000$ inch in the case of cold finished rod, and ± 0.0060 for $1.501 \le d \le 2.000$ inch and ± 0.0080 for $2.001 \le d \le 3.000$ inch in the case of rolled rod.

Most tolerances can be assumed to follow normal distribution for computational convenience. From the design point of view, zero tolerances are desirable. However, from the point of view of manufacturing and production, large tolerances are desirable. As tolerances are reduced, the production costs increase at a much faster rate. As such, the designer has to specify an optimal set of tolerances by compromising between accuracy and production cost. The tolerances obtainable in different machining processes and the corresponding relative costs are indicated in Fig. 7.2. Additional data related to the costs of tolerances can be found in Refs [7.9–7.11].

7.2.2 Assembly of Components

When a number of components are assembled, the variations in their dimensions or geometric parameters are to be considered. Usually, a linear sum of dimensions such as

$$Z = X_1 + X_2 + \cdots + X_N \tag{7.1}$$

is required to determine the overall sizes, a linear combination like

$$Z = X_1 \pm X_2 \pm \cdots \pm X_N \tag{7.2}$$

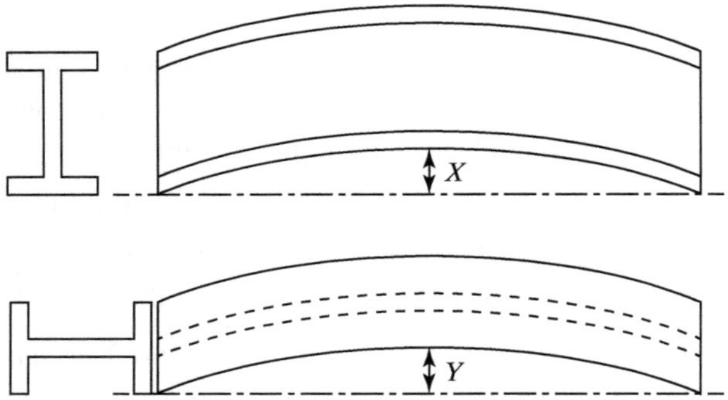

Cutting Tolerances on Length

Beams with a nominal depth of $\leq$24 in	$\pm\frac{3}{8}$ in $\cdots$	when length is $\leq$30 ft
	$+\frac{3}{8}$ in $+\frac{1}{16}$ in for each additional 5 ft or less $\cdots$	when length is $>$30 ft
	$-\frac{3}{8}$ in	
Beams with a nominal depth of $>$24 in	$\pm\frac{1}{2}$ in $\cdots$	when length is $\leq$30 ft
	$+\frac{1}{2}$ in $+\frac{1}{16}$ in for each additional 5 ft or less $\cdots$	when length is $>$30 ft
	$-\frac{1}{2}$ in	

Other Tolerances

Camber and sweep for most sizes: $\frac{1}{8}$ in $\times \left(\frac{\text{total length in ft}}{10}\right)$

Area and weight $\pm 2.5\%$

(a) Cutting tolerances

(i) For sections of all sizes:

Depth (A): $\pm\frac{1}{8}$ in, width (B): $+\frac{1}{4}$ in,

flanges $(T_1 + T_2)$: $-\frac{3}{16}$ in,

total depth (C): $\frac{3}{16}$ in

(ii) Off center of web (E):

$\frac{1}{4}$ in for section sizes $\leq$24 in; $\frac{3}{16}$ in otherwise

(b) Rolling tolerances on cross-sectional dimensions

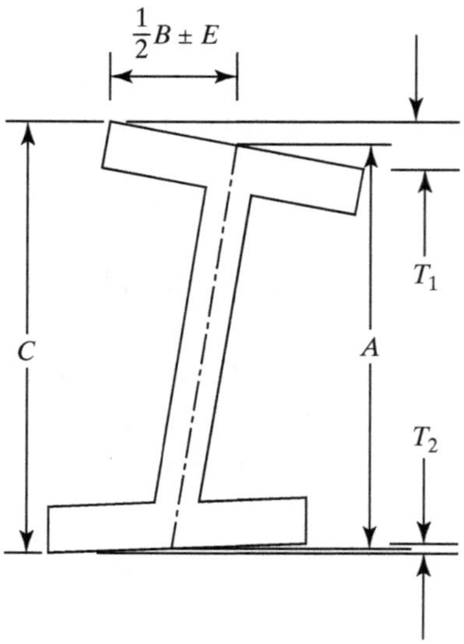

FIGURE 7.1 Tolerances on I-beams.

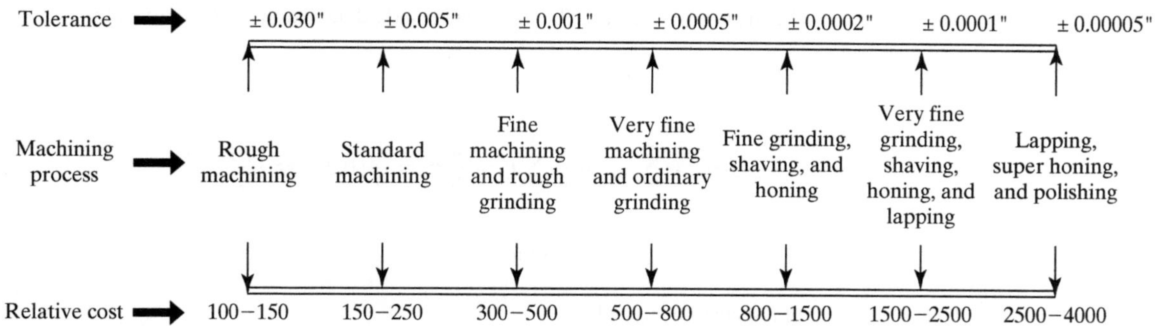

FIGURE 7.2 Relative costs of tolerances.

is required to find the locations, and a difference of dimensions such as

$$Z = X_1 - X_2 \tag{7.3}$$

is needed to find the clearance between two parts. If the individual dimensions $X_1, X_2, \ldots, X_N$ are treated as normally distributed independent random variables, the distribution of the dimension Z will also be normal. The mean and standard deviations of Z can be computed once the means and standard deviations of X_i, $i = 1, 2, \ldots, N$ are known. Even if the individual random variables X_i follow different distributions, Z can be considered to follow normal distribution according to the central limit theorem.

If the overall size of an assembly of N components is given by Eq. (7.1) with the mean value and standard deviation of the dimension of the ith component (X_i) given by $\overline{X}_i$ and σ_{X_i}, respectively, the mean and standard deviations of Z can be determined as

$$\overline{Z} = \overline{X}_1 + \overline{X}_2 + \cdots + \overline{X}_N \tag{7.4}$$

$$\sigma_Z = (\sigma_{X_1}^2 + \sigma_{X_2}^2 + \cdots + \sigma_{X_N}^2)^{1/2} \tag{7.5}$$

If the dimension X_i is specified as $\overline{X}_i \pm t_i$ with $\pm t_i$ representing the three-sigma tolerance band of X_i, the standard deviation of X_i will be $t_i/3$. In this case, the standard deviation of Z can be expressed as

$$\sigma_Z = \frac{1}{3}(t_1^2 + t_2^2 + \cdots + t_N^2)^{1/2} \tag{7.6}$$

EXAMPLE 7.1

Find the statistics of the length of a chain consisting of four links when the length of each link is given by 12 ± 0.2 in.

Solution: The length of the chain can be expressed as

$$l = l_1 + l_2 + l_3 + l_4$$

with $l_i = 12 \pm 0.2$ in $(i = 1, 2, 3, 4)$. The mean and standard deviations of l are given by

$$\bar{l} = \sum_{i=1}^{4} \bar{l}_i = 4\bar{l}_i = 4(12) = 48 \text{ in}$$

$$\sigma_l^2 = \sum_{i=1}^{4} \sigma_{l_i}^2 = \frac{1}{9}\sum_{i=1}^{4} t_i^2 = \frac{4}{9}t_i^2 = \frac{4}{9}(0.2)^2 = \frac{0.16}{9}$$

$$\therefore \sigma_l = \frac{0.4}{3} = 0.1333 \text{ in}$$

Thus, the length of the chain is given by

$$l = \bar{l} \pm 3\sigma_l = 48 \pm 0.4 \text{ in}$$

■

EXAMPLE 7.2

A mechanical assembly involves inserting the pins of part 1 in the holes A and B of part 2 (see Fig. 7.3). The relevant nominal dimensions and the associated tolerances are also indicated in Fig. 7.3. What is

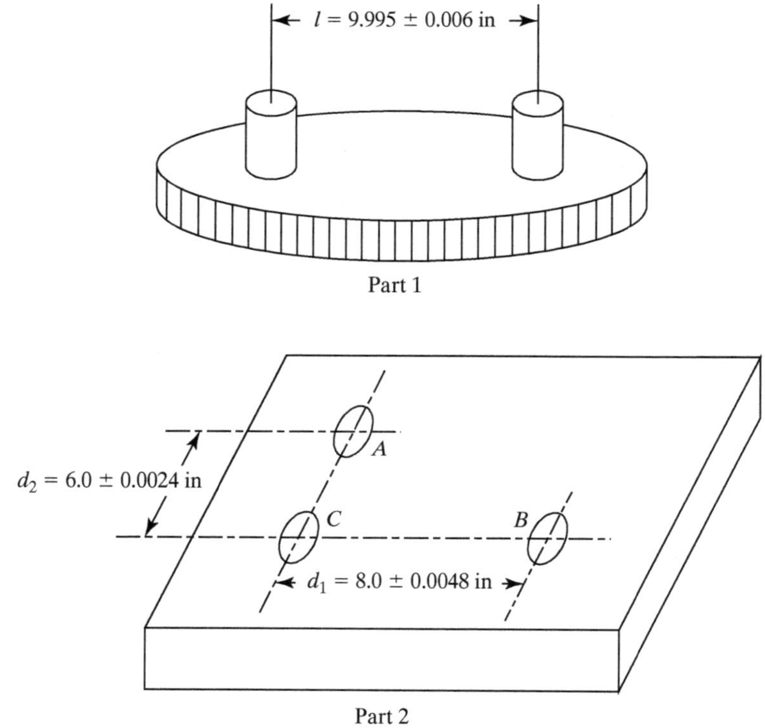

FIGURE 7.3 Two parts to be assembled.

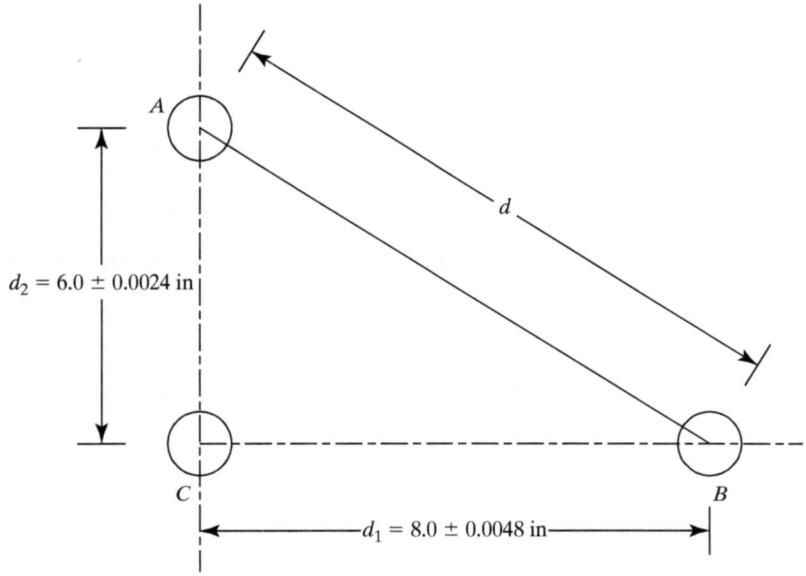

FIGURE 7.4 Tolerances between holes.

the probability of realizing the distance between the pins larger than the distance between the holes A and B in an assembly? Assume that the dimensions follow normal distribution.

Solution: The distance between the holes A and B can be determined from the relation (see Fig. 7.4).

$$d = \sqrt{d_1^2 + d_2^2} \tag{E1}$$

where $d_1 = (8.0 \pm 0.0048)$ in and $d_2 = (6.0 \pm 0.0024)$ in. The mean value of d can be found as

$$\bar{d} = \sqrt{\bar{d}_1^2 + \bar{d}_2^2} \tag{E2}$$

and the standard deviation of d can be determined, using the partial derivative rule, as

$$\sigma_d = \left[\frac{\bar{d}_1^2 \sigma_{d_1}^2 + \bar{d}_2^2 \sigma_{d_2}^2}{\bar{d}_1^2 + \bar{d}_2^2} \right]^{1/2} \tag{E3}$$

Since the dimensions follow normal distribution, the tolerances can be assumed to correspond to three-sigma values and hence the standard deviations of d_1 and d_2 can be found as

$$\sigma_{d_1} = \frac{0.0048}{3} = 0.0016; \quad \sigma_{d_2} = \frac{0.0024}{3} = 0.0008 \tag{E4}$$

Equations (E2) and (E3) yield

$$\bar{d} = \sqrt{8.0^2 + 6.0^2} = 10.0 \text{ in} \tag{E5}$$

$$\sigma_d = \left[\frac{(8.0)^2(0.0016)^2 + (6.0)^2(0.0008)^2}{(8.0)^2 + (6.0)^2}\right]^{1/2}$$
$$= (0.000001869)^{1/2} = 0.001367 \text{ in} \tag{E6}$$

The probability of the distance between the pins (l) larger than the distance between the holes A and $B(d)$ is given by

$$P[l \geq d] = P[l - d \geq 0] \tag{E7}$$

Defining $X = l - d$, Eq. (E7) can be rewritten, in terms of the standard normal variate $z = (X - \overline{X})/\sigma_X$, as

$$P[l \geq d] = P[X \geq 0] = P\left[z = \frac{X - \overline{X}}{\sigma_X} \geq -\frac{\overline{X}}{\sigma_X}\right] \tag{E8}$$

where

$$-\frac{\overline{X}}{\sigma_X} = -\frac{(\bar{l} - \bar{d})}{\sqrt{\sigma_l^2 + \sigma_d^2}} \tag{E9}$$

Since $\bar{l} = 9.995$ in and $\sigma_l = 0.0060/3 = 0.0020$ in, Eq. (E9) gives

$$-\frac{\overline{X}}{\sigma_X} = -\frac{(9.995 - 10.000)}{\sqrt{(0.002)^2 + (0.001367)^2}} = \frac{0.005}{0.002422} = 2.0639 \tag{E10}$$

and hence Eq. (E9) gives the desired probability as

$$P[l \geq d] = P[z \geq 2.0639] = 0.0197 = 1.97\% \tag{E11}$$

■

7.3 Modeling of Material Strength

7.3.1 Statistics of Elastic Properties

Most metals and other engineering materials exhibit variable properties due to random differences in chemistry, heat treatment, and mechanical processing. Static strength is the strength displayed by a material subjected to static loading. The nominal elastic properties of structural steel commonly used in standards, codes, and specifications are Young's modulus in tension and compression (E): 29×10^6 lb/in^2, Poisson's ratio (ν): 0.3, shear modulus (G): $G = \dfrac{E}{2(1 + \nu)} = 0.385\,E$. The statistical characteristics of these properties were investigated experimentally by several researchers [7.12, 7.13]. The coefficient of variation of Young's modulus, Poisson's ratio, and shear modulus of steel are found to be 0.01, 0.025, and 0.042, respectively. The yield stress denotes the principal property that influences structural and mechanical designs. The designer selects the grade of steel to be used for the components to be designed, which gives the corresponding yield stress. The yield stress of steel depends on the chemical composition of the alloy and the method of manufacture. The grade of steel reflects these factors. The data on ASTM A7 and A36 steels, for rolled structural shapes, collected over a time span of 40 years prior to 1957 from several U.S. steel mills yielded a mean value of 33,000 psi with a coefficient of variation of 0.09. The statistics of yield and ultimate strengths of some other materials used in mechanical and structural designs are given in several publications [7.12–7.14]. For example, the mean and standard deviation (in kpsi) are observed to be, respectively, 35.7 and 0.90 for the yield strength and 48.3 and 0.52 for the ultimate strength for C1006 carbon steel (hot rolled sheet). Similarly, for 2024-T6 aluminum alloy sheet, the mean and standard deviation (in kpsi) are found to be, respectively, 50.1 and 2.85 for the yield strength and 63.6 and 2.51 for the ultimate strength. In the case of malleable cast iron (ferritic), the mean and standard deviation (in kpsi) are observed to be, respectively, 34.9 and 1.47 for the yield strength and 53.4 and 2.68 for the ultimate strength [7.7].

7.3.2 Statistical Models for Material Strength

Three statistical models can be used for describing the strength of different materials: the *brittle-material model,* the *plastic model,* and the *fiber-bundle model* [7.15, 7.16]. An ideal brittle material is one that fails whenever a single particle fails. Thus, the strength of the material is governed by the smallest strength of a particle. In an ideal plastic material, yielding starts when the load on a particle reaches its yield capacity. The particle continues to carry the yield load and any additional load is transferred to other particles of the cross section. Thus, the cross section carries its maximum load when all the particles yield. In the fiber-bundle model, the failure of a particle implies that it is no longer capable of carrying the load. Hence the load will be redistributed to other particles. This process continues until all the particles of the body fail.

7.3.3 Model for Brittle Materials

In this model, the material is considered as a chain composed of several (n) elements in series [7.15, 7.16]. The strengths of the individual elements of a series chain are assumed statistically to be independent and follow a common distribution function $F(x)$. The distribution function of the smallest strength in n elements $F_n(x)$ is given by (see Eq. 4.2):

$$F_n(x) = 1 - [1 - F(x)]^n \tag{7.7}$$

which can be rewritten as

$$F_n(x) = 1 - \exp[n \ln\{1 - F(x)\}] \tag{7.8}$$

As n tends to infinity, the distribution function of Eq. (7.8) tends to one of the three asymptotic distributions discussed in Chapter 4. Since the strength is limited by a lower bound, the type-III extremal distribution for the smallest value will be applicable.

Now consider a one-dimensional structure (bar) of length l made of an ideal brittle material subjected to an axial load. Assume that the structure is divided into n equal parts so that the length of each part is $l_0 = l/n$. Let the strengths of these n parts be independent statistically and identically distributed with distribution function $F_0(x)$. Then the distribution function of the strength of the structure can be expressed, using Eqs. (7.7) and (7.8), as

$$F_S(x) = 1 - [1 - F_0(x)]^n = 1 - \exp\left\{\frac{l}{l_0}\ln[1 - F_0(x)]\right\} \tag{7.9}$$

It can be seen from Eq. (7.9) that the size of the structure (l/l_0) appears as a parameter in the distribution function. Equation (7.9) can be extended to a brittle solid subjected to a homogeneous stress state. In this case, the volume of the solid V is divided into several elementary volumes V_0 such as the volume of a standard test specimen. Let the distribution function of the strength of elementary volumes be given by $F_0(x)$. Then the distribution of the strength of the solid can be expressed as

$$F_S(x) = 1 - \exp\left\{\frac{V}{V_0}\ln[1 - F_0(x)]\right\} \tag{7.10}$$

The quantity $\ln[1 - F_0(x)]$ is often approximated as

$$\ln[1 - F_0(x)] \approx -\left(\frac{x - \varepsilon}{\eta - \varepsilon}\right)^{\beta}; \quad x \geq \varepsilon \tag{7.11}$$

where ε is the lower limit on the strength and β and η are positive constants. Typical values of β are 58 for mild steel and 12 for concrete. Using Eq. (7.11), $F_S(x)$ of Eq. (7.10) can be written as

$$F_S(x) = 1 - \exp\left\{-\frac{V}{V_0}\left(\frac{x - \varepsilon}{\eta - \varepsilon}\right)^{\beta}\right\}; \quad x \geq \varepsilon \tag{7.12}$$

which can be identified as a type-III extreme value distribution for the smallest value. The mean and standard deviations of strength S can be expressed as [7.16]

$$\mu_S = \overline{S} = \varepsilon + (\eta - \varepsilon)\,\Gamma\left(1 + \frac{1}{\beta}\right)\left(\frac{V}{V_0}\right)^{-1/\beta} \tag{7.13}$$

$$\sigma_S^2 = (\eta - \varepsilon)^2\left\{\Gamma\left(1 + \frac{2}{\beta}\right) - \Gamma^2\left(1 + \frac{1}{\beta}\right)\right\}\left(\frac{V}{V_0}\right)^{-2/\beta} \tag{7.14}$$

The foregoing analysis was given by Weibull for $\varepsilon = 0$ [7.17]. It can be observed that the mean value of the strength (μ_S) decreases with the volume and for $\varepsilon = 0$ the coefficient of variation is independent of the volume. The size effect is found to be valid for many brittle materials. The decrease in tensile strength with increasing volume is shown qualitatively in Fig. 7.5 for glass threads [7.16].

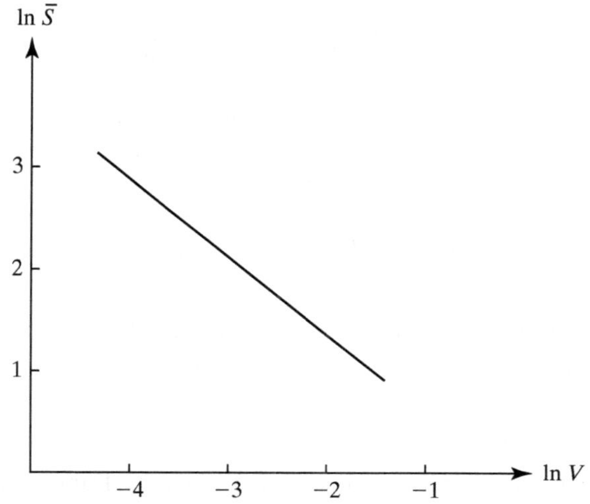

FIGURE 7.5 Decrease of strength with increase in volume.

7.3.4 Model for Plastic Materials

Consider a bar under axial loading. Assume the force-deformation relation for a typical (ith) particle in the cross section of the bar to be ideal plastic or elastic-ideal plastic with a yield strength (force) S_i. The strength of the bar (S) can be expressed as [7.16]

$$S = \sum_{i=1}^{n} S_i \tag{7.15}$$

where the summation is over all the n particles in the cross section. The mean and standard deviation of S are given by

$$\overline{S} = \sum_{i=1}^{n} \overline{S}_i \tag{7.16}$$

$$\sigma_S^2 = \text{var}(S) = \sum_{i=1}^{n}\sum_{j=1}^{n} \text{cov}\,(S_i, S_j) \tag{7.17}$$

If the number of particles n is large, the strength of the bar can be assumed to follow normal distribution according to the central limit theorem. If the yield stress s_y is known to vary as a continuous function within the cross section, Eqs. (7.15) to (7.17) can be rewritten, by replacing summations by integrals, as

$$S = \iint_A s_y(x, y)\, dA \tag{7.18}$$

$$\mu_S = \overline{S} = \iint_A E[\,s_y(x, y)\,]\, dA \tag{7.19}$$

$$\sigma_S^2 = \iint_A \iint_A \operatorname{cov}[s_y(x_i, y_i), s_y(x_j, y_j)]\, dA_i\, dA_j \tag{7.20}$$

If the yield stress is homogeneous, Eqs. (7.19) and (7.20) reduce to [7.16]

$$\mu_S = \overline{S} = \mu_{s_y} A = \overline{s}_y A \tag{7.21}$$

$$\sigma_S^2 = \iint_A \iint_A \operatorname{cov}(x_j - x_i, y_j - y_i)\, dA_i\, dA_j \tag{7.22}$$

where $\mu_{s_y} = \overline{s}_y$ denotes the mean value of strength. It can be seen from Eq. (7.21) that the mean strength per unit area is independent of the area.

7.3.5 Model for Fiber Bundles

The first contribution to fiber-bundle theory was made by Daniels [7.18] in the context of predicting the statistical properties of the combined strength of bundles of threads in textile yarns, given the statistics of the individual threads. In the fiber-bundle model, the cross section is assumed to be composed of ideal fibers in a parallel arrangement. The force-displacement relation for a fiber is such that when a fiber breaks, it will no longer be capable of carrying any load. Thus the failure of a fiber does not necessarily imply failure of the system, but the load is redistributed among the other unbroken fibers which may then be able to carry the increased load. Consider a bundle of n fibers, each with a unit cross-sectional area, as shown in Fig. 7.6(a).

Since the entire bundle supports the load l, each fiber carries a load of l/n. The strengths of all fibers are assumed to be given by the probability distribution, $F(x)$. The bundle is assumed to have failed whenever the fibers $1, 2, \ldots, n$ fail successively in some order. Also, it is known that when j fibers fail $(0 < j < n)$, each of the $(n - j)$ surviving fibers carries a larger load, $l/(n - j)$.

For simplicity, we first derive the probability distribution function for the strength of a bundle consisting of two fibers (see Fig. 7.6(b)). We assume that, as a typical sequence, fiber 1 fails first and fiber 2 fails next. The probability of occurrence of this sequence is given by

$$P(\text{fiber 1 failing first and fiber 2 failing next at a total load} \le l)$$
$$= P\left(\text{strength of fiber 1 } (x_1) \le \frac{l}{2}\right) P(x_1 \le \text{strength of fiber 2 } (x_2) \le l)$$
$$= \int_0^{l/2} f(x_1) \int_{x_1}^{l} f(x_2)\, dx_2\, dx_1 \tag{7.23}$$

where $f(x_i)$ denotes the probability density function of the strength of fiber $i(x_i)$. The distribution function of the bundle strength can be found by noting that the other possible sequence (i.e., fiber 2 failing first and fiber 1 failing next) is statistically identical to the sequence considered in Eq. (7.23). Thus, the distribution function of the bundle strength can be expressed as

$$F_B(l) = 2 \int_0^{l/2} f(x_1) \int_{x_1}^{l} f(x_2)\, dx_2\, dx_1 \tag{7.24}$$

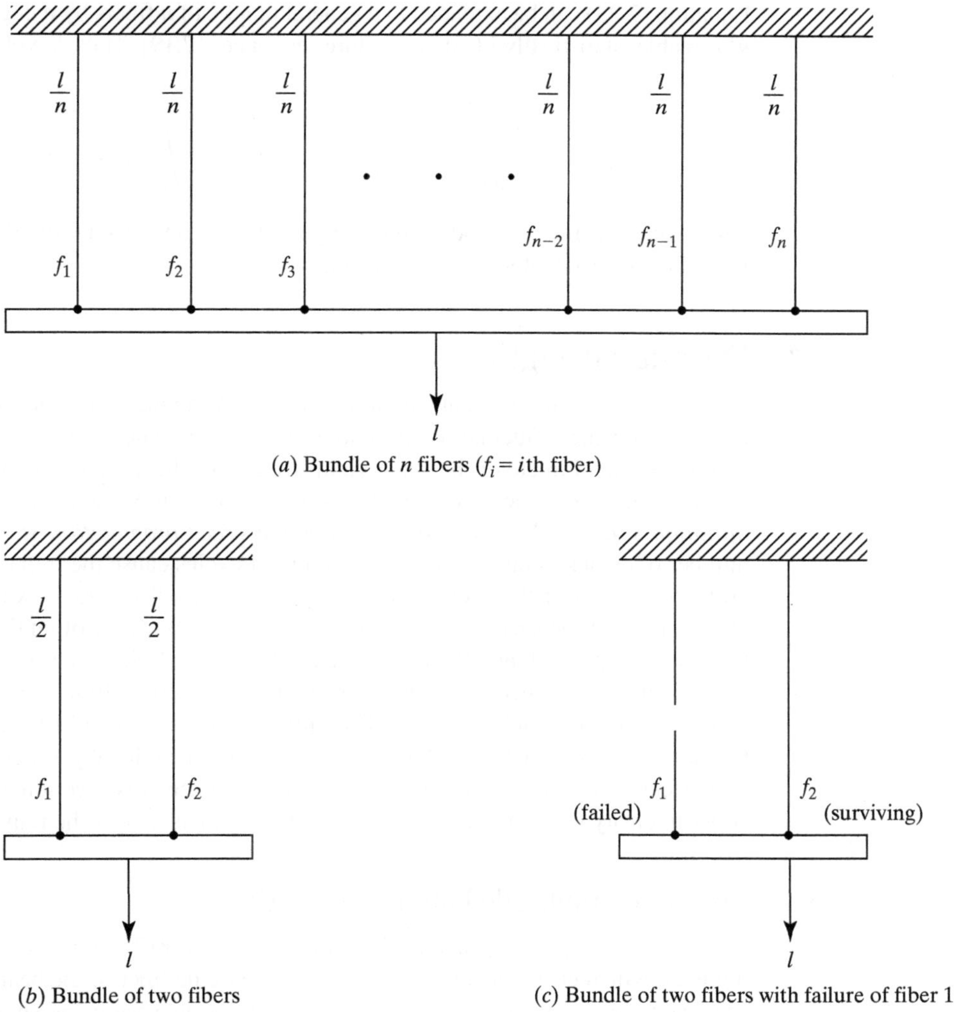

FIGURE 7.6 Bundle of fibers.

which can be rewritten as

$$F_B(l) = 2F(l)F\left(\frac{l}{2}\right) - \left\{F\left(\frac{l}{2}\right)\right\}^2 \tag{7.25}$$

This result can be compared with the corresponding simple parallel system for which

$$F_P(l) = \left\{F\left(\frac{l}{2}\right)\right\}^2 \tag{7.26}$$

Equation (7.23) can be extended to a general bundle of n fibers by recognizing that there are $n!$ possible, statistically identical, failure sequences [7.19]. Thus, we obtain, for a bundle of n fibers

$$F_B(l) = n! \int_0^{l/n} f(x_1) \int_{x_1}^{l/(n-1)} f(x_2) \cdots \int_{x_{n-1}}^{l} f(x_n)\, dx_n \cdots dx_2\, dx_1 \qquad (7.27)$$

Although a general expression cannot be given for the integrated form of Eq. (7.27), recursive formulas can be developed for evaluating $F_B(l)$.

7.4 Fatigue Strength

It is well known that mechanical and structural components fail under the action of fully reversed, repeated, fluctuating, or random loads, even when the peak stress levels are less than the static ultimate or the static yield strength of the material. Such failures are known as *fatigue failures.* There are two stages of fatigue failure: stage one, known as the *crack-initiation stage,* and stage two, known as the *crack-propagation stage.* If N denotes the number of cycles to failure, 60 to 90 percent of cycles cause the initiation of the crack and 10 to 40 percent of the cycles cause the propagation of the crack. A fatigue crack is usually initiated at the surface of a component where stresses are often the highest, imperfections may be present, and the grains are least supported. Surface scratches, inclusions, gas pockets, machining irregularities, changes in cross section, fillets, keyways, and holes are common sites for crack initiation. The initial crack is too small for detection by visual or standard inspection methods. Once the crack is initiated, local geometric-stress concentration effects aid in its growth. At the macro level, as the cross-sectional area is reduced, the stress intensity grows in magnitude until sudden fracture over the remaining area occurs.

7.4.1 Constant-Amplitude Fatigue Strength

In a standard fatigue test such as the rotating-beam test [7.20], a constant-amplitude bending load, which causes a constant stress at the extreme fiber of the beam, is applied to each specimen. As the load varies cyclically, the stress also varies as shown in Fig. 7.7. The number of stress cycles required to cause failure N and the corresponding stress amplitude $S = s_{\max}$ are noted. The test is repeated with different stress amplitudes and the corresponding number of cycles to failure in each case is noted. The data are plotted as cycle life N versus applied tensile stress S on a log–log scale through which a best-fit line is drawn. The resulting graph is known as the S–N diagram (see Fig. 7.8). In this diagram, any ordinate value S_f, just below the failure line is called the fatigue strength corresponding to the number of cycles n of interest. The objective of any design is to assure that the applied load does not cause a stress amplitude larger than the fatigue strength for the material specified for the number of load cycles expected during the life of the component.

It has been observed that failure would not occur at stress levels below the horizontal part of the S–N curve irrespective of the number of cycles applied. The stress corresponding to the horizontal line in Fig. 7.8 is called the endurance limit (S_e). As with most experimental data, a certain amount of scatter of the test-data points about the best–fit S–N line has

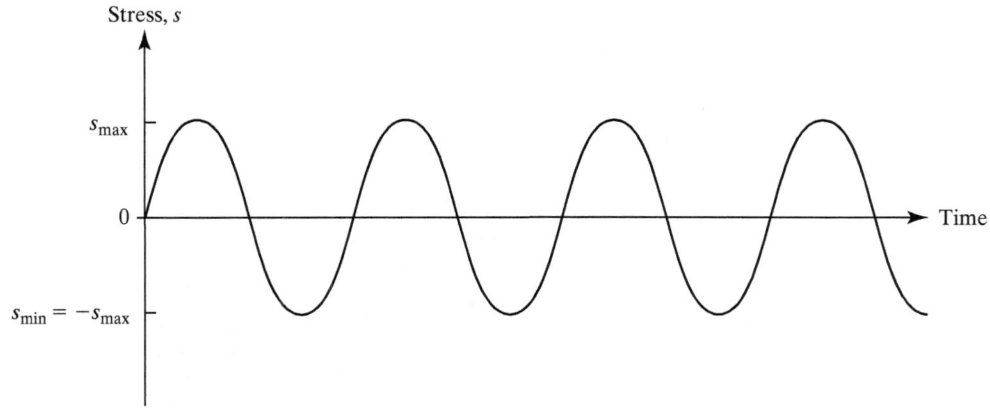

FIGURE 7.7 Cyclic variation of stress with time.

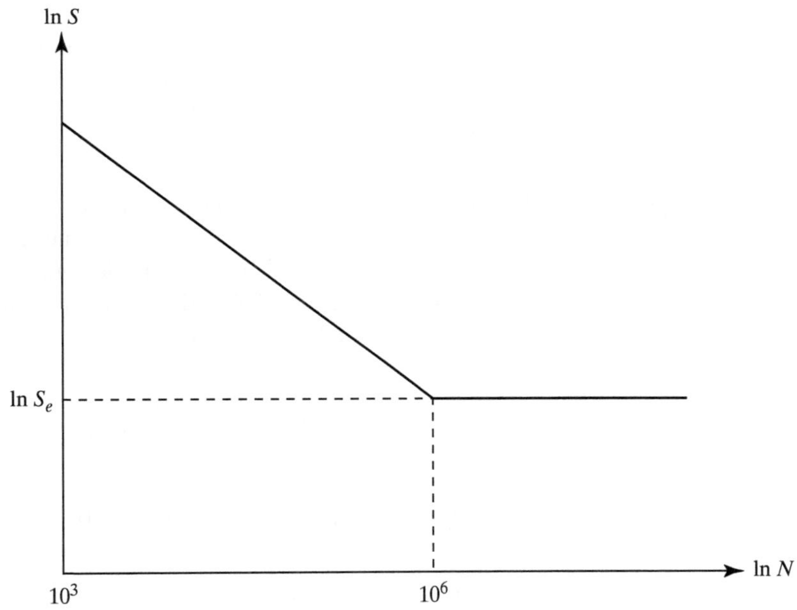

FIGURE 7.8 Endurance limit.

been observed. The statistical fatigue data needed for probabilistic design can be collected in two possible ways as follows.

1. The statistical *S–N* envelope can be developed for a material by plotting the distributions of cycles to failure for a number of stress amplitudes. A typical result, including the scatter in the experimental results, is shown in Fig. 7.9. The number of cycles to failure at a fixed level of maximum bending stress has been found to follow

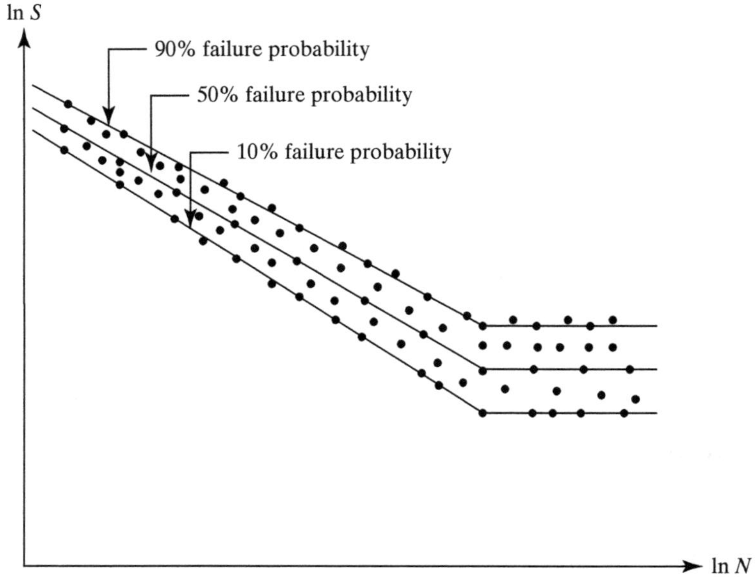

FIGURE 7.9 *S–N* curve with scatter in experimental values.

approximately lognormal distribution. The three-parameter Weibull distribution was also observed to be fitting the fatigue-life data of titanium and steel very closely [7.7, 7.21]. Since the mean value $(\overline{S}_f)$ and standard deviation (σ_{S_f}) are frequently used in engineering design, the values of $\overline{S}_f$ and σ_{S_f} corresponding to Weibull distribution have been expressed in the following form:

$$\overline{S}_f = \varepsilon + (\eta - \varepsilon)\,\Gamma\left(1 + \frac{1}{\beta}\right) \tag{7.28}$$

$$\sigma_{S_f} = \frac{(\overline{S}_f - \varepsilon)}{\beta^{0.926}}; \quad 0.7 < \beta < 10.0 \tag{7.29}$$

where ε, η, and β are the Weibull parameters.

For example, for AISI 1045 steel under constant-amplitude fatigue loading, the values of ε (kpsi), β, and η (kpsi) are found to be, respectively, 79.0, 2.6, and 88.2 at 10^4 cycles, 67.0, 2.75, and 73.0 at 10^5 cycles, and 56.7, 2.85, and 61.65 at 10^6 cycles [7.45]. For 2014 aluminum, the values of ε (kpsi), β, and η (kpsi) are found to be, respectively, 57.85, 1.781, and 63.92 at 10^4 cycles, 48.94, 1.636, and 53.81 at 10^5 cycles, and 41.34, 1.522, and 45.31 at 10^6 cycles [7.46].

2. The statistical *S–N* envelope can be developed by plotting the distributions of stress at failure for a number of fixed-cycle levels. Figure 7.9 shows typical results, including the scatter in the experimental results. The fatigue strength at fixed cycle life was found to follow approximately normal distribution.

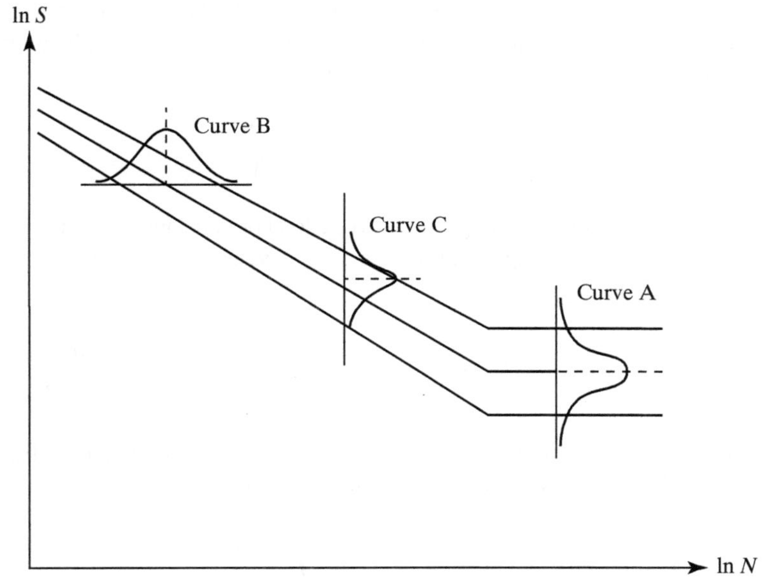

FIGURE 7.10 Distributions related to the S–N curve.

From practical point of view, the three distributions shown in Fig. 7.10 are important. The distribution of the endurance limit (curve A) is useful in the design of mechanical and structural components for infinite life (more than a million cycles of life). The distribution of fatigue life (curve B) is useful in the design of components such as bearings where L_{10}-life is important. The distribution C becomes important when the failure criterion involves fatigue strength at a finite life.

7.4.2 Variable-Amplitude Fatigue Strength

In many engineering applications, the amplitude of fatigue stress varies during the life of a component. First, we shall consider a deterministic fatigue strength model according to Palmgren and Miner's hypothesis [7.22]. Let $N(S)$ denote the number of load cycles with constant stress amplitude S to cause fatigue failure. Then a linear approximation can be obtained from experimental data for typical materials used in structural and machine design between $N(S)$ and S on a log–log scale as shown in Fig. 7.8:

$$\log N = b - \alpha \log S \tag{7.30}$$

or,

$$\log(NS^{\alpha}) = b$$

that is,

$$NS^{\alpha} = \beta \tag{7.31}$$

where the constants α and $\beta = 10^b$ depend only on the material. Actually, a wide scatter prevails in this fatigue law and Eq. (7.31) should be interpreted as representing a mean fatigue curve. Based on this idealized experimental law, Palmgren and Miner's hypothesis states that the percentage or fraction of damage D due to the application of stress amplitude S_i for n_i cycles is accumulated linearly, that is,

$$D = \frac{n_i}{N(S_i)} \tag{7.32}$$

where $N(S_i)$ is the fatigue life with cyclic load of constant amplitude S_i. When stress amplitudes $S_1, S_2, \ldots,$ are applied for $n_1, n_2, \ldots,$ cycles, respectively, the total percentage (fractional) damage is given by

$$D = \frac{n_1}{N(S_1)} + \frac{n_2}{N(S_2)} + \cdots = \sum_i \frac{n_i}{N(S_i)} \tag{7.33}$$

Clearly, fatigue failure is defined by $D = 1$ (100 percent damage). Equation (7.33) can be extended to continuous variations of loading cycles as

$$D = \int_{S=0}^{\infty} \frac{n(S)dS}{N(S)} \tag{7.34}$$

where $n(S)dS$ represents the number of cyclic loadings with amplitude in the interval $(S, S + dS)$ and $N(S)$ is the fatigue life at stress amplitude S.

We now consider the cyclic loading as a stationary narrow band random process and derive an expression for the expected fatigue damage [7.23, 7.47]. Now $n(S)dS$ represents the random number of load cycles with amplitudes in the interval $(S, S + dS)$. From Eq. (7.33), we see that the percentage damage D is also a random variable. The mean or expected value of D is given by

$$E(D) = \int_0^{\infty} \frac{E[n(S)dS]}{N(S)} \tag{7.35}$$

by assuming fatigue life $N(S)$ to be a deterministic or nonrandom function of the amplitude S. In Eq. (7.35), $E(n(S)dS)$ denotes the mean number of load cycles with stress amplitude or peak in the interval $(S, S + dS)$. It can be shown [7.24, 7.47] that the mean number of load cycles is equal to the difference between the mean numbers of up-crossings of the two peaks S and $S + dS$:

$$E[n(S)dS] = (\nu_S^+ - \nu_{S+dS}^+)T = (-d\nu_S^+)T \tag{7.36}$$

where ν_S^+ is the mean up-crossing rate of the stress level S and T is the time duration of load application. The probability density function of the amplitude of the load (S) [7.23] can be derived as

$$f(S)\, dS = -\frac{d\nu_S^+}{\nu_0^+} \tag{7.37}$$

where ν_0^+ is the zero crossing rate. Substitution of Eqs. (7.36) and (7.37) into Eq. (7.35) leads to

$$E(D) = \int_0^\infty \nu_0^+ T \frac{f(S)\,dS}{N(S)} = \nu_0^+ T \int_0^\infty \frac{f(S)\,dS}{N(S)} \tag{7.38}$$

If the stress history $S(t)$ follows normal distribution, then the stress peaks follow Rayleigh's distribution as [7.47]

$$f(S) = \frac{S}{\sigma_S^2} \exp\left(-\frac{S^2}{2\sigma_S^2}\right) \tag{7.39}$$

where σ_S is the standard deviation of the random stress, S. The Palmgren–Miner fatigue law can be written as

$$N(S)S^\alpha = \beta \tag{7.40}$$

By using Eqs. (7.39) and (7.40), Eq. (7.38) can be expressed as [7.23, 7.47]

$$\begin{aligned} E(D) &= \frac{\nu_0^+ T}{\sigma_S^2 \beta} \int_0^\infty S^{\alpha+1} \exp\left(-\frac{S^2}{2\sigma_S^2}\right) dS \\ &= \frac{\nu_0^+ T}{\beta}\left(\sqrt{2}\sigma_S\right)^\alpha \Gamma\left(1 + \frac{\alpha}{2}\right) \end{aligned} \tag{7.41}$$

where

$$\Gamma(1 + x) = \int_0^\infty e^{-y} y^x \, dy \tag{7.42}$$

is the gamma function. Noting that

$$\nu_0^+ = \frac{\omega_0}{2\pi} \tag{7.43}$$

where ω_0 is the frequency of cyclic load variation, the expected fatigue damage in time T can be expressed as

$$E(D) = \frac{\omega_0 T}{2\pi\beta}\left(\sqrt{2}\sigma_S\right)^\alpha \Gamma\left(1 + \frac{\alpha}{2}\right) \tag{7.44}$$

7.5 Modeling of Loads

7.5.1 Introduction

The analysis and design of any machine or structure requires the identification of all important environmental factors that induce loads on the machine or structure, as well as the quantification of those loads. The following types of loads are important in the design of machines and structures:

1. dead loads;
2. live loads;
3. wind loads; and
4. earthquake loads.

7.5.2 Dead Loads

Dead load or gravity load is due to the weight of the machine or structure including all the permanent attachments. The mean values of dead loads can be determined from the specific weights of materials and components. The variability in the dead load is due to the variations in the specific weight of the material and permanent attachments. The dead load is usually described by normal distribution with a coefficient of variation of 0.1. Although normal distribution is often used to describe the dead loads, it is desirable to use a lognormal distribution due to its ability to consider positive loads only.

7.5.3 Live Loads

In structural design, the live load consists of weights of people, furniture and portable fixtures, and equipment. Since these loads vary in time and space, it is very difficult to establish the random variability of these loads. However, several comprehensive statistical surveys have been conducted on live loads [7.25–7.27]. Typical values of the minimum live loads recommended by the 1979 *Uniform Building Code* (*UBC*) are given in Table 7.1. Culver [7.25, 7.27] conducted a survey of live loads of 625 general and clerical offices in private office buildings and observed the mean value and standard deviation of the live load to be 9.0 lb/ft^2 and 5.5 lb/ft^2, respectively. The histogram shows a skew shape and hence a lognormal distribution will be more appropriate than normal distribution to describe the live load.

TABLE 7.1 Minimum live loads recommended by *UBC*†

Nature of Occupancy	Concentrated Load (lb)	Uniform Load†† (lb/ft^2)
Wholesale stores	3000	100
Heavy storage	—	250
School classrooms	1000	40
Residential buildings	0	40
Offices	2000	50
Heavy manufacturing	3000	125
Libraries:		
Reading rooms	1000	60
Stack rooms	1500	125
Hospitals	1000	40
Printing plants (press rooms)	2500	150
Auditoriums (stage areas)	0	125

† Loads specified in Table 7.1 are from Table 23-A of the 1985 *Uniform Building Code* Copyright © 1985, and are reproduced with the written permission of the publishers, the International Conference of Building Officials.

†† See Table 23-A of [7.29] for more details.

EXAMPLE 7.3

If the mean value and the standard deviation of the live load in a heavy manufacturing plant are 125 lb/ft^2 and 62.5 lb/ft^2, respectively, determine the probability of the live load exceeding a value of 150 lb/ft^2. Assume lognormal distribution for the magnitude of the live load.

Solution: The probability density function of the live load L can be expressed as

$$f_L(l) = \frac{1}{l\,\sigma_X\sqrt{2\pi}}\exp\left\{-\frac{1}{2}\left(\frac{\ln l - \overline{X}}{\sigma_X}\right)^2\right\}$$

where

$$\gamma_L = \frac{\sigma_L}{\overline{L}} = \frac{62.5}{125.0} = 0.5$$

$$\sigma_X^2 = \ln(\gamma_L^2 + 1) = \ln(0.5^2 + 1) = 0.223143$$

$$\overline{X} = \ln\overline{L} - \frac{1}{2}\sigma_X^2 = \ln 125.0 - \frac{1}{2}(0.223143) = 4.80341$$

Defining the standard normal variate z_1 corresponding to the specified live load of 150 lb/ft^2 as

$$z_1 = \frac{\ln l - \overline{X}}{\sigma_X} = \frac{\ln 150 - 4.80341}{0.223143} = 0.92864$$

the desired probability can be found as

$$\begin{aligned} P[L > 150] &= P[z > z_1] = P[z > 0.92864] \\ &= 1 - P[z \le 0.92864] = 1 - 0.8238 = 0.1762 \end{aligned}$$

■

7.5.4 Wind Loads

The statistical analysis of the maximum annual wind velocities at several locations in the United States was conducted by Thom [7.30]. He found that type-II extreme value distribution for largest values, also known as the *Frechet distribution,* is the most appropriate one to describe the yearly maximum wind velocity at any location. The corresponding probability density and distribution functions are given by

$$f_V(v) = \left(\frac{\gamma}{\beta}\right)\left(\frac{\beta}{v}\right)^{\gamma+1}\exp\left(-\left\{\frac{\beta}{v}\right\}^{\gamma}\right); \quad v \ge 0, \gamma > 0, \beta > 0 \tag{7.45}$$

$$F_V(v) = P(V \le v) = \exp\left\{-\left(\frac{\beta}{v}\right)^{\gamma}\right\} \tag{7.46}$$

where V is the maximum wind velocity, and γ and β are the shape and scale parameters of the distribution. Thom determined the numerical values of the parameters γ and β for 141 locations in the United States. The mean and standard deviation of the random variable V are given by

$$\overline{V} = \beta\,\Gamma\left(1 - \frac{1}{\gamma}\right) \tag{7.47}$$

$$\sigma_V^2 = \beta^2\left[\Gamma\left(1 - \frac{2}{\gamma}\right) - \Gamma^2\left(1 - \frac{1}{\gamma}\right)\right] \tag{7.48}$$

The probability that the wind velocity will exceed a specified value v in any single year can be found as

$$P(V > v) = 1 - F_V(v) = r(say) \tag{7.49}$$

The return period corresponding to a wind velocity of magnitude v is defined as the number of years elapsed before the recurrence of the wind velocity of magnitude v (see Section 4.7). By denoting the return period as $R(v)$, the probability of the return period exactly equal to i years can be determined as

$$P[R(v) = i] = (1 - r)^{i-1}r \tag{7.50}$$

where r denotes the probability that the wind velocity will exceed v in the ith year and $(1 - r)^{i-1}$ represents the probability that the wind velocity will not exceed v in each of the first $(i - 1)$ years. The mean return period can be found as

$$E[R(v)] = \overline{R} = \sum_{i=1}^{\infty} i\,P[R(v) = i] = \sum_{i=1}^{\infty} i(1 - r)^{i-1}r \tag{7.51}$$

From Eqs. (7.49) and (7.51), we obtain

$$\overline{R} = \frac{1}{1 - F_V(v)}$$

or

$$F_V(v) = 1 - \frac{1}{\overline{R}} \tag{7.52}$$

For any specified mean return period $\overline{R}$ at any particular location, the value of the probability distribution function $F_V(v)$ and the corresponding value of the wind velocity v can be determined.

The building code requirements specify the minimum design loads to be used for buildings by taking into consideration the probabilistic nature of the wind velocity [7.33]. The extreme wind velocity data for several cities in the United States can be found in Ref [7.32]. For example, the mean value, $\overline{V}$, and the coefficient of variation, γ_V, of wind velocity have been observed in Baltimore, MD, Detroit, MI, Austin, TX, and Sacramento, CA, respectively, as (55.9, 0.12), (48.9, 0.14), (45.1, 0.12), and (46.0, 0.22) mph. For the same cities, the mean value and coefficient of variation of the 50-year maximum wind velocity, $(\overline{V}_{50}, \gamma_{V_{50}})$, have been found to be (76.9, 0.09), (69.8, 0.10), (61.9, 0.09), and (77.3, 0.13) mph, respectively.

EXAMPLE 7.4

The annual maximum wind velocities measured at a particular location are given in the following table:

Year	Maximum Wind Velocity (mi/h)
1978	82.4
1979	76.9
1980	70.2
1981	91.8
1982	76.1
1983	72.7
1984	58.6
1985	85.3
1986	92.5
1987	68.5

Determine the parameters of the probability density function assuming type-II extreme value distribution for the largest values to be valid for the maximum wind velocity.

Solution: The mean and standard deviation of the measured annual maximum wind velocity are given by

$$\begin{aligned}\overline{V} &= \frac{1}{10}\sum_{i=1}^{10} v_i \\ &= \frac{1}{10}(82.4 + 76.9 + 70.2 + 91.8 + 76.1 + 72.7 + 58.6 + 85.3 + 92.5 + 68.5) \\ &= 77.5 \text{ mi/h} \qquad \text{(E1)}\end{aligned}$$

$$\begin{aligned}\sigma_V^2 &= \frac{1}{10}\sum_{i=1}^{10} (v_i - \overline{V})^2 \\ &= \frac{1}{10}\left[4.9^2 + 0.6^2 + 7.3^2 + \quad + 14.3^2 + 1.4^2 + 4.8^2 + 18.9^2 + 7.8^2 + 15.0^2 + 9.0^2\right] \\ &= 103.12\,(\text{mi/h})^2 \qquad \text{(E2)}\end{aligned}$$

Equation (E2) gives

$$\sigma_V = 10.1548 \text{ mi/h}$$

and

$$\gamma_V = \frac{\sigma_V}{\overline{V}} = \frac{10.1548}{77.5} = 0.13103 \qquad \text{(E3)}$$

For type-II extremal distribution for the maximum values, Eqs. (7.47) and (7.48) give

$$\gamma_V = \frac{\sigma_V}{\overline{V}} = \frac{\left\{\Gamma\left(1 - \frac{2}{\gamma}\right) - \Gamma^2\left(1 - \frac{1}{\gamma}\right)\right\}^{1/2}}{\Gamma\left(1 - \frac{1}{\gamma}\right)} = 0.13103 \tag{E4}$$

Squaring of Eq. (E4) and rearranging the terms gives

$$1 + \gamma_V^2 = 1 + (0.13102)^2 = 1.01717 = \frac{\Gamma\left(1 - \frac{2}{\gamma}\right)}{\Gamma^2\left(1 - \frac{1}{\gamma}\right)} \tag{E5}$$

Using tables of gamma function [7.50], the solution of Eq. (E5) can be found by trial and error as $\gamma \approx 10$. Equation (7.47) can be used to find β as

$$\beta = \frac{\overline{V}}{\Gamma\left(1 - \frac{1}{\gamma}\right)} = \frac{77.5}{\Gamma(0.9)} = \frac{77.5}{1.0686} = 72.5248$$

■

EXAMPLE 7.5

Find the probability of the maximum wind velocity exceeding a value of 85 mi/h in any year for the data given in Example 7.4. Also, find the return period for maximum velocities greater than 85 mi/h.

Solution: Assuming type-II distribution for the largest values to be valid for the maximum wind velocity V the probability of V, exceeding 85 mi/h is given by

$$P[V > 85 \text{ mi/h}] = 1 - F_V(85) = 1 - \exp\left\{-\left(\frac{\beta}{85}\right)^{\gamma}\right\}$$

$$= 1 - \exp\left\{-\left(\frac{72.5248}{85}\right)^{10}\right\} = 0.18493$$

The return period for maximum wind velocity greater than 85 mi/h can be found from Eq. (7.52) as

$$\overline{R} = \frac{1}{P[V > 85]} = \frac{1}{1 - F_V(85)} = \frac{1}{0.18493} = 5.40745 \text{ years}$$

■

Variation with Height. Wind velocity is found to vary with height. The mean wind velocity decreases at ground level due to the friction forces created by the roughness and other obstructions on the ground level. In addition, the fluctuations around the mean wind velocity will also be larger at ground level because of the same surface roughness. The height at which the frictional effect of the surface roughness is negligible is known as the gradient height H_g. The wind velocity at this height is called the gradient wind velocity V_g. At heights greater than the gradient height, the wind velocity is assumed to be same as the gradient wind velocity. The variation of mean wind velocity with height is described by the equation [7.28, 7.48]

$$V_H = \left(\frac{H}{H_g}\right)^{\frac{1}{n}} V_g \tag{7.53}$$

where V_H is the mean wind velocity at height H above the ground and n is a constant. The values of H_g and n can be taken as 950 ft and 7 for level ground with some obstructions (e.g., airports and farmland with scattered trees and buildings), 1150 ft and 5 for level ground with numerous obstructions of varying sizes (e.g., suburbs with lot sizes of half acre or more), and 1800 ft and 2 for grounds with large obstructions (e.g., centers of large cities).

EXAMPLE 7.6

If the annual maximum wind velocity data of Example 7.4 was obtained at a height of 100 ft in the middle of a city, find the annual maximum wind velocity corresponding to a 25-year return period. Also, find the corresponding annual maximum wind velocity at the gradient height.

Solution: The 25-year return period $\overline{R}$ is given by (see Eq. (7.52))

$$1 - F_V(v) = \frac{1}{\overline{R}} = \frac{1}{25} = 0.04$$

or

$$F_V(v) = 0.96$$

Using Eq. (7.46) and the solution of Example 7.4, we can obtain

$$F_V(v) = \exp\left\{-\left(\frac{72.5248}{v}\right)^{10}\right\} = 0.96$$

This equation gives $v = V_H = 99.8613\,\text{mi/h}$. Since the gradient height and constant n in Eq. (7.53) for middle of city are given by $H_g = 1800$ ft and $n = 2$, the gradient wind velocity can be determined as

$$V_g = V_H\left(\frac{H_g}{H}\right)^{\frac{1}{n}} = (99.8613)\left(\frac{1800}{100}\right)^{0.5} = 423.6756\ \text{mi/h}$$

■

Determination of Wind Load. The wind velocity can be related to the wind pressure acting on any given structure once the geometry of the structure, the vibrational characteristics of the structure, and the density of air are known. In general, the wind pressure load W acting at a particular point on a structure can be expressed as

$$W = cEGC_p V^2 \tag{7.54}$$

where c is a constant, E is an exposure coefficient, G is the gust factor, C_p is the pressure coefficient, and V is the wind velocity. The pressure coefficient, C_p, is used to account for both the building geometry and the site exposure. The gust factor, G, is used to account for the gustiness of the wind (i.e., the dynamic variations of the wind velocity) and the vibrational characteristics of the structure. Methods, based on the time series analysis of the wind velocity data and the theory of structural dynamics, have been developed to compute the gust factor [7.32]. Once the wind pressure is computed using Eq. (7.54), it must be integrated over the surface of the structure to find the total lateral force and moment at any section of the structure. These values can be used directly in the structural design. The mean values of c, E, G, and C_p are given by the codes [7.33]. The coefficients of variation of E, G, and C_p are on the order of 0.16, 0.11, and 0.12, respectively. Normal distribution can be used to describe these parameters [7.34]. As stated earlier, the yearly maximum wind velocity V is described by the type-II extreme value distribution. Thus, the probability distribution of the wind load, given by Eq. (7.54), cannot be expressed in closed form. The Monte Carlo approach can be used to determine the probability characteristics of the wind load.

7.5.5 Earthquake Loads

The discontinuities present in Earth's crust are always subjected to stress due to the natural forces acting on them. A rapid release of this stress produces seismic waves which cause the ground to shake, leading to an earthquake. An earthquake causes vibration of the buildings and the machinery located on the ground. Since the structural/machinery mass times the earthquake-induced acceleration represents the inertia forces acting on the system, the structure/machine must be designed to withstand the stresses created by these inertia forces. Figure 7.11 gives the common terminology used in the study of earthquakes. The focus or hypocenter represents the point of origin (center of initial rupture) of an earthquake. The epicenter denotes a point on the surface of the ground directly above the focus. The focal depth is the depth of the focus beneath the surface of the ground. The intensity

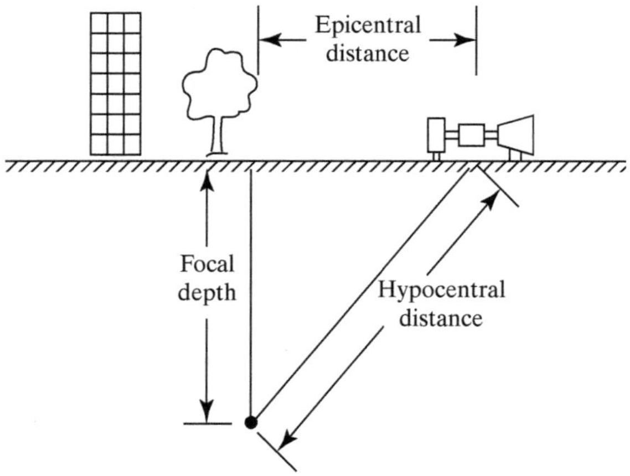

FIGURE 7.11 Terminology used in describing earthquakes.

of an earthquake is commonly measured on the *Richter scale*. A magnitude of M on the Richter scale [7.28, 7.35] is defined as

$$M = \log_{10}\left(\frac{a}{a_0}\right) \tag{7.55}$$

where a denotes the maximum amplitude of the earthquake recorded by an instrument called Wood-Anderson seismograph at a distance of 100 km from the epicenter and a_0 is the reference amplitude (one thousandth of a millimeter). The Richter magnitude [7.28] is related to the energy released by the earthquake W as

$$\log W = 11.8 + 1.5M \tag{7.56}$$

where the energy W is in ergs. A magnitude of M greater than 5 on the Richter scale indicates potential damage to structures. The value of M also gives the length of the slipped fault which, in turn, reflects the size of the area affected by the earthquake. A magnitude of 4 corresponds to a fault length of approximately 1 mile while a magnitude of 9 denotes a 1000 mile slip.

The ground motion during an earthquake is measured by an instrument called a *strong motion accelerograph* placed on the ground at a particular location. A typical graph of the acceleration versus time (also known as an *accelerogram*) recorded by an accelerograph is shown in Fig. 7.12. The accelerograms are generally recorded on photographic paper or film and are digitized for engineering applications. The peak ground acceleration, duration, and frequency content of the earthquake can be obtained from an accelerogram. An accelerogram can be integrated to obtain the time variations of the ground velocity and ground displacement.

Response Spectrum. The response spectrum is a graph between the maximum response of a single degree of freedom oscillator under a base excitation equal to the earthquake

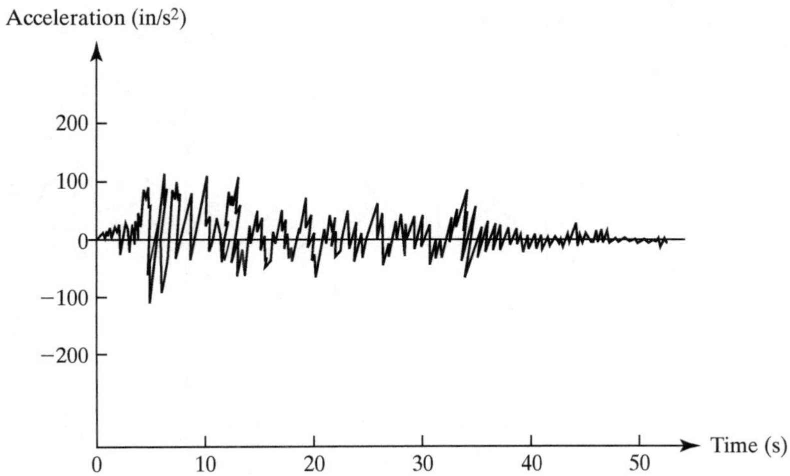

FIGURE 7.12 A typical accelerograph.

acceleration and the natural period (or frequency) of the oscillator. The response spectrum provides the most descriptive representation of the influence of a given earthquake on a structure. The maximum relative displacement of the oscillator for any specified values of the natural period of vibration (τ_n) and damping (ζ) is called the spectral displacement and is represented as $S_d(\tau_n, \zeta)$. Similarly, the maximum relative velocity of the oscillator for any specified values of τ_n and ζ is called the spectral velocity and is denoted as $S_v(\tau_n, \zeta)$. The spectral velocity S_v and the spectral acceleration S_a can be computed from the spectral displacement S_d as

$$S_v(\tau_n, \zeta) = \frac{2\pi}{\tau_n} S_d(\tau_n, \zeta) \tag{7.57}$$

$$S_a(\tau_n, \zeta) = \left(\frac{2\pi}{\tau_n}\right)^2 S_d(\tau_n, \zeta) \tag{7.58}$$

The response spectra are plotted on a four-way logarithmic paper as shown in Fig. 7.13. In this figure, the vertical axis denotes the spectral velocity, the horizontal axis represents the natural time period, the 45° inclined axis going up from left (right) indicates the spectral

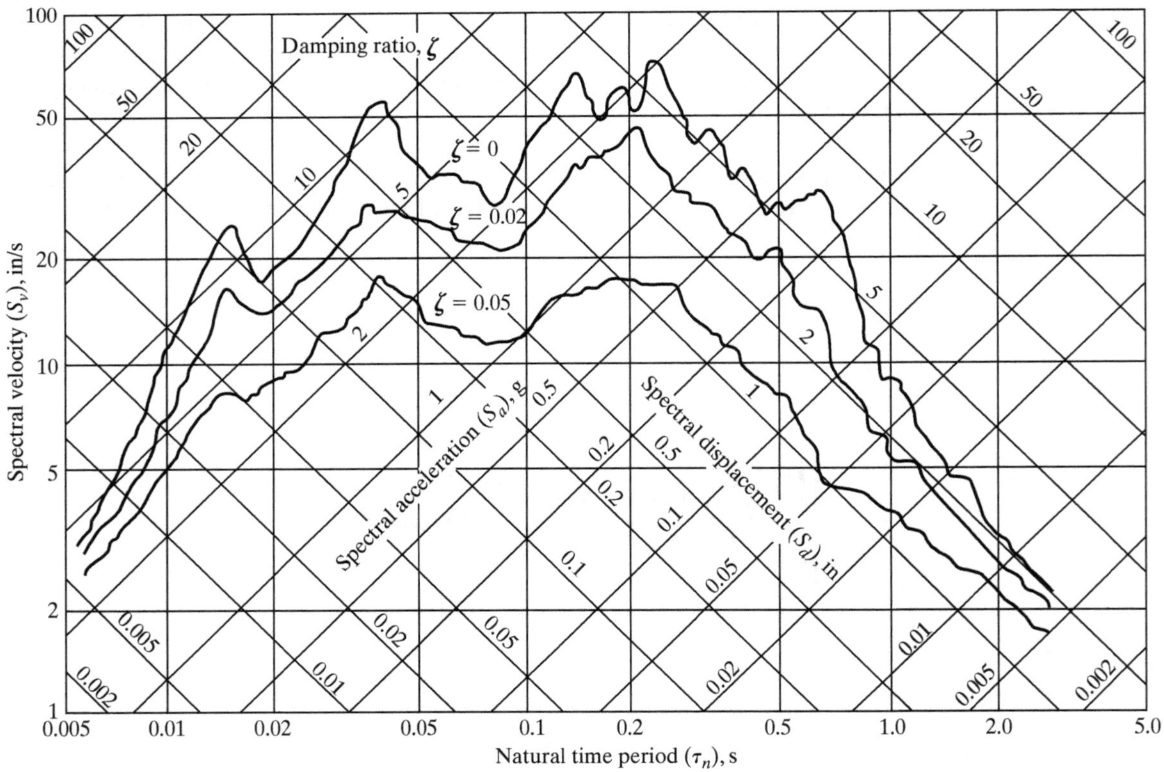

FIGURE 7.13 Response spectrum of an earthquake.

acceleration (displacement). The characteristics of the peak ground motions or the spectral response parameters can be related to the Richter magnitude and the hypocentral distance as [7.28, 7.36]

$$x = \frac{c_1 10^{c_2 M}}{(d + 25)^{c_3}} \tag{7.59}$$

where x = characteristic of peak ground motion; M = Richter magnitude; d = hypocentral distance (km); and c_1, c_2, and c_3 = constants. The values of c_1, c_2, and c_3 are given, respectively, by 472.3, 0.278, and 1.301 for peak ground acceleration; 5.64, 0.401, and 1.202 for peak ground velocity; 0.393, 0.434, and 0.885 for peak ground displacement; 11.0, 0.278, and 1.346 for spectral velocity with 0 percent damping; and 10.09, 0.233, and 1.341 for spectral velocity with 5 percent damping [7.42]. For reliability studies, the peak ground acceleration (or velocity or displacement) can be considered as a random variable. The coefficient of variation is given by 0.548 for peak ground acceleration, 0.969 for peak ground velocity, 0.883 for peak ground displacement, 0.941 for spectral velocity with 0 percent damping, and 0.651 for spectral velocity with 5 percent damping [7.42].

EXAMPLE 7.7

Determine the following for an earthquake with $M = 7$ and $d = 50$ km:

1. The mean and standard deviation of S_v for $\tau_n = 1$ s and $\zeta = 0.05$ using Eq. (7.59).
2. The mean and standard deviation of the corresponding S_a assuming τ_n to be deterministic.
3. The probability of the spectral acceleration S_a exceeding a value of 0.15 g by assuming normal distribution for S_a.

Assume the values of c_1, c_2, and c_3 in Eq. (7.59) for $\tau_n = 1$ s as 0.432, 0.399, and 0.704, respectively, and the coefficient of variation of S_v as 0.703.

Solution:

1. The mean value of the spectral velocity S_v is given by Eq. (7.59) as

$$\bar{x} = \bar{S}_v = \frac{c_1 10^{c_2 M}}{(d + 25)^{c_3}}$$

Using the given values of c_1, c_2, and c_3, we obtain

$$\bar{x} = \bar{S}_v = \frac{0.432(10)^{0.399\times 7}}{(50 + 25)^{0.704}} = \frac{268.2154}{20.8950} = 12.8363 \text{ cm/s}$$

The standard deviation of spectral velocity is given by

$$\sigma_x = \sigma_{S_v} = \gamma_x \bar{x} = 0.703(12.8363) = 9.0239 \text{ cm/s}$$

2. The spectral acceleration S_a is given by

$$S_a = \left(\frac{2\pi}{\tau_n}\right) S_v$$

and hence the mean and standard deviation of S_a can be obtained as

$$\overline{S}_a = \left(\frac{2\pi}{\tau_n}\right)\overline{S}_v = \left(\frac{2\pi}{1}\right)12.8363 = 80.6529\,\text{cm/s}^2$$

$$\sigma_{S_a} = \left(\frac{2\pi}{\tau_n}\right)\sigma_{S_v} = \left(\frac{2\pi}{1}\right)9.0239 = 56.6988\,\text{cm/s}^2$$

3. If the spectral acceleration S_a follows normal distribution, the required probability can be determined as

$$P(S_a > 0.15\,\text{g}) = P(S_a > 147.15\,\text{cm/s}^2) = P\left(\frac{S_a - \overline{S}_a}{\sigma_{S_a}} > \frac{147.15 - 80.6529}{56.6988}\right)$$

$$= P(z > 1.1728) = 1 - P(z \leq 1.1728) = 1 - 0.8796 = 0.1204$$

■

Earthquake Loads. Whenever any structure or machine is subjected to ground acceleration due to an earthquake, all the components of the structure or machine will vibrate. Since the acceleration times the mass gives the inertia load or force, the structure must be designed to withstand the stresses produced by the inertia loads. The equation of motion of the structure, idealized as an undamped single-degree-of-freedom oscillator, subjected to base excitation [7.37] can be expressed as

$$m\,\ddot{y}(t) + k\,x(t) = 0 \tag{7.60}$$

where $m = W/g$ = mass of the structure or oscillator, k = spring stiffness or elasticity of the structure, $x(t)$ = displacement of the mass of the oscillator relative to the base, and $\ddot{y}(t)$ = absolute acceleration of the mass of the oscillator due to the earthquake support motion. From Eq. (7.60), the earthquake inertia force can be expressed as

$$m\,\ddot{y}(t) = -kx(t) \tag{7.61}$$

Thus, the maximum force F is given by k times the maximum relative displacement induced by the earthquake excitation (S_d):

$$F = kS_d(\tau_n, \zeta) = k\left(\frac{\tau_n}{2\pi}\right)^2 S_a(\tau_n, \zeta) \tag{7.62}$$

where S_a is the spectral acceleration and $\tau_n = (2\pi)/\omega_n$ is the natural period of vibration, and $\omega_n = (k/m)^{1/2}$ is the natural frequency of vibration of the oscillator. Equation (7.62) can be rewritten as

$$F = \frac{k}{\omega_n^2} S_a = mS_a$$

or

$$\frac{F}{W} = \frac{S_a}{g} \tag{7.63}$$

This shows that the ratio F/W is equal to the spectral acceleration expressed in units of g.

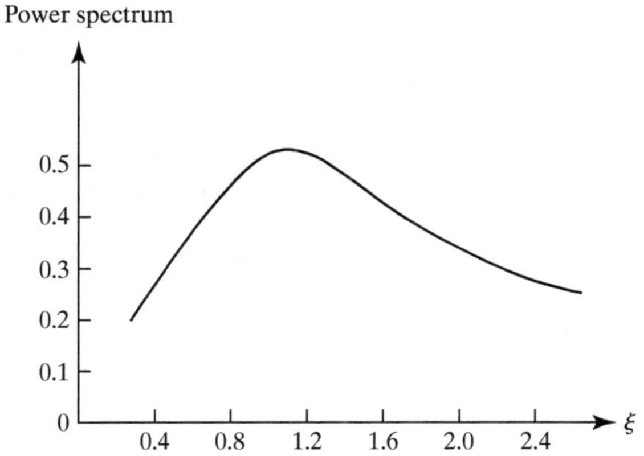

FIGURE 7.14 Power spectrum of ground acceleration.

Power Spectrum of Ground Acceleration. As seen in the last section, the response spectrum and its coefficient of variation can be used to predict the mean value and standard deviation of the earthquake-induced loads in structures. An alternative procedure involves the use of the power spectrum of the ground acceleration. Although the random process techniques are not as well developed, it is possible to apply the power spectral analysis to earthquake analysis.

It has been suggested that an earthquake can be adequately represented by a time history, which consists of a portion of a stationary random process of finite duration, say 15 to 50 seconds. One form of the power spectrum of ground acceleration, $S_a(\omega)$, is given by Tajimi [7.38] as

$$\frac{\omega\, S_a(\omega)}{\sigma_a^2} = \frac{4}{\pi}\left(\frac{H}{1 + 4H^2}\right)\frac{\xi[1 + 4H^2\xi^2]}{[(1 - \xi^2)^2 + 4H^2\xi^2]} \tag{7.64}$$

where a = ground acceleration, σ_a = standard deviation of ground acceleration, $\xi = \omega/\omega_g = \omega/(2\pi f_g)$ = nondimensional frequency, ω = frequency, f_g = characteristic ground frequency (2–3 Hz), and H = nondimensional characteristic damping property of the ground (0.6–0.7). The graph of Eq. (7.64) is shown in Fig. 7.14. The power spectrum can be used to find the variance of the response once the modal characteristics of the structure are known [7.36].

7.6 Numerical Solutions Using MATLAB and Excel

EXAMPLE 7.8

In a mechanical assembly, similar to the one described in Fig. 7.3, the distance between pins 1 and 2 is specified as $l = 9.996 \pm 0.010$ in while the dimensions d_1 and d_2 in part 2 are specified as

$d_1 = 6 \pm 0.003$ in and $d_2 = 8 \pm 0.006$ in. Use MATLAB to find the probability of realizing the distance between the pins larger than the distance between the holes A and B during the assembly process. Assume that all dimensions follow normal distribution.

Solution: Assuming that the tolerances correspond to three-sigma values, we have $\sigma_l = 0.003333$ in, $\sigma_{d_1} = 0.001$ in, and $\sigma_{d_2} = 0.002$ in. The probability of the distance between the pins 1 and 2 (l) larger than the distance between the holes A and B (d) is given by

$$P(l \geq d) = P\left[z = \frac{X - \overline{X}}{\sigma_X} \geq z_1\right] = \Phi(-z_1) \tag{E1}$$

where $X = l - d$ and

$$z_1 = -\frac{\overline{X}}{\sigma_X} = -\frac{\overline{l} - \overline{d}}{\sqrt{\sigma_l^2 + \sigma_d^2}} \tag{E2}$$

The probability indicated by Eq. (E1) is computed using MATLAB. The program and the results are given below.

```
clear all ; clc ;

Lbar = 9.996   ;
D1bar = 6     ;
D2bar = 8     ;

Ls = 0.0033333    ;
D1s = 0.001    ;
D2s = 0.002    ;

Dbar = sqrt( D1bar^2 + D2bar^2 )
Ds = sqrt((D1bar^2)*(D1s^2) + (D2bar^2)*(D2s^2))

z1 = -(Lbar-Dbar)/(sqrt(Ls^2 + Ds^2))

Prob = normcdf(-z1)

Dbar =

     10

Ds =

     0.0171

z1 =

     0.2298

Prob =

     0.4091
```

■

EXAMPLE 7.9

Solve Example 7.8 using Excel.

Solution: The solution procedure is indicated in the following spreadsheet:

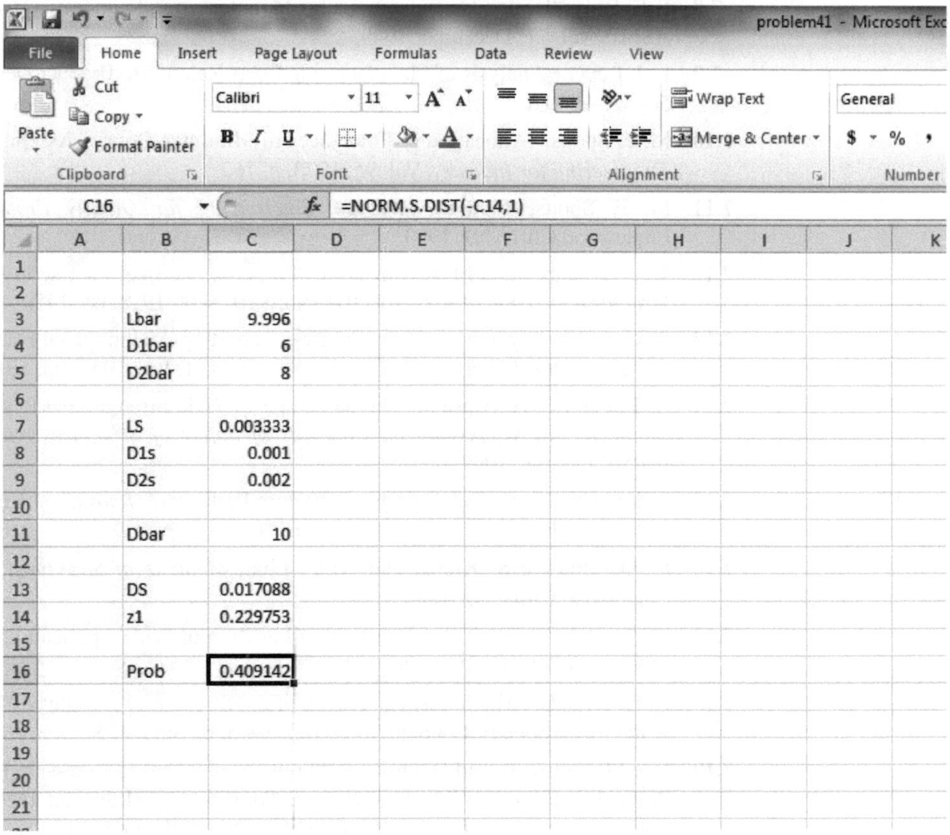

■

REFERENCES AND BIBLIOGRAPHY

7.1 J. Datsko, *Material Properties and Manufacturing Processes,* John Wiley, New York, 1966.

7.2 *Manual of Steel Construction,* 9th ed., American Institute of Steel Construction, Inc., Chicago, IL, 1989.

7.3 G. A. Alpsten, "Variations in Mechanical and Cross-Sectional Properties of Steel," In T. C. Kavanagh (Group Coordinator), *Proceedings of International Conference on Planning and Design of Tall Buildings,* Vol. 1b, *Tall Building Criteria and Loading,* Lehigh University, Bethlehem, PA, August 1972, pp. 755–805.

7.4 J. C. Bralla (ed.), *Handbook of Product Design for Manufacturing,* McGraw-Hill, New York, 1986.

7.5 J. E. Shigley and C. R. Mischke (eds.), *Standard Handbook of Machine Design,* McGraw-Hill, New York, 1986.

7.6 *Metals Handbook,* Vol. 1, T. Lyman (ed.), *Properties and Selection of Metals,* 8th ed., American Society for Metals, Metals Park, OH, 1961.

7.7 E. B. Haugen, *Probabilistic Mechanical Design,* John Wiley, New York, 1980.

7.8 F. E. Bolz, "Design Considerations for Manufacturing Economy," *Mechanical Engineering,* ASME, Vol. 71, December 1949.

7.9 L. J. Bayer, "Analysis of Manufacturing Costs Relative to Product Design," *ASME* paper no. 56–SA–9, 1956.

7.10 M. F. Spotts, "Allocation of Tolerances to Minimize Cost of Assembly," *ASME Journal of Engineering for Industry*, Vol. 95, 1973, p. 762.

7.11 M. F. Spotts, *Dimensioning and Tolerancing for Quality Production,* Prentice-Hall, Englewood Cliffs, NJ, 1983.

7.12 T. V. Galambos and M. K. Ravindra, "Properties of Steel for Use in LRFD," *Journal of the Structural Division, ASCE,* Vol. 104, No. ST9, Sept. 1978, pp. 1459–1468.

7.13 O. C. Julian, "Synopsis of First Progress Report of Committee on Factor of Safety," *Journal of the Structural Division, ASCE,* Vol. 83, No. ST4, July 1957, pp. 1316.1–1316.22.

7.14 L. Tall and G. A. Alpsten, "On the Scatter of Yield Strength and Residual Stresses in Steel Members," in *Symposium on Concepts of Safety of Structures and Methods of Design, IABSE,* London, 1969, pp. 151–163.

7.15 B. Epstein, "Statistical Aspects of Fracture Problems," *Journal of Applied Physics,* Vol. 19, 1948, pp. 140–147.

7.16 H. O. Madsen, S. Krenk, and N. C. Lind, *Methods of Structural Safety,* Prentice-Hall, Englewood Cliffs, NJ, 1986.

7.17 W. Weibull, "A Statistical Distribution Function of Wide Applicability," *Journal of Applied Mechanics,* Vol. 18, 1951, pp. 293–297.

7.18 H. E. Daniels, "The Statistical Theory of the Strength of Bundles of Threads," *Proceedings of the Royal Society,* London, Vol. A183, 1945, pp. 405–435.

7.19 J. F. McCarthy, Jr. and O. Orringer, "Some Approaches to Assessing Failure Probabilities of Redundant Structures," In *Composite Reliability,* ASTM Special Tech. Pub. 580, American Society for Testing Materials, Philadelphia, PA, 1975, pp. 5–31.

7.20 R. E. Little and E. H. Jebe, "Statistical Design of Fatigue Experiments," John Wiley, New York, 1975.

7.21 C. R. Mischke, "Prediction of Stochastic Endurance Strength," *ASME Journal of Vibration, Acoustics, Stress, and Reliability in Design,* Vol. 109, 1987, pp. 113–122.

7.22 J. L. Bogdanoff and F. Kozin, *Probabilistic Models of Cumulative Damage,* John Wiley, New York, 1985.

7.23 T. Y. Yang, *Random Vibration of Structures,* John Wiley, New York, 1986.

7.24 A. Powell, "On the Fatigue Failure of Structures Due to Vibrations Excited by Random Pressure Fields," *Journal of the Acoustical Society of America,* Vol. 30, no. 12, December 1958, pp. 1130–1135.

7.25 C. G. Culver, "Survey Results for Fire Loads and Live Loads in Office Buildings," *NBS Building Science Series Report 85,* Center for Building Technology, National Bureau of Standards, Washington, DC, 1976.

7.26 R. B. Corotis and V. A. Doshi, "Probability Models for Live Load Survey Results," *Journal of the Structural Division, ASCE,* Vol. 103, No. ST6, June 1977, pp. 1257–1274.

7.27 B. Ellingwood and C. Culver, "Analysis of Live Loads in Office Buildings," *Journal of the Structural Division, ASCE,* Vol. 103, No. ST8, August 1977, pp. 1551–1560.

7.28 G. C. Hart, *Uncertainty Analysis, Loads, and Safety in Structural Engineering,* Prentice-Hall, Englewood Cliffs, NJ, 1982.

7.29 "Uniform Building Code," International Conference on Building Officials, Whittier, CA, 1985.

7.30 H. C. S. Thom, "New Distributions of Extreme Winds in the United States," *Journal of the Structural Division, ASCE,* Vol. 94, No. ST7, July 1968, pp. 1787–1802.

7.31 B. Ellingwood, T. V. Galambos, J. G. MacGregor, and C. A. Cornell, "Development of a Probability Based Load Criterion for American National Standard A58," *NBS Special Publication 577,* National Bureau of Standards, Washington, DC, 1980.

7.32 www.itl.nist.gov/div898/winds/daily.htm

7.33 "Building Code Requirements for Minimum Design Loads in Buildings and Other Structures," *ANSI* A58.1, American National Standards Institution, New York, 1972.

7.34 G. I. Schueller, H. Hirtz, and G. Booz, "The Effects of Uncertainties in Wind Load Estimation on Reliability Assessments," *Journal of Wind Engineering and Industry Aerodynamics,* Vol. 14, 1983, pp. 15–26.

7.35 F. Naeim (ed.), *The Seismic Design Handbook,* Van Nostrand Reinhold, New York, 1989.

7.36 P. L. Gould and S. H. Abu-Sitta, *Dynamic Response of Structures to Wind and Earthquake Loading,* John Wiley, New York, 1980.

7.37 S. S. Rao, "Mechanical Vibrations," 5th ed., Prentice Hall, Upper Saddle River, NJ, 2011.

7.38 H. Tajimi, "A Statistical Method of Determining the Maximum Response of a Building Structure During an Earthquake," *Proceedings of the Second World Conference on Earthquake Engineering,* Tokyo, Japan, Vol. 2, 1960, pp. 781–796.

7.39 R. E. Melchers, "Structural Reliability Analysis and Prediction," Ellis Horwood Ltd., Chichester, England, 1987.

7.40 E. Simiu and J. J. Filliben, "Weibull Distributions and Extreme Wind Speeds," *Journal of the Structural Division, ASCE,* Vol. 106, No. ST12, 1980, pp. 491–501.

7.41 W. H. Melbourne, "Probability Distributions Associated with the Wind Loading of Structures," *Civil Engineering Transactions, Institution of Engineers,* Australia, CE, Vol. 19, No. 1, 1977, pp. 58–67.

7.42 R. K. McGuire, "Seismic Structural Response Risk Analysis, Incorporating Peak Response Regressions on Earthquake Magnitude and Distance," *Department of Civil Engineering Research Report R74-51,* MIT, Cambridge, MA, 1974.

7.43 M. Vorlicek and M. Holicky, *Analysis of Dimensional Accuracy of Building Structures,* Elsevier Science Publishing Co. Inc., Amsterdam, The Netherlands, 1989.

7.44 *Aluminum Standards and Data,* Aluminum Association, Inc., Washington, DC, 1990.

7.45 C. Lipson, N. J. Sheth, and R. Disney, "Reliability Prediction—Mechanical Stress/Strength Interference (Ferrous)," Rome Air Development Center, Technical Report No. RADC-TR-66-710, AD 813574, March 1967.

7.46 C. Lipson, N. J. Sheth, R. Disney, and M. Alton, "Reliability Prediction—Mechanical Stress/Strength Interference (Nonferrous)," Rome Air Development Center, Technical Report No. RADC-TR-68-403, AD 856021, February 1969.

7.47 S. H. Crandall and W. D. Mark, *Random Vibration in Mechanical Systems,* Academic Press, New York, 1963.

7.48 A. G. Davenport, "Gust Loading Factors," *Journal of the Structural Division, ASCE,* Vol. 93, No. ST3, June 1967, pp. 11–34.

7.49 R. L. Wiegel (ed.), *Earthquake Engineering,* Prentice-Hall, Englewood Cliffs, NJ, 1970.

7.50 H. T. Davis, *Tables of the Mathematical Functions,* Vol. I, The Principia Press of Trinity University, San Antonio, TX, 1963.

7.51 H. Volkmann, "Ernst Abbe and His Work," *Applied Optics,* Vol. 5, November 1966, pp. 1720–1731.

7.52 *Building Code Requirements for Reinforced Concrete,* ACI 318-95, American Concrete Institute, 1995.

7.53 *Manual of Steel Construction Load & Resistance Factor Design,* 2nd ed., American Institute of Steel Construction, 1994.

7.54 *Minimum Design Loads for Buildings and Other Structures,* American Society of Civil Engineers, ASCE 7-95, New York, NY, 1996.

7.55 A. H-S. Ang and C. A. Cornell, "Reliability Bases of Structural Safety and Design," *Journal of Structural Engineering,* ASCE, Vol. 100, No. ST9, pp. 1755–1769, 1974.

7.56 B. M. Ayyub, "The Nature of Uncertainty in Structural Engineering," In B. M. Ayyub and M. M. Gupta (eds.), *Uncertainty Modeling and Analysis: Theory and Applications,* Elsevier Science, Amsterdam, The Netherlands, 1994.

7.57 B. M. Ayyub and A. Haldar, "Decisions in Construction Operation," *Journal of the Construction Engineering and Management Division,* ASCE, Vol. 111, No. 4, pp. 343–357, 1985.

7.58 *Standards for the Design of Cold-Formed Steel Members in Buildings,* Canadian Standards Association, CSA S-137, 1974.

7.59 C. A. Cornell, "A Probability-Based Structural Code," *Journal of the American Concrete Institute,* Vol. 66, No. 12, pp. 974–985, 1969.

7.60 B. R. Ellingwood, J. G. MacGregor, T. V. Galambos, and C. A. Cornell, "Probability Based Load Criteria: Load Factors and Load Combinations," *Journal of Structural Engineering,* ASCE, Vol. 108, No. ST5, pp. 978–997, 1982.

7.61 G. Fu, "Reliability Models for Assessing Highway Bridge Safety," *Proceedings of International Conference on Structural Safety and Reliability,* Kyoto, Japan, pp. 1883–1888, 1998.

7.62 K. Gurley, M. Tognarelli, and A. Kareem, "Analysis and Simulation Tools for Wind Engineering," *Probabilistic Engineering Mechanics,* Vol. 12, No. 1, pp. 9–31, 1997.

7.63 A. Haldar, "Probabilistic Evaluation of Construction Deficiencies," *Journal of the Construction Engineering Division,* ASCE, Vol. 107, No. CO1, pp. 107–119, 1981.

7.64 A. Haldar, "Probabilistic Evaluation of Welded Structures," *Journal of the Structural Engineering Division,* ASCE, Vol. 108, No. ST9, pp. 1943–1955, 1982.

7.65 A. M. Hosofer and N. C. Lind, "Exact and Invariant Second Moment Code Format," *Journal of the Engineering Mechanics Division,* ASCE, Vol. 100, No. EM1, pp. 111–121, 1974.

7.66 A. Kareem, "Wind Effects on Structures: A Probabilistic Viewpoint," *Probabilistic Engineering Mechanics,* Vol. 2, No. 4, pp. 166–200, 1987.

7.67 Y. K. Wen, *Structural Load Modeling and Combination for Performance and Safety Evaluation,* Elsevier, Amsterdam, The Netherlands, 1990.

7.68 Z. Zhao, A. Haldar, and F. L. Breen, "Fatigue Reliability Updating Through Inspections for Bridges," *Journal of Structural Engineering,* ASCE, Vol. 120, No. 5, pp. 1624–1642, 1994.

REVIEW QUESTIONS

7.1 Give brief answers to the following:

1. What are the primary reasons for the randomness of geometry of machine parts?
2. What are the causes of variation in the strength of a material?
3. What type of loads are of interest to a mechanical designer?
4. Indicate qualitatively the variation of cost with tolerance on a machine component.
5. How are the dimensions of an assembly influenced by the dimensions of individual parts?
6. Name three statistical models that can be used to describe the strength of different types of materials.
7. How is the strength of a brittle material described probabilistically?
8. How can we model the strength of a fiber-reinforced plastic material?
9. How does a fatigue failure occur?
10. What is the significance of *S–N* diagram?
11. What is endurance limit?
12. How can we estimate the damage of a component subjected to a variable amplitude cyclic load?
13. What is the difference between a dead load and a live load?
14. Which probability distribution is commonly used to describe the annual maximum wind velocities?
15. What is the significance of return period in the wind loading on a structure?
16. Describe the variation of wind velocity with height.
17. How is the magnitude of an earthquake defined on the Richter scale?
18. What is an accelerogram and what is its use?
19. What is a response spectrum?
20. Write the equation of motion of a structure, idealized as a single degree of freedom system, subjected to a base excitation.

7.2 Indicate whether each of the following statements is true or false:

1. The tolerance limits are to be specified so as to conform to the capabilities of the manufacturing process.
2. The relative cost of grinding a machine part is much more expensive compared to machining.
3. Very fine machining is more expensive than lapping.

4. The strength of a plastic material follows the weakest-link model.
5. In the fiber-bundle model, whenever a particle fails, the total load will be carried by all the remaining particles.
6. The standard fatigue test is called the rotating beam test.

7.3 Fill in the blank with the proper word(s):

1. The three parameters that influence the analysis/design of any mechanical/structural system are material strength, load, and ______.
2. Tolerance denotes the permissible ______ of a dimension produced.
3. An example of a static load acting on a building floor is ______.
4. If a dimension is represented as $X \pm t$ with a three-sigma tolerance band, the standard deviation of X will be ______.
5. The nominal Young's modulus of steel commonly used in standards and codes is ______ psi.
6. The statistical models used to describe the strength of materials are brittle, plastic, and ______ models.
7. The strength of a brittle material follows the ______ model.
8. In general, the mean value of the strength of a material ______ with the volume.
9. As the size (number of particles) of a solid body increases, its strength can be assumed to follow ______ distribution according to the central limit theorem.
10. The origin of the fiber-bundle theory is the ______ ______ with several threads.
11. Failures due to fluctuating loads are known as ______ failures.
12. The two stages of fatigue failure are the crack initiation stage and the crack ______ stage.
13. The *S–N* diagram in a fatigue test uses ______ scale.
14. The stress value below which fatigue failure will not occur is called the ______ limit.
15. The weights of steel desks and cabinets are called ______ loads.
16. The weights of people and movable furniture are called ______ loads.
17. The intensity of an earthquake is commonly measured on the ______ scale.

7.4 Select the most appropriate answer from the multiple choices given:

1. The tolerance that can be achieved in a milling operation is:
   a. $\pm 0.0002''$ **b.** $\pm 0.001''$ **c.** $\pm 0.01''$
2. The tolerance that can be achieved in spot welding is:
   a. $\pm 0.01''$ **b.** $\pm 0.001''$ **c.** $\pm 0.0001''$
3. Most engineering materials exhibit variable properties due to random differences in their:
   a. Dimensions **b.** Chemistry **c.** Use
4. The nominal Poisson's ratio of steel commonly used in codes and standards is:
   a. 0.33 **b.** 0.25 **c.** 0.30
5. The coefficient of variation of the Young's modulus of structural steel is:
   a. 0.1 **b.** 0.0001 **c.** 0.01
6. An ideal brittle material fails whenever the following fails:
   a. A single particle **b.** All particles **c.** Half of the particles
7. An ideal plastic material fails whenever the following fails:
   a. A single particle **b.** All particles **c.** Half of the particles

8. The strength of a brittle material can be described using the following asymptotic distribution for the smallest value:
   a. Type I b. Type II c. Type III
9. In fatigue failure, the magnitude of stress compared to the yield strength of the material is:
   a. Larger b. Smaller c. Equal
10. In fatigue failure, the crack propagation stage involves:
   a. 90% of cycles b. 70% of cycles c. 20% of cycles
11. In fatigue failure, crack initiation occurs due to:
   a. Surface scratches b. Large static load c. Large volume
12. The probability distribution of the endurance limit is useful for
   a. L_{10}–life design b. Infinite life design c. Finite life design
13. The Palmgren and Miner's hypothesis can consider fatigue damage due to different:
   a. Stress magnitudes b. Strain magnitudes c. Endurance limits
14. If stress variations follow normal distribution, the stress peaks follow the following distribution:
   a. Normal b. Rayleigh c. Uniform
15. The most descriptive representation of the influence of a given earthquake on a structure is provided by:
   a. Power spectrum b. Response spectrum c. Accelerogram
16. The mean wind velocity above the ground level compared to that at the ground level is:
   a. Smaller b. Larger c. Same

7.5 Match the tolerances that can be achieved in different types of manufacturing processes:

1. Turning	**a.** ± 0.00005 in
2. Grinding	**b.** ± 0.02 in
3. Lapping	**c.** $\pm \frac{1}{32}$ in
4. Flame cutting	**d.** ± 0.001 in
5. Sand casting	**e.** ± 0.01 in
6. Heat treatment	**f.** ± 0.0002 in

7.6 Match the following loads with their representations:

1. Dead and live loads	**a.** Frechet distribution (type II for largest values)
2. Wind loads	**b.** Power spectrum
3. Earthquake ground acceleration	**c.** Lognormal distribution

PROBLEMS

7.1 The link lengths of a robot manipulator (Fig. 7.15) are given by

$$l_1 = 10 \pm 0.1 \text{ in}, \quad l_2 = 15 \pm 0.3 \text{ in}, \quad l_3 = 12 \pm 0.2 \text{ in}$$

Find the probability of the end effector P reaching a point located at a distance of 37 in from the base point O.

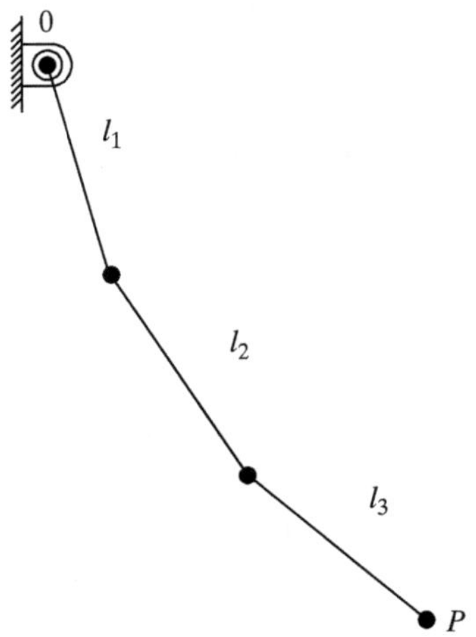

FIGURE 7.15 Links of a robot manipulator.

7.2 The battery compartment (see Fig. 7.16) in an electronic instrument has a length of $d = 1.0 \pm 0.1$ in. The free length of the spring is given by $d_1 = 0.3 \pm 0.05$ in. If the length of the battery is specified as $d_2 = 0.8 \pm 0.05$ in, find the probability of compressing the spring more than 0.15 in. Assume the tolerances to correspond to three-sigma values.

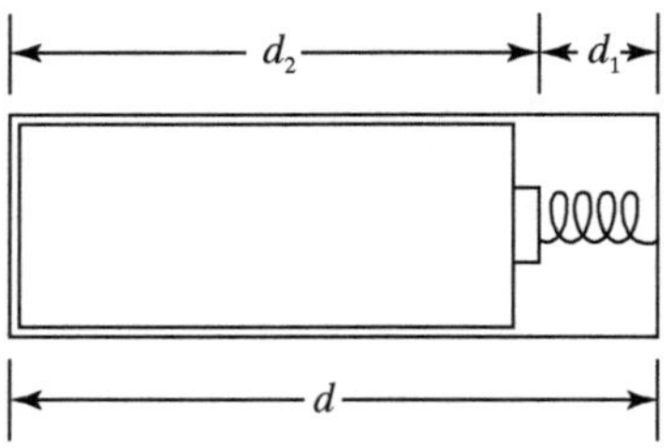

FIGURE 7.16 A battery compartment.

7.3 A packaging box has a length of 20 ± 0.2 in. If the box is to be packed with components of length 4 ± 0.05 in each, determine the probability of accommodating five components in the box without interference. Assume normal distribution for all the dimensions.

7.4 The radial clearance between the rotor and stator of a Francis water turbine is given by 5 ± 1 mm (see Fig. 7.17). The deflection of the shaft, on which the rotor is mounted, can cause a radial motion of 4 ± 1 mm. Determine the probability of rotor touching the stator during the operation of the turbine.

7.5 The joint between two links is formed by first aligning the holes of links 1 and 2 (Fig. 7.18(*a*) and (*b*)) and then inserting the pin shown in Fig. 7.18(*c*). If the dimensions A and a are specified as $A = 1.2 \pm 0.1$ in and $a = 1.1 \pm 0.05$ in, determine the probability of interference between the links 1 and 2. Assume the tolerances to correspond to three-sigma values.

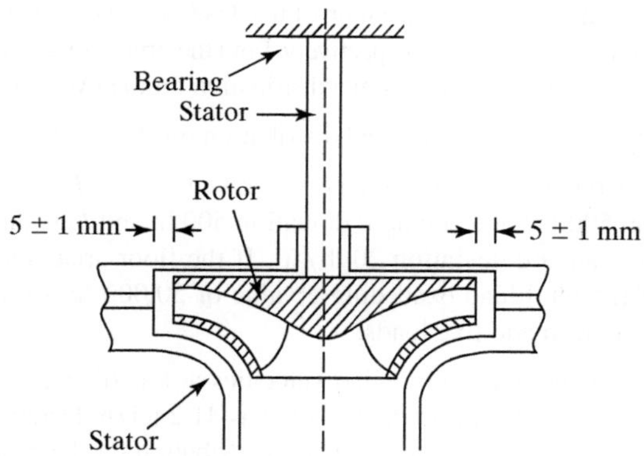

FIGURE 7.17 A Francis water turbine.

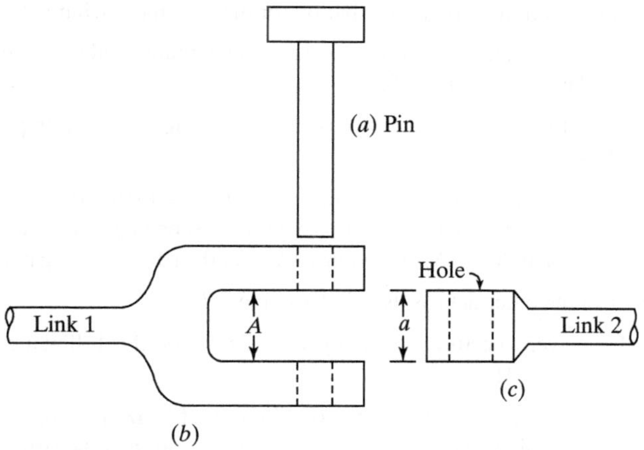

FIGURE 7.18 A joint between two links.

7.6 A load is supported by three identical cables as shown in Fig. 7.19. If the load and strength of each cable follow normal distribution with the parameters $\mu_L = 200$, $\sigma_L = 50$, $\mu_{S_i} = 90$, and $\sigma_{S_i} = 10$, $i = 1, 2, 3$, determine the probability of failure of the system.

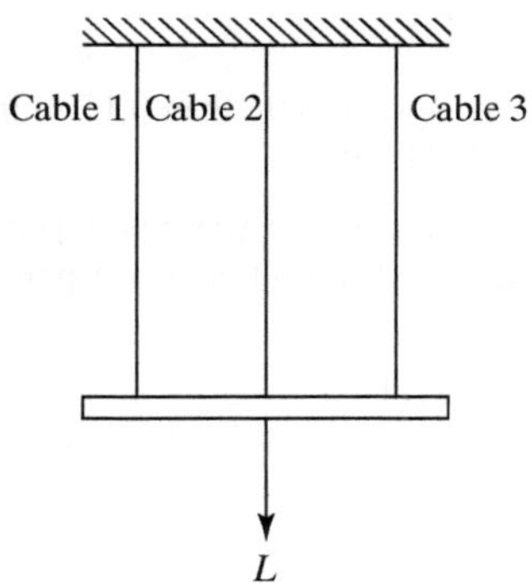

FIGURE 7.19 Load acting on three cables.

7.7 The mean value and the standard deviation of live load in a heavy-manufacturing building are given by 125 lb/ft^2 and 25 lb/ft^2, respectively. Find the probability of the live load exceeding a value of 150 lb/ft^2 assuming lognormal distribution for the live load.

7.8 Solve Problem 7.7 by assuming normal distribution for the live load.

7.9 The total load on the floor of a printing press is given by $L = L_d + L_l$ where L_d is the dead load with mean 5000 lb and standard deviation 500 lb and L_l is the live load with mean 100 lb/ft^2 and standard deviation 20 lb/ft^2. If the floor area is 400 ft^2, determine the probability of the total load exceeding a value of 50,000 lb. Assume normal distribution for both the dead and live loads.

7.10 A record of the annual maximum wind velocities at a particular location gave the mean value as 79.5 mi/h and the standard deviation as 41.2 mi/h. Determine the parameters of the distribution using a type-II extreme value distribution for the maximum wind velocity.

7.11 Using the solution of Problem 7.10, find the probability of the maximum wind velocity exceeding a value of 90 mi/h in any single year. Compare your result with the one obtained by assuming normal distribution for the maximum wind velocity.

7.12 Determine the mean return period for annual maximum wind velocities greater than 90 mi/h using the data of Problem 7.10.

7.13 Determine the following quantities corresponding to an earthquake with $M = 8$ and $d = 100$ km:

1. The mean and standard deviation of S_v for $\tau_n = 1.0$ s and $\zeta = 0.05$ using Eq. (7.59).
2. The mean and standard deviation of the corresponding S_a assuming τ_n to be deterministic.
3. The probability of S_a exceeding a value of 0.1 g by assuming normal distribution for S_a.

Note: Assume other data as used in Example 7.7.

7.14 Plot the power spectrum given by Eq. (7.64) for the following data: $\sigma_a = 10$ in/s^2, $H = 0.6, f_g = 2$ Hz.

7.15 A structure has a natural frequency of vibration of 2 Hz and a damping ratio of $\zeta = 0.02$. Find the spectral displacement and spectral acceleration of the structure corresponding to a spectral velocity of 20 in/s.

7.16 The response spectrum of an earthquake is shown in Fig. 7.13. Find the spectral displacement, spectral velocity, and spectral acceleration of a machine structure whose natural frequency of vibration is 10 Hz and the damping ratio is $\zeta = 0.05$.

MATLAB and Excel Problems

7.17 Find the probability of compressing the spring more than 0.15 in in the battery compartment described in Problem 7.2 using MATLAB.

7.18 Find the annual wind velocity corresponding to a 25-year return period for the wind velocity data in Example 7.4 using MATLAB. Also, find the corresponding annual maximum wind velocity at the gradient height.

7.19 Find the solution of Problem 7.2 using Excel.

7.20 Solve Problem 7.18 using Excel.

CHAPTER 8

Strength-Based Reliability

What You Will Learn

In this chapter, you will learn

- The basic relations for determining the reliability of structural and mechanical components
- The expressions for the reliability (R_0) and probability of failure (P_f) of a structural or mechanical component in terms of the probability distributions of the strength (S) and the load (L)
- The computation of reliability when both the strength and the load follow normal, lognormal, exponential, and extreme value distributions
- A graphical procedure for determining the reliability of components whose strength and load distributions are known experimentally
- The relation between the factor of safety and the reliability
- The limit state equation of systems involving more than two random parameters
- The classification of reliability analysis methods as first-order reliability methods (FORM) and second-order reliability methods (SORM)
- The computation of the reliability index in FORM methods
- The Hasofer-Lind reliability index with multiple normally distributed variables, along with its interpretation as an optimization problem
- The classification of multimember structural and mechanical systems as weakest-link and fail-safe systems
- The computation of reliability and probability of failure as well as the bounds on the probability of failure of weakest-link and fail-safe systems

Siméon Denis Poisson was a French mathematician who was born on June 21, 1781, and died on April 25, 1840. During his childhood, he once suspended himself from a ceiling and started swinging as a pendulum to avoid being bothered by the small animals and bugs that roamed around the floor. The study of the pendulum occupied much of his time in later years. His family wanted him to become a physician. However, during his medical studies, when his first patient died while under his ministrations, he developed a dislike of the profession; he changed to the study of mathematics, joined the Ecole Polytechnic in 1798, and succeeded Fourier as professor there in 1808. He made many contributions to the study of mathematical physics, elasticity of materials, Fourier series, heat, acoustics, and probability. The Poisson's ratio in elasticity and Poisson's density and distribution functions in probability are associated with his name [8.15, 8.16].

8.1 Introduction

This chapter deals with strength-based reliability analysis of structural and mechanical components. There are some differences between electronic and electrical systems and structural and mechanical systems. Most electronic and electrical components and systems deteriorate during use as a result of elevated operating temperature or other similar reasons. In addition, most electronic and electrical systems are produced in large numbers and hence can be assumed to be nominally identical. The existence of a large sample of nominally identical systems enables us to interpret failure probabilities in terms of relative frequencies. Thus, the reliability of an electronic or electrical system is closely associated with the life of the system. The time at which failure occurs will be the primary random variable in these cases. The methods presented in Chapter 6 are applicable to these systems. On the other hand, most structural and mechanical components fail due to extremely small values of strength (S) and/or large values of load (L). Also, many structural and mechanical systems are not mass produced; hence their failure data are very limited. A structural or mechanical component or system is considered to be safe and reliable when its strength or resistance exceeds the value of the load acting on it. This approach is also known as *interference theory.* As indicated in Chapter 1, the overlapping area, primarily in the tail parts, of the probability density or distribution functions of strength and load determines the probability of failure or reliability of the component. Thus, in the face of imprecisely known probability distributions of strength and load (at least in the tail parts), the computation of reliability necessarily involves suitable assumptions and approximations. The expressions for the reliability/probability of failure of a structural or mechanical component are derived in terms of the probability distributions of the strength (S) and the load (L). In certain cases, S and L follow normal, lognormal, exponential, or Weibull distributions, the integral equations can be reduced to a simple form. However, in a more general case, the reliability of the component can be found only by evaluating the integrals numerically. The methods of evaluating reliability of structural and mechanical components and systems are discussed in this chapter.

The exact probability distributions of S and L are not available in most practical cases. Even if their exact distributions are available, the evaluation, in closed form, of the integrals involved in the computation of reliability may not be possible. In most cases, numerical integrations are to be used. To simplify the procedure, sometimes, equivalent normal distributions are used in finding the reliability of a system. In many cases, the strength and load of a system may depend on several random variables of the system such as design variables and preassigned parameters (data) of the system. Also, the calculation of the reliability of the system requires knowledge of the probability distributions of all the random variables involved. In several practical cases, the available information or data may permit the determination of only the mean values, variances, and covariances between the random variables. Under these conditions, the reliability of the system needs to be found (or estimated) using only the first and second moments of the random variables. Several methods have been developed to find the reliability as a function of only the first and second moments of the random variables of the system. Different second-moment methods of evaluating the reliability or probability of failure of a system are presented in this chapter.

Most structural and mechanical systems are composed of multiple members and are subject to several load conditions. These systems may also have multiple failure modes. Two types of models—the weakest-link and fail-safe models—can be used for the reliability analysis of these systems. The weakest-link model, also known as the series model, can be used whenever the failure of any one member causes the failure of the complete system. The fail-safe model, also known as the parallel or redundant model, can be used when the system fails only when all of its members fail. The computation of the reliability/probability of failure of weakest-link and fail-safe systems, as well as the bounds on the probability of failure, is also presented in this chapter.

8.2 General Expression for Reliability

The reliability of a component R is given by

$$R = P(S > L) = P(S - L > 0)$$
$$= \iint f_{S,L}(s, l)\, ds\, dl \tag{8.1}$$

where $f_{S,L}(s, l)$ is the joint density function of S and L, and the double integral has to be evaluated over the region $[(s, l)|_{g(s,l) = s - l > 0}]$ as shown in Fig. 8.1. If S and L are

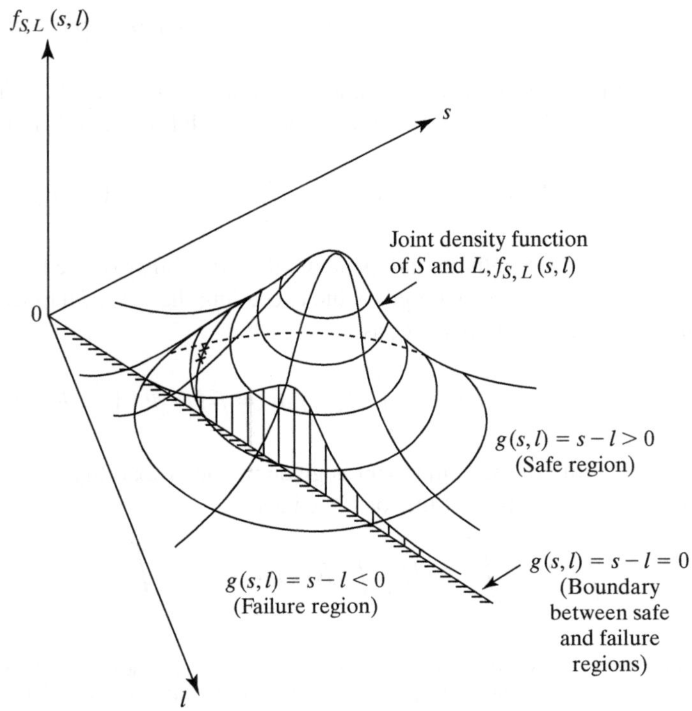

FIGURE 8.1 Regions of integration for determining reliability $(g > 0)$ and probability of failure $(g < 0)$.

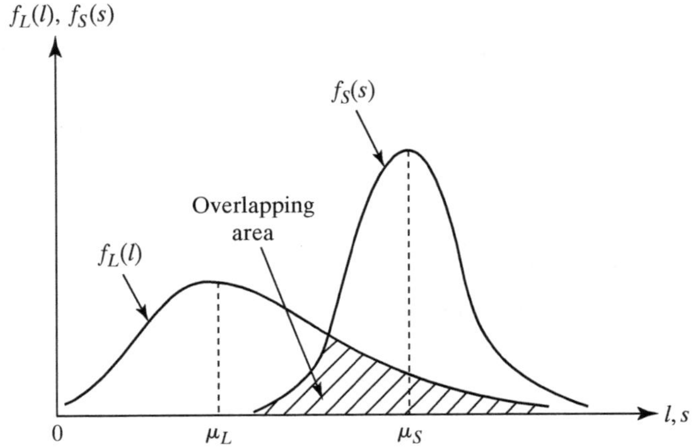

FIGURE 8.2 Interference area between the density functions of S and L.

independent,[1] then the interference area shown in Fig. 8.2, between the probability density functions of S and L, gives a measure of the probability of failure. With reference to Fig. 8.3, the probability of the load (L) falling in a small interval dl around l is equal to the area A_l

$$P\left(l - \frac{dl}{2} \leq L \leq l + \frac{dl}{2}\right) = \text{area } A_l = f_L(l)\ dl \tag{8.2}$$

The probability of strength (S) assuming a value larger than the load l is equal to the shaded area under the density function of the strength (i.e., area A_s in Fig. 8.3):

$$P(S > l) = \text{area } A_s = \int_l^\infty f_S(s)\ ds = 1 - F_S(l) \tag{8.3}$$

If the load and the strength are independent, the probability of the load value falling in the interval dl around l and the strength value exceeding the value l happening simultaneously gives the elemental reliability, dR, as

$$dR = f_L(l)\ dl \int_l^\infty f_S(s)\ ds = f_L(l)\ dl\,[1 - F_S(l)] \tag{8.4}$$

Since the reliability of the component is given by the probability of strength exceeding the load for all possible values of the load, we have

$$R = \int dR = \int_{-\infty}^{\infty} f_L(l) \left[\int_l^\infty f_S(s)\ ds \right] dl = \int_{-\infty}^{\infty} f_L(l)[1 - F_S(l)]\ dl \tag{8.5}$$

[1]The terms *strength* (S) and *load* (L) are used in a general sense and are measured in the same units. For example, in most structural and mechanical designs, S represents the allowable or permissible stress and L denotes the induced or applied stress. The general interpretation of strength and load, with examples to several other engineering situations, is given in Section 8.4. In addition, for computational simplicity in deriving the reliability expressions, the strength and load are considered to be the only two random variables in the system. But, in general, the strength as well as the load will be a function of several other random variables and hence the reliability computation involves multiple random variables. Such problems are considered in Section 8.7.

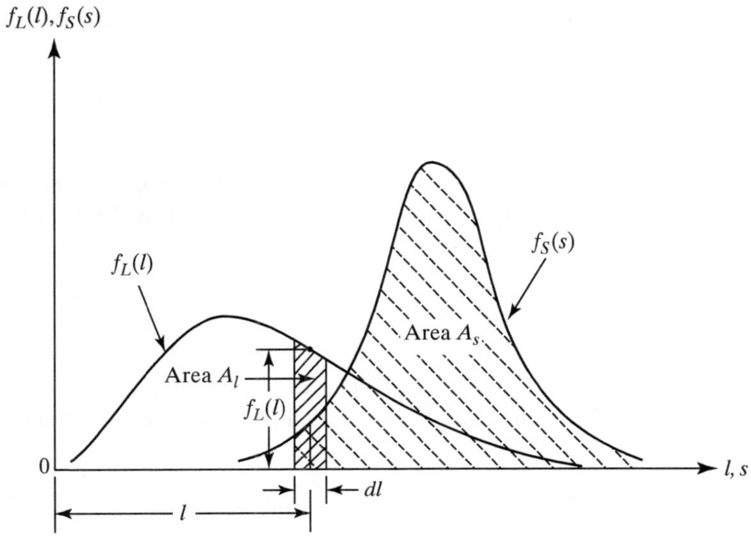

FIGURE 8.3 Probability of realizing strength S larger than the load l.

Alternate Expression for Reliability. The expression for reliability can also be derived by finding the probability of the load assuming a smaller value than the value of the strength. As before, the probability of strength falling in the interval ds around s is given by (see Fig. 8.4)

$$P\left(s - \frac{ds}{2} \leq S \leq s + \frac{ds}{2}\right) = \text{area } A_s' = f_S(s)\, ds \tag{8.6}$$

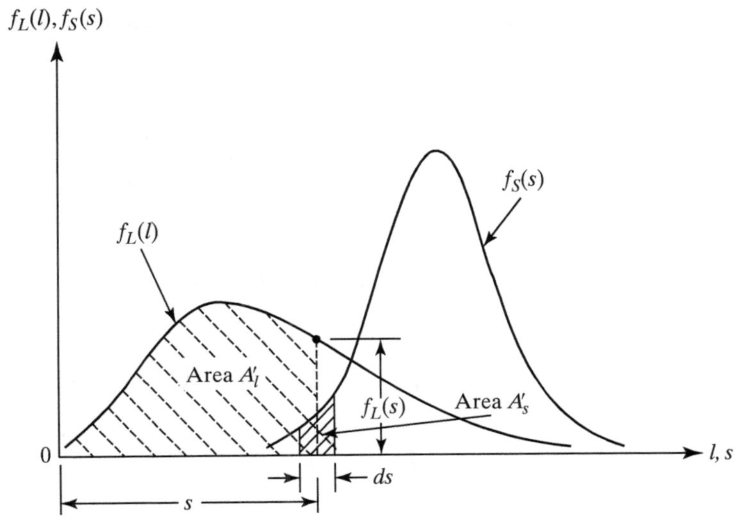

FIGURE 8.4 Probability of realizing load L smaller than the strength s.

The probability of the load assuming a smaller value than s can be obtained as

$$P(L < s) = \text{area } A'_l = \int_{-\infty}^{s} f_L(l)\, dl = F_L(s) \tag{8.7}$$

The probability of strength value falling in the interval ds and the load assuming a value smaller than s occurring simultaneously gives the elemental reliability, dR, as

$$dR = f_S(s)\, ds \int_{-\infty}^{s} f_L(l)\, dl = f_S(s)\, ds\, F_L(s) \tag{8.8}$$

Hence, the reliability of the component can be determined as the probability of load assuming a smaller value than strength for all possible values of strength

$$R = \int dR = \int_{-\infty}^{\infty} f_S(s) \left[\int_{-\infty}^{s} f_L(l)\, dl \right] ds = \int_{-\infty}^{\infty} f_S(s)\, F_L(s)\, ds \tag{8.9}$$

8.3 Expression for Probability of Failure

Just as we can derive an expression for reliability, we can also derive an expression for the probability of failure or unreliability of the component. The probability of failure (P_f) is the probability of strength assuming to a value smaller than that of the load:

$$P_f = P(S \leq L) = \iint f_{S,L}(s, l)\, ds\, dl \tag{8.10}$$

where $f_{S,L}(s, l)$ is the joint density function of S and L, and the double integral has to be evaluated over the region $[(s, l)|_{g(s,l)=s-l\leq 0}]$ as shown in Fig. 8.1. If S and L are independent, Eq. (8.10) can be rewritten as

$$\begin{aligned} P_f &= 1 - P(L \leq S) = 1 - R \\ &= 1 - \int_{-\infty}^{\infty} f_S(s) \left[\int_{-\infty}^{s} f_L(l)\, dl \right] ds \\ &= 1 - \int_{-\infty}^{\infty} f_S(s)\, F_L(s)\, ds \\ &= \int_{-\infty}^{\infty} f_S(s)\, ds - \int_{-\infty}^{\infty} f_S(s)\, F_L(s)\, ds \\ &= \int_{-\infty}^{\infty} [1 - F_L(s)]\, f_S(s)\, ds \end{aligned} \tag{8.11}$$

where $F_L(l)$ is the probability distribution function of L.

Alternate Expression for Probability of Failure. The probability of failure can also be expressed as

$$\begin{aligned}
P_f &= 1 - R = 1 - P(S \geq L) \\
&= 1 - \int_{-\infty}^{\infty} f_L(l)\left[\int_{l}^{\infty} f_S(s)\, ds\right] dl \\
&= 1 - \int_{-\infty}^{\infty} f_L(l)[1 - F_S(l)]\, dl \\
&= 1 - \int_{-\infty}^{\infty} f_L(l)\, dl + \int_{-\infty}^{\infty} f_L(l)\, F_S(l)\, dl
\end{aligned} \tag{8.12}$$

where $F_S(s)$ is the probability distribution function of *S*. Since the integral of the density function from $-\infty$ to ∞ is equal to 1, Eq. (8.12) simplifies to

$$P_f = \int_{-\infty}^{\infty} f_L(l)\, F_S(l)\, dl \tag{8.13}$$

8.4 General Interpretation of Strength and Load

In the reliability and failure probability expressions derived in Sections 8.2 and 8.3, the units of strength and load should be consistent. For example, in the design of a long column subjected to an axial compressive load for safety against buckling, the units of the buckling load and the applied load will be in pounds or Newtons. In the design of a gear, the strength may be the yield stress of the material and the load may be the bending stress induced in the gear tooth during the transmission of power. In this case, the units of strength and load will be in pounds per square inch (psi) or Newtons per square meter (Pascal, Pa). The concept can be used for defining the reliability of nonstructural systems as well. For example, in the hydrological or water resources design, a system is considered reliable if the available supply of water (strength) exceeds the demand or usage of water. In this case, the units of strength and load will be cubic feet or cubic meters. In general, the concepts of strength and load can be interpreted in terms of capacity or supply and requirement or demand, respectively. Some engineering examples and the corresponding strength and load are indicated in Table 8.1.

8.5 Reliability for Known Probability Distributions of *S* and *L*

As indicated in Sections 8.2 and 8.3, the computation of the reliability or probability of failure requires a knowledge of the probability distributions of the strength and load, $f_S(s)$ and $f_L(l)$, if they are statistically independent, or of their joint distribution, $f_{S,L}(s, l)$, if they are correlated. Unfortunately, the exact probability distributions of *S* and *L* are not available in most practical cases. Even if their exact probability distributions are available, the evaluation, in closed form, of the integrals involved in the computation of reliability

TABLE 8.1 Strength and load in some engineering examples

Description of the System	Strength	Load
Machine or structure subjected to static load	Yield strength	Maximum stress induced
Vibrating component in an assembly	Clearance available	Maximum displacement experienced
Maintained system	Target life	Remaining life
Design of a system for fracture	Stress intensity factor	Fracture toughness
Design of a system for fracture	Critical crack size	Growing crack size
Aircraft	Flutter velocity	Flight speed
Machine operating in corrosive atmosphere	Material thickness	Corrosion depth
Vibrating machine or system	Natural frequency	Forcing frequency
Air-conditioning system of a building	Cooling capacity (or power)	Cooling load (or power) required
Machine tool	Horsepower of the machine tool	Power required for the machining operation

(Eqs. (8.1), (8.5), and (8.9)) may not be possible. In most cases, numerical integrations are to be used. However, for certain distributions of strength and load, simple expressions can be derived for the reliability. In this section, we shall consider few typical distributions of S and L to find the reliability of the system.

8.5.1 Reliability When S and L Follow Normal Distribution

If the strength (S) and load (L) are known to follow normal distribution with density functions

$$f_S(s) = \frac{1}{\sigma_S \sqrt{2\pi}} e^{-1/2\{(s-\mu_S)/\sigma_S\}^2}; \quad -\infty < s < \infty \tag{8.14}$$

and

$$f_L(l) = \frac{1}{\sigma_L \sqrt{2\pi}} e^{-1/2\{(l-\mu_L)/\sigma_L\}^2}; \quad -\infty < l < \infty \tag{8.15}$$

where μ_S and μ_L denote the mean values, and σ_S and σ_L represent the standard deviations of the variables S and L, respectively, the reliability of the component can be expressed as

$$R = P(S - L \geq 0) = P(X \geq 0) \tag{8.16}$$

where $X = S - L$ is called the margin of safety or the limit state function of the system. Because S and L are random, the margin of safety X will also be a random variable. Since

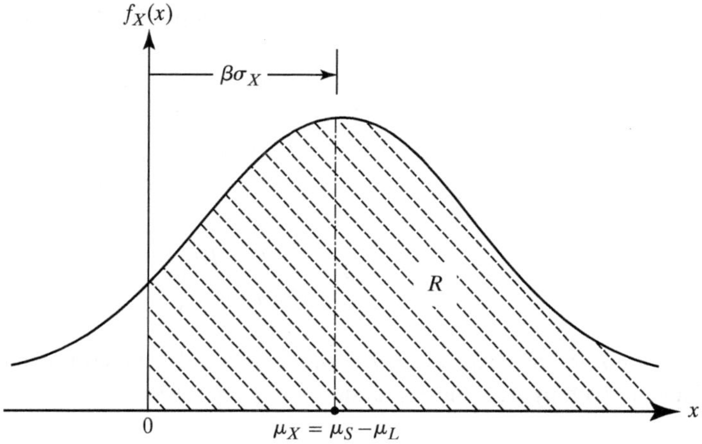

FIGURE 8.5 Probability density function of $X = S - L$.

X is a linear function of the normally distributed random variables S and L, it also follows normal distribution. The probability density function of X (see Fig. 8.5) is given by

$$f_X(x) = \frac{1}{\sigma_X \sqrt{2\pi}} e^{-1/2\{(x-\mu_X)/\sigma_X\}^2}; \quad -\infty < x < \infty \tag{8.17}$$

If S and L are assumed to be independent, the mean and standard deviations of X can be expressed as

$$\mu_X = \mu_S - \mu_L \tag{8.18}$$

$$\sigma_X = (\sigma_S^2 + \sigma_L^2)^{1/2} \tag{8.19}$$

Equation (8.16) can be rewritten as

$$R = P(X \geq 0) = \int_0^\infty f_X(x)\, dx$$

$$= \frac{1}{\sigma_X\sqrt{2\pi}} \int_0^\infty e^{-1/2\{(x-\mu_X)/\sigma_X\}^2}\, dx \tag{8.20}$$

By defining the standard normal variate z as

$$z = \frac{x - \mu_X}{\sigma_X} \tag{8.21}$$

we can express

$$dz = \frac{dx}{\sigma_X} \quad \text{or} \quad dx = \sigma_X\, dz \tag{8.22}$$

The values of z corresponding to the lower and upper limits of the integration of X are given by

$$z_1 = \frac{0 - \mu_X}{\sigma_X} = -\frac{\mu_S - \mu_L}{\sqrt{\sigma_S^2 + \sigma_L^2}} = -\beta \tag{8.23}$$

where

$$\beta = \frac{\mu_S - \mu_L}{\sqrt{\sigma_S^2 + \sigma_L^2}} \tag{8.24}$$

is called the *reliability index* or *safety index.* If S and L are correlated with the correlation coefficient given by $\rho_{S,L}$, the expression for z_1 becomes

$$z_1 = -\beta = -\left(\frac{\mu_S - \mu_L}{\sqrt{\sigma_S^2 + \sigma_L^2 - 2\,\rho_{S,L}\,\sigma_S\,\sigma_L}} \right) \tag{8.25}$$

It can be seen from Eq. (8.23) that the reliability index denotes the distance, $\beta\,\sigma_X$, of the mean value of the margin of safety, $\mu_{g(X)} = \mu_X$, from $g(X) = X = 0$ where $g(X) = X = S - L$. The lower limit of integration in Eq. (8.20), $x = 0$ corresponds to z_1 given by Eq. (8.23), and the upper limit of integration $x = \infty$ corresponds to $z_2 = \infty$.
Thus, Eq. (8.20) can be expressed as

$$R = \frac{1}{\sqrt{2\pi}} \int_{z=z_1}^{\infty} e^{-1/2\,z^2}\, dz = \Phi(\beta) \tag{8.26}$$

Once the value of z_1 is calculated from the known mean and standard deviations of S and L, the corresponding reliability of the component (R) can be determined from Eq. (8.26). The probability of failure of the system can be expressed as (see Problem 8.36):

$$P_f = 1 - R = \Phi(-\beta) \tag{8.27}$$

where $\Phi(-\beta)$ denotes the value of the cumulative distribution function of standard normal variate at $z = -\beta$. Standard normal tables (Appendix A) can be used to find the reliability or probability of failure using Eq. (8.26) or Eq. (8.27), respectively.

8.5.2 Approximate Expressions of Reliability for Normal Distribution

Although the area under the density function of a standard normal variate, given by Eq. (8.26), cannot be evaluated exactly, the area $P_f(z)$ shown in Fig. 8.6 can be approximated as

$$\begin{aligned} P_f(z) = 1 - R(z) = f(z)\,\{&0.31938153\,t - 0.356563782\,t^2 \\ &+ 1.781477937\,t^3 - 1.821255978\,t^4 \\ &+ 1.330274429\,t^5\} + \text{error} \end{aligned} \tag{8.28}$$

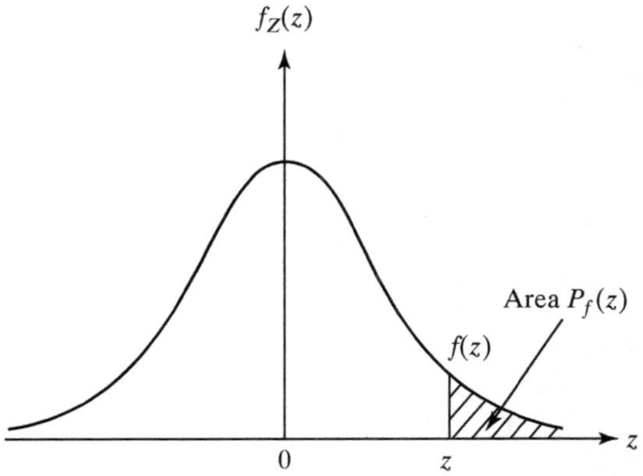

FIGURE 8.6 Area under the probability density function beyond the value of z.

where

$$f(z) = \frac{1}{\sqrt{2\pi}} e^{-1/2 z^2} \tag{8.29}$$

$$t = \frac{1}{(1 + 0.2316419z)} \tag{8.30}$$

and the error is less than 7.5×10^{-8}. Similarly, if $P_f(z) = P_o$ is given, we can find z approximately as

$$z = t - \frac{2.515517 + 0.802853\,t + 0.010328\,t^2}{1 + 1.432788\,t + 0.189269\,t^2 + 0.001308\,t^3} + \text{error} \tag{8.31}$$

where

$$t = \sqrt{\log_e\left(\frac{1}{P_o^2}\right)} \tag{8.32}$$

and the error is less than 4.5×10^{-5}. Many other approximations to the normal distribution function have been reported in the literature [8.41, 8.42].

EXAMPLE 8.1

The torque-transmitting capacity (T) of a cone clutch under uniform pressure [8.8, 8.11] is given by

$$T = \frac{Ff}{3\sin\alpha}\left[\frac{D^3 - d^3}{D^2 - d^2}\right]$$

where D = outer diameter of the cone, d = inner diameter of the cone, α = semi-cone angle, f = coefficient of friction between cup and cone, and F = axial force applied (see Fig. 8.7). For a

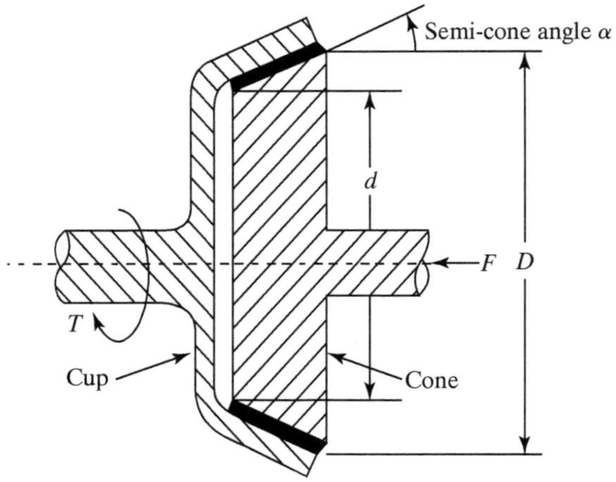

FIGURE 8.7 A cone clutch.

specific clutch, which has molded asbestos on a cast-iron cone, $D = 10$ in, $d = 8$ in, $\alpha = 20°$, and $F = 100$ lb. The coefficient of friction can be assumed to be normally distributed with a mean value of 0.3 and a standard deviation of 0.02. If the torque required to be transmitted by the cone clutch is also normally distributed with mean and standard deviations of 400 lb · in and 40 lb · in, respectively, estimate the reliability of the cone clutch.

Solution: Since the coefficient of friction (f) is the only random variable, the mean and standard deviations of the torque-transmitting capacity (T) of the clutch can be found as

$$\overline{T} = \frac{F\overline{f}}{3\sin\alpha}\left[\frac{D^3 - d^3}{D^2 - d^2}\right] = \frac{100(0.3)}{3\sin 20°}\left[\frac{10^3 - 8^3}{10^2 - 8^2}\right] = 396.33 \text{ lb} \cdot \text{in}$$

$$\sigma_T = \frac{F\,\sigma_f}{3\sin\alpha}\left[\frac{D^3 - d^3}{D^2 - d^2}\right] = \frac{100(0.02)}{3\sin 20°}\left[\frac{10^3 - 8^3}{10^2 - 8^2}\right] = 26.4225 \text{ lb} \cdot \text{in}$$

The required torque to be transmitted (T_r) is given by

$$N(T_r, \sigma_{T_r}) = N(400, 40) \text{ lb} \cdot \text{in}$$

By defining a random variable X as $X = T - T_r$, we find

$$\overline{X} = \overline{T} - \overline{T}_r = 396.33 - 400 = -3.67 \text{ lb} \cdot \text{in}$$

and

$$\sigma_X = \sqrt{\sigma_T^2 + \sigma_{T_r}^2} = \sqrt{26.4225^2 + 40^2} = 47.9390 \text{ lb} \cdot \text{in}$$

Thus, the reliability of the cone clutch can be determined as

$$R = P[X \geq 0] = P\left[z \geq -\frac{\overline{X}}{\sigma_X}\right] = P\left[z \geq \frac{3.67}{47.939}\right] = P[z \geq 0.07655] = 1 - P[z \leq 0.07655]$$

$$= 1 - \Phi(0.07655)$$

From standard normal tables, we find that $R = 1 - 0.53051 = 0.46949$.

■

EXAMPLE 8.2

A mechanical fuse has a specially designed screw head that is required to shear off when the applied torque exceeds a value of 400 in · lb. The fuse is permitted to have a maximum of 100 failures per 50,000 units. Find the torque at which the fuse is to be designed to fail if the standard deviation of the shear strength of the material of the fuse, expressed in terms of the applied torque, is 15 in · lb. The standard deviation of the torque wrenches used to tighten the screw head of the fuse is 10 in · lb of torque. Assume that the applied torque and the torsional strength of the fuse follow normal distribution.

Solution: The permissible probability of failure of the fuse is

$$P_f = \frac{100}{50{,}000} = 0.002$$

From standard normal tables, the value of the probability of failure can be seen to correspond to $\Phi(-2.88) = 0.002$. Let the torque, T, at which the fuse is to be designed to fail (strength) be denoted as $S = T$ and the applied torque or load applied, T_a, be denoted as $L = T_a$. From the given data, $\mu_L = 400$ in-lb, $\sigma_L = 10$ in-lb, and $\sigma_S = 15$ in-lb. If $X = S - L$, $\mu_X = \mu_S - \mu_L = \mu_S - 400$ in-lb and $\sigma_X = \sqrt{\sigma_S^2 + \sigma_L^2} = \sqrt{15^2 + 10^2} = 18.0278$ in-lb. The probability of failure of the fuse can be expressed as

$$P_f = P[X \le 0] = P\left[z \le -\frac{\overline{X}}{\sigma_X}\right] \equiv P[z \le -2.88]$$

which yields $\frac{\overline{X}}{\sigma_X} = 2.88$ or $\overline{X} = 2.88\,\sigma_X$ or $\mu_S - 400 = 2.88(18.0278) = 51.9199$. This gives the mean value of the required torque, $\mu_S = \mu_T$, as 451.9199 in-lb. Thus, the mechanical fuse must be designed to fail at a mean torque of $\mu_S = 451.9199$ in-lb.

■

EXAMPLE 8.3

Find the values of the reliability index corresponding to different values of reliability of a system (in the range $0.5 \le R \le 0.9999999$) for which the strength (S) and load (L) follow normal distribution.

Solution: For practical systems, the reliability index (β) will be positive and hence the reliability (R) of the structure is given by

$$R = \Phi(\beta), \quad \beta = \frac{\mu_X}{\sigma_X} \tag{E1}$$

where $X = S - L$ is the margin of safety and Φ is the cumulative distribution function of the standard normal variate, $N(0, 1)$. For different values of reliability, the corresponding reliability index can be found as

$$\beta = \Phi^{-1}(R) \tag{E2}$$

From standard normal distribution table, the values given in Table 8.2 can be obtained.

TABLE 8.2 Values of reliability index for different reliabilities

R	$P_f = 1 - R$	$\beta = -\Phi^{-1}(R)$
0.5	0.5	0
0.6	0.4	0.253
0.7	0.3	0.524
0.8	0.2	0.842
0.9	0.1	1.282
0.95	0.05	1.645
0.99	0.01	2.326
0.999	0.001	3.090
0.9999	0.0001	3.719
0.99999	0.00001	4.27
0.999999	0.000001	4.75
0.9999999	0.0000001	5.20

■

8.5.3 Reliability When S and L Follow Lognormal Distribution

When the strength and load of a component are known to follow lognormal distribution, we define a new random variable X as the ratio of strength to load[2]

$$X = \frac{S}{L} \tag{8.33}$$

This gives

$$\ln X = \ln S - \ln L \tag{8.34}$$

and $\ln X$ can be seen to be normally distributed since both $\ln S$ and $\ln L$ are normally distributed. The reliability of the component can be expressed as

$$R = P\left(\frac{S}{L} > 1\right) = P(X > 1)$$

$$= \int_1^{\infty} f_X(x)\, dx \tag{8.35}$$

Since $\ln X$ is normally distributed, X follows lognormal distribution. We define a standard normal variate z as

$$z = \frac{\ln x - \mu_{\ln X}}{\sigma_{\ln X}} \tag{8.36}$$

[2]The quantity X can be considered as an alternate form of the safety margin.

where $\mu_{\ln X}$ and $\sigma_{\ln X}$ are the mean and standard deviation of ln *X*, respectively. If Eq. (8.35) is rewritten in terms of z, the new limits of integration can be found as follows:

When

$$x = 1, z = z_1 = \frac{\ln 1 - \mu_{\ln X}}{\sigma_{\ln X}} = \frac{0 - (\mu_{\ln S} - \mu_{\ln L})}{\sqrt{\sigma^2_{\ln S} + \sigma^2_{\ln L}}}$$

When

$$x = \infty, z = z_2 = \infty$$

Thus, Eq. (8.35) can be rewritten in terms of z as

$$R = \int_{z_1}^{\infty} \frac{1}{\sqrt{2\pi}} e^{-1/2z^2}\,dz \tag{8.37}$$

where

$$z_1 = -\frac{(\mu_{\ln S} - \mu_{\ln L})}{\sqrt{\sigma^2_{\ln S} + \sigma^2_{\ln L}}} \tag{8.38}$$

Mean and Standard Deviation of ln *X* in Terms of Mean and Standard Deviation of *X*. If *X* is lognormally distributed, its density function is given by

$$f_X(x) = \frac{1}{\sqrt{2\pi}\,\sigma x} \exp\left\{-\frac{1}{2}\left(\frac{\ln x - \mu}{\sigma}\right)^2\right\}; \quad x > 0 \tag{8.39}$$

where $\mu = E(\ln X) = \mu_{\ln X}$, $\sigma = \sigma_{\ln X}$, and the variable $\ln x$ is normally distributed. Defining $Y = \ln X$, we can find the expected value and the standard deviation of X as follows:

$$E(X) = \int_{-\infty}^{\infty} \frac{1}{\sqrt{2\pi}\,\sigma} \exp\left\{-\frac{1}{2}\left(\frac{\ln x - \mu}{\sigma}\right)^2\right\} dx \tag{8.40}$$

Since $x = e^y$ and $dy = \dfrac{1}{x}dx$ or $dx = x\,dy$, Eq. (8.40) can be rewritten as

$$E(X) = E(e^Y) = \int_{-\infty}^{\infty} \frac{1}{\sqrt{2\pi}\,\sigma} e^y \exp\left\{-\frac{1}{2}\left(\frac{\ln x - \mu}{\sigma}\right)^2\right\} dy \tag{8.41}$$

This equation can be simplified (see Problem 8.31) as:

$$E(X) = \exp\left(\mu + \frac{\sigma^2}{2}\right) \tag{8.42}$$

The computation of σ_X requires the evaluation of $E(X^2)$

$$E(X^2) = E(e^{2y}) = \int_{-\infty}^{\infty} \frac{e^{2y}}{\sqrt{2\pi}\,\sigma} \exp\left\{-\frac{1}{2}\left(\frac{\ln x - \mu}{\sigma}\right)^2\right\} dx \tag{8.43}$$

This equation can be simplified (see Problem 8.31) as:

$$E(X^2) = \exp\{2(\mu + \sigma^2)\} \tag{8.44}$$

Thus, the variance of X is given by

$$\sigma_X^2 = E[\{X - E(X)\}^2] = E(X^2) - [E(X)]^2$$

$$= \exp\{2(\mu + \sigma^2)\} - \exp\left\{2\left(\mu + \frac{\sigma^2}{2}\right)\right\}$$

$$= \exp(2\,\mu + \sigma^2)(\exp(\sigma^2) - 1) \tag{8.45}$$

Equations (8.42) and (8.45) can be solved to find (see Eqs. 3.63 and 3.64):

$$\mu = \ln E(X) - \frac{\sigma^2}{2} = \ln \mu_X - \frac{\sigma^2}{2} \tag{8.46}$$

and

$$\sigma^2 = \ln\left[\frac{\sigma_X^2}{\{E(X)\}^2} + 1\right] = \ln\left[\frac{\sigma_X^2}{\mu_X^2} + 1\right] \tag{8.47}$$

EXAMPLE 8.4

The bending stress developed in a gear tooth is given by the Lewis equation [8.7]

$$s = \frac{F_t\,p_d}{w\,Y}$$

where F_t = tangential load acting on the gear tooth, p_d = diametral pitch $= \pi/p$, p = circular pitch, w = face width, and Y = Lewis form factor (see Fig. 8.8). In an automobile transmission, a gear has full-depth teeth with a pressure angle of 20°, a diametral pitch of unity, a face width of 0.5 in, and a Lewis form factor of 0.3. The tangential load on the gear tooth is lognormally distributed with a mean value of 3720 lb and a coefficient of variation of 0.080645. The yield strength of the material is known to follow lognormal distribution with a mean value of 32,000 (lb/in^2). Find the maximum permissible value of the coefficient of variation of yield strength required in order to have a reliability of 0.9999 for the gear tooth.

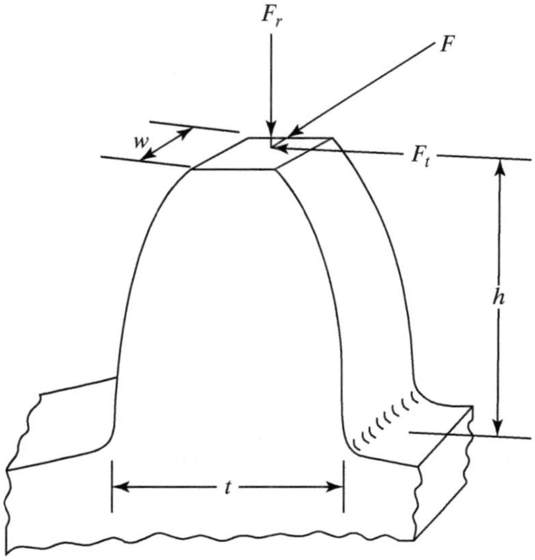

FIGURE 8.8 A gear tooth.

Solution: Since the bending stress induced in the gear tooth $(s \equiv L)$ is given by a constant $[p_d/\{w\,Y\} = 1/\{0.5(0.3)\} = 1/0.15]$ times the tangential load (F_t), L can also be assumed to follow lognormal distribution with a mean value

$$\overline{L} = \overline{s} = \frac{\overline{F}_t}{0.15} = \frac{3720}{0.15} = 24{,}800 \text{ lb/in}^2$$

and a standard deviation

$$\sigma_L = \sigma_s = \frac{1}{0.15}\sigma_{F_t} = \frac{1}{0.15}\gamma_{F_t}\overline{F}_t = \frac{1}{0.15}(0.080645)(3720) = 2000 \text{ lb/in}^2$$

Thus, $E(S) = 32{,}000 \text{ lb/in}^2$, $E(L) = 24{,}800 \text{ lb/in}^2$, $\sigma_L = 2000 \text{ lb/in}^2$, and hence

$$\sigma^2_{\ln L} = \ln\left[\frac{2000^2}{24{,}800^2} + 1\right] = \ln 1.0065 = 0.006479$$

$$\mu_{\ln L} = \ln E(L) - \left\{\frac{\sigma^2_{\ln L}}{2}\right\}$$

$$= 10.118599 - \frac{1}{2}(0.006479) = 10.115360$$

$$\mu_{\ln S} = \ln E(S) - \left\{\frac{\sigma^2_{\ln S}}{2}\right\}$$

$$= 10.3735 - \frac{1}{2}\sigma^2_{\ln S} \tag{E1}$$

From standard normal tables, we find the value of z_1 corresponding to a reliability of 0.9999 as

$$z_1 = -\frac{\mu_{\ln S} - \mu_{\ln L}}{\sqrt{\sigma^2_{\ln S} + \sigma^2_{\ln L}}} = -3.72 \tag{E2}$$

Squaring and rearranging of Eq. (E2) gives

$$\mu^2_{\ln S} - 2\,\mu_{\ln S}\,\mu_{\ln L} + \mu^2_{\ln L} = 13.8384\,\sigma^2_{\ln S} + 13.8384\,\sigma^2_{\ln L} \tag{E3}$$

Equations (E3) and (E1) yield

$$\left(10.3735 - \frac{1}{2}\sigma^2_{\ln S}\right)^2 - 2\left(10.3735 - \frac{1}{2}\sigma^2_{\ln S}\right)(10.11536) + (10.11536)^2$$

$$= 13.8384\,\sigma^2_{\ln S} + 13.8384(0.006479)$$

that is,

$$0.25\,\sigma^4_{\ln S} - 14.0965\,\sigma^2_{\ln S} - 0.02306 = 0 \tag{E4}$$

The roots of Eq. (E4) are

$$\sigma^2_{\ln S} = 0.001636; \quad 56.3876$$

Considering the smaller value, we have

$$\sigma^2_{\ln S} = 0.001636$$

By definition,

$$\sigma^2_{\ln S} = \ln\left[\frac{\sigma^2_S}{\{E(S)\}^2} + 1\right]$$

that is,

$$0.001636 = \ln\left[\frac{\sigma^2_S}{32000^2} + 1\right]$$

that is,

$$\sigma^2_S = (e^{0.001636} - 1)(32000)^2 = (0.001637)(32000)^2 = (1294.8495)^2$$

Thus, the permissible value of the coefficient of variation of yield strength is

$$\gamma_S = \frac{\sigma_S}{E(S)} = \frac{1294.8495}{32000} = 0.040464$$

■

8.5.4 Reliability When S and L Follow Exponential Distribution

When S and L follow exponential distribution, their probability density functions can be expressed as

$$f_S(s) = \frac{1}{\mu_S} e^{-s/\mu_S}; \quad 0 \le s < \infty \tag{8.48}$$

$$f_L(l) = \frac{1}{\mu_L} e^{-l/\mu_L}, \quad 0 \le l < \infty \tag{8.49}$$

where μ_S and μ_L denote the mean values of strength and load, respectively. The reliability of the component can be computed by substituting Eqs. (8.48) and (8.49) into Eq. (8.9) as

$$\begin{aligned}
R &= \int_o^\infty f_S(s) \cdot \left[\int_o^s f_L(l) \cdot dl \right] \cdot ds \\
&= \int_o^\infty \frac{1}{\mu_S} e^{-s/\mu_S} \left[\int_o^s \frac{1}{\mu_L} e^{-l/\mu_L}\, dl \right] ds \\
&= \frac{1}{\mu_S\,\mu_L} \int_o^\infty e^{-s/\mu_S} \left[-\mu_L\, e^{-l/\mu_L} \right] \Big|_o^s \, ds \\
&= -\frac{1}{\mu_S} \int_o^\infty e^{-s/\mu_S} \left[e^{-s/\mu_L} - 1 \right] ds \\
&= -\frac{1}{\mu_S} \left[\int_o^\infty e^{-(1/\mu_S + 1/\mu_L)s}\, ds - \int_o^\infty e^{-s/\mu_S}\, ds \right] \\
&= \frac{1}{\mu_S \left(\dfrac{1}{\mu_S} + \dfrac{1}{\mu_L} \right)} \left[e^{-(1/\mu_S + 1/\mu_L)s} \right] \Big|_o^\infty - \frac{1}{\mu_S} \frac{1}{\left(\dfrac{1}{\mu_S} \right)} \left[e^{-s/\mu_S} \right] \Big|_o^\infty \\
&= \frac{1}{\mu_S \left(\dfrac{1}{\mu_S} + \dfrac{1}{\mu_L} \right)} [0 - 1] - [0 - 1] \\
&= \frac{\mu_S}{\mu_L + \mu_S}
\end{aligned} \tag{8.50}$$

Equation (8.50) can also be written as

$$R = \frac{\lambda_L}{\lambda_L + \lambda_S} \tag{8.51}$$

where

$$\lambda_S = \frac{1}{\mu_S} \tag{8.52}$$

and

$$\lambda_L = \frac{1}{\mu_L} \tag{8.53}$$

EXAMPLE 8.5

The hoop (circumferential) stress developed in a thin-walled cylindrical pressure vessel is given by

$$s_\theta = \frac{pd}{2t}$$

where d = inside diameter, p = internal pressure, and t = wall thickness (see Fig. 8.9). In a particular application, the vessel is subjected to an exponentially distributed pressure with a mean value of 400 lb/in^2. Assuming exponential distribution for the yield strength of the vessel (s_y) with a mean value of 30,000 lb/in^2, determine the reliability of the pressure vessel. Assume the values of d and t to be 72 in and 0.5 in, respectively.

Solution: The mean value of the hoop stress induced in the pressure vessel is given by

$$\bar{s}_\theta = \frac{\bar{p}d}{2t} = \frac{(400)(72)}{(2)(0.5)} = 28{,}800 \text{ lb/in}^2$$

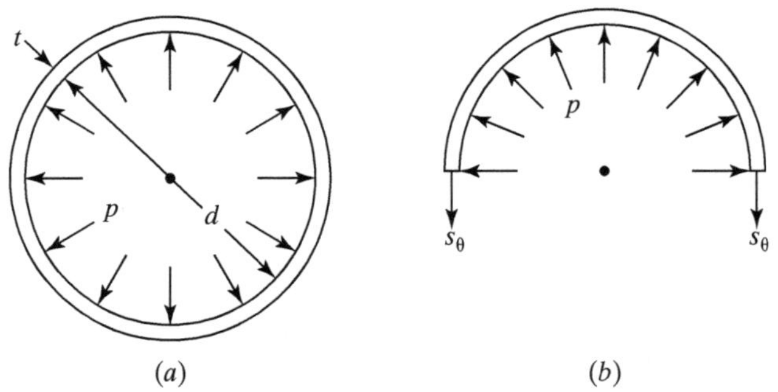

FIGURE 8.9 A thin-walled pressure vessel.

Thus, we have $\overline{S} = \overline{s}_y = 30{,}000\ \text{lb/in}^2$ and $\overline{L} = \overline{s}_\theta = 28{,}800\ \text{lb/in}^2$ and hence the reliability of the pressure vessel is given by Eq. (8.50):

$$R = \frac{\overline{S}}{\overline{L} + \overline{S}} = \frac{30{,}000}{28{,}800 + 30{,}000} = 0.5102$$

■

8.5.5 Reliability When S and L Follow Extreme Value Distributions

As discussed in Chapters 4 and 7, the strength of most materials can be described, based on the weakest-link concept, using extreme value distributions for the smallest value. Similarly, if the load is caused by earthquakes, wind velocities, gusts, landing loads, or ocean waves, it can be described by the extreme value distributions for the largest value. The computation of reliability of a component is considered in this section with extreme value distributions for strength and load. The density functions of extreme value distributions for the smallest value of strength S (see Chapter 4) can be expressed as follows:

Type I

$$f_S(s) = a \exp[a(s - w) - \exp\{a(s - w)\}]; \quad -\infty < s < \infty \tag{8.54}$$

where a and w are the parameters of the distribution.

Type II

$$f_S(s) = \frac{m}{w}\left(\frac{w}{s}\right)^{m+1} \exp\left[-\left(\frac{w}{s}\right)^m\right] \tag{8.55}$$

where $s \leq 0$ and $m > 0$.

Type III

$$f_S(s) = \frac{m}{v - \varepsilon}\left(\frac{s - \varepsilon}{v - \varepsilon}\right)^{m-1} \exp\left[-\left(\frac{s - \varepsilon}{v - \varepsilon}\right)^m\right] \tag{8.56}$$

where $s \geq \varepsilon$, $v > \varepsilon$, and $m > 0$. The density functions of extremal distributions corresponding to the largest value of load L (see Chapter 4) are given by:

Type I

$$f_L(l) = a \exp\{-a(l - w) - \exp[-a(l - w)]\} \tag{8.57}$$

where $-\infty < l < \infty$ and a and w are the parameters of the distribution.

Type II

$$f_L(l) = \frac{m}{w}\left(\frac{w}{l}\right)^{m+1} \exp\left[-\left(\frac{w}{l}\right)^m\right] \tag{8.58}$$

where $l \geq 0$, $w > 0$, and $m > 0$.

Type III

$$f_L(l) = \frac{m}{w - v}\left(\frac{w - l}{w - v}\right)^{m-1} \exp\left[-\left(\frac{w - l}{w - v}\right)^{m}\right] \tag{8.59}$$

where $l \leq w$, $v < w$, and $m > 0$.

The reliability of a component can be determined using Eq. (8.5) for any specified $f_L(l)$ and $f_S(s)$:

$$R = \int_{-\infty}^{\infty} f_L(l)\left[\int_{l}^{\infty} f_S(s) \cdot ds\right] dl \tag{8.60}$$

Equation (8.9) can also be used for this purpose. The procedure is illustrated by considering a typical case in the next section.

8.5.6 When S and L Follow Type-III Extremal Distributions

When S and L follow type-III extremal distributions that correspond to the smallest and the largest values, respectively, we have

$$f_L(l) = \frac{m_1}{w_1 - v_1}\left(\frac{w_1 - l}{w_1 - v_1}\right)^{m_1-1} \exp\left[-\left(\frac{w_1 - l}{w_1 - v_1}\right)^{m_1}\right] \tag{8.61}$$

$$f_S(s) = \frac{m_2}{v_2 - \varepsilon_2}\left(\frac{s - \varepsilon_2}{v_2 - \varepsilon_2}\right)^{m_2-1} \exp\left[-\left(\frac{s - \varepsilon_2}{v_2 - \varepsilon_2}\right)^{m_2}\right] \tag{8.62}$$

with $l \leq w_1$, $v_1 < w_1$, $m_1 > 0$, $s \geq \varepsilon_2$, $v_2 > \varepsilon_2$, and $m_2 > 0$. Thus, the reliability of the system can be evaluated as

$$R = \int_{-\infty}^{\infty} \frac{m_1}{w_1 - v_1}\left(\frac{w_1 - l}{w_1 - v_1}\right)^{m_1-1} \exp\left[-\left(\frac{w_1 - l}{w_1 - v_1}\right)^{m_1}\right] \times \left\{\int_{l}^{\infty} \frac{m_2}{v_2 - \varepsilon_2}\left(\frac{s - \varepsilon_2}{v_2 - \varepsilon_2}\right)^{m_2-1} \exp\left[-\left(\frac{s - \varepsilon_2}{v_2 - \varepsilon_2}\right)^{m_2}\right] ds\right\} dl \tag{8.63}$$

that is,

$$R = \frac{m_1 m_2}{(w_1 - v_1)^{m_1}(v_2 - \varepsilon_2)^{m_2}} \int_{-\infty}^{\infty} (w_1 - l)^{m_1-1} \exp\left[-\left(\frac{w_1 - l}{w_1 - v_1}\right)^{m_1}\right] \times \left\{\int_{l}^{\infty} (s - \varepsilon_2)^{m_2-1} \exp\left[-\left(\frac{s - \varepsilon_2}{v_2 - \varepsilon_2}\right)^{m_2}\right] ds\right\} dl \tag{8.64}$$

The integral in Eq. (8.64) has to be evaluated numerically using a suitable procedure [8.5].

8.5.7 Reliability in Terms of Experimentally Determined Distributions of S and L

When experiments are conducted to find the values of the random variables S and L, several values of S and L can be observed. From these observed values, empirical distribution functions can be determined for describing the random variables S and L. In this case, a graphical procedure can be used for finding the reliability of the component as described by the following steps [8.2, 8.17].

Step 1 Let n experiments be conducted to find the values of the random variable S and let $s_1, s_2, \ldots, s_n$ be the observed values of S. From these values, we can plot the approximate distribution function of S, that is, $F_S(s)$.

Step 2 Similarly, let m experiments be conducted to find the values of the random variable L and let $l_1, l_2, \ldots, l_m$ be the observed values of L. From these values, plot the approximate distribution function of L, that is, $F_L(l)$.

Step 3 Define two variables X and Y as

$$X = \int_0^l f_L(l')\, dl' = F_L(l) \tag{8.65}$$

$$Y = \int_l^\infty f_S(s)\, ds = 1 - F_S(l) \tag{8.66}$$

and find the values of the pairs $(X_1, Y_1), (X_2, Y_2), \ldots$, corresponding to different discrete values of the load $l_1, l_2, \ldots$, respectively, from the results of Steps 1 and 2. Plot a graph of X versus Y from the computed pairs of values $(X_i, Y_i); i = 1, 2, \ldots, m$.

Step 4 According to Eq. (8.5), the reliability of the component is given by

$$R = \int_{-\infty}^\infty f_L(l) \left[\int_l^\infty f_S(s)\, ds \right] dl \tag{8.67}$$

Noting that

$$\int_l^\infty f_S(s)\, ds = Y,$$

$$f_L(l)\, dl = dX,$$

and the range of X is from 0 to 1, Eq. (8.67) can be expressed in terms of X and Y as

$$R = \int_0^1 Y\, dX \tag{8.68}$$

This equation shows that reliability of the component is the same as the area under X versus Y curve. Since the curve X versus Y was plotted in Step 3, the area under the curve (and hence the reliability R) can be found graphically. The procedure is illustrated with the help of the following example.

EXAMPLE 8.6

Several machine tool belt drives were tested experimentally under identical operating (cutting) conditions. The stresses induced in ten belts were found to be 22,300, 11,600, 15,850, 19,900, 13,650, 16,950, 26,750, 12,700, 18,400, and 15,100 lb/in^2. The values of strengths found by testing twelve of the belts used in the machine tool drives were 21,100, 26,950, 19,200, 30,150, 22,050, 24,350, 18,250, 25,700, 23,400, 19,950, 21,600, and 20,500 lb/in^2. Find the reliability of the belt drives.

Solution: First, the estimated cumulative distribution functions for the strength and load (induced stress) are to be determined. For finding the distribution functions of strength, the strength data are arranged in an increasing order and a cumulative probability of (i/n) is assigned to the ith value with n denoting the total number of values observed. A similar procedure is adopted for finding the distribution function of the induced stress. The results are shown in Table 8.3. These results are used to plot the approximate probability distributions shown in Figs. 8.10(a) and (b).

TABLE 8.3 Distribution functions of strength and stress

Strength Data			Stress (Load) Data		
Number	Strength (lb/in^2)	Cumulative Probability	Number	Stress (Load) (lb/in^2)	Cumulative Probability
1	18,250	0.083	1	11,600	0.1
2	19,200	0.167	2	12,700	0.2
3	19,950	0.250	3	13,650	0.3
4	20,500	0.333	4	15,100	0.4
5	21,100	0.417	5	15,850	0.5
6	21,600	0.500	6	16,950	0.6
7	22,050	0.583	7	18,400	0.7
8	23,400	0.667	8	19,900	0.8
9	24,350	0.750	9	22,300	0.9
10	25,700	0.833	10	26,750	1.0
11	26,950	0.917			
12	30,150	1.000			

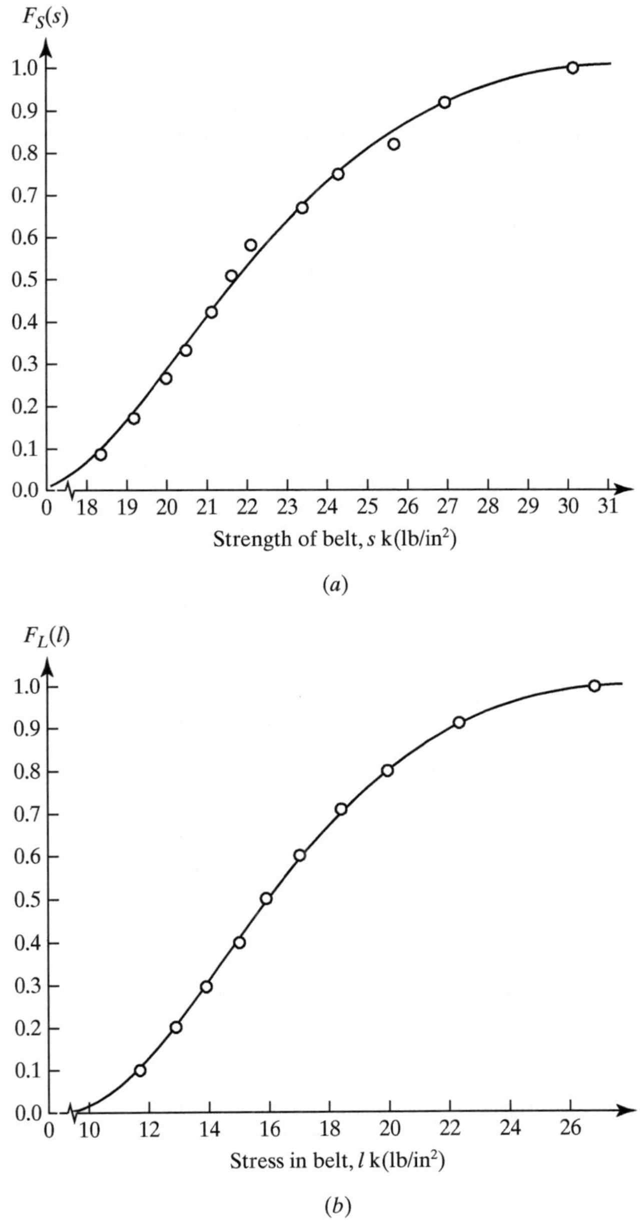

FIGURE 8.10 (a) Approximate distribution function of strength s; (b) approximate distribution function of induced stress l.

TABLE 8.4 Values of X and Y for different settings of the induced stress

Stress, l (lb/in^2)	$X = F_L(l)$ (from Fig. 8.10(*b*))	$Y = 1 - F_S(l)$ (from Fig. 8.10(*a*))
0	0.0	$1 - 0 = 1$
2,000	0.0	$1 - 0 = 1$
4,000	0.0	$1 - 0 = 1$
6,000	0.0	$1 - 0 = 1$
8,000	0.0	$1 - 0 = 1$
10,000	0.02	$1 - 0 = 1$
12,000	0.13	$1 - 0 = 1$
14,000	0.31	$1 - 0 = 1$
16,000	0.50	$1 - 0 = 1$
17,000	0.60	$1 - 0.02 = 0.98$
18,000	0.67	$1 - 0.06 = 0.94$
19,000	0.74	$1 - 0.22 = 0.78$
20,000	0.80	$1 - 0.28 = 0.72$
21,000	0.85	$1 - 0.42 = 0.58$
22,000	0.89	$1 - 0.54 = 0.46$
23,000	0.93	$1 - 0.65 = 0.35$
24,000	0.95	$1 - 0.73 = 0.27$
25,000	0.97	$1 - 0.81 = 0.19$
26,000	0.98	$1 - 0.86 = 0.14$
27,000	1.00	$1 - 0.91 = 0.09$
28,000	1.00	$1 - 0.95 = 0.05$
29,000	1.00	$1 - 0.97 = 0.03$
30,000	1.00	$1 - 0.99 = 0.01$
31,000	1.00	$1 - 1.00 = 0.00$
32,000	1.00	$1 - 1.00 = 0.00$

Once the approximate $F_S(s)$ and $F_L(l)$ are determined, the values of X and Y for different values of the induced stress (l) can be found as indicated in Table 8.4. These results can be used to plot a graph between X and Y as shown in Fig. 8.11. The area under the $X - Y$ curve gives the reliability of the belt drive. This area and, hence, the reliability of the belt drive is found to be approximately 0.8525.

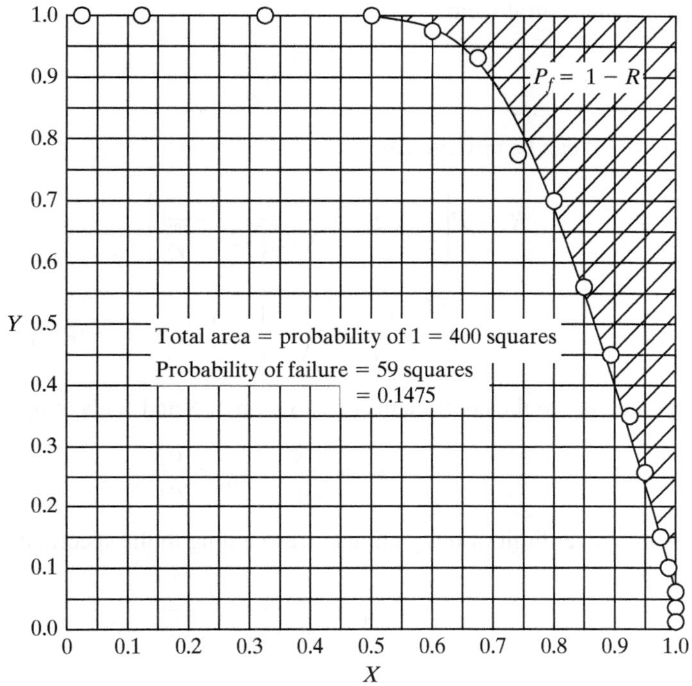

FIGURE 8.11 Reliability as the area under the X-Y curve.

■

8.6 Factor of Safety Corresponding to a Given Reliability

The factor of safety (n) corresponding to any specified reliability can be derived once the probability distribution functions of strength and load are known [8.3]. To illustrate the procedure, we assume that S and L follow normal distribution. Let $X = S - L$ so that the mean and standard deviation of X are given by

$$\mu_X = \mu_S - \mu_L \tag{8.69}$$

$$\sigma_X = \sqrt{\sigma_S^2 + \sigma_L^2} \tag{8.70}$$

where S and L have been assumed to be independent. By defining a standard normal variate z as

$$z = \frac{x - \mu_X}{\sigma_X} \tag{8.71}$$

the reliability can be expressed as

$$R = P[X \geq 0] = P\left[z \geq z_1 = -\frac{\mu_X}{\sigma_X}\right] = P\left[z \geq z_1 = \frac{\mu_L - \mu_S}{\sqrt{\sigma_S^2 + \sigma_L^2}}\right] \tag{8.72}$$

By defining the central factor of safety[3] as

$$n = \frac{\mu_S}{\mu_L} \tag{8.73}$$

Equation (8.72) can be rewritten as

$$R = P\left[z \geq z_1 = \frac{1 - n}{\sqrt{\gamma_S^2 n^2 + \gamma_L^2}} \right]$$

$$= \int_{z_1}^{\infty} \frac{1}{\sqrt{2\pi}} \exp\left(-\frac{1}{2} z^2 \right) dz = 1 - \Phi(z_1) \tag{8.74}$$

where γ_S and γ_L are the coefficients of variation of S and L given by

$$\gamma_S = \frac{\sigma_S}{\mu_S}; \quad \gamma_L = \frac{\sigma_L}{\mu_L} \tag{8.75}$$

and z_1 is the lower limit of integration corresponding to the specified reliability R:

$$z_1 = \frac{1 - n}{\sqrt{\gamma_S^2 n^2 + \gamma_L^2}} \tag{8.76}$$

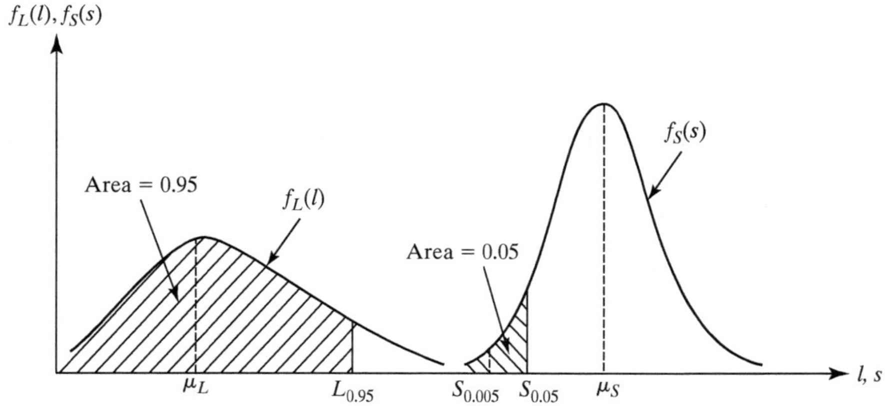

FIGURE 8.12 Significance of $L_{0.95}$, $L_{0.005}$ and $L_{0.05}$.

[3]Other factors of safety that are often used are the characteristic factor of safety, n_k, and the design factor of safety, n^*, defined as [8.4]

$$n_k = \frac{S_{0.05}}{L_{0.95}}$$

$$n^* = \frac{S_{0.005}}{L_{0.95}} \quad \text{(see Fig. 8.12)}$$

For any specified reliability, the value of the lower limit of integration z_1 in Eq. (8.76) can be found, say $z_1 = t$, from the standard normal tables. By setting the expression for z_1 equal to t, we obtain

$$t = \frac{1 - n}{\sqrt{n^2 \gamma_S^2 + \gamma_L^2}}$$

or

$$0 = n^2(1 - t^2\gamma_S^2) - 2n + (1 - t^2\gamma_L^2) \tag{8.77}$$

The solution of Eq. (8.77) gives

$$n = \frac{1 \pm t\sqrt{\gamma_S^2 + \gamma_L^2 - t^2\gamma_S^2\gamma_L^2}}{(1 - t^2\gamma_S^2)} \tag{8.78}$$

Thus, Eq. (8.78) gives the required factor of safety for any specified reliability. We can see an anomaly with Eq. (8.78). By letting $\gamma_L = 0$, we obtain from Eq. (8.78),

$$n = \frac{1}{1 \pm t\gamma_S} \tag{8.79}$$

If we choose the positive sign on the right-hand side of Eq. (8.79) and plot n as a function of γ_S (for any fixed value of reliability), we obtain a curve as shown in Fig. 8.13. It can be seen that the value of n required for achieving the specified reliability R tends to infinity as the product $t\gamma_S$ tends to minus one. The reason for this apparent anomaly can be seen from Fig. 8.14 where the load is assumed to be deterministic $(\gamma_L = 0)$, and the strength

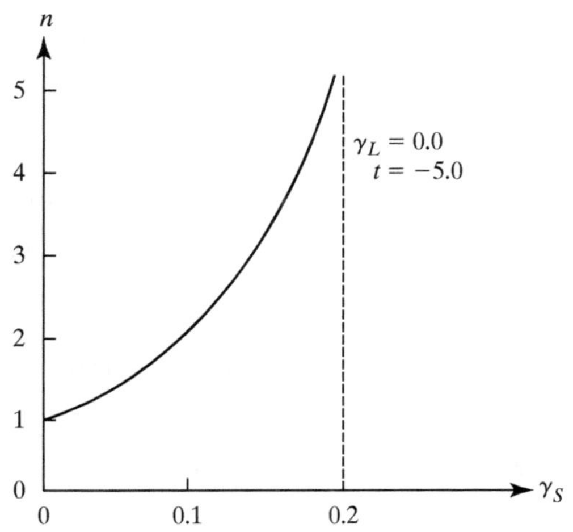

FIGURE 8.13 Factor of safety n versus the coefficient of variation of strength.

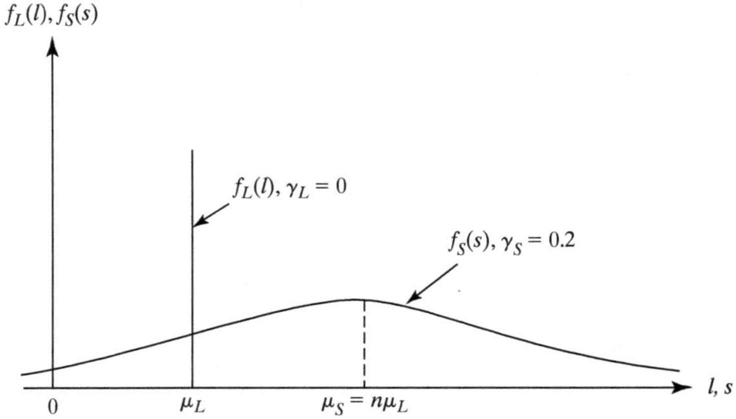

FIGURE 8.14 Probability density functions of L and S.

is assumed to be probabilistic with $\gamma_S = 0.2$. As $t\gamma_S \to 1$ or $|z_1| = |t| \to 5.0$ (i.e., as the required reliability of the component approaches a value of 0.99999971), the distance between μ_L and μ_S needs to approach infinity.

EXAMPLE 8.7

Find the reliability and the corresponding central factor of safety of a system for which $\mu_S = 15000 \text{ lb/in}^2$, $\mu_L = 10000 \text{ lb/in}^2$, $\sigma_S = 3000 \text{ lb/in}^2$, and $\sigma_L = 1000 \text{ lb/in}^2$ if

1. S and L follow normal distribution, and
2. S and L follow lognormal distribution.

Solution:

1. When S and L follow normal distribution,

$$R = P[X \geq 0] = P\left[z \geq z_1 = -\frac{\mu_X}{\sigma_X}\right] = P\left[z \geq z_1 = \frac{\mu_L - \mu_S}{\sqrt{\sigma_S^2 + \sigma_L^2}}\right]$$

Here

$$z_1 = \frac{\mu_L - \mu_S}{(\sigma_L^2 + \sigma_S^2)^{1/2}} = \frac{10{,}000 - 15{,}000}{(1000^2 + 3000^2)^{1/2}} = -\frac{5000}{3162} = -1.5811$$

From standard normal tables, we find that $R = 1 - \Phi(z_1) = 0.94305$. The central factor of safety is given by

$$n = \frac{\mu_S}{\mu_L} = \frac{15{,}000}{10{,}000} = 1.5$$

2. When S and L follow lognormal distribution,

$$\sigma^2_{\ln S} = \ln\left(\frac{\sigma_S^2}{\mu_S^2} + 1\right) = \ln\left(\frac{3000^2}{15{,}000^2} + 1\right)$$

$$= \ln\left(\frac{9}{225} + 1\right) = \ln 1.04 = 0.03922$$

$$\therefore \sigma_{\ln S} = 0.19804$$

$$\mu_{\ln S} = \ln \mu_S - \frac{\sigma^2_{\ln S}}{2} = \ln 15{,}000 - \frac{1}{2}(0.03922)$$

$$= 9.61581 - 0.01961 = 9.59620$$

$$\sigma^2_{\ln L} = \ln\left(\frac{\sigma_L^2}{\mu_L^2} + 1\right) = \ln\left(\frac{1000^2}{10000^2} + 1\right) = \ln\left(\frac{1}{100} + 1\right) = \ln 1.01 = 0.00995$$

$$\therefore \sigma_{\ln L} = 0.09975$$

$$\mu_{\ln L} = \ln \mu_L - \frac{1}{2}\sigma^2_{\ln L} = \ln 10000 - \frac{1}{2}(0.00995)$$

$$= 9.21034 - 0.00497 = 9.20537$$

The reliability of the system can be determined as

$$R = P(z_1 \geq 0)$$

where

$$z_1 = \frac{\mu_{\ln L} - \mu_{\ln S}}{\sqrt{\sigma^2_{\ln L} + \sigma^2_{\ln S}}} = \frac{9.20537 - 9.59620}{\sqrt{0.09975^2 + 0.19804^2}} = -1.762536$$

From standard normal tables, the reliability corresponding to this value of z_1 can be found as

$$R = 1 - \Phi(z_1) = 1 - \Phi(-1.762536) = 1 - 0.0391 = 0.9609$$

■

8.7 Reliability of Systems Involving More Than Two Random Parameters

As shown in Fig. 8.5, β denotes the distance (in terms of the standard deviation σ_g) of the mean value of the limit state function, μ_g, from the origin $g = S - L = 0$. The point (μ_g) indicates the boundary of the failure region. Since β is a direct measure of the reliability of the system, a larger value of β represents a higher reliability.

The strength (S) and load (L) in most practical systems depend on several other parameters such as design variables and preassigned parameters (data) of the system depending on the performance criterion used in defining the reliability/probability of failure of the system. For example, in the case of the cone clutch considered in Example 8.1, the strength S (or the torque-transmitting capacity) depends on the coefficient of friction (f),

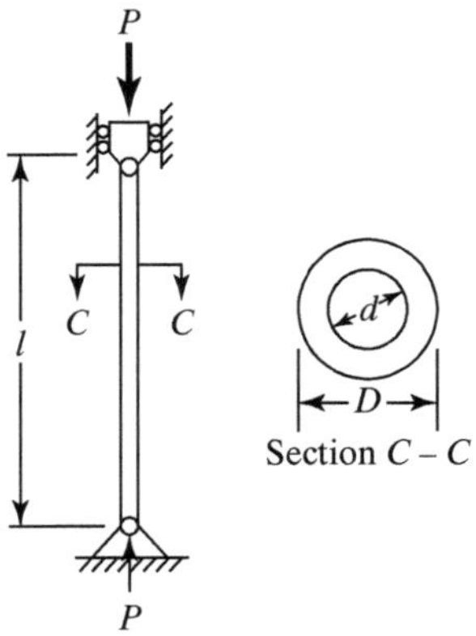

FIGURE 8.15 Pin-connected column.

axial force applied (F), semi-cone angle (α), outer diameter of the cone (D), and the inner diameter of the cone (d). Similarly, for a column with a hollow circular cross section subjected to an axial force, as shown in Fig. 8.15, if the reliability is considered in terms of avoiding buckling (performance criterion), the limit state function (or the margin of safety) can be defined as $Y = g = S - L$ with

$$S = \frac{\pi^2 E I}{l^2} = \frac{\pi^2 E}{l^2} \frac{\pi(D^4 - d^4)}{64}, \quad L = \frac{P}{A} = \frac{P}{\frac{\pi}{4}(D^2 - d^2)} \tag{8.80}$$

where the parameters of the system (column) are given by E = Young's modulus, l = length; D = outer diameter of the column, d = inner diameter of the column, and P = axial force applied (with I denoting the area moment of inertia and A the area of cross section of the column).

Thus, in general, the limit state function, also known as the performance function, of the system can be expressed as

$$Y = g(\vec{X}) = g(X_1, X_2, \ldots, X_n) \tag{8.81}$$

where $X_1, X_2, \ldots,$ and X_n denote the basic parameters or variables of the system which are considered to be random or probabilistic in nature, and $\vec{X} = \{X_1, X_2, \ldots, X_n\}^T$ is the vector of the basic random variables. The function g may be linear or nonlinear in terms of the basic random variables $X_1, X_2, \ldots,$ and X_n. In many practical problems, the function g may not be available as an explicit function of $X_1, X_2, \ldots,$ and X_n. In such cases, g is an implicit function of $X_1, X_2, \ldots,$ and X_n; the value of g can only be evaluated at a known set of values of $X_1, X_2, \ldots,$ and X_n. When the limit state function is zero,

$$Y = g(\vec{X}) = g(X_1, X_2, \ldots, X_n) = 0 \tag{8.82}$$

Equation (8.82) is known as the limit state or failure surface equation that denotes the boundary between the failure and safe regions. If the joint probability density function of the random variables $X_1, X_2, \ldots, X_n$ is known to be

$$f_{X_1, X_2, \ldots, X_n}(x_1, x_2, \ldots, x_n)$$

the probability of failure of the system is given by

$$P_f = \underset{g(\vec{X}) < 0}{\int_{X_1} \int_{X_2} \cdots \int_{X_n}} f_{X_1, X_2, \ldots, X_n}(x_1, x_2, \ldots, x_n)\, dx_1\, dx_2 \ldots dx_n \tag{8.83}$$

and the reliability by

$$R = 1 - P_f = \underset{g(\vec{X}) > 0}{\int_{X_1} \int_{X_2} \cdots \int_{X_n}} f_{X_1, X_2, \ldots, X_n}(x_1, x_2, \ldots, x_n)\, dx_1\, dx_2 \ldots dx_n \tag{8.84}$$

where the multiple integrations are performed over the failure or safe region (in the $\vec{X}$-space) such that $g(\vec{X}) < 0$ in Eq. (8.83) and $g(\vec{X}) > 0$ in Eq. (8.84). In most practical problems, the joint probability density function of the random variables, $f_{X_1, X_2, \ldots, X_n}(x_1, x_2, \ldots, x_n)$, is not known or is too complex to find. If the random variables $X_1, X_2, \ldots,$ and X_n are statistically independent, then the joint probability density function is given by the product of the probability density functions of the individual random variables so that

$$f_{X_1, X_2, \ldots, X_n}(x_1, x_2, \ldots, x_n) = \prod_{i=1}^{n} f_{X_i}(x_i) = f_{X_1}(x_1)\, f_{X_2}(x_2) \ldots f_{X_n}(x_n) \tag{8.85}$$

In several practical cases, the available information may not be adequate to determine the probability distributions of the basic random variables involved. However, the data may be sufficient to estimate the first and second moments of the random variables. This means that only the mean values, variances, and covariances between the random variables are available. Under these conditions, the reliability of the system needs to be found (or estimated) using only the first and second moments of the random variables. This approach is known as the second moment approach. Several methods have been developed to find the reliability as a function of only the first and second moments of the random variables of the system. If the limit state function involves more than two basic random variables and is linear in terms of the random variables, a procedure similar to the one presented in the previous section yields a reasonably accurate value of the reliability or probability of failure of the system.

However, if the limit state function, $g(\vec{X})$, is nonlinear, it can be approximated by using Taylor series expansion about a reference point (vector of random variables). When only the first-order expansion is used, the limit state function will be linear (straight line when only two random variables are involved) as shown in Fig. 8.16.

The procedure used to compute reliability based on a linear approximation of the limit state function is known as the first-order second-moment (FOSM) method, which is also

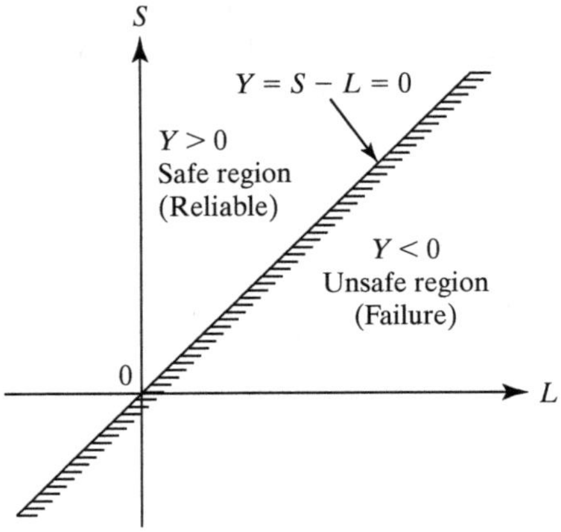

FIGURE 8.16 Significance of the limit state function Y.

called the mean value first-order second-moment (MVFOSM) method. The FOSM method is considered in detail in the next section. When the second-order expansion is used, the limit state function will be a quadratic function. The procedure used to compute the reliability based on a quadratic approximation of the limit state function is known as the second-order second-moment (SOSM) method. If the mean values of the random variables are taken as the reference point, the Taylor's series expansion of the limit state or performance criterion function is given by

$$Y = g(X_1, X_2, \ldots, X_n) \approx g(\overline{X}_1, \overline{X}_2, \ldots, \overline{X}_n) + \sum_{i=1}^{n} \left.\frac{\partial g}{\partial X_i}\right|_m (X_i - \overline{X}_i) + \sum_{i=1}^{n}\sum_{j=1}^{n} \left.\frac{\partial^2 g}{\partial X_i\,\partial X_j}\right|_m (X_i - \overline{X}_i)(X_j - \overline{X}_j) + \ldots \tag{8.86}$$

where the first and second derivatives of g with respect to the random variables $X_i, i = 1, 2, \ldots, n$ are evaluated at the mean values of the random variables, $\overline{X}_i, i = 1, 2, \ldots, n$. The truncation of Eq. (8.86) at the linear terms gives

$$Y = g(X_1, X_2, \ldots, X_n) \approx g(\overline{X}_1, \overline{X}_2, \ldots, \overline{X}_n) + \sum_{i=1}^{n} \left.\frac{\partial g}{\partial X_i}\right|_m (X_i - \overline{X}_i) \tag{8.87}$$

from which the mean value and standard deviation of the limit state function, μ_g and σ_g, can be obtained as

$$\mu_Y = \mu_g = g(\mu_{X_1}, \mu_{X_2}, \ldots, \mu_{X_n}) \tag{8.88}$$

$$\sigma_Y^2 = \sigma_g^2 = \sum_{i=1}^{n}\sum_{j=1}^{n} \left.\frac{\partial g}{\partial X_i}\right|_m \left.\frac{\partial g}{\partial X_j}\right|_m Cov\,(X_i, X_j) \tag{8.89}$$

where $\text{Cov}(X_i, X_j)$ is the covariance of X_i and X_j and the subscript m indicates that the quantity is evaluated at the mean values of the random variables. If the random variables $X_1, X_2, \ldots,$ and X_n are independent (uncorrelated), Eq. (8.89) reduces to

$$\sigma_Y^2 = \sigma_g^2 \approx \sum_{i=1}^{n} \left(\frac{\partial g}{\partial X_i}\bigg|_m \right)^2 Var(X_i) = \sum_{i=1}^{n} \left(\frac{\partial g}{\partial X_i}\bigg|_m \right)^2 \sigma_{X_i}^2 \tag{8.90}$$

The reference point $\vec{X}^*$ or $\vec{X}_m$ used in the Taylor's series expansion of $g(\vec{X})$ is usually chosen as the vector of mean values of the random variables, denoted as $\mu_{\vec{X}}$. However, in general, the reliability (or probability of failure) of the system computed, based on the linearization of the limit state function using Taylor's series expansion about the mean values of the random variables, may not be accurate. In fact, the reliability predicted will be accurate only when the limit state function is linearized about a point, known as the optimal design point or most probable point (MPP) in the random variable space.

The following examples show that the use of different reference points for the linearization of the limit state function or different forms of the limit state function for the same performance criterion predicts different values of the margin of safety and reliability of the system.

EXAMPLE 8.8

The limit state function of a system is given by

$$Y = g(\vec{X}) = X_1 - X_2 X_3 \tag{E1}$$

where the random variables X_1, X_2, and X_3 are statistically independent and normally distributed with mean values given by $\mu_{X_1} = 800$, $\mu_{X_2} = 30$, and $\mu_{X_3} = 20$ and standard deviations $\sigma_{X_1} = 40$, $\sigma_{X_2} = 10$, and $\sigma_{X_3} = 5$. Find the reliability index and the corresponding reliability of the system if the linearization of the limit state function is carried about the following points:

$$\textbf{1.}\ \vec{X}_1^* = \begin{Bmatrix} \mu_{X_1} \\ \mu_{X_2} \\ \mu_{X_3} \end{Bmatrix} = \begin{Bmatrix} 800 \\ 30 \\ 20 \end{Bmatrix}, \quad \text{and} \quad \textbf{2.}\ \vec{X}_2^* = \begin{Bmatrix} \mu_{X_1} \\ \mu_{X_2}/2 \\ \mu_{X_3} \end{Bmatrix} = \begin{Bmatrix} 800 \\ 15 \\ 20 \end{Bmatrix}.$$

Solution:

1. When the linearization of the limit state function is carried about the point $\vec{X}_1^*$, the mean and standard deviation of the safety margin Y, μ_Y and σ_Y, are given by

$$\mu_Y \approx \mu_{X_1} - \mu_{X_2}\mu_{X_3} = 800 - (30)(20) = 200 \tag{E2}$$

$$\sigma_Y^2 \approx \left(\frac{\partial Y}{\partial X_1}\bigg|_{\vec{X}_1^*} \right)^2 \sigma_{X_1}^2 + \left(\frac{\partial Y}{\partial X_2}\bigg|_{\vec{X}_1^*} \right)^2 \sigma_{X_2}^2 + \left(\frac{\partial Y}{\partial X_3}\bigg|_{\vec{X}_1^*} \right)^2 \sigma_{X_3}^2 \tag{E3}$$

Using $\frac{\partial Y}{\partial X_1}\Big|_{\vec{X}_1^*} = 1$, $\frac{\partial Y}{\partial X_2}\Big|_{\vec{X}_1^*} = -X_3\Big|_{\vec{X}_1^*} = -20$, $\frac{\partial Y}{\partial X_3}\Big|_{\vec{X}_1^*} = -X_2\Big|_{\vec{X}_1^*} = -30$, and the known values of σ_{X_i}, $i = 1, 2, 3$ in Eq. (E3), we obtain $\sigma_Y = 253.1798$. This gives $\beta = 0.7899$ and $R = 0.78524$.

2. When the linearization of the limit state function is carried about the point $\vec{X}_2^*$, the mean and standard deviation of the safety margin Y, μ_Y and σ_Y, are given by

$$\mu_Y \approx \mu_{X_1} - (\mu_{X_2}/2)\mu_{X_3} = 800 - (30/2)(20) = 500 \tag{E4}$$

$$\sigma_Y^2 \approx \left(\frac{\partial Y}{\partial X_1}\bigg|_{\vec{X}_2^*}\right)^2 \sigma_{X_1}^2 + \left(\frac{\partial Y}{\partial X_2}\bigg|_{\vec{X}_2^*}\right)^2 \sigma_{X_2}^2 + \left(\frac{\partial Y}{\partial X_3}\bigg|_{\vec{X}_2^*}\right)^2 \sigma_{X_3}^2 \tag{E5}$$

Using $\frac{\partial Y}{\partial X_1}\bigg|_{\vec{X}_2^*} = 1$, $\frac{\partial Y}{\partial X_2}\bigg|_{\vec{X}_2^*} = -X_3\bigg|_{\vec{X}_2^*} = -20$, $\frac{\partial Y}{\partial X_3}\bigg|_{\vec{X}_2^*} = -X_2\bigg|_{\vec{X}_2^*} = -15$, and the known values of σ_{X_i}, $i = 1, 2, 3$ in Eq. (E5), we obtain $\sigma_Y = 217.3131$. This gives $\beta = 2.3008$ and $R = 0.9893$.

The results of this example clearly indicate that the choice of $\vec{X}^*$ affects the estimated values of β and R.

■

EXAMPLE 8.9

The limit state function corresponding to the buckling failure mode of a long column (shown in Fig. 8.15) can be expressed in the following two equivalent forms:

Buckling stress should be larger than the compressive stress due to axial load:

$$g_1(P, E, A, I, l) = \frac{\pi^2 E I}{l^2 A} - \frac{P}{A} \geq 0 \tag{E1}$$

Buckling load should be larger than the axial load applied:

$$g_2(P, E, I, l) = \frac{\pi^2 E I}{l^2} - P \geq 0 \tag{E2}$$

where P = axial load, E = Young's modulus, A = cross-sectional area, I = area moment of inertia, and l = length denote the basic random variables of the column. If all the random variables follow normal distribution with means and standard deviations given by $(\mu_P, \sigma_P) = (1000, 200)$ lb, $(\mu_A, \sigma_A) = (2, 0.1)$ in^2, $(\mu_I, \sigma_I) = (1, 0.05)$ in^4, $(\mu_E, \sigma_E) = (30 \times 10^6, 3 \times 10^6)$ psi, and $(\mu_l, \sigma_l) = (50, 2)$ in, find the safety index corresponding to each of the two limit state functions g_1 and g_2.

Solution: The mean value of the limit state function g_1 is given by

$$\mu_{g_1} = g_1(\mu_P, \mu_E, \mu_A, \mu_I, \mu_l) = \frac{\pi^2(30 \times 10^6)(1.0)}{(50^2)(2.0)} - \frac{1000}{2.0} = 58{,}717.6 \text{ psi}$$

The partial derivatives of g_1 with respect to the random variables, evaluated at the mean values of the random variables, are

$$\frac{\partial g_1}{\partial P}\bigg|_m = -\frac{1}{A}\bigg|_m = -\frac{1}{\mu_A} = -\frac{1}{2} = -0.5$$

$$\frac{\partial g_1}{\partial E}\bigg|_m = \frac{\pi^2 I}{A l^2}\bigg|_m = \frac{\pi^2 \mu_I}{\mu_A \mu_l^2} = \frac{\pi^2(1.0)}{(2.0)(50^2)} = 19.7393 \times 10^{-4}$$

$$\left.\frac{\partial g_1}{\partial A}\right|_m = \left.\left(-\frac{\pi^2 E I}{A^2 l^2} + \frac{P}{A^2}\right)\right|_m = -\frac{\pi^2 \mu_E \mu_I}{\mu_A^2 \mu_l^2} + \frac{\mu_P}{\mu_A^2} = -\frac{\pi^2 (30 \times 10^6)(1.0)}{(2.0^2)(50^2)} + \frac{1000}{(2.0^2)}$$

$$= -29{,}358.9$$

$$\left.\frac{\partial g_1}{\partial l}\right|_m = -\left.\frac{2\pi^2 E I}{A l^3}\right|_m = -\frac{2\pi^2 \mu_E \mu_I}{\mu_A \mu_l^3} = -\frac{2\pi^2 (30 \times 10^6)(1.0)}{(2.0)(50^3)} = -2{,}368.7$$

$$\left.\frac{\partial g_1}{\partial I}\right|_m = \left.\frac{\pi^2 E}{A l^2}\right|_m = \frac{\pi^2 \mu_E}{\mu_A \mu_l^2} = \frac{\pi^2 (30 \times 10^6)}{(2.0)(50^2)} = 59{,}217.9$$

The standard deviation of the limit state function g_1 can be computed as

$$\sigma_{g_1}^2 = (-0.5)^2(200)^2 + (-29{,}358.9)^2(0.1)^2 + (59{,}217.9)^2(0.05)^2$$
$$+ (19.7393 \times 10^{-4})^2(3 \times 10^6)^2 + (-2{,}368.7)^2(2)^2 = 7{,}490.6905 \times 10^4$$

or

$$\sigma_{g_1} = 8{,}654.8775 \text{ psi}$$

The safety index, β_1, corresponding to the limit state function g_1 is given by

$$\beta_1 = \frac{\mu_{g_1}}{\sigma_{g_1}} = \frac{58{,}717.6}{8{,}654.8775} = 6.7843$$

We proceed in a similar manner to compute the safety index corresponding to g_2 by starting with the computation of the mean value of g_2 as

$$\mu_{g_2} = g_2(\mu_P, \mu_E, \mu_I, \mu_l) = \frac{\pi^2 (30 \times 10^6)(1.0)}{(50^2)} - 1000 = 117{,}435.8 \text{ lb}$$

The partial derivatives of g_2 with respect to the random variables, evaluated at the mean values of the random variables, are

$$\left.\frac{\partial g_2}{\partial P}\right|_m = -1$$

$$\left.\frac{\partial g_2}{\partial E}\right|_m = \left.\frac{\pi^2 I}{l^2}\right|_m = \frac{\pi^2 \mu_I}{\mu_l^2} = \frac{\pi^2 (1.0)}{(50^2)} = 0.003948$$

$$\left.\frac{\partial g_2}{\partial l}\right|_m = -\left.\frac{2\pi^2 E I}{l^3}\right|_m = -\frac{2\pi^2 \mu_E \mu_I}{\mu_l^3} = -\frac{2\pi^2 (30 \times 10^6)(1.0)}{(50^3)} = -4{,}737.4323$$

$$\left.\frac{\partial g_2}{\partial I}\right|_m = \left.\frac{\pi^2 E}{l^2}\right|_m = \frac{\pi^2 \mu_E}{\mu_l^2} = \frac{\pi^2 (30 \times 10^6)}{(50^2)} = 11.8436 \times 10^4$$

The standard deviation of the limit state function g_2 can be computed as

$$\sigma_{g_2}^2 = (-1)^2(200)^2 + (11.8436 \times 10^4)^2(0.05)^2$$
$$+ (0.003948)^2(3 \times 10^6)^2 + (-4{,}737.4323)^2(2)^2 = 26{,}516.11 \times 10^4$$

or

$$\sigma_{g_2} = 16{,}283.77 \text{ lb}$$

The safety index, β_2, corresponding to the limit state function g_2 is given by

$$\beta_2 = \frac{\mu_{g_2}}{\sigma_{g_2}} = \frac{117{,}435.8}{16{,}283.77} = 7.2118$$

It can be observed that the two limit state functions corresponding to the same (buckling) failure mode yielded two different values of the safety index.

■

In practice, the integral in Eqs. (8.83) and (8.84) is evaluated using some type of analytical approximation. Several types of approximations have been used to evaluate the probability of failure (and reliability) of the system given by Eqs. (8.83) and (8.84).

As indicated earlier, based on the way in which the limit state function is approximated in terms of the basic random variables, the reliability analysis methods can be classified into different categories as first-order reliability methods (FORM), second-order reliability methods (SORM), and so on. The FORM methods use the first and second moments of the random variables in computing the reliability (or probability of failure) of the system. The FORM method is described in the following section.

8.8 First-Order Second-Moment (FOSM) Method

The first-order second-moment (FOSM) method is also known as the mean value first-order second-moment (MVFOSM) method. As indicated in the previous section, the FOSM method is based on the first-order Taylor's series approximation of the limit state function at the mean values of the random variables and uses only the first and second moments (mean values and covariances) of the random variables. In this case, the mean values and covariances of the limit state function are given by Eqs. (8.88) and (8.89) (or Eq. (8.90) for independent random variables), respectively. The probability of failure of the system can be expressed as

$$P_f = P[Y < 0] = P[S - L < 0] \tag{8.91}$$

In this case, the reliability index β is given by

$$\beta = \frac{\mu_g}{\sigma_g} \tag{8.92}$$

If all the random variables are further assumed to follow normal distribution, the limit state function Y, approximated as a linear function of $X_1, X_2, \ldots,$ and X_n in Eq. (8.87), will also follow normal distribution with mean, μ_Y, and standard deviation, σ_Y, given by Eqs. (8.88) and (8.90). In this case, the normally distributed variable Y can be converted to an equivalent standard normal variate Z as

$$Z = \frac{Y - \mu_Y}{\sigma_Y} \tag{8.93}$$

and the probability of failure of the system can be found as

$$P_f = P\left[Z = \frac{Y - \mu_Y}{\sigma_Y} < \frac{0 - \mu_Y}{\sigma_Y}\right] = 1 - \Phi\left(\frac{\mu_Y}{\sigma_Y}\right) = 1 - \Phi(\beta) \tag{8.94}$$

where Φ is the cumulative distribution function of the standard normal variate at the reliability index, $\beta = \dfrac{\mu_Y}{\sigma_Y}$.

Notes:

1. The reliability index β given by Eq. (8.92) and the probability of failure given by Eq. (8.91) will be accurate only if the basic random variables X_i are independent and normally distributed.
2. In most practical problems, the random variables X_i may not be statistically independent and may not follow normal distribution. In such cases, the safety index β given by Eq. (8.92) and the probability of failure computed with Eq. (8.94) will not be correct. Even if the exact probability distributions (non-normal) of the random variables X_i are available, the present method (MVFOSM) cannot accommodate the information.
3. The MVFOSM method was used to compute system reliability in the early years of development of reliability-based design procedures. In fact, MVFOSM was used in developing standards of reliability-based structural design and criteria by organizations such as Canadian Standard Associations (CSA, 1994), Comite European du Beton (CEB, 1976), and American Institute of Steel Construction (AISC, 1986).
4. The MVFOSM is based on the linearization of the limit state function g. This can introduce significant errors depending on the nonlinearity of the function g.
5. The reliability index, defined in Eq. (8.92), may not maintain a constant value when the safety margin is restated in a different, but equivalent, form. For example, using the safety margin, we can find the probability of failure as

$$P_f = P\,[S - L < 0] \tag{8.95}$$

which can be restated, equivalently, as

$$P_f = P\left[\frac{S}{L} < 1\right] \tag{8.96}$$

The reliability or safety index corresponding to Eqs. (8.95) and (8.96) will be different.

EXAMPLE 8.10

If the strength (S) and load (L) of a structure are independent and follow lognormal distribution with mean values μ_S and μ_L and standard deviations σ_S and σ_L, respectively, derive an expression for the reliability index.

Solution: The safety factor of the structure (Y) is given by

$$Y = \frac{S}{L} \tag{E1}$$

By taking logarithms of Eq. (E1), we have

$$\ln Y = \ln S - \ln L \tag{E2}$$

Because S and L follow lognormal distribution, $\ln S$ and $\ln L$, and hence $\ln Y$ follow normal distribution. The mean and standard deviations of $\ln S$, $\ln L$, and $\ln Y$ can be expressed as

$$\mu_{\ln S} = \ln \mu_S - \frac{1}{2} \ln \left(1 + \frac{\sigma_S^2}{\mu_S^2}\right), \quad \sigma_{\ln S} = \sqrt{\ln \left(1 + \frac{\sigma_S^2}{\mu_S^2}\right)} \tag{E3}$$

$$\mu_{\ln L} = \ln \mu_L - \frac{1}{2} \ln \left(1 + \frac{\sigma_L^2}{\mu_L^2}\right), \quad \sigma_{\ln L} = \sqrt{\ln \left(1 + \frac{\sigma_L^2}{\mu_L^2}\right)} \tag{E4}$$

$$\mu_{\ln Y} = \mu_{\ln S} - \mu_{\ln L}, \quad \sigma_{\ln Y} = \sqrt{\sigma_{\ln S}^2 + \sigma_{\ln L}^2} \tag{E5}$$

The reliability of the structure is given by

$$R = P\left[Y = \frac{S}{L} > 1.0\right] = F_Y(1.0) \tag{E6}$$

where $F_Y(1.0)$ is the cumulative distribution function of Y at 1.0. Because $\ln Y$ follows normal distribution, we can rewrite Eq. (E6) as

$$R = P[\ln Y > \ln 1.0] = P\left[\frac{\ln Y - \mu_{\ln Y}}{\sigma_{\ln Y}} > \frac{\ln 1.0 - \mu_{\ln Y}}{\sigma_{\ln Y}}\right] = P[Z > -\beta]$$

$$= \Phi(\beta) = 1 - \Phi(-\beta) \tag{E7}$$

where the reliability index (β) is defined as

$$\beta = \frac{\mu_{\ln Y}}{\sigma_{\ln Y}} \tag{E8}$$

■

EXAMPLE 8.11

The limit state function of a system is given by

$$g(x_1, x_2) = 25\,x_1^2 + 20\,x_1\,x_2 + 100\,x_2^2 - 10\,x_1 - 20\,x_2 - 2500 \tag{E1}$$

where the basic random variables X_1 and X_2 follow normal distribution with mean values $\mu_{X_1} = 10$ and $\mu_{X_2} = 15$ and standard deviations $\sigma_{X_1} = 5$ and $\sigma_{X_2} = 8$. Determine the safety index β using the mean value FOSM method.

Solution: The mean value and standard deviation of the limit state function can be found as

$$\mu_g = g(\mu_{X_1}, \mu_{X_2}) = 25\,\mu_{X_1}^2 + 20\,\mu_{X_1}\,\mu_{X_2} + 100\,\mu_{X_2}^2 - 10\,\mu_{X_1} - 20\,\mu_{X_2} - 2500$$

$$= 25(10)^2 + 20(10)(15) + 100(15)^2 - 10(10) - 20(15) - 2500 = 25{,}100.0 \tag{E2}$$

$$\sigma_g^2 = \left(\frac{\partial g}{\partial x_1}\bigg|_m \right)^2 \sigma_{X_1}^2 + \left(\frac{\partial g}{\partial x_2}\bigg|_m \right)^2 \sigma_{X_2}^2 \tag{E3}$$

where the partial derivatives of g are evaluated at the mean values of the random variables. The partial derivatives of g appearing in Eq. (E3) are given by

$$\frac{\partial g}{\partial x_1}\bigg|_m = (50\, x_1 + 20\, x_2 - 10)\bigg|_m = 50\, \mu_{X_1} + 20\, \mu_{X_2} - 10 = 50(10) + 20(15) - 10 = 770$$

$$\frac{\partial g}{\partial x_2}\bigg|_m = (20\, x_1 + 200\, x_2 - 20)\bigg|_m = 20\, \mu_{X_1} + 200\, \mu_{X_2} - 20 = 20(10) + 200(15) - 20 = 3{,}180$$

Thus Eq. (E3) gives

$$\sigma_g^2 = (770)^2(5)^2 + (3{,}180)^2(8)^2 = 662.7961 \times 10^6$$

or

$$\sigma_g = 25{,}744.8267 \tag{E4}$$

From Eq. (8.92), the safety index can be found as

$$\beta = \frac{\mu_g}{\sigma_g} = \frac{25{,}100.0}{25{,}744.8267} = 0.9749 \tag{E5}$$

■

8.9 Hasofer-Lind Reliability Index with Two Normally Distributed Variables

In the MVFOSM, the limit state function $g(\vec{X})$ is expanded in Taylor's series about the mean values of the random variables $X_1, X_2, \ldots,$ and X_n. In the Hasofer-Lind method, the limit state function $g(\vec{X})$ is expanded about the *most probable point* (MPP) for failure, also known as the *design point.* To find the MPP, consider the limit state function, $Y = g$, given by

$$g(S, L) = S - L \tag{8.97}$$

where S and L are assumed to follow normal distribution with mean and standard deviations given by (μ_S, σ_S) and (μ_L, σ_L), respectively. In this case, the failure or limit state surface is defined by

$$g(S, L) = S - L = 0 \tag{8.98}$$

and is shown by the diagonal line in the (S, L)-plane in Fig. 8.17. The normally distributed random variables S and L are transformed into standard normal variates $\widetilde{S}$ and $\widetilde{L}$ as

$$\widetilde{S} = \frac{S - \mu_S}{\sigma_S}, \quad \widetilde{L} = \frac{L - \mu_L}{\sigma_L} \tag{8.99}$$

From Eq. (8.99), S and L can be expressed as

$$S = \mu_S + \widetilde{S}\, \sigma_S, \quad L = \mu_L + \widetilde{L}\, \sigma_L \tag{8.100}$$

Thus the limit state surface, given by Eq. (8.98), can be expressed in the $(\widetilde{S}, \widetilde{L})$-coordinate system as

$$g(\widetilde{S}, \widetilde{L}) = S - L = \widetilde{S}\, \sigma_S - \widetilde{L}\, \sigma_L + \mu_S - \mu_L = 0 \tag{8.101}$$

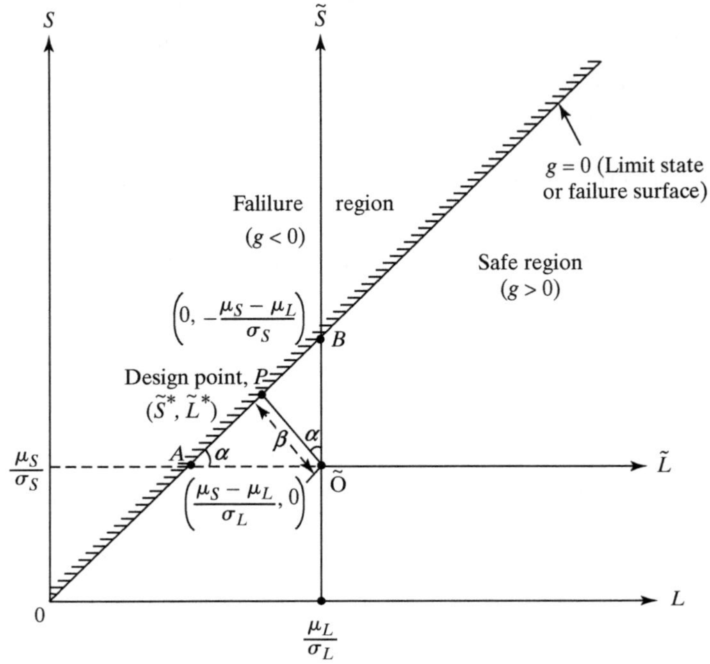

FIGURE 8.17 Limit state surface in $\tilde{S} - \tilde{L}$ plane.

and is shown by the diagonal line in Fig. 8.17. The shortest distance from the origin $\tilde{O}$ to the limit state surface (line) can be found using the relation

From triangle $\tilde{O}PA$:
$$\sin\alpha = \frac{\beta}{\tilde{O}A} = \frac{\beta}{\left(\dfrac{\mu_S - \mu_L}{\sigma_L}\right)} \tag{8.102}$$

From triangle $\tilde{O}PB$:
$$\cos\alpha = \frac{\beta}{\tilde{O}B} = \frac{\beta}{\left(\dfrac{\mu_S - \mu_L}{\sigma_S}\right)} \tag{8.103}$$

Using the relation

$$\sin^2\alpha + \cos^2\alpha = \left(\frac{\beta^2\,\sigma_{\tilde{L}}^2}{(\mu_S - \mu_L)^2}\right) + \left(\frac{\beta^2\,\sigma_{\tilde{S}}^2}{(\mu_S - \mu_L)^2}\right) = 1 \tag{8.104}$$

the reliability index, β, can be expressed as

$$\beta = \frac{\mu_S - \mu_L}{\sqrt{\sigma_{\tilde{S}}^2 + \sigma_{\tilde{L}}^2}} \tag{8.105}$$

Equation (8.105) indicates that, geometrically, the reliability index β is the shortest distance from the origin to the linear limit state or failure surface.

Notes:

1. The point $P(\widetilde{S}^*, \widetilde{L}^*)$ lying on the limit state function $g(\widetilde{S}, \widetilde{L}) = 0$, which corresponds to the shortest distance from the origin $(\widetilde{S} = 0, \widetilde{L} = 0)$, is known as the most probable point (MPP).
2. Although this geometric interpretation of the reliability index β is given by considering a linear limit state surface with only two random variables (S and L), the procedure and interpretation can be extended to any linear limit state surface involving any (n) number of random variables X_i, $i = 1, 2, \ldots, n$.
3. The reliability index given by Eq. (8.105) can be seen to be same as the one defined in MVFOSM (Eq. (8.92)). However, Eq. (8.105) has been derived using a different approach based on geometry. If S and L are independent and follow normal distribution, both the methods will give the same reliability index. However, the two methods will give different values of the reliability index if the number of normally distributed variables is more than two.

8.10 Hasofer-Lind Reliability Index with Several Normally Distributed Variables

The procedure used in the case of a linear limit state function in terms of two normally distributed variables can be extended to the case where the limit state function is in terms of n normally distributed variables X_i, $i = 1, 2, \ldots, n$. Consider a general nonlinear limit state or failure surface given by (Fig. 8.18(a)):

$$g(\vec{X}) = g(X_1, X_2, \ldots, X_n) = 0 \tag{8.106}$$

Let the random variables $X_1, X_2, \ldots$, and X_n be independent (uncorrelated) with the mean values μ_{X_i} and standard deviations σ_{X_i}, $i = 1, 2, \ldots, n$. The variables X_i are transformed into standard normal variates Z_i as

$$Z_i = \frac{X_i - \mu_{X_i}}{\sigma_{X_i}}; \quad i = 1, 2, \ldots, n \tag{8.107}$$

so that

$$\mu_{Z_i} = 0, \quad \sigma_{Z_i} = 1; \quad i = 1, 2, \ldots, n \tag{8.108}$$

Equation (8.107) yields the relations

$$X_i = \sigma_{X_i} Z_i + \mu_{X_i}; \quad i = 1, 2, \ldots, n \tag{8.109}$$

The linear transformation of Eq. (8.107) maps the failure surface $g(\vec{X}) = 0$ in the $\vec{X}$-coordinate system (as shown in Fig. 8.18(b)) into the failure surface $g(\vec{Z}) = 0$ in the $\vec{Z}$-coordinate system as

$$g(\vec{Z}) = g(\sigma_{X_1} Z_1 + \mu_{X_1}, \sigma_{X_2} Z_2 + \mu_{X_2}, \ldots, \sigma_{X_n} Z_n + \mu_{X_n}) = 0 \tag{8.110}$$

(shown in Fig. 8.18(c)). Here, the safe regions are denoted by $g(\vec{X}) > 0$ and $g(\vec{Z}) > 0$, while the failure regions are indicated by $g(\vec{X}) < 0$ and $g(\vec{Z}) < 0$.

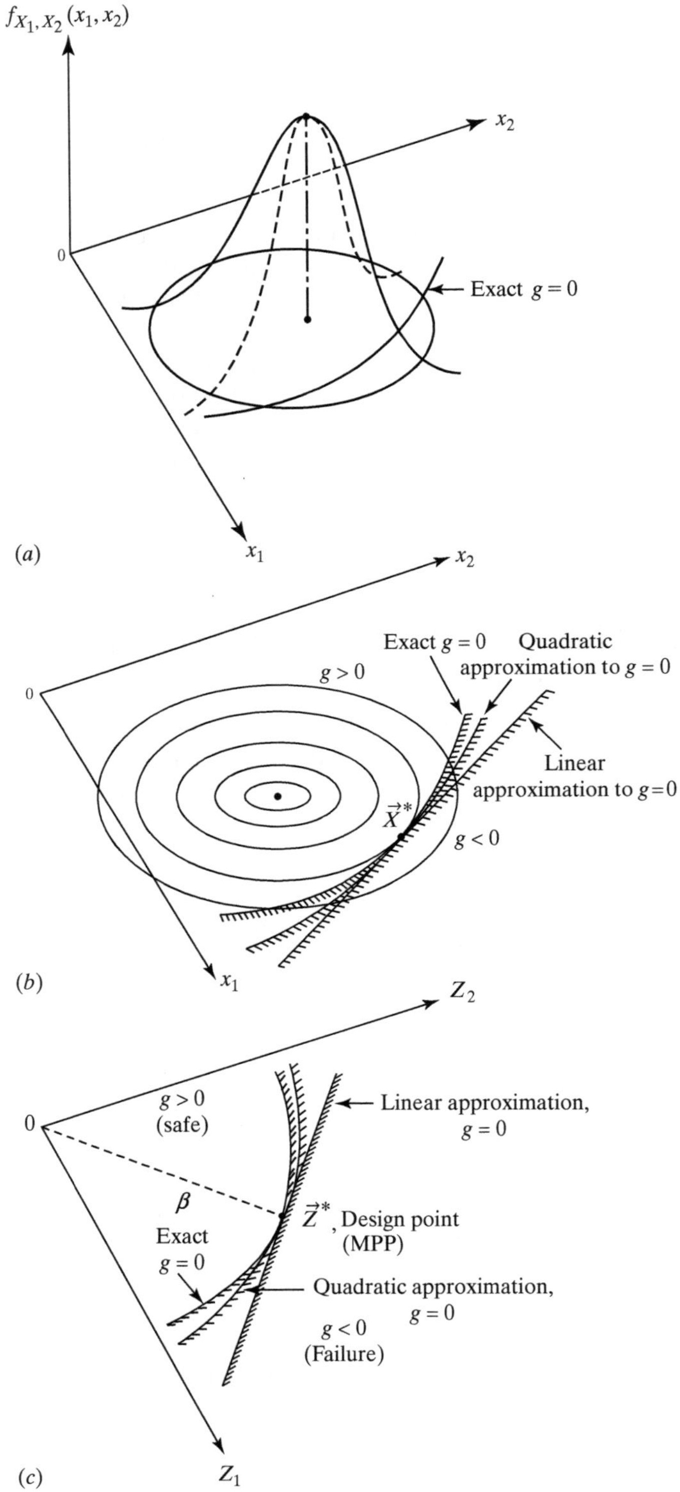

FIGURE 8.18 Failure surface in the $x_1\,x_2$ – and $Z_1\,Z_2$ – planes.

If the limit state surface moves further away from the origin or closer to the origin, the safe region (with $g(\vec{X}) > 0$) will increase or decrease accordingly as indicated in Fig. 8.19(a). Thus, the position of the failure surface (with $g(\vec{X}) = 0$) relative to the origin will determine the reliability (or probability of failure) of the system. The position of the failure surface can be denoted by the minimum distance from the surface $g(\vec{X}) = 0$.

Since the mean values of X_i, μ_{X_i}, are mapped into the origin, $\mu_{Z_i} = 0$, of the $\vec{Z}$-space because of the transformation of Eq. (8.107), the origin of $\vec{Z}$-coordinate system will usually lie in the safe region. In addition, the $\vec{Z}$-coordinate system has an important property, known as rotational symmetry with respect to its second moment or standard deviation. This implies that the distance of any point on the failure surface, $g(\vec{Z}) = 0$, from the origin in the $\vec{Z}$-space is equal to the number of standard deviations from the mean value point in the $\vec{X}$-space to the corresponding point on the failure surface, $g(\vec{X}) = 0$. In fact, the point on the failure surface with minimum distance to the origin denotes the most probable point for failure [8.28]. Thus, the minimum distance of the failure surface from the origin can be used, in some sense, as a measure of reliability of the system (Fig. 8.19(b)).

The distance of any point $\vec{Z} = \{Z_1\, Z_2 \ldots Z_n\}^T$ on the failure surface, $g(\vec{Z}) = 0$, from the origin (in $\vec{Z}$-space) is given by

$$d = \sqrt{Z_1^2 + Z_2^2 + \cdots + Z_n^2} = \vec{Z}^T\vec{Z} \tag{8.111}$$

The point on the failure surface having minimum distance from the origin, $(z_1{}^*, z_2{}^*, \ldots, z_n{}^*)$, can be determined by minimizing the function

$$d = (\vec{Z}^T\vec{Z})^{\frac{1}{2}} \tag{8.112}$$

subject to the constraint

$$g(\vec{Z}) = 0 \tag{8.113}$$

The shortest distance from the origin to the failure surface, $g(\vec{Z}) = 0$, is defined as the Hasofer-Lind safety index, $\beta_{HL} = (\vec{Z}^T\vec{Z})^{\frac{1}{2}}$, and $\vec{Z}$ is found by solving the following optimization problem:

$$\text{Find } \vec{Z} \text{ to minimize } d = (\vec{Z}^T\vec{Z}) \text{ subject to } g(\vec{Z}) = 0 \tag{8.114}$$

The Hasofer-Lind reliability index can be used to find the first-order approximation of the probability of failure of the system as

$$P_f = \Phi(-\beta_{HL}) = 1 - \Phi(\beta_{HL}) \tag{8.115}$$

Notes:

1. The reliability index, as defined by Eq. (8.114), is invariant because its geometric shape and the distance from the origin remain constant irrespective of the form in which the limit state equation is expressed.
2. In practice, the MPP, $\vec{Z}^*$, denotes the design point. It can be seen, from Fig. 8.19, that the nearer the MPP or the design point is to the origin, the higher the probability of failure of the system.

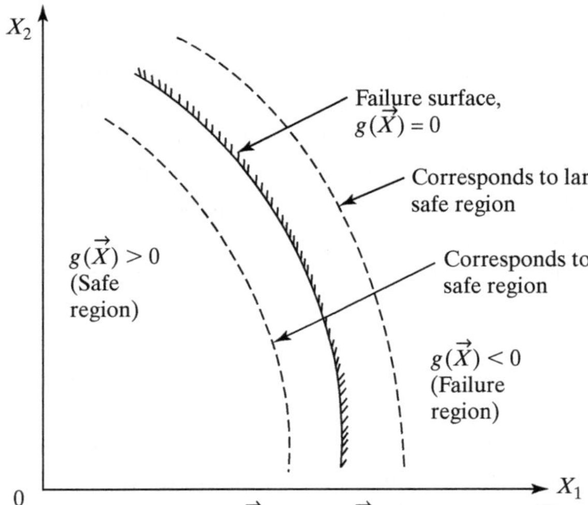

(a) Failure surface in $\vec{X}$ – space, $\vec{X} = \{X_1, X_2, \ldots, X_n\}^T$

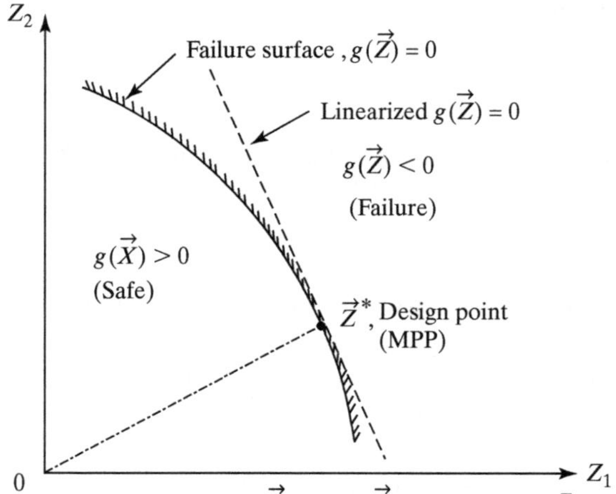

(b) Failure surface in $\vec{Z}$– space, $\vec{Z} = \{Z_1, Z_2, \ldots, Z_n\}^T$

FIGURE 8.19 Failure surface and the most probable point (MPP).

8.11 Reliability of Weakest-Link and Fail-Safe Systems

8.11.1 Introduction

The concept of strength-based reliability considered in the previous sections assumes a single structural element or member for which loading is described by the random variable L and the strength by the random variable S. The reliability and the probability of failure of the member is given by Eqs. (8.1) (or (8.5) or (8.9)) and (8.10) (or (8.11) or (8.13)), respectively. Most machines and structures consist of more than one element or member and are subjected to more than one load condition. In addition, a machine or structure may have more than one possible failure mode. Systems with multiple members can be classified into two categories, namely, weakest-link systems and fail-safe systems. A system is called a *weakest-link system* (also known as a series system) if it fails whenever any of its elements fails. An example of a weakest-link system is the statically determinate truss shown in Fig. 8.20. A system is called a *fail-safe system* (also known as a parallel system) if the failure of a single element will not always result in the failure of the total system, because the remaining elements may be able to sustain the applied loads. An example of a fail-safe system is the statically indeterminate truss shown in Fig. 8.21. In general, there will be more than one way in which a parallel system can fail under a given load. For example, a parallel system may fail due to the failure of different elements or the elements may fail in different orders. All these considerations are important in the reliability analysis of a parallel system.

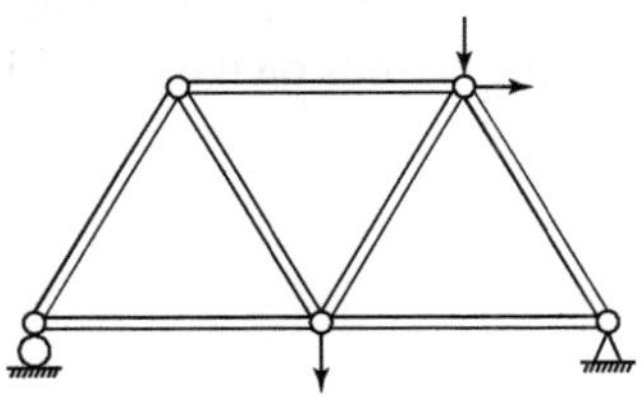

FIGURE 8.20 A statically determinate truss (weakest-link system).

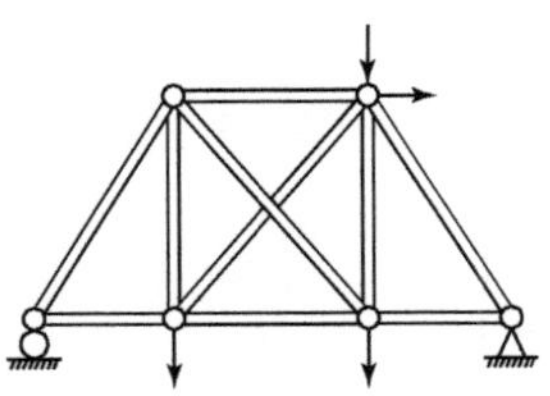

FIGURE 8.21 A statically indeterminate truss (fail-safe system).

8.11.2 Reliability of the Fundamental Problem

A weakest-link system consisting of only a single member of strength S subjected to a load L is known as the fundamental problem. If the strength and load are assumed to be independent random variables, the reliability of the system will be identical to the component reliability considered earlier (in Sections 8.2 and 8.3). Thus, the probability of failure can be expressed as

$$P_f = P(S < L) = \int_{l=0}^{\infty} F_S(l) \cdot f_L(l) \cdot dl \tag{8.116}$$

or, equivalently,

$$P_f = 1 - P(L < S) = 1 - \int_{s=0}^{\infty} F_L(s) \cdot f_S(s) \cdot ds \tag{8.117}$$

If the strength and the load are discrete random variables, Eqs. (8.116) and (8.117) become

$$P_f = \sum_i F_S(l_i) \cdot f_L(l_i) \tag{8.118}$$

and

$$P_f = 1 - \sum_j F_L(s_j) \cdot f_S(s_j) \tag{8.119}$$

where l_i and s_j denote the ith value of the load and the jth value of the strength, respectively.

Bounds on the Probability of Failure. If the strength (S) and the load (L) are represented along two axes as shown in Fig. 8.22, the bounds on the probability of failure can be

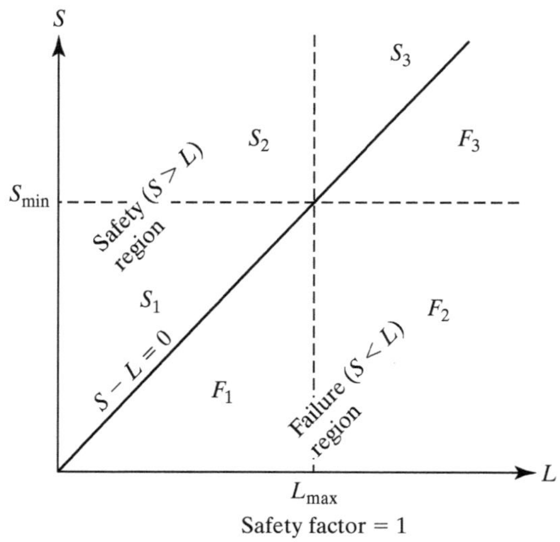

FIGURE 8.22 Bounds on the probability of failure.

derived as follows. The probability of failure is the same as the probability of realizing the strength less than the load. Thus, for a factor of safety of one, we obtain

$$\begin{aligned} P_f &= \text{probability of realizing the point}(S, L) \text{ in areas } (F_1 + F_2 + F_3) \text{ in Fig. 8.22} \\ &= P(S < L < L_{\max};\quad S < S_{\min}) + P(L > L_{\max};\quad S < S_{\min}) \\ &\quad + P(L > L_{\max};\quad S_{\min} < S < L) \end{aligned} \tag{8.120}$$

that is,

$$\begin{aligned} P_f &> \text{probability of realizing the point}(S, L) \text{ in area } F_2 \\ &> P(L > L_{\max};\quad S < S_{\min}) \\ &> p_1 p_2 \end{aligned} \tag{8.121}$$

where

$$p_1 = P(S < S_{\min}) \tag{8.122}$$

$$p_2 = P(L > L_{\max}) \tag{8.123}$$

$L_{\max}$ is the maximum value of load and $S_{\min}$ is the minimum value of strength used in the deterministic design procedure.

Similarly, the probability of survival or reliability can be expressed as

$$\begin{aligned} R_0 &= \text{probability of realizing the point}(S, L) \text{ in areas}(S_1 + S_2 + S_3) \text{ in Fig. 8.22} \\ &= P(L < S < S_{\min};\quad L < L_{\max}) + P(S > S_{\min};\quad L < L_{\max}) \\ &\quad + P(L > L_{\max};\quad L < S < \infty) \end{aligned} \tag{8.124}$$

that is,

$$\begin{aligned} R_0 &> \text{probability of realizing the point}(S, L) \text{ in area } S_2 \\ &> P(S > S_{\min};\quad L < L_{\max}) \\ &> (1 - p_1)(1 - p_2) \end{aligned} \tag{8.125}$$

Since $P_f = 1 - R_0$, the inequality (8.125) can be written as

$$P_f < 1 - (1 - p_1)(1 - p_2) \tag{8.126}$$

Combining the inequalities (8.121) and (8.126), the bounds on the probability of failure can be expressed as

$$(p_1 p_2) < P_f < (p_1 + p_2 - p_1 p_2) \tag{8.127}$$

EXAMPLE 8.12

Find the bounds on the probability of failure of a system for which the strength is found to be less than the design strength once in 10 cases and the load is found to exceed the design load once in 500 cases.

Solution: Here $p_1 = 1/10$ and $p_2 = 1/500$, and hence Eq. (8.127) can be used to find the bounds on the probability of failure as

$$0.0002 < P_f < 0.1018$$

Note that this range is too wide for design purposes and we need to use the upper bound to ensure a conservative design.

■

8.11.3 Reliability of Weakest-Link (or Series) Systems[4]

Statically determinate trusses subjected to a single loading condition, for example, can be idealized as weakest-link structures with several members as shown in Fig. 8.23. If $F_{S_i}(s_i)$ denotes the probability distribution function of the strength of member i, the probability distribution function of the strength of the series system, $F_S(s)$, can be expressed as

$$\begin{aligned} F_S(s) &= P(S \le s) = 1 - P(S > s) = 1 - P[S_1 > s_1 \cap S_2 > s_2 \cap \cdots \cap S_n > s_n] \\ &= 1 - [1 - F_{S_1}(s_1)][1 - F_{S_2}(s_2)] \cdots [1 - F_{S_n}(s_n)] \\ &= 1 - \prod_{i=1}^{n} [1 - F_{S_i}(s_i)] \end{aligned} \tag{8.128}$$

where S denotes the strength of the series system and the strengths of the individual members (S_i) are assumed to be independent. If a load l is applied to the system, the force (or stress) induced in member i will be $a_i l$ where a_i denotes the force (or stress) induced in member i when a unit load is applied to the system. The value of a_i can be determined

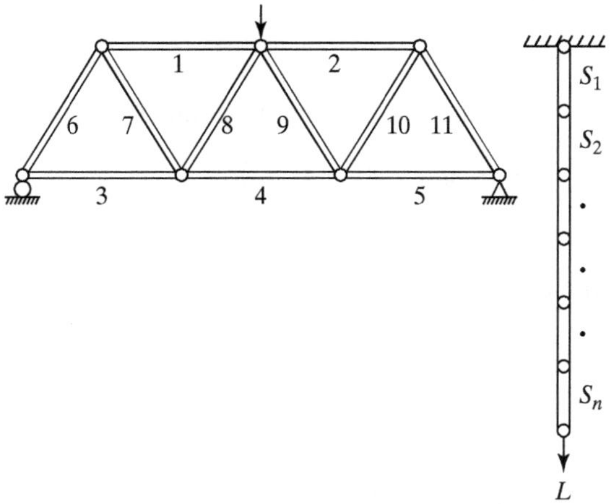

FIGURE 8.23 A weakest-link system with n members and one load.

---

[4]The material presented in this section is valid for any general system. However, the procedures are indicated in the context of structural applications.

by using the methods of structural analysis. In this case, the probability of failure can be found as

$$P_f = 1 - R_0 = 1 - \int_{l=0}^{\infty} [P(S > l)] f_L(l)\, dl$$

$$= 1 - \int_{l=0}^{\infty} \left[\prod_{i=1}^{n} \{1 - F_{S_i}(a_i l)\} \right] f_L(l)\, dl \tag{8.129}$$

where the expression in the square brackets denotes the probability of all members surviving the load l. If the strengths and load are discrete random variables, Eq. (8.129) can be written as

$$P_f = 1 - \sum_{l_j = l_{\min}}^{l_{\max}} \left[\prod_{i=1}^{n} \{1 - F_{S_i}(a_i l_j)\} \right] f_L(l_j) \tag{8.130}$$

where l_j denotes the jth discrete value of the load, and $l_{\min}$ and $l_{\max}$ represent, respectively, the minimum and maximum values of the load, L.

Simple Bounds on the Probability of Failure. Let a variable x_i be defined as follows:

$$x_i = \begin{cases} 1 & \text{if member } i \text{ has survived} \\ 0 & \text{if member } i \text{ has failed} \end{cases}; \quad i = 1, 2, \ldots, n \tag{8.131}$$

The probability of failure of the series system can be expressed as

$$P_f = 1 - R_0 = 1 - P[x_1 = 1 \cap x_2 = 1 \cap \cdots \cap x_n = 1] \tag{8.132}$$

Using the multiplicative rule for the probability of intersection of n events $A_1, A_2, \ldots, A_n$

$$P(A_1 \cap A_2 \cap \cdots \cap A_n) = P(A_1)P(A_2|_{A_1})P(A_3|_{A_1 A_2}) \cdots P(A_n|_{A_1 A_2 \cdots A_{n-1}}) \tag{8.133}$$

Equation (8.132) can be rewritten as

$$P_f = 1 - P(x_1 = 1)\frac{P(x_1 = 1 \cap x_2 = 1)}{P(x_1 = 1)} \cdots \frac{P(x_1 = 1 \cap x_2 = 1 \cap \cdots \cap x_n = 1)}{P(x_1 = 1 \cap x_2 = 1 \cap \cdots \cap x_{n-1} = 1)} \tag{8.134}$$

If the safety margins of the members are positively correlated, it can be shown that the following relationship is valid

$$P(x_1 = 1 \cap x_2 = 1 \cap \cdots \cap x_{k+1} = 1)$$
$$\geq P(x_1 = 1 \cap x_2 = 1 \cap \cdots \cap x_k = 1)P(x_{k+1} = 1); \quad 1 \leq k \leq n - 1 \tag{8.135}$$

Then, Eq. (8.134) can be used to obtain an upper bound on the probability of failure of the system as

$$P_f \leq 1 - \prod_{k=1}^{n} P(x_k = 1) = 1 - \prod_{k=1}^{n} [1 - P(x_k = 0)] \tag{8.136}$$

To obtain a lower bound on the probability of failure, we assume that the member failures are perfectly dependent on one another (correlated). Then the survival of any member implies the survival of all the members, and hence, that of the total system. In this case, the probability of failure of the series system is given by

$$P_f = P_{f_k} = P(x_k = 0) \tag{8.137}$$

Equation (8.137) will be a lower bound to the true value of the probability of failure and hence the bounds on the probability of failure can be expressed as

$$\max_k P(x_k = 0) \le P_f \le 1 - \prod_{k=1}^{n} [1 - P(x_k = 0)]$$

or,

$$\max_k P_{f_k} \le P_f \le 1 - \prod_{k=1}^{n} [1 - P_{f_k}] \tag{8.138}$$

The right-hand side of the inequality (8.138) can be used to obtain another upper bound on the probability of failure of the system as follows:

$$\begin{aligned}
1 - \prod_{k=1}^{n} [1 - P_{f_k}] &= 1 - (1 - P_{f_1})(1 - P_{f_2}) \cdots (1 - P_{f_n}) \\
&= 1 - [1 - (P_{f_1} + P_{f_2} + \cdots + P_{f_n}) + (P_{f_1}P_{f_2} + P_{f_1}P_{f_3} + \cdots) \\
&\qquad - (P_{f_1}P_{f_2}P_{f_3} + \cdots) + \cdots] \\
&= \sum_{k=1}^{n} P_{f_k} - \sum_{j=1}^{n} \sum_{\substack{k=1 \\ j \ne k}}^{n} P_{f_j} P_{f_k} + \cdots
\end{aligned} \tag{8.139}$$

In most practical situations, P_{f_k} are very small compared to unity, and hence the terms involving their products will be negligible. Thus, Eq. (8.139) can be written as

$$1 - \prod_{k=1}^{n} [1 - P_{f_k}] \approx \sum_{k=1}^{n} P_{f_k} \tag{8.140}$$

Since the sum of the terms dropped is non-negative, we have

$$1 - \prod_{k=1}^{n} [1 - P_{f_k}] \le \sum_{k=1}^{n} P_{f_k} \tag{8.141}$$

Ditlevsen Bounds on the Probability of Failure. The bounds given by Eq. (8.138) will be too wide for many practical applications. Tighter bounds, known as *Ditlevsen bounds,* can be derived for series systems as follows [8.35]. Defining a variable y as

$$y = \begin{cases} 1 & \text{if the series system has survived} \\ 0 & \text{if the series system has failed} \end{cases} \tag{8.142}$$

we can express y in terms of $x_1, x_2, \ldots, x_n$ as

$$y = x_1 x_2 \cdots x_n \tag{8.143}$$

By replacing x_n as $1 - (1 - x_n)$, Eq. (8.143) can be rewritten as

$$y = x_1 x_2 \cdots x_{n-1} - x_1 x_2 \cdots x_{n-1}(1 - x_n) \tag{8.144}$$

By proceeding in a similar manner, Eq. (8.144) can be written as

$$y = x_1 - x_1(1 - x_2) - x_1 x_2(1 - x_3) - \cdots x_1 x_2 \cdots x_{n-1}(1 - x_n) \tag{8.145}$$

Noting that the reliability of the system R_0 is same as $E[y]$, Eq. (8.145) can be used to obtain

$$\begin{aligned} P_f &= 1 - R_0 = 1 - E[y] \\ &= E[1 - x_1] + E[x_1(1 - x_2)] + \cdots + E[x_1 x_2 \cdots x_{n-1}(1 - x_n)] \end{aligned} \tag{8.146}$$

where all the terms on the right-hand side of Eq. (8.146) are non-negative. Noting that the inequality

$$x_1 x_2 \cdots x_{k-1} \geq 1 - [(1 - x_1) + (1 - x_2) + \cdots + (1 - x_{k-1})] \tag{8.147}$$

is valid for any $k = 2, 3, \ldots, n$, we obtain

$$E[x_1 x_2 \cdots x_{k-1}(1 - x_k)] \geq E[1 - x_k] - \sum_{i=1}^{k-1} (1 - x_i)(1 - x_k) \tag{8.148}$$

Using Eqs. (8.146) and (8.148), the following inequality can be obtained:

$$P_f \geq P_{f_1} + \sum_{k=2}^{n} \max\left[\left\{P_{f_k} - \sum_{i=1}^{k-1} (P_{f_i} \cap P_{f_k})\right\},\ 0\right] \tag{8.149}$$

Similarly, by noting that

$$x_1 x_2 \cdots x_{i-1} \leq x_k = 1 - (1 - x_k); \quad \text{for } i \geq 2; \quad k < i \tag{8.150}$$

we obtain an upper bound on the probability of failure of the system as

$$P_f \leq \sum_{i=1}^{n} P_{f_i} - \sum_{i=2}^{n} \max_{j<i} P[(1 - x_i) \cap (1 - x_j)] \tag{8.151}$$

Thus, the Ditlevsen bounds are given by

$$P_{f_1} + \sum_{k=2}^{n} \max\left[\left\{P_{f_k} - \sum_{i=1}^{k-1} (P_{f_i} \cap P_{f_k})\right\};\ 0\right] \leq P_f \leq \sum_{i=1}^{n} P_{f_i} - \sum_{i=2}^{n} \max_{j<i} [P_{f_i} \cap P_{f_j}] \tag{8.152}$$

EXAMPLE 8.13

The cross-sectional areas of the members of the two-bar truss shown in Fig. 8.24(*a*) are given by $A_1 = A_2 = A$. The probability mass functions of the strengths of the members and the load acting on the truss are shown in Fig. 8.24(*b*–*d*). Determine the following:

1. Probability of failure of the truss
2. Simple bounds on the probability of failure of the truss
3. Ditlevsen bounds on the probability of failure of the truss

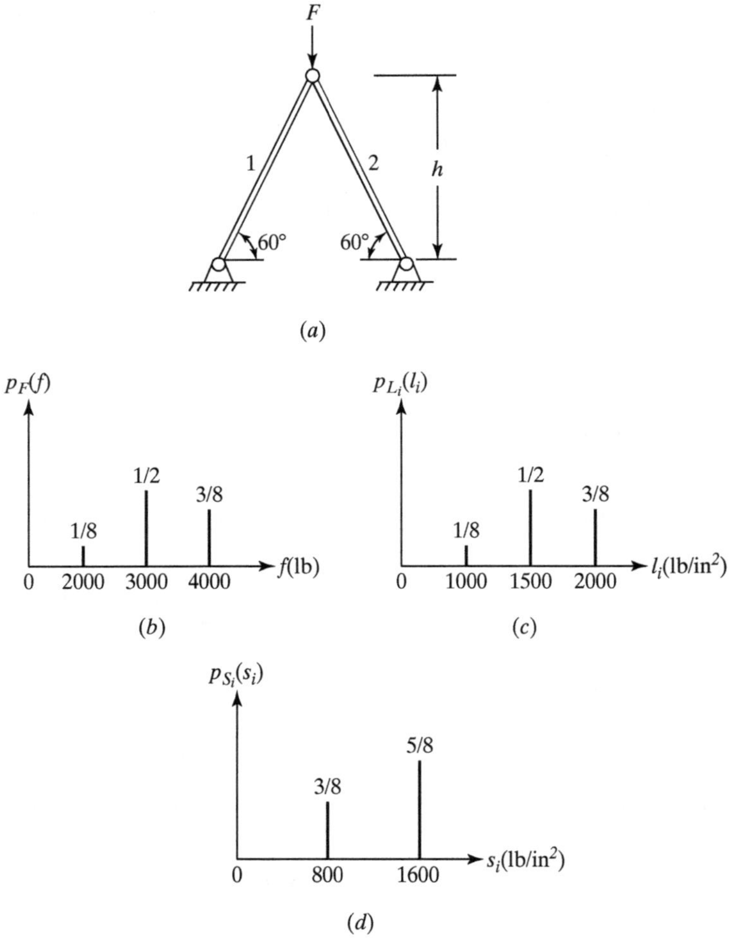

FIGURE 8.24 (*a*) A two-bar truss; (*b*) probability mass function of the load acting on the truss; (*c*) probability mass function of the force in member $i (i = 1, 2)$; and (*d*) probability mass function of strength of member $i (i = 1, 2)$.

Solution:

1. The force in member i is given by $F/2(i = 1, 2)$ and the stress by $F_i/A_i = F/(2A)(i = 1, 2)$. The probability mass function of the forces in the members is shown in Fig. 8.24(*d*). The exact value of the probability of failure of the truss can be computed as:

$$P_f = P_f|_{F=2000}\, P(F = 2000) + P_f|_{F=3000}\, P(F = 3000) + P_f|_{F=4000}\, P(F = 4000) \quad \text{(E1)}$$

If S_i denotes the strength of member $i(i = 1, 2)$, Eq. (E1) gives

$$\begin{aligned} P_f &= \{1 - P(S_1 > 1000, S_2 > 1000)\}\, P(L_i = 1000) \\ &\quad + \{1 - P(S_1 > 1500, S_2 > 1500)\}\, P(L_i = 1500) \\ &\quad + \{1 - P(S_1 > 2000, S_2 > 2000)\}\, P(L_i = 2000) \\ &= \left\{1 - \left(\frac{5}{8}\right)\left(\frac{5}{8}\right)\right\}\frac{1}{8} + \left\{1 - \left(\frac{5}{8}\right)\left(\frac{5}{8}\right)\right\}\frac{1}{2} + \{1 - (0)(0)\}\frac{3}{8} \\ &= \frac{39}{64}\left(\frac{1}{8}\right) + \frac{39}{64}\left(\frac{1}{2}\right) + 1\left(\frac{3}{8}\right) = \frac{387}{512} = 0.7559 \end{aligned}$$

2. The probability of failure of individual members P_{f_i} can be computed as

$$\begin{aligned} P_{f_1} &= P_{f_2} = P(S_1 < L_1) \\ &= \{1 - P(S_1 > 1000)\}P(L_1 = 1000) + \{1 - P(S_1 > 1500)\}P(L_1 = 1500) \\ &\quad + \{1 - P(S_1 > 2000)\}P(L_1 = 2000) \\ &= \left\{1 - \frac{5}{8}\right\}\frac{1}{8} + \left\{1 - \frac{5}{8}\right\}\frac{1}{2} + \{1 - 0\}\frac{3}{8} = \frac{39}{64} = 0.6094 \end{aligned}$$

The simple bounds on the probability of failure, given by Eq. (8.138), are

$$\max\left(\frac{39}{64}, \frac{39}{64}\right) \le P_f \le 1 - \left(1 - \frac{39}{64}\right)\left(1 - \frac{39}{64}\right)$$

that is,

$$\frac{39}{64} < P_f < \frac{3471}{4096}$$

that is,

$$0.6094 < P_f < 0.8474$$

while the exact value of P_f is 0.7559. If the right-hand side of the inequality (8.141) is used, the upper bound on P_f becomes

$$\sum_{k=1}^{2} P_{f_k} = \frac{39}{64} + \frac{39}{64} = 1.2188$$

3. To determine the Ditlevsen bounds, we approximate the value of $P_{f_1} \cap P_{f_2}$ as the product of P_{f_1} and P_{f_2}, which gives a value of $(39/64)^2 = 0.3714$. Thus, the lower bound on P_f can be determined from Eq. (8.149) as $0.6094 + \max[0.6094 - 0.3714, 0] = 0.8474$. Similarly, the upper bound on P_f can be computed from Eq. (8.151) as $0.6094 + 0.6094 - 0.3714 = 0.8474$. It can be seen that the two bounds coincide in this case.

■

8.11.4 Reliability Analysis of Fail-Safe (or Parallel) Systems

A simplified model of a parallel system is shown in Fig. 8.25. A redundant structure with ductile members can be modeled as a parallel system since the structure can only fail when all its members reach their limit states. A redundant system might have several failure modes. In such a case, each of the failure modes can be modeled as a parallel system and all the modes, in turn, can be modeled as a series system to find the reliability of the complete system.

Probability of Failure. The probability of failure of a parallel system consisting of n ductile elements (shown in Fig. 8.25) can be defined as

$$P_f = P(S < L) \tag{8.153}$$

where L is the load and S is the total strength of the system given by

$$S = S_1 + S_2 + \cdots + S_n \tag{8.154}$$

where S_i denotes the strength of member $i (i = 1, 2, \ldots, n)$. The probability of failure of the system can be evaluated using the probability density functions of the strengths of individual members, $f_{S_i}(s_i)$, as follows.

By defining a set of new random variables R_j as

$$R_j = S_1 + S_2 + \cdots + S_j; \quad j = 1, 2, \ldots, n \tag{8.155}$$

the probability distribution function of R_1 can be identified as

$$F_{R_1}(r) = F_{S_1}(r) = \int_{t=0}^{r} f_{S_1}(t)\, dt \tag{8.156}$$

The probability distribution function of R_2 is given by

$$F_{R_2}(r) = P(R_2 \le r) = P(S_1 + S_2 \le r) \tag{8.157}$$

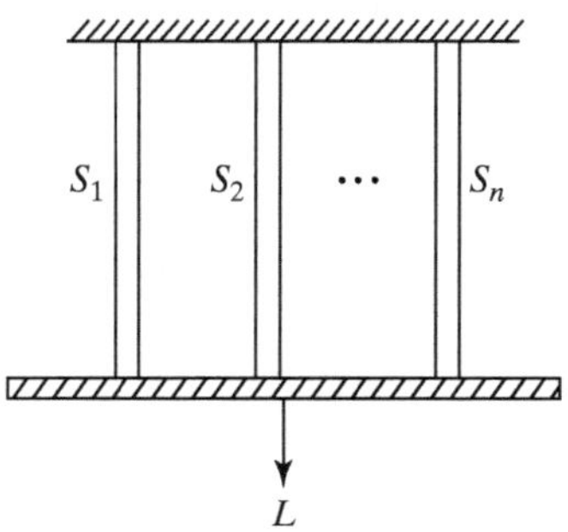

FIGURE 8.25 A parallel system.

Since $S_1 + S_2 \le r$ implies that $S_1 \le r - t$ when $S_2 = t$, Eq. (8.157) can be rewritten as

$$F_{R_2}(r) = \int_{R_2=t=0}^{\infty} F_{R_1}(r - t) f_{S_2}(t)\, dt \tag{8.158}$$

Similarly, the probability distribution function of R_3 is given by

$$F_{R_3}(r) = P(R_3 \le r) = \int_{R_3=t=0}^{\infty} F_{R_2}(r - t) f_{S_3}(t)\, dt \tag{8.159}$$

This procedure can be continued until the probability distribution function of $S = R_n$ is found as

$$F_S(r) = \int_{t=0}^{\infty} F_{R_{n-1}}(r - t) f_{S_n}(t)\, dt \tag{8.160}$$

Once $F_S(r)$ is known, the probability of failure of the system, Eq. (8.153), can be evaluated as

$$P_f = \int_{l=0}^{\infty} F_S(l) f_L(l)\, dl \tag{8.161}$$

and the reliability R_0 as

$$R_0 = 1 - P_f \tag{8.162}$$

Equation (8.161) shows that the probability of failure of the system can be computed by integrating the probability density functions of $S_1, S_2, \ldots,$ and S_n recursively. It can be seen that the recursive integration involves a large amount of computation. Hence, it is desirable to investigate alternate methods.

Bounds on the Probability of Failure. Simple bounds on the probability of failure of a parallel system can be derived by proceeding as in the case of a series system. Let a variable x_i be defined as follows:

$$x_i = \begin{cases} 1 & \text{if member } i \text{ has survived} \\ 0 & \text{if member } i \text{ has failed} \end{cases}; \quad i = 1, 2, \ldots, n \tag{8.163}$$

The probability of failure of the parallel system can be expressed as

$$\begin{aligned} P_f &= P[x_1 = 0 \cap x_2 = 0 \cap \cdots \cap x_n = 0] \\ &= P(x_1 = 0)\, \frac{P(x_1 = 0 \cap x_2 = 0)}{P(x_1 = 0)} \cdots \frac{P(x_1 = 0 \cap x_2 = 0 \cap \cdots \cap x_n = 0)}{P(x_1 = 0 \cap x_2 = 0 \cap \cdots \cap x_{n-1} = 0)} \end{aligned} \tag{8.164}$$

By assuming that

$$P(x_1 = 0 \cap x_2 = 0) \ge P(x_1 = 0) P(x_2 = 0) \tag{8.165}$$

or, in general,

$$P(x_1 = 0 \cap x_2 = 0 \cap \cdots \cap x_i = 0) \geq P(x_1 = 0 \cap x_2 = 0 \cap \cdots \cap x_{i-1} = 0) P(x_i = 0); \quad 2 \leq i \leq n \tag{8.166}$$

we can obtain from Eq. (8.164)

$$P_f \geq \prod_{i=1}^{n} P(x_i = 0) \tag{8.167}$$

The probability of failure of the member that has the lowest value will serve as a simple upper bound on P_f. Thus, the bounds on the probability of failure of the parallel system can be expressed as

$$\prod_{i=1}^{n} P(x_i = 0) \leq P_f \leq \min_{i=1,2,\ldots,n} P(x_i = 0)$$

or

$$\prod_{i=1}^{n} P_{f_i} \leq P_f \leq \min_{i=1,2,\ldots,n} P_{f_i} \tag{8.168}$$

Note that the exact probability of failure P_f will be equal to the lower bound value if there is no correlation between member failures ($\rho_{ij} = 0$ for all i and j with $i \neq j$) and will be equal to the upper bound if there is complete correlation between member failures ($\rho_{ij} = 1$ for all i and j).

8.12 Numerical Solutions Using MATLAB and Excel

EXAMPLE 8.14

Find the reliability of the cone clutch described in Example 8.1 using MATLAB.

Solution: Since the torque-transmitting capacity (T) and the required torque to be transmitted (T_r) follow normal distribution with parameters $T = N(396.33, 26.4225)$ lb-in and $T_r = N(400, 40)$ lb-in, a MATLAB program is developed to find the reliability of the cone clutch. The program and the results are shown below.

```
clear all; clc    ;
T_bar = 396.33    ;
T_div = 26.4225   ;
Tr_bar = 400.00   ;
Tr_div = 40       ;
x_bar = T_bar - Tr_bar
x_div = sqrt(T_div^2 + Tr_div^2)
z = -(x_bar/x_div)
R = 1 - normcdf(z,0,1)
```

```
x_bar =

   -3.6700

x_div =

   47.9390

z =

    0.0766

R =

    0.4695

>>
```

■

EXAMPLE 8.15

The bending stress induced in a gear tooth (L) and the yield strength of the gear material (S) are known to follow lognormal distribution with mean values $\overline{L} = 24{,}800$ psi and $\overline{S} = 32{,}000$ psi and standard deviations $\sigma_L = 2{,}000$ psi and $\sigma_S = 1{,}300$ psi. Find the reliability of the gear tooth using MATLAB.

Solution: A MATLAB program is written to find the mean and standard deviations of the logarithms of the variables L and S and to compute the reliability of the gear tooth. The MATLAB program along with the results obtained from the program are indicated below.

```
clear all;clc;

MS = 32000
ML = 24800
SS = 1300
SL = 2000

SSLOG = sqrt(log((SS^2)/(MS^2) + 1))

SLLOG = sqrt(log((SL^2)/(ML^2) + 1))

MSLOG = log(MS) - (SSLOG^2)/2

MLLOG = log(ML) - (SLLOG^2)/2

z = (MLLOG-MSLOG)/sqrt(SLLOG^2+SSLOG^2)

REL = 1 - normcdf(z,0,1)

MS =

    32000

ML =

    24800

SS =

    1300

SL =

    2000

SSLOG =

  0.040608253187978

SLLOG =

  0.080514499379490

MSLOG =

 10.372666666668374

MLLOG =

 10.115357639847909

z =

 -2.853426054341222

REL =

  0.997837470387882

>>
```

■

EXAMPLE 8.16

Generate the values of the probability of failure (P_f) and the reliability index (β) corresponding to the following values of reliability (R) using MATLAB:

$$R = 0.5, 0.6, 0.7, 0.8, 0.9, 0.95, 0.99, 0.999, 0.9999, 0.99999, 0.999999, 0.9999999$$

Solution: The MATLAB code used and the results are given below:

```
clear all; clc ;

R= [0.5 0.6 0.7 0.8 0.9 0.95 0.99 0.999 0.9999 0.99999
0.999999 0.9999999]

for i =1:length(R)
    pf(i) = 1-R(i)             ;
    beta(i) = -norminv(R(i))   ;

end

pf
beta

R =

    0.5000    0.6000    0.7000    0.8000    0.9000    0.9500    0.9900    0.9990    0.9999 ↙
1.0000    1.0000    1.0000

pf =

    0.5000    0.4000    0.3000    0.2000    0.1000    0.0500    0.0100    0.0010    0.0001 ↙
0.0000    0.0000    0.0000

beta =

            0   -0.2533   -0.5244   -0.8416   -1.2816   -1.6449   -2.3263   -3.0902   -3.7190 ↙
-4.2649   -4.7534   -5.1993

>>
```

■

EXAMPLE 8.17

Find the exact probability of failure and the lower and upper bounds on the probability of failure of the truss described in Example 8.13 and Fig. 8.24 using MATLAB.

Solution: The exact probability of failure can be found using the conditional probabilities and the total probability theorem. The bounds on the probability of failure can be computed using Eqs. (8.138) and (8.141). The MATLAB program and the results given by the program are given below.

```
clear all ; clc ;

PL = [1/8  1/2  3/8 ]  ;
PS = [5/8  5/8]      ;

PS1000 = 5/8      ;
PS1500 = 5/8      ;
PS2000 = 0        ;
```

```
PFC(1) = 1 - PS1000*PS1000
PFC(2) = 1 - PS1500*PS1500
PFC(3) = 1 - PS2000*PS2000

ExactPF = PFC(1)*PL(1) + PFC(2)*PL(2) + PFC(3)*PL(3)

PFM(1) = (1-PS1000)*PL(1) + (1-PS1500)*PL(2) + (1-PS2000)*PL(3)

PFM(2) = PFM(1)
LBPF = max(PFM)
UBPF1 = 1 - (1 -PFM(1))*(1-PFM(2))
UBPF2 = PFM(1) + PFM(2)

PFC =

   0.6094

PFC =

   0.6094    0.6094

PFC =

   0.6094    0.6094    1.0000

ExactPF =

   0.7559

PFM =

   0.6094

PFM =

   0.6094    0.6094

LBPF =

   0.6094

UBPF1 =

   0.8474

UBPF2 =

   1.2188
```

■

EXAMPLE 8.18

Find the reliability of the gear tooth described in Example 8.1 using Excel.

Solution: The computation of the reliability of the gear tooth using Excel is shown in the spreadsheet that follows.

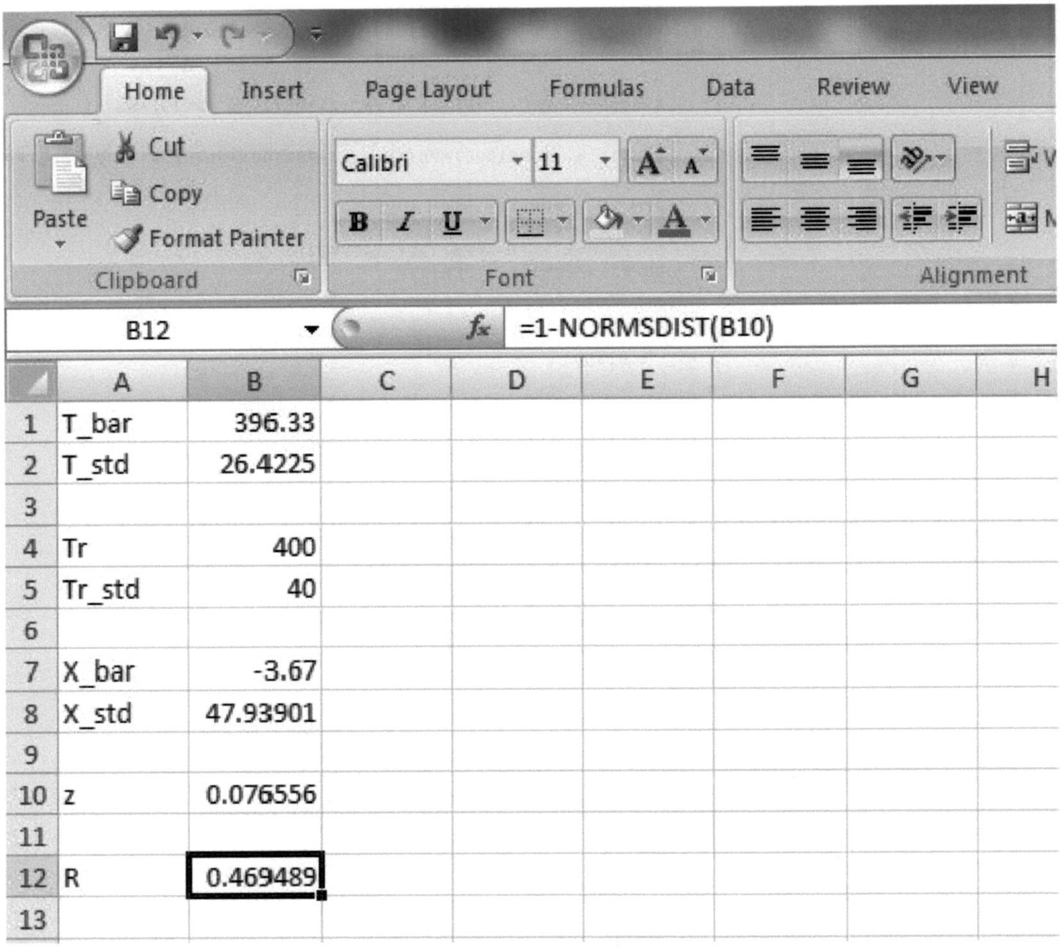

■

EXAMPLE 8.19

Generate the value of the probability of failure (P_f) and the reliability index (β) corresponding to the reliability $R = 0.9999$ using Excel.

Solution: The data and the command used in the Excel spreadsheet to find the results follows.

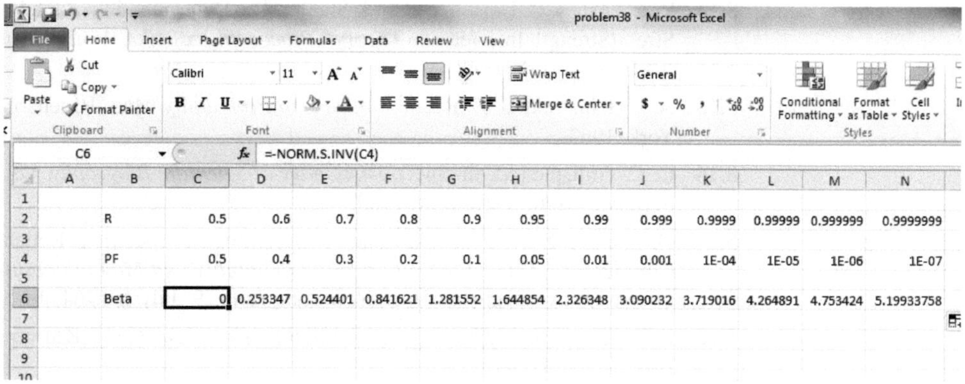

problem38 - Microsoft Excel

C6 f_x =-NORM.S.INV(C4)

	A	B	C	D	E	F	G	H	I	J	K	L	M	N
1														
2		R	0.5	0.6	0.7	0.8	0.9	0.95	0.99	0.999	0.9999	0.99999	0.999999	0.9999999
3														
4		PF	0.5	0.4	0.3	0.2	0.1	0.05	0.01	0.001	1E-04	1E-05	1E-06	1E-07
5														
6		Beta	0	0.253347	0.524401	0.841621	1.281552	1.644854	2.326348	3.090232	3.719016	4.264891	4.753424	5.19933758

■

REFERENCES AND BIBLIOGRAPHY

8.1 P. T. Christensen and M. J. Baker, *Structural Reliability Theory and Its Applications,* Springer-Verlag, Berlin, 1986.

8.2 K. C. Kapur and L. R. Lamberson, *Reliability in Engineering Design,* John Wiley, New York, 1977.

8.3 C. Mischke, "A Method of Relating Factor of Safety and Reliability," *Journal of Engineering for Industry,* Vol. 92, Aug. 1970, pp. 537–542.

8.4 D. Kececioglu and E. B. Haugen, "A Unified Look at Design Safety Factors, Safety Margins and Measures of Reliability," *Annals of Assurance Sciences—Seventh Reliability and Maintainability Conference* 1968, pp. 520–530.

8.5 Singiresu S. Rao, *Applied Numerical Methods for Engineers and Scientists,* Prentice Hall, Upper Saddle River, NJ, 2002.

8.6 G. C. Hart, *Uncertainty Analysis, Loads, and Safety in Structural Engineering,* Prentice-Hall, Englewood Cliffs, NJ, 1982.

8.7 R. C. Juvinall, *Fundamentals of Machine Component Design,* John Wiley, New York, 1983.

8.8 J. E. Shigley and C. R. Mischke, *Mechanical Engineering Design,* 5th ed., McGraw-Hill, New York, 1989.

8.9 E. Sandgren, G. Gim, and K. M. Ragsdell, "Optimal Design of a Class of Welded Structures Based on Design for Latitude," *ASME Journal of Mechanisms, Transmissions, and Automation in Design,* Vol. 107, 1985, pp. 482–487.

8.10 J. H. Faupel and F. E. Fisher, *Engineering Design,* 2nd ed., John Wiley, New York, 1981.

8.11 S. S. Rao, "Application of Complementary Geometric Programming in Mechanical Design," *International Journal of Mechanical Engineering Education,* Vol. 13, 1985, pp. 19–29.

8.12 E. B. Haugen, *Probabilistic Approaches to Design,* John Wiley, New York, 1968.

8.13 E. B. Haugen, *Probabilistic Mechanical Design,* John Wiley, New York, 1980.

8.14 R. L. Disney, C. Lipson, and N. J. Sheth, "The Determination of the Probability of Failure by Stress/Strength Interference Theory," *Proceedings of 1968 Annual Symposium on Reliability,* IEEE, New York, 1968.

8.15 A. C. King and C. B. Read, *Pathways to Probability,* Holt, Rinehart & Winston, Inc., New York, 1963.

8.16 C. C. Gillispie (ed.-in-chief), *Dictionary of Scientific Biography,* Vol. 15, Suppl. I, Charles Scribner's Sons, New York, 1980.

8.17 D. Kececioglu and D. Cormier, "Designing a Specified Reliability Directly into a Component," *Third Annual Aerospace Reliability and Maintainability Conference,* Society of Automotive Engineers, Inc., New York, June 29–July 1, 1964, pp. 546–565.

8.18 I. Elishakoff, *Safety Factors and Reliability: Friends or Foes?*, Kluwer Academic, Publishers, Boston MA, 2004.

8.19 G. I. Schueller (ed.), "A State-of-the-Art Report on Computational Stochastic Mechanics," *Journal of Probabilistic Engineering Mechanics*, Vol. 12, No. 4, pp. 197–313, 1997.

8.20 A. M. Freudenthal, "Safety and Probability of Structural Failure," *ASCE Transactions*, Vol. 121, pp. 1337–1397, 1956.

8.21 H. O. Madsen, S. Krenk, and N. C. Lind, *Methods of Structural Safety*, Prentice-Hall, Englewood Cliffs, NJ, 1986.

8.22 A. H-S. Ang and C. A. Cornell, "Reliability Bases of Structural Safety and Design, *Journal of Structural Engineering*, ASCE, Vol. 100, No. ST9, pp. 1755–1769, 1974.

8.23 B. M. Ayyub and H. McCuen, *Probability, Statistics, & Reliability for Engineers*, CRC Press, Boca Raton FL, 1997.

8.24 X. Chen and N. C. Lind, "Fast Probability Integration by Three-Parameter Normal Tail Approximation," *Structural Safety*, Vol. 1, pp. 269–276, 1983.

8.25 O. Ditlevsen, "Principles of Normal Tail Approximation," *Journal of Engineering Mechanics*, ASCE, Vol. 107, No. EM6, pp. 1191–1208, 1981.

8.26 R. E. Melchers, *Structural Reliability: Analysis and Prediction*, Ellis Horwood Ltd. and Halsted Press, Chichester, England, 1987.

8.27 A. Haldar and S. Mahadevan, *Probability, Reliability, and Statistical Methods in Engineering Design*, John Wiley, New York, 2000.

8.28 A. M. Hasofer and N. C. Lind, "Exact and Invariant Second Moment Code Format," *Journal of the Engineering Mechanics Division*, ASCE, Vol. 100, No. EM1, pp. 111–121, 1974.

8.29 R. Rackwitz, "Practical Probabilistic Approach to Design," *Bulletin No. 112*, Comite European du Beton, Paris, France, 1976.

8.30 H. U. Koyluoglu and S. R. K. Nielsen, "New Approximations for SORM Integrals," *Structural Safety*, Vol. 13, No. 4, pp. 235–246, 1994.

8.31 S.-K. Choi, R. V. Grandhi, and R. A. Canfield, *Reliability-Based Structural Design*, Springer-Verlag, London, 2007.

8.32 M. Shinozuka, "Basic Analysis of Structural Safety," *Journal of the Structural Engineering*, ASCE, Vol. 109, No. 3, pp. 721–749, 1983.

8.33 Singiresu S. Rao, Engineering Optimization: Theory and Practice, 4th ed., John Wiley, Hoboken, NJ, 2009.

8.34 C. A. Cornell, "Bounds on the Reliability of Structural Systems," *Journal of the Structural Division, ASCE*, Vol. 93, 1967, pp. 171–200.

8.35 O. Ditlevsen, "Narrow Reliability Bounds for Structural Systems," *Journal of Structural Mechanics*, Vol. 7, 1979, pp. 453–472.

8.36 F. Moses and J. D. Stevenson, "Reliability-Based Structural Design," *Journal of the Structural Division, ASCE*, Vol. 96, 1970, pp. 221–244.

8.37 R. E. Melchers, "Reliability of Parallel Structural Systems," *Journal of the Structural Division, ASCE*, Vol. 109, 1983, pp. 2651–2665.

8.38 M. Grigoriu and C. Turkstra, "Safety of Structural Systems with Correlated Resistances," *Applied Mathematical Modelling*, Vol. 3, 1979, pp. 130–136.

8.39 R. Nakib and D. M. Frangopal, "RBSA and RBSA-OPT: Two Computer Programs for Structural System Reliability Analysis and Optimization," *Computers & Structures*, Vol. 36, 1990, pp. 13–27.

8.40 F. B. Song, "A Numerical Integration Method for Computing Structural System Reliability," *Computers & Structures*, Vol. 36, 1990, pp. 65–70.

8.41 K. M. Aludaat and M. T. Alodat, "A Note on Approximating the Normal Distribution Function," *Applied Mathematical Sciences*, Vol. 2, No. 9, pp. 425–429, 2008.

8.42 H. Shore, "Simple Approximations for the Inverse Cumulative Function, the Density Function and Loss Integral of the Normal Distribution," *Journal of the Royal Statistical Society*, Vol. 31, No. 2, pp. 108–114, 1982.

REVIEW QUESTIONS

8.1 Give brief answers to the following:

1. What is the major difference between the failure of electronic components and mechanical components?
2. How is the strength-based reliability of a component defined?
3. Derive an expression for the reliability of a component when $f_L(l)$ and $F_S(s)$ are known in discrete form.
4. Find an expression for the reliability of a component when $f_S(s)$ and $F_L(l)$ are known in continuous form.
5. What is a central factor of safety?
6. Define the characteristic factor of safety.
7. How is design factor of safety defined?
8. What type of anomaly is observed in relating the factor of safety to a specified value of reliability?
9. Why is reliability defined as $P(S - L > 0)$ for normally distributed S and L and as $P(S/L > 1)$ for lognormally distributed S and L?
10. What is the justification for assuming S and L as statistically independent?
11. Define the safety margin of a system.
12. Express the standard deviation of $g = S - L$ when S and L are correlated.
13. What is the significance of $\frac{S}{L} = 1$ in the $S-L$ space?
14. What is the reliability of a system with the reliability index $\beta = 0$?
15. Give an example of a system involving more than two random variables in the computation of its reliability.

16. Expand the FORM method.
17. Expand the SORM method.
18. What is Hasofer-Lind safety index?
19. What is the fundamental problem in structural reliability?
20. What is a weakest-link system?
21. Define a fail-safe system.
22. Give two examples of each of the statically determinate and indeterminate structures.

8.2 Indicate whether each of the following statements is true or false:

1. Failure of an electronic/electrical system is not related to the life of the system.
2. When both S and L follow normal distribution, it is convenient to define reliability as the probability of realizing $S - L \geq 0$.
3. The ratio $\frac{S_{0.005}}{L_{0.95}}$ is called the design factor of safety.
4. When S and L follow normal distribution, any specified value of reliability yields a unique value of factor of safety.
5. It is possible that the normal distribution assumption for S and L can lead to infinite factor of safety in some special cases.
6. The safety margin is also known as the limit state function.
7. The units of strength and load can be different in reliability analysis.
8. The strength-based reliability analysis involves only two basic random variables in all cases.
9. The limit state equation, $g(\vec{X}) = 0$, defines the boundary between safe and failure regions on the $\vec{X}$-space.
10. The linearization of the limit state function gives a reasonably good estimate of reliability.
11. The linearization of the limit state function about different points leads to different values of reliability.
12. The FOSM and MVFOSM methods are different.
13. The error involved with MVFOSM method is expected to be more with nonlinear limit state functions.
14. The failure region in the S–L plane is defined by $s - l \leq 0$.
15. There will be a unique way in which a statically indeterminate truss can fail.
16. Ditlevesen bounds are less tight compared to the other type of bounds derived.

8.3 Fill in the blank with the proper word(s):

1. From reliability (R), the failure probability (P_f) can be found as $P_f =$ ______.
2. The ratio $\frac{\mu_S}{\mu_L}$ is called the ______ factor of safety.
3. The ratio $\frac{S_{0.05}}{L_{0.95}}$ is called the ______ factor of safety.
4. With normal variates, the probability of failure of a system can be expressed in terms of the reliability index β as ______.
5. Geometrically, the reliability index β denotes the ______ ______ from origin to the linear limit state function.

6. The nearer the MPP is to the origin, the ______ the probability of failure of the system.
7. The limit state function, approximated at the mean values of the random variables, uses the mean values and ______ of the random variables in the MVFOSM method.
8. The reliability index represents the distance of the ______ of the limit state function g from the origin $g = 0$.
9. The quantity $\frac{S}{L}$ can also be used as a measure of ______ ______ instead of S–L.
10. Systems with multiple members can be classified into ______ categories for the purpose of reliability computation.
11. A weakest-link system is also called a ______ system.
12. A weakest-link system fails whenever any ______ link fails.
13. A fail-safe system ______ ______ fail whenever any one link fails.
14. A fundamental problem of a weakest-link system consists of ______ member(s) and ______ load(s).
15. The reliability index, β, is also called the Hasofer and ______ reliability index.
16. Elastic-plastic structures fail due to the formation of plastic-hinge ______ (s).

8.4 Select the most appropriate answer from the multiple choices given:

1. The primary random variable for describing the failure of an electronic/electrical component is:
   a. Time **b.** Strength **c.** Environment
2. The primary random variable for describing the failure of a mechanical/structural component is:
   a. Time **b.** Strength **c.** Environment
3. When S and L follow normal distribution, the reliability can be defined as $R = \frac{1}{2\pi}\int_{z_1}^{\infty} e^{-\frac{1}{2}z^2}\,dz$ where z_1 is given by:
   a. $-\frac{\mu_S}{\mu_L}$ **b.** $-(\mu_S - \mu_L)$ **c.** $-\frac{(\mu_S - \mu_L)}{\sqrt{\sigma_S^2 + \sigma_L^2}}$
4. When S and L follow lognormal distribution, it is convenient to define reliability as
   a. $P\left(\frac{L}{S} < 1\right)$ **b.** $P\left(\frac{S}{L} > 1\right)$ **c.** $P(S - L \geq 0)$
5. When both S and L follow exponential distribution, the reliability can be found as:
   a. $\frac{\mu_S}{\mu_L}$ **b.** $\frac{\mu_S}{(\mu_S + \mu_L)}$ **c.** $\frac{(\mu_S + \mu_L)}{\mu_S}$
6. The MVFOSM method has been used in developing standards of structural design in:
   a. U.K. **b.** Canada **c.** France
7. In the Hasofer-Lind method, the limit state function is expanded about:
   a. Most probable point
   b. Mean values of random variables
   c. Any arbitrary point

8. The determination of the MPP requires:
a. Linearization about MPP
b. Solution of an optimization problem
c. Use of normal probability tables

9. The reliability of a system is given by:
a. $S - L < 0$ **b.** $S - L = 0$ **c.** $S - L > 0$

10. The reliability index is given by:

a. $\dfrac{\mu_g}{\sigma_g}$ **b.** $1 - \dfrac{\mu_g}{\sigma_g}$ **c.** $\dfrac{\sigma_g}{\mu_g}$

11. The reliability of a system can be expressed as:
a. $\Phi(\beta)$ **b.** $\Phi(-\beta)$ **c.** $\Phi(1 - \beta)$

12. A statically determinate truss can be modeled as a
a. Weakest-link system **b.** Fail-safe system **c.** Complex system

13. A statically indeterminate truss can be modeled as a
a. Weakest-link system **b.** Fail-safe system **c.** Complex system

14. The quantity *S–L* is called
a. Safety index **b.** Safety margin **c.** Safety parameter

15. Reliability index, β, is defined as

a. $-\dfrac{\mu_S - \mu_L}{\sqrt{\sigma_S^2 + \sigma_L^2}}$ **b.** $\mu_S - \mu_L$ **c.** $\dfrac{\mu_S - \mu_L}{\sqrt{\sigma_S^2 + \sigma_L^2}}$

8.5 Match the following:

1. $P\left(l - \dfrac{dl}{2} \leq L \leq l + \dfrac{dl}{2}\right)$	**a.** $\displaystyle\int_l^{\infty} f_S(s)\,ds$
2. $P(L < S)$	**b.** $\displaystyle\int_{-\infty}^{s} f_L(l)\,dl$
3. $P(L < s)$	**c.** $f_L(l)\,dl$
4. $P(S > l)$	**d.** $f_S(s)\,ds$
5. $P\left(s - \dfrac{ds}{2} \leq S \leq s + \dfrac{ds}{2}\right)$	**e.** $\displaystyle\int_{-\infty}^{+\infty} f_S(s)F_L(s)\,ds$

8.6 Match the following reliabilities and the reliability indexes:

1. 0.5	**a.** 3.090
2. 0.7	**b.** 2.326
3. 0.9	**c.** 0.524
4. 0.99	**d.** 1.282
5. 0.999	**e.** 0

8.7 Match the following expressions for reliability and probability of failure:

1. $\int_0^\infty F_S(l)\, f_L(l)\, dl$ — **a.** Probability of failure based on discrete $S < L$

2. $\iint\limits_{s-l>0} f_{S,L}(s,l)\,ds\,dl$ — **b.** Reliability based on discrete $L < S$

3. $\sum_j F_L(s_j) f_S(s_j)$ — **c.** Probability of failure based on continuous $S < L$

4. $\sum_i F_S(l_i) f_L(l_i)$ — **d.** Probability of failure based on continuous $L < S$

5. $1 - \int_0^\infty F_L(s) f_S(s)\,ds$ — **e.** Reliability based on continuous $S > L$

PROBLEMS

8.1 The maximum stress induced near the hole in a rectangular plate of width w and thickness t subjected to an axial load P (see Fig. 8.26) is given by

$$s_{\max} = k_t s_o$$

where $s_o = P/\{(w-d)t\}$ is the nominal stress induced in the cross section A-A and k_t is the stress concentration factor. The mean and standard deviations are given by (for $d/w = 0.2$) $\bar{k}_t = 2.5$ and $\sigma_{k_t} = 0.1$. The yield stress of the plate material, s_y, has mean and standard deviations of $\bar{s}_y = 30{,}000$ lb/in^2 and $\sigma_{s_y} = 1000$ lb/in^2. If $P = 10{,}000$ lb, $w = 10$ in, $d = 2$ in, $t = 0.1$ in, and both k_t and s_y follow normal distribution, find the reliability of the plate.

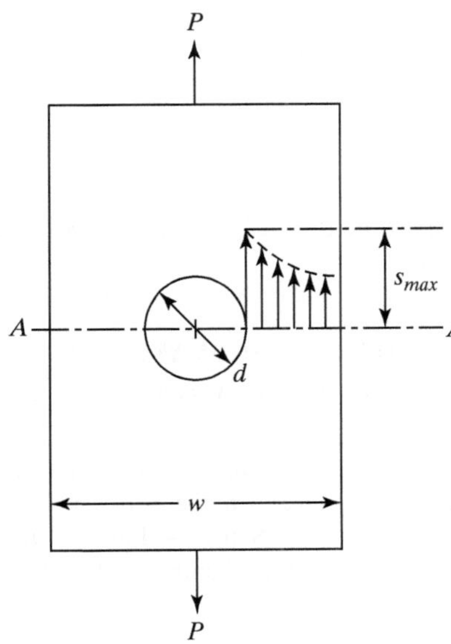

FIGURE 8.26 A rectangular plate with a hole under axial load.

8.2 Find the reliability of the plate described in Problem 8.1 if k_t and s_y are uniformly distributed over the ranges 2.2–2.8 and 27,000–33,000 lb/in^2, respectively.

8.3 The mean and standard deviation of shear strength of a bolted joint are 45 $\text{k(lb/in}^2)$ and 4.2 $\text{k(lb/in}^2)$, respectively. The joint is loaded such that the stress induced has a mean value of 38 $\text{k(lb/in}^2)$ and a standard deviation of 6.4 $\text{k(lb/in}^2)$. Assuming that the shear strength and the induced stress are independent and normally distributed, find the probability of failure of the bolted joint.

8.4 If the internal pressure and the yield strength of the vessel described in Example 8.5 are lognormally distributed, determine the reliability of the pressure vessel. Assume the standard deviation of internal pressure as 50 lb/in^2 and of yield strength as 1,000 lb/in^2.

8.5 Two metal strips are clamped at the ends and are welded as shown in Fig. 8.27. The thermal stress (compressive) that developed in these metal strips due to a uniform rise in temperature during welding is given by

$$s = \alpha E \, \Delta T$$

where α = coefficient of thermal expansion, E = Young's modulus, and ΔT = temperature rise. For steel, $\alpha = 12 \times 10^{-6}/°C$ and $E = 207$ GPa. If the temperature rise during welding is normally distributed with $N(\Delta\overline{T}, \sigma_{\Delta T}) = N(40, 5)°C$, and the permissible compressive stress in steel is also normally distributed with $N(\overline{s}_y, \sigma_{sy}) = N(199.9479, 6.8948)$ MPa, determine the reliability of the steel strips.

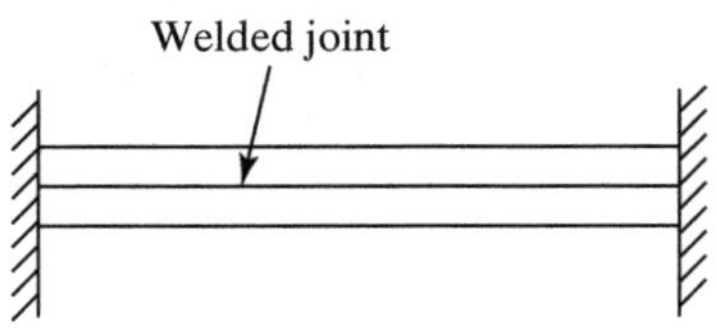

FIGURE 8.27 Welded metal strips.

8.6 Solve Problem 8.5 if the temperature rise, ΔT, and the permissible compressive stress, s_y, follow exponential distributions

$$f(\Delta T) = 0.025 \exp(-0.025 \, \Delta T); \quad \Delta T \geq 0$$
$$f(s_y) = 0.00035 \exp(-0.00035 \, s_y); \quad s_y \geq 0$$

8.7 The knuckle joint shown in Fig. 8.28(a) is subjected to a tensile load F. Although the actual load distribution on the pin AB is unknown, the loads on AB can be approximated as shown in Fig. 8.28(b). For $x = 8$ in, $y = 8$ in, $d = 4$ in, $F = N(60{,}000, 6000)$ lb, and $s_y = N(30{,}000, 2000)$ lb/in^2, determine the reliability of the knuckle joint.

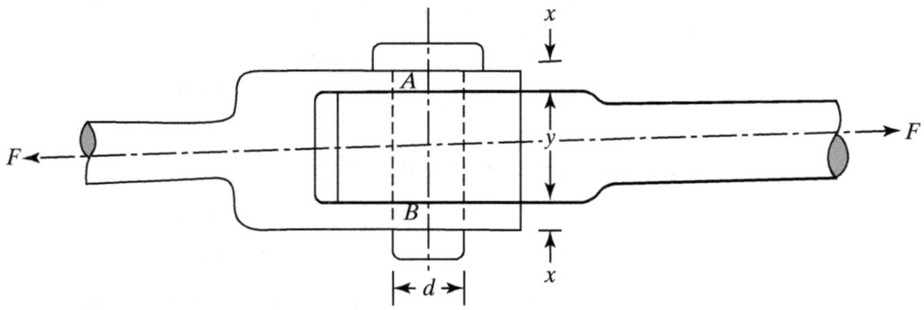

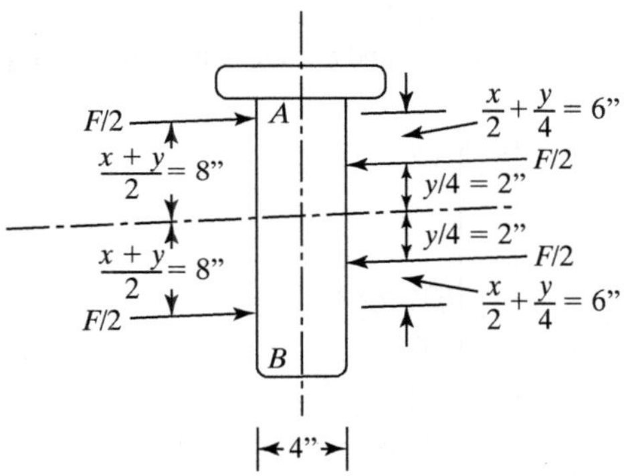

FIGURE 8.28 A knuckle joint.

8.8 If the tensile load and the yield stress are uniformly distributed as $F = U(40{,}000, 80{,}000)$ lb and $s_y = U(20{,}000, 40{,}000)$ lb/in^2, find the reliability of the knuckle joint described in Problem 8.7.

8.9 If both the shear strenth and the stress induced in the bolted joint follow lognormal distribution in Problem 8.3 (with the same means and standard deviations), find the reliability of the bolted joint.

8.10 A bolted joint has a mean fatigue life of 100,000 cycles. If it is subjected to an average of 80,000 load cycles, find the reliability of the joint if both the fatigue life and the number of applied-load cycles follow exponential distribution.

8.11 If the tangential load on the gear tooth is uniformly distributed in the range 4000–6000 lb, and the yield strength of the material is uniformly distributed in the range 28,000–32,000 psi, determine the reliability of the gear tooth described in Example 8.4.

8.12 When a shaft rotates in a guide bearing as shown in Fig. 8.29, the shear stress developed in the lubricant is given by [8.8]

$$\tau = \frac{\pi d \, \mu n}{\varepsilon}$$

and the pressure exerted on the bearing by

$$p = \frac{\pi^2 d \, \mu n}{\varepsilon f}$$

where d = diameter of the shaft, μ = viscosity of the lubricant, n = rotational speed of the shaft, f = coeffcient of friction, and ε = radial clearance. The data are $d = 2$ in, $n = 50$ rad/s, $\varepsilon = 0.002$ in, $f = 0.01$, and the viscosity of the lubricant is a random variable given by $N(4, 0.4)$ reyn. If it is desirable to limit the pressure on the bearing to 200 ± 20 lb/in^2, find the reliability of the bearing.

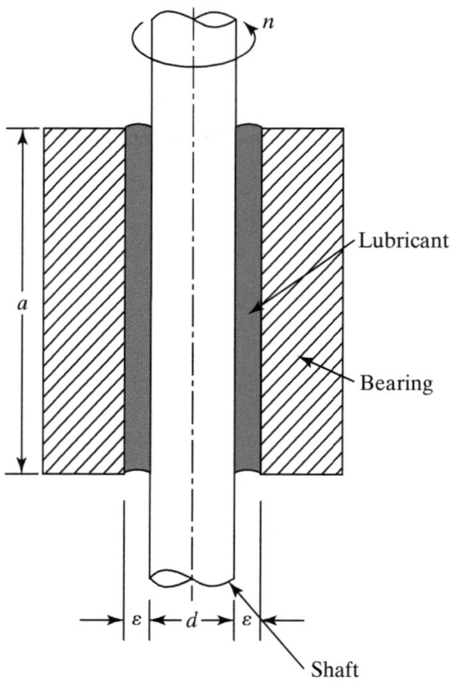

FIGURE 8.29 A shaft in a bearing.

8.13 The spring rate of a helical torsional spring (Fig. 8.30) is given by [8.7]

$$k = \frac{T}{\theta} = \frac{d^4 E}{64 D n}$$

where T = torsional moment applied, θ = angular deflection of the spring, d = wire diameter, D = mean spring diameter, n = number of turns (coils), and E = Young's

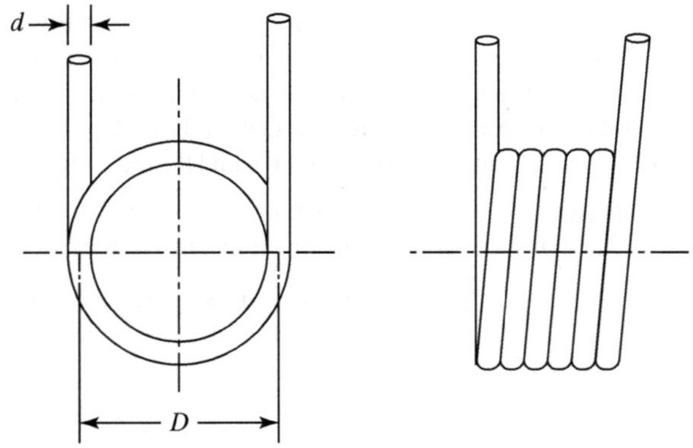

FIGURE 8.30 Torsional spring.

modulus of the material. A torsional spring used in a door hinge is made up of music wire and has $D = 1$ in, and $d = 0.1$ in, and $n = 5.5$. The Young's modulus of music wire is normally distributed as $N(\overline{E}, \sigma_E) = N(30 \times 10^6, 3 \times 10^6)$ lb/in^2. If it is required to achieve a spring rate of 50 $\pm$ 5 lb-in per turn, determine the reliability of the helical spring.

8.14 If the Young's modulus is uniformly distributed in the range 25×10^6 lb/in^2 -35×10^6 lb/in^2, and the spring rate required is also uniformly distributed in the range 40 lb · in/turn–60 lb · in/turn, determine the reliability of the torsional spring discussed in Problem 8.13.

8.15 The shear stress induced in a helical spring τ under compression (see Fig. 8.31) can be expressed as [8.8]

$$\tau = \left(\frac{2C + 1}{2C}\right)\frac{8FD}{\pi d^3}$$

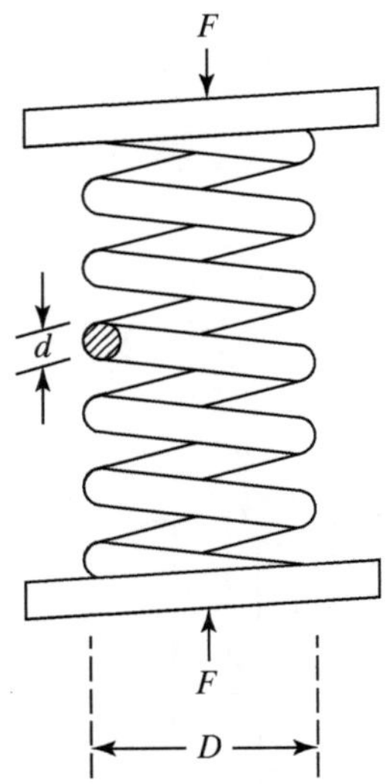

FIGURE 8.31 Compression spring.

where C = spring index = D/d, d = wire diameter, D = mean coil diameter, and F = compressive load. The spring used in a machine suspension (isolation) system has $d = 0.5$ in, $D = 10$ in, and the axial force induced in the spring is uniformly distributed in the range 400–600 lb. If the permissible yield strength in shear of the spring material τ_o is also uniformly distributed in the range 90–110 k(lb/in^2), find the reliability of the compression spring.

8.16 If the probability density functions of the compressive force on the spring and the permissible yield strength in shear of the spring material are given by

$$f(F) = 0.002 \exp(-0.002\, F); \quad F \geq 0$$

and

$$f(\tau_o) = 10^{-5} \exp(-10^{-5}\, \tau_o); \quad \tau_o \geq 0$$

find the reliability of the compression spring described in Problem 8.15.

8.17 A beam, having length L and cross-sectional dimensions t and b, is welded to a fixed support as shown in Fig. 8.32. The weld length is l on both top and bottom surfaces and the beam is required to support a load P. The weld is in the form of a triangle of depth h. The maximum shear stress developed in the weld, τ, is given by [8.9]

$$\tau = \sqrt{(\tau')^2 + 2\tau'\tau'' \cos\theta + (\tau'')^2}$$

where

$$\tau' = \frac{P}{\sqrt{2}\, h\, l}$$

$$\tau'' = \frac{MR}{J} = \left[P\left(L + \frac{l}{2}\right)\right] \frac{\sqrt{\left[\frac{l^2}{4} + \left(\frac{h+t}{2}\right)^2\right]}}{2\left(\frac{hl}{\sqrt{2}}\left[\frac{l^2}{12} + \left(\frac{h+t}{2}\right)^2\right]\right)}$$

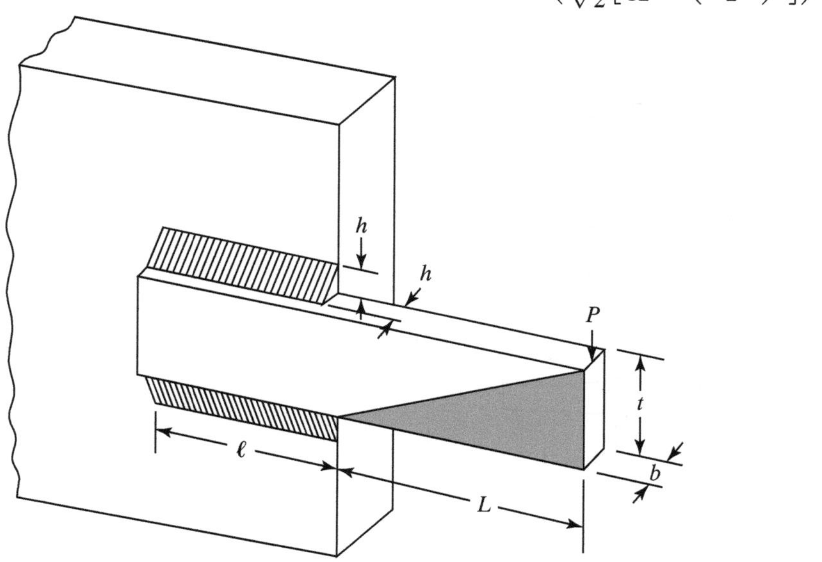

FIGURE 8.32 Welded beam.

and

$$\cos\theta = \frac{l}{2\sqrt{\left[\frac{l^2}{4} + \left(\frac{h+t}{2}\right)^2\right]}}$$

The dimensions of the weld and the beam are $L = 25$ in, $h = 2$ in, $l = 10$ in, $t = 10$ in, $b = 2$ in, and the load is normally distributed with $P = N(500, 50)$ lb. If the design shear stress, τ_o, in the weld is given by $\tau_o = N(\overline{\tau}_o, \sigma_{\tau_o}) = N(50, 10)$ lb/in^2, find the reliability of the welded joint.

8.18 Solve Problem 8.17 by assuming that the load, P, and the design shear stress in the weld, τ_o, are exponentially distributed as

$$f(P) = 0.002 \exp(-0.002\,P); \quad P \geq 0$$

and

$$f(\tau_o) = 0.02 \exp(-0.02\,\tau_o); \quad \tau_o \geq 0$$

8.19 Three plates are jointed by a fillet weld as shown in Fig. 8.33. The maximum principal stress, s_1, and the maximum shear stress, τ, induced in the weld are given by [8.8]

$$s_1 = 1.61803\,\frac{F}{hl}$$

$$\tau = 1.11803\,\frac{F}{hl}$$

In a specific application, $h = 0.1$ in, $l = 10$ in, and $F = N(10{,}000, 2500)$ lb. If the tensile and shear strengths of the weld material are normally distributed with $s_o = N(\overline{s}_o, \sigma_{s_o}) = N(18{,}000, 2500)$ lb/in^2 and $\tau_o = N(\overline{\tau}_o, \sigma_{\tau_o}) = N(12{,}000, 2000)$ lb/in^2, find the reliability of the welded joint.

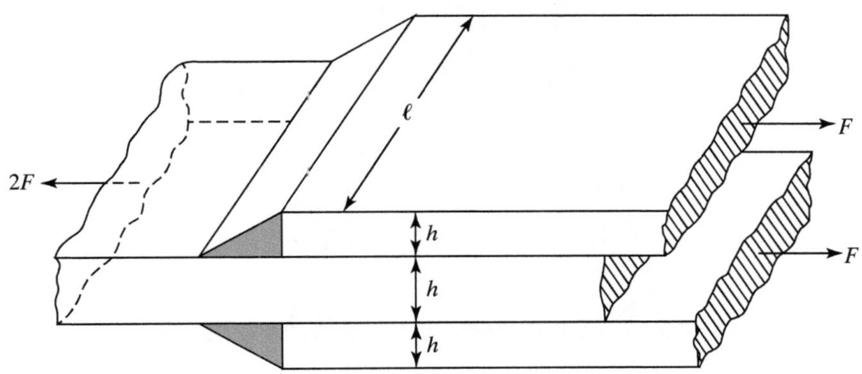

FIGURE 8.33 A fillet weld.

8.20 Solve Problem 8.19 by assuming that the load applied, tensile strength, and shear strength of the weld material follow lognormal distribution with the same values of mean and standard deviation given in Problem 8.19.

8.21 Solve Problem 8.19 by assuming that the load applied, F, the tensile strength of the weld material, s_o, and the shear strength of the weld material, τ_o, follow exponential distributions as

$$f(F) = 0.0001 \exp(-0.0001\, F); \quad F \geq 0$$
$$f(s_o) = 0.000056 \exp(-0.000056\, s_o); \quad s_o \geq 0$$
$$f(\tau_o) = 0.000083 \exp(-0.000083\, \tau_o); \quad \tau_o \geq 0$$

8.22 The torque required, T, to lower a load by a screw jack (see Fig. 8.34) is given by [8.7]

$$T = \frac{Fd_m}{2}\left(\frac{\pi\mu d_m - l}{\pi d_m + \mu l}\right)$$

where F = load, d_m = mean diameter, l = lead (same as the pitch for a single threaded screw jack), and μ = coefficient of friction. It is desirable to have a positive value of T (self-locking condition) to avoid the load coming down by itself without any external effort. The condition for self-locking is

$$T > 0 \quad \text{or} \quad \pi\mu d_m > l$$

The manufacturing tolerances make d_m and l as random variables with $d_m = N(30, 0.05)$ mm and $l = N(8, 0.01)$ mm. Find the value of μ necessary in order to achieve the self-locking condition with a reliability of 0.999.

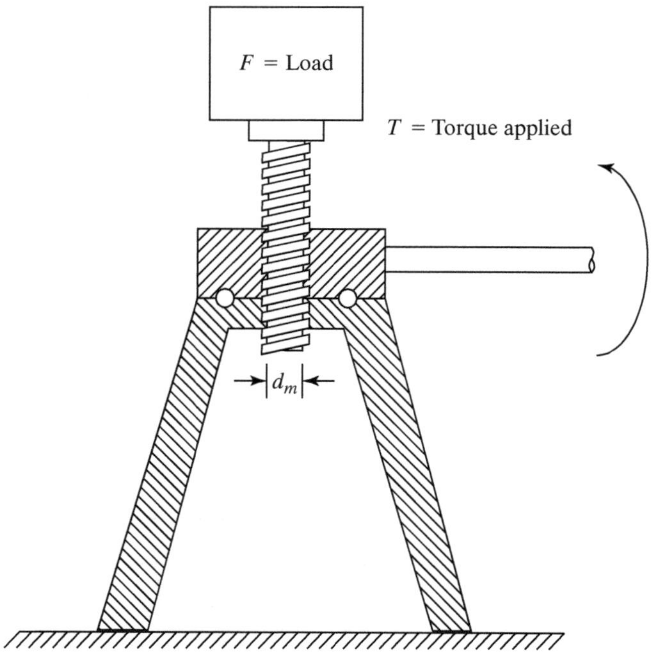

FIGURE 8.34 A screw jack.

8.23 The split piston ring, shown in Fig. 8.35(*a*), is mounted on a piston. The tangential deflection of the split ring under a force F is given by [8.10]

$$\delta = \frac{3\pi F R^3}{EI} + \frac{\pi F R}{EA} + \frac{6\pi F R}{5\,GA}$$

where the three terms on the right-hand side represent the contributions from bending moment, axial load, and transverse shear. A particular cast-iron piston ring has $R = 2$ in, $b = 0.2$ in, $h = 0.3$ in, $E = 18 \times 10^6\ \text{lb/in}^2$, and $G = 7 \times 10^6\ \text{lb/in}^2$. The tangential force applied is a normally distributed random variable with a mean value of 50 lb and a standard deviation of 10 lb. Find the probability of the two ends X and Y overlapping.

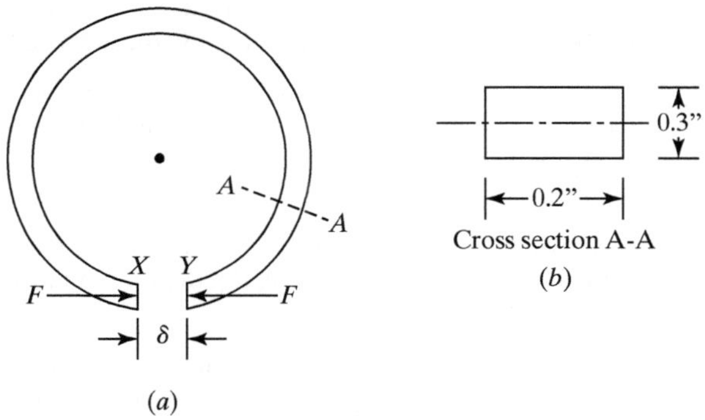

FIGURE 8.35 A split piston ring.

8.24 A connecting rod of length $l = 20$ in and a diameter of 0.5 in is subjected to an axial compressive force as shown in Fig. 8.36(*a*). Assuming the ends as pin joints, the Euler buckling load, P_{cri} [8.10], is given by

$$P_{cri} = \frac{\pi^2 EI}{l^2}$$

where E = Young's modulus and I = moment of inertia of the cross section. The Young's modulus and the compressive load can be assumed to be random variables with $E = N(30 \times 10^6, 3 \times 10^5)\ (\text{lb/in}^2)$ and $P = N(2000, 400)$ lb. Find the reliability of the connecting rod in buckling.

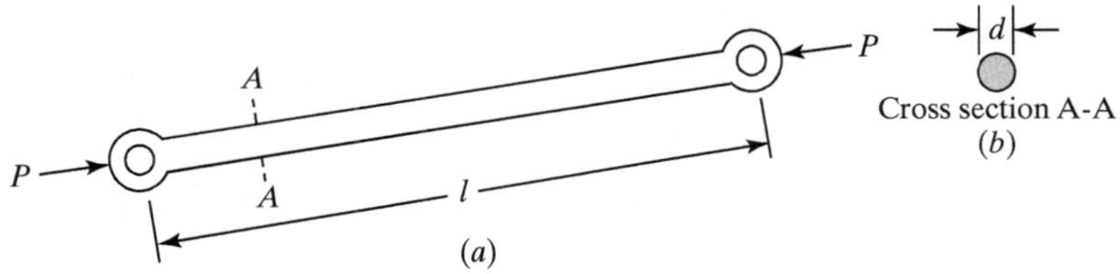

FIGURE 8.36 A connecting rod under axial compression.

8.25 Solve Problem 8.24 by assuming that the Young's modulus and the compressive load on the connecting rod follow uniform distributions over the range $28 \times 10^6 - 32 \times 10^6$ lb/in^2 and 1500–2500 lb, respectively.

8.26 Solve Problem 8.24 for the case where the Young's modulus follows type-I distribution for the smallest value:

$$f(E) = a_1 \exp\left[a_1(E - w_1) - e^{a_1(E-w_1)}\right]; \quad -\infty < E < \infty$$

with $a_1 = 4.2752 \times 10^{-6}$ and $w_1 = 30.1350 \times 10^6$, and the compressive load on the connecting rod follows type-I distribution for the largest value:

$$f(P) = a_2 \exp\left[a_2(P - w_2) - e^{-a_2(P-w_2)}\right]; \quad -\infty < P < \infty$$

with $a_2 = 3.2064 \times 10^{-3}$ and $w_2 = 1820.0464$.

8.27 **a.** Derive an expression for the reliability of a system whose strength is described by the type-III distribution for the smallest value, Eq. (8.56), and load by normal distribution.

b. Using the result of (a), solve Problem 8.24 for the case where the compressive load follows normal distribution with $P = N(2000, 400)$ lb, and the Young's modulus follows Weibull distribution with mean 30×10^6 lb/in^2 and standard deviation 30×10^5 lb/in^2.

8.28 **a.** Derive an expression for the reliability of a system whose strength is described by the type-III distribution for the smallest value, Eq. (8.56), and the load by the type-I distribution for the largest value, Eq. (8.57).

b. Using the result of (a), solve Problem 8.24 for the case where the compressive load follows type-I distribution for the largest value with mean 2000 lb and has a standard deviation of 400 lb, and the Young's modulus follows type-III distribution for the smallest value with mean 30×10^6 lb/in^2 and standard deviation 30×10^5 lb/in^2.

8.29 A steel gear with a central hole of diameter 1.500 in is shrink-fitted over a solid steel shaft of diameter 1.505 in as shown in Fig. 8.37. The gear has a pitch circle diameter of 6 in. The pressure developed at the interface (p) due to shrink-fit is given by [8.8]

$$p = \frac{E\delta}{2R}\left[\frac{(r_o^2 - R^2)}{r_o^2}\right]$$

where E = Young's modulus, δ = radial interference, r_o = outer radius of the outer part, and R = transition (nominal) radius at the interface. If the Young's modulus is normally

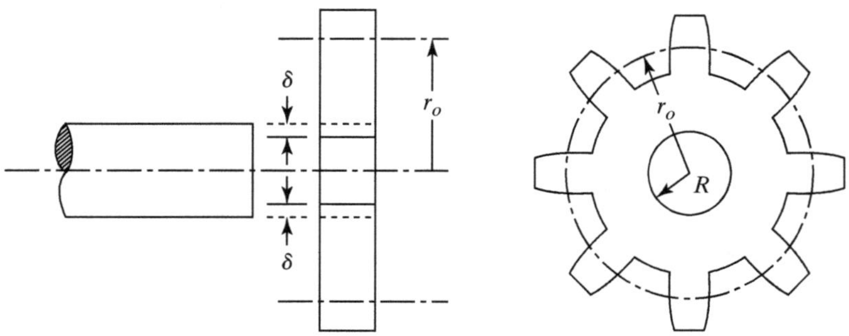

FIGURE 8.37 Shrink fit between gear and shaft.

distributed with $E = N(30 \times 10^6, 3 \times 10^6)$ lb/in^2 and the permissible shrink-fit pressure, p_o, can be taken as $p_o = N(45{,}000, 5000)$ lb/in^2, determine the failure probability of the assembly.

8.30 Experiments conducted to find the stresses induced in power transmitting shafts gave the values 23.4, 28.6, 23.9, 25.2, 25.5, 26.7, 20.3, 27.7, 19.2, 22.8, 23.1, 24.6, 25.9, 26.8, 27.5, 21.7, 24.1, 24.9, 24.8, and 22.6 k(lb/in^2). The experimental values of the strength of the shaft material yielded the values 31.8, 25.2, 28.3, 29.6, 30.1, 32.7, 26.5, 28.9, 30.8, 33.5, 27.4, 29.1, 30.7, 32.2, 29.9, 30.6, 31.7, 31.3, 33.9, and 34.3 k(lb/in^2). Find the reliability of the shafts.

8.31 Derive Eqs. (8.42) and (8.44).

8.32 The following values of compressive load, in lb, have been observed to act on the spring described in Problem 8.15:

400, 420, 440, 460, 470, 480, 490, 500, 510, 520, 540, 550, 570, 580, 590, 610

The yield strength of the spring material in shear, in k(lb/in^2), for different specimens has been observed to be:

99, 100, 101, 103, 105, 106, 108, 110, 95, 97, 98, 91

Determine the reliability of the spring.

8.33 Plot a graph between reliability, R, and the lower limit of integration, z_1, when strength and load follow normal distribution (Eq. 8.26).

8.34 For normally distributed strength and load, plot the variations of n with respect to t, Eq. (8.78), for different values of γ_S and γ_L given by $\gamma_S = 0.1, 0.2$ and $\gamma_L = 0.1, 0.2$.

8.35 For the case of normally distributed strength and load, plot the variation of the probability of failure, P_f, with respect to the factor of safety, n, for the following cases:

1. $\gamma_S = 0.1, \gamma_L = 0.1$
2. $\gamma_S = 0.1, \gamma_L = 0.2$
3. $\gamma_S = 0.2, \gamma_L = 0.1$
4. $\gamma_S = 0.2, \gamma_L = 0.2$

8.36 Derive Eq. (8.27).

8.37 Find the values of the reliability index corresponding to the following values of reliability for a system whose strength (S) and load (L) follow normal distribution:

Reliability = 0.05, 0.15, 0.25, 0.35, 0.45, 0.55, 0.65, 0.75, 0.85, 0.95

8.38 Find the values of reliability corresponding to the following values of the reliability index for a system whose strength (S) and load (L) follow normal distribution:

Reliability index: 0, 0.5, 1, 1.5, 2, 2.5, 3, 3.5, 4, 5, 6

8.39 The limit state function of a system is given by

$$Y = g(\vec{X}) = X_1 - X_2 X_3 \tag{E1}$$

where the random variables X_1, X_2, and X_3 are statistically independent and normally distributed with mean values and standard deviations given, respectively, by μ_{X_i} and $\sigma_{X_i} \equiv \sigma$; $i = 1, 2, 3$. By using the expansion points

a. $X_i = \mu_{X_i}$; $i = 1, 2, 3$ and

b. $X_i = \mu_{X_i}$; $i = 1, 2$, $X_3 = 0.9\,\mu_{X_3}$, show that the two reliability indexes are different.

8.40 The limit state function of a system is given by

$$Y = g(\vec{X}) = X_1 + X_2 - X_3 \tag{E1}$$

where the random variables X_1, X_2, and X_3 are statistically independent and normally distributed with mean values and standard deviations given, respectively, by μ_{X_i} and $\sigma_{X_i} \equiv \sigma;\ i = 1, 2, 3$. By using the expansion points

a. $X_i = \mu_{X_i};\ i = 1, 2, 3$ and

b. $X_i = 0.6\,\mu_{X_i};\ i = 1, 2, 3$, show that the two reliability indexes are different.

8.41 The limit state function of a system is given by

$$Y = g(\vec{X}) = X_1 + X_2 - X_3 \tag{E1}$$

where the random variables X_1, X_2, and X_3 are statistically independent and normally distributed with mean values are given by $\mu_{X_1} = 800$, $\mu_{X_2} = 30$, and $\mu_{X_3} = 20$ and standard deviations by $\sigma_{X_1} = 40$, $\sigma_{X_2} = 10$, and $\sigma_{X_3} = 5$. Find the reliability index and the corresponding reliability of the system if the linearization of the limit state function is carried about the following points:

$$\textbf{a. } \vec{X}_1^{\,*} = \begin{Bmatrix} \mu_{X_1} \\ \mu_{X_2} \\ \mu_{X_3} \end{Bmatrix} = \begin{Bmatrix} 800 \\ 30 \\ 20 \end{Bmatrix}, \quad \text{and} \quad \textbf{b. } \vec{X}_2^{\,*} = \begin{Bmatrix} \mu_{X_1} \\ \mu_{X_2}/2 \\ \mu_{X_3} \end{Bmatrix} = \begin{Bmatrix} 800 \\ 15 \\ 20 \end{Bmatrix}.$$

8.42 The limit state function corresponding to the buckling failure mode of a long column (shown in Fig. 8.15) can be expressed in the following two equivalent forms: Buckling stress should be larger than the compressive stress due to axial load:

$$g_1(P, E, A, I, l) = \frac{\pi^2 E I}{l^2 A} - \frac{P}{A} \geq 0 \tag{E1}$$

Buckling load should be larger than the axial load applied:

$$g_2(P, E, I, l) = \frac{\pi^2 E I}{l^2} - P \geq 0 \tag{E2}$$

where P = axial load, and E = Young's modulus, A = cross-sectional area, I = area moment of inertia, and l = length denote the basic random variables of the column. If all the random variables follow normal distribution with means and standard deviations given by $(\mu_P, \sigma_P) = (1500, 100)$ lb, $(\mu_A, \sigma_A) = (4, 0.2)$ in^2, $(\mu_I, \sigma_I) = (1, 0.1)$ in^4, $(\mu_E, \sigma_E) = (30 \times 10^6, 2 \times 10^6)$ psi, and $(\mu_l, \sigma_l) = (50, 1)$ in, find the safety index corresponding to each of the two limit state functions g_1 and g_2.

8.43 If the strength (S) and load (L) of a structure are independent and follow lognormal distribution with mean values $\mu_S = 500$ and $\mu_L = 400$ and standard deviations $\sigma_S = 100$ and $\sigma_L = 50$, find the value of the reliability index.

8.44 If the strength (S) and load (L) of a structure are independent and follow lognormal distribution with mean values $\mu_S = 300$ and $\mu_L = 200$ and standard deviations $\sigma_S = 80$ and $\sigma_L = 40$, find the value of the reliability index.

8.45 The limit state function of a system is given by

$$g(x_1, x_2) = 25x_1^2 + 20x_1x_2 + 100x_2^2 - 10x_1 - 20x_2 - 2500 \tag{E1}$$

where the basic random variables X_1 and X_2 follow normal distribution with mean values $\mu_{X_1} = 30$ and $\mu_{X_2} = 20$ and standard deviations $\sigma_{X_1} = 10$ and $\sigma_{X_2} = 5$. Determine the safety index β using the mean value FOSM method.

8.46 In a journal bearing, the limit state function is defined in terms of the basic random variables X_1, X_2, and X_3 as

$$g(x_1, x_2, x_3) = X_1 - X_2 - X_3 > 0 \text{ (for safe performance)} \tag{E1}$$

where X_1 is the strength of the bearing, X_2 is the stress induced due to the applied load, and X_3 is the stress induced due to the eccentricity of the rotating shaft or journal. If the mean values, standard deviations, and the correlation coefficients of the random variables are known to be $\mu_{X_1} = 20{,}000$ psi, $\mu_{X_2} = 3{,}000$ psi, $\mu_{X_3} = 10{,}000$ psi, $\sigma_{X_1} = 2{,}000$ psi, $\sigma_{X_2} = 500$ psi, $\sigma_{X_3} = 1{,}000$ psi, $\rho_{X_1, X_2} = 0.2$, $\rho_{X_2, X_3} = 0.9$, and $\rho_{X_3, X_1} = 0.1$, determine the reliability of the bearing assuming that all the random variables follow normal distribution.

8.47 A single member–single load problem has a strength S and load L. The mean value and coefficient of variations of S and L are given by (μ_S, γ_S) and (μ_L, γ_L), respectively. If both S and L follow lognormal distribution, show that the lower limit of the standard normal variate corresponding to a specified reliability R_0 is given by

$$z_1 = \frac{\ln\left\{\left(\frac{\mu_S}{\mu_L}\right)\left(\frac{1+\gamma_S^2}{1+\gamma_L^2}\right)^{1/2}\right\}}{\left\{\ln\left[\left(1+\gamma_S^2\right)\left(1+\gamma_L^2\right)\right]\right\}^{1/2}}$$

where $R_0 = 1 - \Phi(-z_1)$ and Φ is the cumulative (normal) distribution function.

8.48 A chain, consisting of three identical links, carries a load L as shown in Fig. 8.38(*a*). The probability density functions of the strengths of the links and the load are shown in Figs. 8.38(*b*) and (*c*), respectively. Determine (a) the probability of failure of a link, (b) the probability of failure of the chain, and (c) the bounds on the probability of failure of the chain.

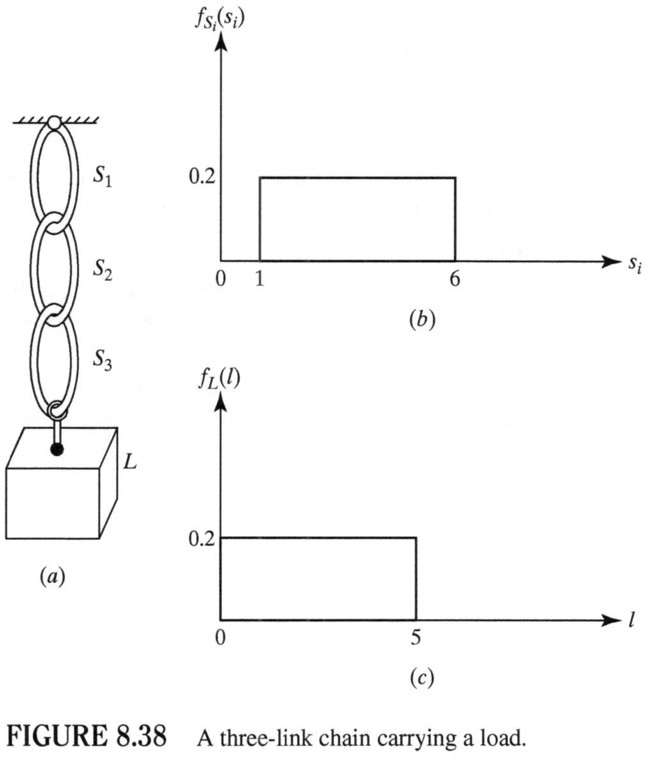

FIGURE 8.38 A three-link chain carrying a load.

MATLAB and Excel Problems

8.49 Find the reliability of a bolted joint when the shear stress induced and the shear strength of the joint follow normal distribution, with parameters as indicated in Problem 8.3, using MATLAB.

8.50 The torque carrying capacity (T) and the required torque to be transmitted (T_r) of a cone clutch (described in Example 8.1) are known to follow lognormal distribution with mean values $\overline{T} = 396.33$ lb-in and $\overline{T}_r = 400$ lb-in and standard deviations $\sigma_T = 26.4225$ lb-in and $\sigma_{T_r} = 40$ lb-in. Find the reliability of the gear tooth using MATLAB.

8.51 Solve Problem 8.49 using Excel.

8.52 Solve Problem 8.50 using Excel.

8.53 Find the reliability of the pressure vessel described in Example 8.5 when the hoop stress induced in the pressure vessel (L) and the yield strength of the material (S) follow exponential distribution with mean values $\overline{L} = 28{,}800$ psi and $\overline{S} = 30{,}000$ psi using Excel.

8.54 The maximum pressure developed in a steam boiler is known to follow normal distribution with a mean value 250 psi and a standard deviation 40 psi. Using MATLAB, find the values of the pressure developed in the boiler that correspond to the following probabilities (i.e., the values of the cumulative distribution function): 0.1, 0.2, 0.3, 0.4, 0.5, 0.6, 0.7, 0.8, 0.9, 0.99.

8.55 In the cone clutch described in Example 8.1 (shown in Fig. 8.7), the axial force applied (F), coefficient of friction between the cup and the cone (f), semi-cone angle (α), outer diameter of the cone (D), and the inner diameter of the cone (d) are known to follow normal distribution with parameters given by $F = N(100, 10)$ lb, $f = N(0.3, 0.02)$, $\alpha = N(20, 2)^0$, $D = \mathrm{N}(10, 0.02)$ in, and $d = N(8, 0.02)$ in. The torque transmitted by the cone clutch (T) is given by

$$T = \frac{Ff}{3 \sin \alpha}\left(\frac{D^3 - d^3}{D^2 - d^2}\right)$$

If the torque required to be transmitted by the cone clutch (T_r) is given by $T_r = N(400, 40)$ lb-in, find the reliability of the cone clutch using MATLAB.

8.56 Solve Problem 8.54 using Excel.

8.57 **a.** Write a MATLAB program to find the bounds on the probability of failure of a single member subjected to several loads based on Eq. (8.129). Assume that the data are available in the form of discrete probabilities.

b. Assuming that the data given in Figs. 8.24(c) and (d) are valid for two loads and one member strength, use the MATLAB program developed in part (a) to find the bounds on the probability of failure of the member.

CHAPTER 9

Design of Mechanical Components and Systems

What You Will Learn

In this chapter, you will learn

- The reliability-based design of mechanical components and systems
- That the geometry, material properties, and external loads acting on the component or system are treated as random variables and are assumed to follow normal distribution for computational simplicity
- The use of the limit state equation in the computation of strength-based reliability of single components and multi-component systems
- The design of components/systems using the expression of reliability for any specified value of reliability
- The reliability-based design of a variety of mechanical components and systems, including the design of a connecting rod, a helical spring, the fatigue design of machine components, a compound gear train, a cam-follower system, and a four-bar function generating mechanism
- The possibility of coupling the reliability analysis procedure with optimization procedures for finding the optimum design of a mechanical or structural component/system

William Sealy Gosset was an Englishman who is more popularly known under the pen name of "Student." He was born in Canterbury on 13 June, 1876, and died in Beaconsfield on 16 October, 1937. He worked as a consulting statistician for a famous Irish brewery. Originally the brewery did not permit him to publish his research work fearing that the stockholders might not like to approve spending money on research. Later, when the company permitted him to publish his research work on the condition that he use a pen name, Gosset used the name "Student." He was one of the first who used the theory of probability to statistical data relating the brewing methods, characteristics of the raw materials used, and the quality of the finished product. The experiments he conducted using small samples have led to the *t*-distribution (or Student's *t*-distribution) and *t*-test. His works include such diverse topics as testing of cereals, agronomy, and mathematics. He toured the barley fields before harvest and suggested methods of improving the crops using principles of statistics [9.14, 9.15].

9.1 Introduction

In this chapter, we shall consider the reliability-based design of mechanical components and systems. The geometry, material properties, and external loads are treated as random variables. In most cases, the random variables are assumed to follow normal distribution for computational simplicity. The assumption of normal distribution for the behavior or response parameters can be justified on the basis of central limit theorem. When the induced stress (or load) and the strength of a component follow normal distribution, the reliability can be determined, in terms of the standard normal variate, as

$$R_0 = \frac{1}{\sqrt{2\pi}} \int_{z_1}^{\infty} e^{-1/2\, z^2}\, dz \tag{9.1}$$

where the lower limit of integration is given by

$$z_1 = -\left(\frac{\mu_S - \mu_L}{\sqrt{\sigma_S^2 + \sigma_L^2}} \right) \tag{9.2}$$

with μ_S and σ_S (μ_L and σ_L) denoting, respectively, the mean value and standard deviation of the strength (load). If the induced stress (load) and the strength follow lognormal distribution, the lower limit of integration in Eq. (9.1) can be computed as

$$z_1 = -\left(\frac{\mu_{\ln S} - \mu_{\ln L}}{\sqrt{\sigma_{\ln S}^2 + \sigma_{\ln L}^2}} \right) \tag{9.3}$$

The designs of a connecting rod, a pressure vessel, a helical spring, fatigue design of machine components, and a compound gear train, and the reliability analysis of cam-follower systems and function generating mechanisms are considered as specific examples.

9.2 Design of Mechanical Components

EXAMPLE 9.1

A connecting rod of length l and diameter d is subjected to an axial compressive force P. The Euler buckling load of the connecting rod, P_c, assuming the ends to be pin-connected, is given by [9.1]

$$P_c = \frac{\pi^2 E I}{l^2}$$

where E = Young's modulus and I = moment of inertia of the cross section. The parameters are given by $l = N(20, 2)$ in, $d = N(\bar{d}, 0.1\bar{d})$ in, $P = N(2000, 200)$ lb, and $E = N(30 \times 10^6, 3 \times 10^6)$ lb/in^2. Design the connecting rod to achieve a reliability of 0.99 against buckling.

Solution: The specified reliability of 0.99 corresponds to a lower limit of $z_1 = -2.32635$ in Eq. (9.1) so that

$$z_1 = -2.32635 = -\left(\frac{\overline{P}_c - \overline{P}}{\sqrt{\sigma_{P_c}^2 + \sigma_P^2}} \right) \tag{E1}$$

where $\overline{P} = 2000$ lb and $\sigma_P = 200$ lb. Using the relation

$$I = \frac{\pi d^4}{64} \tag{E2}$$

the buckling load can be expressed as

$$P_c = \frac{\pi^3 E d^4}{64\, l^2} \tag{E3}$$

so that the mean value of P_c is given by

$$\overline{P}_c = \frac{\pi^3 \overline{E}\, \overline{d}^4}{64\, \overline{l}^2} = \frac{\pi^3 (30 \times 10^6) \overline{d}^4}{64 (20)^2} = 36{,}331.1091\, \overline{d}^4 \tag{E4}$$

The partial derivative rule can be used to find the standard deviation of P_c as

$$\sigma_{P_c}^2 = \left[\left(\frac{\partial P_c}{\partial E} \right)^2 \sigma_{\overline{E}}^2 + \left(\frac{\partial P_c}{\partial d} \right)^2 \sigma_{\overline{d}}^2 + \left(\frac{\partial P_c}{\partial l} \right)^2 \sigma_{\overline{l}}^2 \right] \tag{E5}$$

where

$$\frac{\partial P_c}{\partial E} = \frac{P_c}{E}, \quad \frac{\partial P_c}{\partial d} = 4 \frac{P_c}{d}, \quad \text{and} \quad \frac{\partial P_c}{\partial l} = -2 \frac{P_c}{l}$$

Thus Eq. (E5) gives

$$\begin{aligned} \gamma_{P_c}^2 &= \gamma_{\overline{E}}^2 + 16\, \gamma_{\overline{d}}^2 + 4 \gamma_{\overline{l}}^2 \\ &= (0.1)^2 + 16(0.1)^2 + 4(0.1)^2 = 21(0.1)^2 = (0.45825757)^2 \end{aligned}$$

that is,

$$\sigma_{P_c} = \gamma_{P_c} \overline{P}_c = 16{,}649.0058\, \overline{d}^4 \tag{E6}$$

Substitution of Eqs. (E5) and (E6), along with the known values of $\overline{P}$ and σ_P, into Eq. (E1) gives

$$-2.32635 = -\left(\frac{36331.1091\, \overline{d}^4 - 2000}{\sqrt{(16649.0058\, \overline{d}^4)^2 + (200)^2}} \right) \tag{E7}$$

Squaring both sides of Eq. (E7) leads to a quadratic equation in $\overline{d}^4$

$$1.80365827 \times 10^8 \overline{d}^8 + 1.45324436 \times 10^8 \overline{d}^4 - 378.3496 \times 10^4 = 0 \tag{E8}$$

The solution of Eq. (E8) gives

$$\begin{aligned} \overline{d}^4 &= \frac{-1.45324436 \times 10^8 \pm \sqrt{(1.45324436 \times 10^8)^2 + 4(1.80365827 \times 10^8)(0.03783496 \times 10^8)}}{2(1.80365827 \times 10^8)} \\ &= 0.0252439096; \quad -0.83096436 \end{aligned} \tag{E9}$$

which leads to $\overline{d}^2 = 0.15888332$ or $\overline{d} = 0.3986$ in.

■

EXAMPLE 9.2

A pressure vessel, in the form of a torus (see Fig. 9.1), is to be designed for a scientific experiment. The cross section of the cylinder has a radius of $r = N(50, 1)$ in, while the ring has a radius of $R = N(200, 2)$ in. The expected pressure in the vessel is $p = N(500, 50)$ lb/in^2. Find the

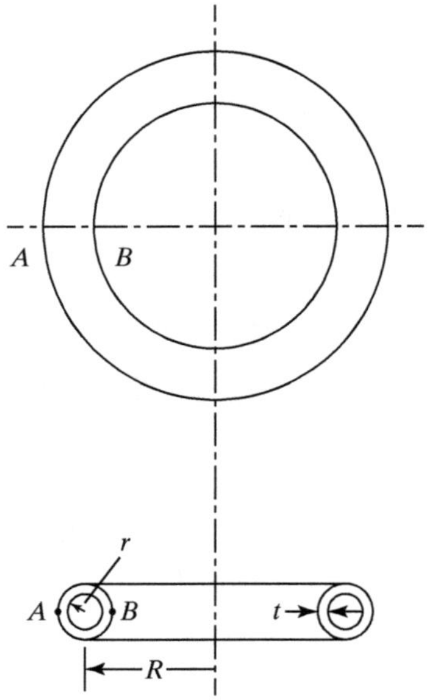

FIGURE 9.1 A pressure vessel in the form of a torus.

thickness of the cylinder $t = N(\bar{t}, 0.1\bar{t})$ in to achieve a reliability of 0.9999997 when the strength of the material is given by $S = N(30000, 2000)$ lb/in^2.

Solution: The stress induced around the rim of the cross section of the torus [9.2] is given by

$$s_\phi = \frac{pr}{2t}\left(1 + \frac{R}{R \pm r}\right) \tag{E1}$$

where the plus and minus signs in Eq. (E1) correspond to the stresses induced at points A and B, respectively, in Fig. 9.1. The larger stress induced (at point B) will be

$$s_\phi = \frac{pr}{2t}\left(1 + \frac{R}{R - r}\right) \tag{E2}$$

The stress induced in the circumferential direction of the cross section is given by [9.2]

$$s_\theta = \frac{p\,r}{2\,t} \tag{E3}$$

Since the reliability of 0.9999997 corresponds to $z_1 = -5$, Eq. (9.2) can be written, by considering the larger induced stress, s_ϕ, as

$$z_1 = -5.0 = -\frac{\overline{S} - \overline{s}_\phi}{\sqrt{\sigma_S^2 + \sigma_{s_\phi}^2}} \tag{E4}$$

where $\overline{S} = 30{,}000$ lb/in^2, $\sigma_S = 2000$ lb/in^2, and $\overline{s}_\phi$ is, from Eq. (E2),

$$\bar{s}_\phi = \frac{\bar{p}\bar{r}}{2\bar{t}}\left(1 + \frac{\bar{R}}{\bar{R} - \bar{r}}\right)$$

$$= \frac{500(50)}{2\bar{t}}\left(1 + \frac{200}{200 - 50}\right) = \frac{29166.6667}{\bar{t}} \tag{E5}$$

The standard deviation of s_ϕ can be found, using the partial derivative rule, as

$$\sigma_{s_\phi}^2 = \left(\frac{\partial s_\phi}{\partial p}\right)^2 \sigma_p^2 + \left(\frac{\partial s_\phi}{\partial r}\right)^2 \sigma_r^2 + \left(\frac{\partial s_\phi}{\partial t}\right)^2 \sigma_t^2 + \left(\frac{\partial s_\phi}{\partial R}\right)^2 \sigma_R^2 \tag{E6}$$

where the partial derivatives, evaluated at the mean values of the random variables, can be determined as

$$\frac{\partial s_\phi}{\partial p} = \frac{s_\phi}{p} = \frac{29{,}166.6667}{\bar{t}(500)} = \frac{58.3333}{\bar{t}}$$

$$\frac{\partial s_\phi}{\partial t} = -\frac{s_\phi}{t} = -\left(\frac{29{,}166.6667}{\bar{t}^2}\right)$$

$$\frac{\partial s_\phi}{\partial r} = \frac{s_\phi}{r} + \frac{p\,r\,R}{2t\,(R - r)^2}$$

$$= \frac{29{,}166.6667}{\bar{t}\,(50)} + \frac{500\,(50)\,(200)}{2\,\bar{t}\,(200 - 50)^2} = \frac{694.4444}{\bar{t}}$$

$$\frac{\partial s_\phi}{\partial R} = \frac{p\,r}{2\,t\,(R - r)}\left(1 - \frac{R}{R - r}\right)$$

$$= \frac{500\,(50)}{2\,\bar{t}\,(200 - 50)}\left(1 - \frac{200}{200 - 50}\right) = -\frac{27.775}{\bar{t}}$$

Thus, Eq. (E6) gives

$$\sigma_{s_\phi}^2 = \left(\frac{58.3333}{\bar{t}}\right)^2 (50)^2 + \left(\frac{694.4444}{\bar{t}}\right)^2 (1)^2 + \left(-\frac{29166.6667}{\bar{t}^2}\right)^2 (0.1\bar{t})^2 + \left(-\frac{27.775}{\bar{t}}\right)^2 (2)^2$$

$$= \frac{17.499219 \times 10^6}{\bar{t}^2}$$

that is,

$$\sigma_{s_\phi} = \frac{4183.207}{\bar{t}}$$

Substitution of the known quantities into Eq. (E4) yields

$$-5 = -\frac{30{,}000 - \left(\dfrac{29{,}166.6667}{\bar{t}}\right)}{\sqrt{(2000)^2 + \left(\dfrac{4183.207}{\bar{t}}\right)^2}} \tag{E7}$$

Squaring and rearranging Eq. (E7) results in

$$\frac{413.213971 \times 10^6}{\overline{t}^2} - \frac{1750.0 \times 10^6}{\overline{t}} + 800 \times 10^6 = 0$$

that is,

$$\overline{t}^2(800) - 1750\,\overline{t} + 413.213971 = 0 \tag{E8}$$

The solution of Eq. (E8) gives $\overline{t} = 1.9182$ in and 0.2693 in. Selecting the smaller value, the required mean thickness of the pressure vessel can be taken as $\overline{t} = 0.2693$ in.

■

EXAMPLE 9.3

The spring rate of a helical torsional spring [9.3] is given by

$$k = \frac{T}{\theta} = \frac{d^4 E}{64\,D\,n}$$

where T = torque applied, θ = angular deflection of the spring, d = wire diameter, D = mean spring coil diameter, n = number of turns, and E = Young's modulus of the material. The torsional spring used for a door hinge is composed of music wire and has the following probabilistic parameters: $D = N(1, 0.1)$ in, $n = N(5.5, 0.55)$, and $E = N(30 \times 10^6, 3 \times 10^6)$ lb/in^2. Find the mean value of the wire diameter if a spring rate of 50 ± 5 lb · in per turn is to be achieved with a reliability of 0.99. Assume the coefficient of variation of the wire diameter as 0.1.

Solution: If K denotes the required stiffness and k reflects the actual stiffness of the spring, the application of Eq. (9.2) for the specified reliability of R_o gives

$$z_1 = -2.3265 = -\frac{\overline{K} - \overline{k}}{\sqrt{\sigma_K^2 + \sigma_k^2}} \tag{E1}$$

where $\overline{K} = 50$ and $\sigma_K = \Delta K/3 = 5/3 = 1.6667$. The mean value of k is given by

$$\overline{k} = \frac{\overline{d}^4 \overline{E}}{64 \overline{D}\,\overline{n}} = \frac{\overline{d}^4(30 \times 10^6)}{64(1)(5.5)} = 85{,}227.2727\,\overline{d}^4 \tag{E2}$$

The standard deviation of k can be obtained, from the partial derivative rule, as

$$\sigma_k^2 = \left[\left(\frac{\partial k}{\partial E}\right)^2 \sigma_E^2 + \left(\frac{\partial k}{\partial d}\right)^2 \sigma_d^2 + \left(\frac{\partial k}{\partial D}\right)^2 \sigma_D^2 + \left(\frac{\partial k}{\partial n}\right)^2 \sigma_n^2\right] \tag{E3}$$

where

$$\left(\frac{\partial k}{\partial E}\right)^2 \sigma_E^2 = \left(\frac{k}{E}\right)^2 \sigma_E^2 = k^2 \gamma_E^2$$

$$\left(\frac{\partial k}{\partial d}\right)^2 \sigma_d^2 = \left(\frac{4d^3 E}{64\,D\,n}\right)^2 \sigma_d^2 = \left[4\left(\frac{k}{d}\right)\right]^2 \sigma_d^2 = 16\,k^2 \gamma_d^2$$

$$\left(\frac{\partial k}{\partial D}\right)^2 \sigma_D^2 = \left(\frac{-d^4 E}{64\,D^2\,n}\right)^2 \sigma_D^2 = \left(-\frac{k}{D}\right)^2 \sigma_D^2 = k^2 \gamma_D^2$$

and

$$\left(\frac{\partial k}{\partial n}\right)^2 \sigma_n^2 = \left(\frac{-d^4 E}{64 D n^2}\right)^2 \sigma_n^2 = \left(-\frac{k}{n}\right)^2 \sigma_n^2 = k^2 \gamma_n^2$$

Thus, Eq. (E3) can be rewritten as

$$\gamma_k^2 = \gamma_E^2 + 16\,\gamma_d^2 + \gamma_D^2 + \gamma_n^2 = (0.1)^2 + 16\,(0.1)^2 + (0.1)^2 + (0.1)^2 = 19\,(0.1)^2$$

that is,

$$\gamma_k = 0.4358899$$

and hence

$$\sigma_k = \gamma_k \bar{k} = 0\,.4358899\,\bar{k} = 37{,}149.7074\,\bar{d}^{\,4} \tag{E4}$$

Substitution of Eqs. (E2) and (E4) into (E1) yields

$$-2.3265 = -\frac{50 - 85227.2727\bar{d}^4}{\sqrt{(1.6667)^2 + (37149.7074\bar{d}^4)^2}} \tag{E5}$$

Squaring Eq. (E5) and rearranging the terms gives

$$2.062456 \times 10^8\,\bar{d}^8 + 852.27273 \times 10^4\,\bar{d}^4 - 2484.964338 = 0 \tag{E6}$$

The solution of Eq. (E6) gives

$$\bar{d}^4 = \frac{-852.27273 \times 10^4 \pm \sqrt{72.636881 \times 10^{12} + 2.050052 \times 10^{12}}}{4.124912 \times 10^8}$$

$$= 0.02895405 \times 10^{-2}; \quad -4.16127362 \times 10^{-2}$$

This gives $\bar{d}^2 = 0.0170158894$ or $\bar{d} = 0.1304$ in.

■

9.3 Fatigue Design

A fatigue failure is one in which the component fails gradually when subjected to repeated loads that induce stresses that are often below the yield strength of the material. If the induced repeated stress is below the endurance strength of the material, theoretically the component will not fail no matter how many cycles are repeated. High-cycle fatigue design is concerned with designing the component for infinite life. Thus, in the limiting case (deterministic design procedure), the induced stress is equated to the endurance strength of the component.

The endurance strength of the component is determined using the fatigue-strength modification factors suggested by Marin [9.3–9.5] as follows:

$$S_e = k_a\,k_b\,k_c\,k_d\,k_e\,k_f\,S_e' \tag{9.4}$$

where S_e = fatigue strength at the critical location of the component in the geometry and condition of use, k_a = surface-condition modification factor, k_b = size modification factor, k_c = load modification factor, k_d = temperature modification factor, k_e = stress-concentration and notch-sensitivity modification factor, k_f = miscellaneous-effects modification factor, and S'_e = endurance limit of the laboratory specimen, which is often estimated from the ultimate tensile strength of the material (S_{ut}), as

$$S'_e = \begin{cases} 0.504\, S_{ut} & \text{if} \quad S_{ut} \le 200\ \text{k(lb/in}^2) \\ 100\ \text{k(lb/in)}^2 & \text{if} \quad S_{ut} > 200\ \text{k(lb/in}^2) \end{cases} \tag{9.5}$$

9.3.1 Deterministic Design Procedure

The modification factors k_a, k_b, k_c, k_d, k_e, and k_f and the ultimate tensile strength S_{ut} are treated as constants [9.3–9.5]. The surface modification factor, k_a, is given by

$$k_a = a S_{ut}^b \tag{9.6}$$

where $a = 16.45$ and $b = -0.7427$ for hot-rolled components, $a = 39.9$ and $b = -0.995$ for as-forged components, $a = 2.7$ and $b = -0.2653$ for machined surfaces, and $a = 1.34$ and $b = -0.0848$ for ground surfaces when S_{ut} is expressed in k(lb/in^2). The size modification factor, k_b, is expressed as

$$k_b = \begin{cases} \left(\dfrac{d}{0.3}\right)^{-0.1133}; & \text{for bending and torsion loads; } 0.11\ \text{in} \le d \le 2\ \text{in} \\ 1; & \text{for axial loads} \end{cases} \tag{9.7}$$

where d is the diameter (or equivalent diameter) of the component in inches. The load modification factor, k_c, is given by

$$k_c = \begin{cases} 0.923; & \text{for axial loads} \qquad (\text{when } S_{ut} \le 220\ \text{k(lb/in}^2)) \\ 1; & \text{for bending loads} \\ 0.583; & \text{for torsional loads} \end{cases} \tag{9.8}$$

The temperature modification factor, k_d, can be expressed as

$$k_d = \begin{cases} 1; & \text{when operating temperature is 70°F} \\ 1.018; & \text{when operating temperature is 400°F} \end{cases} \tag{9.9}$$

The stress-concentration and notch-sensitivity modification factor, k_e, is given by

$$k_e = \frac{1}{K_F} \tag{9.10}$$

where K_F is the fatigue-strength reduction factor whose value depends on the type and size of the stress concentration present in the component. The value of K_F [9.4] can be computed as

$$K_F = \frac{K_t}{1 + \dfrac{2}{\sqrt{r}}\left(\dfrac{K_t - 1}{K_t}\right)\sqrt{a}} \tag{9.11}$$

where

$$\sqrt{a} = \begin{cases} \dfrac{5}{S_{ut}}; & \text{for a transverse hole} \\ \dfrac{4}{S_{ut}}; & \text{for a shoulder} \\ \dfrac{3}{S_{ut}}; & \text{for a groove} \end{cases} \tag{9.12}$$

K_t is the geometric stress concentration factor [9.6], and r is the notch radius. The value of the miscellaneous-effects modification factor, k_f, depends on factors such as the residual stresses present in the component and the nature of corrosive atmosphere in which the component operates. The deterministic design procedure is illustrated by working through the following example.

EXAMPLE 9.4

The machined steel shaft shown in Fig. 9.2(a) is subjected to a force $F = 3000$ lb. The material has an ultimate tensile strength of 80 k(lb/in^2). Determine the factor of safety of the shaft. The geometric stress concentration factor at the shoulder C is known to be 1.75.

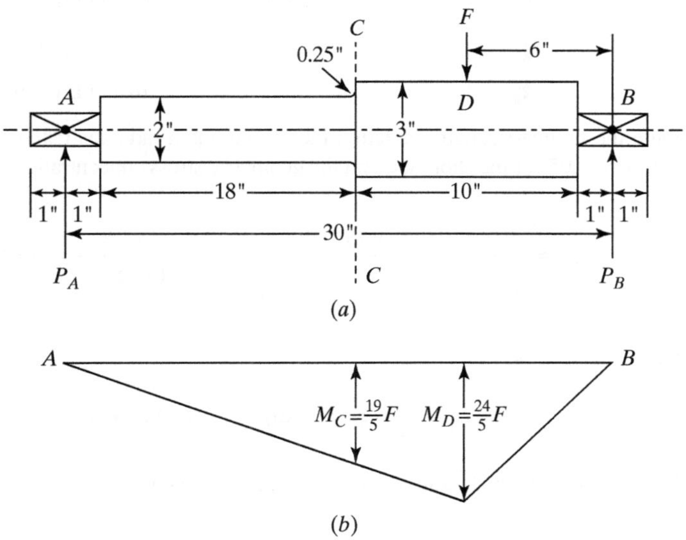

FIGURE 9.2 (a) Machined steel shaft; and (b) bending-moment diagram.

Solution: The reactions at the bearings can be determined as $P_A = (1/5)\,\mathrm{F}$ and $P_B = (4/5)\,\mathrm{F}$. The bending-moment diagram is shown in Fig. 9.2(*b*). The critical section can be seen to be the section C. The bending moment at section C, M_c, is given by

$$M_C = 19P_A = \frac{19}{5}F = \frac{19}{5}(3000) = 11{,}400 \text{ lb} \cdot \text{in}$$

The stress induced at C, s_C, is given by

$$s_C = \frac{M_C c}{I}$$

where

$$\frac{I}{c} = \frac{\frac{\pi}{64}d^4}{\frac{d}{2}} = \frac{\pi d^3}{32} = \frac{\pi}{32}(2)^3 = \frac{\pi}{4}\text{ in}^3$$

Thus

$$s_C = \frac{4M_C}{\pi} = \frac{4(11{,}400)}{\pi}$$
$$= 14{,}514.896868 \text{ lb/in}^2$$

The endurance strength of the component at section C can be computed using Eq. (9.4) with

$$S_e' = 0.504(80) = 40.32\text{ k}(1\text{b/in}^2) \quad (\text{using Eq. 9.5})$$
$$k_a = 2.7(80)^{-0.2653} = 0.844256 \quad (\text{using Eq. 9.6})$$
$$k_b = \left(\frac{2}{0.3}\right)^{-0.1133} = 0.806587 \quad (\text{using Eq. 9.7})$$
$$k_c = 1 \quad (\text{using Eq. 9.8})$$
$$k_d = 1 \quad (\text{using Eq. 9.9})$$

To compute the fatigue-strength reduction factor, the notch radius is taken as $r = 0.25$ in, $\sqrt{a}$ as $\sqrt{a} = 4/80 = 0.05$ for the shoulder, and the geometric stress concentration factor K_t as $K_t = 1.75$. This gives

$$K_F = \frac{1.75}{1 + \frac{2}{\sqrt{0.25}}\left(\frac{1.75 - 1}{1.75}\right)(0.05)} = \frac{1.75}{1.085714} = 1.611842$$

and hence

$$k_e = \frac{1}{K_F} = 0.620408 \quad (\text{using Eq. 9.10})$$

The miscellaneous-effects factor, k_f, is taken as 1. This gives the endurance strength of the component as

$$S_e = (0.844256)(0.806587)(1)(1)(0.620408)(1)(40.32) = 17.034257\text{ k}(1\text{b/in}^2)$$

Thus the factor of safety, n, is given by

$$n = \frac{S_e}{s_C} = \frac{17.034257}{14.514897} = 1.173571$$

■

9.3.2 Probabilistic Design Procedure

The same approach used in the deterministic procedure is used in the probabilistic case, also, except that both the induced stress and the endurance strength of the component are assumed to be probabilistic. Thus, the probability distribution functions of the induced stress and the endurance strength play a role in determining the reliability of the component by using the interference theory. Mischke [9.4] proposed the use of a stochastic form of Marin factors to express the endurance strength in probabilistic form as

$$S_e = k_a\, k_b\, k_c\, k_d\, k_e\, k_f\, \phi\, \overline{S}_{ut} \tag{9.13}$$

where S_e, k_a, k_b, k_c, k_d, k_e, and k_f are treated as random variables; ϕ is an empirical random proportionality factor; and $\overline{S}_{ut}$ is the deterministic, mean ultimate tensile strength of the material. The random surface modification factor is given by:

$$k_a = a\, \overline{S}_{ut}^{b}(1, \gamma_{k_a}) \tag{9.14}$$

where a and b are same as those given for Eq. (9.6) and γ_{k_a} is the coefficient of variation of k_a given by 0.098 for hot-rolled components, 0.078 for as-forged components, 0.131 for ground surfaces, and 0.06 for machined surfaces. The size factor, k_b, is considered as a deterministic quantity so that the value given by Eq. (9.7) is taken as its mean value and its coefficient of variation γ_{k_b} is taken as zero. The loading modification factor, k_c, is given by

$$k_c = \begin{cases} (1, 0); & \text{for bending loads} \\ 0.774(1, 0.163); & \text{for axial loads} \\ 0.583(1, 0.123); & \text{for torsional loads} \end{cases} \tag{9.15}$$

The temperature and miscellaneous-effects factors, k_d and k_f, depend on the details of the particular application and cannot be expressed in a general form. The stress-concentration and notch-sensitivity modification factor, k_e, can be expressed as

$$k_e = \frac{1}{K_F} \tag{9.16}$$

where

$$K_F = \frac{K_t(1, \gamma_t)}{1 + \dfrac{2}{\sqrt{r}}\left(\dfrac{K_t - 1}{K_t}\right)\sqrt{a}} = \overline{K}_F(1, \gamma_F) \tag{9.17}$$

with

$$\sqrt{a} = \begin{cases} \dfrac{5}{\overline{S}_{ut}}; & \text{for a transverse hole} \\ \dfrac{4}{\overline{S}_{ut}}; & \text{for a shoulder} \\ \dfrac{3}{\overline{S}_{ut}}; & \text{for a groove} \end{cases} \tag{9.18}$$

γ_t = coefficient of variation of K_t, γ_F = coefficient of variation of K_F

$$\gamma_F = \begin{cases} 0.11; & \text{for a transverse hole} \\ 0.08; & \text{for a shoulder} \\ 0.13; & \text{for a groove} \end{cases} \tag{9.19}$$

r = notch radius, and $\overline{K}_F$ = mean value of K_F. The random proportionality factor ϕ is given by

$$\phi = \begin{cases} 0.504\ (1, 0.146); & \text{for bending loads} \\ 0.390\ (1, 0.309); & \text{for axial loads} \\ 0.294\ (1, 0.269); & \text{for torsional loads} \end{cases} \tag{9.20}$$

Since the endurance strength, S_e, given by Eq. (9.13) is a product of several random variables, its mean value and standard deviation can be found using the partial derivative rule. The results can be expressed as

$$\overline{S}_e = \overline{k}_a\, \overline{k}_b\, \overline{k}_c\, \overline{k}_d\, \overline{k}_e\, \overline{k}_f \overline{\phi}\, \overline{S}_{ut} \tag{9.21}$$

and

$$\gamma_{S_e}^2 = \gamma_{k_a}^2 + \gamma_{k_b}^2 + \gamma_{k_c}^2 + \gamma_{k_d}^2 + \gamma_{k_e}^2 + \gamma_{k_f}^2 + \gamma_{\phi}^2 \tag{9.22}$$

where a bar over a symbol indicates its mean value and γ denotes the coefficient of variation.

EXAMPLE 9.5

The machined steel shaft shown in Fig. 9.2(a) is subjected to a random force $F = N(3000, 300)$ 1b. The material has a mean ultimate tensile strength of 80 k(1b/in^2). The geometric stress concentration factor is known to be 1.75. Determine the reliability of the shaft assuming normal distribution for the induced and permissible stresses.

Solution: From the bending-moment diagram shown in Fig. 9.2(b), the mean and standard deviation of the bending moment at section C, M_C, can be expressed as

$$M_C = 19P_A = \frac{19}{5}F = \frac{19}{5}(3000, 300)\ \text{1b}\cdot\text{in} = (11400, 1140)\ \text{1b}\cdot\text{in}$$

The stress induced at C, s_C, is given by

$$s_C = \frac{M_C c}{I}$$

where

$$\frac{I}{c} = \frac{\pi d^3}{32} = \frac{\pi(2)^3}{32} = \frac{\pi}{4}\,\text{in}^3$$

so that

$$s_C = \frac{4M_C}{\pi} = \frac{4(11400, 1140)}{\pi}$$

$$= 14514.896868\ (1, 0.1)\ \text{1b/in}^2$$

The endurance strength of the component at section C can be computed using Eq. (9.13) with:

$$k_a = a\overline{S}_{ut}^{\,b}(1, \gamma_{k_a}) = 2.7(80)^{-0.2653}(1, 0.06) = 0.844256(1, 0.06)$$

$$k_b = \left(\frac{d}{0.3}\right)^{-0.1133}(1, 0) = \left(\frac{2}{0.3}\right)^{-0.1133}(1, 0) = 0.806587(1, 0)$$

$$k_c = (1, 0)$$

$$k_d = (1, 0)$$

$$k_f = (1, 0)$$

since no specific information is known about the operating temperature and other effects. Here, $K_t = 1.75$, $r = 0.25$ in, and $\sqrt{a} = 4/\overline{S}_{ut} = 4/80 = 0.05$ and hence the mean value of K_F can be determined using Eq. (9.17) as

$$\overline{K}_F = \frac{K_t}{1 + \dfrac{2}{\sqrt{r}}\left(\dfrac{K_t - 1}{K_t}\right)\sqrt{a}} = \frac{1.75}{1.0 + \dfrac{2}{\sqrt{0.25}}\left(\dfrac{1.75 - 1}{1.75}\right)(0.05)} = 1.611842$$

and hence $K_F = 1.611842(1, 0.08)$ with

$$k_e = \frac{1}{K_F} = \frac{1}{1.611842(1, 0.08)}$$

The random proportionality factor ϕ is given by

$$\phi = 0.504(1, 0.146)$$

Thus the endurance strength of the component, given by Eq. (9.13), is

$$S_e = [0.844256(1, 0.06)][0.806587(1, 0)][(1, 0)][(1, 0)]\left[\frac{1}{1.611842\ (1, 0.08)}\right]$$

$$\times\ [(1, 0)][0.504(1, 0.146)][80] \qquad \text{(E1)}$$

Equation (E1) gives the mean value of S_e as

$$\overline{S}_e = (0.844256)(0.806587)(1)(1)\left(\frac{1}{1.611842}\right)(1)(0.504)(80) = 17.034241 \text{ k}(1\text{b}/\text{in}^2)$$

The coefficient of variation of S_e can be determined using Eq. (9.22) as

$$\gamma_{S_e}^2 = (0.06)^2 + (0)^2 + (0)^2 + (0)^2 + (0.08)^2 + (0)^2 + (0.146)^2$$
$$= 0.0036 + 0.0064 + 0.021316 + 0.031316 = (0.176963)^2$$

from which the standard deviation of S_e can be found as

$$\sigma_{S_e} = 0.176963(17.034241) = 3.014435 \text{ k}(1\text{b}/\text{in}^2)$$

If s_C and S_e follow normal distribution, the lower limit of integration, z_1, can be found as

$$z_1 = -\frac{\overline{S}_e - \overline{s}_C}{\sqrt{\sigma_{S_e}^2 + \sigma_{s_C}^2}}$$
$$= -\frac{17.034241 - 14.514897}{\sqrt{(3.014435)^2 + (1.451490)^2}} = -\frac{2.519344}{\sqrt{11.193642}} = -0.753012$$

From standard normal tables, this value of z_1 can be seen to correspond to a reliability of 0.77427. ■

EXAMPLE 9.6

Find the reliability of the shaft described in Example 9.5 by assuming lognormal distribution for both the induced stress and the endurance strength of the component.

Solution: The mean and standard deviations of the induced stress and the endurance strengths have been found, in Example 9.5, to be

$$\overline{s}_C = 14.514897 \text{ k}(\text{lb}/\text{in}^2), \quad \sigma_{s_C} = 1.451490 \text{ k}(1\text{b}/\text{in}^2)$$
$$\overline{S}_e = 17.034241 \text{ k}(1\text{b}/\text{in}^2), \quad \sigma_{S_e} = 3.014430 \text{ k}(1\text{b}/\text{in}^2)$$

Since s_C and S_e follow lognormal distributions, $\ln s_C$ and $\ln S_e$ follow normal distributions with

$$\mu_{\ln s_C} \approx \ln \overline{s}_C - \frac{1}{2}\left(\frac{\sigma_{s_C}}{\overline{s}_C}\right)^2 = \ln 14.514897 - \frac{1}{2}\left(\frac{1.451490}{14.514897}\right)^2 = 2.670175$$

$$\sigma_{\ln s_C} = \frac{\sigma_{s_C}}{\overline{s}_C} = \frac{1.451490}{14.514897} = 0.1$$

$$\mu_{\ln S_e} \approx \ln \overline{S}_e - \frac{1}{2}\left(\frac{\sigma_{S_e}}{\overline{S}_e}\right)^2 = \ln 17.034241 - \frac{1}{2}\left(\frac{3.014430}{17.034241}\right)^2 = 2.819567$$

and

$$\sigma_{\ln S_e} = \frac{\sigma_{S_e}}{\overline{S}_e} = \frac{3.014430}{17.034241} = 0.176963$$

The lower limit of integration, z_1, in Eq. (9.3), can be found as

$$z_1 = -\frac{\mu_{\ln S_e} - \mu_{\ln s_C}}{\sqrt{\sigma^2_{\ln S_e} + \sigma^2_{\ln s_C}}} = -\frac{2.819567 - 2.670175}{\sqrt{(0.176963)^2 + (0.1)^2}}$$

$$= -\frac{0.149392}{\sqrt{0.0413159}} = -0.734968$$

From standard normal tables, this value of z_1 can be seen to correspond to a reliability of 0.7688. ■

9.4 Design of Mechanical Systems

9.4.1 Reliability-Based Design of Gear Trains

Introduction. A gear train is composed of several gears mounted on different shafts. Different gears can be brought into mesh by shifting the positions of certain gears. Depending on the number of teeth on the various meshing gear pairs, different speeds can be obtained at the output shaft for a single input speed. Gear trains find application in automobiles, machine tools, helicopters, and a variety of industrial machinery. A typical gear train is shown in Fig. 9.3. The safe performance of a gear train depends on its overall reliability, which in turn depends on the probability of failure of each of the gear pairs in the drive in different failure modes. The problem, thus, is the design of a multicomponent transmission for a specified overall probability of failure [9.8, 9.9].

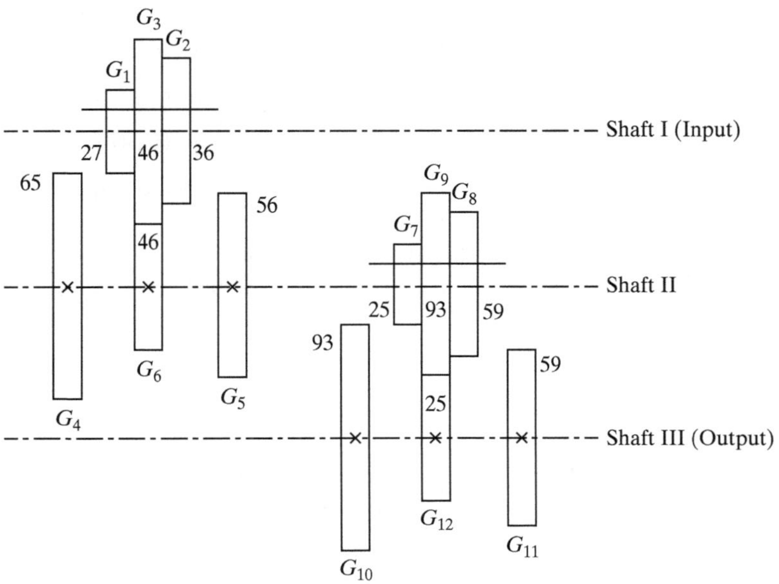

FIGURE 9.3 A typical gear train (nine-speed gear train).

Assumptions. The following assumptions are made in the analysis:

1. Two failure modes, namely, the bending and surface-wear failure modes, are considered for each gear pair.
2. The gear train is idealized as a weakest-link kinematic chain, a concept analogous to the series system.
3. The layout of the gears, the number of teeth on different gears, the module, and the interconnection of the various gear pairs of the gear train are known. The interconnection of the gear pairs of the gear train are known. The interconnection of the gear pairs and the output shaft speeds obtainable for the gear train of Fig. 9.3 are shown in Fig. 9.4.
4. The power transmitted by all the gear pairs is the same.
5. The power transmitted, the speeds of the input shaft, the allowable and induced stresses in bending and surface wear, the center distances of the gear pairs, and the face widths of the gear pairs are random variables and are statistically independent.
6. All the random variables follow normal distribution.

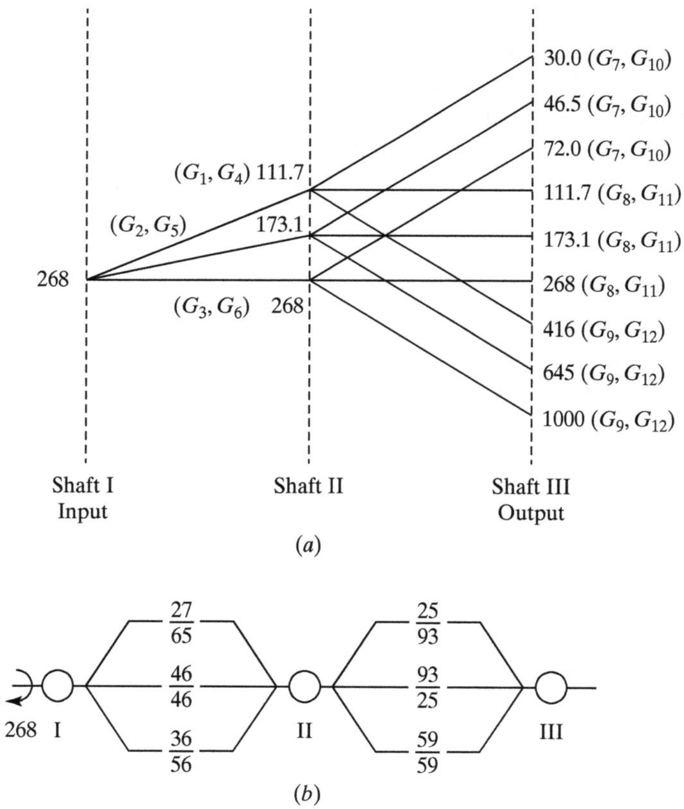

FIGURE 9.4 (*a*) Interconnection of the gear pairs; and (*b*) the output shaft speeds.

7. The mean values of the face widths of the gear pairs are the design variables.
8. A design is considered to be safe and adequate if the probability of failure of the gear train is less than or equal to a specified small quantity in each of the two failure modes.

Design Equations. If two meshing gears (see Fig. 9.5) transmit a power of p (watts) at a wheel speed[1] of n_w (rpm), the torque acting on the gear teeth, M_t (N-mm), is given by

$$M_t = \frac{7.1324 \times 10^9 p}{n_w} \tag{9.23}$$

The induced bending stress, $s_b(\text{N/mm}^2)$, assuming the tooth to be a cantilever beam with end load [9.8, 9.9], can be computed as

$$s_b = \frac{\beta M_t}{At} \tag{9.24}$$

where

$$\beta = \frac{K_c K_d(i + 1)}{imy \cos \alpha} \tag{9.25}$$

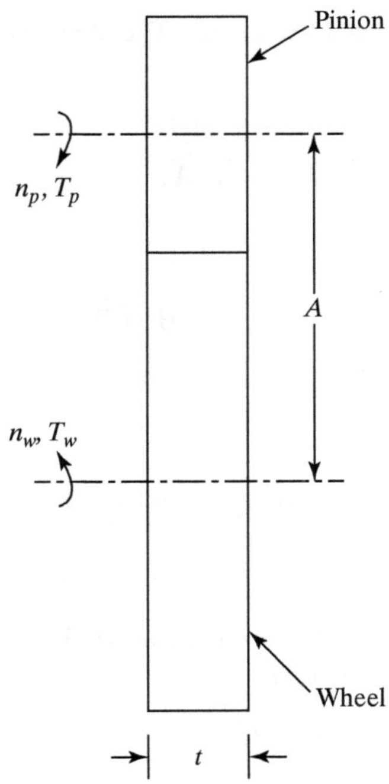

FIGURE 9.5 A gear pair in mesh.

[1]The gear having a larger diameter in a gear pair is called the wheel and the one having a smaller diameter is called the pinion. If both the gears have the same diameter, any one can be called the wheel.

A is the center distance between the gears (mm), t is the face width (mm), K_c is the stress concentration factor, K_d is the dynamic load factor, $i = T_w/T_p = n_p/n_w$ is the ratio of speeds (transmission ratio), T_w is the number of teeth on the wheel, T_p is the number of teeth on the pinion, n_p is the speed of the pinion (rpm), m is the module (mm), α is the pressure angle, and y is the Lewis form factor given by

$$y = 0.52\left(1 + \frac{20}{T_w}\right) \tag{9.26}$$

The induced wear stress, s_w, $(\mathrm{N/mm^2})$, [9.8, 9.9] is given by

$$s_w = \frac{\gamma}{A}\sqrt{\frac{M_t}{t}} \tag{9.27}$$

where

$$\gamma = 0.1848\left(\frac{i+1}{i}\right)\left\{\frac{(i+1)EK_cK_d}{\sin\,2\,\alpha}\right\}^{1/2} \tag{9.28}$$

and E is the Young's modulus of the material of the gear pair.

Mean and Standard Deviations of Induced Stresses. The mean and standard deviation of s_b are given by

$$\bar{s}_b = 7.1324 \times 10^9 \frac{\beta\,\bar{p}}{\bar{n}_w \bar{A}\,\bar{t}} \tag{9.29}$$

and

$$\sigma_{s_b} = \frac{1}{\bar{A}\,\bar{t}}\left[\frac{\beta^2\,\overline{M}_t^2(\bar{t}^2\sigma_A^2 + \bar{A}^2\sigma_t^2) + \beta^2\bar{A}^2\bar{t}^2\sigma_{M_t}^2}{\bar{A}^2\bar{t}^2 + \bar{t}^2\sigma_A^2 + \bar{A}^2\sigma_t^2}\right]^{1/2} \tag{9.30}$$

where

$$\sigma_{M_t} = \frac{7.1324 \times 10^9}{\bar{n}_w}\left[\frac{\overline{P}^2\sigma_{n_w}^2 + \bar{n}_w^2\sigma_P^2}{\bar{n}_w^2 + \sigma_{n_w}^2}\right]^{1/2} \tag{9.31}$$

and the partial derivative rule has been used in deriving Eqs. (9.30) and (9.31). Similarly, the mean and standard deviation of s_w are given by

$$\bar{s}_w = \frac{\gamma}{\bar{A}}\sqrt{\frac{\overline{M}_t}{\bar{t}}} \tag{9.32}$$

and

$$\sigma_{s_w} = \frac{1}{\bar{A}\sqrt{\bar{t}}}\left[\frac{4\overline{M}_t^2\gamma^2\bar{t}^2\sigma_A^2 + \gamma^2\bar{A}^2\overline{M}_t^2\sigma_t^2 + \bar{A}^2\bar{t}^2\gamma^2\sigma_{M_t}^2}{4\overline{M}_t\bar{A}^2\bar{t}^2 + 4\overline{M}_t\bar{t}\sigma_A^2 + \overline{M}_t\bar{A}^2\sigma_t^2}\right]^{1/2} \tag{9.33}$$

Probabilistic Model. For the gear train shown in Figs. 9.3 and 9.4, two gear pairs are in mesh while delivering any particular output speed. For example, the gear pairs (G_1, G_4) and (G_9, G_{12}) are in mesh for obtaining an output speed of 416 rpm when the input speed is 268 rpm. To use the design equations, we consider a typical ith gear pair in jth output speed ($i = 1, 2$ and $j = 1, 2, \ldots, 9$ for the gear train of Fig. 9.3). This gear pair is considered as a single link in a series kinematic chain consisting of a total of k links ($k = 2$ for the gear train of Fig. 9.3). The probability of failure of the gear train, P_f, can be expressed, using the weakest-link model, as

$$P_f = 1 - \prod_{i=1}^{k} (1 - P_{fi}) \tag{9.34}$$

where P_{fi} is the probability of failure of the ith link (gear pair). Since two failure modes are considered for each gear pair, the overall probabilities of failure of the gear train while transmitting the jth output speed can be computed as

$$P_{f_{bj}} = 1 - \prod_{i=1}^{k} (1 - P_{f_{bij}}); \quad j = 1, 2, \ldots, N \tag{9.35}$$

$$P_{f_{wj}} = 1 - \prod_{i=1}^{k} (1 - P_{f_{wij}}); \quad j = 1, 2, \ldots, N \tag{9.36}$$

where $P_{f_{bj}}(P_{f_{wj}})$ is the probability of failure of the gear train in bending mode (surface-wear mode), $P_{f_{bij}}$ $(P_{f_{wij}})$ is the probability of failure of the ith gear pair in delivering the jth output shaft speed, and N is the total number of speeds obtainable at the output shaft. The probabilities of failure of all the gear pairs, $P_{f_{bij}}$ and $P_{f_{wij}}$, are assumed to be same in the present analysis.

If we consider the ith gear pair in jth output shaft speed as a single link being acted upon by the bending stress, the probability of failure of the gear pair (link) is given by

$$P_{f_b} = P_{f_{bij}} = \int_{-\infty}^{\infty} F_{S_b}(r) f_{s_b}(r)\, dr \tag{9.37}$$

where $f_{s_b}(r)$ is the probability density function of the induced stress, s_b, and $F_{S_b}(r)$ is the probability distribution function of the permissible bending stress. Since s_b and S_b are assumed to follow normal distribution with means $(\bar{s}_b, \bar{S}_b)$ and standard deviations $(\sigma_{s_b}, \sigma_{S_b})$, the probability of failure can also be expressed as

$$P_{f_{bij}} = P(\bar{S}_b - \bar{s}_b < 0) \tag{9.38}$$

with the corresponding upper limit of integration given by

$$z_{1b} = -\left[\frac{\bar{S}_b - \bar{s}_b}{\sqrt{\sigma_{S_b}^2 + \sigma_{s_b}^2}} \right] \tag{9.39}$$

The value of z_{1b} can be found from standard normal distribution tables for any specified probability of failure $P_{f_{bij}}$. Similarly, the upper limit of integration corresponding to the probability of failure in surface wear $P_{f_{wij}}$ is given by

$$z_{1w} = -\left[\frac{\overline{S}_w - \overline{s}_w}{\sqrt{\sigma_{S_w}^2 + \sigma_{s_w}^2}}\right] \tag{9.40}$$

Design Procedure. Assuming that the speed of the input shaft, the number of teeth on different gears, the power transmitted, the allowable stresses, the allowable probabilities of failure of the gear train, and other factors involved in the various equations are known as design data, the following procedure can be adopted for designing the gear train for bending strength [9.8, 9.9].

1. From the known $P_{f_b} = P_{f_{bj}}$, compute the probability of failure of any gear pair at any speed by assuming equal probabilities of failure for all gear pairs.
2. Find the upper limit of integration in Eq. (9.39) from standard normal tables.
3. Calculate the coefficients of variation of torque and induced bending stress using Eqs. (9.31) and (9.30) as

$$\gamma_{M_t}^2 = \frac{\gamma_p^2 + \gamma_{n_w}^2}{1 + \gamma_{n_w}^2} \tag{9.41}$$

and

$$\gamma_{s_b}^2 = \frac{\gamma_t^2 + \gamma_A^2 + \gamma_{M_t}^2}{1 + \gamma_A^2 + \gamma_t^2} \tag{9.42}$$

4. Find the mean value of the induced stress as

$$\overline{s}_b = \frac{\overline{S}_b \pm [-\overline{S}_b^2(1 - z_{1b}^2\gamma_{S_b}^2)(1 - z_{1b}^2\gamma_{s_b}^2) + \overline{S}_b^2]^{1/2}}{(1 - z_{1b}^2 \cdot \gamma_{s_b}^2)} \tag{9.43}$$

5. Using the smaller value of $\overline{s}_b$ given by Eq. (9.43), compute the mean value of the face width of the particular gear pair using Eq. (9.29). From this, find the standard deviation of t as

$$\sigma_t = \gamma_t \overline{t} \tag{9.44}$$

6. Repeat steps 3 through 5 until the thicknesses of different gear pairs are calculated for all the speeds.
7. Determine the design thickness $\overline{t}$ of any gear pair as the largest value out of all the speeds in which the gear pair participates.

A similar procedure can be adopted for the design of the gear train for surface-wear strength. The relevant equations in this case are

$$\gamma_{s_w}^2 = \frac{\gamma_t^2 + \gamma_{M_t}^2 + 4\gamma_A^2}{4 + \gamma_t^2 + 4\gamma_A^2} \tag{9.45}$$

$$\bar{s}_w = \frac{\bar{S}_w \pm \left[-\bar{S}_w^2 (1 - z_{1w}^2 \gamma_{S_w}^2)(1 - z_{1w}^2 \gamma_{s_w}^2) + \bar{S}_w^2 \right]^{1/2}}{(1 - z_{1w}^2 \gamma_{s_w}^2)} \tag{9.46}$$

and

$$\bar{t} = \left(\frac{\gamma \sqrt{\bar{M}_t}}{\bar{A} \bar{s}_w} \right)^2 \tag{9.47}$$

Numerical Results. The reliability-based design of the three shaft, nine-speed lathe gear train shown in Figs. 9.3 and 9.4 is illustrated using the following data:

Mean power, 7.457 kW
Mean input shaft speed, 268 rpm
Gear tooth profile, 20° involute type
Module of gears, 4 mm
Material of gears, hardened steel
Mean allowable stress in bending, 2.4525×10^8 N/m^2
Mean allowable stress in surface wear, 17.1675×10^8 N/m^2
Density of the material, 7.60275×10^4 N/m^3
Stress concentration factor, 1.5
Dynamic load factor, 1.1

The results obtained for an allowable probability of failure of $P_{f_b} = P_{f_w} = 0.2 \times 10^{-5}$ for different values of the coefficient of variation of the input shaft speed are shown in Table 9.1. The results given by the deterministic procedure are also shown in the table. It can be seen that the thicknesses of all the gear pairs except that of the sixth one are governed by the surface-wear strength. The deterministic design gives the weight of the gear train as 1264.0185 N, whereas the probabilistic design gives a varying weight of 1236.8448 N to 1305.2205 N as the value of γ_{n_w} changes from 0.01 to 0.10. Thus the factor γ_{n_w}, which is not even mentioned in the deterministic procedure, can be seen to influence the weight by 5.33 percent. A similar behavior can be expected in respect of the other coefficients of variation like γ_p, γ_t, γ_{S_b}, and γ_{S_w}. It is to be noted that the weight of the gear train is obtained by adding the weights of the individual gears without considering the weights of the shafts and bearings.

9.5 Reliability Analysis of Mechanical Systems

9.5.1 Cam-Follower Systems

Introduction. It is well known that the limitations of the manufacturing machines, poor workmanship, inaccuracies in assembly, and clearances in the bearings of kinematic pairs cause errors that affect the kinematic accuracy, dynamic performance, noise level, and life of cam mechanisms. In the case of internal combustion engines with valves,

TABLE 9.1 Face widths of gear pairs

Gear pair	(G_1, G_4)	(G_3, G_6)	(G_2, G_5)	(G_7, G_{10})	(G_9, G_{12})	(G_8, G_{11})
Speed of wheel, n_w	111.7 rpm	268.0 rpm	173.1 rpm	30.0* rpm	416.0* rpm	111.7* rpm
Probabilistic design with $P_{fb} = P_{fw} = 2 \times 10^{-6}$, $\gamma_{n_w} = \gamma_p = 0.1$, $\gamma_A = \gamma_t = 0.01$, $\gamma_{S_b} = \gamma_{S_w} = 0.1$, and $E = 20.601 \times 10^{11}$ N/m^2						
Bending strength only considered	2.936 cm	1.576 cm	2.206 cm	8.222 cm	0.593 cm	3.159 cm
Surface wear strength only considered	3.266 cm	1.595 cm	2.194 cm	8.228 cm	0.593 cm	2.327 cm
Probabilistic design with $P_{fb} = P_{fw} = 2 \times 10^{-6}$, $\gamma_{n_w} = 0.01$, $\gamma_p = 0.1$, $\gamma_A = \gamma_t = 0.01$, $\gamma_{S_b} = \gamma_{S_w} = 0.1$, and $E = 20.601 \times 10^{11}$ N/m^2						
Bending strength only considered	2.733 cm	1.467 cm	2.053 cm	7.654 cm	0.552 cm	2.940 cm
Surface wear strength only considered	3.104 cm	1.516 cm	2.086 cm	7.821 cm	0.564 cm	2.211 cm
Deterministic design with factor of safety = 2.0						
Bending strength only considered	2.586 cm	1.388 cm	1.943 cm	7.242 cm	0.522 cm	2.782 cm
Surface wear strength only considered	3.216 cm	1.571 cm	2.161 cm	8.104 cm	0.584 cm	2.291 cm
Weight of gear train: Probabilistic design: 1305.221 N for $\gamma_{n_w} = 0.10$ and 1236.845 N for $\gamma_{n_w} = 0.01$ Deterministic design: 1264.019 N						

*Speed requiring the largest value of face width.

the lift of a valve, which is governed by a cam-follower system, should be accurately known so as to clearly define the closing and opening of the valves. Otherwise, it may result in the inefficient use of the fuel. Thus, the analysis and control of errors in cam mechanisms is extremely important to achieve the desired level of accuracy. In this section, the effect of errors in the geometrical parameters of the cam-follower system on the kinematic response is studied. The errors are treated as probabilistic in nature. The disc cam with translating roller-ended follower system shown in Fig. 9.6 [9.10] is considered for illustration.

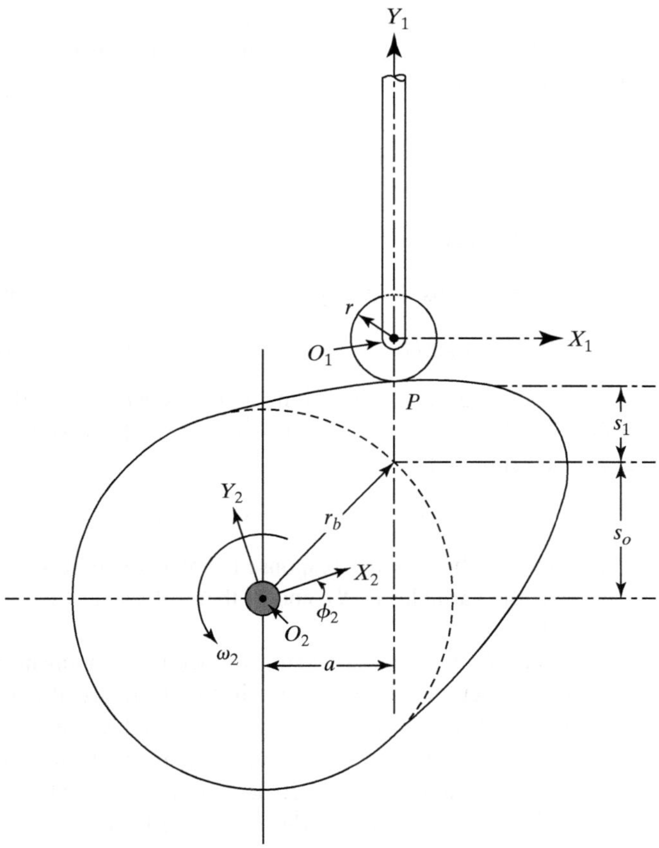

FIGURE 9.6 Disc cam with translating-roller follower.

Kinematic Response of the Cam-Follower System. For the cam-follower mechanism shown in Fig. 9.6, the condition of contact [9.10] can be expressed as:

$$E \sin \theta + F \cos \theta + G = 0 \tag{9.48}$$

with

$$E = s_1' - a$$

$$F = s_1 + s_o$$

$$G = 0$$

$$s_1' = \frac{ds_1}{d\phi_2}$$

$$s_o = \sqrt{r_b^2 - a^2}$$

s_1 = displacement of the follower, a = offset or eccentricity, r_b = radius of the base circle of the cam, ϕ_2 = cam-rotation angle, and θ = parameter of the follower surface. Equation (9.48) can be used to express θ as

$$\theta = \tan^{-1}\left(\frac{s_1 + s_o}{-s_1' + a}\right) \tag{9.49}$$

The cam profile [9.10–9.12] is defined by

$$X_2 = -r\cos(\theta + \phi_2) + a\cos\phi_2 + (s_1 + s_o)\sin\phi_2 \tag{9.50}$$

$$Y_2 = -r\sin(\theta + \phi_2) - a\sin\phi_2 + (s_1 + s_o)\cos\phi_2 \tag{9.51}$$

where X_2 and Y_2 denote the coordinates of the point of contact, P, and r indicates the radius of the roller (see Fig. 9.6). Using Eqs. (9.50) and (9.51), the displacement or response of the follower can be expressed as

$$s_1 = X_2\sin\phi_2 + Y_2\cos\phi_2 - r\sin\theta - s_o \tag{9.52}$$

This equation shows that the follower response is not an explicit function of the geometric parameters, because the variables θ, X_2, and Y_2 themselves are functions of s_1 as given by Eqs. (9.49) to (9.51).

Here the errors in X_2, Y_2, r, r_b, and a are assumed to be independent random variables following normal distribution. The variations in the geometrical parameters r and r_b are treated as random variables because of the nature of manufacturing tolerances. Some error is bound to occur in mounting the cam-follower system, and hence the offset or eccentricity a is assumed to be a random variable. To account for the manufacturing errors, the tolerances in the coordinates of the cam profile are assumed to be random variables. Although these coordinates change with the cam-rotation angle, the tolerances ΔX and ΔY on the coordinates X_2 and Y_2 will be the same for all the cam-rotation angles. By defining a vector of random parameters as

$$\mathbf{X} = \begin{Bmatrix} x_1 \\ x_2 \\ x_3 \\ x_4 \\ x_5 \end{Bmatrix} = \begin{Bmatrix} \Delta X \\ \Delta Y \\ \Delta r \\ \Delta r_b \\ \Delta a \end{Bmatrix}$$

the standard deviation of the kinematic response s_1 can be obtained by applying the partial derivative rule as

$$(\sigma_{s_1})_j = \left[\sum_{i=1}^{5}\left(\frac{\partial s_1}{\partial x_i}\right)^2\bigg|_{\mathbf{X}} \sigma_{x_i}^2\right]^{1/2} \tag{9.53}$$

where $(\sigma_{s_1})_j$ = standard deviation of s_1 at $\phi_2 = (\phi_2)_j$ and $\mathbf{X}$ = mean design vector = $(0.0, 0.0, 0.0, 0.0, 0.0)^T$. The partial derivatives required for the computation of $(\sigma_{s_1})_j$ can be found as

$$\frac{\partial s_1}{\partial x_1} = \sin \phi_2$$

$$\frac{\partial s_1}{\partial x_2} = \cos \phi_2$$

$$\frac{\partial s_1}{\partial x_3} = -\sin \theta$$

$$\frac{\partial s_1}{\partial x_4} = -r \cos \theta \left(\frac{\partial \theta}{\partial r_b} \right) - \left(\frac{\partial s_o}{\partial r_b} \right)$$

$$\frac{\partial s_1}{\partial x_5} = -r \cos \theta \left(\frac{\partial \theta}{\partial a} \right) - \left(\frac{\partial s_o}{\partial a} \right)$$

$$\frac{\partial \theta}{\partial r_b} = \frac{\left(\dfrac{\partial s_o}{\partial r_b} \right)(s_1' - a)}{[(s_1' - a)^2 + (s_1 + s_o)^2]}$$

$$\frac{\partial \theta}{\partial a} = \frac{\left(\dfrac{\partial s_o}{\partial a} \right)(s_1' - a)}{[(s_1' - a)^2 + (s_1 + s_o)^2]}$$

$$\frac{\partial s_o}{\partial r_b} = \frac{r_b}{\sqrt{r_b^2 - a^2}}$$

$$\frac{\partial s_o}{\partial a} = \frac{-a}{\sqrt{r_b^2 - a^2}}$$

Mean and Standard Deviation of the Kinematic Response. The various steps involved in the computation of the mean and standard deviation of the follower response are now outlined [9.10]:

1. Divide the cam angle for total lift, ϕ_o, into a convenient number of intervals, $n - 1$. The discrete values of ϕ_2 are labeled as $(\phi_2)_j$, $j = 1, 2, \ldots, n$. Set $j = 1$.
2. Assume a suitable lift curve. The cycloidal motion is assumed in this work. The lift s_1 corresponding to any cam angle ϕ_2, according to cycloidal motion, is given by

$$s_1 = \frac{L_o}{\pi} \left[\frac{\pi \phi_2}{\phi_o} - \frac{1}{2} \sin \left(\frac{2\pi \phi_2}{\phi_o} \right) \right] \tag{9.54}$$

and hence

$$s_1' = \frac{ds_1}{d\phi_2} = \left(\frac{L_0}{\phi_0}\right)\left[1 - \cos\left(\frac{2\pi\phi_2}{\phi_0}\right)\right] \tag{9.55}$$

Calculate the values of s_1 and s_1' for a particular value of $\phi_2 = (\phi_2)_j$ and substitute them into Eq. (9.49) to find the corresponding value of θ at $\phi_2 = (\phi_2)_j$.

3. Determine the mean lift $(s_1)_j$ and standard deviation $(\sigma_{s_1})_j$ using Eqs. (9.52) and (9.53).
4. Set new j = current $j + 1$, and repeat steps 2 through 4 until the angle $(\phi_2)_n$ is considered.

Numerical Results. The tolerances for all the geometrical parameters are assumed to be normally distributed with $\pm 3\sigma$ band. Hence, if the mean value of one of the parameters is r with a standard deviation σ_r, then the allowable value will be $(r \pm 3\sigma_r)$. Thus $\sigma_r = \Delta r/3 = x_3/3$ where Δr is the tolerance given to r. Similarly, $\sigma_{r_b} = \Delta r_b/3 = x_4/3$, $\sigma_a = \Delta a/3 = x_5/3$, $\sigma_{X_2} = \Delta X/3 = x_1/3$, and $\sigma_{Y_2} = \Delta Y/3 = x_2/3$. The following data are assumed for the numerical computation: $\bar{r} = 1.0$, $\bar{r}_b = 2.5$, $\bar{a} = 1.2$, and total angle of rise $= \phi_0 = 150°$. The standard deviation of the response is computed for two different sets of errors—one given by

$$\mathbf{X}_1 = \begin{Bmatrix} 0.010 \\ 0.010 \\ 0.003 \\ 0.006 \\ 0.003 \end{Bmatrix}$$

and the other by

$$\mathbf{X}_2 = \begin{Bmatrix} 0.1847 \\ 0.1843 \\ 0.0299 \\ 0.0731 \\ 0.0357 \end{Bmatrix}$$

The numerical results, along with the mean kinematic response, are shown in Table 9.2. Since normal distribution is assumed for all the random variables, the actual response of the cam-follower system is expected to fall in the range (mean response ± 3 standard deviations) 99.73 percent of the time. This can be interpreted as a measure of reliability of the system.

9.5.2 Four-Bar Mechanisms

A probabilistic model of linkage mechanisms, considering tolerances on the link lengths and clearances in the hinges, is developed in this section. Consider the planar four-bar mechanism shown in Fig. 9.7. In this mechanism, l_1, l_2, l_3, and l_4 are the link lengths, θ is the input angle, and ϕ is the output angle. Let this mechanism be used for generating

TABLE 9.2 Standard deviation of the kinematic response of the cam-follower system

Angle ϕ_2 (Degrees)	Mean Kinematic Response	Standard Deviation Corresponding to X_1	Standard Deviation Corresponding to X_2
0	0.0000	0.004051	0.06521
10	0.00773	0.004074	0.06742
20	0.06023	0.004138	0.06795
30	0.1945	0.004194	0.06844
40	0.4335	0.004209	0.06860
50	0.7820	0.004232	0.06886
60	1.226	0.004237	0.06952
70	1.734	0.004230	0.06890
80	2.266	0.004219	0.06880
90	2.774	0.004209	0.06869
100	3.218	0.004201	0.06860
110	3.566	0.004197	0.06854
120	3.805	0.004196	0.06852
130	3.940	0.004192	0.06847
140	3.992	0.004187	0.06840
150	4.000	0.004184	0.06837

Source: S. S. Rao and S. S. Gavane, "Analysis of Synthesis of Mechanical Error in Cam-Follower Systems," *Journal of Mechanical Design*, ASME, Vol. 104, January 1982.

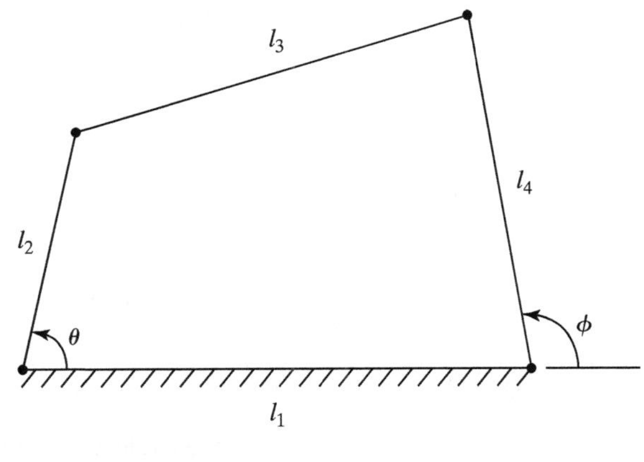

FIGURE 9.7 A planar four-bar mechanism.

a function, $\phi = \phi(\theta)$. For an arbitrary choice of the link lengths, the output angle ϕ may vary with θ as $\phi = \phi_s(\theta)$, instead of $\phi(\theta)$ as shown in Fig. 9.8. The difference, $|\phi(\theta) - \phi_s(\theta)|$, is known as the structural error. In practical mechanisms, the play or

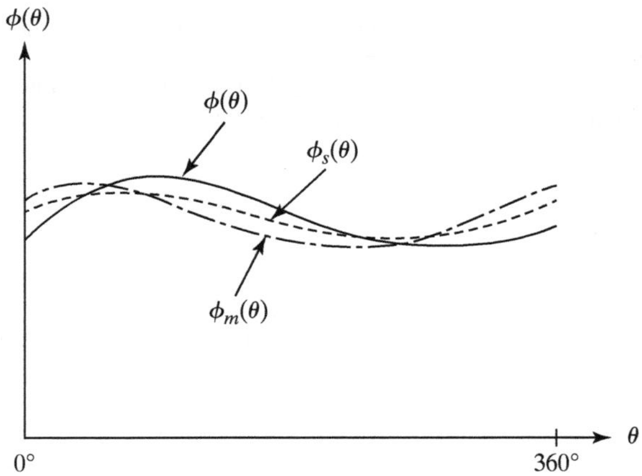

FIGURE 9.8 Generalized output angle—deterministic analysis.

clearances in the joints and tolerances on the link lengths of the mechanism cause an additional error, known as the mechanical error, and the output angle may vary with θ as $\phi = \phi_m(\theta)$. Depending on the values of the clearances in the joints and the tolerances on the link lengths, the mechanical error may be much larger than the structural error. Since the tolerances and clearances are probabilistic in nature, a probabilistic model is necessary for the analysis of the four-bar mechanism.

Deterministic Analysis. The input–output relation of the ideal mechanism [9.13] shown in Fig. 9.7 can be expressed as:

$$\phi_a(\theta) = 2\tan^{-1}\left(\frac{A \pm D}{B + C}\right) \tag{9.56}$$

where

$$A = \sin\theta \tag{9.57}$$

$$B = \cos\theta - \frac{l_1}{l_2} \tag{9.58}$$

$$C = \frac{l_1^2 + l_2^2 - l_3^2 + l_4^2}{2 l_2 l_4} - \frac{l_1}{l_4}\cos\theta \tag{9.59}$$

$$D = \sqrt{A^2 + B^2 - C^2} \tag{9.60}$$

and ϕ_a is the actually generated value of the output angle corresponding to the input angle θ. The two values of ϕ_a given by Eq. (9.56) correspond to the two possible configurations of the linkage for any specific value of the input angle θ as shown in Fig. 9.9.

Probabilistic Analysis. The line diagram of a four-bar linkage showing the effect of clearances and tolerances in an exaggerated way is given in Fig. 9.10. If only the tolerances are considered, the random link length, R_i, of the ith link lies anywhere in the range

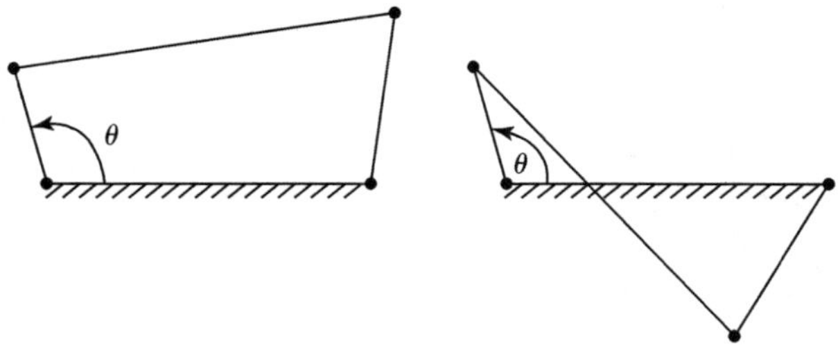

FIGURE 9.9 Two possible configurations of the linkage.

$(l_i \pm \Delta l_i)$, where l_i is the nominal link length and Δl_i is the tolerance. The axis of the pin may lie anywhere in the clearance circle of radius r_{ij}, where r_{ij} is the radial clearance in the hinge between jth and ith links. Let us fix four sets of rectangular coordinate systems $X_{ij}Y_{ij}$ at the hinges so that the X_{ij} is coincident with the center line of the link ending in the race, Fig. 9.11 [9.13]. When the effects of both tolerances and clearances are considered, the equivalent linkage is shown by the dashed lines in Fig. 9.10 [9.13]. The link lengths of the equivalent mechanism are given by

$$R_{i1}^2 = (R_i + x_{ij})^2 + y_{ij}^2 \approx (R_i + x_{ij})^2;$$
$$j = i + 1 \text{ for } i = 1, 2, 3; \qquad j = 1 \text{ for } i = 4 \tag{9.61}$$

as $R_i + x_{ij} \gg y_{ij}$ and hence

$$R_{i1} \approx R_i + x_{ij} \tag{9.62}$$

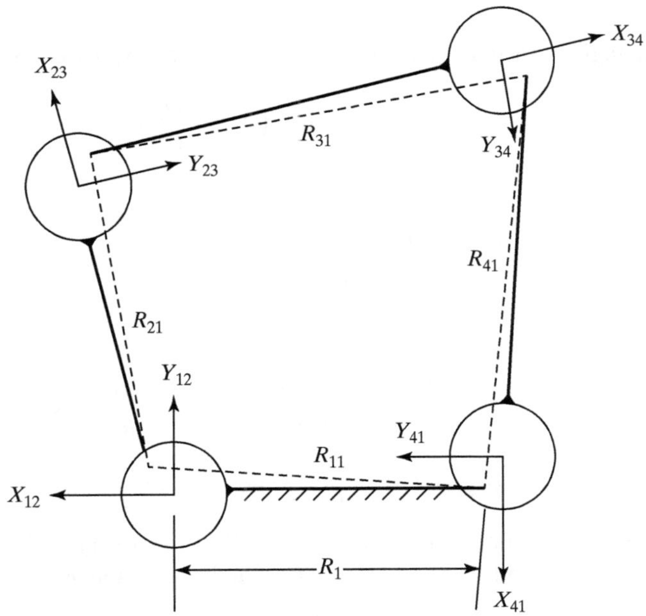

FIGURE 9.10 A four-bar linkage with tolerances and clearances.

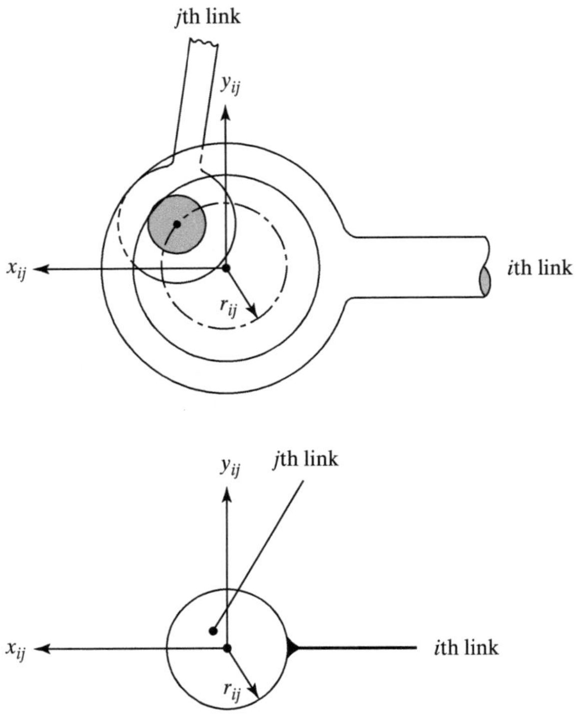

FIGURE 9.11 Clearance in the joint between two links.

The output angle of the equivalent linkage [9.13] can be expressed as:

$$\phi = 2 \tan^{-1}\left(\frac{A' \pm D'}{B' + C'}\right) \tag{9.63}$$

where

$$A' = \sin\theta \tag{9.64}$$

$$B' = \cos\theta - \frac{R_{11}}{R_{21}} \tag{9.65}$$

$$C' = \frac{R_{11}^2 + R_{21}^2 - R_{31}^2 + R_{41}^2}{2R_{21}R_{41}} - \frac{R_{11}}{R_{41}}\cos\theta \tag{9.66}$$

$$D' = \sqrt{A'^2 + B'^2 - C'^2} \tag{9.67}$$

By defining a set of random variables V_i as $V_1 = R_1$, $V_2 = R_2$, $V_3 = R_3$, $V_4 = R_4$, $V_5 = x_{12}$, $V_6 = x_{23}$, $V_7 = x_{34}$, and $V_8 = x_{41}$, Eq. (9.63) can be rewritten as

$$\phi = \phi(\theta; V_i, i = 1, 2, \dots, 8) \tag{9.68}$$

The actual output angle, in the presence of random tolerances and clearances, will fall in, say, $\pm 3\sigma_\phi$ band with a 99.73 percent probability if ϕ follows normal distribution. This is indicated in Fig. 9.12. To find the probability of generating $\phi(\theta)$ within a specified range of ϕ, we need to know the probability distribution function of ϕ, which is a difficult task. An approximate probability analysis can be conducted by finding the mean and standard deviation of ϕ using the partial derivative rule. For this, the link lengths are assumed to follow normal distribution with

$$\overline{V}_i = E(V_i) = l_i; \quad \sigma_{V_i} = \left(\frac{\Delta l_i}{3}\right); \quad i = 1, 2, 3, 4 \tag{9.69}$$

For the location of the pin center in the clearance circle, we assume a bivariate normal probability density function with independent random variables (x_{ij}, y_{ij})

$$f(x_{ij}, y_{ij}) = \frac{1}{2\pi\sigma_{x_{ij}}\sigma_{y_{ij}}} \exp\left\{-\frac{1}{2}\left(\frac{x_{ij} - \overline{x}_{ij}}{\sigma_{x_{ij}}}\right)^2 - \frac{1}{2}\left(\frac{y_{ij} - \overline{y}_{ij}}{\sigma_{y_{ij}}}\right)^2\right\} \tag{9.70}$$

In the present case, $\overline{x}_{ij} = \overline{y}_{ij} = 0$, $\sigma_{x_{ij}} = \sigma_{y_{ij}} = r_{ij}/3$, and hence Eq. (9.70) reduces to

$$f(x_{ij}, y_{ij}) = \frac{9}{2\pi r_{ij}^2} \exp\left\{-\frac{9(x_{ij}^2 + y_{ij}^2)}{2r_{ij}^2}\right\} \tag{9.71}$$

By using the partial derivative rule, the mean and standard deviations of ϕ can be determined, from Eq. (9.68), as

$$\overline{\phi} = \phi(\theta; \overline{V}_i, i = 1, 2, \ldots, 8) \tag{9.72}$$

$$\sigma_\phi = \left\{\sum_{i=1}^{8}\left(\frac{\partial\phi}{\partial V_i}\bigg|_{\overline{V}_1, \overline{V}_2, \ldots, \overline{V}_8}\right)^2 \sigma_{V_i}^2\right\}^{1/2} \tag{9.73}$$

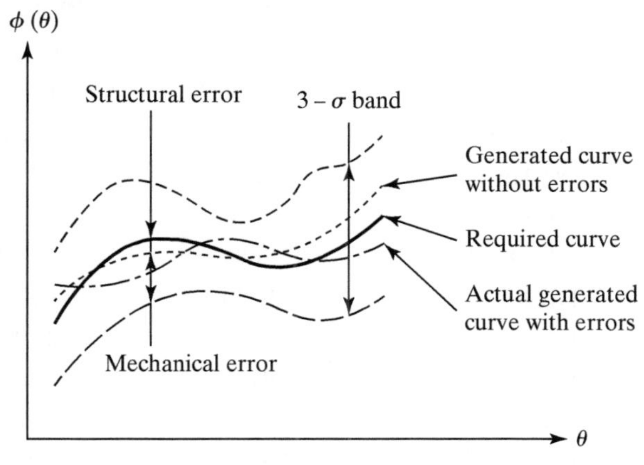

FIGURE 9.12 Generated output angle.

Once $\overline{\phi}$ and σ_ϕ are known, the $\pm 3\sigma_\phi$ band of the output angle ϕ about the mean value $\overline{\phi}$ can be constructed. Since all V_i are assumed to follow normal distribution, ϕ also follows normal distribution in view of the central limit theorem. Thus, the actually generated output angle ϕ is guaranteed to fall in the band 99.73 percent of the time.

Numerical Results. The $\pm 3\sigma_\phi$ band of the output angle is determined using the following data:

Function generated: $y = \sin x$

Range of x: $0° \le x \le 90°$, x is proportional to θ, y is proportional to ϕ

Range of the input angle: $\Delta\theta = 90°$

Range of the output angle: $\Delta\phi = 90°$

Nominal link lengths: $l_1 = 1, l_2 = 1.9, l_3 = 2.7, l_4 = 0.85$

Tolerances: $\Delta l_i = 0.0002; i = 1, 2, 3, 4$

Clearances: $\Delta r_{ij} = 0.0002; ij = 12, 23, 34, 41$

The range of the input angle $(\Delta\theta)$ is divided into 10 equal parts for computational convenience so that θ_1 and θ_{11} indicate the starting and the final angular positions of the input link. θ_1 is assumed to be 2.0283 radians. The required and mean-generated angular positions of the output link, the structural error, the 3 σ_ϕ value of the mechanical error at various positions of the input link are given in Table 9.3. Since ϕ follows normal distribution,

TABLE 9.3 $3\sigma_\phi$–values for the four-bar mechanism

Input Angle (i)	Angular Position of the Output Link (Degrees): Required Value (ϕ_{required})	Angular Position of the Output Link (Degrees): Mean-Generated Value $(\overline{\phi})$	Structural Error (Degrees)	Mechanical Error (Degrees) $3-\sigma$ Value
1	43.4510	43.4510	0.0000	3.2946
2	57.5301	56.5167	−1.0134	3.1956
3	71.2625	68.3578	−2.9047	3.1601
4	84.3101	79.0859	−4.2242	3.1630
5	96.3516	88.7510	−7.6006	3.1851
6	107.0905	97.3676	−9.7229	3.2104
7	116.2624	104.9340	−11.3284	3.2260
8	123.6415	111.4473	−12.1942	3.2232
9	129.0460	116.9146	−12.1314	3.1989
10	132.3429	121.3578	−10.9851	3.1563
11	133.4509	124.8113	−8.6396	3.1045

Source: Reprinted with permission from S. S. Rao and C.P. Reddy, "Mechanism Design by Chance Constrained Programming," *Mechanism and Machine Theory*, Vol. 14, copyright 1979, Pergamon Press PLC, pp. 413–424.

the probability of the actual value of ϕ falling in the band (mean-generated value $\pm 3\sigma_\phi$) at any value of θ will be 99.73 percent.

9.6 Numerical Solutions Using MATLAB and Excel

EXAMPLE 9.7

Use MATLAB to find the mean value of the wire diameter of the spring considered in Example 9.3 to achieve a reliability of 0.99.

Solution: The lower limit of integration in the standard normal probability density function corresponding to the required reliability of 0.99 is given by $z_1 = -2.3263$. The equation governing the mean value of the wire diameter is given by

$$z_1 = \frac{50 - 85227.2727\,\bar{d}^4}{\sqrt{(1.6667)^2 + (37149.7074\,\bar{d}^4)^2}} \tag{E1}$$

By squaring both sides of Eq. (E1), by substituting the value of z_1, and simplifying results in

$$2.0624847 \times 10^8\,\bar{d}^8 + 426.13636 \times 10^4\,\bar{d}^4 - 2484.964392 = 0 \tag{E2}$$

The solution of this quadratic equation in $\bar{d}^4$ yields the (real) solution

$$\bar{d}^4 = 5.6754825 \times 10^{-4} \quad \text{or} \quad \bar{d} = 0.154348 \text{ in} \tag{E3}$$

A MATLAB program is written to find the mean wire diameter of the spring. The program and the results are shown below:

```
clear all ; clc ;

reliab = 0.99     ;
z1 = norminv(reliab)

aa = (z1^2)*(37149.7074^2) - (85227.2727^2)
bb = 4.2613636 * 10^6
cc = (z1^2)*(1.6667^2)-(50^2)

disc = (bb^2) - 4*aa*cc

d4_p = (-bb + sqrt(disc))/(2*aa)
d4_n = (-bb - sqrt(disc))/(2*aa)

d = d4_p^0.25
z1 =

   2.3263

aa =

   2.0527e+008

bb =

   4.2614e+006
```

```
cc =

  -2.4850e+003

disc =

   2.0200e+013

d4_p =

   5.6762e-004

d4_n =

  -0.0213

d =

   0.1544
```

■

EXAMPLE 9.8

Use Excel to find the mean value of the wire diameter of the spring considered in Example 9.3 to achieve a reliability of 0.99.

Solution: The lower limit of integration in the standard normal probability density function corresponding to the required reliability of 0.99 is given by $z_1 = -2.3263$. The solution of the resulting quadratic equation in $\overline{d}^4$, Eq. (E2) of Example 9.7, is given by $\overline{d} = 0.154348$ in. The Excel computation of the solution is shown in the following spreadsheet.

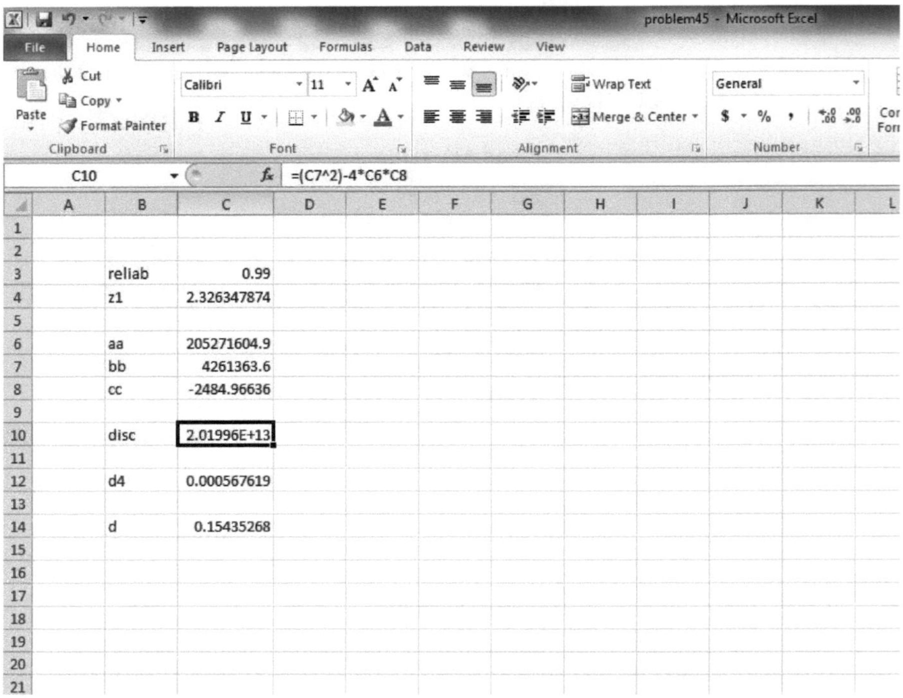

■

REFERENCES AND BIBLIOGRAPHY

9.1 R. W. Fitzgerald, *Mechanics of Materials*, 2nd ed., Addison-Wesley, Reading, MA., 1982.

9.2 J. H. Faupel and F. E. Fisher, *Engineering Design*, 2nd ed., John Wiley, New York, 1981.

9.3 J. E. Shigley and C. R. Mischke, *Mechanical Engineering Design*, 5th ed., McGraw-Hill, New York, 1989.

9.4 C. R. Mischke, "Prediction of Stochastic Endurance Strength," *ASME Journal of Vibration, Acoustics, Stress, and Reliability in Design*, Vol. 109, 1987, pp. 113–122.

9.5 J. Marin, *Mechanical Behavior of Engineering Materials*, Prentice-Hall, Englewood Cliffs, NJ, 1962.

9.6 R. E. Peterson, "Design Factors for Stress Concentration," *Machine Design*, Vol. 23, No. 2, February. 1951, p. 169; No. 3, March 1951, p. 161; No. 5, May 1951, p. 159; No. 6, June 1951, p. 173; No. 7, July 1951, p. 155.

9.7 R. C. Juvinall and K. M. Marshek, *Fundamentals of Machine Component Design*, 2nd ed., John Wiley, New York, 1991.

9.8 S. S. Rao, "A Probabilistic Approach to the Design of Gear Trains," *International Journal of Machine Tool Design and Research*, Vol. 14, 1974, pp. 267–278.

9.9 S. S. Rao and G. Das, "Reliability Based Optimum Design of Gear Trains," *ASME Journal of Mechanisms, Transmissions, and Automation in Design*, Vol. 106, 1984, pp. 17–22.

9.10 S. S. Rao and S. S. Gavane, "Analysis and Synthesis of Mechanical Error in Cam-Follower Systems," *Journal of Mechanical Design, Transactions of ASME*, Vol. 104, 1982 pp. 52–62.

9.11 S. S. Rao, "Error Analysis of Cam-Follower Systems: A Probabilistic Approach," *Proceedings of the Institution of Mechanical Engineers, London*, Part C, *Mechanical Engineering Science*, Vol. 198C, 1984, pp. 155–162.

9.12 H. R. Kim and W. R. Newcombe, "Stochastic Error Analysis in Cam Mechanisms," *Mechanism and Machine Theory*, Vol. 13, 1978, pp. 631–641.

9.13 S. S. Rao and C. P. Reddy, "Mechanism Design by Chance Constrained Programming," *Mechanism and Machine Theory*, Vol. 14, 1979, pp. 413–424.

9.14 A. C. King and C. B. Read, "Pathways to Probability," Holt, Rinehart and Winston, Inc., New York, 1963.

9.15 C. C. Gillispie (ed.-in-chief), *Dictionary of Scientific Biography*, Vol. 5, Charles Scribner's Sons, New York, 1980.

REVIEW QUESTIONS

9.1 Give brief answers to the following:

1. What is the purpose of Marin factors in fatigue design of machine components?
2. What is meant by endurance strength of a material?
3. How is the ultimate tensile strength related to the endurance strength of a material?
4. What is a fatigue failure?
5. What is the significance of three-sigma band of the performance of a mechanism?
6. What type of probabilistic model is useful for a multispeed gear train?

7. Give two examples of mechanical systems where the weakest-link concept is useful for probabilistic modeling.
8. Give two examples of mechanical systems where the fail-safe concept is useful for probabilistic modeling.
9. Suggest a method of including the probabilities of failure of shafts and bearings in the reliability analysis of a gear box.
10. What is the difference between structural error and mechanical error in a linkage mechanism?
11. What are the sources of error in a cam-follower mechanism?
12. What is the influence of error in the kinematic response of a cam-follower mechanism?

9.2 Indicate whether each of the following statements is true or false:

1. Gear trains are used in helicopters.
2. The purpose of a gear train is to provide multiple output speeds for a single input speed.
3. For any input angle, a four-bar mechanism will have more than two possible configurations.

9.3 Fill in the blank with the proper word(s):

1. The error between the desired output angle $\phi(\theta)$ and the generated output angle $\phi_s(\theta)$ for the input angle θ of a four-bar mechanism with specified configuration is called ________ error.
2. The justification for assuming normal distribution for the response parameters is the ________ theorem.
3. The lower limit of integration, z_1, for computing reliability with normal distribution assumption is ________.
4. A gear train can be modeled as a ________ kinematic chain.
5. Two common failure modes of a gear are bending and ________.
6. In any mechanism, the mechanical error is caused by ________ on link lengths.

9.4 Select the most appropriate answer from the multiple choices given:

1. In the case of lognormally distributed strength and load, the lower limit of integration, z_1, for computing reliability is:

   a. $-\dfrac{(\mu_S - \mu_L)}{\sqrt{\sigma_S^2 + \sigma_L^2}}$ **b.** $-\dfrac{(\mu_{\ln S} - \mu_{\ln L})}{\sqrt{\sigma_{\ln S}^2 + \sigma_{\ln L}^2}}$ **c.** $-\ln\left\{\dfrac{(\mu_S - \mu_L)}{\sqrt{\sigma_S^2 + \sigma_L^2}}\right\}$

2. The endurance limit of the laboratory specimen (S'_e) is often estimated from:
   a. Yield strength **b.** Nominal strength **c.** Ultimate strength

3. The miscellaneous-effects modification factor (k_f) in fatigue design is used for:
   a. Combined bending and twisting loads
   b. Corrosive atmosphere
   c. Hot-rolled components

4. The purpose of a cam, in general, is to provide
   a. Desired form of output displacement
   b. Sinusoidal form of output displacement
   c. Triangular form of output displacement

5. Noise level of a cam mechanism increases due to
   a. Clearances in bearings
   b. Radius of the bearings
   c. Distance between bearings

6. Marin fatigue strength reduction factors are used to compute:
   a. Yield strength
   b. Endurance strength
   c. Ultimate strength

9.5 Match the following in the context of fatigue design:

1. k_a	**a.** Temperature modification factor
2. k_b	**b.** Load modification factor
3. k_c	**c.** Miscellaneous-effects modification factor
4. k_d	**d.** Stress-concentration modification factor
5. k_e	**e.** Size modification factor
6. k_f	**f.** Surface-condition modification factor

PROBLEMS

9.1 A cold-rolled steel rectangular plate, having a central hole, is subjected to a completely reversed axial load of $F = 1$ kip as shown in Fig. 9.13. If the ultimate tensile strength of the material is 90 k(lb/in^2) and the geometric stress concentration factor is 2.42, find the factor of safety against fatigue failure.

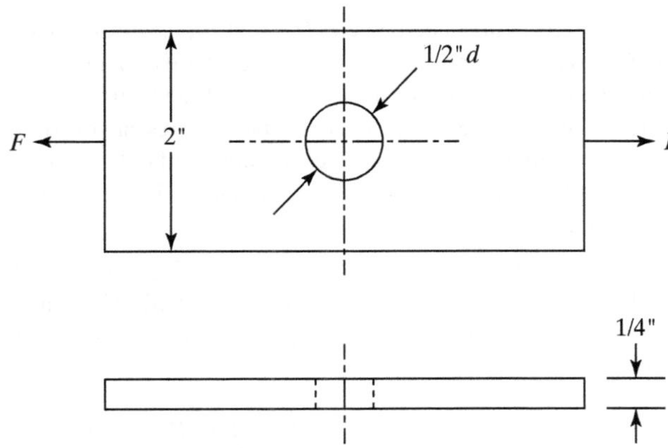

FIGURE 9.13 Rectangular plate with a hole subjected to axial load.

9.2 For the rectangular steel plate described in Problem 9.1, the reversed axial load is given by $F = N(1.0, 0.08)$ kip. If the mean ultimate tensile strength of the material is 90 k(lb/in^2), determine the reliability of the plate by assuming normal distribution for both the induced stress and the endurance strength of the component.

9.3 For the rectangular steel plate described in Problem 9.1, the reversed axial load is given by $F = N(1.0, 0.08)$ kip. If the mean ultimate tensile strength of the material is 90 k(lb/in^2), estimate the reliability of the plate by assuming lognormal distribution for both the induced stress and the endurance strength of the component.

9.4 A thin-walled steel spherical shell of radius 40 in is supported along the equator as shown in Fig. 9.14. It is filled with a liquid having a specific weight of $w = N(62.4, 3.12)$ lb/ft^3. The shell material has a yield stress of $s_y = N(20{,}000, 2000)$ lb/in^2. Determine the shell wall thickness for a reliability of 0.99.

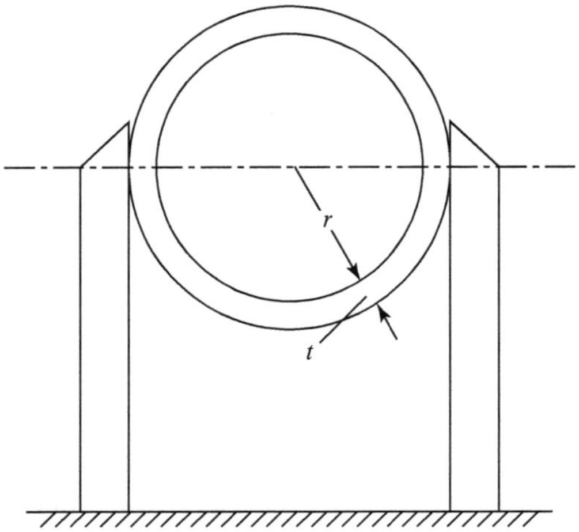

FIGURE 9.14 A thin-walled spherical shell.

9.5 A thin-walled steel cylinder of diameter $D = N(24, 0.5)$ in and thickness $t = N(0.1, 0.01)$ in is reinforced by a layer of closely wound steel wire of diameter $d = N(\bar{d}, 0.1\,\bar{d})$ in. The wire is wound with an initial tension of $p_1 = N(10{,}000, 1000)$ lb/in^2 while the cylinder is empty. The cylinder is then subjected to an internal pressure of $p_2 = N(1000, 100)$ lb/in^2. Find the mean wire diameter to achieve a reliability of 0.999 assuming that the stress in the cylinder in the longitudinal direction is negligible (as in the case of a section of a pipe with expansion joints).

9.6 A steel gear with a central hole of radius $R = N(1.5, 0.001)$ in is shrink-fitted over a solid steel shaft of radius $(1.5 + \delta)$ in as shown in Fig. 8.37. The gear has a pitch circle radius of $r_o = N(6, 0.1)$ in. The pressure developed at the interface, p, due to shrink-fit [9.3] is given by

$$p = E\frac{\delta}{R}\left[\frac{(r_o^2 - R^2)\,(R^2 - r_i^2)}{2\,R^2(r_o^2 - r_i^2)}\right]$$

where E = Young's modulus, δ = radial interference, r_o = outer radius of the outer part, r_i = inner radius of the inner part, and R = transition (nominal) radius at the interface. If the Young's modulus is normally distributed with $E = N(30 \times 10^6, 3 \times 10^6)$ lb/in^2 and the permissible shrink-fit pressure, p_o, can be taken as $p_o = N(45{,}000, 4500)$ lb/in^2, determine the radial interference δ to achieve a reliability of 0.99.

9.7 The bending stress developed in a gear tooth is given by the Lewis equation [9.7]

$$s = \frac{F_t p_d}{wY}$$

where F_t = tangential load acting on the gear tooth, p_d = diametral pitch = π/p, p = circular pitch, w = face width, and Y = Lewis form factor:

$$Y = \frac{p_d\, t^2}{6\, h}$$

where t = root thickness and h = radial depth of the tooth (see Fig. 8.8). In an automobile transmission, a gear has full-depth teeth with a pressure angle of 20°, a diametral pitch of $p_d = N(1, 0.05)$, a face width of $w = N(\overline{w}, 0.1\,\overline{w})$ in, a Lewis form factor of $Y = N(0.322, 0.0322)$ and is subjected to a tangential load of $F_t = N(5000, 500)$ lb. If the yield strength of the material is given by $s_y = N(30{,}000, 3000)$ lb/in^2, find the mean face width of the gear tooth for a reliability of 0.99.

9.8 The torque-transmitting capacity of a cone clutch under uniform pressure [9.3] is given by

$$T = \frac{Ff}{3 \sin \alpha}\left[\frac{D^3 - d^3}{D^2 - d^2}\right]$$

where D = outer diameter of the cone, d = inner diameter of the cone, α = semi-cone angle, f = coefficient of friction between cup and cone, and F = axial force applied (see Fig. 8.7). For a specific clutch, which has molded asbestos on a cast-iron cone, $d = N(8, 0.08)$ in, $\alpha = N(20°, 2°)$, $F = N(100, 10)$ lb, and $f = N(0.3, 0.03)$. If the torque required to be transmitted by the cone clutch is also normally distributed with mean and standard deviations of 400 lb · in and 40 lb · in, respectively, find the outer diameter of the cone (D) to achieve a reliability of 0.99. Assume that the outer diameter of the cone is given by $D = N(\overline{D}, 0.05\overline{D})$ in.

9.9 The cantilever beam shown in Fig. 9.15 is subjected to an alternating force $F = N(400, 40)$ lb. The beam is machined from AISI 1040 steel and heat-treated with an ultimate tensile strength of $S_{ut} = N(85, 8.5)$ k(lb/in^2). Find the factor of safety based on a deterministic approach.

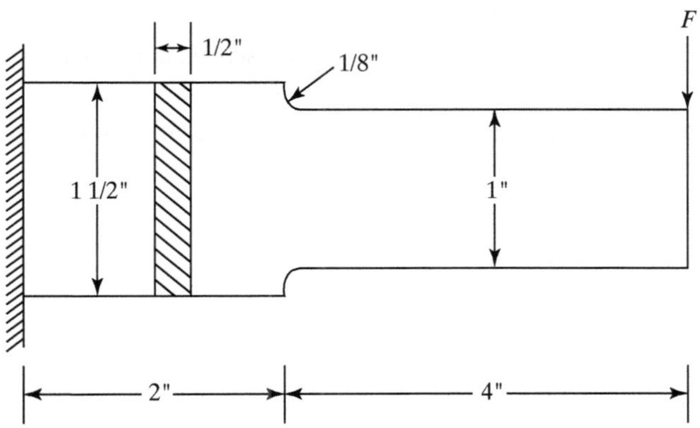

FIGURE 9.15 A cantilever beam.

9.10 Find the reliability of the beam described in Problem 9.9 by assuming that the induced stress and the strength of the material follow normal distribution.

9.11 Find the reliability of the beam described in Problem 9.9 by assuming that the induced stress and the strength of the material follow lognormal distribution.

9.12 The steel shaft shown in Fig. 9.16 has a transverse hole of diameter a and is subjected to an alternating torque of $T = N(30{,}000, 3000)$ lb · in. The mean ultimate tensile strength of the material is 150 k(lb/in^2). The geometric stress concentration factor at the hole is 1.62. If the diameter of the shaft is $D = N(\overline{D}, 0.1\overline{D})$ in, and that of the hole is $a = N(0.25, 0.025)$ in, determine the mean diameter of the shaft, $\overline{D}$, for a factor of safety of 2.

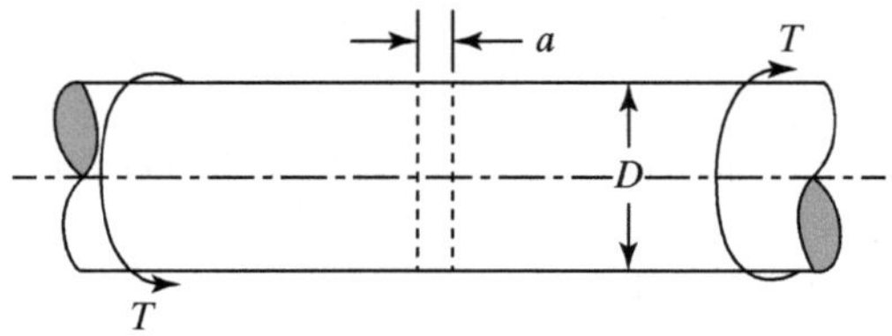

FIGURE 9.16 A shaft subjected to alternating torque.

9.13 Find the mean diameter of the shaft, $\overline{D}$, described in Problem 9.12 to achieve a reliability of 0.99 by assuming that the induced stress and the strength of the material follow normal distribution.

9.14 Find the mean diameter of the shaft described in Problem 9.12 to achieve a reliability of 0.99 by assuming that the induced stress and the strength of the material follow lognormal distribution.

MATLAB and Excel Problems

9.15 Find the thickness of the pressure vessel considered in Example 9.2 to achieve a reliability of 0.9999997 using MATLAB.

9.16 Use MATLAB to determine the mean face width of the gear tooth described in Problem 9.7 to achieve a reliability of 0.99.

9.17 Find the thickness of the pressure vessel described in Example 9.2 to achieve a reliability of 0.9999997 using Excel.

9.18 Use Excel to find the mean face width of the gear tooth considered in Problem 9.7 to achieve a reliability of 0.99.

CHAPTER 10

Monte Carlo Simulation

What You Will Learn

In this chapter, you will learn

- That the Monte Carlo method, based on the use of random numbers, is a powerful technique for the uncertainty analysis of complex engineering problems
- The methods of generating random numbers that follow a specified probability distribution using uniformly distributed random numbers
- The method of generating sets of random numbers that follow a specified joint probability distribution
- The generation of random numbers that represent a set a correlated normal variables
- The computation of reliability of a system using a large number of randomly generated trial systems
- The estimation of the error involved with the computed reliability of a system in terms of the number of trial systems used in the computation
- The use of the Monte Carlo method in the estimation of the reliability of a straight-line generating four-bar mechanism

Monte Carlo is a town in Monaco with a population of 35,986 as of 2011 and is located about one mile north of Monaco Ville, the capital of the country. The name of the town was given by Prince Charles III. Although the citizens of Monaco are not admitted to the casino, the town is very famous for its luxurious casino, which was founded in 1856 by the prince himself. Besides being famous as a gambling place, Monte Carlo is one of the most fashionable seaside resorts on the French Riviera. Two world famous automobile races are held annually in this town—the International Monaco Grand Prix and the Monte Carlo Rally. The name *Monte Carlo method* dates back to about 1944 when von Neumann and Ulam introduced it as a code name for their secret work on neutron diffusion problems, encountered in their work on the atomic bomb, at the Los Alamos Scientific Laboratory. Probably the name was chosen because roulette is one of the simplest tools available for generating random numbers and the roulette is traditionally associated with the casino town, Monte Carlo [10.14, 10.15].

10.1 Introduction

Monte Carlo simulation is a powerful engineering tool that can be used for the statistical analysis of uncertainty in engineering problems. It is particularly useful for complex problems in which several random variables are related through nonlinear equations. The Monte Carlo analysis can be considered as an experiment performed on a computer rather than performed in an engineering laboratory. If a system parameter is known to follow certain probability distribution, the performance of the system is studied by considering several possible values of the parameter, each following the specified probability distribution. In the context of structural reliability, for example, the Monte Carlo method involves constructing trial structures on a computer according to generated random numbers and determining the percentage of structures that fail [10.1]. Of course, a large number of trial structures are needed if high confidence is required at small failure probability levels.

Since physical experimentation on real systems is a time-consuming and costly affair, this experimentation on trial systems generated with random numbers has proved to be very convenient. The Monte Carlo approach, thus, requires considerable calculations and is useful only for complex interrelated systems or for the verification of other analyses. Also, the solutions derived from Monte Carlo simulations cannot provide generalizations or extrapolations about the system behavior. The term *Monte Carlo* is associated with the city of Monte Carlo in Monaco, which is well known for gambling, especially the roulette wheel. Since the simulation process involves the generation and use of random or chance variables, it has been called the *Monte Carlo simulation*.

10.2 Generation of Random Numbers

The essential feature common to all Monte Carlo computations is that we need to substitute for each random variable a corresponding set of numbers having the statistical properties of that random variable. The numbers that we substitute are called random numbers, on the grounds that they could well have been produced by chance by a suitable random process. Special devices may be used for generating random variables. For example, a fair coin can be tossed for a discrete random variable having two equally likely possible values. Uniformly distributed random numbers within a specified range may be generated by spinning a frictionless disc or wheel with the range subdivided equally on the circumference of the disc. Before digital computers became popular, published tables of random numbers [10.2] were used for hand computations. These tables were generated by physical processes which are, as far as one can tell, random in the strict sense, and have been successfully subjected to a number of statistical tests. For Monte Carlo simulation using digital computers, it will be necessary to generate automatically the requisite random numbers with specified probability distributions. This is accomplished by first generating a uniformly distributed random number between 0 and 1 and then using a suitable transformation to obtain the corresponding random number following the specified probability distribution. For the standard uniform random variable, U, the density and distribution functions are given by (see Figs. 10.1(a) and (b)):

$$f_U(u) = \begin{cases} 0; & u < 0 \\ 1; & 0 \le u \le 1 \\ 0; & u > 1 \end{cases} \tag{10.1}$$

and

$$F_U(u) = \begin{cases} 0; & u < 0 \\ u; & 0 \le u \le 1 \\ 1; & u > 1 \end{cases} \tag{10.2}$$

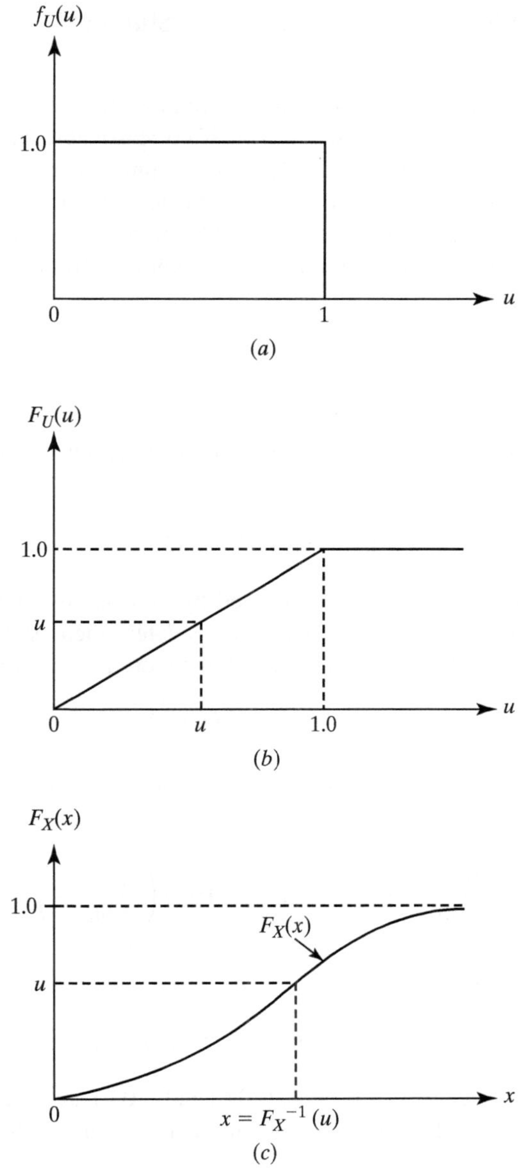

FIGURE 10.1 (*a*) Probability density function of standard uniform variate U; (*b*) probability distribution function of standard uniform variate U; and (*c*) probability distribution function of a general random variable X.

Let X be a random variable with the distribution function $F_X(x)$. Then the value of x corresponding to a given value of the cumulative probability, say, u, is given by

$$x = F_X^{-1}(u) \tag{10.3}$$

where F_X^{-1} denotes the inverse function of F_X. The significance of Eq. (10.3) can be seen in Fig. 10.1(*c*).

10.2.1 Generation of Random Numbers Following Standard Uniform Distribution

It is most convenient to calculate a sequence of random numbers, one at a time, by a completely specified rule that is devised such that no reasonable statistical test will detect any significant departure from randomness. Such numbers are called *pseudo-random numbers* [10.3]. The main advantage of a specified rule is that the random sequence can be reproduced for purposes of computational verification. The pseudo-random numbers u_i, following standard uniform distribution, can be computed from a sequence of positive integers x_i using the relation

$$u_i = \frac{x_i}{m} \tag{10.4}$$

where m is a suitable positive integer. A pseudo-random sequence that can be generated by a recurrence relation is given by

$$x_{i+1} = ax_i (\text{mod } m) \tag{10.5}$$

where x_{i+1} is the remainder when ax_i is divided by m, a is any integer between 0 and $m - 1$, and m is a large integer whose magnitude is determined by the limitations of the computer used. Equation (10.5) can be expressed in a more general form as

$$x_{i+1} = (ax_i + b)(\text{mod } m) \tag{10.6}$$

where a and b are integers between 0 and $m - 1$. Thus, if n_i denotes the integer part of the ratio

$$\left(\frac{ax_i + b}{m}\right) \quad \text{i.e.,} \quad n_i = \text{integer}\left(\frac{ax_i + b}{m}\right) \tag{10.7}$$

then the remainder x_{i+1} is given by

$$x_{i+1} = ax_i + b - m\, n_i \tag{10.8}$$

The sequence of numbers generated by Eqs. (10.6) and (10.4) are repetitive after, at most, m numbers. This means that the numbers are periodic. For example, the recurrence relation

$$x_{i+1} = (7x_i + 1)(\text{mod } 15) \tag{10.9}$$

with $x_0 = 1$ generates the sequence 1, 8, 12, 10, 11, 3, 7, 5, 6, 13, 2, 0; 1, 8, ... The period of this pseudo-random sequence can be seen to be twelve. Obviously, the period must be selected to be equal to or larger than the number of random numbers needed in a given problem. The uniformly distributed random numbers corresponding to the above pseudo-random sequence can be found by dividing the numbers by $m = 15$ (i.e., by using Eq. 10.4) as 0.066667, 0.533333, 0.800000, 0.666667, 0.733333, 0.20000, 0.4666667, 0.333333, 0.400000, 0.866667, 0.133333, 0.000000; 0.066667, 0.533333, ... It can be shown that for large values of m, the numbers generated using the deterministic formula, Eq. (10.6), are uniformly distributed and statistically independent [10.4]. Representative values are $a = 2^7 + 1$, $b = 1$, and $m = 2^{35}$ for binary computers [10.11].

10.2.2 Random Variables with Nonuniform Distribution

As stated earlier, if u denotes a particular value of a uniform random variate, the corresponding random number, x, following the distribution function $F_X(x)$ can be found as

$$x = F_X^{-1}(u) \tag{10.10}$$

The method of determining random numbers using Eq. (10.10) is known as the inverse-transformation method. If the function F can be inverted analytically (as in the cases of exponential and Weibull distributions), the application of Eq. (10.6) is simple and straightforward. On the other hand, if the probability distribution function cannot be inverted analytically (as in the cases of normal and lognormal distributions), other methods such as the composition method and the method of functions of random variables have to be used.

EXAMPLE 10.1

If $u_i, i = 1, 2, \ldots, N$ denote a set of standard uniformly distributed random numbers, find the corresponding numbers x_i of a random variable X, which is uniformly distributed in the range a to b.

Solution: The probability density and distribution functions of X are given by

$$f_X(x) = \begin{cases} \dfrac{1}{b-a}; & a \le x \le b \\ 0; & \text{elsewhere} \end{cases} \tag{E1}$$

and

$$F_X(x) = \int_a^x f_X(x')\,dx' = \begin{cases} \dfrac{x-a}{b-a}; & a \le x \le b \\ 0; & \text{elsewhere} \end{cases} \tag{E2}$$

Since u_i is uniformly distributed in $(0, 1)$, we set

$$u_i = F_X(x_i) = \frac{x_i - a}{b - a} \tag{E3}$$

which can be inverted to obtain

$$x_i = a + (b - a)u_i; \quad i = 1, 2, \ldots, N \tag{E4}$$

■

EXAMPLE 10.2

If $u_i, i = 1, 2, \ldots, N$ represent a set of standard uniformly distributed random numbers, find the corresponding numbers x_i of a random variable X that follows exponential distribution.

Solution: The probability density and distribution functions of X are given by

$$f_X(x) = \begin{cases} \dfrac{1}{\alpha} e^{-(x/\alpha)}; & 0 \le x < \infty, \alpha > 0 \\ 0; & \text{elsewhere} \end{cases} \tag{E1}$$

$$F_X(x) = \int_o^x f_X(x')\, dx' = \frac{1}{\alpha}\int_o^x e^{-(x'/\alpha)}\, dx' = 1 - e^{-(x/\alpha)} \tag{E2}$$

Equating the standard uniform number u_i to $F_X(x_i)$,

$$u_i = F_X(x_i) = 1 - e^{-(x_i/\alpha)} \tag{E3}$$

Equation (E3) can be inverted by first rewriting it as

$$1 - u_i = e^{-(x_i/\alpha)} \tag{E4}$$

and then taking logarithms on both sides of Eq. (E4) to obtain

$$x_i = -\alpha \ln(1 - u_i); \quad i = 1, 2, \ldots, N \tag{E5}$$

Since $(1 - u_i)$ also can be considered as a standard uniform random number, Eq. (E5) can be rewritten as

$$x_i = -\alpha \ln u_i; \quad i = 1, 2, \ldots, N \tag{E6}$$

■

EXAMPLE 10.3

If $u_i, i = 1, 2, \ldots, N$ indicate a set of standard uniformly distributed random numbers, find the corresponding numbers x_i of a random variable X that follows Weibull distribution (with two parameters).

Solution: The probability density and distribution functions of X are given by

$$f_X(x) = \begin{cases} \dfrac{\alpha}{\beta^\alpha} x^{\alpha-1} e^{-(x/\beta)^\alpha}; & 0 \le x < \infty \\ 0; & \text{elsewhere} \end{cases} \tag{E1}$$

and

$$F_X(x) = \begin{cases} 1 - e^{-(x/\beta)^\alpha}; & 0 \le x < \infty \\ 0; & \text{elsewhere} \end{cases} \tag{E2}$$

Equating u_i to $F_X(x_i)$,

$$u_i = 1 - e^{-(x_i/\beta)^\alpha} \tag{E3}$$

we can solve for x_i to obtain

$$x_i = \beta[-\ln(1 - u_i)]^{1>a}; \quad i = 1, 2, \ldots, N \tag{E4}$$

Since $(1 - u_i)$ can also be considered as a standard uniform number, Eq. (E4) can be rewritten as

$$x_i = \beta\,[-\ln u_i]^{1/\alpha}; \quad i = 1, 2, \ldots, N \tag{E5}$$

■

EXAMPLE 10.4

If u_i denotes a standard uniformly distributed random number, determine the corresponding number x_i of a random variable X that follows normal distribution.

Solution: If X follows normal distribution with mean μ and standard deviation σ, the corresponding standard normal variate Z is given by

$$Z = \frac{x - \mu}{\sigma} \tag{E1}$$

with

$$f_Z(z) = \frac{1}{\sqrt{2\pi}} e^{-z^2/2} \tag{E2}$$

The standard normal tables list the values z versus $F_Z(z)$. Corresponding to the given uniform number $u_i = F_Z(z_i)$, we can find the value of z_i from standard normal tables. Once z_i is known, the random number x_i can be determined from Eq. (E1) as

$$x_i = \mu + \sigma\, z_i \tag{E3}$$

Instead of using standard normal tables, z_i corresponding to $u_i = F_Z(z_i)$ can be determined by making use of the approximate formula given by Eq. (8.28). Another approximate method, presented by Tocher [10.5], is based on the approximation

$$e^{z^2/2} \approx \frac{2e^{-kz}}{(1 + e^{-kz})^2}; \quad z > 0; \quad k = \sqrt{\frac{8}{\pi}} \tag{E4}$$

The integral of $e^{z^2/2}$ gives

$$F_Z(z) = \int \frac{1}{\sqrt{2\pi}} e^{z^2/2}\, dz = \frac{1}{\sqrt{2\pi}} \int \frac{2e^{-kz}}{(1 + e^{-kz})^2}\, dz$$

$$= \frac{2}{1 + e^{-kz}} - 1 \tag{E5}$$

Equating $F_Z(z)$, given by Eq. (E5), to u_i and inverting the relation yields

$$z = \frac{1}{k} \ln\left(\frac{1 + u_i}{1 - u_i}\right) \tag{E6}$$

The desired random number z_i can be obtained by assigning randomly a sign to z in Eq. (E6).

■

EXAMPLE 10.5

The cumulative probability distribution function of a random variable X is known at a finite number of points from experimental results as shown in Fig. 10.2. Generate a random number x_k corresponding to the uniformly distributed number u_k.

Solution: Since the distribution function of X, $F_X(x)$, is not known in the form of an expression, we use a linear interpolation technique to generate the random number x_i. Let the given uniformly distributed number $u_k \equiv F_k$ fall in between the known discrete distribution function values F_i and F_{i+1}, which correspond to the values x_i and x_{i+1}, respectively, of the random variable X. Then the linear interpolation of F_k between F_i and F_{i+1} gives

$$\frac{F_k - F_i}{x_k - x_i} = \frac{F_{i+1} - F_i}{x_{i+1} - x_i}; \quad x_i \le x_k \le x_{i+1} \tag{E1}$$

By equating F_k to u_k, Eq. (E1) gives

$$x_k = x_i + (u_k - F_i)\left(\frac{x_{i+1} - x_i}{F_{i+1} - F_i}\right)$$

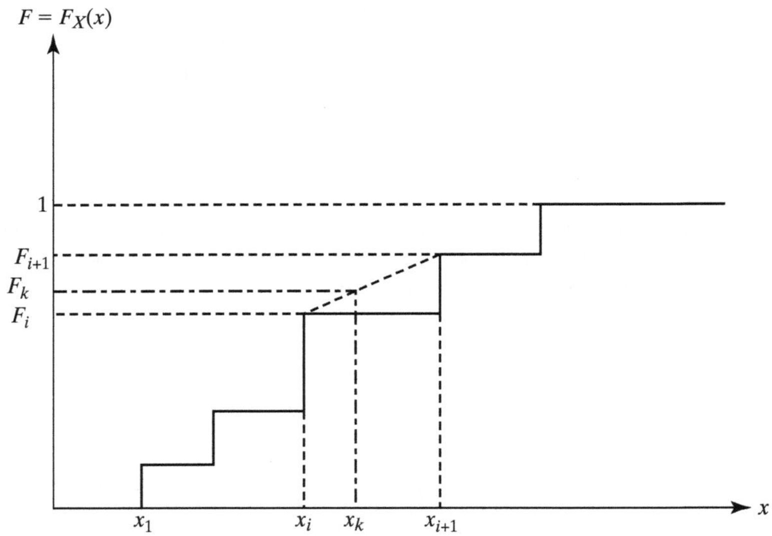

FIGURE 10.2 Cumulative distribution function of the random variable X.

■

10.2.3 Generation of Discrete Random Variables

Let X be a discrete random variable that can take $n + 1$ distinct values with the probability mass function $p_X(x_i) = P(X = x_i), i = 0, 1, \ldots, n$. The cumulative distribution function of X is given by $F_X(x_i) = P(X \le x_i) = \sum_{j=0}^{i} p_X(x_j)$. The standard uniformly distributed random numbers can be used to generate random numbers following the discrete distribution $F_X(x)$. Let u be a standard uniformly distributed number. The corresponding discrete random number x_i can be determined by identifying x_{i-1} and x_i such that

$$p_X(x_o) + p_X(x_1) + \cdots + p_X(x_{i-1}) < u \le p_X(x_o) + p_X(x_1) + \cdots + p_X(x_i)$$

that is

$$F_X(x_{i-1}) < u \le F_X(x_i) \tag{10.11}$$

It can be seen that the generation of discrete random numbers by this method requires the determination of the cumulative distribution function, in general, for all the possible values of the random variable. Each time a number u is generated, we need to search for the corresponding x_i which satisfies Eq. (10.11). This method is basically the inverse-transformation procedure requiring, in most cases, a numerical search to determine the inverse of the distribution function.

EXAMPLE 10.6

Generate random integers following uniform distribution between the integers a and b.

Solution: If u_i denotes a standard (continuous) uniformly distributed number in $(0, 1)$, we can set

$$u_i = F_X(x_i) = \frac{x_i - a}{b - a} \tag{E1}$$

and get

$$x_i = a + (b - a)u_i \tag{E2}$$

Since x_i is restricted to take only integer values, we can use the relation

$$x_i = a + [u_i(b - a) + 1] \tag{E3}$$

where the expression in brackets, $[.]$, denotes the largest integer value of $[u_i\,(b - a) + 1]$. For example, if $u_i = 0.2892$, $a = 1$, and $b = 15$, we obtain

$$\begin{aligned} x_i &= 1 + [0.2892(15 - 1) + 1] = 1 + [4.0488 + 1] \\ &= 1 + [5.0488] = 1 + 5 = 6 \end{aligned}$$

■

EXAMPLE 10.7

Generate random numbers following binomial distribution with $n = 5$ and $p = 0.4$ corresponding to the standard uniformly distributed numbers 0.75300 and 0.22015.

Solution: The probability mass function of X is given by

$$p_X(x_i) = \binom{5}{i}(0.4)^i(0.6)^{5-i}; \quad i = 0, 1, 2, 3, 4, 5 \tag{E1}$$

and the cumulative distribution function by

$$F_X(x_i) = \sum_{j=0}^{i}\binom{5}{j}(0.4)^j(0.6)^{5-j}; \quad i = 0, 1, 2, 3, 4, 5 \tag{E2}$$

The numerical values given by Eqs. (E1) and (E2) are shown in Table 10.1. The probability mass and distribution functions are shown graphically in Fig. 10.3. By equating the first standard uniform random number to the distribution function as $u_1 = 0.75300 = F_X(x_1)$, we find that u_1 is bounded by $F_X(2)$ and $F_X(3)$. Hence, the value of x_1 is given by 3. Similarly, corresponding to the second standard uniform random number, we have $u_2 = 0.22015 = F_X(x_2)$. Since u_2 is bounded by $F_X(0)$ and $F_X(1)$, we choose the value of x_2 as 1.

TABLE 10.1 Probability mass and distribution functions of X

i	$p_X(i)$	$F_X(i)$
0	0.07776	0.07776
1	0.2592	0.33696
2	0.3456	0.68256
3	0.2304	0.91296
4	0.0768	0.98976
5	0.01024	1.00000

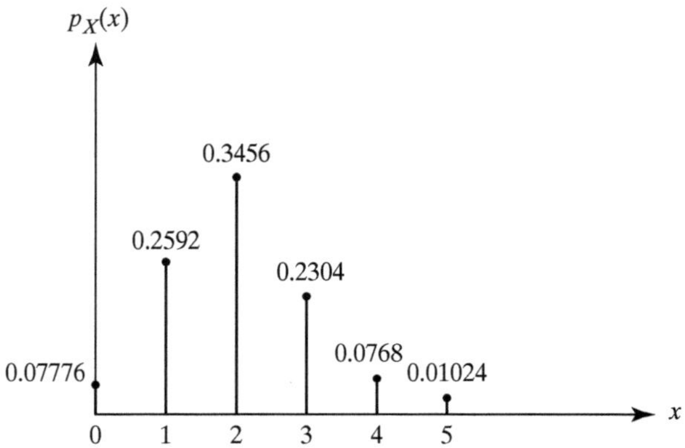

(*a*)

FIGURE 10.3 Probability mass and distribution functions of X.

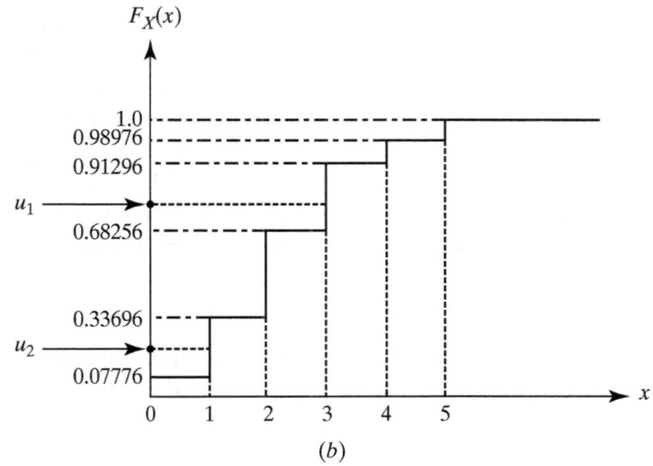

FIGURE 10.3 Probability mass and distribution functions of X. (*Continued*)

■

10.3 Generation of Jointly Distributed Random Numbers

10.3.1 Independent Random Variables

Let $X_1, X_2, \ldots, X_n$ be a set of n random variables. If these variables are statistically independent, then their joint density and distribution functions can be expressed as

$$f_{X_1,\ldots,X_n}(x_1, \ldots, x_n) = \prod_{i=1}^{n} f_{X_i}(x_i) \tag{10.12}$$

$$F_{X_1,\ldots,X_n}(x_1, \ldots, x_n) = \prod_{i=1}^{n} F_{X_i}(x_i) \tag{10.13}$$

where $f_{X_i}(x_i)$ and $F_{X_i}(x_i)$ denote the marginal (individual) probability density and distribution functions of X_i. In this case, the random numbers for each variable can be generated separately and independently of one another using the procedures described earlier.

10.3.2 Dependent Random Variables

For a set of n correlated (dependent) random variables $X_1, X_2, \ldots, X_n$, the joint probability density and distribution functions can be expressed as

$$f_{X_1,\ldots,X_n}(x_1, \ldots, x_n) = f_{X_1}(x_1) \cdot f_{X_2}(x_2|_{x_1}) \cdots f_{X_n}(x_n|_{x_1,\ldots,x_{n-1}}) \tag{10.14}$$

$$F_{X_1,\ldots,X_n}(x_1, \ldots, x_n) = F_{X_1}(x_1) \cdot F_{X_2}(x_2|_{x_1}) \cdots F_{X_n}(x_n|_{x_1,\ldots,x_{n-1}}) \tag{10.15}$$

where $f_{X_1}(x_1)$ and $F_{X_1}(x_1)$ are the marginal density and distribution functions of X_1, and $f_{X_i}(x_i|_{x_1, x_2, \ldots, x_{i-1}})$ and $F_{X_i}(x_i|_{x_1, x_2, \ldots, x_{i-1}})$ are the conditional density and distribution functions of X_i given $X_1 = x_1, X_2 = x_2, \ldots, X_{i-1} = x_{i-1}$. Since the random numbers are dependent, we cannot directly use the set of uniformly distributed random numbers to generate the random numbers corresponding to $X_1, X_2, \ldots, X_n$. However, the following procedure can be used to generate random numbers for $X_1, X_2, \ldots, X_n$.

Let $(u_1, \ldots, u_n)$ denote a set of uniformly distributed random numbers. Then the random number x_1 corresponding to X_1 can be determined from u_1 as

$$x_1 = F_{X_1}^{-1}(u_1) \tag{10.16}$$

With the value of x_1 known, the conditional distribution function $F_{X_2}(x_2|_{x_1})$ becomes a function only of x_2, and hence it can be inverted to find x_2 as

$$x_2 = F_{X_2}^{-1}(u_2|_{x_1}) \tag{10.17}$$

In general, we can generate x_i with the known values of $x_1, x_2, \ldots, x_{i-1}$ as

$$x_i = F_{X_i}^{-1}(u_i|_{x_1, x_2, \ldots, x_{i-1}}) \tag{10.18}$$

This recursive procedure can be continued until the last number x_n is generated as

$$x_n = F_{X_n}^{-1}(u_n|_{x_1, x_2, \ldots, x_{n-1}}) \tag{10.19}$$

It can be seen that the method requires the inversion of marginal and conditional distribution functions as indicated in Eqs. (10.16) to (10.19).

EXAMPLE 10.8

The contact pressure, p_{max}, developed between two cylinders [10.6] under a load F (see Fig. 10.4) is given by

$$p_{max} = \frac{2F}{\pi \varepsilon a} \tag{E1}$$

where 2 ε denotes the width of contact after deformation:

$$\varepsilon = \sqrt{\frac{4F(1 - \nu^2) d_1 d_2}{\pi E a (d_1 + d_2)}} \tag{E2}$$

where F = applied load, d_1 and d_2 = diameters of cylinders, ν = Poisson's ratio of cylinders, E = Young's modulus of cylinders, and a = axial length of cylinders. Assume the load and axial length of cylinders to be independently normally distributed random variables with $(\overline{F}, \sigma_F) = (10{,}000, 2000)$ N and $(\overline{a}, \sigma_a) = (0.2, 0.02)$ m. If 0.41277, 0.64903, 0.39797, 0.26741, 0.25185, 0.04836, 0.94539, 0.89573, 0.75300, and 0.22015 denote the uniformly distributed random numbers, determine the mean and standard deviation of the contact pressure developed. Assume $\nu = 0.3$, $E = 2.06 \times 10^{11}$ N/m^2, $d_1 = 0.1$ m, and $d_2 = 0.25$ m.

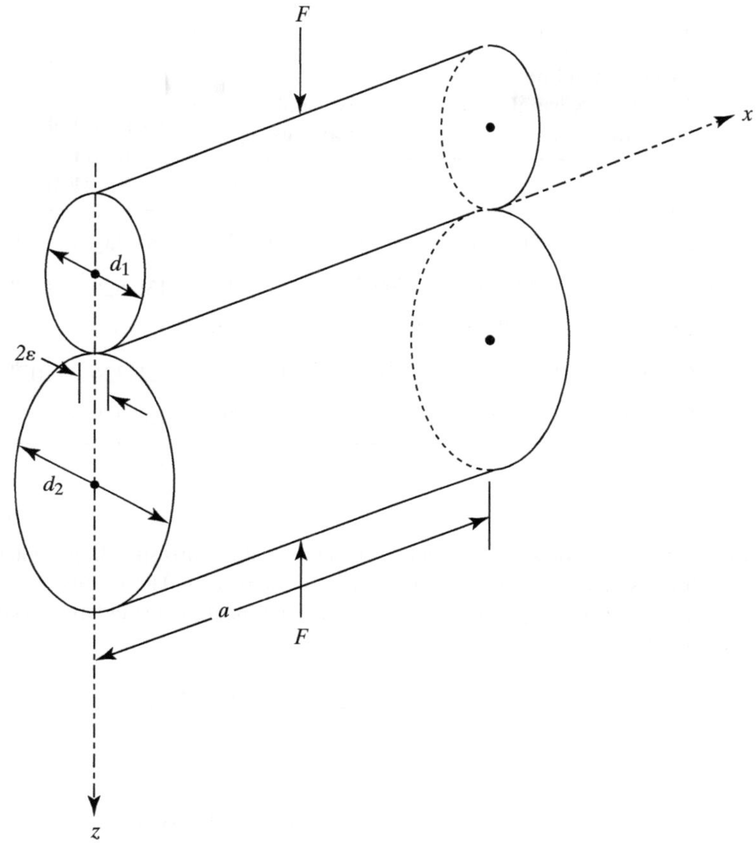

FIGURE 10.4 Two cylinders under contact stress.

Solution: Since there are two random variables in the problem, we use the first five random numbers for generating the values of F and the remaining five numbers for generating the values of a. By using the definition of the standard normal distribution

$$F_Z(z) = \frac{1}{\sqrt{2\pi}} \int_{-\infty}^{z} e^{-\frac{1}{2}u^2} du$$

and standard normal tables, we find for the uniformly distributed number 0.41277, the corresponding standard normally distributed number x is given by

$$F_Z(z) = 0.41277 \rightarrow z = -0.220425$$

If z_1 and z_2 denote the standard normally distributed numbers for F and a, we have by definition,

$$z_1 = \frac{F - \overline{F}}{\sigma_F} \quad \text{and} \quad z_2 = \frac{a - \overline{a}}{\sigma_a}$$

The sample values of F and a can be determined as

$$F = \overline{F} + \sigma_F z_1 = (10{,}000 + 2000\, z_1)\ \text{N}$$
$$a = \overline{a} + \sigma_a z_2 = (0.2 + 0.02\, z_2)\ \text{m}$$

TABLE 10.2 Sample values of contact pressure developed between cylinders

Standard Uniformly Distributed Numbers		Corresponding Standard Normal Numbers		Normal Variates		Sample Values of ε(m) Eq. (E2)	Sample Value of p_{max}(N/m^2) Eq. (E1)
u_1	u_2	z_1	z_2	F(N)	a(m)		
0.41277	0.04836	−0.220425	−1.660967	9559.149	0.166781	151.745×10^{-6}	240.4574×10^6
0.64903	0.94539	0.382703	1.601708	10765.406	0.232034	136.526×10^{-6}	216.3419×10^6
0.39797	0.89573	−0.258605	1.257590	9482.790	0.225152	130.079×10^{-6}	206.1255×10^6
0.26741	0.75300	−0.620665	0.683961	8758.670	0.213679	128.326×10^{-6}	203.3478×10^6
0.25185	0.22015	−0.668679	−0.771687	8662.642	0.184566	137.318×10^{-6}	217.5958×10^6

Once the sample values of F and a are known, the corresponding contact pressure between the cylinders can be computed using Eqs, (E1) and (E2). The results are shown in Table 10.2. From the values of $p_{\max}$ given in Table 10.2, the mean and standard deviations of $p_{\max}$ can be determined as

$$\overline{p}_{\max} = 216.7737 \times 10^6 \,\text{N/m}^2$$

and

$$\sigma_{p\max} = 13.0799 \times 10^6 \,\text{N/m}^2$$

■

10.3.3 Generation of Correlated Normal Random Variables

Let the random variables $X_1, X_2, \ldots, X_n$ be correlated and follow normal distribution. Their joint probability density function is defined by the means, variances, and covariances of the random variables. We assume that the mean values $(\overline{X}_i)$, variances $(\sigma_i^2 = \sigma_{X_i}^2)$, and convariances $(\sigma_{ij} = \sigma_{X_i X_j})$ of the variables X_i are known, that is, the vector of mean values

$$\overline{X} = \begin{Bmatrix} \overline{X}_1 \\ \overline{X}_2 \\ \cdot \\ \cdot \\ \cdot \\ \overline{X}_n \end{Bmatrix} \tag{10.20}$$

and the covariance matrix

$$[V_X] = \begin{bmatrix} \sigma_1^2 & \sigma_{12} & \cdots & \sigma_{1n} \\ \sigma_{12} & \sigma_2^2 & \cdots & \sigma_{2n} \\ \cdot & \cdot & \cdots & \cdot \\ \cdot & \cdot & \cdots & \cdot \\ \cdot & \cdot & \cdots & \cdot \\ \sigma_{n1} & \sigma_{n2} & \cdots & \sigma_n^2 \end{bmatrix} \tag{10.21}$$

are known. To generate the required set of correlated normally distributed random numbers X_i, $i = 1, 2, \ldots, n$, we first generate a set of n statistically independent normally distributed random numbers W_i, $i = 1, 2, \ldots, n$ whose mean values $\overline{W}_i$ and variances $\sigma_{W_i}^2$ are unknown at this time. Express the desired correlated random variables X_i as linear functions of the independent random variables W_i [10.10] as

$$\begin{aligned} X_i &= a_{i1}\, W_1 + a_{i2}\, W_2 + \cdots + a_{in}\, W_n \\ &= \sum_{j=1}^{n} a_{ij}\, W_j; \quad i = 1, 2, \ldots, n \end{aligned}$$

or

$$\underset{n\times 1}{X} = \underset{n\times n}{[a]}\ \underset{n\times 1}{W} \tag{10.22}$$

where the elements of the matrix $[a]$ are also unknown at this stage. Since the random variables X_i are expressed as linear functions of W_j, the mean values and standard deviations of X_i can be expressed as

$$\overline{\text{X}} = [a]\,\overline{\text{W}} \tag{10.23}$$

and

$$[V_X] = [a][V_W][a]^T \tag{10.24}$$

where $\overline{\text{X}}$ and $\overline{\text{W}}$ are the vectors of mean values of X_i and W_j, and $[V_X]$ and $[V_W]$ are the covariance matrices of X_i and W_j, respectively. In Eqs. (10.23) and (10.24), the vector $\overline{X}$ and the matrix $[V_X]$ are known from the characteristics of the random variables to be generated, and the vector $\overline{\text{W}}$ and the matrices $[a]$ and $[V_W]$ are to be determined. The Choleski decomposition method [10.7] can be used for this purpose.

In the Choleski decomposition method, a symmetric matrix $[S]$ of order $n \times n$ is decomposed into the form of a matrix product as

$$[S] = [L][D][L]^T \tag{10.25}$$

where $[L]$ is a lower triangular matrix of order $n \times n$ that contains ones on the main diagonal, and $[D]$ is a diagonal matrix of order $n \times n$. The elements of $[S]$ and $[D]$ [10.7, 10.10] are given by

$$\begin{aligned} D_{11} &= S_{11} \\ L_{ii} &= 1; \quad i = 1, 2, \ldots, n \end{aligned}$$

$$L_{j1} = \frac{S_{1j}}{D_{11}}; \quad j \geq 2$$

$$D_{ii} = S_{ii} - \sum_{j=1}^{i-1} L_{ij}^2 D_{jj}; \quad i \geq 2$$

$$L_{ji} = \frac{1}{D_{ii}}\left[S_{ij} - \sum_{k=1}^{i-1} L_{ik} L_{jk} D_{kk}\right]; \quad i \geq 2; \quad j \geq i+1 \tag{10.26}$$

By comparing Eqs. (10.24) and (10.25), we find that

$$[S] = [V_X] \tag{10.27}$$

$$[a] = [L] \tag{10.28}$$

and

$$[V_W] = [D] \tag{10.29}$$

From Eqs. (10.26) and (10.28), the elements a_{ij} and hence the linear relationships between X_i and W_j are completely known. Similarly, Eqs. (10.26) and (10.29) define the variances of W_j. Further, since the covariance matrix $[V_W] = [D]$ is diagonal, the random variables W_j are ensured to be independent. Finally, the mean values of W_j can be determined from Eq. (10.23) as

$$\overline{\mathrm{W}} = [a]^{-1}\overline{\mathrm{X}} \tag{10.30}$$

It can be observed that the random variables X_i will be normally distributed since they are obtained from a linear combination of the normal random variables, W_j.

The procedure described for generating the correlated normal random numbers is indicated as a flow diagram in Fig. 10.5.

EXAMPLE 10.9

The torque transmitted by a plate clutch with a single pair of friction surfaces, shown in Fig. 10.6, can be expressed as [10.6]

$$T = \frac{Ff}{4}(D + d) \text{ for uniform wear condition} \tag{E1}$$

and

$$T = \frac{Ff}{3}\left(\frac{D^3 - d^3}{D^2 - d^2}\right) \text{ for uniform pressure condition} \tag{E2}$$

where F = axial force applied = $(\pi p_a d/2)(D - d)$, p_a = maximum pressure induced, f = coefficient of friction, D = outer diameter, and d = inner diameter. If $\overline{D}$ = 250 mm, $\overline{d}$ = 200 mm, σ_D = 12.5 mm, σ_d = 15 mm, $\rho_{D,d}$ = 0.4, f = 0.25, and p_a = 750 kPa, compute the torque transmitted by the clutch using the Monte Carlo simulation. Use the following uniformly distributed random

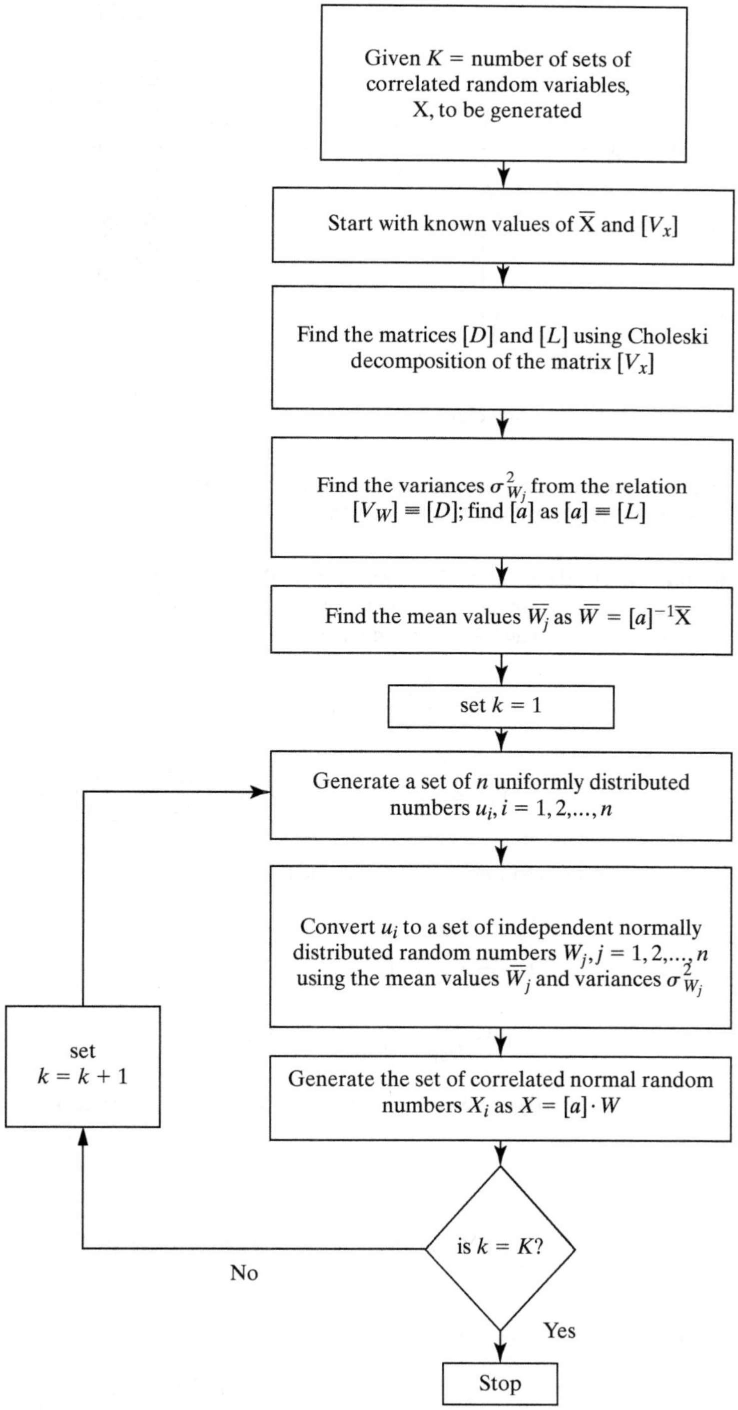

FIGURE 10.5 Procedure for generating correlated normal random variables.

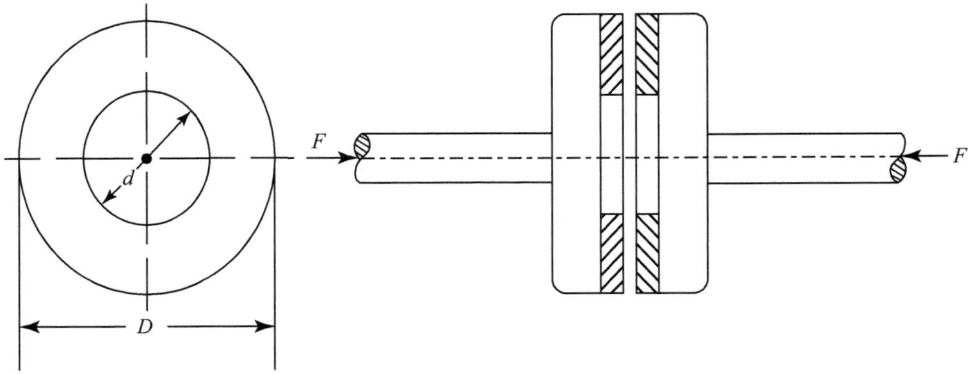

FIGURE 10.6 A plate clutch.

numbers to generate the correlated random variables: 0.41277, 0.64903, 0.39797, 0.26741, 0.25185, 0.04836, 0.94539, 0.89573, 0.75300, 0.22015.

Solution: Given data: $\overline{D} = 250$ mm, $\overline{d} = 200$ mm, $\sigma_D = 12.5$ mm, $\sigma_d = 15$ mm, $\rho_{D,d} = 0.4 = \sigma_{D,d}/(\sigma_D, \sigma_d)$, $\sigma_{D,d} = (0.4)(12.5)(15) = 75$ mm^2.

Defining $X_1 = D$ and $X_2 = d$, we have

$$\overline{X} = \begin{Bmatrix} \overline{X}_1 \\ \overline{X}_2 \end{Bmatrix} = \begin{Bmatrix} 250 \\ 200 \end{Bmatrix} \text{ mm}$$

$$[V_X] = \begin{bmatrix} \sigma_{X_1}^2 & \sigma_{X_1,X_2} \\ \sigma_{X_1,X_2} & \sigma_{X_2}^2 \end{bmatrix} = \begin{bmatrix} 156.25 & 75 \\ 75 & 225 \end{bmatrix} \text{mm}^2$$

The Choleski decomposition of $[S] \equiv [V_X]$ leads to

$$D_{11} = S_{11} = 156.25$$

$$L_{11} = L_{22} = 1$$

$$L_{21} = \frac{S_{12}}{D_{11}} = \frac{75}{156.25} = 0.48$$

$$D_{22} = S_{22} - L_{21}^2 D_{11} = 225 - (0.48)^2(156.25) = 189$$

$$[a] \equiv [L] = \begin{bmatrix} 1 & 0 \\ 0.48 & 1 \end{bmatrix}$$

$$[V_W] \equiv [D] = \begin{bmatrix} 156.25 & 0 \\ 0 & 189 \end{bmatrix} = \begin{bmatrix} 12.5^2 & 0 \\ 0 & 13.7477^2 \end{bmatrix}$$

The vector $\overline{\text{W}}$ is determined as

$$\overline{W} = [a]^{-1}\overline{X} = \begin{bmatrix} 1 & 0 \\ -0.48 & 1 \end{bmatrix} \begin{Bmatrix} 250 \\ 200 \end{Bmatrix} = \begin{Bmatrix} 250 \\ 80 \end{Bmatrix} \text{ mm}$$

TABLE 10.3 Sample values of torque transmitted by the plate clutch

						Torque Transmitted	
						Uniform Wear	Uniform Pressure
u_i	z_i	D(m)	u_i	z_i	d(m)	kN − m	kN − m
0.41277	−0.220425	0.24724	0.04836	−1.660967	0.17584	0.39114	0.39486
0.64903	0.382703	0.25478	0.94539	1.601708	0.22432	0.24107	0.24140
0.39797	−0.258605	0.24677	0.89573	1.257590	0.21574	0.22796	0.22830
0.26741	−0.620665	0.24224	0.75300	0.683961	0.20568	0.24801	0.24857
0.25185	−0.668679	0.24164	0.22015	−0.771687	0.18538	0.32794	0.32984

For this given uniformly distributed random number u_i, the corresponding normally distributed numbers w_i are determined. The correlated random numbers x_i are found as

$$X = [a]W = \begin{Bmatrix} w_1 \\ 0.48w_1 + w_2 \end{Bmatrix} = \begin{Bmatrix} D \\ d \end{Bmatrix}$$

These values are shown in Table 10.3. Once the values of d and D are known, the torque transmitted by the clutch can be determined using Eqs. (E1) and (E2).

■

10.4 Computation of Reliability

Due to the random nature of parameters influencing the performance of an engineering system, there is a small probability of not meeting the specified performance requirements. The Monte Carlo method can be used to estimate the reliability or probability of failure of the system. In this method, a uniformly distributed random number is used to generate a sample value of each of the random parameters of the system. These parameters are then used to predict the performance of the system. The sample system's performance is compared with the specified performance requirement. The sample is considered to be a successful one if the specified performance requirement is met; otherwise, it is considered to be a failed one. After generating the predetermined number of sample systems, the overall reliability of the system is computed as

$$\begin{pmatrix} \text{Reliability} \\ \text{of the system} \end{pmatrix} = \left(\frac{\text{Number of successful sample systems}}{\text{Total number of sample systems generated}} \right)$$

This procedure is shown as a flow diagram in Fig. 10.7.

10.4.1 Sample Size and Error in Simulation

The random numbers generated in the Monte Carlo simulation can be used to estimate the reliability (or probability of failure) of an engineering system. It is important to know the error involved in the estimated failure probability. It is equally important to know how

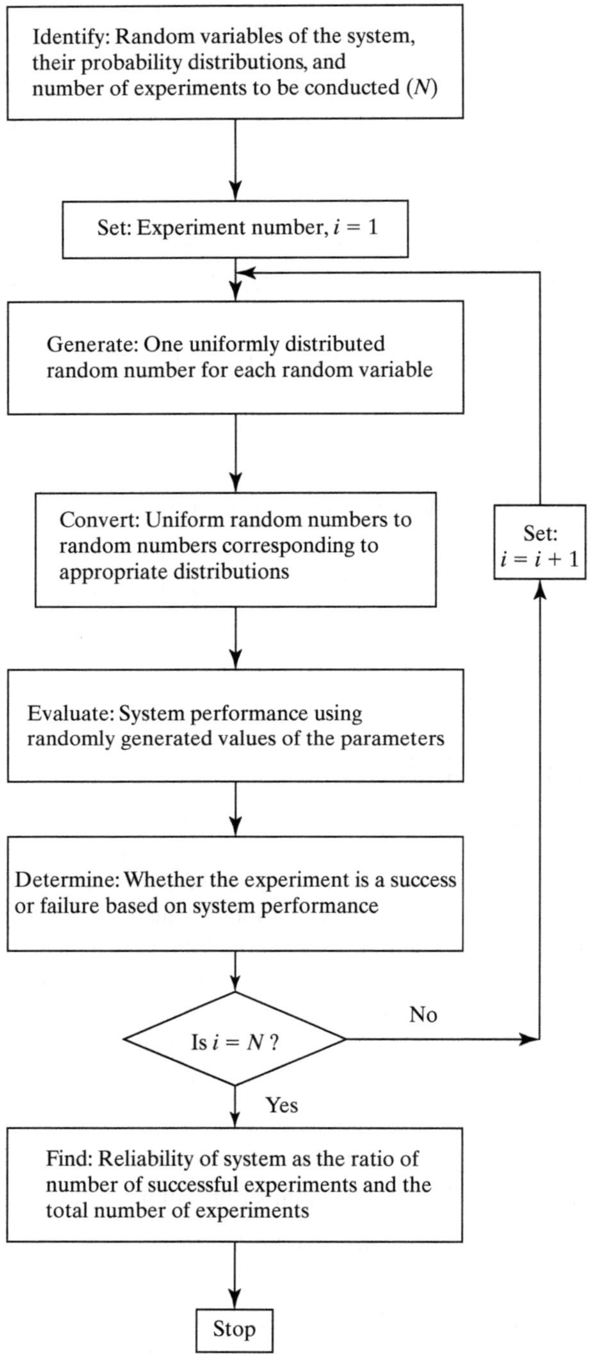

FIGURE 10.7 Computation of reliability using Monte Carlo simulation.

many simulations (i.e., sample size) are required to achieve a specified accuracy. The percent error involved with the estimated probability P_f [10.8] has been found to be

$$\text{Percent error} = 200 \cdot \sqrt{\frac{1 - P_f}{n \cdot P_f}} \tag{10.31}$$

where n is the number of simulations (sample size) used in estimating P_f. There is a 95 percent chance that the percent error in the estimated value of P_f will be less than that given by Eq. (10.31). For example, if 1,000 simulations were performed in finding an estimated probability of failure of 0.1, Eq. (10.31) will give a 18 percent error. This means that there is a 95 percent chance that the actual failure probability will be within 0.1 ± 0.018. Equation (10.31) can also be used to find the sample size required for achieving a specified accuracy in the estimated probability of failure. For example, if an accuracy of 0.1 ± 0.01 is desired for the probability of failure, Eq. (10.31) gives the required number of simulations (n) as 3600.

10.4.2 Example: Reliability Analysis of a Straight-Line Mechanism

The four-bar straight-line mechanism shown in Fig. 10.8 is considered to illustrate the Monte Carlo method for computing reliability [10.9]. The reliability of this function generating mechanism can be defined as the probability of generating the desired function with a specific accuracy. The mechanism has six design parameters, namely, r_1, r_2, r_3, r_4, r_5, and θ. The nominal values of the parameters are assumed as $r_1 = 0.963$, $r_2 = 0.764$, $r_3 = 0.528$, $r_4 = 1.815$, $r_5 = 0.778$, and $\theta = -89.65°$. These parameters are assumed to be random in nature, and the reliability of the mechanism is computed for normal, Weibull, and beta distributions. To calculate the reliability of the mechanism, trial values for the six design parameters are generated using random numbers.

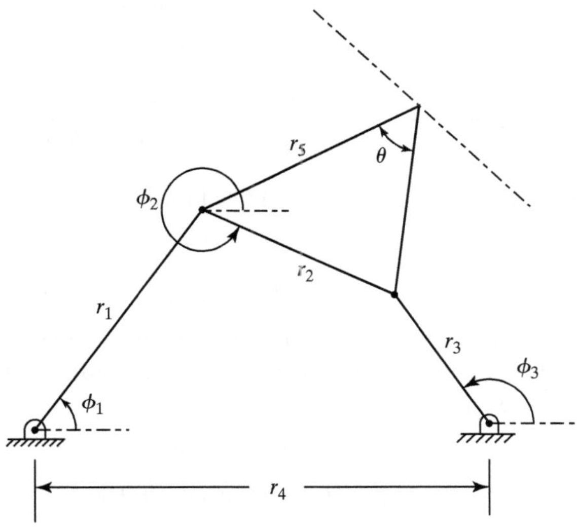

FIGURE 10.8 Four-bar straight line mechanism.

These values specify one mechanism in the sample. The input link length is r_1. Positions of the coupler point are computed by considering the vector loop equations of the mechanism

$$\begin{aligned} r_1 \cos\phi_1 + r_2 \cos\phi_2 - r_3 \cos\phi_3 - r_4 &= 0 \\ r_1 \sin\phi_1 + r_2 \sin\phi_2 - r_3 \sin\phi_3 &= 0 \end{aligned} \tag{10.32}$$

For a given mechanism design, the parameters r_1, r_2, r_3, and r_4 are known. The input angle ϕ_1 is assumed to be deterministic. For a specified value of ϕ_1, Eqs. (10.32) are solved for the values of the coupler angle, ϕ_2, and the follower angle, ϕ_3, using the Newton-Raphson method. The vector loop equations become the residual functions

$$\begin{aligned} f_1 &= r_1 \cos\phi_1 + r_2 \cos\phi_2 - r_3 \cos\phi_3 - r_4 \\ f_2 &= r_1 \sin\phi_1 + r_2 \sin\phi_2 - r_3 \sin\phi_3 \end{aligned} \tag{10.33}$$

The Jacobian matrix for these equations is given by

$$[A] = \begin{bmatrix} \dfrac{\partial f_1}{\partial \phi_2} & \dfrac{\partial f_1}{\partial \phi_3} \\ \dfrac{\partial f_2}{\partial \phi_2} & \dfrac{\partial f_2}{\partial \phi_3} \end{bmatrix} = \begin{bmatrix} -r_2 \sin\phi_2 & r_3 \sin\phi_3 \\ r_2 \cos\phi_2 & -r_3 \cos\phi_3 \end{bmatrix} \tag{10.34}$$

Given initial guesses for the angles ϕ_2 and ϕ_3, the Newton-Raphson algorithm consists of iteratively solving the following matrix equation for the correction vector $\overrightarrow{\Delta\phi}$:

$$[A^i]\overrightarrow{\Delta\phi} = \begin{bmatrix} -r_2 \sin\phi_2 & r_3 \sin\phi_3 \\ r_2 \cos\phi_2 & -r_3 \cos\phi_3 \end{bmatrix}^i \begin{bmatrix} \Delta\phi_2 \\ \Delta\phi_3 \end{bmatrix}^i = \begin{bmatrix} -f_1 \\ -f_2 \end{bmatrix}^i \tag{10.35}$$

where the superscript i indicates the iteration number. For this system, the solution can be written explicitly by Cramer's rule as

$$\begin{aligned} \Delta\phi_2 &= \frac{f_1 \cos\phi_3 + f_2 \sin\phi_3}{r_2(\sin\phi_2 \cos\phi_3 - \cos\phi_2 \sin\phi_3)} \\ \Delta\phi_3 &= \frac{f_2 \sin\phi_2 + f_1 \cos\phi_2}{r_3(\sin\phi_2 \cos\phi_3 - \cos\phi_2 \sin\phi_3)} \end{aligned} \tag{10.36}$$

The updated values of the angles are then computed from

$$\begin{aligned} \phi_2^{i+1} &= \phi_2^i + \Delta\phi_2 \\ \phi_3^{i+1} &= \phi_3^i + \Delta\phi_3 \end{aligned} \tag{10.37}$$

The convergence criteria for a solution can be checked on both the residual vector and the correction vector. Once the angles are known for a given input angle, the coupler point position can be computed as

$$\begin{aligned} x_c &= r_1 \cos\phi_1 + r_5 \cos(\phi_2 + \theta) \\ y_c &= r_1 \sin\phi_1 + r_5 \sin(\phi_2 + \theta) \end{aligned} \tag{10.38}$$

This procedure is followed over the input angle range of $-14°$ to $42°$ in increments of two degrees. The resulting coupler points (x_c, y_c) are then fit to a line using least-squares analysis. Finally, the perpendicular distance from the desired straight line to each of the coupler

TABLE 10.4 Results of Monte Carlo simulation of the mechanism

	Probability Distribution								
	Normal		Weibull			Beta			
Parameter	μ	σ	b	h	l	q	r	l	u
Input link	0.963	0.005	3.0	0.964	0.95	3.0	3.0	0.953	0.973
Coupler	0.764	0.005	3.0	0.765	0.75	3.0	3.0	0.754	0.774
Follower	0.528	0.005	3.0	0.529	0.51	3.0	3.0	0.518	0.538
Ground link	1.815	0.005	3.0	1.816	1.80	3.0	3.0	1.805	1.825
Coupler radius	0.778	0.005	3.0	0.779	0.76	3.0	3.0	0.768	0.788
Coupler angle	−1.565	0.01	3.0	−1.564	−1.57	3.0	3.0	−1.575	−1.555
Reliability	0.9514			0.9190				0.9873	

point positions is computed. If any such distance exceeds the specified error tolerance, the mechanism is declared unreliable. Also, if the Newton-Raphson iteration fails to converge after a maximum number of steps for any input angle, the mechanism is declared unreliable.

The results of reliability computations are given in Table 10.4. For each simulation, 10,000 trials were used with an error tolerance of 0.05. Plots of the density functions of three random parameters are given in Figs. 10.9–10.11. The simulation results indicate that all distributions produce reliabilities over 0.9. The beta distributions yield

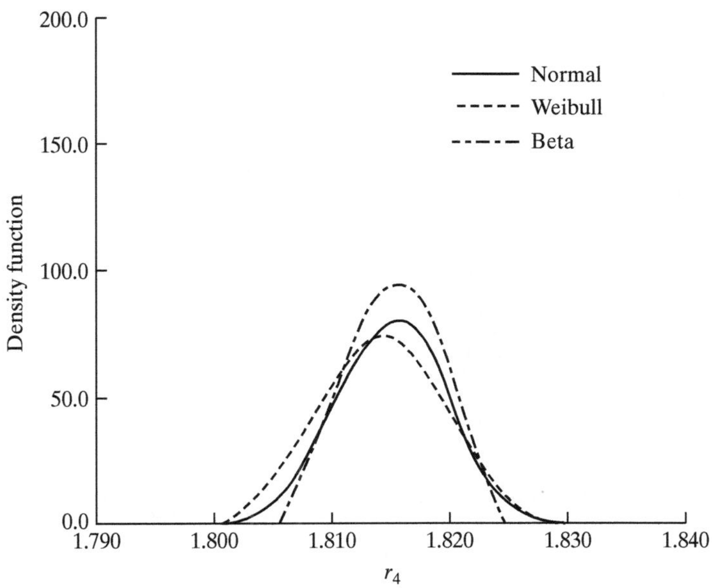

FIGURE 10.9 Distributions for ground link. (Reprinted with permission from R. H. Crawford and S. S. Rao, "Probabilistic Analysis of Function Generating Mechanisms," *Journal of Mechanisms, Transmissions, and Automation in Design*, ASME, Vol. 111, December 1989.)

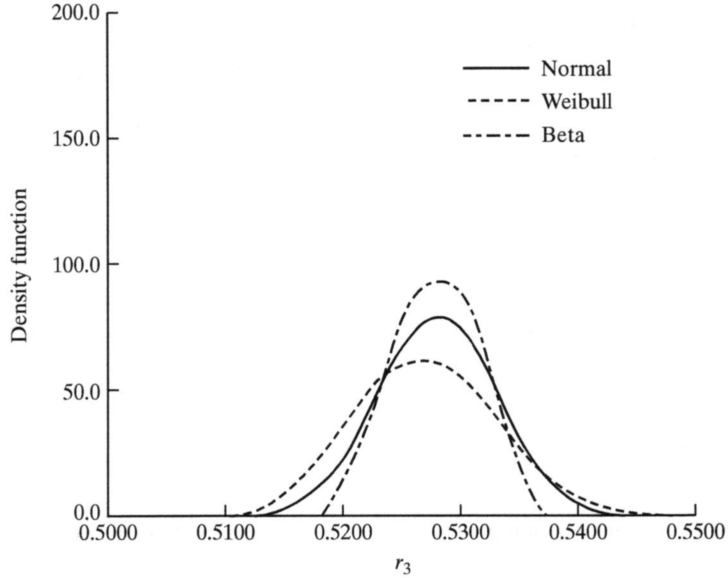

FIGURE 10.10 Distributions for follower. (Reprinted with permission from R.H. Crawford and S.S. Rao, "Probabilistic Analysis of Function Generating Mechanisms," *Journal of Mechanisms, Transmissions, and Automation in Design*, ASME, Vol. 111, December 1989.)

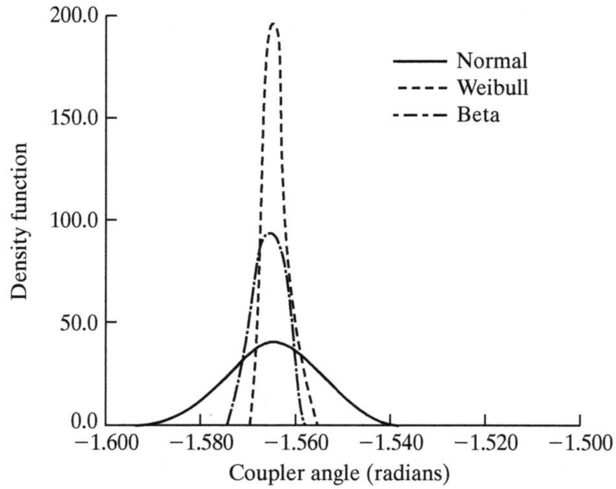

FIGURE 10.11 Distributions for coupler angle. (Reprinted with permission from R.H. Crawford and S.S. Rao, "Probabilistic Analysis of Function Generating Mechanisms," *Journal of Mechanisms, Transmissions, and Automation in Design*, ASME, Vol. 111, December 1989.)

the highest reliability because they are bounded at both ends and have the smallest dispersion about the mean. The reliability computed from the normal distributions was somewhat lower due to the wider dispersions, which can be seen in Figs. 10.9–10.11. The Weibull distributions yielded the smallest reliability because these distributions have slightly different means. Using Eq. (10.31), the probability of failure, according to normal distribution, has been found to be 0.0486 using 10,000 simulations. Equation (10.31) gives the percent error as 8.8489. Thus, it is likely that the actual probability of failure will be within 0.0486 $\pm$ 0.0043 or the actual reliability will be within 0.9514 $\pm$ 0.0043.

10.5 Numerical Solutions Using MATLAB and Excel

EXAMPLE 10.10

Generate five normally distributed random numbers for each of the diameters D and d of the clutch plate shown in Fig. 10.6 assuming zero correlation between D and d. Develop a MATLAB program to compute the torque transmitted by the clutch according to uniform wear and uniform pressure conditions, given, respectively, by Eqs. (E1) and E2) of Example 10.9.

Solution: The following procedure is used to develop the required MATLAB program.

1. Specify the known data of the mean values and standard deviations of D and d.
2. Generate ten uniformly distributed random numbers, $u_i(\mathrm{i} = 1, 2, \ldots, 10)$, each lying in the interval 0 to 1.
3. Convert the uniformly distributed random numbers to the corresponding standard normally distributed variables, $z_i(i = 1, 2, \ldots, 10)$.
4. Use the standard normally distributed random variables 1 to 5 and 6 to 10 to the normally distributed diameters D and d, respectively, as

$$D_i = \mu_D + z_i \sigma_D\,; \quad i = 1, 2, \ldots, 5$$

$$d_i = \mu_d + z_{i+5}\, \sigma_d\,; \quad i = 1, 2, \ldots, 5$$

5. Find the random axial force applied on the clutch plate (F_i) using the relation

$$F_i = \frac{\pi p_a d_i}{2}(D_i - d_i); \quad i = 1, 2, \ldots, 5$$

6. Determine the random torque transmitted by the clutch as

$$T_{1i} = \frac{F_i f}{4}(D_i + d_i); \quad \text{uniform wear condition}$$

$$T_{2i} = \frac{F_i f}{3}\left(\frac{D_i^3 - d_i^3}{D_i^2 - d_i^2}\right); \quad \text{uniform pressure condition}$$

The MATLAB program developed and the results obtained are given below.

```
clear all;clc;

Dbar = 0.25 ;  SD = 0.0125 ; dbar = 0.2 ; Sd = 0.015 ;

RHODd = 0.4 ; SDd = 75 *10^-6 ; f = 0.25 ;Pa = 750 ;

% __________(b)______

u = rand(10,1)'

z = norminv(u,0,1)

%__________(d)______

for i = 1:5
    W(i) = Dbar + SD * z(i)    ;
end

for i = 6:10

    W(i) = dbar + Sd * z(i)    ;
end

W

% __________(e)__________
D(1:5) = W(1:5)

for i = 1:5
    d(i) = 0.0*W(i) + W(i+5)    ;
end

d
% __________(f)______

for i =1:5
    F(i) = (0.5*pi*Pa) * d(i) * (D(i)-d(i))    ;
end

F

% __________(g)______

for j =1:5
    T1(j) = 0.25 * F(j) * f * (D(j)+d(j))    ;
    T2(j) = F(j)* f * (D(j)^3 - d(j)^3)/(3*D(j)^2 - d(j)^2))    ;
end

T1

T2

u=  0.8147  0.9058   0.1270   0.9134  0.6324   0.0975   0.2785  0.5469  0.9575  0.9649

z=  0.8954  1.3153  -1.1408   1.3618  0.3381  -1.2957  -0.5873  0.1178  1.7225  1.8105

w=  0.2612  0.2664   0.2357   0.2670  0.2542   0.1806   0.1912  0.2018  0.2258  0.2272

D=  0.2612  0.2664   0.2357   0.2670  0.2542

d=  0.1806  0.1912   0.2018   0.2258  0.2272

F= 17.1515 16.9495   8.0756  10.9579  7.2441

T1= 0.4735  0.4848   0.2208   0.3375  0.2179

T2= 0.4788  0.4892   0.2213   0.3383  0.2182
```

■

EXAMPLE 10.11

Using uniformly distributed random numbers, generate random variables that follow exponential distribution with parameter $\alpha = 1000$ (mean life of a component, in hours) using Excel.

Solution: The computations involved in the Excel sheet are shown below.

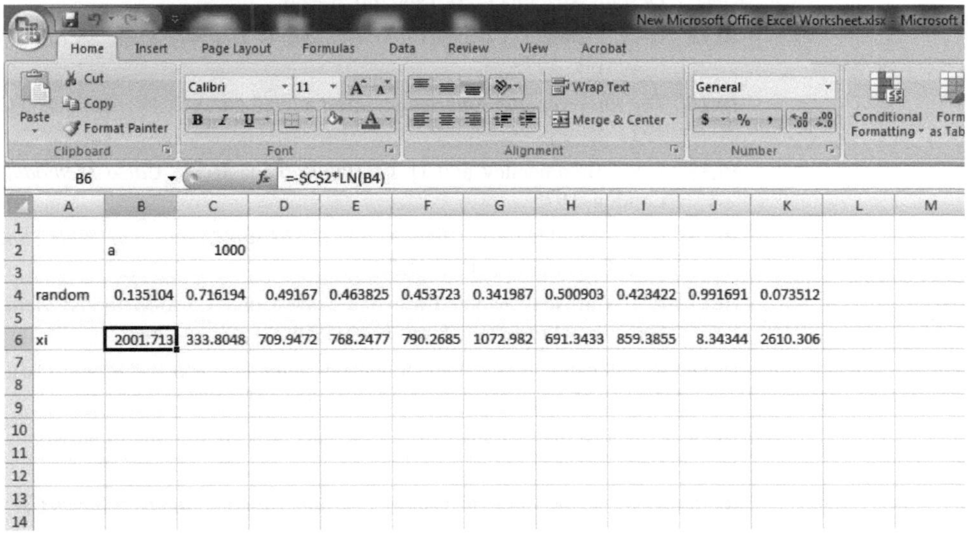

■

REFERENCES AND BIBLIOGRAPHY

10.1 R. F. Warner and A. P. Kabaila, "Monte Carlo Study of Structural Safety," *ASCE Journal of the Structural Division*, Vol. 94, No. ST12, 1968, pp. 2847–2859.

10.2 Rand Corporation, *A Million Random Digits with 100,000 Normal Deviates*, Free Press, Glencoe, IL, 1955.

10.3 F. Neelamkavil, *Computer Simulation and Modelling*, John Wiley, Chichester, 1987.

10.4 D. E. Knuth, *The Art of Computer Programming: Seminumerical Algorithms*, Vol. 2, Addison-Wesley, Reading, MA, 1969.

10.5 K. D. Tocher, *The Art of Simulation*, English Universities Press, London, 1963.

10.6 J. E. Shigley and C. R. Mischke, *Mechanical Engineering Design*, 5th ed., McGraw-Hill, New York, 1989.

10.7 S. S. Rao, *The Finite Method in Engineering*, 5th ed., Butterworth-Heinemann, Elsevier, Burlington, MA, 2011.

10.8 M. L. Shooman, *Probabilistic Reliability: An Engineering Approach*, McGraw-Hill, New York, 1968.

10.9 R. H. Crawford and S. S. Rao, "Probabilistic Analysis of Function Generating Mechanisms," *ASME Journal of Mechanisms, Transmissions, and Automation in Design*, Vol. 111, December 1989, pp. 479–481.

10.10 G. C. Hart, *Uncertainty Analysis, Loads, and Safety in Structural Engineering*, Prentice-Hall, Englewood Cliffs, NJ, 1982.

10.11 A. H. S. Ang and W. H. Tang, *Probability Concepts in Engineering Planning and Design*, Vol. II: *Decision, Risk, and Reliability*, John Wiley, New York, 1984.

10.12 M. F. Spotts, "Predicting Length of Assemblies with Monte Carlo Simulation," *Machine Design*, November, 20, 1980, pp. 84–88.

10.13 G. S. Fishman, *Concepts and Methods in Discrete Event Digital Simulation*, John Wiley, New York, 1973.

10.14 *The Encyclopedia Americana*, international ed., Vol. 19, Americana Corporation, New York, 1971.

10.15 J. M. Hammersley and D. C. Handscomb, *Monte Carlo Methods*, Methuen & Co. Ltd., London, 1975.

10.16 B. A. Berg, *Markov Chain Monte Carlo Simulations and Their Statistical Analysis*, World Scientific, Hackensack, NJ, 2004.

10.17 R. E. Caflisch, "Monte Carlo and quasi-Monte Carlo Methods," *Acta Numerica*, Cambridge University Press, Vol. 7, pp. 1–49, 1998.

10.18 R. Eckhardt, Stan Ulam, John von Neumann, and the Monte Carlo Method, *Los Alamos Science*, Special Issue, Vol. 15, pp. 131–137, 1987.

10.19 G. S. Fishman, *Monte Carlo: Concepts, Algorithms, and Applications*, Springer-Verlag, New York, 1995.

10.20 M. H. Kalos and P. A. Whitlock, *Monte Carlo Methods*, Wiley-VCH, 2008.

10.21 D. P. Kroese, T. Taimre, and Z. I. Botev, *Handbook of Monte Carlo Methods*, Wiley, Hoboken, NJ, 2011.

10.22 N. Metropolis, "The Beginning of the Monte Carlo Method," *Los Alamos Science*, Special Issue, Vol. 15, pp. 125–130, 1987.

10.23 C. P. Robert and G. Casella, *Monte Carlo Statistical Methods*, 2nd ed., Springer-Verlag, New York, 2004.

10.24 D. Vose, *Risk Analysis: A Quantitative Guide*, 2nd ed., Wiley, New York, 2000.

10.25 I. M. Sobol, *A Primer for the Monte Carlo Method*, CRC Press, Boca Raton, FL, 1994.

REVIEW QUESTIONS

10.1 Give brief answers to the following:

1. What is Monte Carlo simulation?
2. List five physical processes which can be used to generate random numbers?
3. What is a uniformly distributed random number?
4. How do you generate a random variable following the distribution function $F_X(x)$ using a uniformly distributed random number?
5. What is a pseudo-random number?

6. State the characteristics of a good random number generator.
7. How do you generate pseudo-random numbers?
8. How do you generate a discrete random number?
9. How do you find the reliability of a system using Monte Carlo simulation?
10. Generate the numbers given by the formula

$$X_{i+1} = (8X_i + 1)(\bmod 14)$$
$$\text{with } X_1 = 2.$$

10.2 Indicate whether the following statement is true or false:

1. Uniformly distributed random numbers in the range 0 to 1 can be used to generate random numbers that follow any specified probability distribution.
2. Monte Carlo method was invented by John von Neumann, Stanislaw Ulam, and Nicholas Metropolis in the 1940s.

10.3 Fill in the blank with the proper word(s):

1. Monte Carlo analysis is similar to conducting an experiment on ______.
2. The process of tossing a fair coin can be used to generate a discrete random variable with ______ possible values.
3. The probability density function of a uniformly distributed random variable in the range 0 to 1 is given by ______.
4. The probability distribution of a uniformly distributed random variable in the range 0 to 1 is ______.
5. The pseudo random numbers given by the formula $x_{i+1} = (9x_i + 2)(\bmod 15)$ using $x_0 = 4$ are ______.
6. Using Monte Carlo simulation, the reliability of a system can be determined as the ratio of the number of ______ sample systems and the total number of sample systems generated.

10.4 Select the most appropriate answer from the multiple choices given:

1. Monte Carlo simulation can be used for the analysis of engineering systems involving:
   a. Complex parameters **b.** Statistical parameters **c.** Lengthy analysis

2. The city of Monte Carlo is famous for:
   a. Gambling **b.** Horse racing **c.** Car racing

3. If the probability distribution of X is $F_X(x)$, the value of X that has a specific value of $F_X(x) = u$ can be generated as:
   a. $x = u$ **b.** $x = F_X^{-1}(u)$ **c.** $x = F_X(u)$

4. Pseudo-random numbers are generated using a:
   a. Random process **b.** Wheel **c.** Formula

5. The percent error involved in the probability of failure P_f estimated using n samples is given by

$$\text{Percent error} = 200\sqrt{\frac{1 - P_f}{n\,P_f}}$$

   The percent error corresponding to $P_f = 0.001$ and $n = 10000$ is:
   a. 63.21% **b.** 13.21% **c.** 42.31%

PROBLEMS

10.1 The probability density function of a random variable X is given by

$$f_X(x) = \begin{cases} \dfrac{1}{x^2}; & x \ge 1 \\ 0; & x < 1 \end{cases}$$

Generate the random numbers x_i corresponding to X in terms of a set of standard uniformly distributed numbers u_i, $i = 1, 2, \ldots, N$.

10.2 The probability distribution function of a random variable X is given by

$$F_X(x) = 1 - \left(1 - \frac{x}{k}\right)^2; \quad 0 \le x \le k$$

Generate the random numbers x_i corresponding to X in terms of a set of standard uniformly distributed numbers u_i, $i = 1, 2, \ldots, N$.

10.3 If u_i, $i = 1, 2, \ldots, N$ denote a set of standard uniformly distributed random numbers, find the corresponding numbers x_i of a random variable X which follows the following triangular distribution (see Fig. 10.12):

$$f_X(x) = \begin{cases} k\left(\dfrac{x-a}{b-a}\right); & a \le x \le b \\ k\left(\dfrac{c-x}{c-b}\right); & b \le x \le c \end{cases}$$

$$\text{where } k = \left(\frac{2}{c-a}\right)$$

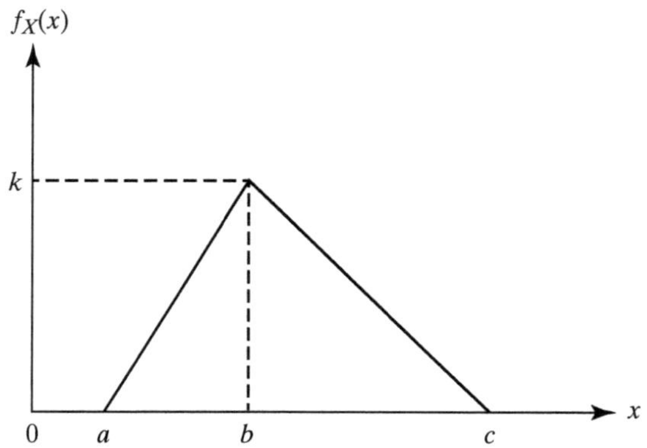

FIGURE 10.12 Triangular density function of X.

10.4 Consider a set of five standard uniformly distributed random numbers given in the first row of Table 10.5. Generate the corresponding numbers x_i of a lognormally distributed random variable X

$$f_X(x) = \frac{1}{x\sigma\sqrt{2\pi}} \exp\left[-\frac{(\ln x - \mu)^2}{2\sigma^2}\right]$$

with $\sigma = 1.0$ and $\mu = 2.0$.

TABLE 10.5 Standard uniformly distributed random numbers

Row 1	0.52478	0.22835	0.33307	0.73842	0.67277
Row 2	0.32880	0.76457	0.94489	0.82597	0.04836
Row 3	0.58815	0.01919	0.22225	0.38562	0.45731
Row 4	0.91743	0.99315	0.70350	0.78240	0.22015
Row 5	0.66047	0.26430	0.22415	0.98215	0.10413
Row 6	0.54380	0.10492	0.59665	0.42368	0.15138
Row 7	0.71899	0.68860	0.33415	0.72559	0.19902
Row 8	0.40024	0.74215	0.93857	0.04988	0.24389
Row 9	0.22094	0.89237	0.41999	0.12790	0.87990
Row 10	0.77646	0.33177	0.62684	0.34119	0.09212

10.5 The stress induced in a wire rope (σ) when it is bent around a sheave (Fig. 10.13) is approximately given by

$$\sigma = E\frac{d_w}{D}$$

where E = Young's modulus of the wire, d_w = diameter of the wire, and D = diameter of the sheave. If d_w and D follow normal distribution with $\bar{d}_w = 2$ mm, $\bar{D} = 200$ mm,

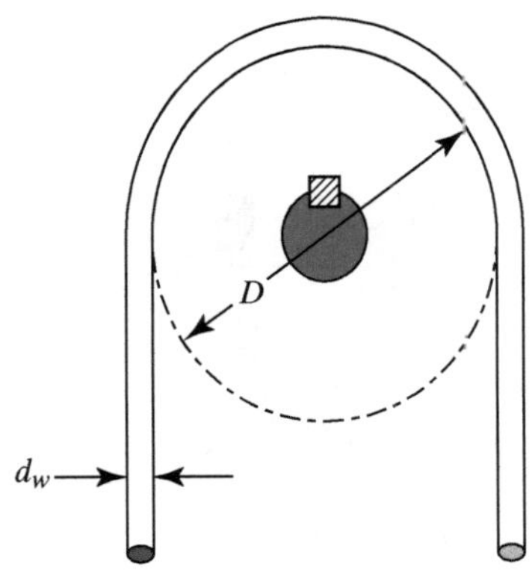

FIGURE 10.13 A wire rope around a sheave.

$\sigma_{d_w} = 0.2$ mm, $\sigma_D = 15$ mm, $\rho_{d_w, D} = 0.3$, and $E = 2.07 \times 10^{11}$ Pa, find the mean and standard deviations of the stress induced in the wire using the Monte Carlo simulation. Use the uniformly distributed random numbers given in the first two rows of Table 10.5 to generate the correlated random numbers. Compare the results with those given by the partial derivative rule.

10.6 The bearing pressure (p) induced by a loaded wire rope on the sheave is given by

$$p = 2\frac{F}{d_r D}$$

where F = tensile force in the rope, d_r = diameter of the rope, and D = diameter of the sheave over which the wire rope is mounted.

a. If the mean values of d_r and D are 25 mm and 200 mm and the variances are 6.25 mm^2 and 225 mm^2, respectively, with $\rho_{d_r, D} = 0.25$ and $F = 4000$ N, compute the mean and standard deviations of the bearing pressure on the sheave using the Monte Carlo simulation. Use the uniformly distributed random numbers given in rows 3 and 4 of Table 10.5 to generate the correlated variables. Assume that d_r and D follow normal distribution.

b. Find the mean and standard deviations of p using the partial derivative rule.

10.7 The length of a roller chain (L) mounted over two sprockets (see Fig. 10.14) is approximately given by

$$L = 2C + \left(\frac{N_1 + N_2}{2}\right)p + \left(\frac{N_2 - N_1}{2\pi}\right)^2 \frac{p^2}{C}$$

where p = chain pitch, C = center distance, N_1 = number of teeth on the small sprocket, and N_2 = number of teeth on the large sprocket. The following data is known for a particular application

$$N_1 = 20; N_2 = 30; \overline{C} = 30 \text{ in}; \sigma_C = 0.5 \text{ in}; \overline{p} = 0.625 \text{ in}; \sigma_p = 0.04 \text{ in}; \rho_{p,C} = 0.$$

Find the mean and standard deviations of the length of the chain using the uniformly distributed random numbers given in rows 5 and 6 of Table 10.5.

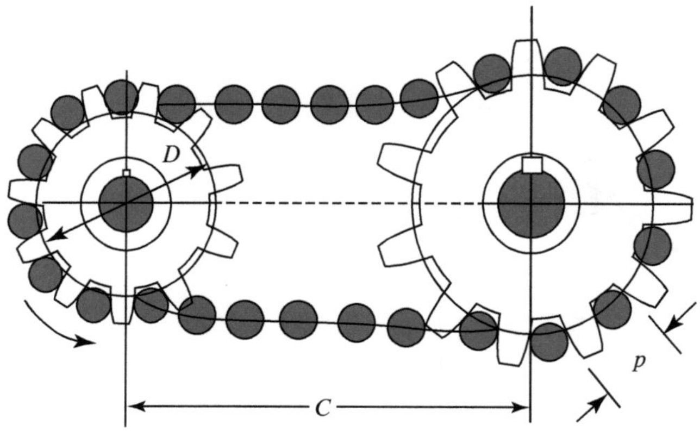

FIGURE 10.14 A roller chain mounted over two sprockets.

10.8 The actuating force (F) necessary to apply the brake shown in Fig. 5.12 is given by [10.6]

$$F = \frac{M_n - M_f}{a}$$

where

$$M_n = p_a trb\left(\frac{\theta_2}{2} - \frac{1}{4}\sin 2\theta_2\right)$$

$$M_f = p_a trf\left(r - r\cos\theta_2 - \frac{b}{2}\sin^2\theta_2\right)$$

where p_a = maximum pressure exerted on the brake lining, t = face width of brake shoes, r = inner radius of the brake drum, f = coefficient of friction, a = linear distance between the brake shoe hinge and the line of application of the actuating force, b = radius of the brake shoe, and θ_2 = angular distance between the brake shoe hinge and the point of application of the actuating force. For the brake shown in Fig. 5.12, the data is given by $\bar{b} = 120$ mm, $\sigma_b = 4$ mm, $\bar{p}_a = 10^6$ Pa, $\sigma_{p_a} = 10^5$ Pa, $\bar{t} = 30$ mm, $\sigma_t = 1$ mm, $\bar{r} = 150$ mm, $\sigma_r = 10$ mm, $\bar{f} = 0.35$, $\sigma_f = 0.035$, $\bar{a} = 200$ mm, $\sigma_a = 10$ mm, $\bar{\theta}_2 = 120°$, and $\sigma_{\theta_2} = 3°$, Assuming all parameters to be independent, estimate the mean and standard deviations of the actuating force needed to apply the brake using Monte Carlo simulation. Use five random numbers for each parameter starting from the first row in Table 10.5.

10.9 The energy absorbed by a clutch or brake is given by

$$E = \frac{J_1 J_2(\omega_1 - \omega_2)^2}{2(J_1 + J_2)}$$

where J_1 and J_2 are the mass moments of inertia and ω_1 and ω_2 are the angular velocities (see Fig. 10.15). If $\bar{J}_1 = 1.0$ kg-m^2, $\bar{J}_2 = 0.5$ kg-m^2, $\sigma_{J_1} = \sigma_{J_2} = 0.075$ kg-m^2, $\rho_{J_1, J_2} = 0.5$, $\bar{\omega}_1 = 200$ rpm, $\bar{\omega}_2 = 1000$ rpm, $\sigma_{\omega_1} = \sigma_{\omega_2} = 25$ rpm, and $\rho_{\omega_1, \omega_2} = 0.4$, estimate the mean and standard deviations of the energy absorbed by the clutch using Monte Carlo simulation. Use the following uniformly distributed random numbers to generate the correlated variables:

For J_1 and J_2: Numbers given in rows 2 and 3 of Table 10.5.

For ω_1 and ω_2: Numbers given in rows 4 and 5 of Table 10.5.

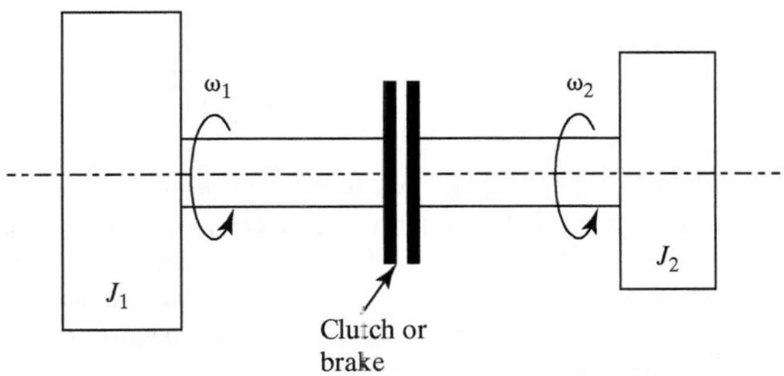

FIGURE 10.15 A clutch/brake system.

10.10 The horsepower transmitted by a belt drive (P) can be expressed as

$$P = \frac{K_p K_v F_a V}{16{,}500\, K_s}$$

where K_p = pulley correction factor, K_v = velocity correction factor, F_a = allowable tension in belt (lb), V = linear velocity of belt (ft/min), and K_s = service factor. The data for a flat leather belt drive is: $\overline{K}_p = 0.6$, $\sigma_{K_p} = 0.06$, $\overline{K}_v = 0.9$, $\sigma_{K_v} = 0.09$, $\overline{V} = 3000$ ft/min, $\sigma_v = 100$ ft/min, $\overline{F}_a = 80$ lb, $\sigma_{F_a} = 4$ lb, $\overline{K}_s = 1.1$, and $\sigma_{K_s} = 0.1$. Assuming that all the random variables are independent, estimate the mean and standard deviations of the horsepower transmitted by the belt drive using the Monte Carlo simulation method. Use five uniformly distributed numbers, starting from row 1 of Table 10.5, to simulate each random variable.

10.11 The contact pressure (p) developed at the interface when a steel gear with a central hole is shrink-fitted over a hollow steel shaft (see Fig. 10.16) is given by

$$p = \frac{E\delta}{R}\left[\frac{(r_o^2 - R^2)(R^2 - r_i^2)}{2R^2(r_o^2 - r_i^2)}\right]$$

where E = Young's modulus of steel, δ = total radial interference, r_o = pitch circle radius of the gear, r_i = inner radius of the shaft, and R = interface radius. If $E = 2.07 \times 10^{11}\ \text{N/m}^2$, $\delta = 1$ mm, $R = 300$ mm, $\overline{r}_o = 350$ mm, $\overline{r}_i = 250$ mm, $\sigma_{r_o} = 20$ mm, $\sigma_{r_i} = 15$ mm, $\rho_{r_o, r_i} = 0.3$, determine the mean value and the standard deviation of the contact pressure using the Monte Carlo simulation procedure. Use the uniformly distributed random numbers given in rows 4 and 5 of Table 10.5 to generate the correlated random variables.

10.12 The maximum stresses induced at the inner and outer surfaces [10.6] of a crane hook (see Fig. 10.17) are given by

$$\sigma_i = \frac{W}{A} + \frac{Mc_i}{Aer_i}; \quad \sigma_o = \frac{W}{A} - \frac{Mc_o}{Aer_o}$$

where W = weight lifted, A = cross-sectional area, M = bending moment induced, c_i = distance from neutral axis to inner fiber, c_o = distance from neutral axis to outer fiber, e = distance from centroidal axis to neutral axis, r_i = radius of inner fiber, and r_o = radius of outer fiber, with the radius of neutral axis, for the rectangular cross section shown in Fig. 10.17, is given by

$$r_n = \frac{w}{\ln\left(\dfrac{r_o}{r_i}\right)}$$

and the radius of the centroidal axis is given by $R = (r_o + r_i)/2$, so that $e = R - r_n$. The following data is applicable for a specific crane hook: $\overline{r}_i = 2$ in, $\overline{r}_o = 6$ in, $w = 4$ in, $t = 3/4$ in, $\overline{W} = 5000$ lb, $\sigma_{r_i} = 0.1$ in, $\sigma_{r_o} = 0.5$ in, $\sigma_W = 1000$ lb, and $\rho_{r_i, r_o} = 0.5$. Estimate the mean and standard deviations of the stresses induced σ_i and σ_o using Monte Carlo simulation. Use the uniformly distributed random numbers given in the last three rows of Table 10.5 to generate the random variables.

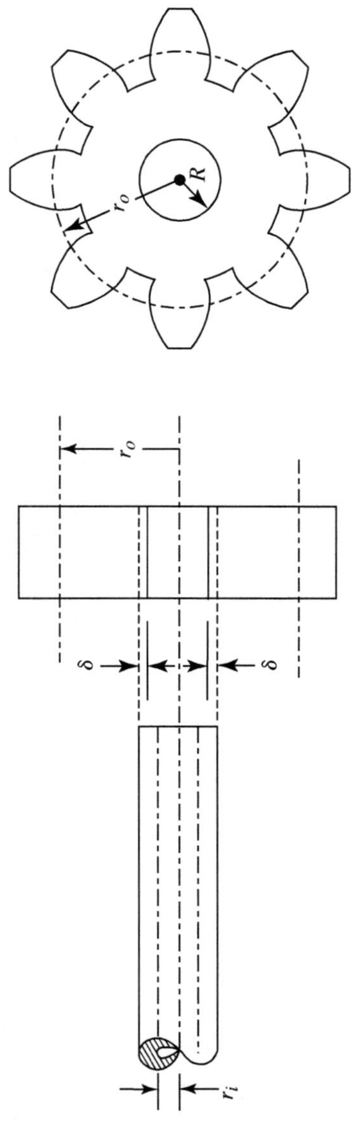

FIGURE 10.16 Shrink fit.

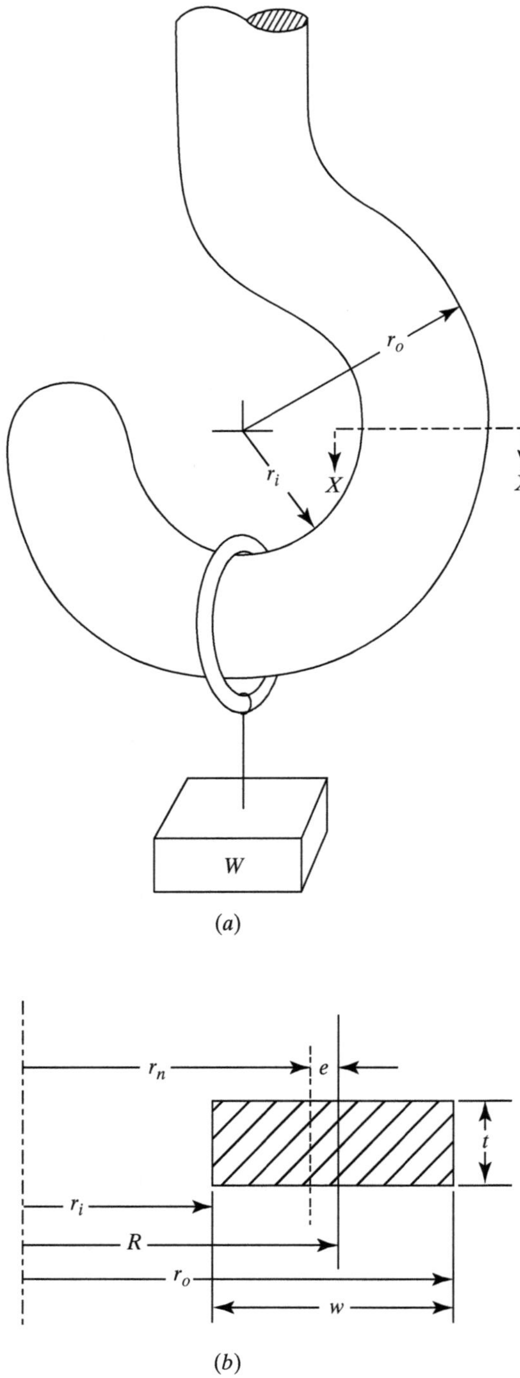

FIGURE 10.17 (*a*) Crane hook; and (*b*) Section X–X.

10.13 The maximum contact pressure (p_{max}) developed between the two spheres [10.6] shown in Fig. 10.18 is given by

$$p_{max} = \frac{3P}{2\pi\varepsilon^2}$$

with

$$\varepsilon = \left\{ \frac{3P}{4} \left[\frac{\left(\frac{1 - \nu_1^2}{E_1}\right) + \left(\frac{1 - \nu_2^2}{E_2}\right)}{\frac{1}{r_1} + \frac{1}{r_2}} \right] \right\}^{1/3}$$

where ν = Poisson's ratio, E = Young's modulus, r = radius of sphere, ε = radius of the contact area, P = load applied, and the subscripts 1 and 2 refer to the spheres 1 and 2, respectively. Find the mean and standard deviations of p_{max} using Monte Carlo simulation for the following data: $\overline{P} = 5000$ lb, $\sigma_P = 500$ lb, $\overline{\nu}_1 = 0.29$, $\sigma_{\nu_1} = 0.029$, $\overline{E}_1 = 30 \times 10^6$ lb/in^2, $\sigma_{E_1} = 3 \times 10^6$ lb/in^2, $\rho_{\nu_1, E_1} = 0.3$, $\overline{\nu}_2 = 0.32$, $\sigma_{\nu_2} = 0.016$, $\overline{E}_2 = 15 \times 10^6$ lb/in^2, $\sigma_{E_2} = 0.75 \times 10^6$ lb/in^2, $\rho_{\nu_2, E_2} = 0.2$, $r_1 = 2$ in, $r_2 = 3$ in. Use five uniformly distributed random numbers starting from row 1 of Table 10.5 to generate each of the random variables.

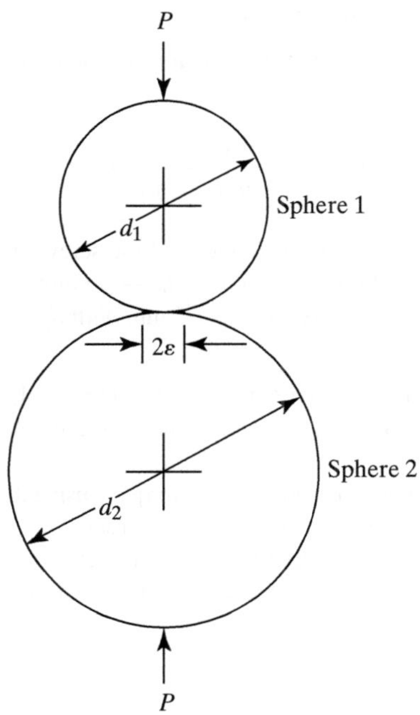

FIGURE 10.18 Two spheres under contact pressure.

10.14 The maximum contact pressure ($p_{\max}$) developed between two cylinders made of different materials (see Fig. 10.4) is given by [10.6]

$$p_{\max} = \frac{2F}{\pi \varepsilon a}$$

with

$$\varepsilon = \left\{ \frac{2F}{\pi a} \left[\frac{\left(\frac{1 - v_1^2}{E_1} \right) + \left(\frac{1 - v_2^2}{E_2} \right)}{\frac{1}{d_1} + \frac{1}{d_2}} \right] \right\}^{1/2}$$

where v = Poisson's ratio, E = Young's modulus, d = diameter of cylinder, ε = semi-width of contact, a = axial length of contact, F = load applied, and the subscripts 1 and 2 refer to the cylinders 1 and 2, respectively. The data for a specific case follow:

$$\overline{F} = 5000 \text{ lb}; \quad \sigma_F = 500 \text{ lb}; \quad \overline{v}_1 = 0.29; \quad \overline{E}_1 = 30 \times 10^6 \text{ lb/in}^2;$$

$$\sigma_{v_1} = 0.029; \quad \sigma_{E_1} = 3 \times 10^6 \text{ lb/in}^2; \quad \rho_{v_1, E_1} = 0.3; \quad \overline{v}_2 = 0.32;$$

$$\overline{E}_2 = 15 \times 10^6 \text{ lb/in}^2; \quad \sigma_{v_2} = 0.016; \quad \sigma_{E_2} = 0.75 \times 10^6 \text{ lb/in}^2;$$

$$\rho_{v_2, E_2} = 0.2; \quad a = 20 \text{ in}; \quad d_1 = 4 \text{ in}; \quad d_2 = 6 \text{ in}.$$

Use five uniformly distributed random numbers starting from row 3 of Table 10.5 to generate each of the random variables and estimate the mean value and standard deviation of $p_{\max}$ using the Monte Carlo simulation method.

10.15 The torque required (T) to raise the load in a square threaded screw jack (Fig. 10.19) is given by [10.6]

$$T = \frac{F\, d_m}{2} \left(\frac{\pi\, \mu\, d_m + e}{\mu\, d_m - \mu e} \right)$$

where F = load being raised, d_m = mean diameter of the screw, e = lead of the threads, and μ = coefficient of friction. In a power screw, $\overline{d}_m = 30$ mm, $\sigma_{d_m} = 1$ mm, $\overline{\mu} = 0.1$, $\sigma_\mu = 0.01$, $\overline{F} = 5000$ N, $\sigma_F = 1000$ N, and $e = 5$ mm with d_m, μ, and F following uniform distributions:

a. Use the uniformly distributed random numbers given in rows 4, 5, and 6 of Table 10.5 to generate the random variables d_m, μ, and F, and find the mean and standard deviations of the torque, T.

b. Find the mean value and standard deviations of the torque T using the partial derivative rule.

c. Discuss why the answers in parts (a) and (b) are different.

d. Assuming that instead of using just five values in part (a), we use a very large number of (infinite) values for each random variable, do you expect the results of parts (a) and (b) to be different? If so, why?

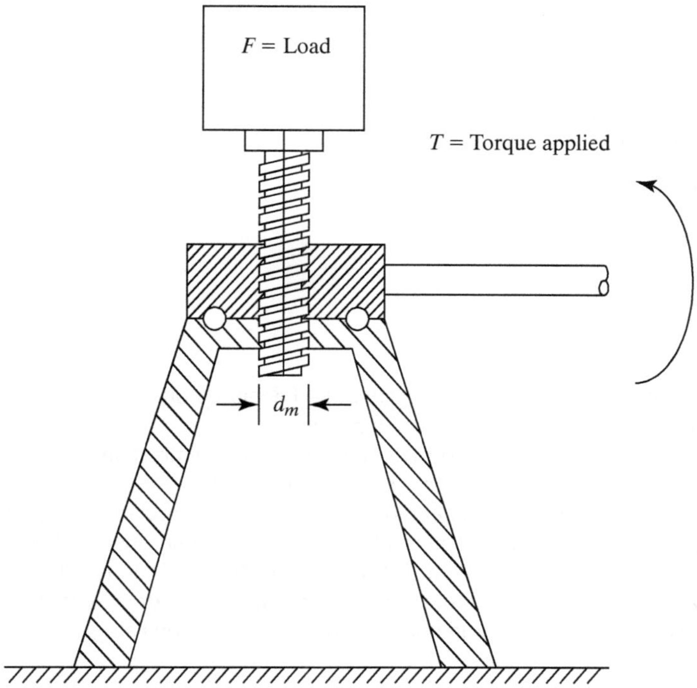

FIGURE 10.19 A screw jack.

MATLAB and Excel Problems

10.16 Using a uniformly distributed random number lying in the interval 0 to 1, find the corresponding number x of a random variable X that follows Weibull distribution with parameters $\alpha = 0.5$ and $\beta = 6000$ using MATLAB.†

10.17 When two cylinders are in contact with each other as shown in Fig. 10.4, the contact pressure is given by Eq. (E1) of Example 10.8. The load, F, and the length of the cylinder, a, are known to follow normal distribution with mean values and standard deviations given by $\mu_F = 10{,}000$ N, $\sigma_F = 2{,}000$ N, $\mu_a = 0.2$ m, and $\sigma_a = 0.02$ m. Using MATLAB, generate 20 random values for each of F and a, use them to find 20 random values of the contact pressure, and compute the corresponding mean value and standard deviation of the contact pressure. Assume the values of the other parameters as $v = 0.3$, $E = 2.06 \times 10^{11}$ N/m^2, $d_1 = 0.1$ m, and $d_2 = 0.25$ m.

10.18 Using Excel, generate five random numbers following Weibull distribution with parameters $\alpha = 0.5$ and $\beta = 6000$.†

---

†α denotes the shape parameter and β indicates the characteristic value.

CHAPTER 11

Reliability-Based Optimum Design

What You Will Learn

In this chapter, you will learn

- The application of optimization techniques to reliability-based design problems
- The importance of the reliability-based optimization in practice
- An introduction to optimization and the basic terminologies used in optimization
- The classification of optimization problems and their solution methods
- The formulation of strength-based mechanical and structural reliability problems as optimization problems
- The graphical method of solving two-variable optimization problems with an illustrative example
- An outline of the Lagrange multiplier method and the penalty function method for the solution of reliability-based optimization problems subject to equality and inequality constraints, respectively
- The computational details of dynamic programming for the solution of multistage optimization problems with an illustrative example

Joseph Louis Lagrange was a French mathematician who was born on January 25, 1736. He was the youngest and the only one among his parents' 11 children to survive beyond infancy. He studied at the Royal Artillery School in Turin, Italy, and became a professor of mathematics there at the age of 18. He left for Berlin in 1766 and remained there until 1786. Later he went to Paris, became a professor at the newly established Ecole Normale, and taught there until his death on April 10, 1813. Lagrange was very brilliant and invented the calculus of variations. He solved the vibrating-string problem and explained the phenomena of beats, echos, and compound sounds. He contributed to the theory of numbers and differential equations and developed the principles of virtual velocities and least action. The method he presented for finding the extremum solution of a multivariable function in the presence of constraints is known as the Lagrange multiplier method and plays an important role in the literature of optimization [11.10, 11.11].

11.1 Introduction

Optimization plays an important role in system design. In the design of any component or system, we will be interested either in maximizing the reliability subject to a constraint on the cost or in minimizing the cost with a restriction on the reliability of the system. In structural and mechanical design problems, the strength-based reliability becomes important. In this chapter, the standard optimization problem is stated first and then a variety of reliability-based design problems are formulated as optimization problems. The various techniques available for the solution of optimum-design problems are summarized. The computational details of some of the commonly used optimization techniques are given along with illustrative examples. The basic principles of graphical techniques, calculus-based methods, and nonlinear programming-based strategies are explained. The illustrative examples deal with the reliability-based optimum design of machine and structural components and the optimal-reliability allocation in multicomponent systems.

11.2 Optimization Problem

Optimization is the process of obtaining the best result under given circumstances. The conventional design procedures aim at finding an adequate design that merely satisfies the functional requirements within the confines of the existing limitations. The optimum-design procedures seek to find the design variables in such a way that the design will be the best one (among the many adequate designs) that satisfies all of the limitations and restrictions placed on it.

A standard optimization problem can be stated in the following form:

$$\text{Find X} = \begin{Bmatrix} x_1 \\ x_2 \\ \cdot \\ \cdot \\ \cdot \\ x_n \end{Bmatrix}$$

which minimizes $f(\text{X})$

subject to

$$g_j(\text{X}) \leq 0; \quad j = 1, 2, \ldots, m$$

and

$$l_k(\text{X}) = 0; \quad k = 1, 2, \ldots, p \tag{11.1}$$

where x_i = ith design variable $(i = 1, 2, \ldots, n)$, X = vector of design variables, f = objective or criterion function, g_j = inequality-constrained function, l_k = equality-constrained function, n = number of design variables, m = number of inequality constraints, and p = number of equality constraints. The problem stated in Eq. (11.1) is also known as a mathematical programming problem. Several methods are available for solving the problem stated in Eq. (11.1). However, no single method can solve all types of optimization problems efficiently. Depending on the type of optimization problem, only certain methods can be used for the efficient solution of Eq. (11.1). The classification of optimization problems is given in Table 11.1.

TABLE 11.1 Classification of optimization problems

Type of Problem	Characteristics of Problem in Eq. (11.1): Nature of Design Variables (x_i)	Nature of the Objective Function $f(\mathbf{X})$	Nature of the Constraint Functions $g_j(\mathbf{X})$ and $l_j(\mathbf{X})$	Solution Method(s) Available	Name(s) of the Person(s) Associated with the Problem
1. Linear programming problem	Real	Linear function of x_i	Linear functions of x_i	Simplex method	Dantzig
2. Integer linear programming problem	Integers	Linear function of x_i	Linear functions of x_i	Cutting-plate method Branch-bound method Zero-one programming method	Gomory, Land and Doig, Balas
3. Quadratic programming problem	Real	Quadratic function of x_i	Linear functions of x_i	Modified simplex method	Wolfe
4. Nonlinear programming problem	Real	General nonlinear function of x_i	General nonlinear functions of x_i	Penalty function methods (SUMT), Feasible-directions method	Fiacco and McCormick, Zoutendijk
5. Geometric programming problem	Real	Posynomial in x_i	Posynomials in x_i	Method based on arithmetic-geometric inequality	Duffin, Zener, and Peterson
6. Dynamic programming problem	Real or integers	Staged nonlinear function of x_i	Staged nonlinear functions of x_i	Multistage optimization based on the principle of optimality	Bellman
7. Calculus of variations	Real functions, $x_i(t)$ where t is some parameter	Nonlinear function of x_i in the form of an integral	Nonlinear functions of x_i in the form of integrals	Calculus of variations	Bernoulli, Euler, Lagrange, and Weirstrass

As indicated in Table 11.1, if one or more of the functions f, g_j, and l_k are nonlinear in terms of the design variables x_i, the problem stated in Eq. (11.1) is called a constrained-nonlinear programming problem. In most reliability-based optimum-design problems, many of the functions f, g_j, and l_k are nonlinear, and hence we shall study nonlinear optimization techniques in this chapter. We shall consider four techniques. The first one is a graphical method that can be used when the number of design variables (unknowns) is one or two. The second technique is a calculus-based method, called the Lagrange multiplier method, which can be used to solve problems containing only equality constraints. The third technique is termed the sequential unconstrained minimization technique (SUMT) or the penalty function method in which the solution of a general optimization problem, as expressed in Eq. (11.1), is found by solving a series of unconstrained minimization problems. The fourth method is known as the dynamic-programming approach that can be used to solve problems whose physical structure is in series form. We also note that the maximization of $f(\mathrm{X})$ can be achieved by the minimization of $-f(\mathrm{X})$.

11.3 Formulation of Optimization Problems

11.3.1 Reliability Allocation Problems

Reliability allocation is the process of assigning reliabilities to individual components so as to attain a specified overall reliability for the system. The specified overall system reliability can be achieved in several ways. However, we will be interested in minimizing a criterion or objective function such as cost, weight, or volume while achieving the overall system reliability over a specified length of time, T. Thus, the problem can be formulated in different ways as illustrated by the following examples.

EXAMPLE 11.1

Formulate a reliability allocation problem using the minimization of the total cost as the objective function.

Solution: Assuming the cost of the ith component to be a function of its reliability as $c_i = c_i(R_i)$, the optimization problem can be stated as follows:

$$\text{Find } \mathrm{X} = \begin{Bmatrix} R_1 \\ R_2 \\ \cdot \\ \cdot \\ \cdot \\ R_n \end{Bmatrix}$$

$$\text{which minimizes } f(\mathrm{X}) = \sum_{i=1}^{n} c_i(R_i) \tag{E1a}$$

subject to

$$R_{\text{system}}(R_1, R_2, \ldots, R_n) \geq R_{\text{specified}} \tag{E1b}$$

$$0 \leq R_i \leq 1; \quad i = 1, 2, \ldots, n \tag{E1c}$$

where R_i indicates the reliability of ith component and n denotes the number of components. As seen earlier, the expression for system reliability, R_{system}, depends on the type of system. For example, for a series system,

$$R_{\text{system}}(t) = \prod_{i=1}^{n} R_i(t) \tag{E2}$$

If the failure rates of the components (λ_i) are constant, Eq. (E2) becomes

$$R_{\text{system}}(t) = \prod_{i=1}^{n} e^{-\lambda_i t} = e^{-\lambda_{\text{system}} t} \equiv e^{-(\lambda_1 + \lambda_2 + \cdots + \lambda_n)t} \tag{E3}$$

In this case, the reliability constraint, Eq. (E1b), can be expressed in equivalent form as

$$\sum_{i=1}^{n} \lambda_i - \lambda_{\text{specified}} \leq 0 \tag{E4}$$

where $\lambda_{\text{specified}}$ is the failure rate corresponding to the specified reliability of the system.

■

EXAMPLE 11.2

Formulate a reliability-allocation problem using the maximization of system reliability as the objective function.

Solution: In this case, the problem can be stated as follows:

$$\text{Find } \mathbf{X} = \begin{Bmatrix} R_1 \\ R_2 \\ \cdot \\ \cdot \\ \cdot \\ R_n \end{Bmatrix}$$

which maximizes

$$f(\mathbf{X}) = R_{\text{systems}}(R_1, R_2, \ldots, R_n)$$

subject to

$$c_{\text{system}} = \sum_{i=1}^{n} c_i(R_i) \leq c_{\text{specified}}$$

and

$$0 \leq R_i \leq 1; \quad i = 1, 2, \ldots, n$$

where c_i is the cost of the ith component and $c_{\text{specified}}$ is the maximum specified cost of the system.

■

EXAMPLE 11.3

Formulate the reliability allocation problem of maximizing the system reliability for the parallel–series system shown in Fig. 11.1.

Solution: In this case, the optimization problem can be formulated as follows. Assuming that all the components of stage i are identical $(i = 1, 2, \ldots, N)$, we have $R_1, R_2, \ldots, R_N$ as design variables. Thus, the problem can be formulated as

$$\text{Find } \mathbf{X} = \begin{Bmatrix} R_1 \\ R_2 \\ \cdot \\ \cdot \\ \cdot \\ R_N \end{Bmatrix}$$

which maximizes

$$R_{\text{system}}(\mathbf{X}) = \prod_{i=1}^{N} \left[1 - (1 - R_i)^{n_i}\right]$$

subject to

$$\sum_{i=1}^{N} c_i n_i \le c_o$$

where c_i is the cost of one component of stage i and c_o is the maximum permissible cost of the system.

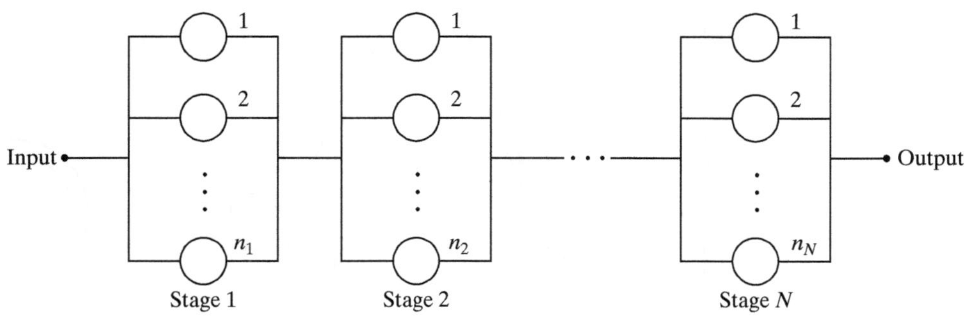

FIGURE 11.1 Parallel–series system.

■

11.3.2 Structural and Mechanical Design Problems

The following examples illustrate the formulation of reliability-based structural and mechanical design problems as optimization problems.

EXAMPLE 11.4

A two-bar truss, subjected to the loads F_v and F_h, is shown in Fig. 11.2. The members have a tubular section with mean diameter d and thickness t. Each member has two failure modes, namely, yielding and buckling. Formulate the problem of finding the mean values of d and t to minimize the weight of the truss subject to constraints on the reliability of the members in yielding and buckling. The data of the problem are given in the following list:

$F_v = N(50{,}000, 5000)$ lb
$F_h = N(30{,}000, 3000)$ lb
$t = N(\bar{t}, 0.01\ \bar{t})$ in
$d = N(\bar{d}, 0.01\bar{d})$ in
s_y = yield stress = $N(30{,}000, 3000)$ lb/in^2
E = Young's modulus = $N(3, 0.3) \times 10^7$ lb/in^2
$B = N(20, 1)$ in
$H = N(30, 1)$ in
$\rho = N(0.3, 0.03)$ lb/in^3
Minimum acceptable reliability = 0.9973
Permissible ranges of $\bar{d}$ and $\bar{t}$ are 1.0 in to 5.0 in and 0.1 in to 1.0 in, respectively

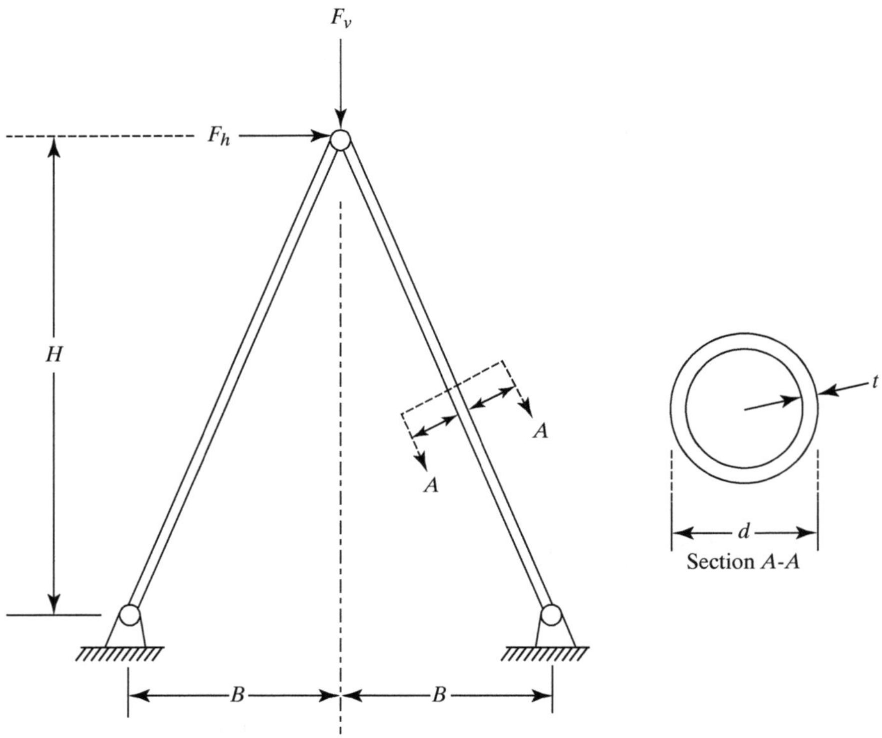

FIGURE 11.2 Two-bar truss.

Solution: Assuming that the stress induced in a member and the Euler buckling stress follow normal distribution, the minimum acceptable reliability in each failure mode can be seen to correspond to a value of $z_1 = -3.0$. Thus, the reliability constraints can be expressed as

$$z_1 = -\frac{\bar{s}_y - \bar{s}_c}{\sqrt{\sigma_{s_y}^2 + \sigma_{s_c}^2}} \tag{E1}$$

and

$$\bar{z}_2 = -\frac{\bar{s}_b - \bar{s}_c}{\sqrt{\sigma_{s_b}^2 + \sigma_{s_c}^2}} \tag{E2}$$

where s_c is the compressive stress induced in a member and s_b is the buckling stress of the member

$$s_c = \frac{\sqrt{(B^2 + H^2)}}{2\pi\, dt}\left(\frac{F_v}{H} + \frac{F_h}{B}\right) \tag{E3}$$

$$s_b = \frac{\pi^2 EI}{l^2 \pi\, dt} = \frac{\pi^2 E(d^2 + t^2)}{8(B^2 + H^2)} \tag{E4}$$

where I = area moment of inertia and l = length of the bar. The mean values of s_c and s_b are given by

$$\bar{s}_c = \frac{\sqrt{(\bar{B}^2 + \bar{H}^2)}}{2\pi\, \bar{d}\bar{t}}\left(\frac{\bar{F}_v}{\bar{H}} + \frac{\bar{F}_h}{\bar{B}}\right) \tag{E5}$$

$$\bar{s}_b = \frac{\pi^2 \bar{E}(\bar{d}^2 + \bar{t}^2)}{8(\bar{B}^2 + \bar{H}^2)} \tag{E6}$$

The standard deviations of s_c and s_b can be found using the partial derivative rule as

$$\sigma_{s_c}^2 = \left(\frac{\partial s_c}{\partial B}\right)^2 \sigma_B^2 + \left(\frac{\partial s_c}{\partial H}\right)^2 \sigma_H^2 + \left(\frac{\partial s_c}{\partial F_v}\right)^2 \sigma_{F_v}^2 + \left(\frac{\partial s_c}{\partial F_h}\right)^2 \sigma_{F_h}^2 + \left(\frac{\partial s_c}{\partial d}\right)^2 \sigma_d^2 + \left(\frac{\partial s_c}{\partial t}\right)^2 \sigma_t^2 \tag{E7}$$

$$\sigma_{s_b}^2 = \left(\frac{\partial s_b}{\partial E}\right)^2 \sigma_E^2 + \left(\frac{\partial s_b}{\partial d}\right)^2 \sigma_d^2 + \left(\frac{\partial s_b}{\partial t}\right)^2 \sigma_t^2 + \left(\frac{\partial s_b}{\partial B}\right)^2 \sigma_B^2 + \left(\frac{\partial s_b}{\partial H}\right)^2 \sigma_H^2 \tag{E8}$$

where the partial derivatives are to be evaluated at the mean values of the random variables. The partial derivatives are given by

$$\frac{\partial s_c}{\partial B} = (B^2 + H^2)^{-1/2}\frac{B}{2\pi\, dt}\left(\frac{F_v}{H} + \frac{F_h}{B}\right) - \frac{\sqrt{B^2 + H^2}}{2\pi\, dt}(F_h B^{-2}) \tag{E9}$$

$$\frac{\partial s_c}{\partial H} = (B^2 + H^2)^{-1/2}\frac{H}{2\pi\, dt}\left(\frac{F_v}{H} + \frac{F_h}{B}\right) - \frac{\sqrt{B^2 + H^2}}{2\pi\, dt}(F_v H^{-2}) \tag{E10}$$

$$\frac{\partial s_c}{\partial F_v} = \frac{\sqrt{B^2 + H^2}}{2\pi\, dtH} \tag{E11}$$

$$\frac{\partial s_c}{\partial F_h} = \frac{\sqrt{B^2 + H^2}}{2\pi\, dtB} \tag{E12}$$

$$\frac{\partial s_c}{\partial d} = -\frac{\sqrt{B^2 + H^2}}{2\pi\, d^2 t}\left(\frac{F_v}{H} + \frac{F_h}{B}\right) \tag{E13}$$

$$\frac{\partial s_c}{\partial t} = -\frac{\sqrt{B^2 + H^2}}{2\pi\, dt^2}\left(\frac{F_v}{H} + \frac{F_h}{B}\right) \tag{E14}$$

$$\frac{\partial s_b}{\partial E} = \frac{\pi^2(d^2 + t^2)}{8(B^2 + H^2)} \tag{E15}$$

$$\frac{\partial s_b}{\partial d} = \frac{\pi^2 E(2d)}{8(B^2 + H^2)} \tag{E16}$$

$$\frac{\partial s_b}{\partial t} = \frac{\pi^2 E(2t)}{8(B^2 + H^2)} \tag{E17}$$

$$\frac{\partial s_b}{\partial B} = -\frac{\pi^2 E(d^2 + t^2)}{64(B^2 + H^2)^2}(2B) \tag{E18}$$

$$\frac{\partial s_b}{\partial H} = -\frac{\pi^2 E(d^2 + t^2)}{64(B^2 + H^2)^2}(2H) \tag{E19}$$

The weight of the truss (W) is given by

$$W = 2\rho\pi\, dt\sqrt{B^2 + H^2} \tag{E20}$$

By treating the mean values of d and t as design variables and the mean value of the weight as the objective function, the optimization problem can be stated as follows:

$$\text{Find } \mathbf{X} = \begin{Bmatrix} x_1 \\ x_2 \end{Bmatrix} = \begin{Bmatrix} \overline{d} \\ \overline{t} \end{Bmatrix}$$

which minimizes

$$f(\mathbf{X}) = \overline{W} = 2\overline{\rho}\pi\, \overline{d}\, \overline{t}\sqrt{\overline{B}^2 + \overline{H}^2}$$

subject to

$$z_1 \le -\frac{\overline{s}_y - \overline{s}_c}{\sqrt{\sigma_{s_y}^2 + \sigma_{s_c}^2}} \tag{E21}$$

$$z_1 \le -\frac{\overline{s}_b - \overline{s}_c}{\sqrt{\sigma_{s_b}^2 + \sigma_{s_c}^2}}$$

$$0.1 \le \overline{t} \le 1.0$$

$$1.0 \le \overline{d} \le 5.0$$

■

EXAMPLE 11.5

The single-shoe brake, shown in Fig. 11.3, is to be designed to have a braking capacity (torque) of $T_c \pm \Delta T_c = (350 \pm 35)$ lb · in. The applied force F, the coefficient of friction between the brake drum and the brake shoe f, the dimensions a, b, c, and r are known to follow normal distributions as $F = N(50, 5)$ lb, $f = N(0.3, 0.03)$, $a = N(10, 1)$ in, $b = N(\overline{b}, \sigma_b)$ in, $c = N(20, 2)$ in, and $r = N(\overline{r}, \sigma_r)$ in. Find the mean values of b and r to maximize the reliability without exceeding a mean pressure of 120 lb/in^2 on the brake pad. The brake shoe has a contact area of $\overline{A} \pm \Delta A = 4 \pm 0.4$ in^2. Assume that the coefficient of variation of b and r is 0.1 and the tolerances given for the braking capacity and the brake-shoe area correspond to 3σ values. Formulate the optimization problem including the following constraints:

$$\frac{\overline{a}}{2} \leq \overline{r} \leq \overline{a} \leq \overline{b} \leq \overline{c}$$

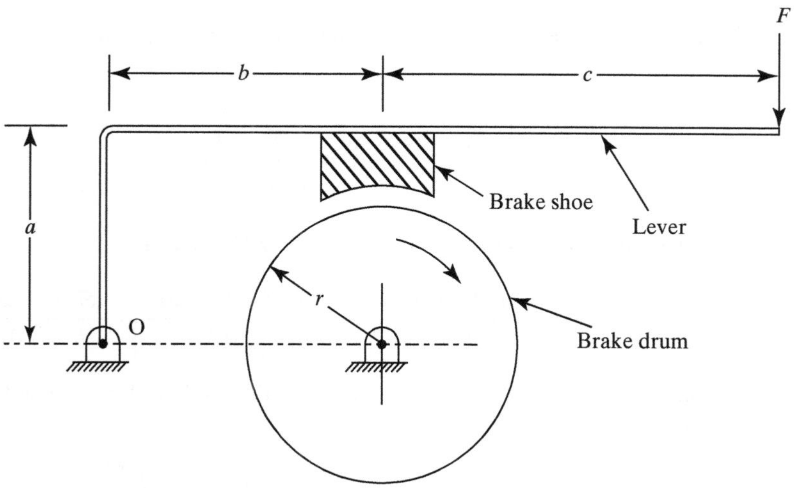

FIGURE 11.3 Single-shoe brake.

Solution: The free-body diagram of the lever and the brake shoe is shown in Fig. 11.4. Taking moments about the pivot point of the lever, O, we obtain

$$F(b + c) + rfN - bN = 0 \quad \text{i.e.,} \quad N = \frac{F(b + c)}{(b - fr)} \tag{E1}$$

where N and fN denote, respectively, the normal and tangential forces acting on the brake shoe. The torque acting on the brake drum (T_b) is given by

$$T_b = fNr = \frac{frF(b + c)}{(b - fr)} \tag{E2}$$

Assuming normal distribution for all the quantities, the reliability of the brake can be expressed as (by substituting T_b for load L and T_c for strength S in Eq. 8.23)

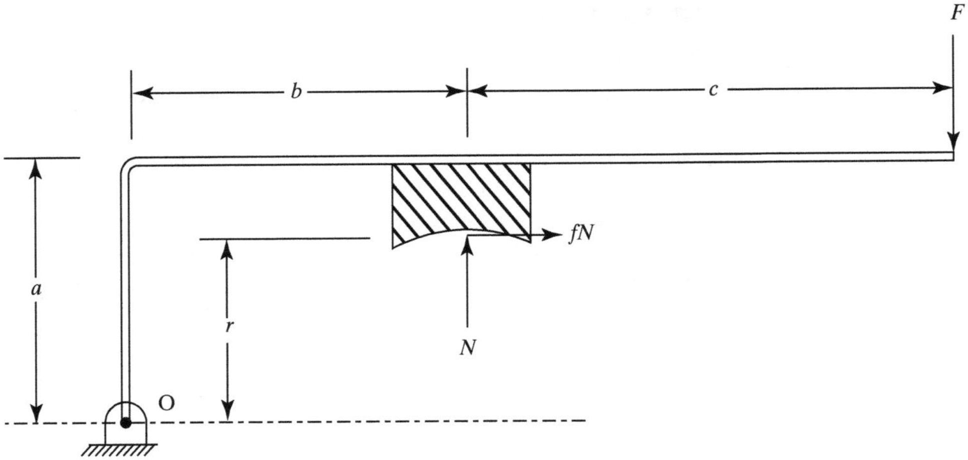

FIGURE 11.4 Free-body diagram of the lever and the brake shoe.

$$z_1 = \frac{\overline{T}_b - \overline{T}_c}{\sqrt{\sigma_{T_b}^2 + \sigma_{T_c}^2}} \tag{E3}$$

The mean values of T_c and T_b are given by $\overline{T}_c = 350$ lb · in and

$$\overline{T}_b = \frac{\overline{f}\,\overline{r}\,\overline{F}(\overline{b} + \overline{c})}{(\overline{b} - \overline{f}\,\overline{r})} = \frac{0.3\overline{r}(50)(\overline{b} + 20)}{(\overline{b} - 0.3\overline{r})} = \frac{15\overline{r}(\overline{b} + 20)}{(\overline{b} - 0.3\overline{r})} \tag{E4}$$

The standard deviation of T_c is $\sigma_{T_c} = 35/3 = 11.67$ lb · in and σ_{T_b} can be found using the partial derivative rule as

$$\sigma_{T_b}^2 = \left(\frac{\partial T_b}{\partial f}\right)^2 \sigma_f^2 + \left(\frac{\partial T_b}{\partial r}\right)^2 \sigma_r^2 + \left(\frac{\partial T_b}{\partial F}\right)^2 \sigma_F^2 + \left(\frac{\partial T_b}{\partial b}\right)^2 \sigma_b^2 + \left(\frac{\partial T_b}{\partial c}\right)^2 \sigma_c^2 \tag{E5}$$

where the partial derivatives, evaluated at the mean values of the random variables, give

$$\frac{\partial T_b}{\partial f} = \frac{rF(b + c)}{(b - fr)} - \frac{frF(b + c)}{(b - fr)^2}(-r) = \frac{50\overline{r}(\overline{b} + 20)}{(\overline{b} - 0.3\overline{r})} + \frac{15\overline{r}^2(\overline{b} + 20)}{(\overline{b} - 0.3\overline{r})^2} \tag{E6}$$

$$\frac{\partial T_b}{\partial r} = \frac{fF(b + c)}{(b - fr)} - \frac{frF(b + c)}{(b - fr)^2}(-f) = \frac{15(\overline{b} + 20)}{(\overline{b} - 0.3\overline{r})} + \frac{4.5\overline{r}(\overline{b} + 20)}{(\overline{b} - 0.3\overline{r})^2} \tag{E7}$$

$$\frac{\partial T_b}{\partial F} = \frac{fr(b + c)}{(b - fr)} = \frac{0.3\overline{r}(\overline{b} + 20)}{(\overline{b} - 0.3\overline{r})} \tag{E8}$$

$$\frac{\partial T_b}{\partial b} = \frac{frF}{(b - fr)} - \frac{frF(b + c)}{(b - fr)^2}$$

$$= \frac{15\overline{r}}{(\overline{b} - 0.3\overline{r})} - \frac{15\overline{r}(\overline{b} + 20)}{(\overline{b} - 0.3\overline{r})^2} \tag{E9}$$

$$\frac{\partial T_b}{\partial c} = \frac{frF}{(b - fr)} = \frac{15\overline{r}}{(\overline{b} - 0.3\overline{r})} \tag{E10}$$

Substitution of Eqs. (E6) to (E10) into (E5), along with $\sigma_f = 0.03$, $\sigma_r = 0.1\overline{r}$, $\sigma_F = 5$, $\sigma_b = 0.1\overline{b}$, and $\sigma_c = 2.0$, gives

$$\sigma^2_{T_b} = 2\left(\frac{1.5\overline{r}(\overline{b} + 20)}{(\overline{b} - 0.3\overline{r})} + \frac{0.45\overline{r}^2(\overline{b} + 20)}{(\overline{b} - 0.3\overline{r})^2}\right)^2 + \left(\frac{1.5\overline{r}(\overline{b} + 20)}{(\overline{b} - 0.3\overline{r})}\right)^2$$

$$+ \left(\frac{1.5\overline{r}\overline{b}}{(\overline{b} - 0.3\overline{r})} - \frac{1.5\overline{r}\overline{b}(\overline{b} + 20)}{(\overline{b} - 0.3\overline{r})^2}\right)^2 + \left(\frac{30\overline{r}}{\overline{b} - 0.3\overline{r}}\right)^2 \tag{E11}$$

The pressure on the brake pad (p) is given by

$$p = \frac{N}{A} = \frac{F(b + c)}{(b - fr)A} \tag{E12}$$

where A is the area of the brake shoe. Hence, the mean value of pressure is given by

$$\overline{p} = \frac{\overline{F}(\overline{b} + \overline{c})}{(\overline{b} - \overline{f}\,\overline{r})\overline{A}} = \frac{50(\overline{b} + 20)}{(\overline{b} - 0.3\overline{r})4} \tag{E13}$$

The optimization problem can be stated as follows:

Find X which minimizes

$$f(\mathbf{X}) = z_1 = \frac{\overline{T}_b - 350}{\sqrt{\sigma^2_{T_b} + 136.1111}}$$

subject to

$$\begin{aligned} \frac{12.5(\overline{b} + 20)}{(\overline{b} - 0.3\overline{r})} - 120 &\le 0 \\ 5 - \overline{r} &\le 0 \\ \overline{r} - 10 &\le 0 \\ 10 - \overline{b} &\le 0 \\ \overline{b} - 20 &\le 0 \end{aligned} \tag{E14}$$

where $\overline{T}_b$ and σ_{T_b} are given by Eqs. (E4) and (E11), respectively.

■

11.4 Solution Techniques

11.4.1 Graphical-Optimization Method

Consider the optimization problem

Find X which minimizes $f(\mathrm{X})$

subject to the constraints

$$g_j(\mathrm{X}) \leq 0, j = 1, 2, \ldots, m \tag{11.2}$$

If each of the n design variables, x_i, is represented along a coordinate axis, the resulting n-dimensional cartesian space is known as the *design space*. Each point in this n-dimensional design space is called a design point and represents either a possible (feasible) or an impossible (infeasible) solution to the design problem. A *feasible design* satisfies all the constraints, while an *infeasible design* violates one or more of the constraints. The feasible and infeasible design points are separated by a surface, known as the *constraint surface*. To understand the concept of *constraint surface*, consider the inequality constraint $g_j(\mathrm{X}) \leq 0$. The set of design points X that satisfy the equation $g_j(\mathrm{X}) = 0$ forms a hypersurface in the design space and is called a constraint surface. The constraint surface divides the design space into two parts: one in which $g_j(\mathrm{X}) < 0$, and the other in which $g_j(\mathrm{X}) > 0$. Thus, the points lying on the hypersurface will satisfy the constraint $g_j(\mathrm{X})$ critically, whereas the points lying in the region where $g_j(\mathrm{X}) > 0$ are infeasible or unacceptable, and the points lying in the region where $g_j(\mathrm{X}) < 0$ are feasible or acceptable. When the constraint surfaces corresponding to all the constraints $g_j(\mathrm{X}) \leq 0, j = 1, 2, \ldots, m$ are considered, they collectively divide the design space into two regions: one called the feasible space in which every design point X corresponds to a value of $g_j(\mathrm{X})$ less than or equal to zero, and the other called the infeasible space in which every design point X gives a value of $g_j(\mathrm{X})$ greater than zero for some j.

When the objective function is plotted with $f(\mathrm{X}) = c =$ constant, the resulting graph forms a hypersurface in the design space. If several graphs are plotted with different values of the constant c, the resulting surfaces are known as the *objective-function surfaces*. Once the objective-function surfaces are known, the optimum solution of the problem can be identified as the design point in the feasible space that gives the minimum value of the objective function.

The constraint surfaces, objective-function surfaces, along with the optimum solution of a typical two-variable optimization problem, are shown in Fig. 11.5. It can be noted that as the number of design variables exceeds two or three, the constraints and objective-function surfaces become complex even for visualization and the problem has to be solved purely as a mathematical problem. Example 11.6 illustrates the graphical optimization-procedure.

EXAMPLE 11.6

Two components in series are shown in Fig. 11.6. The cost of the first component is given by $100R_1^2$ where R_1 denotes its reliability. Similarly, the cost of the second component is given by $200R_2^2$ where R_2 represents its reliability. Find the reliabilities of the components to minimize the total cost. The system reliability as well as the individual component reliabilities are required to be at least 0.7.

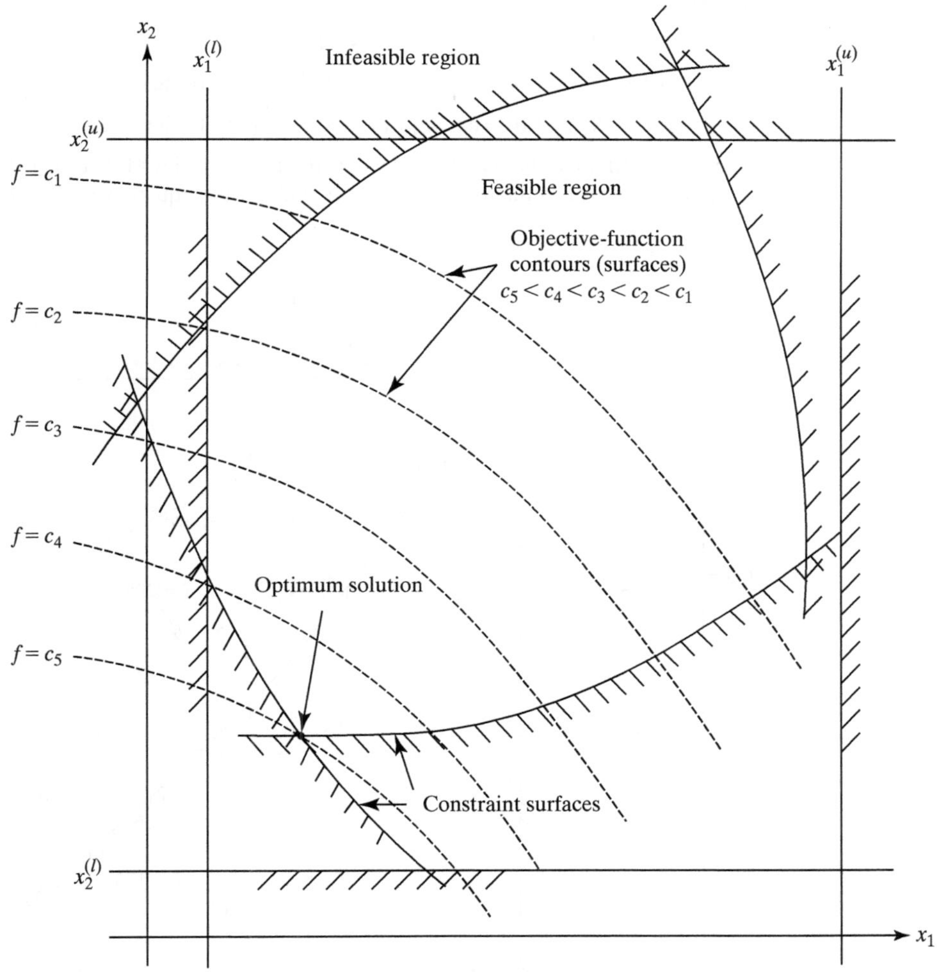

FIGURE 11.5 Graphical optimization procedure.

FIGURE 11.6 Two components in series.

Solution: The design variables are the reliabilities of the components so that

$$\mathbf{X} = \begin{Bmatrix} x_1 \\ x_2 \end{Bmatrix} = \begin{Bmatrix} R_1 \\ R_2 \end{Bmatrix}$$

The objective is to minimize the total cost

$$f(\mathbf{X}) = 100R_1^2 + 200R_2^2$$

The reliability constraints can be stated as

$$R_1R_2 \geq 0.7 \quad \text{i.e.;} \quad g_1(\text{X}) = 0.7 - R_1R_2 \leq 0$$

$$0.7 \leq R_1 \leq 1.0 \quad \text{i.e.;} \quad g_2(\text{X}) = 0.7 - R_1 \leq 0 \quad \text{and} \quad g_3(\text{X}) = R_1 - 1.0 \leq 0$$

$$0.7 \leq R_2 \leq 1.0 \quad \text{i.e.;} \quad g_4(\text{X}) = 0.7 - R_2 \leq 0 \quad \text{and} \quad g_5(\text{X}) = R_2 - 1.0 \leq 0$$

The lower and upper bounds (i.e., the constraint boundaries of g_2 to g_5) are shown in Fig. 11.7. To plot the overall reliability constraint boundary, the equation $g_1 = R_1R_2 - 0.7 = 0$ is to be

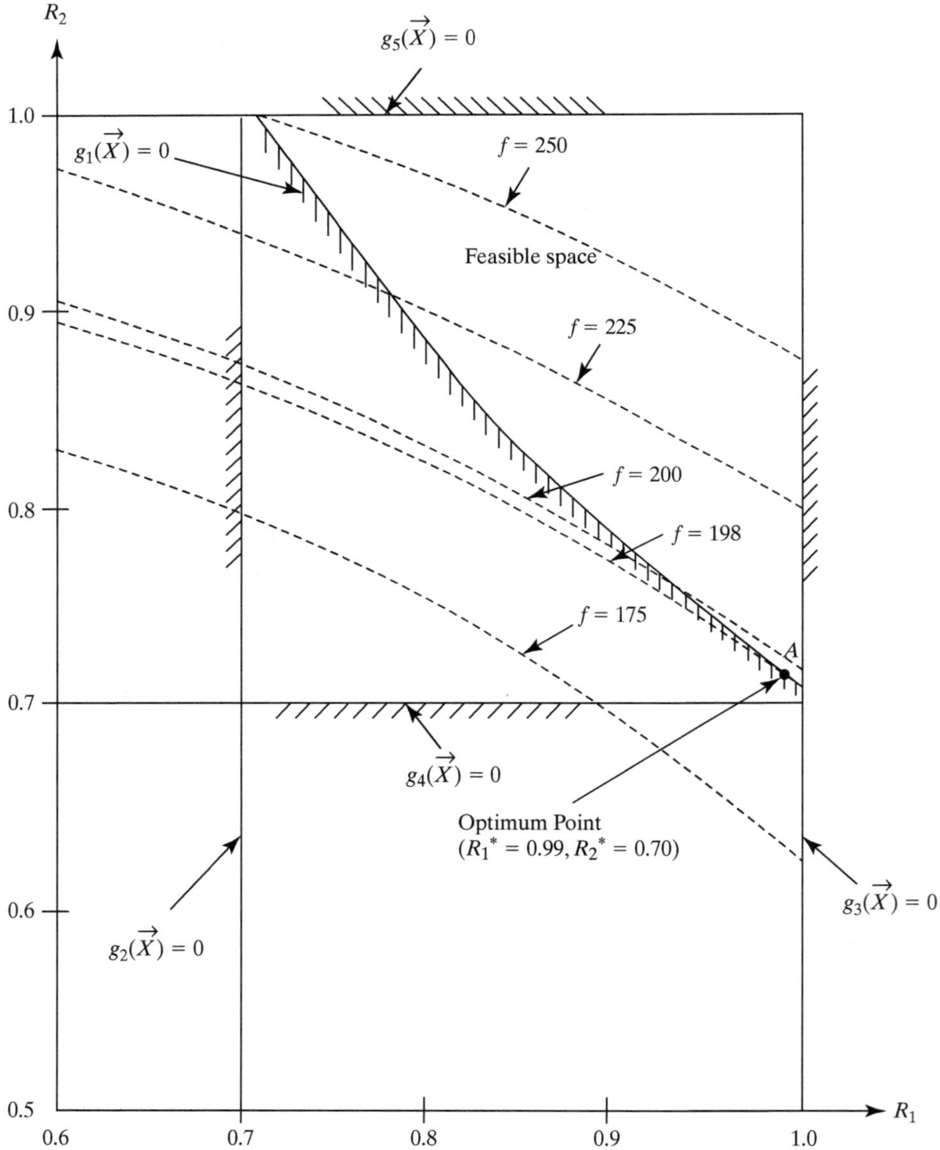

FIGURE 11.7 Graphical solution of Example 11.6.

plotted. For this, a series of values are given to R_1 and the corresponding values of R_2 are obtained as $R_2 = 0.7/R_1$

R_1	0.6	0.65	0.7	0.75	0.8	0.85	0.9	0.95	1.0
R_2	1.167	1.077	1.000	0.933	0.875	0.824	0.778	0.737	0.700

Next, the contours of the objective function,

$$f(X) = 100R_1^2 + 200R_2^2 = c$$

are to be plotted for different values of c. The points on the contours are found as indicated.

$$\text{For } 100R_1^2 + 200R_2^2 = 250; \quad R_2 = \sqrt{\frac{250 - 100R_1^2}{200}}$$

R_1	0.6	0.65	0.7	0.75	0.8	0.85	0.9	0.95	1.0
R_2	1.034	1.019	1.003	0.984	0.964	0.943	0.919	0.894	0.866

$$\text{For } 100R_1^2 + 200R_2^2 = 225; \quad R_2 = \sqrt{\frac{225 - 100R_1^2}{200}}$$

R_1	0.6	0.65	0.7	0.75	0.8	0.85	0.9	0.95	1.0
R_2	0.972	0.956	0.938	0.919	0.897	0.874	0.849	0.821	0.791

$$\text{For } 100R_1^2 + 200R_2^2 = 200; \quad R_2 = \sqrt{\frac{200 - 100R_1^2}{200}}$$

R_1	0.6	0.65	0.7	0.75	0.8	0.85	0.9	0.95	1.0
R_2	0.906	0.888	0.869	0.848	0.825	0.799	0.771	0.741	0.707

$$\text{For } 100R_1^2 + 200R_2^2 = 198; \quad R_2 = \sqrt{\frac{198 - 100R_1^2}{200}}$$

R_1	0.6	0.65	0.7	0.75	0.8	0.85	0.9	0.95	1.0
R_2	0.900	0.882	0.863	0.842	0.819	0.793	0.765	0.734	0.700

$$\text{For } 100R_1^2 + 200R_2^2 = 175; \quad R_2 = \sqrt{\frac{175 - 100R_1^2}{200}}$$

R_1	0.6	0.65	0.7	0.75	0.8	0.85	0.9	0.95	1.0
R_2	0.834	0.815	0.794	0.771	0.745	0.717	0.686	0.651	0.612

These contours are shown in Fig. 11.7 and it can be observed that the objective-function value cannot be reduced below a value of 198 without violating the overall reliability constraint. The optimum point can be seen to be the point of tangency between the curves $g_1 = 0$ and $f = 198$ (point A). Thus, the optimum solution is given by $R_1^* = 0.99$, $R_2^* = 0.70$ with $f_{\min} = 198$.

■

11.4.2 Lagrange Multiplier Method

The Lagrange multiplier method is an analytical method that can be used to find the minimum of a multivariable function in the presence of equality constraints. Thus, the problem to be solved can be stated as

$$\text{Find } \mathbf{X} = \begin{Bmatrix} x_1 \\ x_2 \\ \cdot \\ \cdot \\ \cdot \\ x_n \end{Bmatrix}$$

which minimizes

$$f(\mathbf{X})$$

subject to

$$l_j(\mathbf{X}) = 0; \quad j = 1, 2, \ldots, p \tag{11.3}$$

The solution procedure involves construction of a new function, called the Lagrange function, $L(\mathbf{X}, \boldsymbol{\lambda})$, as

$$L(\mathbf{X}, \boldsymbol{\lambda}) = f(\mathbf{X}) + \sum_{j=1}^{p} \lambda_j l_j(\mathbf{X}) \tag{11.4}$$

where λ_j are unknown constants, called the Lagrange multipliers. Notice that there are as many Lagrange multipliers as there are constraints. It has been proved [11.1] that the solution of the problem stated in Eq. (11.3) can be obtained by finding the unconstrained minimum of the function $L(\mathbf{X}, \boldsymbol{\lambda})$. The necessary conditions for the minimum of $L(\mathbf{X}, \boldsymbol{\lambda})$ are

$$\frac{\partial L}{\partial x_i}(\mathbf{X}, \boldsymbol{\lambda}) = 0; \quad i = 1, 2, \ldots, n$$

$$\frac{\partial L}{\partial \lambda_j}(\mathbf{X}, \boldsymbol{\lambda}) = 0; \quad j = 1, 2, \ldots, p \tag{11.5}$$

The number of unknowns is $n + p$ (n design variables x_i and p Lagrange multipliers λ_j) and the number of available equations, Eq. (11.5), is also $n + p$. Thus, the solution of the $(n + p)$ simultaneous nonlinear equations, Eq. (11.5), gives $\mathbf{X}^*$ and $\boldsymbol{\lambda}^*$. Although this method appears to be simple and straightforward, the solution of the $(n + p)$ simultaneous nonlinear equations, Eq. (11.5), may be tedious. Further, if the functions f and l_j are not available in explicit form in terms of x_i, it will be extremely difficult to solve Eq. (11.5). Hence, a suitable numerical method of optimization, such as the penalty function method discussed in the next section, is to be used for the solution of the optimization problem. The Lagrange multiplier method is illustrated with Examples 11.7 and 11.8.

EXAMPLE 11.7

Two components are in parallel as shown in Fig. 11.8. It is desired to find the component reliabilities R_1 and R_2 to minimize the total cost of the components while achieving a system reliability of 0.7. Assume that the costs of the components 1 and 2 are given by $100R_1$ and $200R_2$, respectively.

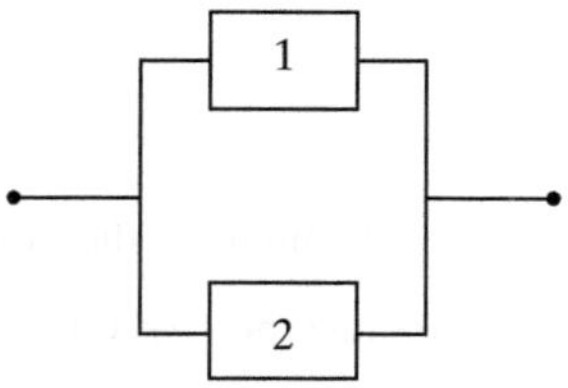

FIGURE 11.8 Two components in parallel.

Solution: The problem can be stated as follows:

$$\text{Find } \mathbf{X} = \begin{Bmatrix} x_1 \\ x_2 \end{Bmatrix} = \begin{Bmatrix} R_1 \\ R_2 \end{Bmatrix}$$

to minimize $f(\mathbf{X}) = 100R_1 + 200R_2$

subject to

$$l(\mathbf{X}) = R_1 + R_2 - R_1R_2 - 0.7 = 0$$

The Lagrange function can be formulated as

$$L = 100R_1 + 200R_2 + \lambda(R_1 + R_2 - R_1R_2 - 0.7) \tag{E1}$$

where λ is the Lagrange multiplier. The necessary conditions of optimality give

$$\frac{\partial L}{\partial R_1} = 100 + \lambda(1 - R_2) = 0 \quad \text{i.e.;} \quad \lambda = -\frac{100}{1 - R_2} \tag{E2}$$

$$\frac{\partial L}{\partial R_2} = 200 + \lambda(1 - R_1) = 0 \quad \text{i.e.;} \quad \lambda = -\frac{200}{1 - R_1} \tag{E3}$$

$$\frac{\partial L}{\partial \lambda} = R_1 + R_2 - R_1R_2 - 0.7 = 0 \tag{E4}$$

Equating (E2) and (E3), we obtain

$$-\lambda = \frac{100}{1 - R_2} = \frac{200}{1 - R_1} \quad \text{i.e.;} \quad R_1 = 2R_2 - 1 \tag{E5}$$

Substitution of Eq. (E5) into (E4) gives

$$(2R_2 - 1) + R_2 - R_2(2R_2 - 1) = 0.7 \quad \text{i.e.;} \quad 2R_2^2 - 4R_2 + 1.7 = 0 \tag{E6}$$

The solution of Eq. (E6) gives

$$R_2 = \frac{4 \pm \sqrt{16 - 13.6}}{4} = 0.6127; \quad 1.3873 \tag{E7}$$

Thus, the physically realizable optimum solution is given by $R_1^* = 0.2254$ and $R_2^* = 0.6127$ with $f_{\min} = 145.08$.

■

EXAMPLE 11.8

Find the component failure rates $\lambda_i, i = 1, 2, \ldots, n$ that minimize the cost (objective function) $f = \sum_{i=1}^{n} f_i(\lambda_i)$ subject to the constraint $\sum_{i=1}^{n} \lambda_i = \lambda_0$. Assume the cost functions to be proportional to the component MTBF (mean time between failures), that is, $f_i(\lambda_i) = a_i/\lambda_i$ where a_i is a constant.

Solution: The Lagrangian function can be constructed as

$$L(\lambda_1, \ldots, \lambda_n; \alpha) = \sum_{i=1}^{n} \frac{a_i}{\lambda_i} + \alpha\left(\sum_{i=1}^{n} \lambda_i - \lambda_0\right) \tag{E1}$$

where α is the Lagrange multiplier. The necessary conditions for the extremum of L give

$$\frac{\partial L}{\partial \lambda_i} = -\frac{a_i}{\lambda_i^2} + \alpha = 0; \quad i = 1, 2, \ldots, n$$

that is,

$$\lambda_i = \left(\frac{a_i}{\alpha}\right)^{1/2}; \quad i = 1, 2, \ldots, n \tag{E2}$$

$$\frac{\partial L}{\partial \alpha} = \sum_{i=1}^{n} \lambda_i - \lambda_0 = 0$$

that is,

$$\sum_{i=1}^{n} \left(\frac{a_i}{\alpha}\right)^{1/2} = \lambda_0$$

that is,

$$\alpha^{1/2} = \frac{1}{\lambda_0} \sum_{i=1}^{n} \sqrt{a_i} \tag{E3}$$

Thus, the optimum values of the design variables are given by

$$\lambda_i = \frac{\lambda_0 \sqrt{a_i}}{\sum_{j=1}^{n} \sqrt{a_j}}; \quad i = 1, 2, \ldots, n \tag{E4}$$

and the optimum value of the objective function by

$$f_{\min} = \sum_{i=1}^{n} \frac{a_i}{\lambda_i} = \frac{1}{\lambda_0}\left(\sum_{i=1}^{n} \sqrt{a_i}\right)^2$$

■

11.4.3 Penalty Function Method (SUMT)

In this method, a new function, $\phi_k(\mathrm{X}, r_k)$, is defined as

$$\phi_k(\mathbf{X}, r_k) = f(\mathbf{X}) + r_k \sum_{j=1}^{m} \frac{1}{g_j(\mathbf{X})} - \frac{1}{\sqrt{r_k}} \sum_{j=1}^{p} l_j^2(\mathbf{X}) \tag{11.6}$$

where r_k is a positive constant, known as the penalty parameter. It has been proved [11.1] that the optimum solution of the problem stated in Eq. (11.6) can be obtained by minimizing the function ϕ_k for a sequence of decreasing values of r_k. The value of r_k is initially chosen as a small positive value, say 1.0, and then decreased toward zero after each unconstrained minimization. The starting point for minimization is assumed to be feasible with respect to the inequality constraints. The iterative procedure involved can be summarized as follows:

1. Start with an initial feasible point $\mathbf{X}_1$ satisfying all the inequality constraints with strict inequality sign, that is, $g_j(\mathbf{X}_1) < 0$ for $j = 1, 2, \ldots, m$, and an initial value of $r_1 > 0$. Set $k = 1$.
2. Minimize $\phi_k(\mathbf{X}, r_k)$ by using an unconstrained minimization method (described below) and obtain the solution $\mathbf{X}_k^*$.
3. Test whether $\mathbf{X}_k^*$ is the optimum solution of the original problem of Eq. (11.6). If $\mathbf{X}_k^*$ is found to be optimum, terminate the process; otherwise, go to the next step.
4. Find the value of the next penalty parameter, r_{k+1}, as $r_{k+1} = c.\, r_k$ where $c < 1$ (usually taken as 0.1).
5. Take the new starting point as $\mathbf{X}_1 = \mathbf{X}_k^*$, set the new value of $k = k + 1$, and go to step 2.

The convergence of the procedure as r_k tends to zero is illustrated graphically in Fig. 11.9 for the simple problem

$$\text{Find } \mathbf{X} = \{x_1\}$$

which minimizes

$$f(\mathbf{X}) = a x_1$$

subject to

$$g_1(\mathbf{X}) = b - x_1 \le 0 \tag{11.7}$$

Unconstrained Minimization. It can be seen that the penalty function method requires the solution of an unconstrained minimization problem for each value of the penalty parameter r_k. Although several unconstrained minimization methods have been developed in the literature [11.1], we shall consider one of the popular and efficient methods, known

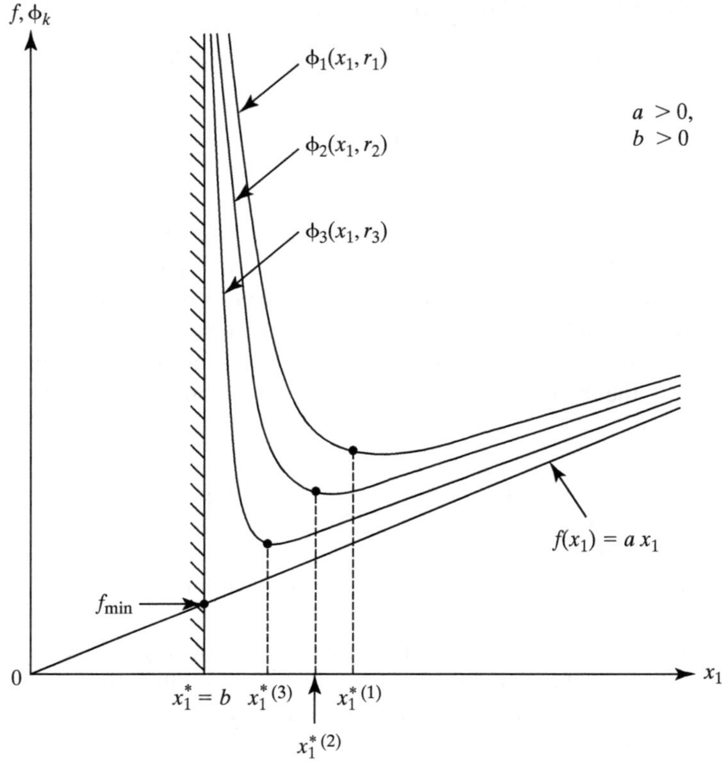

FIGURE 11.9 Convergence of penalty function method.

as the Davidon-Fletcher-Powell (DFP) method, in this section. For convenience of notation, we state the problem to be solved as follows:

$$\text{Find } \mathbf{X} = \begin{Bmatrix} x_1 \\ x_2 \\ \cdot \\ \cdot \\ \cdot \\ x_n \end{Bmatrix} \tag{11.8}$$

which minimizes $f(\mathbf{X})$.

The solution procedure can be summarized through the following steps:

1. Start with any initial vector $\mathbf{X}_1$ and an initial positive definite-symmetric matrix $[H_1]$. Usually $[H_1]$ is taken as the identity matrix, $[I]$.

2. Find the first search direction $\mathbf{S}_1$ as

$$\mathbf{S}_1 = -[H_1]\nabla f_1 \tag{11.9}$$

where $\nabla f_1 = \nabla f(\mathbf{X}_1)$ is the gradient vector evaluated at $\mathbf{X}_1$. Set the iteration number as $i = 1$.

3. Compute the new design vector

$$\mathbf{X}_{i+1} = \mathbf{X}_i + \alpha_i^* \mathbf{S}_i \tag{11.10}$$

where α_i^* is the minimizing step length along the search direction $\mathbf{S}_i$, that is, α_i^* is chosen so that $f(\mathbf{X}_i + \alpha_i \mathbf{S}_i) = f(\alpha_i)$ attains a minimum at $\alpha_i = \alpha_i^*$. A method of finding α_i^* is described in the next section.

4. Test for convergence of the process. If the magnitude of the gradient of the function at $\mathbf{X}_{i+1}$ is zero, that is, if $||\nabla f(X_{i+1})|| \approx 0$, set $\mathbf{X}^* = \mathbf{X}_{\text{opt}} \approx \mathbf{X}_{i+1}$ and stop the process. Otherwise, go to the next step.

5. Compute the new matrix $[H_{i+1}]$ as

$$[H_{i+1}] = [H_i] + [M_i] + [N_i] \tag{11.11}$$

where

$$[M_i] = \alpha_i^* \frac{\mathbf{S}_i \mathbf{S}_i^T}{\mathbf{S}_i^T \mathbf{Y}_i} \tag{11.12}$$

$$[N_i] = -\frac{([H_i]\mathbf{Y}_i)([H_i]\mathbf{Y}_i)^T}{\mathbf{Y}_i^T [H_i] \mathbf{Y}_i} \tag{11.13}$$

and

$$\mathbf{Y}_i = \nabla f(\mathbf{X}_{i+1}) - \nabla f(\mathbf{X}_i) \tag{11.14}$$

6. Compute the new search direction as

$$\mathbf{S}_{i+1} = -[H_{i+1}]\nabla f_{i+1} \tag{11.15}$$

7. Set the new iteration number as $i = i + 1$, and repeat step 3 onwards until the convergence in step 4 is satisfied.

One-dimensional Minimization. The unconstrained-minimization method just presented requires the computation of α_i^* in finding an improved vector $\mathbf{X}_{i+1}$ (Eq. 11.10). Since only one variable α_i^* is to be found by minimizing the function $f(\mathbf{X}_i + \alpha_i \mathbf{S}_i) = f(\alpha_i)$, the problem is called a one-dimensional minimization problem. One of the popular methods, known as the cubic-interpretation method, is considered in this section. This method makes use of the derivative of f with respect to α, which can be computed as (by neglecting the subscript i for simplicity)

$$f'\Big|_{\alpha} = \frac{df}{d\alpha}\Big|_{\alpha} = \frac{d}{d\alpha} f(\mathbf{X} + \alpha\mathbf{S}) = \mathbf{S}^T \nabla f(\mathbf{X} + \alpha\mathbf{S}) \tag{11.16}$$

In this method, the lower and upper bounds on the optimal step length, α^*, are determined first. Always $A = \alpha = 0$ can be taken as a lower bound on α^*. At $A = \alpha = 0$,

$$f'\big|_{\alpha=0} = \mathbf{S}^T\, \nabla f(\mathbf{X}) < 0 \tag{11.17}$$

because $\mathbf{S}$ is assumed to be a descent direction. The upper bound on α^*, B, is taken as the first value among 1, 2, 4, 8, . . . , at which $f'\big|_B = \mathbf{S}^T\, \nabla f(\mathbf{X} + B\mathbf{S})$ is non-negative. Since f' changes sign between $\alpha = A$ and $\alpha = B$, the minimum of f lies between A and B.

Since the exact functional relation between f and α is not known in most practical problems, we approximate $f(\alpha)$ by a cubic equation $h(\alpha) = a + b\alpha + c\alpha^2 + d\alpha^3$ (a, b, c, and d are unknown constants) between $\alpha = A$ and $\alpha = B$. This is the reason the method is called the cubic-interpretation method. By using the relations $h(\alpha = A) = f(\alpha = A)$, $h'(\alpha = A) = f'(\alpha = A)$, $h(\alpha = B) = f(\alpha = B)$, and $h'(\alpha = B) = f'(\alpha = B)$, the constants a, b, c, and d can be determined. Then the minimum of $h(\alpha)$, obtained by solving the equation,

$$h'(\underset{\sim}{\alpha}^*) = b + 2c\underset{\sim}{\alpha}^* + 3d\underset{\sim}{\alpha}^{*2} = 0, \tag{11.18}$$

is tested for convergence to see whether $\underset{\sim}{\alpha}^*$ can be used as an approximation for the minimum of $f(\alpha)$, that is, $\alpha^* \approx \underset{\sim}{\alpha}^*$. If the convergence test is satisfied, the procedure is terminated by taking $\alpha^* = \underset{\sim}{\alpha}^*$. Otherwise, the best two points out of $\alpha = A$, $\alpha = B$, and $\alpha = \alpha^*$ are used to fit another cubic equation. This procedure is continued until convergence is achieved.

EXAMPLE 11.9

Determine the component reliabilities of the system shown in Fig. 11.10 to maximize the system reliability subject to an upper bound on the total cost of the system. The cost of component $i(c_i)$ is given by $c_i = k_i(R_i)^{\alpha_i}$ where R_i = reliability of component i, $k_1 = k_3 = 50$, $k_2 = 200$, $k_4 = 75$, and $\alpha_i = 0.5(i = 1, 2, 3, 4)$. The maximum permissible cost of the system is 550.

Solution: The system reliability (R_s) can be found as $R_s = 1 - P_{fs}$ where P_{fs} is the probability of failure of the system given by

$$P_{fs} = P_{fs}\Big|_{\substack{\text{component 2}\\ \text{operating}}} \cdot P\begin{pmatrix}\text{component 2}\\ \text{operating}\end{pmatrix} + P_{fs}\Big|_{\substack{\text{component 2}\\ \text{failed}}} \cdot P\begin{pmatrix}\text{component 2}\\ \text{failed}\end{pmatrix}$$

$$= \{(1 - R_3)(1 - R_4)\}^2 \cdot R_2 + \{1 - R_1[1 - (1 - R_3)(1 - R_4)]\}^2 \cdot (1 - R_2) \tag{E1}$$

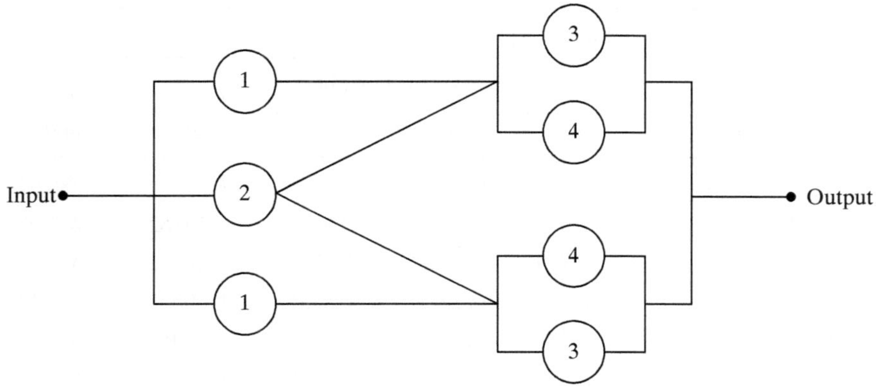

FIGURE 11.10 System considered in Example 11.9.

The optimization problem can be stated as follows:

$$\text{Find } \mathbf{X} = \begin{Bmatrix} R_1 \\ R_2 \\ R_3 \\ R_4 \end{Bmatrix}$$

which maximizes

$$R_s = 1 - P_{fs} = 1 - R_2\{(1-R_3)(1-R_4)\}^2 - (1-R_2)\{1 - R_1[1-(1-R_3)(1-R_4)]\}^2 \quad \text{(E2)}$$

subject to

$$2c_1 + c_2 + 2c_3 + 2c_4 \le c_{\max}$$

that is,

$$100R_1^{0.5} + 200R_2^{0.5} + 100R_3^{0.5} + 150R_4^{0.5} - 550 \le 0 \quad \text{(E3)}$$

and

$$0 \le R_i \le 1; \quad i = 1, 2, 3, 4 \quad \text{(E4)}$$

This problem has been solved using the interior penalty function method and the results are shown in Table 11.2. The optimum solution can be seen to be

$$\mathbf{X}^* = \begin{Bmatrix} R_1^* \\ R_2^* \\ R_3^* \\ R_4^* \end{Bmatrix} = \begin{Bmatrix} 0.99538 \\ 0.70189 \\ 0.99682 \\ 0.95339 \end{Bmatrix}$$

$$R_s|_{\max} = 0.999993$$

$$\text{and total cost} = 513.63.$$

TABLE 11.2 Solution of the optimization problem

Iteration k	Value of r_k	Design Variables R_1	R_2	R_3	R_4	Total Cost	Objective Function (R_s)
—	—	0.5	0.5	0.5	0.5	388.911	0.773438
1	0.05	0.65098	0.49455	0.58076	0.54162	407.929	0.8681115
2	0.005	0.82769	0.57822	0.73809	0.68962	453.536	0.971967
3	0.0005	0.90923	0.65037	0.84353	0.80631	483.181	0.994507
4	0.00005	0.95428	0.70478	0.89284	0.87983	500.779	0.998889
5	0.000005	0.97722	0.70210	0.91760	0.92442	506.450	0.999725
6	0.0000005	0.99303	0.70305	0.91986	0.92687	507.667	0.999927
7	0.00000005	0.99655	0.70103	0.97837	0.94596	512.136	0.999993
8	0.000000005	0.99550	0.70188	0.99669	0.95339	513.628	0.999993
9	0.0000000005	0.99538	0.70189	0.99682	0.95339	513.630	0.999993

■

EXAMPLE 11.10

Optimum Design of a Machine Tool Gear Train

The problem of the design of the nine-speed machine tool gear train shown in Fig. 9.3 is to be formulated as an optimization problem. The gear train consists of a cluster gear set on the input shaft as shown in Fig. 9.3. The intermediate shaft carries three fixed gears and a cluster gear set. The output shaft carries three fixed gears. Depending on the gear pair in contact, the intermediate shaft will have three different speeds. For each intermediate shaft speed, three speeds can be realized at the output shaft. Thus nine speeds are possible for the output shaft. The speed diagram of the gear train is shown in Fig. 9.4. A gear tooth may fail due to excessive bending stress or surface-wear stress. The mass of the gear train is to be minimized by treating the face widths of the mating gear pairs as design variables. Formulate the design problem as an optimization problem.

Solution: The reliability analysis of this problem was considered in Chapter 9. The optimization problem is formulated first in a deterministic format and then in a probabilistic format.

Deterministic Formulation. The face widths of the gear pairs are treated as design variables and the optimum-design problem is stated as a nonlinear programming problem as follows:

$$\text{Find } \boldsymbol{X} = \begin{Bmatrix} x_1 \\ x_2 \\ \cdot \\ \cdot \\ \cdot \\ x_6 \end{Bmatrix} \tag{E1}$$

which minimizes the mass of the gear train

$$f(\boldsymbol{X}) = \sum_{i=1}^{12} \rho \frac{\pi D_i^2}{4} b_i \tag{E2}$$

subject to

$$S_{b_{ij}} - S_b \le 0; \quad i = 1, 2, \ldots, k, \quad j = 1, 2, \ldots, l \tag{E3}$$

$$S_{w_{ij}} - S_w \le 0; \quad i = 1, 2, \ldots, k, \quad j = 1, 2, \ldots, l \tag{E4}$$

$$x_i^{(l)} \le x_i; \quad i = 1, 2, \ldots, 6 \tag{E5}$$

where $x_i = b_i$ = face width of ith gear pair (mm), ρ = density of the material (kg/mm^3), D_i = pitch circle diameter of ith gear (mm), $x_i^{(l)}$ = lower bound on the face width of ith gear (mm), k = number of gears in engagement while delivering any particular output speed, l = total number of output shaft speeds (9), $S_{b_{ij}}(S_{w_{ij}})$ = bending (surface-wear) stress induced in ith gear at jth output shaft speed (N/mm^2), and $S_b(S_w)$ = permissible bending (surface-wear) stress for material (N/mm^2).

The stresses induced in a gear, $S_{b_{ij}}$ and $S_{w_{ij}}$, can be expressed as indicated in Eqs. (9.24) to (9.27).

Probabilistic Formulation. To formulate the optimum-design problem in the probabilistic format, the power transmitted, the input shaft speed, the allowable and induced stresses, the center distances of the gear pairs, and the face widths of the gears are treated as normally distributed random variables. The mean values of the face widths are considered as the design variables. A linear combination of the mean value and standard deviation of the mass of the gear train is considered as the objective for minimization. Two failure modes, namely the bending and surface-wear modes, are considered in the design process. The reliability of the gear train is required to be greater than or equal to a specified quantity in bending and surface-wear failure modes. The gear train is modeled as a weakest-kinematic chain in both the failure modes as indicated in Chapter 9. The optimization problem can be stated as follows:

$$\text{Find } \boldsymbol{X} = \begin{Bmatrix} x_1 \\ x_2 \\ \cdot \\ \cdot \\ \cdot \\ x_6 \end{Bmatrix} \tag{E6}$$

which minimizes

$$f(\boldsymbol{X}) = C_1 \bar{f}(\boldsymbol{X}) + C_2 \sigma_f(\boldsymbol{X}) \tag{E7}$$

subject to

$$R_{bj} \ge R_s; \quad j = 1, 2, \ldots, l \tag{E8}$$

$$R_{wj} \ge R_s; \quad j = 1, 2, \ldots, l \tag{E9}$$

$$x_i \ge x_i(l); \quad i = 1, 2, \ldots, 6 \tag{E10}$$

where $\bar{f}$ and σ_f are the mean and standard deviations of the mass of the gearbox, C_1 and C_2 are constants denoting the relative importances of the mean value and standard deviations of the mass during minimization, $R_{bj}\,(R_{wj})$ is the reliability of the gear train in jth speed in bending (surface-wear) failure mode, and R_s is the specified reliability. Since the weakest-link hypothesis is used, the reliabilities R_{bj} and R_{wj} can be expressed as

$$R_{bj} = \prod_{i=1}^{k} R_{b_{ij}} \tag{E11}$$

$$R_{wj} = \prod_{i=1}^{k} R_{w_{ij}} \tag{E12}$$

where $R_{b_{ij}}(R_{w_{ij}})$ indicates the reliability of ith gear in jth speed in bending (surface-wear) failure mode.

If the strength (S) and the load acting (s) on a typical link of a weakest-link chain are assumed to be independent normally distributed random variables, the reliability of the link (R_0) can be determined as follows. A new random variable u is defined as

$$u = S - s \tag{E13}$$

which implies that u is also normally distributed. If z denotes the standard normal variate

$$z = \frac{u - \overline{u}}{\sigma_u} \tag{E14}$$

the normality relation becomes

$$\int_{-\infty}^{\infty} \frac{1}{\sqrt{2\pi}} \exp\left(\frac{-t^2}{2}\right) dt = 1 \tag{E15}$$

and the reliability of the link (R_0) can be expressed as

$$R_0 = \frac{1}{\sqrt{2\pi}} \int_{-(\overline{u}/\sigma_u)}^{\infty} \exp\left(\frac{-z^2}{2}\right) dz = \int_{z_1}^{\infty} \frac{1}{\sqrt{2\pi}} \exp\left(\frac{-z^2}{2}\right) dz \tag{E16}$$

where the lower limit of integration can be expressed in terms of the expected values ($\overline{S}$ and $\overline{s}$) and the standard deviations (σ_S and σ_s) of the random variables S and s as

$$z_1 = -\frac{\overline{u}}{\sigma_u} = -\frac{(\overline{S} - \overline{s})}{\sqrt{\sigma_S^2 + \sigma_s^2}} \tag{E17}$$

If the reliability of the gear pair is known, the normal probability tables provide the value of the lower limit of integration, $z_1 = z_{10}$. By assuming equal reliabilities for all the gears participating in the jth output speed, z_{10} can be determined from the relation

$$(R_0)^{1/k} = \frac{1}{\sqrt{2\pi}} \int_{z_{10}}^{\infty} \exp\left(\frac{-z^2}{2}\right) dz \tag{E18}$$

Thus, the constraints (E8) and (E9) can be rewritten as

$$(z_{1b})_{ij} \le z_{10} \tag{E19}$$

$$(z_{1w})_{ij} \le z_{10} \tag{E20}$$

where

$$(z_{1b})_{ij} = -\frac{(\overline{S}_b - \overline{s}_{b_{ij}})}{\sqrt{\sigma_{S_b}^2 + \sigma_{s_{b_{ij}}}^2}} \tag{E21}$$

and

$$(z_{1w})_{ij} = -\frac{(\overline{S}_w - \overline{s}_{w_{ij}})}{\sqrt{\sigma_{S_w}^2 + \sigma_{s_{w_{ij}}}^2}} \tag{E22}$$

The mean and standard deviations of f, $s_{b_{ij}}$, and $s_{w_{ij}}$ can be computed using the procedure of Section 5.5.5.

■

The optimization problems have been solved using the penalty function method. The following data are assumed for the deterministic design calculations: p = power = 7.457 kW, m = module = 4 mm, $\rho = 7.75 \times 10^{-6}$ kg/mm^3, K_c = stress concentration factor = 1.5, K_d = dynamic load factor = 1.1, $x_i^{(l)} = 5$ mm for $i = 1$ to 6, $E = 2.0601 \times 10^6$ N/mm^2, input shaft speed = 268 rpm, gear tooth profile = involute type, pressure angle = 20°, material of the gears = case-hardened steel, and strength of the material = 245.25 N/mm^2 in bending and 1716.75 N/mm^2 in the surface wear. The values of l and k are 9 and 4, respectively. The factor of safety (FS) is applied to the material strength and the permissible stresses are takes as $S_b = (245.25/\text{FS})$ and $S_w = (1716.75/\text{FS})$. The results obtained with deterministic optimization procedure with FS = 1 are given in Table 11.3. The minimum mass of the gear train can be seen to be 85.901 kg.

The data of the deterministic problem are taken as the mean values for the probabilistic problem. The coefficients of variation of all the random variables are assumed to be the same for simplicity. Further, the values of C_1 and C_2 are chosen such that the contributions from the mean value and the standard deviation of the mass are the same at the starting design point, that is, $C_1 f(\boldsymbol{X}_s) = C_2 \sigma_f(\boldsymbol{X}_s)$, where $\boldsymbol{X}_s$ denotes the starting design vector. The reliability-based optimum-design results obtained with three sets of values of C_1 and

TABLE 11.3 Results of deterministic optimization (with FS = 1)

Quantity	Initial Design	Optimum Design
Design Variables		
x_1(mm)	60	17.339
x_2(mm)	40	12.527
x_3(mm)	40	9.311
x_4(mm)	150	48.563
x_5(mm)	60	13.044
x_6(mm)	60	18.655
Objective Function		
$f(\boldsymbol{X})$, kg	293.0505	85.9010

TABLE 11.4 Results of reliability-based optimization (with $R_0 = 0.9999$); Initial design: $\bar{x}_1 = \bar{x}_5 = \bar{x}_6 = 60$ mm, $\bar{x}_2 = \bar{x}_3 = 40$ mm, $\bar{x}_4 = 130$ mm, $\bar{f} = 274.9867$ kg

Quantity	**With Coefficient of Variation of All Random Variables = 0.01**		
Design Variables	$C_1 = 1, C_2 = 7.524$	$C_1 = 1, C_2 = 0$	$C_1 = 0, C_2 = 7.514$
x_1(mm)	19.127	19.121	19.147
x_2(mm)	10.266	10.269	12.575
x_3(mm)	13.820	13.820	13.975
x_4(mm)	53.550	53.550	53.553
x_5(mm)	14.382	14.383	14.383
x_6(mm)	19.640	19.642	19.657
Objective Function			
$f(\boldsymbol{X})$, kg	104.530	94.097	10.453
Quantity	**With Coefficient of Variation of All Random Variables = 0.10**		
Design Variables	$C_1 = 1, C_2 = 7.524$	$C_1 = 1, C_2 = 0$	$C_1 = 0, C_2 = 7.514$
x_1(mm)	50.415	50.416	50.416
x_2(mm)	24.635	24.632	24.805
x_3(mm)	32.896	32.905	32.915
x_4(mm)	127.014	127.017	127.014
x_5(mm)	34.123	34.115	34.116
x_6(mm)	42.711	42.717	42.712
Objective Function			
$f(\boldsymbol{X})$, kg	470.722	223.171	247.560

Source: S. S. Rao and G. Das, "Reliability Based Optimum Design of Gear Trains," *ASME Journal of Mechanisms, Transmissions, and Automation in Design*, Vol. 106, 1984, with permission from ASME.

C_2 and two different values of the coefficient of variation of the random variables are given in Table 11.4. It can be observed that all combinations of C_1 and C_2 yield essentially the same optimum-design vector. The mean value of the mass ($\bar{f}$ with $C_1 = 1$ and $C_2 = 0$) increased by more than 200 percent when the coefficient of variation of all the random variables is increased from 0.01 to 0.10. This indicates that the optimum value of the objective functions increases at a rapid rate as the uncertainty in the design parameters increases.

11.4.4 Dynamic Programming

The dynamic programming technique, developed by Richard Bellman in the early 1950s, solves a multistage decision problem as a sequence of single-stage decision problems [11.1, 11.3, 11.6]. In most cases, the solution of a set of N single-stage optimization problems is much easier compared to the solution of a N-stage optimization problem. An additional advantage of dynamic programming is that the method can handle discrete variables, nonconvex, continuous, and nondifferentiable functions without any difficulty. The main disadvantage of dynamic programming is that it suffers from the so-called "curse of dimensionality," which necessitates a large storage for large problems. In spite of this disadvantage, the method is suitable for solving a wide range of decision-making problems, including those arising in the field of reliability. An example of a multistage decision problem is shown in Fig. 11.11 where, for convenience, the stages $n, n-1, \ldots, i, \ldots, 2, 1$, are labeled in the decreasing order. The input state of stage i is assumed to be s_{i+1} and the output state as s_i. The design parameter of stage i is denoted as x_i and its contribution to the objective function as $f_i = f_i(\mathbf{X})$.

The problem is to find $\mathbf{X} = (x_1, x_2, \ldots, x_n)^T$ which minimizes the objective function

$$F(\mathbf{X}) = f_1(\mathbf{X}) + f_2(\mathbf{X}) + \cdots + f_n(\mathbf{X})$$

subject to

$$s_i = s_i(s_{i+1}, x_i); \quad i = 1, 2, \ldots, n \tag{11.19}$$

and other constraints specified on x_i and $s_i (i = 1, 2, \ldots, n)$.

The dynamic programming makes use of the concept of suboptimization and the principle of optimality in solving the multistage decision problem. The basic procedure can be stated as follows. The last component (the component labeled 1 in Fig. 11.11) is used first for suboptimization. If the input state of this component, s_2, is specified, x_1 must be chosen to minimize $f_1(x_1, s_2)$. Denoting the minimum value of f_1 as

$$F_1^*(s_2) = \min_{x_1} \left[f_1(x_1, s_2) \right] \tag{11.20}$$

This equation gives the minimum value of f_1 as a function of the input state s_2 in a parametric form. Next, we consider the last two components (components labeled 2 and 1 in Fig. 11.11) together for suboptimization. Here we choose x_1 and x_2 such that

$$F_2(s_3) = f_2(x_2, s_3) + f_1(x_1, s_2) \tag{11.21}$$

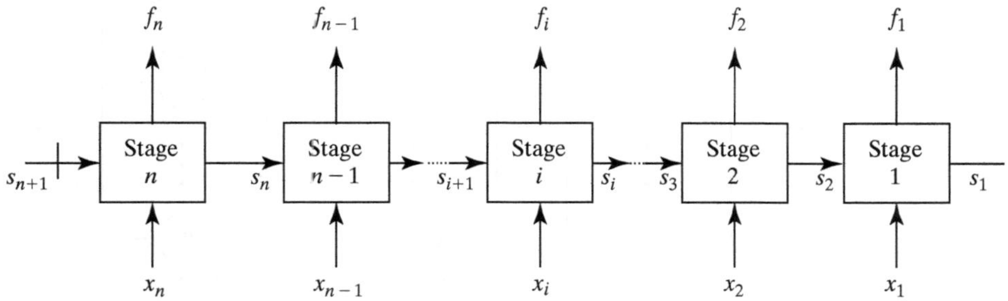

FIGURE 11.11 Multistage decision problem.

is minimized for any specified value of s_3 to obtain

$$F_2^*(s_3) = \min_{x_1, x_2} \left[f_2(x_2, s_3) + f_1(x_1, s_2) \right] \tag{11.22}$$

Since s_2 can be computed once s_3 and x_2 are specified, Eq. (11.22) can be rewritten, with the help of Eq. (11.20), as

$$F_2^*(s_3) = \min_{x_2} \left[f_2(x_2, s_3) + F_1^*(s_2) \right] \tag{11.23}$$

Thus F_2^* denotes what is known as the *optimal policy* for the two-stage subproblem. This process of suboptimization can be continued by considering one more component each time until all the n components are considered in the nth suboptimization problem. In general, the ith suboptimization problem can be stated as

$$F_i^*(s_{i+1}) = \min_{x_i} \left[f_i(x_i, s_{i+1}) + F_{i-1}^*(s_i) \right] \tag{11.24}$$

In the last (nth) suboptimization problem, the input state, s_{n+1}, is usually specified and hence F_n^* need to be found just for one input, namely, s_{n+1}. Also note that only one design variable is involved at any given stage; hence the suboptimization is expected to be very simple. The multistage optimization procedure of dynamic programming is illustrated with the following example.

EXAMPLE 11.11

A series–parallel system is composed of three types of components in three stages as shown in Fig. 11.12. The type, number, reliability, and cost of components used in the various stages are given Table 11.5. If the minimum required reliability of the system is 0.99, and the permissible total cost is 40, determine the number of components to be used in stages 1, 2, and 3 for maximizing the system reliability.

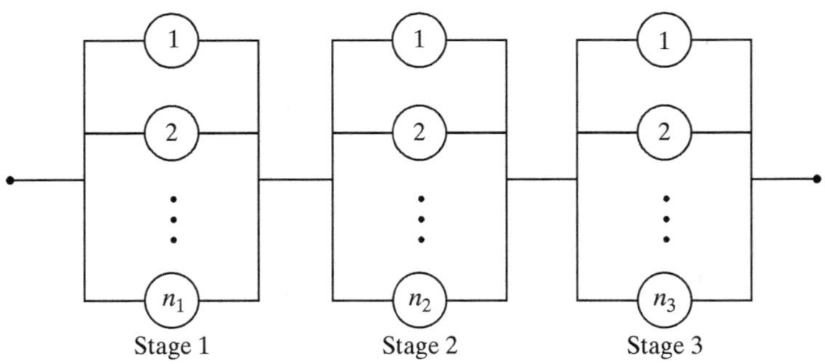

FIGURE 11.12 Three-stage problem.

TABLE 11.5 Details of components used in various stages

Component	Stage $i = 1$	Stage $i = 2$	Stage $i = 3$
Type of component	1	2	3
Number of components	n_1	n_2	n_3
Reliability of each component (R_i)	0.9	0.7	0.8
Cost of each component (c_i)	4	2	3

Solution: The reliability of the system can be expressed as

$$R_0 = \prod_{i=1}^{3} \{1 - (1 - R_i)^{n_i}\} \tag{E1}$$

Thus the optimization problem involves finding

$$\mathbf{X} = \begin{Bmatrix} n_1 \\ n_2 \\ n_3 \end{Bmatrix}$$

to maximize $R_0(\mathbf{X})$ given by Eq. (E1) subject to the constraints

$$R_0 \geq 0.99$$

and

$$\sum_{i=1}^{3} c_i \, n_i \leq 40$$

Since the three stages are in series, the minimum required overall system reliability (0.99) dictates that each stage should have a reliability of at least 0.99. Since the reliability of stage i (with n_i components in parallel) is given by

$$1 - (1 - R_i)^{n_i} \tag{E2}$$

the reliability of each stage with different number of components can be obtained as indicated in Table 11.6.

The results shown in Table 11.6 indicate that the minimum number of components required to achieve the specified overall reliability will be 2, 4, and 3 for stages 1, 2, and 3, respectively. The cost of these components will be $2 \times 4 + 4 \times 2 + 3 \times 3 = 25$. In order to apply the dynamic programming procedure, we first consider the suboptimization of stage 1. Since the minimum values of n_3, n_2, and n_1 have to be 3, 4, and 2, respectively, with an associated cost of 25, we investigate all possible values of cost between 26 and 40 (in increments of 1) and find the additional number of type-1 components that can be added in stage 1 (with the additional money available) to improve the overall reliability of the system. The results are shown in Table 11.7. For example, a total cost of 26 implies that only $26 - 25 = 1$ is available to procure additional type-1 components. Since each component of type 1 costs 4 units of money, the value of n_1 cannot be increased from 2. This situation persists until the total cost increases to 29 units at which time n_1 can be increased from 2 to 3 by spending $29 - 25 = 4$ units of money to procure an additional component of type 1.

TABLE 11.6 Reliability of each stage for different number of components

Number of Components in the Stage (n_i)	Stage 1	Stage 2	Stage 3
1	0.9	0.7	0.8
2	0.99	0.91	0.96
3	0.999	0.973	0.992
4	0.9999	0.9919	0.9984
5	0.99999	0.99757	0.99968
6	—	0.999271	0.999936
7	—	0.9997813	0.9999872
8	—	0.9999344	0.9999975
9	—	0.9999804	—
10	—	0.9999942	—
11	—	0.9999983	—

TABLE 11.7 Result of suboptimization of Stage 1

Total Money (cost)	R_{03} Reliability of Stage 3 ($n_3 = 3$ fixed)	R_{02} Reliability of Stage 2 ($n_2 = 4$ fixed)	Value of n_1 ($n_1 = 2$ minimum)	R_{01} Reliability of Stage 1	R_0 Overall System Reliability
26	0.992	0.9919	2	0.99	0.9741251
27	0.992	0.9919	2	0.99	0.9741251
28	0.992	0.9919	2	0.99	0.9741251
29	0.992	0.9919	3	0.999	0.9829808
30	0.992	0.9919	3	0.999	0.9829808
31	0.992	0.9919	3	0.999	0.9829808
32	0.992	0.9919	3	0.999	0.9829808
33	0.992	0.9919	4	0.9999	0.9838664
34	0.992	0.9919	4	0.9999	0.9838664
35	0.992	0.9919	4	0.9999	0.9838664
36	0.992	0.9919	4	0.9999	0.9838664
37	0.992	0.9919	5	0.99999	0.9839550
38	0.992	0.9919	5	0.99999	0.9839550
39	0.992	0.9919	5	0.99999	0.9839550
40	0.992	0.9919	5	0.99999	0.9839550

Next, we consider the suboptimization of stages 1 and 2 together by fixing the value of n_3 at 3. By varying the total cost from 26 to 40 in increments of 1, we find the additional number of components of type 1 and 2 (beyond $n_1 = 2$ and $n_2 = 4$) that can be procured with the money available after meeting the cost of 25 units for obtaining the minimum number of components $n_1 = 2$, $n_2 = 4$, and $n_3 = 3$. The results are summarized in Table 11.8. For example, when the total money available (cost) is 31 units, there will be $31 - 25 = 6$ units available for procuring additional components beyond $n_1 = 2$ and $n_2 = 4$. Since each component of type 1 and type 2 costs 4 and 2 units, respectively, we have $n_1 = 3$ and $n_2 = 5$, or $n_1 = 2$ and $n_2 = 7$.

TABLE 11.8 Result of suboptimization of Stages 1 and 2

Total Money	R_{03} Reliability of Stage 3 ($n_3 = 3$ fixed)	Value of n_2 ($n_2 = 4$ minimum)	R_{02} Reliability of Stage 2	Value of n_1 ($n_1 = 2$ minimum)	R_{01} Reliability of Stage 1	R_0 = Overall System Reliability
25	0.992	4	0.9919	2	0.99	0.9741251
26	0.992	4	0.9919	2	0.99	0.9741251
27	0.992	5	0.99757	2	0.99	0.9796935
28	0.992	5	0.99757	2	0.99	0.9796935
29	0.992	6	0.999271	2	0.99	0.9813640
29	0.992	4	0.9919	3	0.999	0.9829808
30	0.992	6	0.999271	2	0.99	0.9813640
30	0.992	4	0.9919	3	0.999	0.9829808
31	0.992	7	0.9997813	2	0.99	0.9818651
31	0.992	5	0.99757	3	0.999	0.9885998
32	0.992	7	0.9997813	2	0.99	0.9818651
32	0.992	5	0.99757	3	0.999	0.9885998
33	0.992	8	0.9999344	2	0.99	0.9820155
33	0.992	6	0.999271	3	0.999	0.9902855
33	0.992	4	0.9919	4	0.9999	0.9838664
34	0.992	8	0.9999344	2	0.99	0.9820155
34	0.992	6	0.999271	3	0.999	0.9902855
34	0.992	4	0.9919	4	0.9999	0.9838664
35	0.992	9	0.9999804	2	0.99	0.9820606
35	0.992	7	0.9997813	3	0.999	0.9907912
35	0.992	5	0.99757	4	0.9999	0.9894904
36	0.992	9	0.9999804	2	0.99	0.9820606
36	0.992	7	0.9997813	3	0.999	0.9907912
36	0.992	5	0.99757	4	0.9999	0.9894904
37	0.992	10	0.9999942	2	0.99	0.9820742

(Continued)

TABLE 11.8 Result of suboptimization of Stages 1 and 2 *(Continued)*

Total Money	R_{03} Reliability of Stage 3 ($n_3 = 3$ fixed)	Value of n_2 ($n_2 = 4$ minimum)	R_{02} Reliability of Stage 2	Value of n_1 ($n_1 = 2$ minimum)	R_{01} Reliability of Stage 1	R_0 = Overall System Reliability
37	0.992	8	0.9999344	3	0.999	0.9909429
37	0.992	6	0.999271	4	0.9999	0.9911776
37	0.992	4	0.9919	5	0.99999	0.9839549
38	0.992	10	0.9999942	2	0.99	0.9820742
38	0.992	8	0.9999344	3	0.999	0.9909429
38	0.992	6	0.999271	4	0.9999	0.9911776
38	0.992	4	0.9919	5	0.99999	0.9839549
39	0.992	11	0.9999983	2	0.99	0.9820783
39	0.992	9	0.9999804	3	0.999	0.9909885
39	0.992	7	0.9997813	4	0.9999	0.9916838
39	0.992	5	0.99757	5	0.99999	0.9895795
40	0.992	11	0.9999983	2	0.99	0.9820783
40	0.992	9	0.9999804	3	0.999	0.9909885
40	0.992	7	0.9997813	4	0.9999	0.9916838
40	0.992	5	0.99757	5	0.99999	0.9895795

Finally, the optimization of stages 1, 2, and 3 is considered. Since the total permissible cost is 40 units, the additional money available, namely, $40 - 25 = 15$ units, is used to procure additional components beyond $n_1 = 2$, $n_2 = 4$, and $n_3 = 3$. For example, if n_3 is increased from 3 to 6 (by spending $3 \times 3 = 9$ units of money), there remains $15 - 9 = 6$ units of money to procure additional components of type 1 and type 2. This means that we can choose either $n_2 = 5, n_1 = 3$ or $n_2 = 7, n_1 = 2$. All these possibilities, along with their corresponding system reliabilities, are summarized in Table 11.9. From this table, the optimum solution of the problem can be identified as $n_1 = 3$, $R_{01} = 0.999$, $n_2 = 6$, $R_{02} = 0.999271$, $n_3 = 5$, $R_{03} = 0.99968$, and $R_0 = 0.99795228$. ■

11.5 Numerical Solutions Using MATLAB

The solution of reliability-based optimum-design problems can be found using the programs or m-files of the *Optimization Toolbox* of MATLAB. The basic information necessary for using the various programs can be found in the user's guide of the Optimization Toolbox. The program **`fmincon`** can be used for the minimization of a nonlinear function of several variables subject to inequality and equality constraints. If the optimization problem is unconstrained, the program **`fminunc`** can be used. Some of the optimization problems formulated in this chapter are solved using the program **`fmincon`** in the following examples.

TABLE 11.9 Result of optimization of Stages 1, 2, and 3

Value of n_3 ($n_3 = 3$ minimum)	R_{03} = Reliability of Stage 3	Value of n_2 ($n_2 = 4$ minimum)	R_{02} = Reliability of Stage 2	Value of n_1 ($n_1 = 2$ minimum)	R_{01} = Reliability of Stage 1	R_0 = Overall System Reliability
8	0.9999975	4	0.9919	2	0.99	0.98197854
7	0.9999872	5	0.99757	2	0.99	0.98758167
6	0.999936	7	0.9997813	2	0.99	0.98972011
6	0.999936	5	0.99757	3	0.999	0.99650866
5	0.99968	8	0.9999344	2	0.99	0.98961824
5	0.99968	6	0.999271	3	0.999	0.99795228
5	0.99968	4	0.9919	4	0.9999	0.99148339
4	0.9984	10	0.9999942	2	0.99	0.98841023
4	0.9984	8	0.9999344	3	0.999	0.99733615
4	0.9984	6	0.999271	4	0.9999	0.99757236
4	0.9984	4	0.9919	5	0.99999	0.99030304
3	0.992	11	0.9999983	2	0.99	0.98207825
3	0.992	9	0.9999804	3	0.999	0.99098861
3	0.992	7	0.9997813	4	0.9999	0.99168384
3	0.992	5	0.99757	5	0.99999	0.98957950

EXAMPLE 11.12

Find the solution of Example 11.6 using the MATLAB program **fmincon**.

Solution: The optimum-design problem can be stated as follows:

$$\text{Find } \vec{X} = \begin{Bmatrix} x_1 \\ x_2 \end{Bmatrix} = \begin{Bmatrix} R_1 \\ R_2 \end{Bmatrix}$$

to minimize

$$f(\vec{X}) = 100x_1^2 + 200x_2^2$$

subject to

$$g_1(\vec{X}) = 0.7 - x_1 x_2 \leq 0$$
$$g_2(\vec{X}) = x_1 - 1.0 \leq 0$$
$$g_3(\vec{X}) = x_2 - 1.0 \leq 0$$

By using the initial design vector $\vec{X}_0 = \begin{Bmatrix} 0.8 \\ 0.8 \end{Bmatrix}$, the following program is developed for the solution of the problem. The program and the results are given below.

```
clear all;clc;close all;format short

x0=[0.8 0.8];
Lbnd=[    ] ;
Ubnd=[    ] ;

options=optimset('display','iter');
[xopt,fval,exitflag,output]=fmincon(@objfun3,x0,[],[],[],[],Lbnd,Ubnd,
@consfun3,options)

end

function ff=objfun3(x)
x
ff=100*x(1)^2+200*x(2)^2   ;

end

function [c,ceq]=consfun3(x)

c(1)=0.7-x(1)*x(2)  ;
c(2)=x(1)-1  ;
c(3)=x(2)-1  ;
ceq=[];

end

    Results:

x=  0.8000  0.8000
x=  0.8000  0.8000
x=  0.8000  0.8000

                    Max    Line search Directional First-order
Iter  F-count  f(x)  constraint  steplength derivative  optimality     Procedure
   0  3        192      0.06                                           Infeasible start point

x=  1.0000  0.6750
x=  1.0000  0.6750
x=  1.0000  0.6750
```

```
Iter  F-count     f(x)  constraint  steplength  derivative  optimality    Procedure
   1  6      191.125   0.025             1       6.25              130

x=  0.9990  0.7006
x=  0.9990  0.7006
x=  0.9990  0.7006

Iter  F-count     f(x)  constraint  steplength  derivative  optimality    Procedure
   2  9      197.99    2.465e-005        1          7              2.47

x=  0.9950  0.7035
x=  0.9950  0.7035
x=  0.9950  0.7035

Iter  F-count     f(x)  constraint  steplength  derivative  optimality    Procedure
   3  12     197.987   1.176e-005        1       0.000294        0.006

x=  0.9950  0.7035
x=  0.9950  0.7035
x=  0.9950  0.7035

Iter  F-count     f(x)  constraint  steplength  derivative  optimality    Procedure
   4  15     197.99   -5.344e-012        1       0.00333     8.22e-006

x=  0.9950  0.7035
x=  0.9950  0.7035
x=  0.9950  0.7035

Iter  F-count     f(x)  constraint  steplength  derivative  optimality    Procedure
   5  18     197.99   -1.11e-016         1      -1.51e-009   1.87e-006   Hessian modified

Optimization terminated:first-order optimality measure less than options.TolFun
and maximum constraint violation is less than options.TolCon.
Active inequalities(to within options.TolCon=1e-006):
lower  upper  ineqlin  ineqnonlin
                 1

xopt=    0.9950  0.7035
fval= 197.9899
exitflag= 1

output=
     iterations:5
     funcCount:18
  lssteplength:1
     stepsize:3.6169e-009
     algorithm:'medium-scale:SQP,Quasi-Newton,line-search'
  firstorderopt:2.6936e-007
constrviolation:-1.1102e-016
     message:[1x143char]
```

■

EXAMPLE 11.13

Find the solution of Example 11.9 using the MATLAB program **fmincon**.

Solution: Noting that the maximization of the reliability of the system, $R_s = 1 - P_{fs}$, is equivalent to minimizing the probability of failure of the system, P_{fs}, the optimum-design problem is formulated as the minimization of $f = P_{fs}$ as follows:

$$\text{Find } \vec{X} = \begin{Bmatrix} x_1 \\ x_2 \\ x_3 \\ x_4 \end{Bmatrix} = \begin{Bmatrix} R_1 \\ R_2 \\ R_3 \\ R_4 \end{Bmatrix}$$

to minimize

$$f(\vec{X}) = x_2[(1-x_3)(1-x_4)]^2 - (1-x_2)[1-x_1\{1-(1-x_3)(1-x_4)\}]^2$$

$$0 \le x_i \le 1;\ i = 1, 2, 3, 4$$

where

$$c_i = k_i(x_i)^{\alpha_i};\ i = 1, 2, 3, 4$$

$$k_1 = k_3 = 50,\ k_2 = 200,\ k_4 = 75;\quad \alpha_i = 0.5,\ i = 1, 2, 3, 4$$

By using the initial design vector $\vec{X}_0 = \{0.5\ 0.5\ 0.5\ 0.5\}^T$, the following program is developed for the solution of the problem. The program and the results are given below.

```
clear all;clc;close all;
format long

x0=[0.5  0.5   0.5  0.5];
Lbnd=[0  0  0  0] ;
Ubnd=[1  1  1  1] ;

options=optimset('display','iter') ;
[xopt,fval,exitflag,output]=fmincon(@objfun3,x0,[],[],[],[],Lbnd,Ubnd,
@consfun3,options)

end

function ff=objfun3(x)
x  ;
ff = x(2)*((1-x(3))*(1-x(4)))^2 + (1-x(2))*(1-x(1)*(1-(1-x(3))*(1-x
(4))))^2  ;
end

function [c,ceq]=consfun3(x)

kk(1)=50; kk(2)=200; kk(3)=50; kk(4)=75;
af(1)=0.5; af(2)=af(1); af(3)=af(1); af(4)=af(1) ;

for ii=1:4
    cc(ii)=kk(ii)*x(ii)^af(ii)   ;
end

cmax=550 ;

c(1)=2*cc(1)+cc(2)+2*cc(3)+2*cc(4)-cmax ;
ceq=[] ;

end
```

			Max	Line search	Directional	First-order	
Iter	F-count	f(x)	constraint	steplength	derivative	optimality	Procedure
0	5	0.226563	-0.5				
1	10	0.00293137	-0.03125	1	-0.0261	0.0254	
2	15	0.0014759	-0.004067	1	-0.0011	0.0149	
3	20	0.000765756	0	1	-0.00055	0.0128	Hessian modified
4	25	0.000211249	0	1	-0.000321	0.0035	Hessian modified
5	30	7.2942e-005	0	1	-8.89e-005	0.00158	Hessian modified
6	35	2.3128e-005	0	1	-3.08e-005	0.000667	Hessian modified
7	40	7.57902e-006	0	1	-9.75e-006	0.000289	Hessian modified
8	45	2.45239e-006	0	1	-3.2e-006	0.000125	Hessian modified
9	50	7.97416e-007	0	1	-1.03e-006	7.15e-005	Hessian modified

```
Optimization terminated:magnitude of directional derivative in search
direction less than 2*options.TolFun and maximum constraint violation
is less than options.TolCon.
Active inequalities(to within options.TolCon=1e-006):
lower  upper  ineqlin  ineqnonlin
      1

x= 0.500000000000000  0.500000000000000  0.500000000000000  0.500000000000000
ff= 0.226562500000000
c= -1.610912703473989e+002

x= 1.000000000000000  0.838245018634754  0.970117207028541  0.970117221929702
ff= 7.974148350552899e-007
c=-20.652235043463747

                          Max  Line search Directional First-order
Iter  F-count            f(x)  constraint  steplength  derivative  optimality Procedure
   9  50 7.97416e-007  0        1       -1.03e-006   7.15e-005     Hessian modified

Optimization terminated:magnitude of directional derivative in search
direction less than 2*options.TolFun and maximum constraint violation
is less than options.TolCon.
Active inequalities(to within options.TolCon=1e-006):
lower  upper  ineqlin  ineqnonlin
         1

xopt= 1.000000000000000  0.838245018634754  0.970117207028541  0.970117207028541

fval= 7.974156303233959e-007

output=
    iterations:9
    funcCount:50
  lssteplength:1
    stepsize:0.013707901349106
    algorithm:'medium-scale:SQP,Quasi-Newton,line-search'
  firstorderopt:5.336953917378651e-005
constrviolation:0
    message:[1x172char]
```

■

EXAMPLE 11.14

Find the solution of the problem formulated in Example 11.4 using the MATLAB program **fmincon**.

Solution: The optimization problem is given by Eq. (E.21) of Example 11.4. By using the initial design vector as $\vec{X}_0 = \begin{Bmatrix} 0.5 \\ 2.5 \end{Bmatrix}$, the MATLAB program developed for the solution of the optimization problem and the results are given below.

```
clear all;clc;close all;warning off;
format long

x0=[0.5 2.5];
Lbnd=[0.1 1 ] ;
Ubnd=[1 5] ;

options=optimset('display','iter');
[xopt,fval,exitflag,output]=fmincon(@objfun3,x0,[],[],[],[],Lbnd,Ubnd,
@consfun3,options)

end
```

```
function ff=objfun3(x)
x;

B=20 ;  H=30 ;  density=0.3 ;

ff=2*density*pi*x(1)*x(2)*sqrt(B^2+H^2)   ;

end

function [c,ceq]=consfun3(x)

Fv=50000 ; Fh=30000 ; E=3*10^7 ; B=20   ; H=30 ; Sy=30000 ;
rB=1   ;  rH=1 ;  rFv=5000   ; rFh=3000   ; rE=0.3*10^7 ; rSy=3000 ;

rd=0.01*x(1)   ; rt=0.01*x(2)    ;

DSCB=((B^2+H^2)^(-1/2))*B*(Fv/H+Fh/B)/(2*pi*x(1)*x(2))-sqrt(B^2+H^2)*Fh/
(2*pi*x(1)*x(2)*B*B)    ;

DSCH=((B^2+H^2)^(-1/2))*H*(Fv/H+Fh/B)/(2*pi*x(1)*x(2))-sqrt(B^2+H^2)*Fv/
(2*pi*x(1)*x(2)*H*H)       ;

DSCFV=sqrt(B^2+H^2)/(2*pi*x(1)*x(2)*H)    ;

DSCFH=sqrt(B^2+H^2)/(2*pi*x(1)*x(2)*B)     ;

DSCD=-sqrt(B^2+H^2)*(Fv/H+Fh/B)/(2*pi*x(1)*x(1)*x(2))      ;

DSCT=-sqrt(B^2+H^2)*(Fv/H+Fh/B)/(2*pi*x(1)*x(2)*x(2))     ;

DSBE=pi*pi*(x(1)^2+x(2)^2)/(8*(B^2+H^2))     ;

DSBD=pi*pi*E*2*x(1)/(8*(B^2+H^2))     ;

DSBT=pi*pi*E*2*x(2)/(8*(B^2+H^2))      ;

DSBB=-pi*pi*E*(x(1)^2+x(2)^2)*2*B/(64*(B^2+H^2)*(B^2+H^2))      ;

DSBH=-pi*pi*E*(x(1)^2+x(2)^2)*2*H/(64*(B^2+H^2)*(B^2+H^2))     ;

SDSC2=(DSCB^2)*(rB^2)+(DSCH^2)*(rH^2)+(DSCFV^2)*(rFv^2)+(DSCFH^2)*(rFh^
2)+(DSCD^2)*(rd^2)+(DSCT^2)*(rt^2)          ;

SDSB2=(DSBE^2)*(rE^2)+(DSBD^2)*(rd^2)+(DSBT^2)*(rt^2)+(DSBB^2)*(rB^2)+
(DSBH^2)*(rH^2)         ;

Sc=(sqrt(B^2+H^2))*(Fv/H+Fh/B)/(2*pi*x(1)*x(2))     ;

Sb=pi*pi*E*(x(1)^2+x(2)^2)/(8*(B^2+H^2))     ;

c(1)=-3+(Sy-Sc)/(sqrt(rSy^2+SDSC2))       ;

c(2)=-3+(Sb-Sc)/(sqrt(SDSB2+SDSC2))     ;

% c(3)=0.1-x(2)    ;
% c(4)=x(2)-1      ;
%
% c(5)=1-x(1)      ;
% x(6)=x(1)-5      ;

ceq=[] ;

end
```

```
(1)Iteration 0
x= 0.500000000000000  2.500000000000000
```

```
                   Max  Line search Directional First-order
Iter  F-count    f(x)    constraint  steplength  derivative  optimality  Procedure
   0      3   84.9538    6.013                                         Infeasible start point

(2)Iteration 1
x=-0.739588308577973 0.160411691422027
x=-0.119794154288987 1.330205845711013
x= 0.190102922855507 1.915102922855507
x= 0.190102937756668 1.915102922855507
x= 0.190102922855507 1.915102951392764

Iter  F-count    f(x)    constraint  steplength  derivative  optimality  Procedure
   1      8   24.7431    1.883         0.25        -192          130     infeasible

(3)Iteration 2
x= 0.100000000000000  1.363029844999214
x= 0.100000014901161  1.363029844999214
x= 0.100000000000000  1.363029865309941

Iter  F-count    f(x)    constraint  steplength  derivative  optimality  Procedure
   2     11     9.26357   0             1          -12.1         13.8

(4)Iteration 3
x= 0.100000000000000  1.000000000000000
x= 0.100000014901161  1.000000000000000
x= 0.100000000000000  1.000000014901161

Iter  F-count    f(x)    constraint  steplength  derivative  optimality  Procedure
   3     14     6.7963    0             1          -2.47         12.4

(5)Iteration 4
x= 0.100000000000000  1.000000000000000
x= 0.100000014901161  1.000000000000000
x= 0.100000000000000  1.000000014901161

Iter  F-count    f(x)    constraint  steplength  derivative  optimality  Procedure
   4     17     6.7963    0             1           0             0   Hessian modified twice

Optimization terminated:first-order optimality measure less
than options.TolFun and maximum constraint violation is less
than options.TolCon.
Active inequalities(to within options.TolCon=1e-006):
lower  upper  ineqlin  ineqnonlin
   1
   2

xopt= 0.100000000000000  1.000000000000000
fval= 6.796304039483387

exitflag= 1


output=
    iterations:4
    funcCount:17
  lssteplength:1
    stepsize:0
    algorithm:'medium-scale:SQP,Quasi-Newton,line-search'
  firstorderopt:0
constrviolation:0
    message:[1x144char]
```

■

REFERENCES AND BIBLIOGRAPHY

11.1 Singiresu S. Rao, *Engineering Optimization: Theory and Practice,* 4th ed., John Wiley, Hoboken, NJ, 2009.

11.2 S. S. Rao and G. Das, "Reliability Based Optimum Design of Gear Trains," *ASME Journal of Mechanisms, Transmissions, and Automation in Design,* Vol. 106, 1984, pp. 17–22.

11.3 F. A. Tillman, C. L. Hwang, and W. Kuo, *Optimization of Systems Reliability,* Marcel Dekker Inc., New York, 1980.

11.4 K. B. Misra and M. D. Ljubojevic, "Optimal Reliability Design of System: A New Look," *IEEE Transactions on Reliability,* Vol. R-22, 1973, pp. 255–258.

11.5 R. M. Burton and G. T. Howard, "Optimal Design for System Reliability and Maintainability," *IEEE Transactions on Reliability,* Vol. R-20, 1971, pp. 56–60.

11.6 R. E. Bellman and S. E. Dreyfus, *Applied Dynamic Programming,* Princeton University Press, Princeton, NJ, 1962.

11.7 R. C. Juvinall and K. M. Marshek, *Fundamentals of Machine Component Design,* 4th ed., John Wiley, Hoboken, NJ, 2006.

11.8 S. S. Rao, "Minimum Cost Design of Concrete Beams with a Reliability-Based Constraint," *Building Science*, Vol. 8, 1973, pp. 33–38.

11.9 A. Nilson, D. Darwin, and C. Dolan, *Design of Concrete Structures,* 14th ed., McGraw-Hill, New York, 2009.

11.10 A. C. King and C. B. Read, *Pathways to Probability,* Holt, Rinehart & Winston, Inc., New York, 1963.

11.11 C. C. Gillispie (ed.-in-chief), *Dictionary of Scientific Biography,* Vol. 7, Charles Scribner's Sons, New York, 1980.

REVIEW QUESTIONS

11.1 Give brief answers to the following:

1. Define a mathematical programming problem.
2. What is the difference between a linear programming problem and a nonlinear programming problem?
3. What is an integer programming problem?
4. What is the Lagrange multiplier method?
5. What is a SUMT method?
6. What is the role of the penalty parameter in the penalty function method?
7. How is the minimum of a function $f(x)$ found in the cubic interpolation method?
8. What is the role of Davidon-Fletcher-Powell method in the solution of a constrained optimization problem?
9. What is a multistage decision process?
10. State a typical reliability allocation problem as an optimization problem.

11.2 Indicate whether each of the following statements is true or false:

1. An optimization problem can be called a mathematical programming problem.
2. An optimization problem is called a nonlinear programming problem if all the functions f, g_j, and l_k are nonlinear.
3. Graphical optimization method cannot be used if the number of design variables is more than two.
4. The Lagrange multiplier method can be used to solve optimization problems with inequality constraints.
5. A point in the design space can be infeasible.
6. A feasible point $\vec{X}$ is one with all $g_j(\vec{X}) < 0$.
7. An infeasible design point $\vec{X}$ is one with all $g_j(\vec{X}) > 0$.
8. At the optimum-design point $\vec{X}^*$, one or more $g_j(\vec{X}^*) = 0$ always.
9. One Lagrange multiplier is associated with each equality constraint in the Lagrange multiplier method.
10. The starting design point $\vec{X}_0$ is required to be feasible with all $g_j(\vec{X}_0) < 0$ in the SUMT method.
11. The value of the penalty parameter is increased sequentially in the SUMT method.

11.3 Fill in the blank with the proper word(s):

1. Optimization is the process of obtaining the ______ result under given circumstances.
2. The SUMT method requires the solution of a series of ______ optimization problems.
3. A minimization problem can be solved as a maximization problem by changing the ______ of the objective function.
4. The space in which each axis represents one of the design variables is known as ______ space.
5. A constraint surface divides the design space into ______ parts.
6. A design point $\vec{X}$ with one or more $g_j(\vec{X}) = 0$ is called a ______ design.
7. An objective-function surface is one on which every design point will have the same ______ function value.
8. The ______ interpolation method finds the solution of one-dimensional minimization problem.
9. The unconstrained-minimization method discussed in the SUMT method is called the ______ method.
10. Dynamic programming problem is called the ______ decision problem.

11.4 Select the most appropriate answer from the multiple choices given:

1. Any optimization problem can be solved efficiently using:
   a. One prespecified optimization method
   b. Any optimization method
   c. An optimization method that is not known in advance
2. Dynamic programming approach can be used to solve optimization problems whose physical structure is in:
   a. Parallel form **b.** Serial form **c.** Series–parallel form
3. The reliability of a parallel–series system with n_i components in parallel in the ith subsystem of the N subsystems in series is given by:
   a. $\prod_{i=1}^{N}\{1 - (1 - R_i)^{n_i}\}$ **b.** $\prod_{i=1}^{N}(1 - R_i)^{n_i}$ **c.** $(1 - R_i)^{n_i N}$

11.5 Match the following:

1. Linear programming	**a.** Minimizes a nonlinear objective function in one variable
2. Nonlinear programming	**b.** Minimizes a linear objective function subject to linear constraints
3. Davidon-Fletcher-Powell method	**c.** Minimizes an objective function in the form of a sum of several terms
4. Cubic interpolation method	**d.** Minimizes a nonlinear objective function subject to nonlinear constraints
5. Dynamic programming problem	**e.** Minimizes a nonlinear objective function with no constraints

PROBLEMS

11.1 Consider a system consisting of N stages in series with n_j components in parallel in stage j; $j = 1, 2, \ldots, N$. Each component of stage j costs c_j, weighs w_j, and has a reliability of R_j.

1. Formulate the problem of maximizing the overall reliability subject to constraints on the cost and weight of the system.
2. Solve the problem formulated in (1) for the following data:

Stage j	R_j	c_j	w_j
1	0.9	5	3
2	0.7	6	1
3	0.8	8	2

Maximum cost permitted: 100. Maximum weight permitted: 50.

11.2 A system consists of N stages in parallel with n_j components in series in stage j; $j = 1, 2, \ldots, N$. Each component of stage j costs c_j, weighs w_j, and has a reliability of R_j.

1. Formulate the problem of maximizing the reliability of the system subject to constraints on the cost and weight of the system.
2. Solve the problem formulated in (1) for the following data:

Stage j	R_j	c_j	w_j
1	0.9	5	3
2	0.7	6	1
3	0.8	8	2

Maximum cost permitted: 100. Maximum weight permitted: 50.

11.3 The shear stress (τ) induced in a helical spring under a compressive load F [11.7] is given by

$$\tau = \left(\frac{2C + 1}{2C}\right)\frac{8FD}{\pi d^3}$$

where $C = D/d$ = spring index, d = wire diameter, and D = mean coil diameter. The stiffness of the spring k is given by

$$k = \frac{d^4 G}{8D^3 N}$$

where G = shear modulus and N = number of turns (coils). A spring is to be designed for a machine suspension system to achieve a stiffness of 10 ± 0.2 lb/in. The spring material has a permissible (yield) strength in shear of 90 ± 10 k(lb/in^2), a specific weight of 0.28 ± 0.028 lb/in^3, and a shear modulus of $11.5 \times 10^6 \pm 0.5 \times 10^6$ lb/in^2. Formulate the problem of finding $\overline{d}$, $\overline{D}$, and $\overline{N}$ to minimize the mean weight of the spring to achieve a reliability greater than or equal to 0.99 in strength as well as stiffness of the spring. The axial load on the spring is given by 400 ± 100 lb. Assume that all the variables follow normal distribution and the tolerances, where specified, correspond to 3σ-band. Also, assume that the coefficient of variation of d, D, and N is 0.1.

11.4 Solve the reliability-based spring optimization problem described in Problem 11.3 for $\overline{C} = \overline{D}/\overline{d} = 10$ using a graphical procedure.

11.5 A hollow circular shaft is to be designed for minimum weight to achieve a reliability of 0.999 when subjected to a torque, $T = N(1{,}000{,}000, 10{,}000)$ lb · in. The permissible shear stress, τ_0, of the material is given by $\tau_0 = N(50{,}000, 5000)$ lb/in^2, while the maximum induced stress, τ, is given by

$$\tau = \frac{T r_0}{J}$$

where r_0 is the outer radius of the shaft and J is the polar moment of inertia of the cross section. The manufacturing tolerances on the inner and outer radii of the shaft are to be specified as ± 0.045 in. The length of the shaft is given by 50 ± 1 in and the specific weight of the material by 0.3 ± 0.03 lb/in^3. Formulate the design problem and solve it using a graphical procedure. Assume normal distribution for all the random variables and 3σ values for the specified tolerances.

11.6 A flywheel is to be designed for a forging press to store an energy of $10{,}000 \pm 1000$ J. The minimum and maximum operating speeds of the flywheel are 600 ± 60 and 800 ± 80 rpm, respectively. The kinetic energy (E) of the flywheel is given by $E = 1/2\, I \omega^2$ where I is the mass moment of inertia and ω is the rotational speed of the flywheel. By idealizing the flywheel as a rotating annular disk with inner radius r_i, outer radius r_0, and thickness t, the mass moment of inertia can be expressed as

$$I = \frac{\pi (r_0^4 - r_i^4) t \rho}{2}$$

where ρ is the density of the material. The tangential and radial stresses developed in the flywheel at a radius r, $r_i \le r \le r_0$, are given by

$$s_t = \rho \omega^2 \left(\frac{3 + v}{8}\right)\left(r_i^2 + r_0^2 + \frac{r_i^2 r_0^2}{r^2} - \frac{1 + 3v}{3 + v} r^2\right)$$

$$s_r = \rho \omega^2 \left(\frac{3 + v}{8}\right)\left(r_i^2 + r_0^2 - \frac{r_i^2 r_0^2}{r^2} - r^2\right)$$

where ν is the Poisson's ratio. Formulate the problem of determining $\overline{r}_0, \overline{r}_i$, and $\overline{t}$ of the flywheel to minimize the mass while achieving a reliability of 0.99 in both strength and energy storage. The following data are applicable to the design problem: $\rho = N(7700, 770)$ kg/m^3, $s_y = N(30{,}000, 3000)$ lb/in^2, and $\nu = N(0.3, 0.03)$. Assume that all the variables follow normal distribution and the tolerances, where specified, represent the 3σ values. Use a coefficient of variation of 0.1 for r_i, r_0, and t.

11.7 Formulate the flywheel design problem described in Problem 11.6 with $\overline{r}_i$ and $\overline{r}_0$ as design variables assuming the value of $\overline{t}$ as $\overline{t} = 0.5\overline{r}_i$. Solve the resulting problem using a graphical procedure.

11.8 An electronic navigation system consists of n components in series, each following an exponential distribution as:

$$f_{T_i}(t_i) = \lambda_i e^{-\lambda_i t_i}; \quad i = 1, 2, \ldots, n; \quad \lambda_i > 0; \quad t > 0$$

where T_i denotes the time to failure and λ_i represents the failure rate of component i. The cost of the ith component is given by (a_i/λ_i^2) where a_i is a constant. Find the values of $\lambda_1, \lambda_2, \ldots, \lambda_n$ to maximize the reliability of the system without exceeding a total cost of b using the Lagrange multiplier method.

11.9 The preload on each of the eight equally spaced 1/4-in. UNF bolts in the gasketed cylinder head shown in Fig. 11.13 is given by 800 ± 80 lb. The cylinder, cylinder head, and the bolts are all made of steel with an Young's modulus of $(30 \pm 3) \times 10^6$ lb/in^2. The proof strength of bolts is known to be 50 ± 5 k(lb/in^2). The gasket has a thickness of t_2 with a Young's modulus of 250 ± 25 lb/in^2. The stress induced in the bolt (s_b) is given by

$$s_b = \frac{CF}{A_t} + \frac{F_i}{A_t}$$

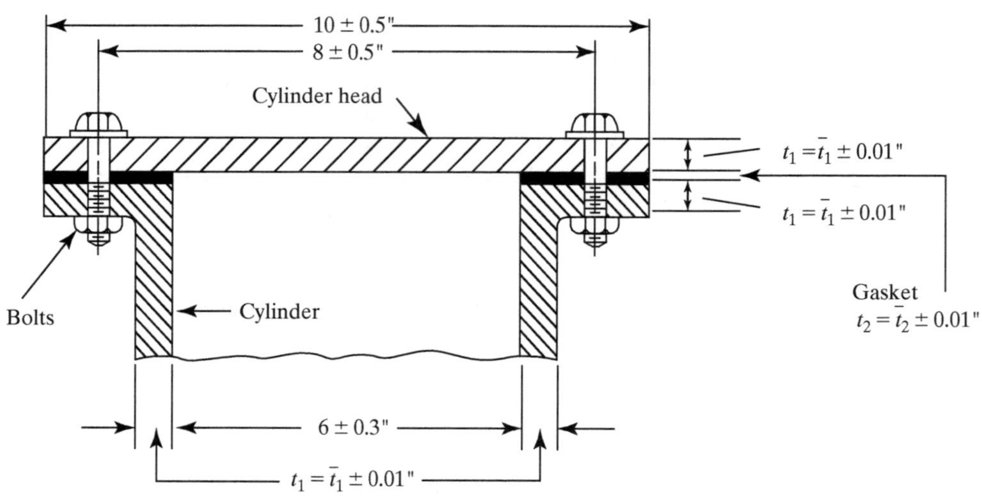

FIGURE 11.13 Bolted cylinder head.

where

$$C = \frac{k_b}{k_b + k_g}$$

is the joint constant, k_b is the stiffness of the bolt, k_g is the stiffness of the gasket, F = tensile load acting on one bolted joint, F_i = preload on the joint, and A_t is the tensile area of the bolt which is equal to 0.0364 in^2 for the 1/4 in UNF bolt. The stiffness of the bolt and gasket are given by

$$k_b = \frac{A_b E_b}{l_b}; \quad k_g = \frac{A_g E_g}{l_g}$$

where $A_b(A_g)$ = area of cross section, $E_b(E_g)$ = Young's modulus, and $l_b(l_g)$ = length of bolt (gasket) in the axial direction of the cylinder. The internal pressure of the cylinder is given by 500 $\pm$ 50 lb/in^2. Formulate the problem of finding $\bar{t}_1$ and $\bar{t}_2$ to minimize the mass of the cylinder head while achieving a reliability of 0.999 for the bolt strength. The bounds on $\bar{t}_i$ are specified as

$$0.5 \text{ in} \leq \bar{t}_i \leq 1.5 \text{ in}; \quad i = 1, 2$$

11.10 Find the solution of Problem 11.9 using a graphical procedure.

11.11 The reliability of a system, consisting of n stages in series with x_j redundant units in each stage $j (j = 1, 2, \ldots, n)$, is given by

$$R_0 = \prod_{j=1}^{n} R_j = \prod_{j=1}^{n} \{1 - (1 - r_j)^{x_j}\} \tag{E1}$$

where r_j is the reliability of a unit in stage j. The cost of each unit in stage j, c_j, is given by

$$c_j = a_j e^{b_j/(1-r_j)} \tag{E2}$$

and the total cost of the system is restricted as

$$\sum_{j=1}^{n} c_j x_j \leq C \tag{E3}$$

where C is a constant. By eliminating x_j using Eq. (E1), Eq. (E3) takes the form

$$\sum_{j=1}^{n} a_j \frac{\ln(1 - R_j)}{(1 - r_j)} e^{b_j/(1-r_j)} \leq C \tag{E4}$$

Using the Lagrange multiplier method, find the values of R_j and r_j which maximizes R_0 subject to the cost constraint of Eq. (E4) for a 4-stage system with the following data [11.4]:

Stage j	**1**	**2**	**3**	**4**
a_j	1.0	3.5	2.0	5.0
b_j	0.30	0.55	0.40	0.65

Assume the value of C as 300 and discuss the practicality of the solution found.

11.12 Minimize $f(x) = 2x^2 - 12x + 5$ subject to $x \leq 2$ using the interior penalty function method. Plot the contours of the ϕ-function for different values of r_k and identify the optimum point in the graph.

11.13 Find the solution of the following optimization problem using a graphical procedure:

$$\text{Maximize } f(x_1, x_2) = 3x_1 + x_2$$

subject to

$$3x_1 + 2x_2 \geq 6$$
$$x_1 + x_2 \leq 8$$
$$2x_1 - 4x_2 \leq 2$$
$$-4x_1 + 3x_2 \leq 12$$
$$x_1 \geq 0; \quad x_2 \geq 0$$

11.14 A parallel system, consisting of two components with constant failure rates of λ_1 and λ_2, is to be optimized for minimum cost. The costs of the components are given by $c_1 = 100/\lambda_1$ and $c_2 = 200/\lambda_2$. Find the values of λ_1 and λ_2 to minimize the total cost while achieving a reliability of 0.99 at 100 hours.

11.15 Solve the problem formulated in Example 11.4 using a graphical procedure.

11.16 Solve the problem formulated in Example 11.5 using a graphical procedure.

11.17 Formulate the problem of minimum cost design of a rectangular under-reinforced concrete beam which can carry a bending moment M with a reliability of 0.99. The costs of concrete (c_c), steel (c_s), and form work (c_f) are given by 0.01 dollar/in^3, 0.20 dollar/in^3, and 0.1 dollar/in^2 of surface area. The bending moment acting on the beam is normally distributed with a mean value $\overline{M} = 10^6$ lb · in and a standard deviation of $\sigma_M = 10^5$ lb · in. The strengths of concrete (s_c) and steel (s_s) are normally distributed:

$$s_c = N(1500, 300) \text{ lb/in}^2; \quad s_s = N(30{,}000, 3000) \text{ lb/in}^2$$

Treat the width (w) and depth (d) of the beam and the reinforcing steel area (A_s) as deterministic design variables (see Fig. 11.14). The resisting moment (M_r) of a reinforced beam section [11.8, 11.9] is given by

$$M_r = A_s\, s_s\left(d - 0.59\frac{A_s\, s_s}{s_c w}\right)$$

and the balanced-steel area, which can be used as an upper bound on A_s to ensure under-reinforcement of the beam [11.9], is given by

$$A_s^{(b)} = 0.542\frac{s_c w d}{s_s}\left(\frac{6117}{6117 + s_s}\right)$$

11.18 Solve the concrete beam design problem formulated in Problem 11.17 using a graphical method assuming $w = 0.5d$.

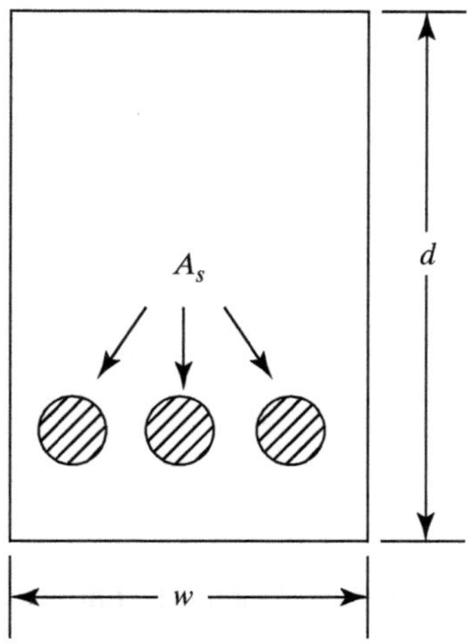

FIGURE 11.14 Reinforced concrete beam section.

MATLAB Problems

11.19 Find the reliability of the series system shown in Fig. 11.6 for minimum overall cost if the costs of the first and second components are given by $50\, R_1^3$ and $100 \exp(R_2^3)$, respectively. The system reliability as well as the individual component reliabilities are required to be at least 0.75.

11.20 Find the reliability of the parallel system shown in Fig. 11.8 for minimum overall cost if the costs of the first and second components are given by $50\, R_1^3$ and $100 \exp(R_2^3)$, respectively. The system reliability as well as the individual component reliabilities are required to be at least 0.75.

11.21 Determine the component reliabilities of the system shown in Fig. 11.10 to minimize the overall cost. The reliability of the system is required to be at least 0.9 and the reliabilities of each of the components is required to be at least 0.75. The component costs (c_i) are given by $c_i = k_i R_i^{\alpha_i}$ where R_i is the reliability of component $i (i = 1, 2, 3, 4)$ with $k_1 = k_3 = 50, k_2 = 200, k_4 = 75$, and $\alpha_i = 0.5 (i = 1, 2, 3, 4)$.

CHAPTER 12

Failure Modes, Event-Tree, and Fault-Tree Analyses

What You Will Learn

In this chapter, you will learn

- That the safety analysis of a system can be conducted using two types of methods, namely, the qualitative and the quantitative methods
- That the failure modes and effects analysis and the event-tree analysis are qualitative methods that help in understanding the logical structure of the various failure modes of the system and their interrelationship
- That the reliability analysis and the fault-tree analysis are quantitative methods that help in estimating the probability of occurrence of certain accidents (or failures) using the available component failure data and the corresponding repair times
- That the failure modes and effects analysis helps in identifying, during the conceptual and initial design phases, different failures and modes of failure that occur at the component, subsystem and system levels and evaluating the consequences of these failures
- That the event-tree analysis uses binary logic and considers each event to have either happened (components worked successfully) or not (components failed to work successfully)
- That the fault-tree analysis involves a graphical representation of the relationship between certain specific events and the ultimate undesirable event which, in turn, helps in studying the cause-and-effect relations among the components of the system

John von Neumann was born in Budapest, Hungary, on December 28, 1903. He received his B.A. in 1925 in Chemical Engineering from the Zurich Institute and his Ph.D. in 1926 in Mathematics from the University of Budapest. He emigrated to the United States in 1930 and taught at Princeton University for the rest of his career. Von Neumann is best known for his work in game theory, quantum mechanics, and computing. He worked on the government's atomic bomb project at Los Alamos Scientific Laboratory during World War II. At that time, the scientists at Los Alamos wanted to find the distance the neutrons travel through various materials in connection with shielding and other important aspects. Although the physicists had most of the data, such as the average distance a neutron of a given speed would travel in a given material before it collides with an atomic nucleus and the probability of a neutron bouncing off instead of being absorbed by the nucleus, they found it to be an extremely complicated problem beyond the reach of theoretical calculations; the experimental determination by trial and error would have been expensive, time-consuming, and hazardous. To that crisis, von Neumann, along with Stanislas Ulam, suggested a remarkably simple, workable answer based on the use of a roulette wheel. When the method was revived for the secret work at Los Alamos, von Neumann gave it the code name "Monte Carlo." The method was found to be so successful on neutron diffusion problems that it later was applied to numerous other fields. Von Neumann was appointed as a member of the Atomic Energy Commission in 1955 and died in Washington, D.C., on February 8, 1957 [12.19 – 12.21].

12.1 Introduction

The failure-mode, event-tree, and the fault-tree analyses fall under the category of "design and safety review techniques." Although the considerations of reliability are included in the design stage, it is desirable to conduct a review of the complete system from safety point of view. The failure of a complex system may not necessarily always be due to the failure of its components, but sometimes due to the faults of operators and maintenance crews. Thus, the consideration of human errors is also important in the safety analysis. Some system failures may simply imply inconvenience and loss of productivity, while some others imply loss of human life. For example, the breakdown of a household refrigerator may not cause a serious loss, whereas the breakdown of an artificial respirator in a hospital emergency room might cause a loss of human life.

Several design and safety review methods have been developed to understand as to how accidents occur, how their probabilities can be estimated, and how to reduce the probability of their occurrences. These methods aid in the systematic analysis of a system for safety, in identifying potential weak spots that might lead to a safety hazard, and in highlighting areas that need special attention in the design, manufacture, operation, and maintenance of the system. In most cases, the design review might reveal that nearly all aspects of the design are satisfactory. However, in some cases, it might reveal certain deficiencies of design from the point of view of system safety. It is desirable to discover any possible design deficiencies at preliminary design stages, even at the expense of considerable effort, rather than having to modify the design at a later stage or live with the consequences of a defective design. Thus, a well-planned and well-conducted design review will prove to be extremely cost-effective in the design and development of complex engineering systems. Among the various methods available for the design and safety review of complex systems, the failure modes and effects analysis, the event-tree analysis, and the fault-tree analysis are particularly popular in industry.

12.2 System-Safety Analysis

Both qualitative and quantitative methods are available for system-safety analysis. The qualitative methods help in understanding the logical structure of the various failure modes of a system and their interrelationship. This, in turn, might lead to clues as to how the likelihood of certain accidents can be reduced or eliminated. The quantitative methods, on the other hand, make use of the available component failure data, estimates of repair times and human errors, and predict the probabilities of occurrence of certain accidents. In practice, the type of system-safety analysis to be used depends on the complexity of the system, the availability of failure data, and the degree to which human factors are involved. The various types of system-safety analysis procedures used for the identification and quantification of safety and risk of a system are shown in Fig. 12.1. As shown in this figure, all the available methods can be broadly classified into inductive and deductive techniques.

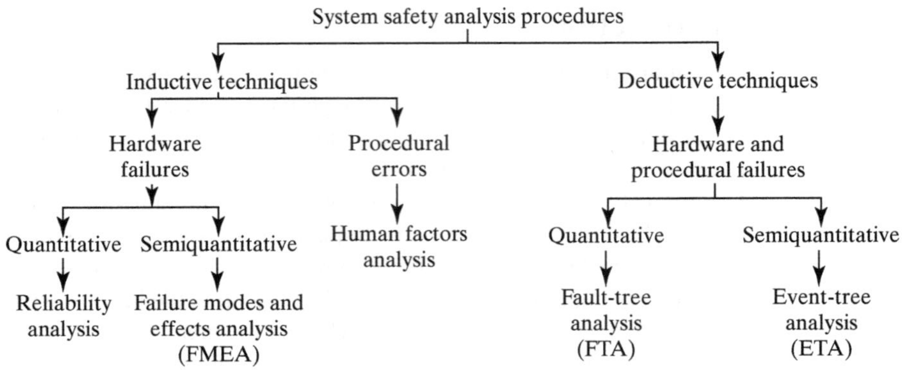

FIGURE 12.1 Basic types of system-safety analysis.

In the inductive procedures, the analysis begins at the component level, identifies the failure modes of each component, and establishes the effect of each component failure on the overall system. Thus, the inductive analysis works its way up in the system through higher assemblies until the complete system is analyzed. The failure modes and reliability analyses belong to the category of inductive hardware-failure analysis techniques. The failure-modes analysis is used to examine each component of the system and find how various modes of failure affect the system. This technique is mostly subjective due to the qualitative nature of the analysis. An extension of this procedure takes human factors into account and results in a semiquantitative method. The reliability analysis deals with the methods of finding the probability of survival of the system over a specified period under prescribed operating conditions.

In the deductive techniques, the analysis starts with an enumeration of the potential hazards and works down through the system to identify the system hardware failures or human errors which could have caused the potential hazards. The fault-tree and event-tree analyses belong to deductive failure-analysis techniques. In the fault-tree analysis, a TOP event, such as an explosion, fire, or release of toxic material which signifies the total failure of the system, is identified first and then all the possible causes (precursor events or basic causes) are identified. This method of analysis is probably the most powerful and accurate technique for the quantification of risk [12.1]. The event-tree analysis is a logical procedure, based on a form of binary logic, in which an event is deemed to have either happened or not, or a component is deemed to have either worked properly or not. A major limitation of the method is that it does not consider partially degraded components. This method uses a logic that is essentially the reverse of that used in fault-tree analysis. Given the occurrence of a particular type of failure (called the TOP event), a fault tree identifies the various combinations and sequences of other failures that lead to the TOP event. On the other hand, in the event-tree analysis, starting with some initiating event, the consequences of that event are followed through a large number of possible chains assigning to each path a probability of occurrence. The end result of an event-tree analysis is a long list of possible consequences of an initiating event with probabilities associated with each consequence.

12.3 Failure Modes and Effects Analysis (FMEA)

The *failure modes and effects analysis* (FMEA) is the most widely used analysis procedure in practice at the initial stages of system development. The purpose is to identify the different failures and modes of failure that can occur at the component, subsystem, and system levels and to evaluate the consequences of these failures. It involves an analysis of the system to determine the effect of component or subsystem failure (1) on the overall performance of the system and (2) on the ability to meet performance requirements or objectives. The FMEA is usually performed during the conceptual and initial design phases of the system in order to assure that all possible failure modes have been considered and that proper provisions have been made to eliminate all the potential failures. Sometimes, a criticality analysis is also made along with the FMEA. In this case, an index related to the criticality or severity of the failure is also included. The analysis is then called a *failure modes, effects, and criticality analysis* (FMECA). A comprehensive FMECA for a complex system is a sizable task. However, this procedure is a very effective reliability design-review technique and is mandatory for many government-funded projects. US MIL-STD-1629 [12.18] describes in detail how FMECA must be conducted, documented, and presented for military contracts. It is also cited in several other reliability program standards in USA and other countries.

The FMECA aids in producing block-diagram reliability analysis and diagnostic charts for repair purposes. In fact, the work involved in generating diagnostic procedures and user-maintenance handbooks can be eased considerably if FMECA details are known to the designer. The level of detail to which FMECA is performed should be based on the purpose for which it is performed. For example, the feasibility study for a large system may consider only the functional subsystems with much of the detailed design yet to be performed. Then the initial FMECA will consider only the failures associated with the subsystems or blocks. As the detailed design progresses, the FMECA will be updated, which aids in the identification of detailed features of the system. Most FMECAs deal with single failure modes. If several critical failure modes are identified, their simultaneous effect will be studied separately. The procedure involved in FMECA can be summarized as follows:

1. Identify all potential failure modes of the system.
2. Relate the causes, effects, and hazards of each mode of failure.
3. Prioritize the modes relative to their probabilities of occurrence, failure criticality (or severity), and detection capability.
4. Provide suitable follow-up or corrective actions for each type of failure mode.

The FMECA procedure can be implemented systematically using a standardized FMECA form. Although the details of the form should reflect the specific needs and characteristics of the particular system being analyzed, a typical FMECA form is shown in Fig. 12.2. In general, the following information is included in a FMECA form:

1. System. Name of the system or component under consideration.
2. Analyst. Name of the person conducting the analysis.
3. Function. Broad description of the function of the system.

System: Function/Mission				Identification Number: Date: Analyst:								
Modes of failure				Failure effects					Criticality	Corrective action		
Serial number	Failure mode	Cause	Frequency of occurence	Local effects	Overall effects	Failure detection method	Degree of severity	Probability of detection	Priority of risk	Design modification	Design verification	Remarks

FIGURE 12.2 Typical FMECA form.

4. Modes of failure. The modes in which the system can fail.
5. Cause of failure. The possible reason for the failure of the system.
6. Frequency of occurrence. The frequency of occurrence of this type of failure on a scale of 1 to 100 with 1 denoting a very rare occurrence and 100 denoting a very frequent occurrence.
7. Effects of failure. The local and overall consequences of failure, including possible loss or damage to humans and property.
8. Failure detection method. The method(s) of detecting/isolating the failure.
9. Degree of severity. The expected degree of severity of the failure on a scale of 1 to 100 with 1 indicating almost no loss and 100 denoting a major loss to humans and property.
10. Probability of detection. The chance of detecting the failure before any damage is caused, on a scale of 1 to 100 with 1 denoting sure detection and 100 indicating impossible detection.
11. Priority of risk. The relative importance of the failure for corrective action on a scale of 1 to 100 with 1 denoting least priority and 100 indicating highest priority. This number can be obtained by multiplying the numbers in steps 6, 9, and 10 and dividing the result by 10,000.
12. Design modification. The action taken or the design modification made to reduce the priority of risk in step 11.
13. Design verification. The procedure to be used to verify the action taken or design modification made in step 12.

12.4 Event-Tree Analysis

An event tree is a graphical representation of all possible events in a system. It is called a "tree" because the graphical representation gradually expands in the form of a tree with more and more branches created as the number of events increases. It is based on binary logic in which an event is assumed to have either happened or not, or a component is assumed to have either worked properly or not. A major limitation of this analysis is that components that are partially degraded cannot be considered as such. This method proceeds essentially in reverse order compared to that used in the fault-tree analysis [12.2]. Given that a particular TOP event (fault) occurred, a fault tree identifies the various combinations and sequences of other failures that might have led to the TOP event. The event-tree analysis was originally devised to assess the safety of nuclear reactors.

In an event-tree analysis, we begin with some initiating event and follow the consequences of that event through a large number of possible paths (chains) until a final outcome is encountered. By assigning to each path a probability of occurrence, the end result is a long list of possible consequences of the initiating event in which each consequence has a certain probability associated with it. Example 12.1 is given to illustrate the procedure.

EXAMPLE 12.1

A room consists of two lightbulbs operated by a single switch as shown in Fig. 12.3. Considering the initial event as the "room without electric light," develop the event tree of the system.

Solution: In this case, we can proceed to trace a series of events by following through Fig. 12.3. Each time a component operates successfully, the corresponding path in the event tree is constructed upwards as shown in Fig. 12.4. Failures of components are indicated using descending paths. We start with two branches at the left of the event tree—one ("up") corresponding to the presence of light in the room and the other ("down") corresponding to the absence of light in the room. For the branch corresponding to "no light in the room," we notice that the fuse must function properly for the electric power to reach the switch. Thus, we get two branches: "up" for fuse functioning and "down" for fuse nonoperating (blown off). The next set of branches are produced by the switch being in operating (closed) or failed (open) condition. Finally, we produce branches for bulbs 1 and 2 being in good or burnt condition. After completing the event tree as shown in Fig. 12.4, we can label the final outcome with letters denoting which failures produced each branch. The second branch, with final event labeled A, results only from the initial event, "room without electric light" (event A), with all components operating in the desired condition. The third branch represents event A followed by the failed condition of bulb 2 (event E) and thus is labeled AE. From the event tree, we can also determine the physical outcome of each output event (percent of light in room corresponding to the event). For example, the second branch results in the operation of both the bulbs (100 percent light), the third and fourth branches represent the operation of one bulb only (50 percent light), and the remaining branches correspond to no light at all in the room (0 percent light). Figure 12.4, thus, shows how an event tree resulting from an initially defined event results in several different consequences.

Once the event tree is constructed, we can compute the probability of each of the final events. If the components are assumed to be statistically independent, the probability of each outcome is simply the product of probabilities of each event in the sequence. Note that in Fig. 12.4, an outcome such as AE implicitly includes the events $\overline{B}$, $\overline{C}$, and $\overline{D}$ where, for example, $\overline{B}$ denotes the complement

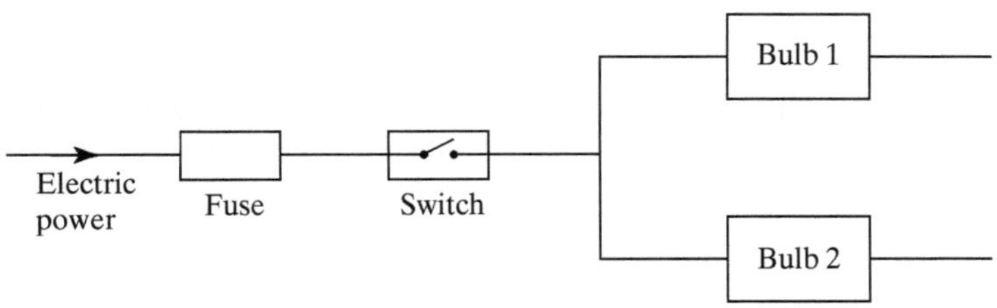

FIGURE 12.3 Two light bulbs connected through a fuse and a switch.

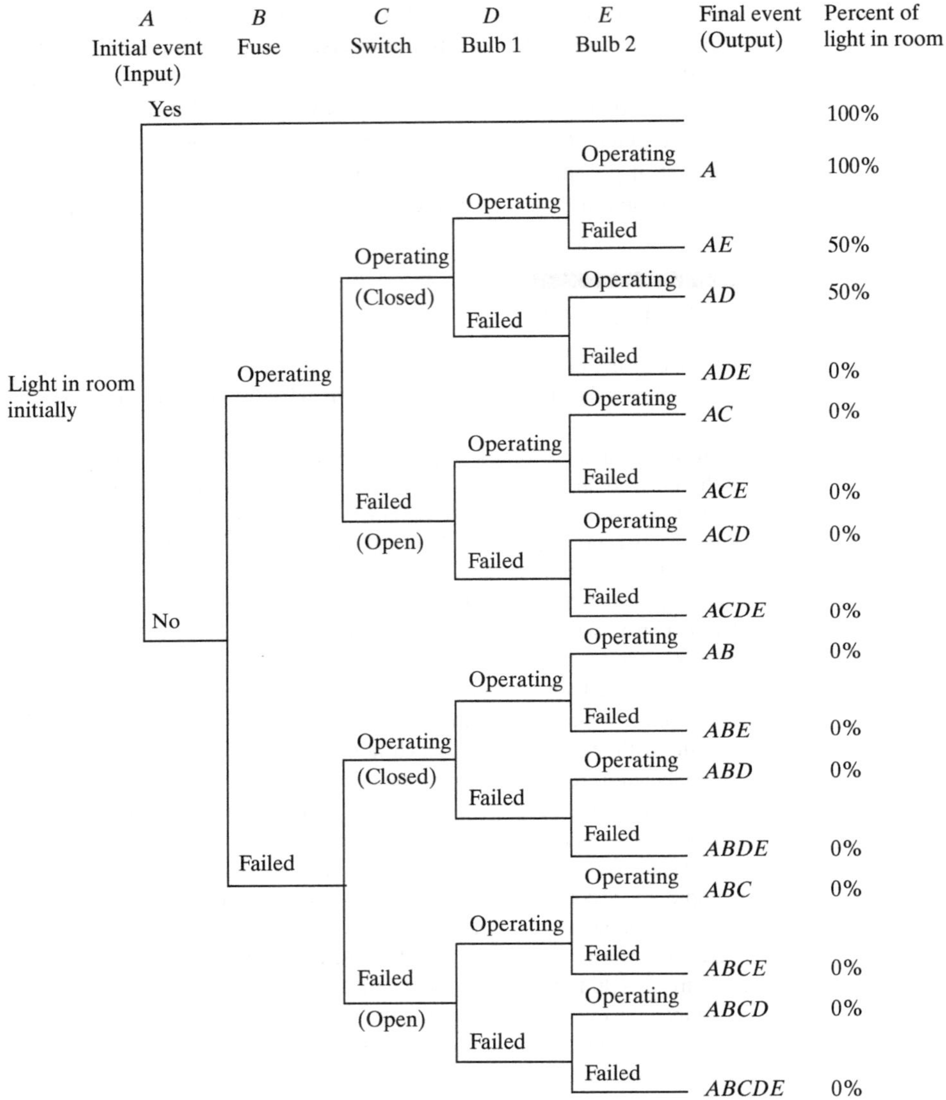

FIGURE 12.4 Event tree for the system depicted in Fig. 12.3.

of event B ($\overline{B}$ indicates the event "fuse in good condition," while B indicates the event "fuse blown off"). Thus, the probability of the event AE (that is, the occurrence of "no electric light in the room" followed by the "failure of bulb 2") can be determined as

$$P_{AE} = P_A P_{\overline{B}} P_{\overline{C}} P_{\overline{D}} P_E \tag{12.1}$$

where P_X denotes the probability of occurrence of the event X. It can be seen that the probability of realizing "50% light in the room" can be found by adding P_{AE} and P_{AD}.

■

In many cases, the construction of the event tree can be simplified further [12.11]. These situations occur when a specific event leads to a final outcome regardless of any subsequent events. For example, if the fuse is in nonoperating condition in Example 12.1, the room will not have any light, irrespective of the conditions of the switch and the bulbs. Thus, it is unnecessary to proceed any further in the chain of events, and hence the branch corresponding to "failed fuse" can be immediately terminated. A similar reduction occurs if the switch fails, eliminating the need to check the conditions of bulbs 1 and 2. The resulting event tree, shown in Fig. 12.5, now has only six branches for the "no electric light in the room" situation, rather than the 16 such outcomes indicated in Fig. 12.4.

The event tree thus provides a pictorial definition of the different possible sequences of events when a major failure occurs. Because of the tree-like appearance, each sequence has separate end points and it is practical to analyze the importance of the sequences by supplementing the diagram with a simple table. In practice, the probability of occurrence of the various paths of the event tree are not quantifiable during the synthesis of a system, but the approach provides a quick means of comparing different system configurations. A more detailed examination of the subsystems would be necessary to identify subsystem failure modes and effects. Such a detailed examination could be achieved by extending the event tree methodology to the fault-tree analysis.

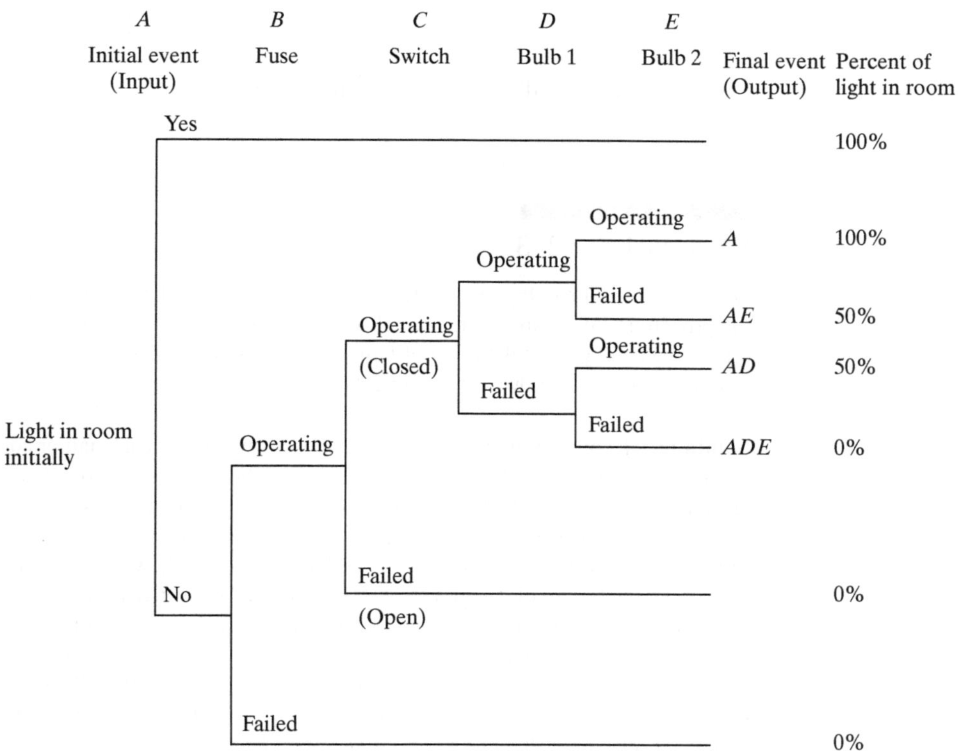

FIGURE 12.5 Reduced event tree for the system depicted in Fig. 12.3.

EXAMPLE 12.2

Consider a diesel power generating system in a hospital (see Fig. 12.6). When normal power supply fails, a "loss of power detector unit" switches on and starts three diesel power generators. Each diesel power generator satisfies one-third of the needs of the hospital. Develop the event tree of the system.

Solution: When the initiating event occurs, that is, when the normal power supply fails, each of the "loss of power detector unit" and the "diesel power generator" may function or fail. The consideration of operation and failure of each component leads to the event tree shown in Fig. 12.7.

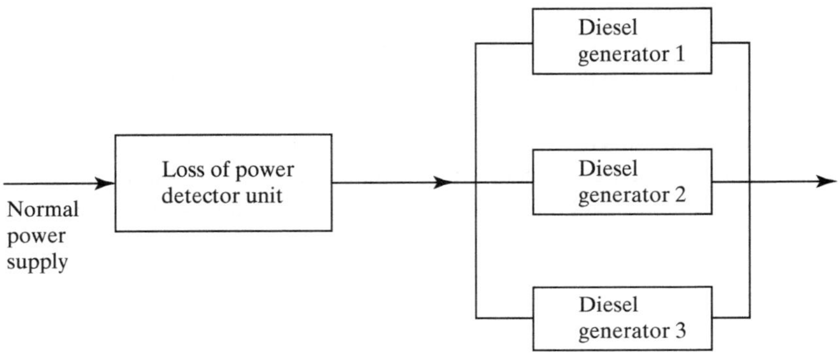

FIGURE 12.6 A diesel power generating system.

■

EXAMPLE 12.3

Determine the probability of occurrence of each outcome of the system described in Example 12.2 if the probability of failure of the normal power supply is 0.01, the probability of failure of the detector unit switching on in the event of failure of normal power supply is 0.001, and the probability of failure of each diesel power generator is 0.05.

Solution: The probabilities of occurrence of each branch (for each unit or component) are indicated in parentheses in Fig. 12.7. For example, the probabilities of normal power supply in operating (good) and failed conditions are 0.99 and 0.01, respectively; hence these values are shown in parentheses along the respective branches. Similarly, the probabilities of the "loss of power detector unit" in operating and failed conditions are, respectively, 0.999 and 0.001, and are shown along the corresponding branches. Once the probabilities of occurrence of all branches are indicated, the probability of occurrence of any final outcome (event) can be obtained by multiplying all the probabilities of the individual branches that lead to the particular event. For example, the probability of occurrence of event #5 is given by $P(\overline{A}) \cdot P(B) \cdot P(C) \cdot P(\overline{D})P(\overline{E}) = (0.01)(0.999)(0.95)(0.05)(0.05) = 0.00002373$. Similar calculations yield results as shown in Table 12.1. The fractional power supply available for each outcome is also indicated in Fig. 12.7 and Table 12.1.

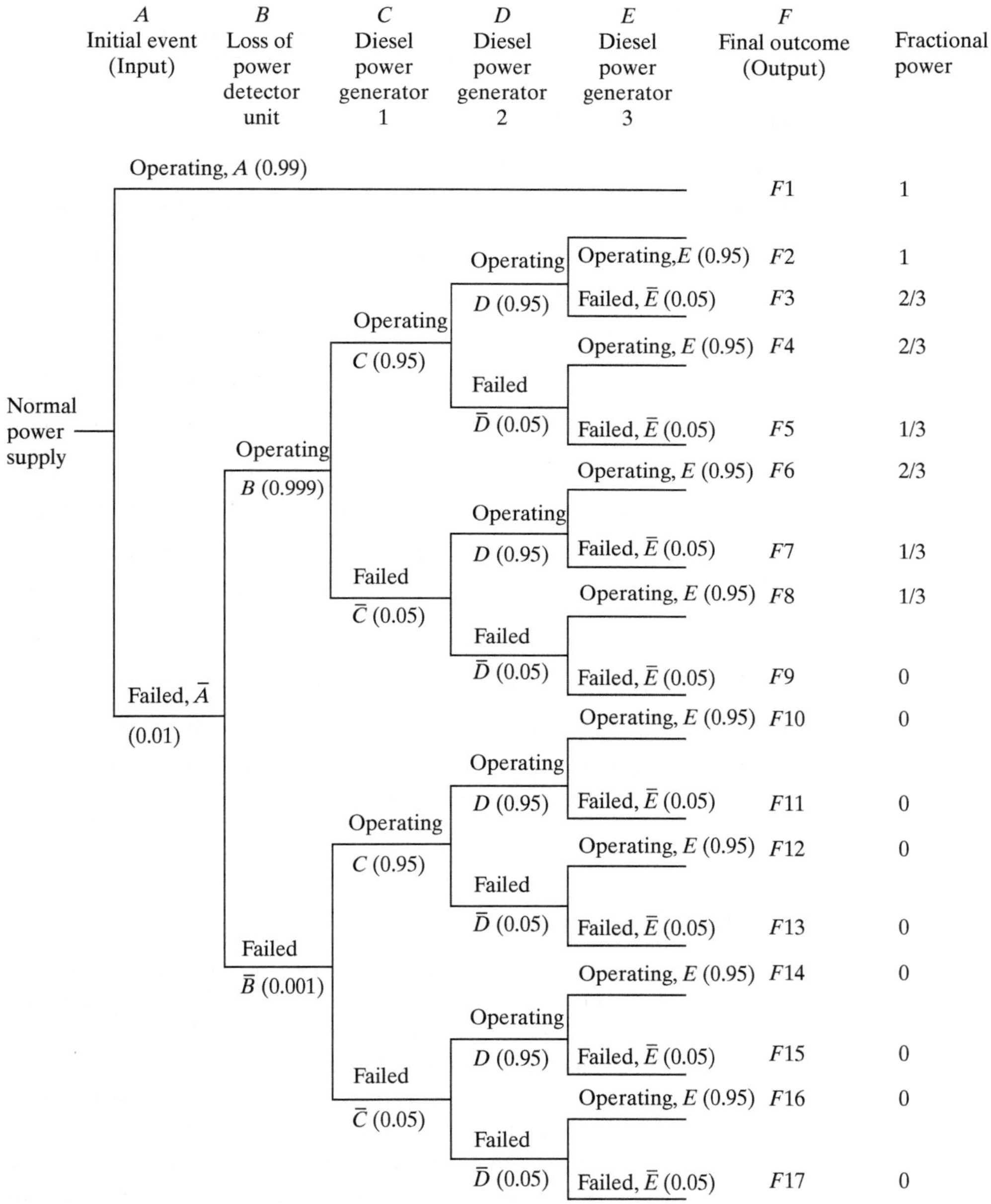

FIGURE 12.7 Event tree of the system in Example 12.2.

TABLE 12.1 Probabilities of realizing different levels of power

$P(F1)$	$= P(A)$		$= 0.99$
$P(F2)$	$= P(\bar{A})P(B)P(C)P(D)P(E)$	$= (0.01)(0.999)(0.95)(0.95)(0.95)$	$= 8.5652 \times 10^{-3}$
$P(F3)$	$= P(\bar{A})P(B)P(C)P(D)P(\bar{E})$	$= (0.01)(0.999)(0.95)(0.95)(0.05)$	$= 4.5080 \times 10^{-4}$
$P(F4)$	$= P(\bar{A})P(B)P(C)P(\bar{D})P(E)$	$= (0.01)(0.999)(0.95)(0.05)(0.95)$	$= 4.5080 \times 10^{-4}$
$P(F5)$	$= P(\bar{A})P(B)P(C)P(\bar{D})P(\bar{E})$	$= (0.01)(0.999)(0.95)(0.05)(0.05)$	$= 2.3726 \times 10^{-5}$
$P(F6)$	$= P(\bar{A})P(B)P(\bar{C})P(D)P(E)$	$= (0.01)(0.999)(0.05)(0.95)(0.95)$	$= 4.5080 \times 10^{-4}$
$P(F7)$	$= P(\bar{A})P(B)P(\bar{C})P(D)P(\bar{E})$	$= (0.01)(0.999)(0.05)(0.95)(0.05)$	$= 2.3726 \times 10^{-5}$
$P(F8)$	$= P(\bar{A})P(B)P(\bar{C})P(\bar{D})P(E)$	$= (0.01)(0.999)(0.05)(0.05)(0.95)$	$= 2.3726 \times 10^{-5}$
$P(F9)$	$= P(\bar{A})P(B)P(\bar{C})P(\bar{D})P(\bar{E})$	$= (0.01)(0.999)(0.05)(0.05)(0.05)$	$= 1.2487 \times 10^{-6}$
$P(F10)$	$= P(\bar{A})P(\bar{B})P(C)P(D)P(E)$	$= (0.01)(0.001)(0.95)(0.95)(0.95)$	$= 8.5737 \times 10^{-6}$
$P(F11)$	$= P(\bar{A})P(\bar{B})P(C)P(D)P(\bar{E})$	$= (0.01)(0.001)(0.95)(0.95)(0.05)$	$= 4.5125 \times 10^{-7}$
$P(F12)$	$= P(\bar{A})P(\bar{B})P(C)P(\bar{D})P(E)$	$= (0.01)(0.001)(0.95)(0.05)(0.95)$	$= 4.5125 \times 10^{-7}$
$P(F13)$	$= P(\bar{A})P(\bar{B})P(C)P(\bar{D})P(\bar{E})$	$= (0.01)(0.001)(0.95)(0.05)(0.05)$	$= 2.3750 \times 10^{-8}$
$P(F14)$	$= P(\bar{A})P(\bar{B})P(\bar{C})P(D)P(E)$	$= (0.01)(0.001)(0.05)(0.95)(0.95)$	$= 4.5125 \times 10^{-7}$
$P(F15)$	$= P(\bar{A})P(\bar{B})P(\bar{C})P(D)P(\bar{E})$	$= (0.01)(0.001)(0.05)(0.95)(0.05)$	$= 2.3750 \times 10^{-8}$
$P(F16)$	$= P(\bar{A})P(\bar{B})P(\bar{C})P(\bar{D})P(E)$	$= (0.01)(0.001)(0.05)(0.05)(0.95)$	$= 2.3750 \times 10^{-8}$
$P(F17)$	$= P(\bar{A})P(\bar{B})P(\bar{C})P(\bar{D})P(\bar{E})$	$= (0.01)(0.001)(0.05)(0.05)(0.05)$	$= 1.2500 \times 10^{-9}$

■

12.5 Fault-Tree Analysis (FTA)

The fault-tree techniques have received a widespread attention in the reliability and safety analysis of complex systems. The techniques were extensively used in aerospace and nuclear industries [12.3, 12.11]. The importance of fault-tree analysis in the nuclear reactor design is discussed at length in the Reactor Safety Study [12.4]. The main reason for the widespread use of fault-tree analysis in nuclear and aerospace industries is due to the concern for human safety. The fault-tree analysis procedure was first introduced at Bell Telephone Laboratories in connection with the safety analysis of the Minuteman missile launch control system in 1962. Subsequently, the method was developed at the Boeing Company in the mid-1960s [12.5]. Since that time, fault trees have been widely used to investigate the reliability and safety of complex and large systems and for diagnostic applications. A comprehensive bibliography on the fault-tree analysis and its applications is presented in Ref. [12.6].

12.5.1 Concept

While the purpose of the failure modes and effects analysis is to identify the various possible failure modes and effects, the fault-tree analysis aims at developing the structure from which simple logical relationships can be used to express the probabilistic relationships among the various events that lead to the failure of the system. In fact, the FTA is usually preceded by a FMEA in which the design, operation, and environment of the system are evaluated and the cause-and-effect relationships leading to the failure of the system are identified. Thus, FMEA becomes an essential step for understanding the behavior of the system without which the fault-tree analysis and the reliability analysis cannot be performed.

A fault tree is a graphical representation of the relationship between certain specific events and the ultimate undesired event. Thus, we must know how the system works before constructing a fault tree. The functioning of a system is usually described by a function or flow diagram in which the flow of information, materials, signals, and other services is indicated. For example, for a steam power plant, the flow diagram shows the flow of all inputs affecting the power plant such as fuel (from receipt to the emission of flue gases from the smokestack) or water (from the intake to the final discharge). The flow diagram is then used to identify the various functional sequences from input to output. Next, a logic diagram is prepared by translating the functional relationships into logical relationships among the various components of the system. Once the logical relationships among the components of the system are established, the cause-and-effect relationships in the operation of the system can be studied, and the fault tree of the system can be developed.

12.5.2 Procedure

The fault-tree analysis starts with the identification of the TOP events, usually the set of most serious system failure events that could occur. Next, the events which contribute directly to the TOP events are identified and connected to the TOP events by logical links. This process is continued until the lowest level or most basic events are reached. The fault tree is constructed using the symbols shown in Fig. 12.8. The two basic symbols involved are AND and OR gates. For an AND gate, the output will be present only if all of the inputs are present. For an OR gate, the output will be present if one or more of the input events are present. Among the other symbols used in the fault-tree analysis, the rectangle is simply used to label an intermediate event in a tree. For instance, if the output from a specific gate is present, a rectangle might be used to describe what the event would physically represent. A rectangle is also used to denote the TOP event from which the construction of the entire fault tree starts. The triangles are used to indicate the transfer of parts of the fault tree to other locations. A circle is used to specify a "primary input event," usually the failure of a specific component, whenever the quantitative behavior of that component is known sufficiently well so that no further analysis of it is required.

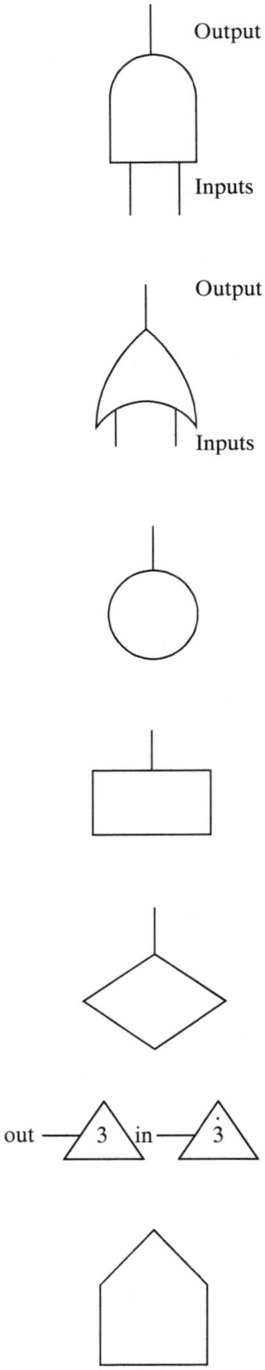

AND gate
The output event is generated only if all the input events are present simultaneously.

OR gate
The output event is generated if any one or more of the input events are present.

Basic failure
A basic fault or event caused by a component or sub-assembly for which a probability can be assigned (from known empirical data).

Intermediate event
A fault or event caused by a combination of other events via a logic gate.

Undeveloped event
A fault that is not subdivided into basic events due to lack of information or importance. The event must be expanded or developed later.

Transfer events
An entire part of the tree is transferred to other locations of the tree.

Basic event
A basic event that is a normal occurrence while the system is operating.

FIGURE 12.8 Important symbols used in the construction of fault trees.

The following points aid in the construction of fault trees [12.11]:

1. The first step in the fault tree analysis is to define a suitable TOP event that constitutes a serious system failure.
2. Usually several different, but equivalent, fault trees can be constructed for a given system. Also, different TOP events lead to different fault trees.
3. For any specified TOP event, each possible event is examined to see whether it can, either alone or in conjunction with some other event(s), cause the TOP event.
4. The primary events that lead to the TOP event and the secondary events that cause each of the primary events are determined. The procedure is continued until all the basic failures are identified.
5. The set of events that are all required to produce an event of interest are connected to AND gates.
6. The set of events that can individually produce an event of interest are connected to OR gates.

EXAMPLE 12.4

Construct a fault tree for the series–parallel system shown in Fig. 12.9.

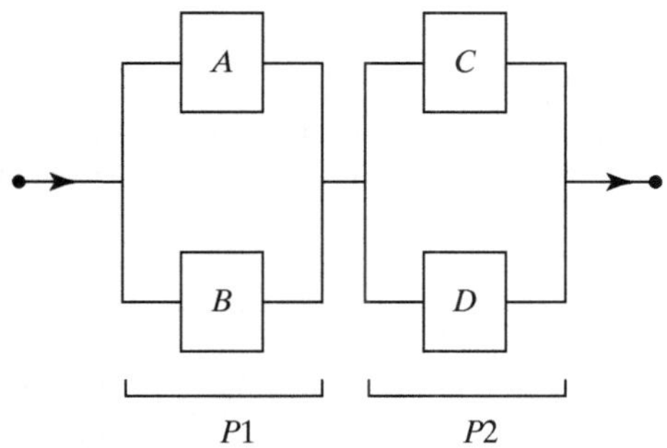

FIGURE 12.9 A series–parallel system.

Solution: The TOP (or the most undesirable) event is the failure of the system. When the system fails, either one or both of the parallel units $P1$ and $P2$ must have failed. Thus the units $P1$ and $P2$ are connected to the TOP event through an OR gate. The unit $P1$ ($P2$) will fail only when both the components A and B (C and D) fail. Hence, the components A and B (C and D) are connected to an AND gate as shown in Fig. 12.10. Note that a bar over a letter is used to indicate the failure of the corresponding component. The complete fault tree is also shown in Fig. 12.10.

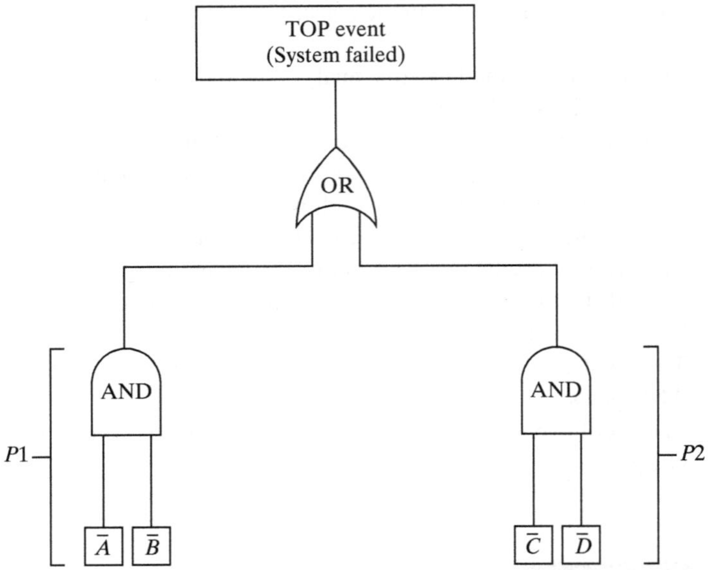

FIGURE 12.10 Fault tree of the series-parallel system shown in Fig. 12.9.

■

EXAMPLE 12.5

Consider the system shown in Fig. 12.11, which consists of three valves, a pump, a pipeline, and a tank to collect water pumped from the pond. Construct a fault tree corresponding to the TOP event: "no flow of water into the tank" [12.11].

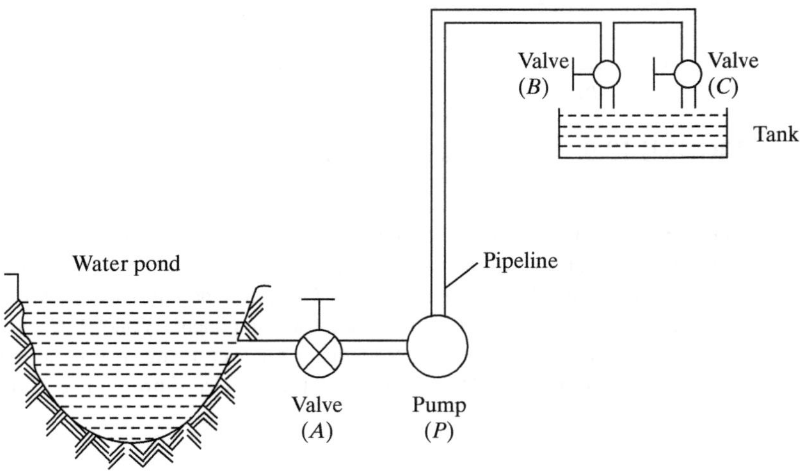

FIGURE 12.11 A water pumping system.

Solution: For simplicity, we will ignore the possibility of loss of electric power (to run the pump), lack of water in the pond, or failure of the pipeline. To construct the fault tree, we first note that there are two primary events, namely, "both valves *B* and *C* closed (failed)" and "no flow of water to valves *B* and *C*," either one of which could cause the TOP event. Thus, an OR gate is introduced with the two stated primary events as inputs and the TOP event as the output (Fig. 12.12). Since the primary event, "no flow of water to valves *B* and *C*," in Fig. 12.12 can be caused by the occurrence of any of the secondary events "valve *A* closed (failed)" and "pump *P* failed," an OR gate is used to connect these primary and secondary events as shown in Fig. 12.13. In addition, the event "pump *P* failed" can be caused by either "pump not started (human failure)" or "pump broken" and hence an OR gate is used to connect them as indicated in Fig. 12.13. By assuming the events "valve *A* closed" and "pump *P* broken" as basic failures with known failure rates, these events are denoted by

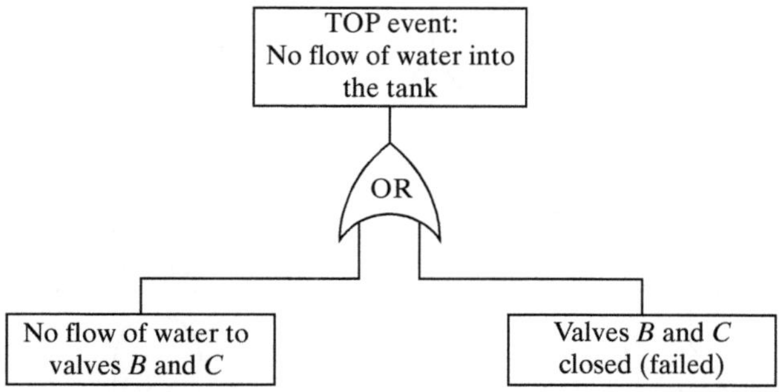

FIGURE 12.12 Connection of primary events and the TOP event through an OR gate.

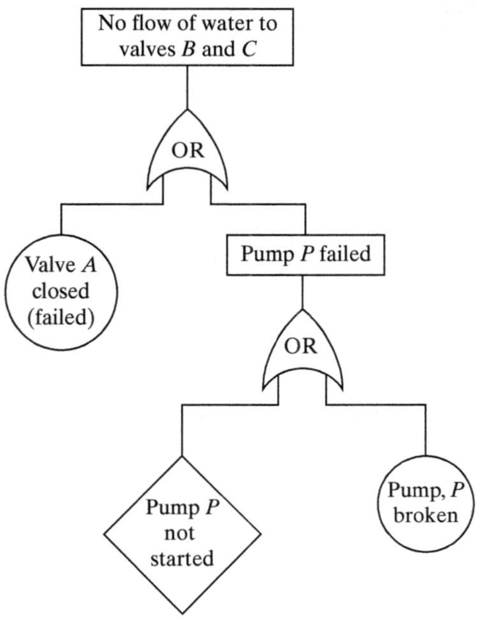

FIGURE 12.13 Use of OR gate at different levels.

circles. The event "pump not started" is assumed to be an undeveloped event (basic events causing this event are unknown), hence is represented by a diamond-shaped block, and this branch of the fault tree is terminated. Next, the branch from the primary event, "valves B and C closed (failed)," of Fig. 12.12 is developed. This primary event can be caused by the simultaneous occurrence of the secondary events, "valve B closed (failed)" and "valve C closed (failed)," and thus an AND gate is to be introduced to connect the primary and secondary events. Further, if the events "valve B closed" and "valve C closed" are assumed to be basic failures with known failure rates, this branch of the fault tree can be terminated by using circles to represent the basic failures. The complete fault tree, with each branch ending with basic or undeveloped failures, is shown in Fig. 12.14.

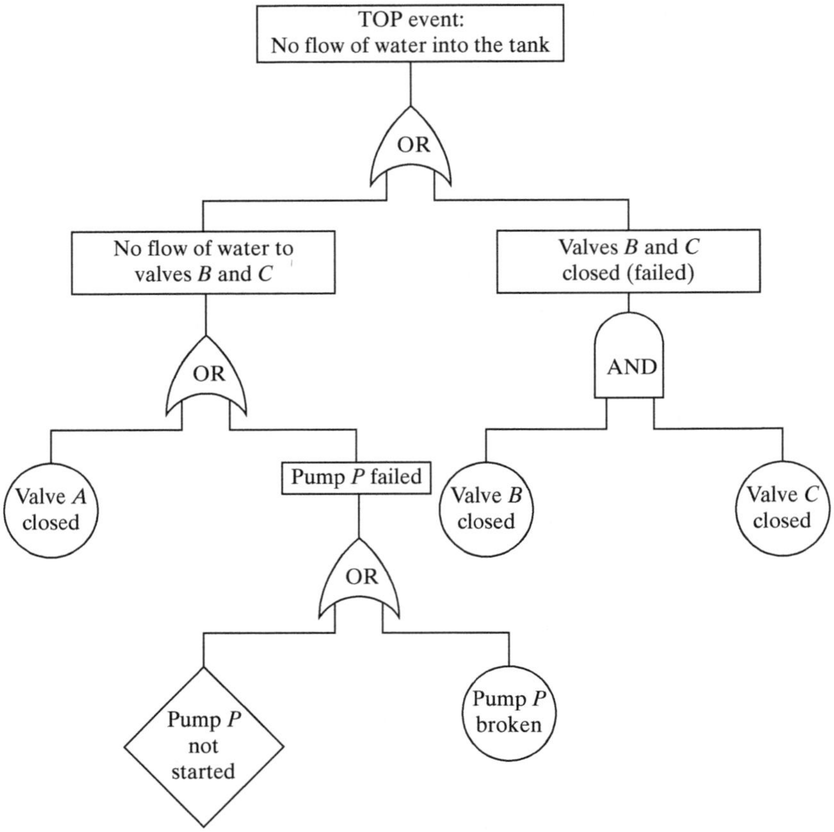

FIGURE 12.14 Fault tree of the water pumping system.

■

EXAMPLE 12.6

Consider a room containing two lightbulbs operated by a single switch (see Fig. 12.3). By assuming the "room without electric light" as the undesirable (TOP) event, construct a fault tree of the system.

Solution: The TOP event can occur due to the occurrence of any of the three input events, namely, "bulbs burnt out," "switch failed to close," and "no electricity." This solution can be represented by a three-input OR gate. The occurrence of the events "bulbs burnt out" and "no electricity" can be investigated further. The event "bulbs burnt out" occurs when both the bulbs are burnt out and so the situation can be represented by a two-input AND gate. The event "no electricity" occurs when electric power is cut off or the fuse is blown, and the situation can be represented by a two-input OR gate. The resulting fault tree is shown in Fig. 12.15.

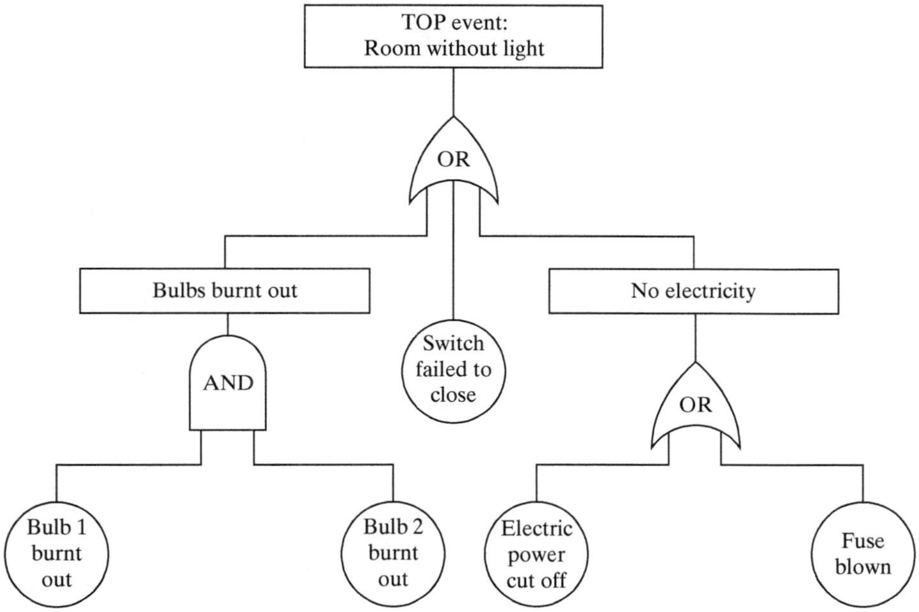

FIGURE 12.15 Construction of a fault tree.

■

EXAMPLE 12.7

A clutch is a mechanical device that provides for the transmission of power (and motion) from one component, called the driving member, to another, called the driven member, when they are engaged. By disengaging the clutch, the two members can be separated thereby cutting off the transmission of power to the driven member. For example, in an automobile, the power from the engine can be transmitted (or cut off) to the wheels by engaging (or disengaging) the clutch. The cone clutch, shown in Fig. 12.16, can be engaged by making the cone (driving member) come into contact with the cup (driven member). The friction at the contacting conical surface causes the two members rotate at the same speed while transmitting the power. Construct a fault tree diagram corresponding to the TOP event, "failure of the clutch to disengage (to cut off transmission of power to the wheels)."

Solution: The major components of the system are indicated in Fig. 12.16. The following failures can be associated with the various components.

1. Shifting mechanism (intended to move the cone away from the cup): Failed
2. Cup and cone: Got locked

3. Friction material: Excessive wear
4. Compression spring: Coil broken
5. Shaft 1: Broken; bearings (not shown in Fig. 12.16) failed

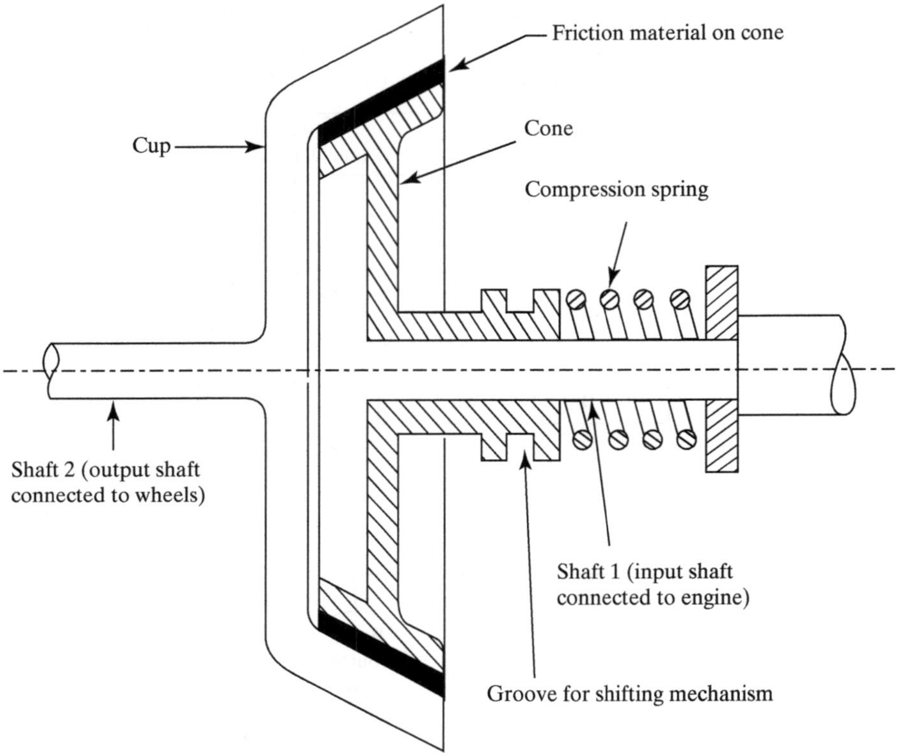

FIGURE 12.16 A cone clutch.

The failure to cut off power to the wheels when required can cause an accident in the case of an automobile and hence is considered as the TOP event. By considering the various modes of failure of the components, the fault tree can be developed as shown in Fig. 12.17.

■

12.6 Minimal Cut-Sets

As defined in Section 6.13.3, a cut-set is a group of primary events whose occurrence causes a system to fail, that is, the occurrence of the TOP event. A minimal cut-set is a set of minimum number of primary events that produces the TOP event if and only if all the events of the set occur. This also implies that there are no extraneous events of the set, whose failures do not contribute to the TOP event. Enumerating the minimal cut-sets is an essential step in evaluating the system reliability. Interpretation of the minimal cut-sets gives a number of qualitative results such as the weak points of the system, false redundancies, or the effect of a given component on the overall system reliability. Different methods

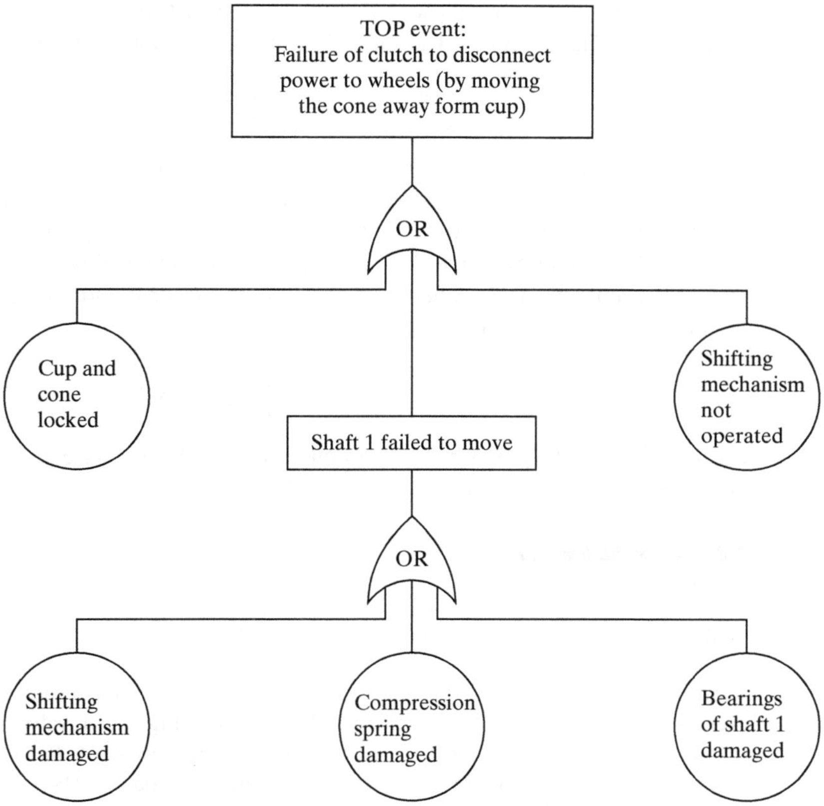

FIGURE 12.17 Fault tree of the clutch.

can be used for determining the minimal cut-sets [12.7]. In one of the methods, known as the inductive method, the minimal cuts are found directly by combining significant failures of the system components. In another procedure, known as the fault tree approach, the minimal cut-set is derived directly from the fault tree. A third alternative is to proceed automatically from the reliability block diagram (or successful paths).

In the fault-tree approach, the minimal cut-sets are determined by converting the fault tree to a boolean expression [12.11]. For this, a boolean variable is assigned to each basic event. The output event of an AND gate is assigned a boolean variable equal to the minimum of the boolean variables for the input events. The output event of an OR gate is assigned a boolean variable equal to the maximum of the input event variables. Finally, a boolean expression is obtained for the undesirable final event as a function of the boolean variables associated with each basic event. It will be convenient to define a "boolean variable" for a typical event A, denoted as X_A, such that $X_A = 1$ if the event A has occurred (true) and $X_A = 0$ if the event A has not occurred (false).

Consider an AND gate with n input events and one output event A. If the n events $1, 2, \ldots, n$ have the boolean variables $X_1, X_2, \ldots, X_n$, the output event A is defined as the product of these n events such that the event A will occur if and only if all the n input events occur. That is, $X_A = 1$ if and only if all $X_i = 1$, and hence we can write

$$X_A = \prod_{i=1}^{n} X_i \tag{12.2}$$

Since X_A is expressed as the product of zeros and ones, X_A will be either zero or one. Alternately, X_A can be defined as the minimum value of the X_is. Next, consider an OR gate with n input events and one output event B. Here the output event B is defined as the union of the n input events 1, 2, . . . , n such that the event B will occur if one or more of the input events occur. In other words, B will not occur if and only if all the input events do not occur. Thus we can write.

$$X_B = 1 - \prod_{i=1}^{n} (1 - X_i) \tag{12.3}$$

It can be seen that X_B will be equal to one if any of the X_is are equal to one. Thus, X_B can also be expressed as the maximum value of the X_is.

EXAMPLE 12.8

Find the minimal cut-sets for the fault tree shown in Fig. 12.14.

Solution: It can be seen that if all indicator variables of a cut-set are true $(X_i = 1)$, then the TOP event is true. It can be seen that for the fault tree shown in Fig. 12.14, valve A by itself forms a cut-set and valves B and C together form another cut-set. Hence, valves A, B, and C also form a cut-set. Several other combinations are also possible. Since no extraneous members of the set, whose failures do not contribute to the TOP event, can be present in a minimal cut-set, "failure of valve A plus pump P broken," for example, is not a minimal cut-set. The reason is that even if the pump is not broken, the failure of valve A by itself will cause the TOP event. On the other hand, valves B and C together form a minimal cut-set, since the simultaneous failure of both the components is needed to produce the TOP event. Thus, we find that the minimal cut-sets of Fig. 12.14 are (valve A), (pump P not started), (pump P broken), (valve B, valve C). These can even be identified from the way the tree is constructed. However, as the size of the fault tree increases, the number of minimal cut-sets increases at a rapid rate, and we may not be able to identify the minimal cut-sets by inspection.

■

12.6.1 Probability of the TOP Event

Once the minimal cut-sets are found, we can use them for a quantitative analysis of a fault tree. By definition, any minimal cut set (MCS) can lead to the TOP event. Thus, we can reconstruct the tree by connecting the TOP event to an OR gate, with each of the minimal cut-sets represented as inputs to the OR gate (see Fig. 12.18). Also, the occurrence of a minimal cut-set implies the failure of all the components (basic failures) of the set. Thus, the individual components of each minimal cut-set are to be connected through an AND gate. Hence, in terms of minimal cut-sets, any fault tree can be represented as shown in Fig. 12.19 [12.11]. For a typical minimal cut-set (labeled as MCS_i) consisting of s components, we can write

$$X_{\text{MCS}_i} = \prod_{j=1}^{s} X_j^{(i)} \tag{12.4}$$

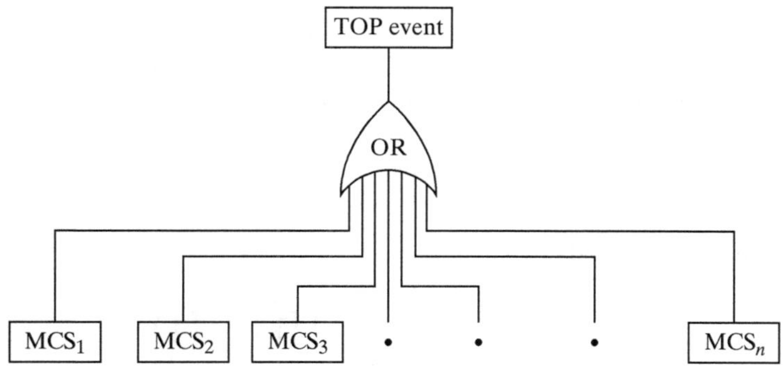

FIGURE 12.18 Connection of minimal cut sets to the TOP event.

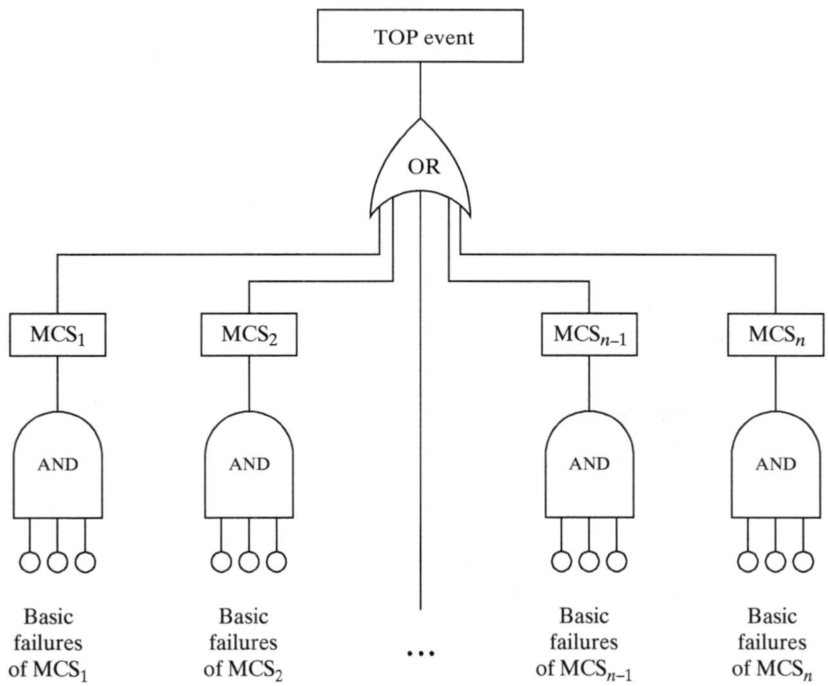

FIGURE 12.19 Fault tree representation using minimal cut sets.

In Eq. (12.4), $X_j^{(i)}$ is the boolean variable for the jth component (or basic failure) of the ith minimal cut-set and all the components of MCS_i are assumed to be independent. The probability of the occurrence of MCS_i is given by

$$P_{\text{MCS}_i} = \prod_{j=1}^{s} P_j^{(i)} \tag{12.5}$$

where $P_j^{(i)}$ is the probability of failure of the jth component in ith minimal cut-set (also equal to the expected value of $X_j^{(i)}$). Thus, the boolean variable corresponding to the TOP event can be expressed, using OR for minimal cut-sets, as

$$X_{\text{TOP}} = 1 - \prod_{i=1}^{n} (1 - X_{\text{MCS}_i}) \tag{12.6}$$

Equations (12.6) and (12.4) yield

$$X_{\text{TOP}} = 1 - \prod_{i=1}^{n} \left(1 - \prod_{j=1}^{s} X_j^{(i)}\right) \tag{12.7}$$

where the component j of $\text{MCS}_i(X_j^{(i)})$ is assumed to be the same as the component l of $\text{MCS}_k(X_l^{(k)})$. It is to be noted that even if the individual components are independent, the minimal cut-sets may not be independent always. Particularly, when several cut-sets share the same components, there will be interdependence between the minimal cut-sets. This dependence can be accounted by using the following boolean reduction relationship

$$X_j^{(i)} X_l^{(k)} = X_j^{(i)} \tag{12.8}$$

This implies that in evaluating the double products such as

$$\prod_{i=1}^{n} \prod_{j=1}^{s} X_j^{(i)}$$

we need to multiply all component probabilities only once without considering any duplicate entry.

EXAMPLE 12.9

Find the probability of occurrence of the TOP event for the system represented by Fig. 12.14.

Solution: For Fig. 12.14, the minimal cut-sets were found in Example 12.8 to be (valve A closed), (pump P broken), (pump P not started), and (valves B and C closed). Thus we find that all the cut-sets are independent; and hence we can write, without reductions,

$$X_{\text{TOP}} = 1 - (1 - X_A)(1 - X_{P1})(1 - X_{P2})(1 - X_B X_C)$$

and

$$P_{\text{TOP}} = 1 - (1 - P_A)(1 - P_{P1})(1 - P_{P2})(1 - P_B P_C)$$

where P_A, P_B, and P_C denote the probability of failure of valves A, B, and C, respectively, P_{P1} represents the probability of occurrence of the event "pump broken," and P_{P2} indicates the probability of occurrence of the event "pump not started."

■

The fault-tree analysis and the method of finding the system reliability in mechanical design is illustrated through Example 12.10.

EXAMPLE 12.10

The gear train shown in Fig. 12.20 consists of a cluster gear set mounted on the input shaft [12.8]. The intermediate shaft carries three fixed gears and a cluster gear set. The output shaft carries three fixed gears. Each shaft is mounted on two bearings. Depending on the gear pair in contact, the intermediate shaft will have three different speeds. For each intermediate shaft speed, three speeds can be realized at the output shaft. Thus the system has three shafts, two clusters, six gears, and six bearings and delivers nine output speeds for any particular input speed.

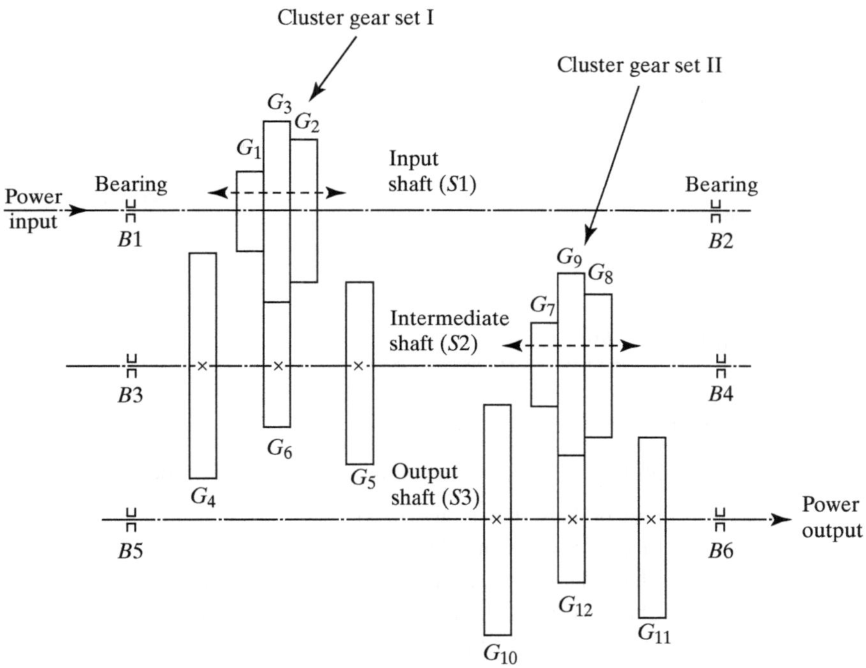

FIGURE 12.20 Layout of the gear train (Gi = ith gear, Bi = ith bearing, Si = ith shaft). (Reprinted with permission from S. S. Rao, "A Probabilistic Approach to the Design of Gear Trains," *International Journal of Machine Tool Design and Research,* Vol. 14, copyright 1974, Pergamon Press PLC, pp. 267–278.)

1. Develop a fault tree of the system corresponding to the TOP event: "no output speed from the gear train."
2. Find the failure rate of the TOP event corresponding to the failure rates of the individual components given in Table 12.2 [the notation used in Table 12.2 is introduced in the solution of part (a)].
3. Find the reliability of the gear train when the component reliabilities are given as follows:

$$R_1 = \text{reliability of each bearing} = 0.9999$$
$$R_2 = \text{reliability of each gear in bending failure mode} = 0.9998$$
$$R_3 = \text{reliability of each gear in surface-wear failure mode} = 0.9998$$
$$R_4 = \text{reliability of each cluster gear set} = 0.9999$$

$$R_5 = \text{reliability of each shaft} = 0.99999$$
$$R_{PE1} = \text{reliability of fuse} = 0.999$$
$$R_{PE2} = \text{reliability of electric switch} = 0.9999$$

TABLE 12.2 Failure rates of the components

Component (Event)	Failure Rate in $hr^{-1}(\lambda)$
*PE*1	10^{-3}
*PE*2	2×10^{-4}
*PE*3	10^{-5}
*PE*4	10^{-5}
*PE*5	2×10^{-6}
*PE*6	2×10^{-6}
*PE*7	2×10^{-6}
*PE*10	10^{-4}
*PE*11	10^{-5}
*PE*12 (*B*1 to *B*6)	10^{-5}

Solution:

1. Fault-tree construction: The basic events can be identified [12.9, 12.10] as follows:

   1. Fuse of the system blows (*PE*1)
   2. Electric power switch fails (*PE*2)
   3. Cluster I sticks to zero position, that is, no output (*PE*3)
   4. Cluster II sticks to zero position, that is, no output (*PE*4)
   5. Shaft I breaks (*PE*5)
   6. Shaft II breaks (*PE*6)
   7. Shaft III breaks (*PE*7)
   8. Gear group I fails, that is, *G*1 or *G*4, *G*2 or *G*5, *G*3 or *G*6 fails (*PE*8)
   9. Gear group II fails, that is, *G*7 or *G*10, *G*8 or *G*11, *G*9 or *G*12 fails (*PE*9)

   The event *PE*8 can be induced by wear-out failure (*PE*10); bending or overload failure (*PE*11), or bearing failure, that is, breakage or misalignment of *B*1, *B*2, *B*3, or *B*4 (*PE*12). Similarly the event *PE*9 can be induced by wear-out failure; bending or overload failure; or bearing failure; that is, breakage or misalignment of *B*3, *B*4, *B*5, or *B*6. The connectivity of the components of the gearbox is shown in Fig. 12.21 and the corresponding fault tree is shown in Fig. 12.22.

2. Failure-rate analysis: It is assumed that component failure times are exponentially distributed with constant failure rates and the components (basic events) are not repaired. For the failure rates given in Table 12.2, the failure rates of intermediate events can be computed as follows:

$$\lambda_a = \text{bearing failure} = \text{OR}\,(B1, B2, B3, B4) = 4 \times 10^{-5}$$

$$\lambda_b = \text{failure rate of } G1 \text{ to } G12 = \text{OR}\,(\lambda_a, \lambda_{10}, \lambda_{11}) = 4 \times 10^{-5} + 10^{-4} + 10^{-5} = 1.5 \times 10^{-4}$$

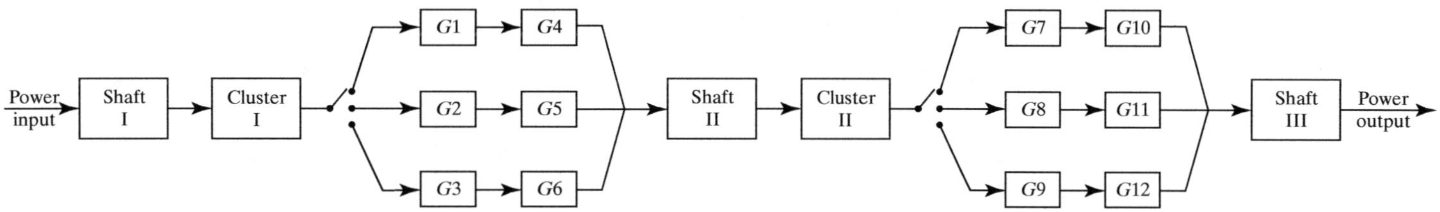

FIGURE 12.21 Connectivity of components of the gear train. (With permission from T. S. Pan and S. S. Rao, "Fault Tree Approach for the Reliability Analysis of Gear Trains," *Journal of Mechanisms, Transmissions, and Automation in Design,* ASME, Vol. 110, 1988.)

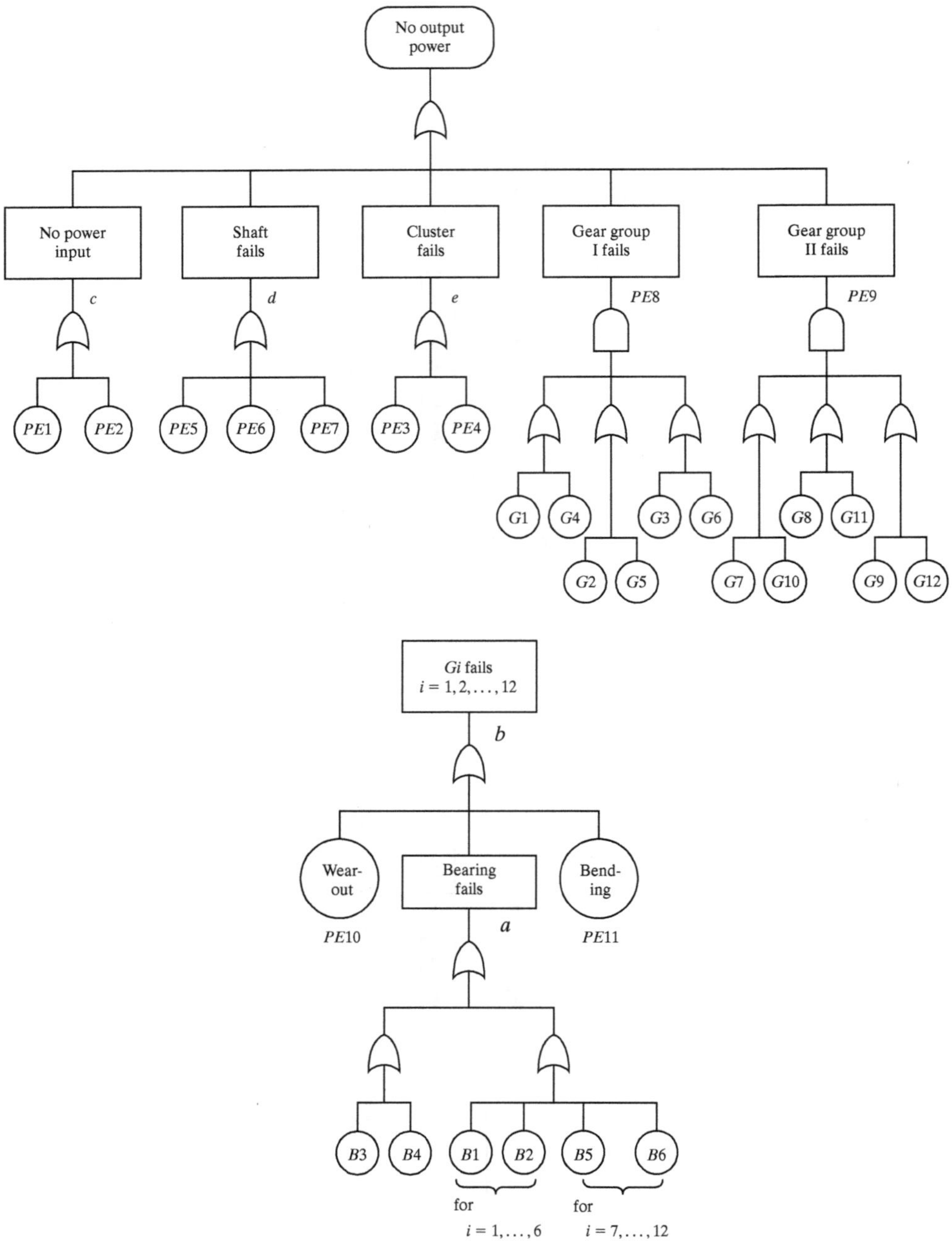

FIGURE 12.22 Fault tree of the gear train. (With permission from T. S. Pan and S. S. Rao, "Fault Tree Approach for the Reliability Analysis of Gear Trains," *Journal of Mechanisms, Transmissions, and Automation in Design,* ASME, Vol. 110, 1988.)

$$\lambda\,(\text{any two pair of gears}) = \lambda_{\text{pair}} = \text{OR}\,(\lambda_{b1}, \lambda_{b2}) = 3 \times 10^{-4}$$

$$\lambda_8 = \lambda_9 = \text{AND}\,(\lambda_{\text{pair}}, \lambda_{\text{pair}}, \lambda_{\text{pair}}) = \frac{3 \cdot \lambda_{\text{pair}} \cdot (Z - 1)}{Z^3 - 1} = 2.848 \times 10^{-4}$$

where

$$Z = \{1 - \exp(-\lambda_{\text{pair}} \cdot t)\}^{-1} = 1.0524(t = 10{,}000\text{ hr})$$

$$\lambda_c = \lambda_1 + \lambda_2 = 1.2 \times 10^{-3}$$

$$\lambda_d = \lambda_5 + \lambda_6 + \lambda_7 = 6 \times 10^{-6}$$

$$\lambda_e = \lambda_3 + \lambda_4 = 2 \times 10^{-5}$$

$$\lambda_{\text{TOP}} = \text{OR}\,(\lambda_c, \lambda_d, \lambda_e, \lambda_8, \lambda_9) = \sum \lambda_i = 1.7956 \times 10^{-3}\text{ hr}^{-1}$$

Thus, the total system failure rate is $1.7956 \times 10^{-3}/\text{hr}$. This implies that on the average there is one failure per 588.6 hr. It is to be noted that the major failure has been contributed by the "fuse of the system blows."

3. Reliability analysis: The AND gate in a fault tree represents a weakest link or series system and the OR gate denotes a fail-safe or parallel system. If the reliabilities of the components of the gear train are known, then the total system (gear train) reliability can be determined. From the fault-tree analysis and the expressions for the reliabilities of weakest-link and fail-safe systems, the overall reliability of the gear train can be obtained as

$$R_{\text{gear train}} = R_{PE8}R_{PE9}R_eR_dR_c = \{1 - (1 - R_b^2)^3\}^2 R_4^2 R_5^3 R_{PE1} R_{PE2}$$
$$= \{1 - (1 - R_2^2 R_3^2 R_1^3)^3\}^2 R_4^2 R_5^3 R_{PE1} R_{PE2}$$

Thus, the failure probability of the gear train is given by $P_{f\text{system}} = 1 - R_{\text{system}}$. The reliabilities of the intermediate events can be computed as

$$R_a = R_1^4,\ R_b = R_aR_2R_3 = R_1^4R_2R_3, \quad R_{PE9} = R_{PE8} = 1 - (1 - R_b^2)^3,$$
$$R_e = R_4^2, \quad R_d = R_5^3 \quad \text{and} \quad R_c = R_{PE1} \cdot R_{PE2}$$

For the given component reliabilities,

$$R_1 = R_4 = 0.9999, \quad R_2 = R_3 = 0.9998, \quad R_5 = 0.99999, \quad R_{PE1} = 0.999, \quad R_{PE2} = 0.9999$$

the reliability of the gear train can be found as

$$R_{\text{system}} = R_{\text{gear train}} = 0.99867$$

and hence the probability of failure is given by

$$P_{f\text{system}} = 0.00133$$

■

REFERENCES AND BIBLIOGRAPHY

12.1 H. M. N. Raafat, "The Quantification of Risk in System Design," *Journal of Engineering for Industry,* Vol. 105, 1983, pp. 223–233.

12.2 S. Levine and W. E. Vesely, "Important Event-Tree and Fault-Tree Considerations in the Reactor Safety Study," *IEEE Transactions on Reliability,* R-25, 1976, pp. 132–139.

12.3 H. R. Roberts, W. E. Vesely, D. F. Haast, and F. F. Goldberg, "Fault Tree Handbook," U.S. Nuclear Regulatory Commission, NURE G-0492, 1981.

12.4 "Reactor Safety Study: An Assessment of Accident Risks in U.S. Commercial Nuclear Power Plants," U.S. Atomic Energy Commission, DRAFT WASH-1400, Washington, D.C., August 1974.

12.5 R. E. Barlow and H. E. Lambert, "Introduction to Fault-Tree Analysis," in *Reliability and Fault-Tree Analysis,* SIAM, Philadelphia PA, 1975, pp. 7–37.

12.6 B. S. Dhillon and C. Singh, "Bibliography of Literature on Fault Trees," *Microelectronics and Reliability,* Vol. 17, 1978, pp. 501–503.

12.7 J. B. Fussell, and W. E. Vesely, "A New Methodology for Obtaining Cut Sets for Fault Trees," *Transactions of American Nuclear Society,* Vol. 15, 1972, pp. 262–263.

12.8 S. S. Rao, "A Probabilistic Approach to the Design of Gear Trains," *International Journal of Machine Tool Design and Research,* Vol. 14, 1974, pp. 267–278.

12.9 T. S. Pan and S. S. Rao, "Fault Tree Approach for the Reliability Analysis of Gear Trains: Part I—Basic Theory, Part II—Example Applications," In S. S. Rao (ed.), *Advances in Design Automation—* 1987, Publication no. DE-Vol-10-1, ASME, New York, 1987.

12.10 T. S. Pan and S. S. Rao, "Fault Tree Approach for the Reliability Analysis of Gear Trains," *ASME Journal of Mechanisms, Transmissions and Automation in Design,* Vol. 110, 1988, pp. 348–353.

12.11 S. L. Salem, G. E. Apostolakis, and D. Okrent, "A Computer-Oriented Approach to Fault-Tree Construction," Report no. UCLA–ENG-7635 and NSF/RA-760320, UCLA, Los Angeles, CA, April 1976.

12.12 T. W. Yellman, "Event-Sequence Analysis," *Proceedings of the Annual Reliability and Maintainability Symposium,* Washington, D.C., January 28–30, 1975, IEEE, New York, 1975, pp. 286–291.

12.13 R. J. Schroder, "Fault Trees for Reliability Analysis," *Proceedings of the Annual Symposium on Reliability,* Los Angeles, CA, February 3–5, IEEE, New York, 1970, pp. 198–205.

12.14 W. F. Stoecker, *Design of Thermal Systems,* 3rd ed., McGraw-Hill, New York, 1989.

12.15 J. E. Shigley and C. R. Mischke, "Mechanical Engineering Design," 5th ed., McGraw-Hill, New York, 1989.

12.16 P. L. Crown, "Design Effective Failure Mode and Effect Analysis," *Proceedings of the Annual Symposium on Reliability,* Chicago, IL, January 21–23, 1969, IEEE, New York 1969, pp. 514–521.

12.17 H. M. N. Raafat, "Reliability in Design and Failure to Safety," *Engineering,* Vol. 219, 1979, pp. 1425–1429.

12.18 MIL-STD-1629, "Procedures for Performing a Failure Mode, Effects and Criticality Analysis," version 1629A, November 24, 1980.

12.19 D. Tarwater (ed.), *The Bicentennial Tribute to American Mathematics 1776–1976,* The Mathematical Association of America, Washington D.C., 1977.

12.20 D. Abbott (ed.), *The Biographical Dictionary of Scientists: Mathematicians,* Peter Bedrick Books, New York, 1986.

12.21 J. von Neumann, *The Computer and the Brain,* Yale University Press, New Haven, CT, 1958.

REVIEW QUESTIONS

12.1 Give brief answers to the following:

1. What is the purpose of FMEA?
2. What is the purpose of a design and safety review technique?
3. What is the difference between the inductive and deductive methods of system-safety analysis?
4. What is a fault tree?
5. What is the difference between an event tree and a fault tree?
6. What is meant by "TOP event"?
7. Explain the significance of AND and OR gates in a fault tree.
8. What meaning is associated with a triangle and a rectangle in a fault tree?
9. How is the minimal cut-set theory used in the fault-tree analysis?
10. How was the fault-tree analysis developed originally?

12.2 Indicate whether each of the following statements is true or false:

1. Failure of a complex system is basically due to the failure of its components.
2. Design reviews can reveal deficiencies of design from the point of view of system safety.
3. In inductive methods, the analysis begins at the component level.
4. In deductive methods, the analysis starts at the component level.
5. The fault-tree analysis is a logical procedure based on a form of binary logic.
6. The logic used in the fault-tree and event-tree analyses is the same.
7. The FMECA procedure is implemented using a standard form.

12.3 Fill in the blank with the proper word(s):

1. The failure-mode, event-tree, and the fault-tree analyses are collectively called ______ techniques.
2. Human errors can lead to a catastrophic ______ of the system.

3. The event-tree analysis does not consider partially ______ components.
4. The failure modes and effects analysis is ______ ______ used at the initial stages of system development.
5. FMECA is ______ for many government-funded projects.
6. An event tree is a ______ representation of all possible events in a system.
7. Event-tree analysis was originally developed to assess the safety of ______ ______.
8. Fault-tree analysis was first developed at ______ ______ laboratories.
9. A cut-set is a group of primary events whose occurrence causes the system to ______.

12.4 Select the most appropriate answer from the multiple choices given:

1. System-safety analysis includes:
   a. Qualitative methods
   b. Quantitative methods
   c. Both qualitative and quantitative methods
2. The failure modes and effects analysis is:
   a. Objective
   b. Subjective
   c. Mathematical
3. The fault-tree analysis is extensively used in:
   a. Automobile industry
   b. Machine tool industry
   c. Aerospace industry
4. The following indicates the minimum number of primary events that produces the TOP event if and only if all the events of the set occur:
   a. Cut-set
   b. Minimal cut-set
   c. Primary cut-sets
5. TOP event of a system denotes:
   a. Serious failure
   b. Any failure
   c. Rare failure

12.5 Match the following system–safety analysis procedures:

1. Reliability analysis	**a.** Deductive and semiquantitative
2. Failure modes and effects analysis	**b.** Inductive and quantitative
3. Fault-tree analysis	**c.** Inductive and semiquantitative
4. Event-tree analysis	**d.** Deductive and quantitative

12.6 Match the following symbols used in fault-tree analysis:

1.

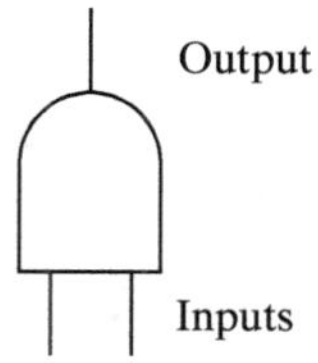

a. Intermediate event

2.

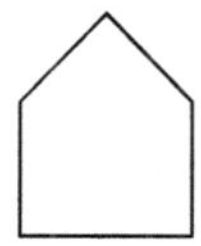

b. Basic failure

3.

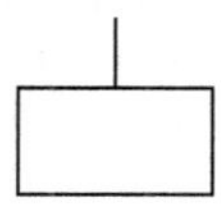

c. AND gate

4.

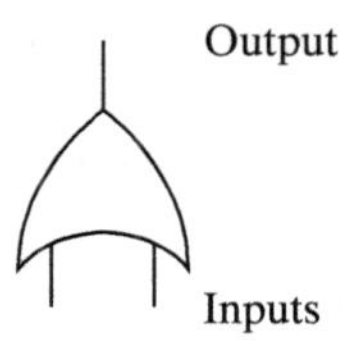

d. Basic event

5.

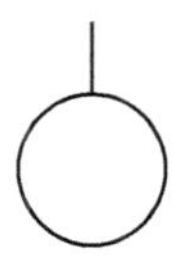

e. OR gate

PROBLEMS

12.1 Construct a fault tree for the flashlight shown in Fig. 12.23 for the TOP event, “no light from the flashlight.” Consider the following basic events: (i) end cap failure (loose or falls off), (ii) retaining spring failure (breaks or falls off from the cap), (iii) battery failure (one or more batteries discharge or die), (iv) switch failure (not moved to “on” position, sticks, or loose contact), (v) bulb failure (burnt or loose contact at base), (vi) reflector failure (falls off), and (vii) lens failure (covered by another object).

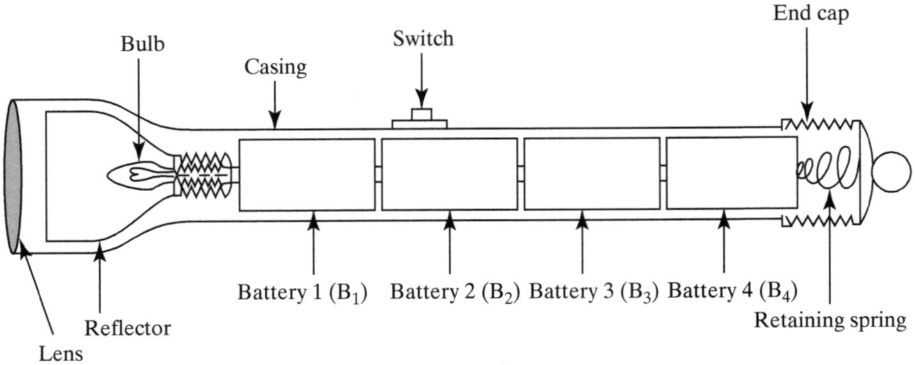

FIGURE 12.23 Components of a flashlight.

12.2 Consider a subsystem of a thermal power plant as shown in Fig. 12.24. It consists of a feed water pump, two boilers, and two steam turbines. The operator is required to turn on the boiler B_1 to start the turbine T_1. If T_1 fails to start, the operator is expected to turn off the boiler B_1 and turn on the boiler B_2 to start the turbine T_2. Construct the event tree for the system by considering the following possibilities: (i) the operator may initially forget to turn on the boiler B_1, (ii) If T_1 does not start, the operator may fail to turn off B_1 or fail to turn on B_2, (iii) when the operator turns on, the boilers B_1 and B_2 may fail to start (damaged), (iv) when steam is received from the respective boilers, the turbines T_1 and T_2 may fail to start (damaged), (v) if both turbines are made to start accidentally, the feed water pump may fail (while trying to supply adequate quantity of water to both the boilers simultaneously).

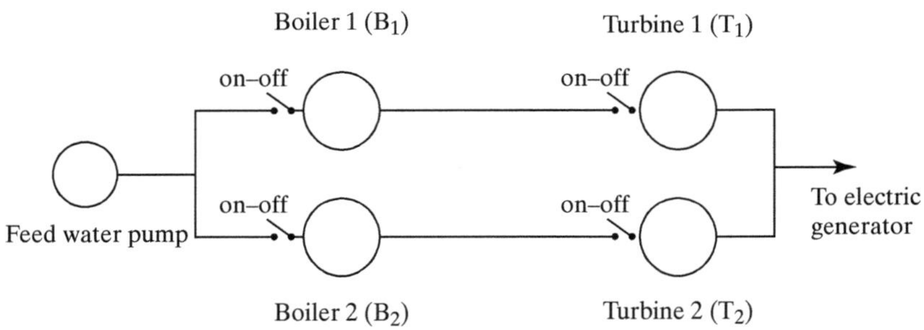

FIGURE 12.24 Subsystem of a thermal power plant.

12.3 Construct a fault tree for the system described in Problem 12.2 assuming the TOP event as "no input to electric generator."

12.4 Figure 12.25 shows an air-driven emergency hydraulic pump that is connected to a manifold through two check valves. The pump is normally in a standby condition and is used to test it before each mission. The check valves are intended to prevent the pump, when de-energized, from being back-driven by pressure in the manifold. Back-driving of the pump causes a hazardous situation. Construct an event tree of the system.

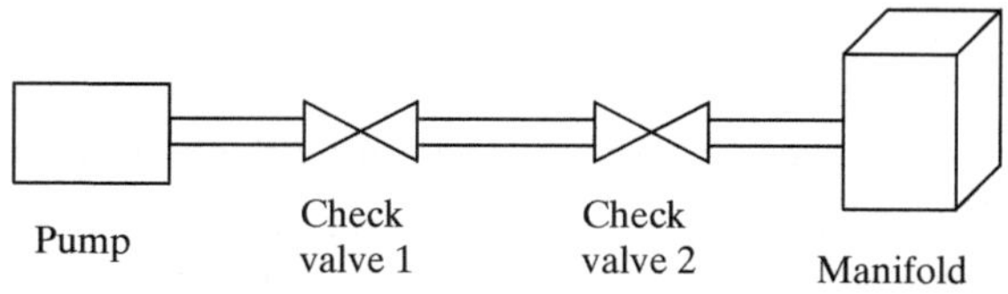

FIGURE 12.25 An emergency hydraulic pump.

12.5 Figure 12.26 shows a typical reactor coolant system that is required to pump coolant from a reservoir in the case of "loss of coolant accident (LOCA)." The coolant from the reservoir is controlled by valve 1 and is pumped through two parallel segments via the valves 2 and 3. Considering LOCA as the initial event, develop the event tree of the system.

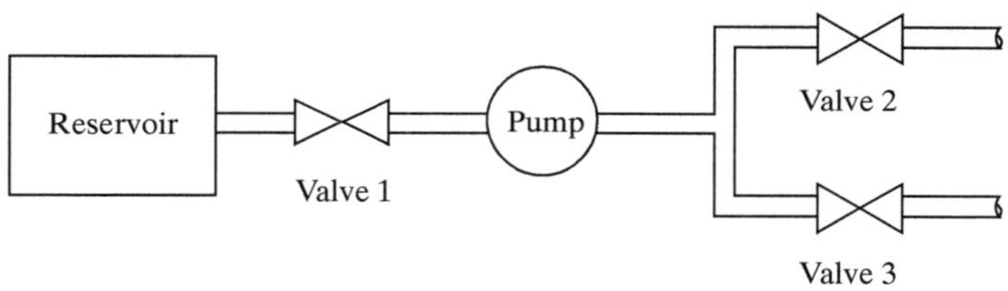

FIGURE 12.26 A typical reactor coolant system.

12.6 Construct a fault tree for the reactor coolant system described in Problem 12.5 considering "no flow of coolant" as the TOP event.

12.7 Figure 12.27 shows a mechanical power transmission system, which transmits power from the motor to shaft 2. Considering "no motion of shaft 2" as the TOP event, develop a

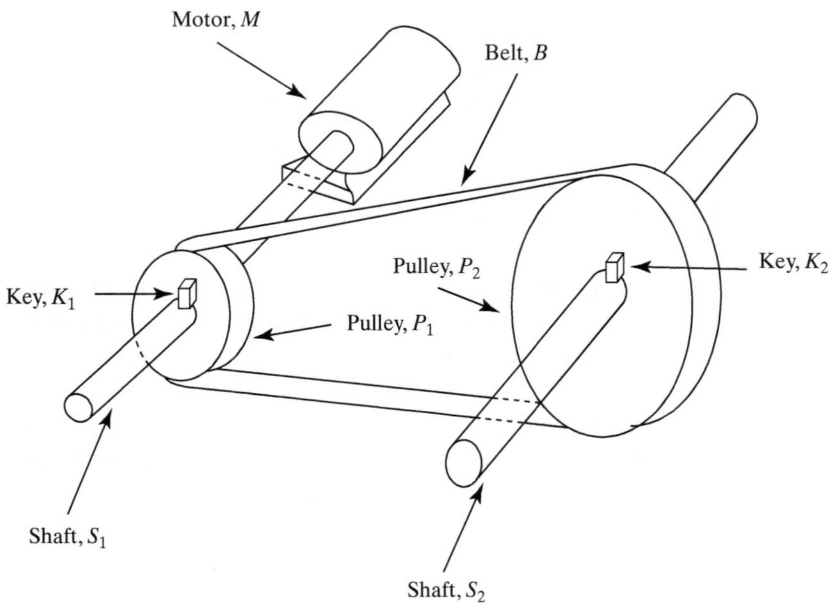

FIGURE 12.27 A mechanical power transmission system.

fault tree of the system. Assume that the motor, keys, shafts, pulleys, and the belt can fail individually.

12.8 Develop an event tree for the mechanical transmission system described in Problem 12.7.

12.9 A schematic diagram of a thermal power plant, consisting of a heating system, feedwater pump, boiler, turbine, and generator, is shown in Fig. 12.28. Assuming the TOP event as "no electric power output," construct a fault tree of the system.

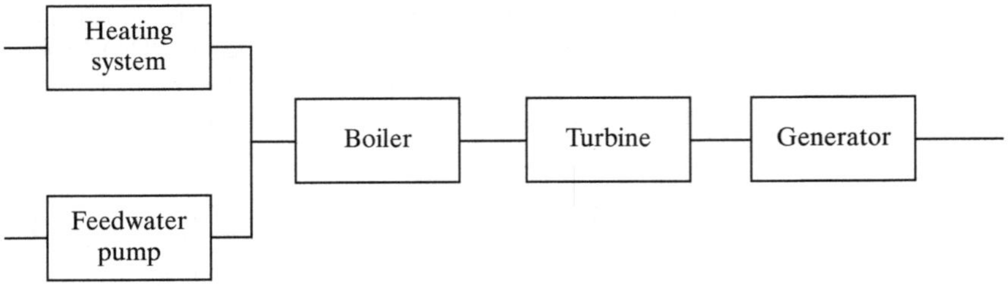

FIGURE 12.28 A thermal power plant.

12.10 Construct a fault tree for the series–parallel network shown in Fig. 12.29.

12.11 Develop an event tree for the series–parallel network shown in Fig. 12.29.

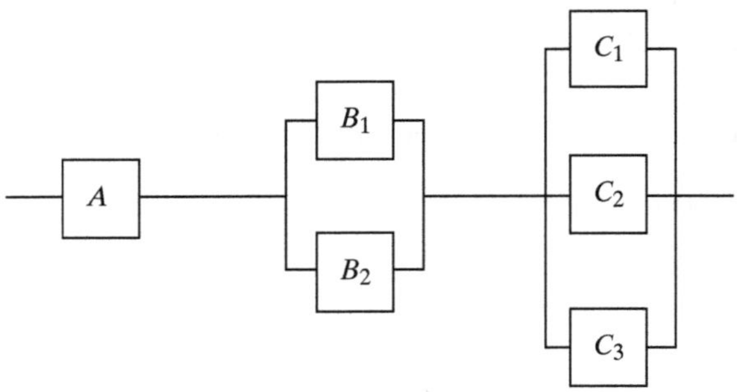

FIGURE 12.29 A series-parallel network.

12.12 An electric overhead traveling crane is shown in Fig. 12.30. The signal from the operator is transmitted to the electric motor, located in the trolley, which lifts the load through the rope and hook. Assuming that "the load is not lifted" when the operator gives the signal, construct a fault tree of the system.

12.13 Develop an event tree for the overhead traveling crane system described in Problem 12.12.

12.14 The various stages of the cement manufacturing process are shown schematically in Fig. 12.31. Develop an event tree of the system.

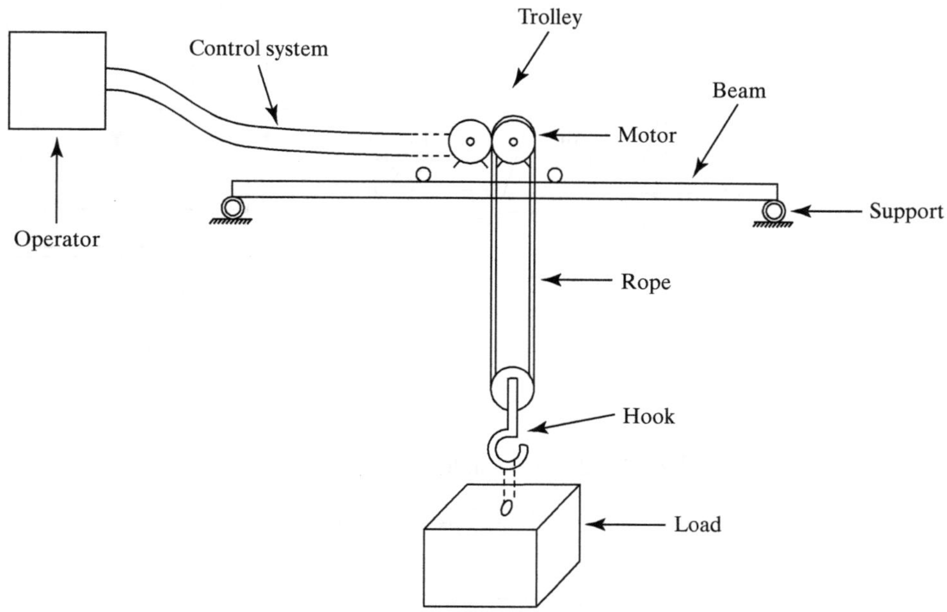

FIGURE 12.30 Electric overhead traveling crane.

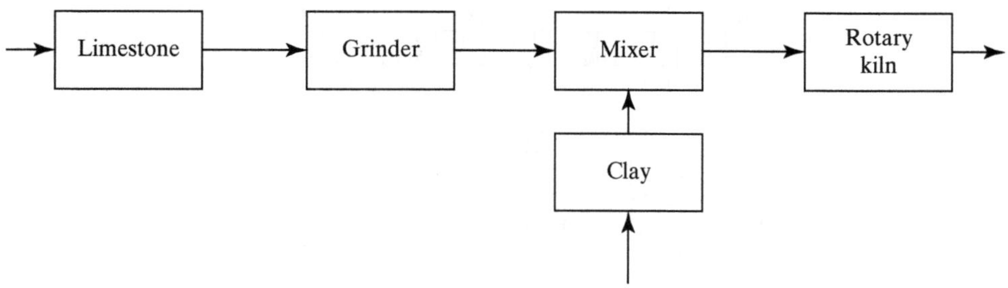

FIGURE 12.31 Cement manufacturing process.

12.15 Figure 12.32 shows the arrangement of a solar air heater [12.14]. Assuming the component failures as (a) solar energy may not be available, (b) air blower failure, (c) insulation failure, (d) absorbing sheet failure, and (e) transparent sheet failure, develop the event tree of the system.

12.16 If the TOP event is "no hot air output" for the solar air heater considered in Problem 12.15, construct a fault tree of the system.

12.17 A mechanical press consists of four bearings, a motor, three gears, two screws, and a flat plate as shown in Fig. 12.33. If the TOP event is "object is not pressed," construct a fault tree of the system.

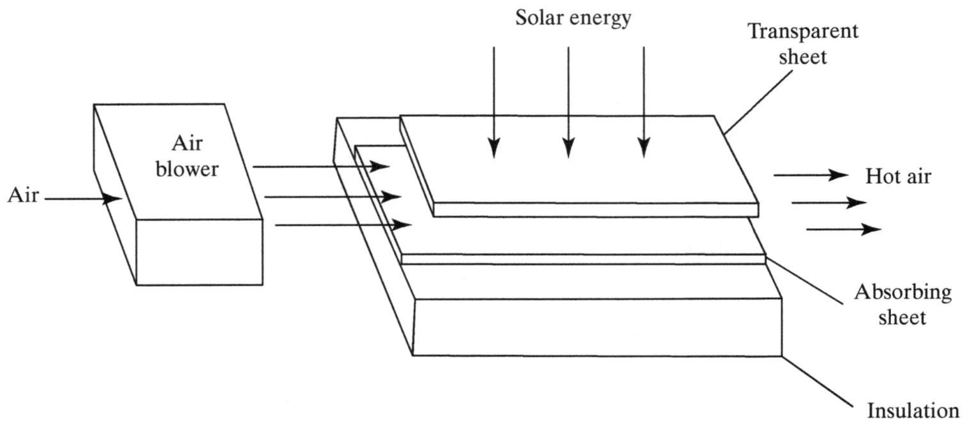

FIGURE 12.32 Solar air heater.

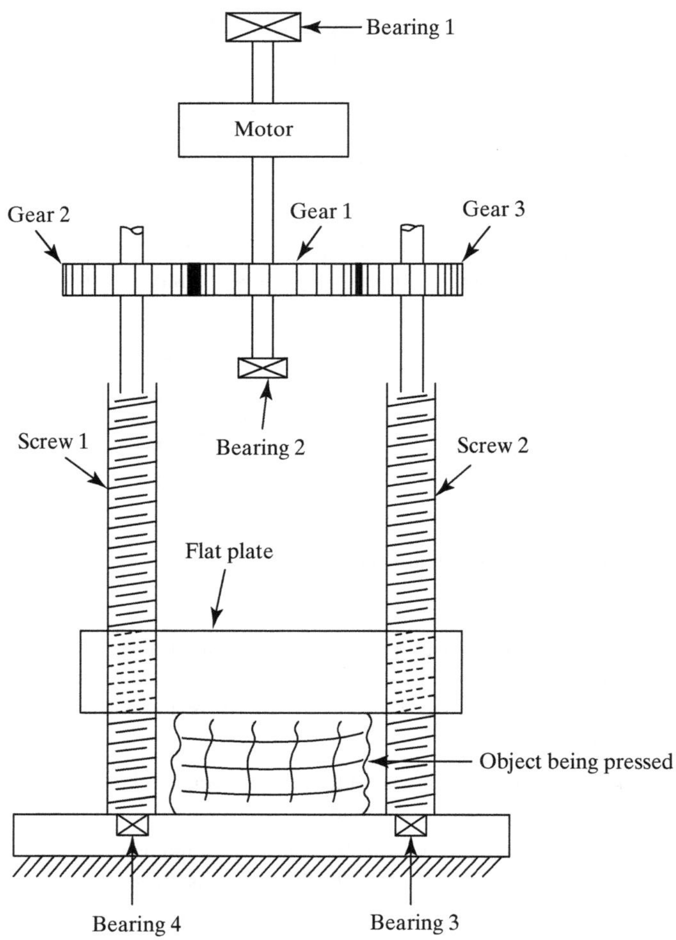

FIGURE 12.33 Mechanical press.

12.18 Construct an event tree for the mechanical press considered in Problem 12.17.

12.19 A compound gear train consists of eight bearings, four shafts, and eight gears as shown in Fig. 12.34. Assuming that a gear can fail either due to excessive bending or surface-wear stress, a shaft can fail either due to excessive stress or deflection, and a bearing can fail due to excessive contact stress, misalignment, or lack of lubricant, construct an event-tree diagram for the system.

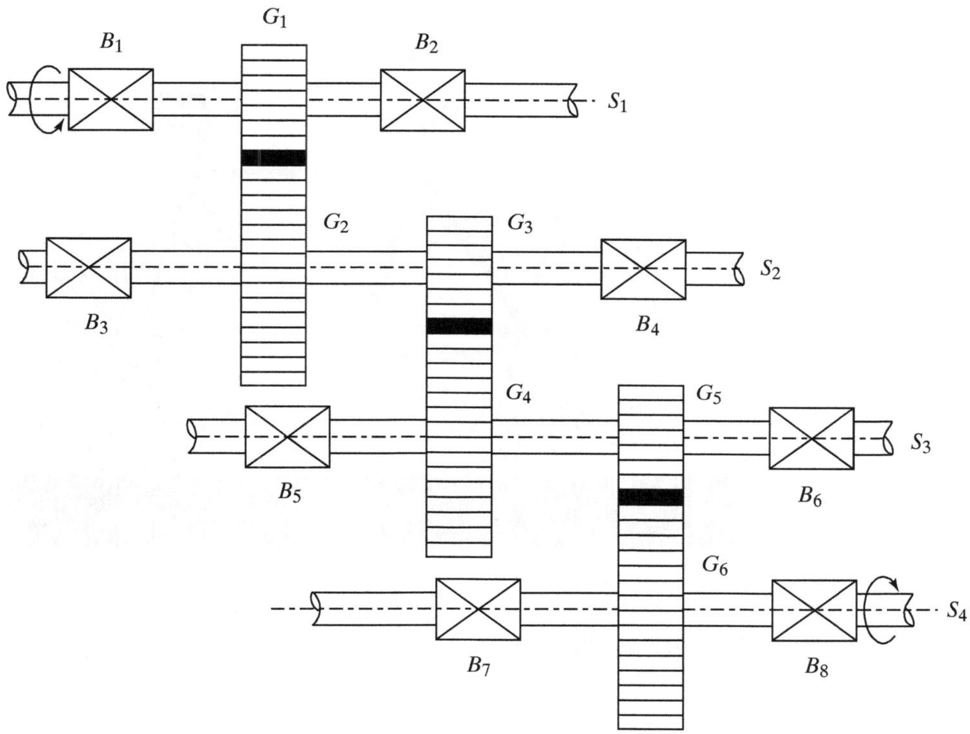

FIGURE 12.34 A compound gear train (B_i = ith bearing, G_i = ith gear, and S_i = ith shaft).

12.20 For the compound gear train described in Problem 12.19, determine a fault tree corresponding to the TOP event, "no power output at shaft S_4."

12.21 For the bicycle shown in Fig. 12.35, construct a fault tree corresponding to the TOP event: "the bicycle is unfit for riding." Consider the following basic events: failure of the seat, failure of the frame, failure of the wheels, failure of the tires, failure of the chain, failure of the sprocket, failure of the handle, failure of the foot pedals.

12.22 Figure 12.36 depicts a scissor jack. Construct a fault tree corresponding to the TOP event "load is not lifted" by considering the possibility of failure of each component/joint indicated in Fig. 12.36.

FIGURE 12.35 A bicycle.

12.23 Construct the event tree and determine the probabilities of occurrence of each of the final outcomes for the flashlight described in Problem 12.1. Assume the probability of occurrence of each of the events (i) to (v) indicated in Problem 12.1 as 0.01, and the initiating event as "pushing the switch of the flashlight on."

12.24 Find the probabilities of occurrence of each of the output events of the event tree of Problem 12.19 assuming that the probabilities of failure of each of the bearings, shafts, and gears are given by 0.97, 0.98, and 0.99, respectively.

12.25 Determine the probabilities of occurrence of each of the outcomes of the event tree considered in Problem 12.15 assuming the probabilities of occurrence of each of the events (a) to (e) as 0.05.

12.26 If the probability of occurrence of each of the events stated in (i) to (v) in Problem 12.2 is 0.99, determine the probability of occurrence of the TOP event: "no input to electric generator."

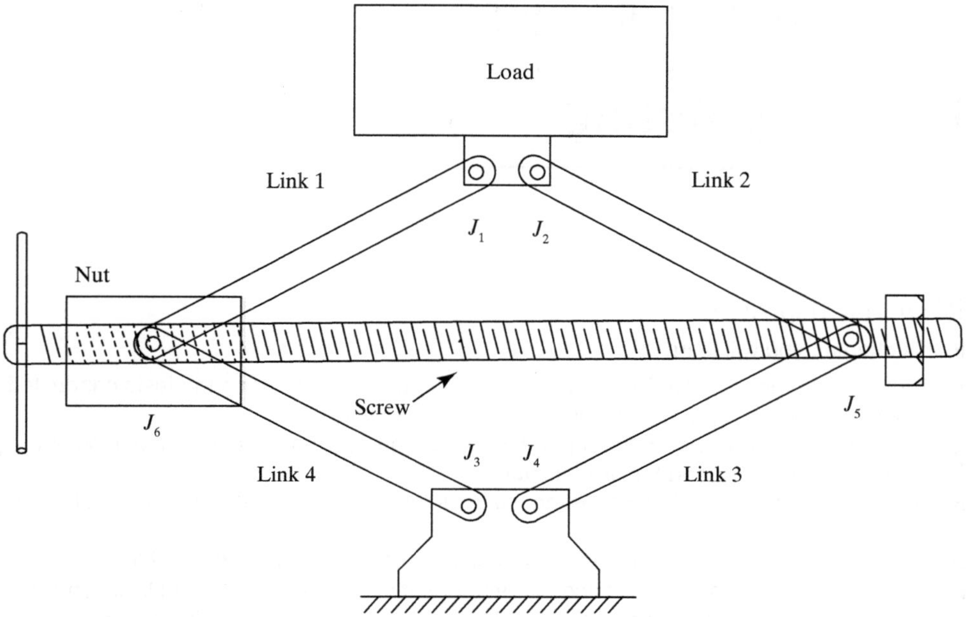

FIGURE 12.36 A scissor jack.

12.27 If the probabilities of failure of the valves 1, 2, and 3 are 0.01, 0.02, and 0.03, respectively, and the probability of failure of the pump is 0.04, find the reliability of the system shown in Fig. 12.26.

12.28 Find the probability of failure of the system (i.e., the occurrence of the event, "no motion of shaft 2") described in Problem 12.7 assuming that the probability of failure of each of the components is 0.01.

12.29 If the failure rates of the heating system, feedwater pump, boiler, turbine, and generator are 2×10^{-6}, 5×10^{-7}, 4×10^{-5}, 8×10^{-6}, and 1×10^{-7} per hour, respectively, in the thermal power plant described in Problem 12.9, find (a) the failure rate of the system and (b) the reliability of the system at 1000 hours.

12.30 Find the reliability of the mechanical press described in Problem 12.17 if the reliabilities of the bearing, motor, gear, and screw are 0.99, 0.95, 0.98, and 0.92, respectively.

CHAPTER 13

Reliability Testing

What You Will Learn

In this chapter, you will learn

- The different types of reliability testing and their objectives, including the tests conducted during the design, manufacturing, and operating stages of a product or system
- The analysis of individual and grouped failure data and the numerical computation of the probability density, distribution, reliability, and hazard functions
- The methods of accelerated life testing, such as testing until partial failure, magnified loading, and sudden-death testing
- The sequential life testing and the method of establishing accept and reject lines
- The concepts of statistical inference, parameter estimation, and maximum likelihood method
- The computation of confidence limits and level of significance and confidence intervals on the mean and standard deviation of a normal variable
- The plotting of reliability data using least-squares technique
- An outline of the method of fitting different probability distributions such as exponential, normal, lognormal, and Weibull distributions to experimental data, as well as the use of normal and Weibull probability papers

Andrey Kolmogorov, born on 25 April, 1903, in Tambov, Imperial Russia, is a well-known mathematician who contributed to the theory of probability as well as topology, intuitionistic logic, and classical mechanics. In probability theory, the Borel–Kolmogorov paradox and in statistics, the Kolmogorov–Smirnov test are well known. The Kolmogorov–Smirnov test is a goodness-of-fit test that can be used to decide if a sample came from a population with a specific distribution. The two-sample Kolmogorov–Smirnov test is used to determine if two data sets differ significantly without making any assumption about the distribution of data. Kolmogorov graduated from Moscow State University in 1925, continued his graduate studies under the supervision of Nikolai Luzin, and received his Ph.D. from the Moscow State University in 1929. In 1931, he became a Professor at Moscow University and published his pioneering work, *About the Analytical Methods of Probability Theory*, (in German) in the same year. In 1933, he published the book *Foundations of the Theory of Probability*, which laid the modern axiomatic foundations of probability theory. In 1935, he became the first chairman of the department of probability at Moscow State University and was elected as academician of the USSR Academy of Sciences in 1939. At Moscow State University, he served, as Dean of Mechanics and Mathematics. Kolmogorov received several prizes, including the Stalin Prize in 1941, the Lenin Prize in 1965, the Wolf Prize in 1980, and the Lobachevsky Prize in 1987. He died on 20 October, 1987, in Moscow, USSR.

13.1 Introduction

Reliability tests are intended to find whether a system can operate satisfactorily for a specified period of time under prescribed operating conditions. Different types of reliability tests are conducted at various stages of the life cycle of a system as indicated in the following list:

1. *Tests conducted during the design stage.* At the end of the design stage, prototypes are built and tested to analyze the failure modes and reliability of the system. The results of the tests are used to check whether the system is behaving as intended. The results are also used to modify or redesign the system for improved reliability.
2. *Tests conducted during the construction (or manufacturing) stage.* Before the system is put into service, qualification and acceptance tests are conducted to prove that the design standards of reliability are met. In acceptance tests, the system or component is tested to determine whether it should be accepted or rejected (on an individual unit or lot basis). Based on the results, better quality control methods can be used to reduce the defects in construction or manufacture. In qualification tests, the system is tested to determine whether it truly qualifies for its intended application. In some cases, the manufacturer has to demonstrate the reliability to fulfill the contractual requirements. The contract might include an incentive clause whereby the profit margin is readjusted either upwards or downwards depending on the value of the demonstrated reliability compared to the originally specified value.
3. *Tests conducted during the operating stage.* Tests are conducted during the operating stage of the system to find the failure rate and reliability of the system. These results are used to verify the reliability analysis conducted previously and also to find whether any modifications are needed in the design and/or operating procedures to improve the reliability. These tests are also useful to establish proper maintenance and parts replacement schedules. In addition, these tests can be used to determine reasonable warranty policies that will result in improved sales while assuring a reasonable profit.

As stated in Chapter 6, the life of a system can be divided into three phases, based on the failure rate, as follows:

1. Infant-mortality phase (also known as burn-in phase) during which a high failure rate is observed due to manufacturing defects.
2. Operating-life phase during which a constant failure rate is observed due to random failures.
3. Aging or wear-out phase during which a high failure rate is observed due to mechanical wear of the components.

Although some reliability tests are conducted during the burn-in phase, most of the reliability tests are conducted during the useful life of a system, since this is the period of interest to the customer.

13.1.1 Objectives of Reliability Tests

Depending on the objectives, reliability tests can be classified into four categories as follows:

1. *Longevity tests.* The objective is to find the length of the useful life of the system that has a constant failure rate.

2. *MTBF tests.* The objective is to determine the mean time between failures of the system.
3. *Operating life tests.* The objective is to find the ability of the system to perform without failure for a prescribed minimum time period.
4. *Reliability margin tests.* The objective is to establish the margin of safety between the extreme operating conditions of the system.

13.1.2 Details of a Reliability Test

In the case of consumer-oriented products and systems, the reliability requirements are related to the cost of the product/system. For military equipment and systems, the reliability requirements are usually specified in the military specifications. Before a test is conducted, the inputs and outputs of the system must be clearly defined, including their nominal values and permissible deviations or tolerances. In multicomponent systems, the failure criteria must be defined clearly. Sometimes, failures may have to be classified into major and minor ones, catastrophic and non-catastrophic ones, and so on. The sample size or the number of units (components or systems) tested and the duration of the test must be determined from the required confidence level and the limitations of cost and time involved in conducting the test. Although a statistically correct sample size can be determined in all the cases, the following sample sizes are commonly used:

1. For testing individual components, 10 to several hundred units.
2. For testing subsystems, a few to 20 or 30 units.
3. For testing complete systems, 1 to 10 units. The duration of a test depends on the component or system and its failure characteristics. It may range from few hundred hours to several thousands of hours.

13.2 Analysis of Failure Time

When experiments are conducted to find the failure time of a component or system, the data may be collected in two ways: as individual data or as grouped data. When individual data are collected, the failure times of individual components are noted. When grouped data are collected, the number of failures in different time intervals is recorded.

13.2.1 Analysis of Individual Failure Data

To analyze the individual failure data, consider a test involving n components. Let the failure times of the components be arranged in an increasing order as $t_1 \le t_2 \le \cdots \le t_i \le \cdots \le t_n$ where t_i denotes the failure time of component i and is also known as the ith rank statistic of the test. The cumulative distribution function, $F(t)$, by definition, represents the probability of realizing the failure time of the component less than or equal to t. Thus, $F(t_i)$ denotes the fraction of components failed in time t_i

$$F(t_i) = \frac{i}{n} \tag{13.1}$$

It has been found that $F(t_i)$ given by Eq. (13.1) provides reasonably accurate values of the distribution function when n is large. However, Eq. (13.1) does not give very accurate estimates of $F(t)$ when n is small, of the order of 10 or less. In such cases, Eq. (13.1) indicates that all the components fail for values of t larger than t_n. However, if another test is conducted with a larger number of units (of the order of $5n$), it is likely that some components may fail at times larger than t_n. Hence other formulas have been suggested for estimating the cumulative distribution function, $F(t)$, as follows:

$$F(t_i) = \frac{i}{n+1} \tag{13.2}$$

$$F(t_i) = \frac{i-0.5}{n} \tag{13.3}$$

$$F(t_i) = \frac{(0.5)^{1/n}(2i-n-1)+n-1}{n-1} \tag{13.4}$$

$$F(t_i) = \frac{i-0.3}{n+0.4} \tag{13.5}$$

where the right-hand sides of Eqs. (13.2) and (13.5) are called the *mean rank* and the *median rank*, respectively. Equation (13.1) implies that when a large number of components are tested, equal number of components fail in each of the time intervals, $t_{i+1} - t_i$; $i = 1, 2, \ldots, n-1$. Equation (13.2), on the other hand, denotes that the number of components failing beyond time t_n is same as the number of failures occurring in each of the time intervals, $t_{i+1} - t_i$; $i = 1, 2, \ldots, n-1$. Equations (13.3) to (13.5) are based on other, more advanced, statistical arguments [13.1–13.3]. If Eq. (13.1) is used, the reliability of the component at time $t = t_i$, $R(t_i)$, can be expressed as

$$R(t_i) = 1 - F(t_i) = \frac{n-i}{n} \tag{13.6}$$

This indicates that i components have failed and $(n-i)$ survived at $t = t_i$. By using a forward-finite difference formula, the probability density function of the failure time can be expressed as

$$f(t) = \frac{F(t_{i+1}) - F(t_i)}{t_{i+1} - t_i} = \frac{\frac{i+1}{n} - \frac{i}{n}}{t_{i+1} - t_i} = \frac{1}{n(t_{i+1} - t_i)};$$

$$t_i \le t \le t_{i+1}; \quad i = 1, 2, \ldots, n-1 \tag{13.7}$$

The failure rate or hazard function of the component can be estimated as (see Eq. 6.11)

$$h(t) = \frac{f(t)}{R(t)} = \frac{1}{(n-i)(t_{i+1} - t_i)}; \quad t_i \le t \le t_{i+1}; \quad i = 1, 2, \ldots, n-1 \tag{13.8}$$

EXAMPLE 13.1

The failure times of ten automobile brakes are observed to be 43,500, 52,000, 63,500, 72,000, 84,500, 93,500, 101,000, 111,500, 116,000, and 123,500 miles of operation. Plot the probability density, probability distribution, reliability, and the hazard functions of the failure time of the brakes.

Solution: Equations (13.1), (13.7), (13.6), and (13.8) can be used to find the probability distribution, probability density, reliability, and hazard functions, respectively, of the life of brakes. The computations are indicated in Table 13.1. These results are shown graphically in Figs. 13.1(*a*)–(*c*).

TABLE 13.1 Computation of the hazard function from failure data

i	t_i (miles)	$t_{i+1} - t_i$ (miles)	$F(t_i) = \frac{i}{n}$	$R(t_i) = \frac{n-i}{n}$	$f(t) = \frac{1}{n(t_{i+1} - t_i)}$ ($\times 10^{-6}$)	$h(t) = \frac{f(t)}{R(t)}$ ($\times 10^{-6}$)
0	0	43,500	0	1.0	2.2989	2.2989
1	43,500	8,500	0.1	0.9	11.7647	13.0719
2	52,000	11,500	0.2	0.8	8.6957	10.8696
3	63,500	8,500	0.3	0.7	11.7647	16.8067
4	72,000	12,500	0.4	0.6	8.0000	13.3333
5	84,500	9,000	0.5	0.5	11.1111	22.2222
6	93,500	7,500	0.6	0.4	13.3333	33.3332
7	101,000	10,500	0.7	0.3	9.5238	31.7460
8	111,500	4,500	0.8	0.2	22.2222	111.1110
9	116,000	7,500	0.9	0.1	13.3333	133.3330
10	123,500	—	1.0	0.0	—	—

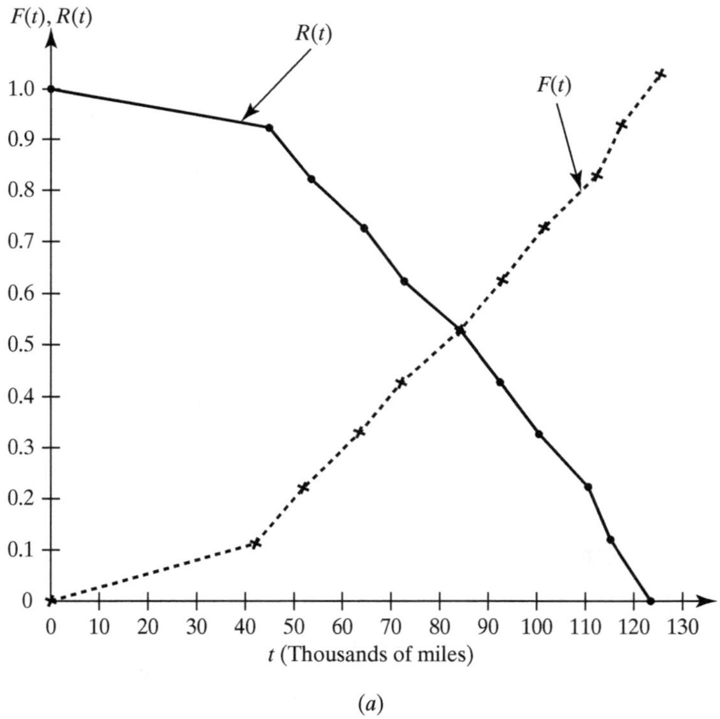

FIGURE 13.1 (*a*) Variations of $F_T(t)$ and $R(t)$ with time.

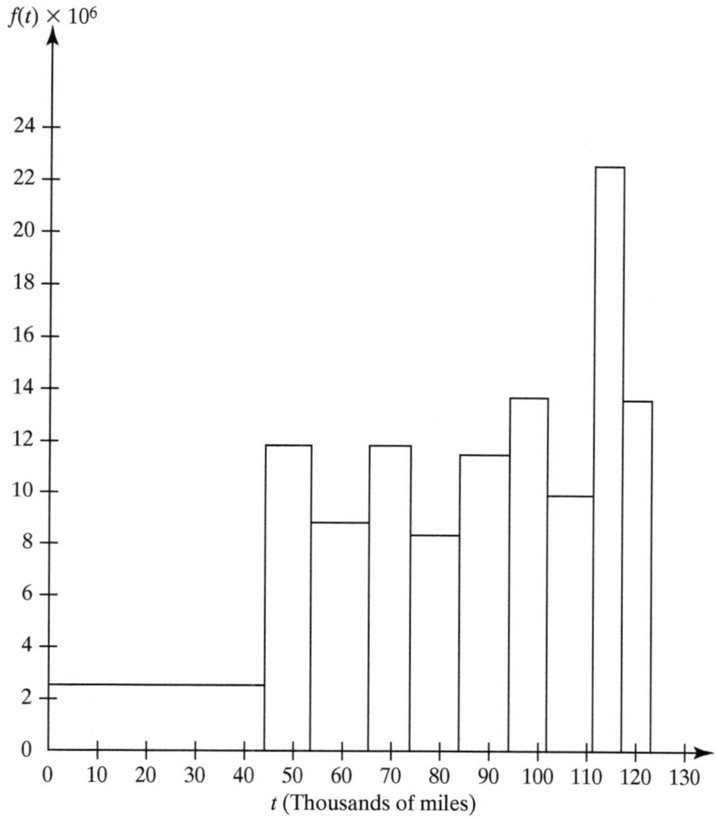

(b)

FIGURE 13.1 (b) Probability density function, $f_T(t)$.

■

13.2.2 Analysis of Grouped Failure Data

Let a test be started with n identical components or systems. Let the number of components surviving at times $t_1, t_2, \ldots,$ and t_k be denoted as $n_1, n_2, \ldots,$ and n_k, respectively $(n_1 \geq n_2 \geq \cdots \geq n_k)$. Thus, the failure times of individual components are not available; only the time intervals in which groups of components failed are available. Then the probability distribution and reliability functions at time $t = t_i$ can be expressed as

$$F(t_i) = \frac{n - n_i}{n}; \quad i = 1, 2, \ldots, k \tag{13.9}$$

$$R(t_i) = \frac{n_i}{n}; \quad i = 1, 2, \ldots, k \tag{13.10}$$

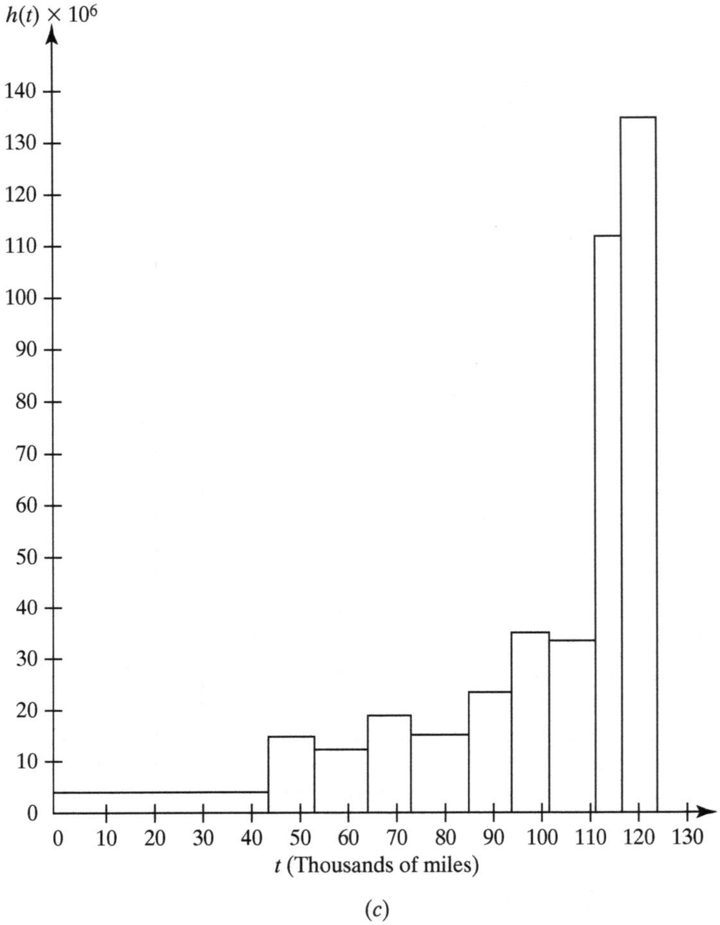

FIGURE 13.1 (*c*) Hazard function, $h(t)$.

The probability density function of the failure time can be determined using a forward difference formula as:

$$f(t) = \frac{F(t_{i+1}) - F(t_i)}{t_{i+1} - t_i} = \frac{(n - n_{i+1}) - (n - n_i)}{n(t_{i+1} - t_i)} = \frac{n_i - n_{i+1}}{n(t_{i+1} - t_i)};$$

$$t_i < t < t_{i+1}; \quad i = 1, 2, \ldots, k - 1 \tag{13.11}$$

The hazard function or the failure rate of the component can be estimated as

$$h(t) = \frac{f(t)}{R(t)} = \frac{n_i - n_{i+1}}{n_i(t_{i+1} - t_i)}; \quad t_i < t < t_{i+1}; \quad i = 1, 2, \ldots, k - 1 \tag{13.12}$$

13.3 Accelerated Life Testing

Usually, life testing is expensive in terms of both money and time. If a component or system is highly reliable, the testing procedure takes a long period of time before observing the failure. Hence, it is desirable to speed up the testing procedure for collecting the failure data. Accelerated testing procedures aim at reducing the time required for life testing by using one of the following strategies:

1. By conducting a test only until a part of the sample fails.
2. By using a magnified load.
3. By using sudden-death testing.

These methods are described below in detail.

13.3.1 Testing Until Partial Failure

In this method, a large number of units are tested simultaneously until the failure of a fraction of the units is observed. Let the failure times of the units be expressed as $t_1 \leq t_2 \leq \cdots \leq t_i \leq \cdots$ where t_i denotes the failure time of the ith component. If we want to observe the failure times of m components, there are two approaches. One approach is to start with a sample of m components and conduct the test until all the components fail. The other approach is to start with a larger sample of n components $(n > m)$ and continue the test until the first m components fail. If the average MTBF is to be estimated from the test results, both the approaches yield approximately the same value of MTBF. However, in general, the approach using the larger sample requires a smaller testing time. This can be seen intuitively as follows. Let there be one component, say A, with an extremely large value of failure time (life) among the m components. In the first approach, where we observe m failures out of m components, we need to conduct the test until the component A fails. However, in the second approach, we need not wait until the component A fails; some other component out of the remaining $(n - m)$ components fails before the component A does. Thus, the overall testing time can be reduced by starting with a larger sample of n components.

The expected waiting time to observe the mth failure in a sample of m units [13.4, 13.5], $E(t_{m,\,m})$, is given by

$$E(t_{m,\,m}) = \theta \sum_{i=1}^{m} \frac{1}{(m - i + 1)} \tag{13.13}$$

where θ is the MTBF of the unit. Similarly, the expected waiting time to observe the mth failure in a sample of n units can be expressed as

$$E(t_{m,\,n}) = \theta \sum_{i=1}^{m} \frac{1}{(n - i + 1)} \tag{13.14}$$

The saving in time realized in using a sample of n units instead of m units for observing m failures can be seen from the ratio

$$\text{Fractional time} = \frac{E(t_{m,n})}{E(t_{m,m})} = \frac{\sum_{i=1}^{m} \frac{1}{(n-i+1)}}{\sum_{i=1}^{m} \frac{1}{(m-i+1)}} \tag{13.15}$$

This ratio is computed for different combinations of m and n and the results are given in Table 13.2. As an example, consider a test involving the operation of five units. Let the waiting period be 100 hours to observe the failure of all the five units. If we use a sample of 10 units, we need to wait, on the average, only $0.2828(100) = 28.28$ hours to observe the failure of the first five units.

If n units are tested until m failures are observed, the $(n - m)$ unfailed units may still be useful for normal application. Even if the units are not suitable for further use, the savings in time realized in testing are well worth the cost of the additional units.

13.3.2 Magnified Loading

Most accelerated tests involve application of magnified stress levels (such as temperature, voltage, or load) as opposed to the normal use values in order to increase the rate of failure and reduce the testing time. In many cases, the use of magnified stress levels also results

TABLE 13.2 Ratio of expected waiting times to observe the mth failure in samples of size n and m

	Value of n							
Value of m	**5**	**10**	**15**	**20**	**25**	**30**	**40**	**50**
1	0.2000	0.1000	0.0667	0.0500	0.0400	0.0333	0.0250	0.0200
2	0.3000	0.1407	0.0921	0.0684	0.0544	0.0452	0.0338	0.0269
3	0.4273	0.1833	0.1173	0.0863	0.0683	0.0565	0.0420	0.0334
4	0.6160	0.2299	0.1432	0.1042	0.0819	0.0675	0.0499	0.0396
5	1.0	0.2828	0.1705	0.1224	0.0956	0.0784	0.0577	0.0457
6	—	0.3452	0.1997	0.1413	0.1095	0.0894	0.0654	0.0516
7	—	0.4226	0.2315	0.1611	0.1237	0.1005	0.0732	0.0575
8	—	0.5258	0.2669	0.1820	0.1385	0.1119	0.0810	0.0634
9	—	0.6819	0.3069	0.2043	0.1538	0.1236	0.0888	0.0694
10	—	1.0	0.3533	0.2283	0.1699	0.1356	0.0968	0.0753

in a fewer number of units required for test. When magnified loads are used, the most important consideration is to find the correlation between the two (normal and magnified) load conditions. This requires a knowledge of the failure characteristics of the unit under normal and magnified load conditions. It is to be noted that the failure mode should not be changed with the intensified stress level. Then, by extrapolating the results suitably to the "normal use" conditions, we can obtain reasonably accurate estimates of the life of the component under the "normal use" conditions.

A practical example where accelerated testing has proved to be useful is in the testing of impregnated paper-based capacitors. For these capacitors, the mean life (θ) is related to the applied DC voltage (V) as [13.5]

$$\theta = \frac{k}{V^p} \tag{13.16}$$

where k and p are constants. A value of $p = 5$ has been found to be applicable for this case. Let the average life of a capacitor be 10 years under normal use conditions in which the applied voltage is V_1. Then the nominal life of capacitors will be

$$\theta_1 = \frac{k}{V_1^5} = 10 \text{ years} \tag{13.17}$$

When the capacitors are subjected to a DC voltage that is twice the normal value, the average life will be

$$\theta_2 = \frac{k}{(2V_1)^5} \tag{13.18}$$

This gives

$$\frac{\theta_1}{\theta_2} = \frac{10}{\theta_2} = \frac{(2V_1)^5}{V_1^5} = 32 \tag{13.19}$$

which yields $\theta_2 = 3.75$ months, a substantial reduction of mean life from the nominal value of 10 years.

Another application of magnified loading is in fatigue testing. In *fatigue testing*, the number of cycles (N) which causes failure is determined under a series of constant stress amplitudes (S). From this data, the standard S–N curve is plotted for the material of the specimens. It is well known that in practice very few components are subjected to constant-amplitude loading during their entire lives. Hence, fatigue tests involving variable amplitude of the stress are developed. If the amplitude of the stress is increased, the fatigue life of the specimen will be smaller. Thus, the results obtained from a series of fatigue tests with different stress amplitudes can be plotted as a straight line on a log-stress (S) and log-fatigue life (N) graph as shown in Fig. 13.2. For example, for the AISI 1045 steel, the fatigue strength relationship between the stress and the number of cycles is given by $SN^{0.085} = 154.0$ where S is in k(lb/in^2) and N is in cycles. In general, for most structural steels, the S–N relationship [13.6] can be expressed as

$$SN^a = b \tag{13.20}$$

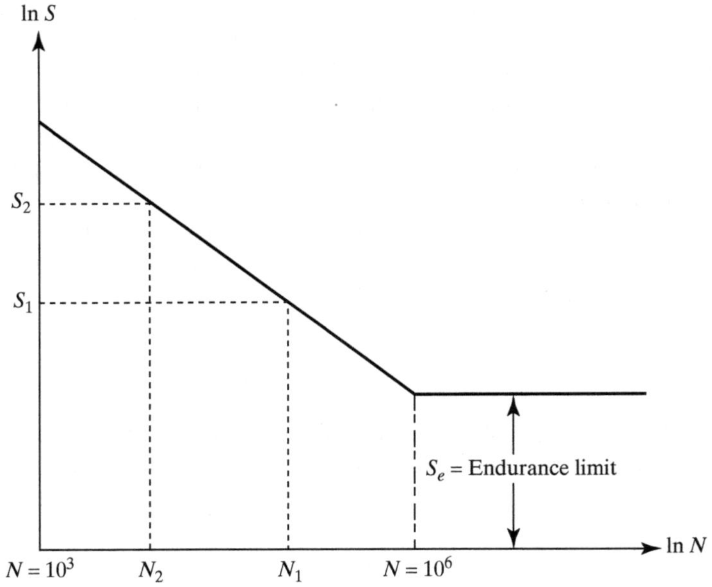

FIGURE 13.2 Stress (S) – fatigue life (N) diagram (log-log scale).

where a and b are constants. Equation (13.20) can be used to predict the fatigue life (N_2) at a stress intensity of S_2 once the fatigue life (N_1) at the stress intensity S_1 is known, as

$$S_1 N_1^a = b = S_2 N_2^a$$

or

$$N_2 = N_1 \left(\frac{S_1}{S_2} \right)^{1/a} \tag{13.21}$$

Let the stress intensity be doubled for a component made up of AISI 1045 steel for which $a = 0.085$. Then the number of cycles to failure (fatigue life) is reduced by a factor of

$$\frac{N_1}{N_2} = \left(\frac{S_2}{S_1} \right)^{1/a} = 2^{1/0.085} = 2^{11.7647} = 3480.68$$

When acceleration is to be achieved by increasing the temperature, the Arrhenius and Eyring models are commonly used.

Arrhenius Model. According to the Arrhenius model, the rate of a chemical reaction or failure mechanism (process) is given by the empirical formula [13.1, 13.7],

$$\text{Rate of reaction} = c e^{-a/(k\,T)} \tag{13.22}$$

where a is a constant known as the activation energy, k is known as the Boltzman's constant, T is the ambient temperature (in degrees K), and c is a normalization constant. In Eq. (13.22), the value of a depends on the specific reaction or the failure model, while the value of k is given by 86.3×10^{-6} eV/°K. Usually, the life of a component is assumed to be inversely proportional to the rate of the failure process, and hence

$$L = c_2 e^{a/(kT)} \tag{13.23}$$

where L represents the life of the component such as the minimum or average life and c_2 is a constant. Thus, if T_u and L_u denote the temperature and life during normal use and T_a and L_a indicate the temperature and life during the accelerated testing, Eq. (13.23) yields

$$\frac{L_a}{L_u} = e^{a/\{k(1/T_a - 1/T_u)\}} \tag{13.24}$$

The acceleration factor, A_f, achieved during the accelerated testing is given by

$$A_f = \frac{L_u}{L_a} = e^{a/\{k(1/T_u - 1/T_a)\}} \tag{13.25}$$

Eyring Model. According to the Eyring model [13.1, 13.7], the rate of reaction is given by

$$\text{Rate of reaction} = cTe^{-a/(kT)} \tag{13.26}$$

The Eyring model is more general and can be used when stresses other than temperature are involved. While the Arrhenius model is based on an empirical equation, the Eyring model has a theoretical basis and can be derived from chemical-reaction rate theory and quantum mechanics. However, the Arrhenius model is more frequently used in the reliability testing of several types of components such as microcircuits.

13.3.3 Sudden-death Testing

When each component is relatively inexpensive, the sudden-death testing can be used to reduce the overall testing time. If each component is relatively expensive, this method is not suitable. In this method, a large number of components are tested by dividing them into several groups with five or six components in each group. For example, if 60 components are available for testing, they are randomly divided into 10 groups of 6 components each. Then the components in each group are tested simultaneously until one component fails. As soon as the first component fails, the remaining components of the group are removed or withdrawn from the test (a sudden death to these components). This procedure is continued until all the 10 groups have been tested. Thus the procedure gives ten numbers, each representing the smallest value of life in a random sample size of six. If we use the median ranking procedure, an estimate of the true fraction of failures occurring in time t_i is given by (see Eq. (13.5))

$$F(t_i) = \frac{i - 0.3}{n + 0.4} \tag{13.27}$$

where i denotes the number of failures observed in a sample of size n in time t_i. Since $i = 1$ and $n = 6$ in the present case, we have $F(t_i) = (1 - 0.3)/(6 + 0.4) = 0.1094$. Thus, the 10 failure times represent the lives of the weakest 10.94 percent of the population. The Weibull distribution can be used to describe the extremal distribution corresponding to the smallest lives of the components. Hence the 10 observed lives, when plotted on the Weibull graph, lie on a straight line as shown in Fig. 13.3. This straight line denotes the distribution of the lives of the weakest 10.94 percent of the population instead of that of the lives of the entire population. The median or 50 percent life corresponding to this data can be identified

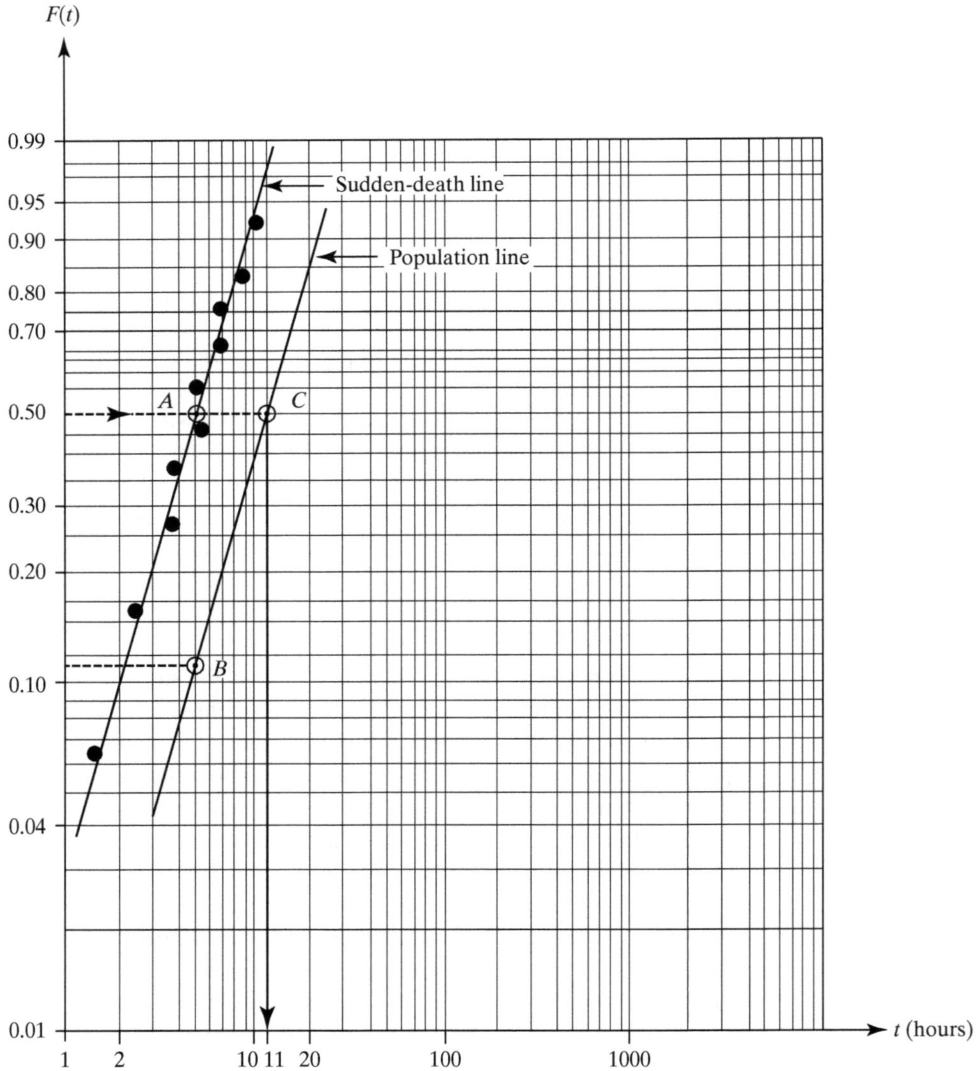

FIGURE 13.3 Weibull probability paper.

(shown as point A in Fig. 13.3). The vertical line drawn through the point A intersects the 10.94 percent line at point B as shown in Fig. 13.3. The straight line drawn through point B, parallel to the sudden-death line, can be taken as the population line to denote the lives of the population. This assumes that the slopes of the sudden-death line and the population line are the same. The characteristics of the population can be estimated from the population line. For example, the median or 50 percent life of the population can be found from the life corresponding to point C in Fig. 13.3.

EXAMPLE 13.2

Sixty switches are randomly divided into 10 equal groups and each group is subjected to sudden-death testing. As soon as the first unit in each group fails, the remaining units of the group are withdrawn from the test. The failure times observed for the 10 failed units, arranged in an increasing order, are given by 1.5, 2.2, 3.5, 3.9, 5.0, 5.2, 6.4, 6.8, 8.2, and 10.1 hours. Find the median life of the population of switches.

Solution: Since each failure is the first failure observed in a sample of six units, it corresponds to a median rank of $(i - 0.3)/(n + 0.4) = (1 - 0.3)/(6 + 0.4) = 0.1094$. Hence the 10 failure times observed correspond to the lives of the weakest 10.94 percent of the population of switches. These lives are plotted on a Weibull paper using the median ranks for a sample of 10 ($n = 10$). The calculations are indicated in Table 13.3 and the points $(t_i, F(t_i))$; $i = 1, 2, \ldots, 10$ are plotted in Fig. 13.3. The median life of the 10 failed units, corresponding to point A, can be seen to be $\approx$ 5.0 hours. The point of intersection of the horizontal line at $F(t) = 0.1094$ and the vertical line through the point A lies on the population line. The population line is drawn parallel to the sudden-death line. The median or 50 percent life of the population of switches can be identified from point C as $\approx$ 11 hours.

TABLE 13.3 Computation of probability distribution function

i	t_i (hours)	$F(t_i) = \dfrac{i - 0.3}{n + 0.4} = \dfrac{i - 0.3}{10 + 0.4}$
1	1.5	0.0673
2	2.2	0.1635
3	3.5	0.2596
4	3.9	0.3558
5	5.0	0.4519
6	5.2	0.5481
7	6.4	0.6442
8	6.8	0.7404
9	8.2	0.8365
10	10.1	0.9327

■

13.4 Sequential Life Testing

The sequential life testing procedure was developed by Abraham Wald during World War II [13.8]. It has been proved that sequential testing requires fewer units to be tested compared to the other testing methods. In sequential life testing, the sample size is not fixed. Components from a lot are examined one by one. Criteria are established so that each component may be classified as good or bad (defective). After testing each component, the accumulated data is reviewed and a decision is made to accept the lot, reject the lot, or continue testing. Thus, the primary aim of this testing procedure is to determine whether the component meets a reliability goal rather than to estimate the MTBF.

To see how a sequential testing scheme works, assume that we are interested in the reliability of a certain type of electric motor and the manufacturer has provided a large lot of these motors. Let the desired reliability of the motor (required by the customer) be R_u. However, we may be willing to accept the lot even if it has a somewhat lower reliability, R_l, provided that there is a very high probability of rejecting a lot if its reliability is less than R_l. It can be seen that sequential testing involves risks to both the manufacturer and the customer. Since the acceptance of the entire lot is based on the results of a random sample, there is a certain probability of rejecting a good lot or accepting a bad lot. The probability of rejecting the lot whose reliability is equal to or greater than R_u is known as the *producer's risk* and is denoted as α. Similarly, the probability of accepting the lot whose reliability is less than R_l is called the *consumer's risk* and is denoted as β.

In sequential testing, the quantities R_u, R_l, α, and β are specified at the beginning. Now consider a graph in which the ordinate represents the cumulative number of failures and the abscissa the cumulative number of successes as shown in Fig. 13.4. In this figure, two lines, known as the "accept" and "reject" lines, are plotted. The equations for these lines [13.5, 13.9] are given by

$$\text{Accept line:}\quad f\ln\left(\frac{1-R_l}{1-R_u}\right)+s\ln\left(\frac{R_l}{R_u}\right)=\ln\left(\frac{1-\beta}{\alpha}\right) \tag{13.28}$$

$$\text{Reject line:}\quad f\ln\left(\frac{1-R_l}{1-R_u}\right)+s\ln\left(\frac{R_l}{R_u}\right)=\ln\left(\frac{\beta}{1-\alpha}\right) \tag{13.29}$$

where f and s denote the cumulative number of failures and successes, respectively. In the electric motor example, let us assume the following values for the parameters: $R_u = 0.95$, $R_l = 0.85$, $\alpha = 0.05$, and $\beta = 0.10$. For these values, Eqs. (13.28) and (13.29) reduce to

$$\text{Accept line:}\quad 1.0986f - 0.1113\,s = 2.8904 \tag{13.30}$$

and

$$\text{Reject line:}\quad 1.0986f - 0.1113\,s = -2.2513 \tag{13.31}$$

These lines are shown in Fig. 13.4. During testing, the results are continuously plotted in between the accept and reject lines. The results of two hypothetical tests are shown by dotted lines in Fig. 13.4. The dotted line corresponding to test 1 indicates that the first 5 units were observed to be good, next 1 unit bad, next 15 units good, next 1 bad, and

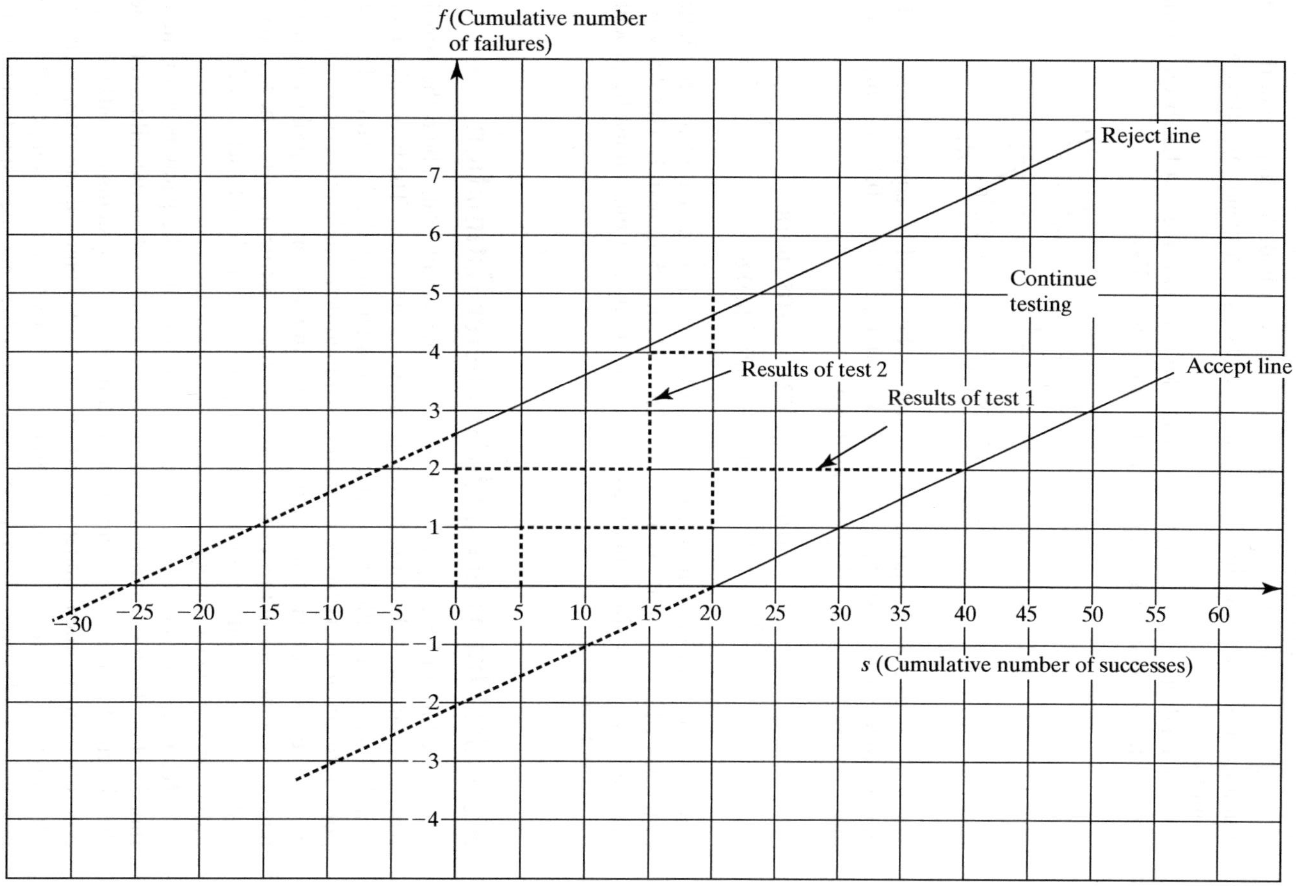

FIGURE 13.4 Construction of accept and reject lines.

next 20 good. At this stage, the dotted line is found to cross the accept line and hence the test is terminated by accepting the entire lot of motors. The dotted line corresponding to test 2 denotes that the first 2 units were observed to be bad, next 15 units good, next 2 units bad, next 5 units good, and the next unit was bad. At this stage, the dotted line is found to cross the reject line and hence the test is terminated by rejecting the entire lot of motors.

Although the exact sample size is not known until the test has been terminated, it is possible to estimate the average sampling number (ASN) as

$$\text{ASN} = \frac{(1-\alpha)\ln\left(\frac{\beta}{1-\alpha}\right) + \alpha\ln\left(\frac{1-\beta}{\alpha}\right)}{(1-R_u)\ln\left(\frac{1-R_l}{1-R_u}\right) + R_u\ln\left(\frac{R_l}{R_u}\right)} \tag{13.32}$$

The ASN gives the average number of units to be tested in sequential testing. For the particular values of $R_l = 0.85$, $R_u = 0.95$, $\alpha = 0.05$, and $\beta = 0.10$ assumed for the electric motor example, Eq. (13.32) gives a value of

$$\text{ASN} = \frac{(0.95)\ln 0.1053 + (0.05)\ln 18}{(0.15)\ln 3 + (0.95)\ln 0.8947} = 39.2712$$

This number can be compared with the test results indicated by the dotted lines in Fig. 13.4 where 42 units were used in test 1 and 25 in test 2. The ASN is useful for planning purposes since, obviously, certain number of units must be procured to start the sequential test.

13.5 Statistical Inference and Parameter Estimation

The procedure of drawing conclusions from the sample data about the population from which the sample is drawn is known as *statistical inference*. The two major types of statistical inference are called parameter estimation and hypothesis testing. The *parameter or point estimation* refers to the method of estimating a single numerical value for a parameter of the distribution from n sample values. In many practical problems, we need to determine whether a statement about some parameter is true or false. The statement is known as the hypothesis and the process of determining the truth or falseness of the hypothesis is known as *hypothesis testing*.

Reliability tests can be used to estimate not only the parameters of the distribution function but also the parameters related to reliability such as the failure rate and MTBF of the system. In each test, a number of components (or systems or units) are operated in a controlled environment under prescribed conditions until all the components have failed. This requires a clear definition of failure, especially for a multicomponent system. As soon as a failure is observed, the time to failure is recorded and the test is terminated after all the components or units have failed. In some cases, the test is terminated after the failure of a predetermined number of components or units.

13.5.1 Maximum-likelihood Method

Several methods are available for estimating the parameters of a distribution. One of the popular methods is the maximum-likelihood estimation method. Consider a random variable X which follows the probability density function $f(x; \theta)$ where θ is the parameter to be estimated (such as the mean or standard deviation in normal distribution and the mean in exponential distribution). Let the sample values from the tests be given by $\{x_1, x_2, \ldots, x_n\}$. These values are assumed to be drawn from a population described by the probability density function $f(x; \theta)$. Then we will be interested in finding the probability of obtaining the specific sample $\{x_1, x_2, \ldots, x_n\}$ from all possible samples of size n from the population. In other words, among the various possible values of θ, we want to find the particular value $\hat{\theta}$ which maximizes the likelihood of obtaining the set of observations $\{x_1, x_2, \ldots, x_n\}$. The maximum-likelihood method of estimation is intended to find the value of $\hat{\theta}$.

The probability of obtaining the ith sample value in the interval $x_i - dx_i/2$ and $x_i + dx_i/2$ can be assumed to be $f(x_i; \theta)dx_i$ (from the definition of the probability density function). Assuming the sample values to be independent, the joint probability of finding the first sample value in the range $x_1 - dx_1/2$ and $x_1 + dx_1/2$, the second sample value in the range $x_2 - dx_2/2$ and $x_2 + dx_2/2, \ldots,$ and the nth sample value in the range $x_n - dx_n/2$ and $x_n + dx_n/2$ is given by

$$[f(x_1; \theta)dx_1][f(x_2; \theta)dx_2] \cdots [f(x_n; \theta)dx_n] = \prod_{i=1}^{n} f(x_i; \theta)dx_i$$

$$= L(\theta)dx_1\, dx_2 \cdots dx_n \tag{13.33}$$

where the function L given by

$$L(\theta) = \prod_{i=1}^{n} f(x_i; \theta) \tag{13.34}$$

is called the likelihood function of the sample. The maximum of the function $L(\theta)$ can be determined by setting the partial derivative of L with respect to θ equal to zero and solving the resulting equation for $\theta = \hat{\theta}$. Since most probability density functions $f(x; \theta)$ involve exponential terms, it is more convenient to maximize the natural logarithm of the likelihood function to find $\hat{\theta}$. Thus Eq. (13.34) gives

$$\ln L(\theta) = \ln \prod_{i=1}^{n} f(x_i; \theta) = \sum_{i=1}^{n} \ln f(x_i; \theta) \tag{13.35}$$

By differentiating Eq. (13.35), we obtain

$$\frac{\partial \ln L(\theta)}{\partial \theta} = 0 \tag{13.36}$$

the solution of which gives the desired maximum-likelihood estimate $\hat{\theta}$.

This procedure can be used for estimating the parameters of any distribution. It can also be generalized to distributions involving two or more parameters.

EXAMPLE 13.3

A reliability test is conducted within the useful life period of a system for which the failure time (T) follows the exponential distribution

$$f_T(t;\theta) = \frac{1}{\theta} e^{-t/\theta}; \quad t > 0, \theta > 0 \tag{E1}$$

where θ indicates the MTBF. Find the maximum-likelihood estimate of the MTBF based on n sample values.

Solution: Let the sample values from the test be $t_1, t_2, \ldots, t_n$ with $t_1 \le t_2 \le \cdots \le t_n$. The corresponding likelihood function is given by

$$L(\theta) = \left(\frac{1}{\theta} e^{-t_1/\theta}\right)\left(\frac{1}{\theta} e^{-t_2/\theta}\right) \cdots \left(\frac{1}{\theta} e^{-t_n/\theta}\right) = \frac{1}{\theta^n} e^{-(1/\theta)\sum_{i=1}^{n} t_i} \tag{E2}$$

so that

$$\ln L(\theta) = -n \ln \theta - \frac{1}{\theta}\sum_{i=1}^{n} t_i \tag{E3}$$

For the maximum of $\ln L(\theta)$,

$$\frac{\partial \ln L(\theta)}{\partial \theta} = -\frac{n}{\theta} + \frac{1}{\theta^2}\sum_{i=1}^{n} t_i = 0 \tag{E4}$$

which yields the value of θ as

$$\theta = \hat{\theta} = \frac{1}{n}\sum_{i=1}^{n} t_i \tag{E5}$$

This value of $\hat{\theta}$ represents the maximum-likelihood estimate of the MTBF available from the observed data.

■

EXAMPLE 13.4

The sample values $x_1, x_2, \ldots, x_n$ have been observed for a random variable X following normal distribution with unknown mean μ and standard deviation σ. Determine the maximum-likelihood estimators of μ and σ.

Solution: The probability density function is

$$f_X(x) = \frac{1}{\sqrt{2\pi}\sigma} e^{-1/2\{(x-\mu)/\sigma\}^2} \tag{E1}$$

where $\mu = \theta_1$ and $\sigma = \theta_2$ are the unknown parameters of the distribution. The likelihood function, $L(\theta_1, \theta_2)$, can be defined as

$$L(\theta_1, \theta_2) = \prod_{i=1}^{n} \frac{1}{\sqrt{2\pi}\sigma} e^{-1/2\{(x_i-\mu)/\sigma\}^2} = \frac{1}{\left(\sqrt{2\pi}\sigma\right)^n} \prod_{i=1}^{n} e^{-1/2\{(x_i-\mu)/\sigma\}^2} \tag{E2}$$

so that

$$\ln L(\theta_1, \theta_2) = -n \ln(\sqrt{2\pi}\,\sigma) - \frac{1}{2\sigma^2}\sum_{i=1}^{n}(x_i - \mu)^2 \tag{E3}$$

The conditions for maximizing $\ln L(\theta_1, \theta_2)$ are given by

$$\frac{\partial L(\theta_1, \theta_2)}{\partial \mu} = \frac{1}{\sigma^2}\sum_{i=1}^{n}(x_i - \mu) = 0 \tag{E4}$$

$$\frac{\partial L(\theta_1, \theta_2)}{\partial \sigma} = -\frac{n}{\sqrt{2\pi}\,\sigma}(\sqrt{2\pi}) - \frac{1}{2}\frac{(-2)}{\sigma^3}\sum_{i=1}^{n}(x_i - \mu)^2 = 0 \tag{E5}$$

The solution of Eqs. (E4) and (E5) gives the maximum-likelihood estimators of μ and σ, namely $\hat{\mu}$ and $\hat{\sigma}$, as

$$\mu = \hat{\mu} = \frac{1}{n}\sum_{i=1}^{n} x_i \tag{E6}$$

$$\sigma^2 = \hat{\sigma}^2 = \frac{1}{n}\sum_{i=1}^{n}(x_i - \mu)^2 \tag{E7}$$

■

13.6 Confidence Intervals

Theoretically, an infinite number of sample values have to be collected to find the true value of a parameter such as the mean value of a random quantity. However, this is not possible in practice due to the limitations of cost and time. Hence, representative samples are tested, and the value of the parameter (such as the mean of the population or universe) is estimated from the sample values. Since only a finite sample size is used, the sample mean, for example, will not be equal to the population mean. However, the difference between the sample mean $(\overline{X})$ and the population mean (μ_X) decreases as the sample size increases. Thus, an infinite sample size is needed to obtain 100 percent assurance that the sample mean is equal to the population mean. The statistical sampling techniques can be used to find from the sample values, the range, known as the confidence interval, in which the population mean lies.

As stated earlier, the sample mean is an example of a point estimate of a population parameter. In addition to obtaining the point estimate, we would also like to find the accuracy of this estimate. For this the concept of confidence intervals is used. The *confidence interval* indicates the range of values within which the true value of the point estimate lies. If the point estimate happens to be the sample mean, the confidence interval indicates the range in which the true mean of the population lies. In fact, a probability can be associated with any confidence interval. For example, we might say that "the true population mean of yield strength of steel lies between 25,000 lb/in^2 and 35,000 lb/in^2 with a probability of 99%." The lower and upper bound values of the confidence interval are known as the *confidence limits*. The probability associated with the confidence interval is called the *confidence level*. In the preceding example, the confidence limits are 25,000 $1b/in^2$ and 35,000 $1b/in^2$, and the confidence level is 99 percent. Once the confidence level is stated for a parameter such as the mean value as p, it indicates that the probability of the true population mean lying in the confidence interval is p. Thus, there is a probability of $(1 - p)$ that the real mean does not lie within the given confidence interval. This quantity is known as the *level of significance* so that

$$\text{Level of significance} = 1 - \text{Confidence level} \tag{13.37}$$

In the case of the preceding example of the yield strength of steel, the level of significance is equal to $1 - 0.99 = 0.01$. Figure 13.5 illustrates the concepts of confidence interval, confidence level, and the level of significance. It can be seen that the confidence level is equal to the area under the probability density function between the lower confidence limit (L) and the upper confidence limit (U)

$$P(L \le x \le U) = 1 - \alpha \tag{13.38}$$

For a symmetric probability density function, we can see that

$$L = -x_{(\alpha/2)} \quad \text{and} \quad U = x_{(\alpha/2)} \tag{13.39}$$

In some applications, a one-sided confidence interval is used as indicated in Fig. 13.6. For example, in reliability testing, there is no need to specify an upper bound on the mean failure time. The larger the mean failure time, the better it will be from reliability point of view. In such a case, the confidence level denotes the area under the probability density

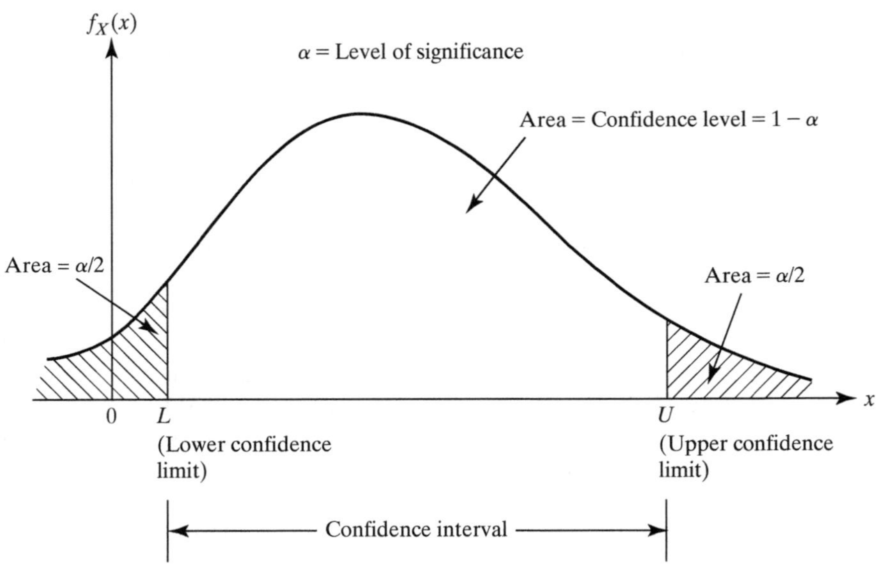

FIGURE 13.5 Two-sided confidence interval.

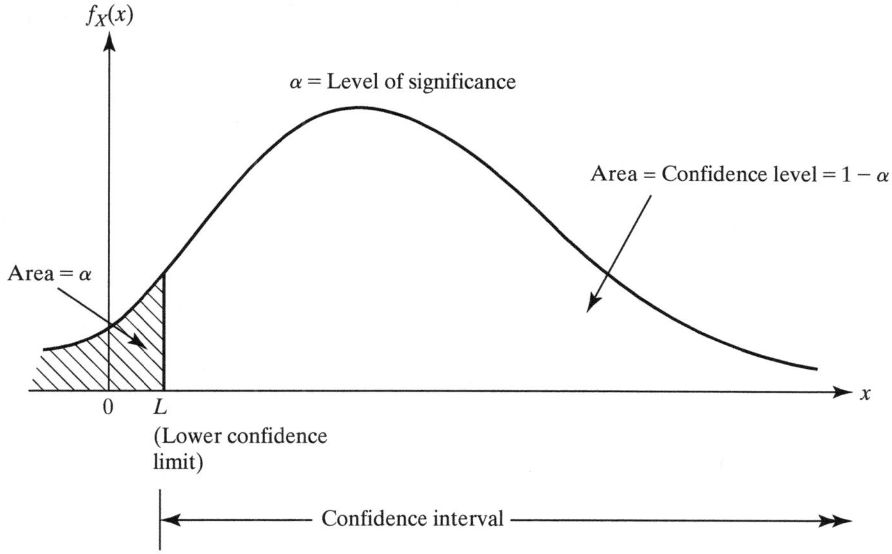

FIGURE 13.6 One-sided confidence interval.

curve from the lower limit while the level of significance represents the area under the curve up to the lower limit

$$P(x \geq L) = 1 - \alpha \tag{13.40}$$

13.6.1 Confidence Interval on the Mean of a Normal Random Variable of Known Standard Deviation

Let a normally distributed random variable X have an unknown population mean μ_X and a known standard deviation σ_X. Let $x_1, x_2, \ldots, x_n$ denote a random sample of size n, which gives the sample mean as $\overline{X}$. It has been proved from central limit theorem [13.10, 13.11] that the sample mean $\overline{X}$ itself is a random variable that has an approximate normal distribution with mean μ_X (population mean) and standard deviation $\sigma_X/\sqrt{n}$ (σ_X is the population standard deviation)

$$\mu_{\overline{X}} = \mu_X \tag{13.41}$$

and

$$\sigma_{\overline{X}} = \frac{\sigma_X}{\sqrt{n}} \tag{13.42}$$

Thus, the random variable $z = (\overline{X} - \mu_X)/(\sigma_X/\sqrt{n})$ will be a standard normal variate with mean 0 and standard deviation 1. The 100 $(1 - \alpha)$ percent confidence interval on the mean μ_X is given by

$$\overline{X} \pm z_{(\alpha/2)} \frac{\sigma_X}{\sqrt{n}}$$

where $z_{(\alpha/2)}$ is the 100 $(\alpha/2)$ percentage point of the normal distribution. This implies that

$$P\left[-z_{(\alpha/2)} \leq \frac{\overline{X} - \mu_X}{\left(\dfrac{\sigma_X}{\sqrt{n}}\right)} \leq z_{(\alpha/2)}\right] = 1 - \alpha \tag{13.43}$$

Rearranging the terms inside the square brackets in Eq. (13.43) leads to

$$P\left[\left(\overline{X} - z_{(\alpha/2)}\frac{\sigma_X}{\sqrt{n}}\right) \leq \mu_X \leq \left(\overline{X} + z_{(\alpha/2)}\frac{\sigma_X}{\sqrt{n}}\right)\right] = 1 - \alpha \tag{13.44}$$

where the lower and upper confidence limits, L and U, can be identified as

$$L = \overline{X} - z_{(\alpha/2)}\frac{\sigma_X}{\sqrt{n}} \tag{13.45}$$

$$U = \overline{X} + z_{(\alpha/2)}\frac{\sigma_X}{\sqrt{n}} \tag{13.46}$$

EXAMPLE 13.5

A normally distributed random variable X has a known standard deviation, σ_X. Determine the sample size n needed so that the error involved in estimating μ_X by $\overline{X}$ will be less than a specified amount δ with 100 $(1 - \alpha)$ percent confidence.

Solution: From Eq. (13.44), we find that

$$z_{(\alpha/2)}\frac{\sigma_X}{\sqrt{n}} = \delta$$

and hence the required sample size n will be

$$n = \left(\frac{z_{(\alpha/2)}\sigma_X}{\delta}\right)^2$$

■

13.6.2 Confidence Interval on the Mean of a Normal Random Variable of Unknown Standard Deviation

Let X be the random variable with unknown mean μ_X and unknown standard deviation σ_X. Let $x_1, x_2, \ldots, x_n$ denote a random sample of size n which gives the sample mean as $\overline{X}$ and the sample standard deviation as s_X. It has been proved that the distribution of the random variable $(\overline{X} - \mu_X)/(s_X/\sqrt{n})$ is a t-distribution with $(n - 1)$ degrees of freedom [13.11, 13.12]. Note that this random variable is similar to the variable $(\overline{X} - \mu_X)/(\sigma_X/\sqrt{n})$ defined in the preceding section. Since the population standard deviation σ_X is not known, it has been replaced by the sample standard deviation s_X in the present case.

The probability density function of a random variable T following t-distribution, also known as student's t-distribution, with n degrees of freedom [13.11, 13.12] is given by

$$f_T(t) = \frac{\Gamma\left(\frac{n+1}{2}\right)}{\Gamma\left(\frac{n}{2}\right)\sqrt{n\pi}}\left(1 + \frac{t^2}{n}\right)^{-(n+1)/2}; \quad -\infty < t < \infty \tag{13.47}$$

The values of the distribution function for different values of t and n are given in Appendix B.

The $100(1 - \alpha)$ percent confidence interval of μ_X is given by

$$\overline{X} \pm t_{(\alpha/2;\, n-1)}\frac{s_X}{\sqrt{n}}$$

where $t_{(\alpha/2;\, n-1)}$ is the $100(\alpha/2)$ percent point of the t-distribution with $n - 1$ degrees of freedom. This implies that

$$P\left[-t_{(\alpha/2;\, n-1)} \le \frac{\overline{X} - \mu_X}{\left(\frac{s_X}{\sqrt{n}}\right)} \le t_{(\alpha/2;\, n-1)}\right] = 1 - \alpha \tag{13.48}$$

or

$$P\left[\overline{X} - t_{(\alpha/2;\, n-1)}\frac{s_X}{\sqrt{n}} \le \mu_X \le \overline{X} + t_{(\alpha/2;\, n-1)}\frac{s_X}{\sqrt{n}}\right] = 1 - \alpha \tag{13.49}$$

with the lower and upper confidence limits given by

$$L = \overline{X} - t_{(\alpha/2;\, n-1)}\frac{s_X}{\sqrt{n}} \tag{13.50}$$

and

$$U = \overline{X} + t_{(\alpha/2;\, n-1)}\frac{s_X}{\sqrt{n}} \tag{13.51}$$

EXAMPLE 13.6

The internal diameters of 10 internal combustion engine cylinders have been found to be 120.5, 118.4, 125.7, 116.9, 119.2, 123.3, 121.5, 119.8, 120.9, and 122.6 mm. If the diameters follow normal distribution, determine the 95 percent confidence interval for the population mean μ_X.

Solution: Here $\alpha = 0.05$, $n = 10$, and the observed sample mean is

$$\overline{X} = \frac{1}{10}(120.5 + 118.4 + \cdots + 122.6) = 120.88 \text{ mm}$$

and the observed sample standard deviation is

$$s_X = \left\{\frac{1}{9}\left[(120.5 - 120.88)^2 + (118.4 - 120.88)^2 + \cdots + (122.6 - 120.88)^2\right]\right\}^{1/2}$$

$$= \left(\frac{58.556}{9}\right)^{1/2} = 2.5507 \text{ mm}$$

From Appendix B, we find that $t_{(9;\,0.025)} = 2.262$. From Eqs. (13.50) and (13.51), we can obtain the confidence limits on the mean μ_X as

$$L = \overline{X} - t_{(\alpha/2;\,n-1)}\frac{s_X}{\sqrt{n}} = 120.88 - \left(\frac{2.262 \times 2.5507}{\sqrt{10}}\right) = 119.0555 \text{ mm}$$

$$U = \overline{X} + t_{(\alpha/2;\,n-1)}\frac{s_X}{\sqrt{n}} = 120.88 + \left(\frac{2.262 \times 2.5507}{\sqrt{10}}\right) = 122.7045 \text{ mm}$$

■

13.6.3 Confidence Interval on the Standard Deviation of a Normal Random Variable with Unknown Mean

Let X be a normally distributed random variable with unknown mean μ_X and unknown standard deviation σ_X. Let $\{x_1, x_2, \ldots, x_n\}$ denote a random sample of size n which gives the sample mean as $\overline{X}$ and the sample standard deviation as s_X

$$\overline{X} = \frac{1}{n}\left(\sum_{i=1}^{n} x_i\right) \tag{13.52}$$

$$s_X = \left\{\frac{1}{n-1}\sum_{i=1}^{n}\left(x_i - \overline{X}\right)^2\right\}^{1/2} \tag{13.53}$$

If we consider the random variable Y as

$$Y = \frac{(n-1)s_X^2}{\sigma_X^2} \tag{13.54}$$

then Y can be shown to follow a chi-square distribution with $n - 1$ degrees of freedom [13.11, 13.12]. For time-dependent processes with constant-failure rates, the chi-square distribution is used for assessing the confidence on the estimated reliability. Let x_i, $i = 1, 2, \ldots, n$ be n independent values or observations of a random variable, $N(\mu, \sigma)$. Then the variates

$$z_i = \frac{x_i - \mu}{\sigma}; \quad i = 1, 2, \ldots, n \tag{13.55}$$

will also be normally distributed with $N(0, 1)$. Let Y (i.e, χ^2) be a random variable denoting the sum of the squares of the z_i

$$Y = z_1^2 + z_2^2 + \cdots + z_n^2 \tag{13.56}$$

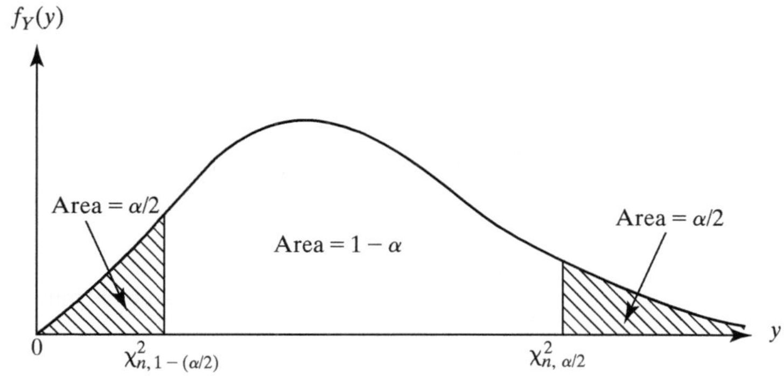

FIGURE 13.7 $100(1-\alpha)$ percent confidence limits for Y with n degrees of freedom.

Then Y follows the χ^2-distribution function with n degrees of freedom given by [13.11]

$$f_{\chi^2(n)}(y) = \frac{y^{\{(n/2)-1\}}e^{-y/2}}{2^{\frac{n}{2}}\Gamma\left(\frac{n}{2}\right)}; \quad y \geq 0 \tag{13.57}$$

The mean and variance of Y are given by

$$\mu_Y = n; \quad \sigma_Y^2 = 2n \tag{13.58}$$

The values of the χ^2-distribution function up to 20 degrees of freedom are given in Appendix C. By letting $\chi^2_{(n,\alpha/2)}$ be the value such that

$$P\left[Y > \chi^2_{(n,\alpha/2)}\right] = \frac{\alpha}{2} \tag{13.59}$$

with n degrees of freedom, we can obtain the confidence interval for σ_X^2 as (see Fig. 13.7)

$$P\left[\chi^2_{(n-1,1-(\alpha/2))} \leq Y \leq \chi^2_{(n-1,\alpha/2)}\right] = 1 - \alpha \tag{13.60}$$

This equation can be rewritten, after substituting for Y, as

$$P\left[\frac{(n-1)s_X^2}{\chi^2_{(n-1,\alpha/2)}} \leq \sigma_X^2 \leq \frac{(n-1)s_X^2}{\chi^2_{(n-1,1-(\alpha/2))}}\right] = 1 - \alpha \tag{13.61}$$

EXAMPLE 13.7

Determine the 95 percent confidence interval for the standard deviation σ_X for the data given in Example 13.6.

Solution: The observed sample standard deviation is $s_X = 2.5507$ mm. The values of $\chi^2_{(9,0.975)}$ and $\chi^2_{(9,0.025)}$ can be obtained from Appendix C as

$$\chi^2_{(9,0.975)} = 2.700; \quad \chi^2_{(9,0.025)} = 19.023$$

Equation (13.61) gives for $n = 10$ and $\alpha = 0.05$; the lower and the upper confidence limits on the standard deviation σ_X are

$$\frac{\sqrt{n-1}\, s_X}{\sqrt{\chi^2_{(n-1,\,\alpha/2)}}} = \frac{\sqrt{9}\, 2.5507}{\sqrt{19.023}} = 1.7544 \text{ mm}$$

and

$$\frac{\sqrt{n-1}\, s_X}{\sqrt{\chi_{(n-1,1-(\alpha/2))}}} = \frac{\sqrt{9}\, 2.5507}{\sqrt{2.7}} = 4.6569 \text{ mm}$$

■

13.7 Plotting of Reliability Data

The analytical equations for various probability distributions were given in Chapter 3. When experimental reliability data is available, it is desirable to have a simple graphical procedure to find a suitable distribution and also to estimate the parameters of the distribution. The methods presented in this section will aid in the rapid analysis of failure time data and estimation of the parameters of the appropriate probability distribution. The use of complete-, single-censored, or multicensored data is discussed. The data plotting techniques are illustrated for the exponential, normal, lognormal, and Weibull distributions.

13.7.1 Least-Squares Technique

When experiments are conducted to find the dependency of a quantity y on an independent variable x, several sets of data points are obtained as $(x_i, y_i), i = 1, 2, \ldots, n$ where n denotes the total number of data points. If x and y are linearly related, we can fit a straight line, also known as the *regression line*, using the least-squares technique. In the *least-squares method*, the sum of the squares of the deviations of the values of y predicted by the straight line from the observed values of y is minimized. To find the least-squares fit, let the equation of the regression line be

$$y = ax + b \tag{13.62}$$

where a and b are constants. By using the experimental points $(x_i, y_i), i = 1, 2, \ldots, n$, the constants a and b can be determined as (see Section 4.9.1)

$$a = \frac{n \sum_{i=1}^{n} x_i y_i - \sum_{i=1}^{n} x_i \sum_{i=1}^{n} y_i}{n \sum_{i=1}^{n} x_i^2 - \left(\sum_{i=1}^{n} x_i \right)^2} \tag{13.63}$$

and

$$b = \frac{n \sum_{i=1}^{n} x_i^2 \sum_{i=1}^{n} y_i - n \sum_{i=1}^{n} x_i \sum_{i=1}^{n} x_i y_i}{n \sum_{i=1}^{n} x_i^2 - \left(\sum_{i=1}^{n} x_i \right)^2} \tag{13.64}$$

13.7.2 Linear Rectification

The least-squares technique can be used to estimate the parameters of many distributions by using a procedure known as linear rectification. *Linear rectification* involves converting a given nonlinear equation into a linear form by suitably changing the variables. Let the relationship between the variables x and y be given by

$$y = \alpha x^{\beta} \tag{13.65}$$

where α and β are constants. By taking logarithms of both sides of Eq. (13.65), we get

$$\ln y = \ln \alpha + \beta \ln x \tag{13.66}$$

which represents a straight line between the variables $\ln y$ and $\ln x$. Thus the nonlinear equation (13.65) has been converted to an equivalent linear form in Eq. (13.66) using linear rectification. The concept of linear rectification is very useful because it allows us to determine in many cases the parameters of a distribution from experimental values.

13.7.3 Plotting Positions

When probability distributions are to be fitted to experimental data, we need to select the plotting positions of the data points. For specificness, let the observed data represent the failure times of a system or unit. If n units are tested, the observed failure times, for convenience, can be arranged in an increasing order as

$$t_1 \le t_2 \le \cdots \le t_i \le \cdots \le t_n$$

where t_i are called the *rank statistics* of the test. The cumulative distribution function, $F(t)$, which denotes the probability of realizing the failure time of the unit less than or equal to t, can be expressed as

$$F(t_i) = \frac{i}{n} \tag{13.67}$$

To plot the test results in the form of $F(t)$ versus t, we plot the points $\{t_i, F(t_i)\}$ for $i = 1, 2, \ldots, n$. The value of $F(t_i)$ given by Eq. (13.67) thus determines the plotting (ordinate) position of the ith data point, t_i. As stated earlier (in Section 13.2.1), other formulas, given by Eqs. (13.2) to (13.5), can also be used for $F(t_i)$. Equation (13.5) can be used to obtain more accurate plotting, although Eq. (13.2) is used in some cases, for simplicity, to plot the experimental data.

13.7.4 Exponential Distribution

When failure time data are available, it is usual to try to fit exponential distribution first. The main reason is that exponential distribution fits best if the system has a constant failure rate. In addition, exponential distribution is one of the simplest distributions involving only one parameter. For this case, the cumulative distribution function is given by

$$F(t) = 1 - e^{-\lambda t} \tag{13.68}$$

which can be rewritten as

$$1 - F(t) = e^{-\lambda t} \tag{13.69}$$

or

$$\ln \frac{1}{1 - F(t)} = \lambda t \tag{13.70}$$

This indicates that $\ln \{1/(1 - F(t))\}$ varies linearly with time. Thus the probability paper for exponential distribution is constructed by plotting the values of $[1/(1 - F(t))]$ on a logarithmic scale against the values of t on a linear scale (semilog paper). The procedure is illustrated through Example 13.8.

EXAMPLE 13.8

The failure times of 15 blenders are observed to be 30, 50, 60, 100, 130, 150, 190, 250, 270, 310, 380, 410, 530, 630, and 850 hours of operation. Plot the data on an exponential probability paper and determine the failure rate of the blenders.

Solution: The data are plotted twice, first by using Eq. (13.2) and then Eq. (13.5) to estimate the cumulative distribution function. The calculations are shown in Table 13.4. The resulting graphs are shown in Fig. 13.8. The solid line corresponds to Eq. (13.2) while the dotted line corresponds to Eq. (13.5). From Eq. (13.70), the failure rate, λ, can be seen to be the slope of

TABLE 13.4 Computation of cumulative distribution function

Data Point i	Failure Time t_i (Hours)	$F(t_i) = \frac{i}{n+1}$	$\frac{1}{1 - F(t_i)}$	$F(t_i) = \frac{i - 0.3}{n + 0.4}$	$\frac{1}{1 - F(t_i)}$
1	30	0.0625	1.0667	0.04545	1.04762
2	50	0.1250	1.1429	0.11039	1.12409
3	60	0.1875	1.2308	0.17532	1.21260
4	100	0.2500	1.3333	0.24026	1.31624
5	130	0.3125	1.4545	0.30519	1.43925
6	150	0.3750	1.6000	0.37013	1.58763
7	190	0.4375	1.7778	0.43506	1.77011
8	250	0.5000	2.0000	0.50000	2.00000
9	270	0.5625	2.2857	0.56493	2.29851
10	310	0.6250	2.6667	0.62987	2.70175
11	380	0.6875	3.2000	0.69481	3.27665
12	410	0.7500	4.0000	0.75974	4.16216
13	530	0.8125	5.3333	0.82468	5.70370
14	630	0.8750	8.0000	0.88961	9.05882
15	850	0.9375	16.0000	0.95454	22.00000

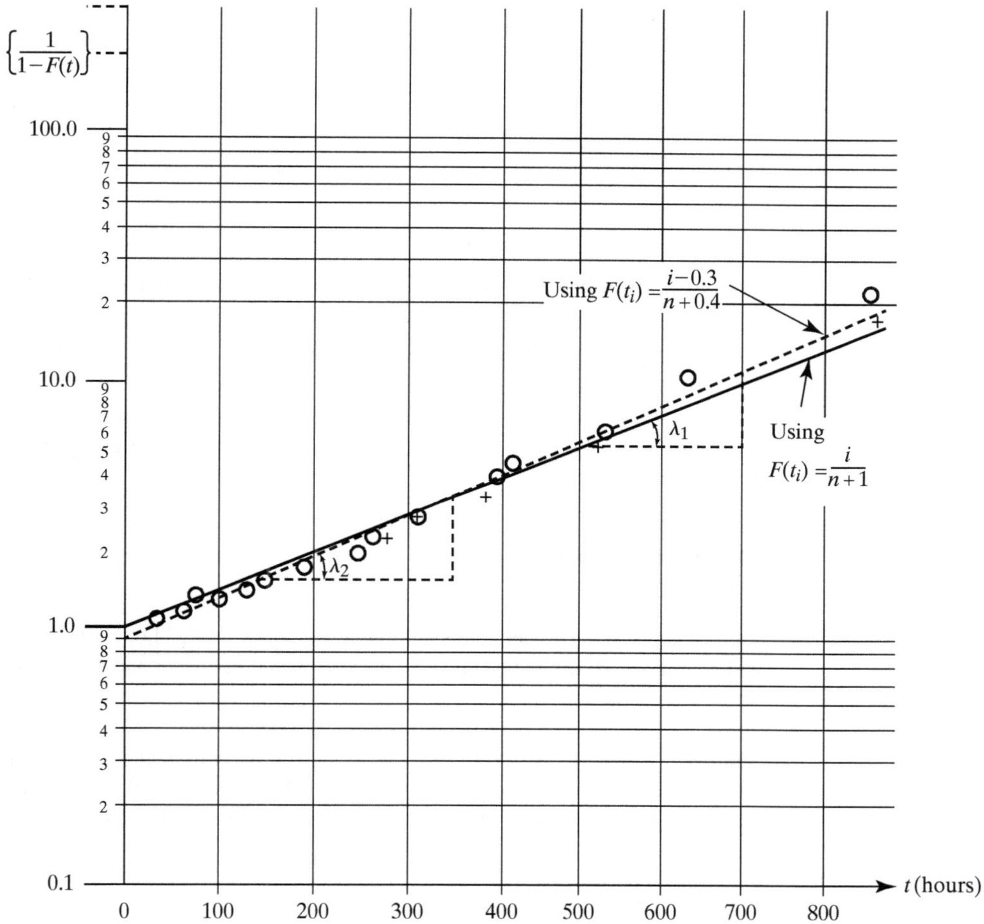

FIGURE 13.8 Plotting of data on exponential probability paper.

the straight line drawn between $\ln[1/(1 - F(t))]$ and t. Thus, Fig. 13.8 gives the value of λ as $\lambda_1 = (\ln 9.2 - \ln 5.0)/(700 - 500) = 0.003049$ when $F(t_i) = \{i/(N + 1)\}$ is used and $\lambda_2 = (\ln 3.0 - \ln 1.5)/(350 - 150) = 0.003466$ when $F(t_i) = (i - 0.3)/(n + 0.4)$ is used. ■

13.7.5 Normal Distribution

When the variable t is normally distributed, we define the standard normal variate z as

$$z = \frac{t - \mu}{\sigma} \tag{13.71}$$

where μ and σ are, respectively, the mean and standard deviations of t. The cumulative distribution function is given by

$$F(t) = \Phi(z) = \Phi\left(\frac{t - \mu}{\sigma}\right) \tag{13.72}$$

and its inverse by

$$\frac{t-\mu}{\sigma} = \Phi^{-1}(F(t)) \tag{13.73}$$

or

$$t = \sigma\Phi^{-1}(F(t)) + \mu \tag{13.74}$$

The normal probability paper is based on Eq. (13.74) where t is plotted on the horizontal axis and $\Phi^{-1}(F(t))$ is plotted on the vertical axis. Thus, if t follows normal distribution, the data will fall on a straight line. Since $F(t) = 0.5$ and $\Phi^{-1}(0.5) = 0$ when $t = \mu$, the value of μ can be determined as the value of t for which $F(t) = 0.5$. Similarly, $F(t) = 0.8413$ and $\Phi^{-1}(0.8413) = 1.0$ when $t = \mu + \sigma$, the value of σ can be determined as the horizontal distance between the points corresponding to $F = 0.5$ and $F = 0.8413$.

EXAMPLE 13.9

The strengths of 10 welded joints are found to be 35.3, 31.8, 34.0, 39.9, 35.8, 41.2, 33.5, 37.1, 36.4, and 38.6 $k(\text{lb/in}^2)$. Plot the data on a normal probability paper and determine the parameters of the distribution from the graph.

Solution: The strengths are arranged in an increasing order and the ith strength value x_i is plotted against the plotting position $F(x_i) = i/(n+1)$ with $n = 10$. The necessary calculations are shown in Table 13.5. The resulting graph is shown in Fig. 13.9. The mean value of the strength

TABLE 13.5 Computation of distribution function, $F(x_i)$

i	$F(x_i) = \frac{i}{n+1}$	x_i
1	0.0909	31.8
2	0.1818	33.5
3	0.2727	34.0
4	0.3636	35.3
5	0.4545	35.8
6	0.5455	36.4
7	0.6364	37.1
8	0.7273	38.6
9	0.8182	39.9
10	0.9091	41.2

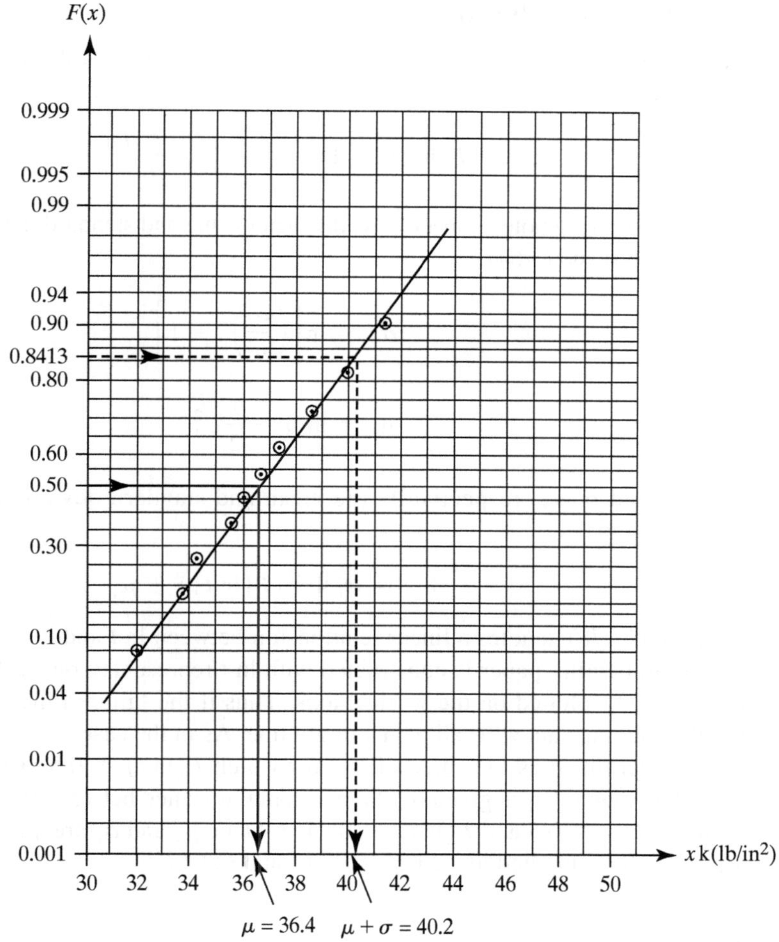

FIGURE 13.9 Normal probability paper.

of the welded joint, corresponding to a value of $\Phi^{-1}(0.5) = 0.0$ or $F(x) = 0.5$, can be seen to be $\mu = 36.4\,\mathrm{k(lb/in^2)}$. Similarly, the value of x corresponding to $\Phi^{-1}(0.8413) = 1.0$ or $F(t) = 0.8413$ gives $\mu + \sigma = 40.2\,\mathrm{k(lb/in^2)}$. Since the value of μ has been found to be $36.4\,\mathrm{k(lb/in^2)}$, we find the value of σ as $\sigma = (\sigma + \mu) - \mu = 40.2 - 36.4 = 3.8\,\mathrm{k(lb/in^2)}$. These values can be compared with the sample values of

$$\overline{X} = \frac{1}{n}\sum_{i=1}^{n} x_i = \frac{1}{10}\sum_{i=1}^{10} x_i = 36.36\,\mathrm{k(lb/in^2)}$$

and

$$s_X = \left\{\frac{1}{n}\sum_{i=1}^{n}(\overline{X} - x_i)^2\right\}^{1/2} = \left\{\frac{1}{10}\sum_{i=1}^{10}(36.36 - x_i)^2\right\}^{1/2} = 2.7840\,\mathrm{k(lb/in^2)}$$

■

13.7.6 Lognormal Distribution

If the failure time, t, follows lognormal distribution, then $\ln t$ follows normal distribution and hence

$$F(\ln t) = \Phi(z) = \Phi\left(\frac{\ln t - \mu_1}{\sigma_1}\right) \tag{13.75}$$

where μ_1 and σ_1 denote, respectively, the mean value and standard deviation of $\ln t$, given by Eqs. (3.63) and (3.64)

$$\sigma_1^2 = \ln\left[\left(\frac{\sigma}{\mu}\right)^2 + 1\right] \tag{13.76}$$

$$\mu_1 = \ln \mu - \frac{1}{2}\sigma_1^2 \tag{13.77}$$

where μ and σ are, respectively, the mean and standard deviation of t. Inverting of Eq. (13.75) gives

$$\ln t = \sigma_1 \, \Phi^{-1}(F(\ln t)) + \mu_1 \tag{13.78}$$

Equation (13.78) denotes a linear relationship between $\ln t$ and $\Phi^{-1}(F(\ln t))$. The lognormal probability paper is constructed with $\ln t$ represented on the horizontal axis and $\Phi^{-1}(F(\ln t))$ plotted on the vertical axis. Thus if the failure time t follows lognormal distribution, the data will fall on a straight line. As in the case of normal distribution, μ_1 can be determined as the value of $\ln t$ for which $F(\ln t) = 0.5$ or $\Phi^{-1}(F(\ln t)) = 0$. Also, the value of σ_1 is given by the horizontal distance between the points corresponding to $F(\ln t) = 0.5$ and $F(\ln t) = 0.8413$. Once μ_1 and σ_1 are known, the values of μ and σ can be computed using Eqs. (13.76) and (13.77).

13.7.7 Weibull Distribution

The three-parameter Weibull distribution function for the failure time t is given by

$$F(t) = 1 - \exp\left\{-\left(\frac{t - t_0}{\eta - t_0}\right)^\beta\right\}; \quad t \geq t_0, \beta > 0, \eta > t_0 \tag{13.79}$$

where t_0 = minimum possible value of $t (t \geq 0)$, η = characteristic parameter $(\eta \geq t_0)$, and β = shape parameter $(\beta > 0)$ denote the three parameters of the distribution. If the minimum life t_0 is taken as zero, Eq. (13.79) becomes a two-parameter Weibull distribution

$$F(t) = 1 - e^{-(t/\eta)^\beta} \tag{13.80}$$

Equation (13.79) can be rewritten as

$$1 - F(t) = \exp\left\{-\left(\frac{t - t_0}{\eta - t_0}\right)^\beta\right\} \tag{13.81}$$

Taking logarithms on both sides of Eq. (13.81) gives

$$\ln\left(\frac{1}{1 - F(t)}\right) = \left\{-\left(\frac{t - t_0}{\eta - t_0}\right)^{\beta}\right\} \tag{13.82}$$

Taking logarithms again, Eq. (13.82) yields

$$\ln\ln\left(\frac{1}{1 - F(t)}\right) = \beta \ln(t - t_0) - \beta \ln(\eta - t_0) \tag{13.83}$$

which can be seen to be of the form $y = cx + d$ where the ordinate (y) is $\ln\ln[1/(1 - F(t))]$ and the abscissa (x) is $\ln(t - t_0)$. The resulting graph paper is called the Weibull probability paper. If the failure time t follows Weibull distribution, the data points will fall on a straight line.

To determine the parameters of the Weibull distribution, first we consider the case of a two-parameter distribution for which $t_0 = 0$. For this case, Eq. (13.83) reduces to

$$\ln\ln\left(\frac{1}{1 - F(t)}\right) = \beta \ln t - \beta \ln \eta \tag{13.84}$$

where β and η are the unknown parameters. When the data is plotted as a straight line on the Weibull probability paper, the slope of the line gives the value of β. To find the value of η, we note that when the left-hand side of Eq. (13.84) is zero, we obtain $\beta(\ln t - \ln \eta) = 0$ or $t = \eta$. Since the left-hand side of Eq. (13.84) can be seen to be zero when $F(t) = 0.632$, the value of t corresponding to $F(t) = 0.632$ in the graph is taken as the value of η.

If the minimum failure time t_0 is not zero, we need to use the three-parameter Weibull distribution. In such a case, the values of t_0, β, and η can be estimated from the experimental data by using the following two-step procedure:

Step 1 Plot the data on a Weibull paper assuming that $t_0 = 0$. Since $t_0 \neq 0$ in reality, the data points fall on a curve instead of a straight line. Draw the best curve possible, using eye judgment, through the data points as shown in Fig. 13.10. Then select three points A_1, A_2, and A_3 on the curve with coordinates (X_1, Y_1), (X_2, Y_2), and (X_3, Y_3), respectively, such that they are equally spaced along the Y-axis:

$$Y_2 - Y_1 = Y_3 - Y_2 \tag{13.85}$$

Since the ordinate of Fig. 13.10 is defined as $Y = \ln\ln[1/(1 - F(t))]$, Eq. (13.85) gives

$$\ln\ln\left(\frac{1}{1 - F(t_2)}\right) - \ln\ln\left(\frac{1}{1 - F(t_1)}\right)$$
$$= \ln\ln\left(\frac{1}{1 - F(t_3)}\right) - \ln\ln\left(\frac{1}{1 - F(t_2)}\right) \tag{13.86}$$

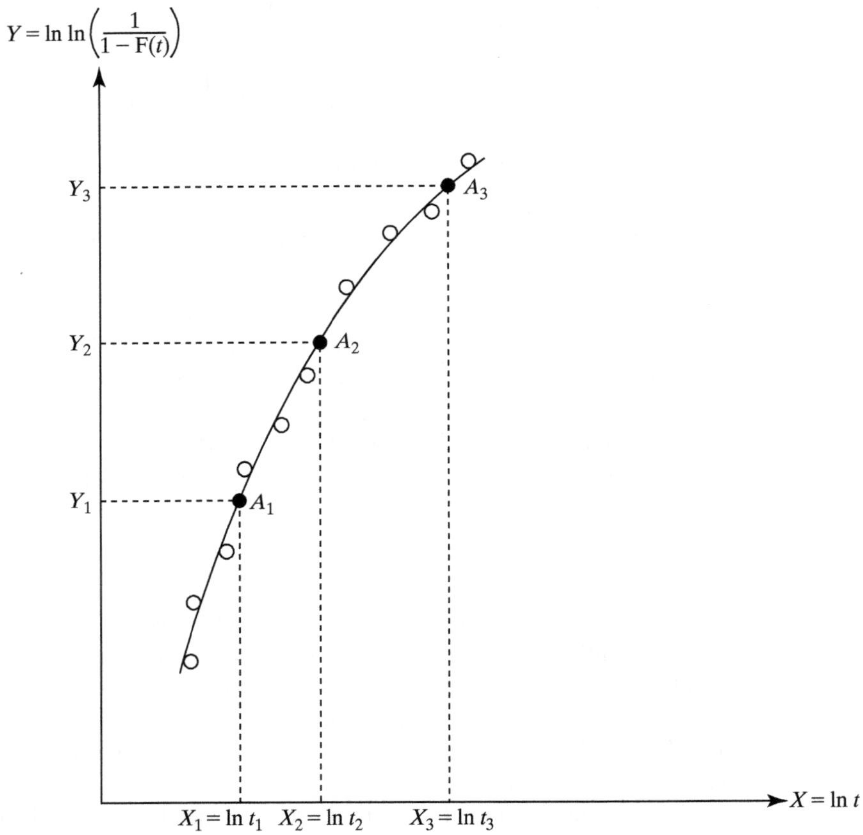

FIGURE 13.10 Plotting of data on Weibull probability paper.

Noting that the point A_i corresponds to $t_i (i = 1, 2, 3)$, Eq. (13.83) can be used to rewrite Eq. (13.86) in equivalent form as

$$\begin{aligned}[\beta \ln(t_2 - t_0) - \beta \ln(\eta - t_0)] - [\beta \ln(t_1 - t_0) - \beta \ln(\eta - t_0)] \\ = [\beta \ln(t_3 - t_0) - \beta \ln(\eta - t_0)] - [\beta \ln(t_2 - t_0) - \beta \ln(\eta - t_0)]\end{aligned} \tag{13.87}$$

Equation (13.87) can be rewritten as

$$\ln\left(\frac{t_2 - t_0}{\eta - t_0}\right) - \ln\left(\frac{t_1 - t_0}{\eta - t_0}\right) = \ln\left(\frac{t_3 - t_0}{\eta - t_0}\right) - \ln\left(\frac{t_2 - t_0}{\eta - t_0}\right)$$

or

$$\ln\left(\frac{t_2 - t_0}{t_1 - t_0}\right) = \ln\left(\frac{t_3 - t_0}{t_2 - t_0}\right) \tag{13.88}$$

By taking the inverse logarithms, Eq. (13.88) gives

$$(t_2 - t_0)^2 = (t_1 - t_0)(t_3 - t_0) \tag{13.89}$$

Equation (13.89) can be solved to obtain t_0 as

$$t_0 = \left(\frac{t_1 t_3 - t_2^2}{t_1 + t_3 - 2t_2} \right) \tag{13.90}$$

Step 2 Once t_0 is known, the data are replotted with the values of $\ln(t_i - t_0)$ as abscissa and the corresponding values of $\ln \ln [1/(1 - F(t_i))]$ as ordinates for $i = 1, 2, \ldots, n$. Since the data now falls on a straight line, the parameters β and η can be estimated from the graph as described for the case of two-parameter Weibull distribution.

EXAMPLE 13.10

The failure times of eight mechanical fuses are found to be 46, 85, 115, 140, 170, 205, 270, and 360 hours. Plot the data on a two-parameter Weibull probability paper and estimate the parameters of the distribution.

Solution: The failure times t_i are plotted using the plotting positions $F(t_i) = (i - 0.3)/(n + 0.4)$ indicated in Table 13.6. The data points $\{t_i, F(t_i)\}$ are plotted on the Weibull probability paper as shown in Fig. 13.11. A straight line is drawn through the data points and the parameters of the distribution are computed as follows. The value of $\ln t$ (on the abscissa) corresponding to a value of $F(t) = 0.632$ gives the characteristic parameter as $\eta = \ln 202 = 5.3083$. The slope of the Weibull line gives the shape parameter β as

$$\beta = \frac{F(t^{(2)}) - F(t^{(1)})}{\ln t^{(2)} - \ln t^{(1)}} = \frac{0.95 - 0.07}{\ln 400 - \ln 40} = 0.3822$$

Thus the two-parameter Weibull distribution corresponding to the given failure time data is given by

$$F(t) = 1 - \exp\left\{ -\left(\frac{t}{5.3083} \right)^{0.3822} \right\}$$

TABLE 13.6 Computation of distribution function

Data Point (i)	$F(t_i) = \dfrac{i - 0.3}{n + 0.4}$	Failure Time t_i (Hours)
1	0.08333	46
2	0.20238	85
3	0.32143	115
4	0.44048	140
5	0.55952	170
6	0.67857	205
7	0.79762	270
8	0.91667	360

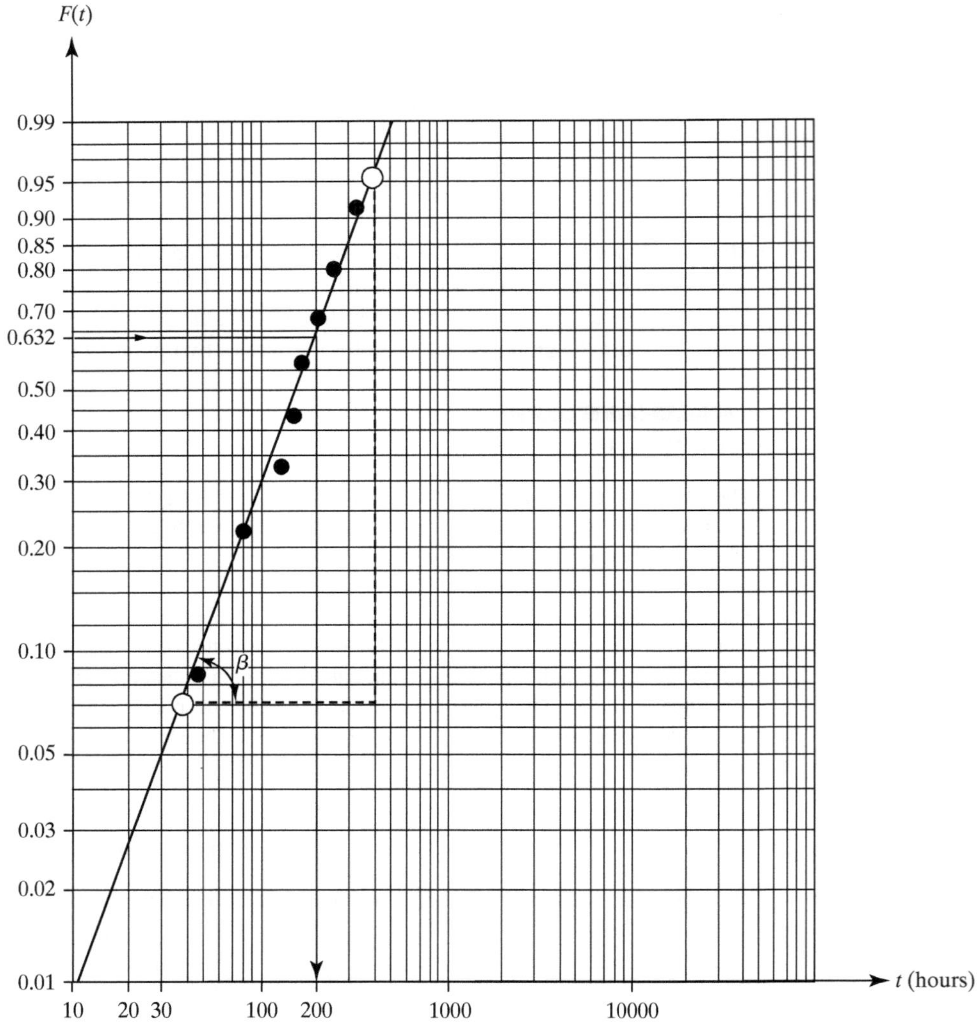

FIGURE 13.11 Two-parameter Weibull probability paper.

■

13.8 Numerical Solutions Using MATLAB

13.8.1 Parameter Estimation and Confidence Intervals

While estimating the parameters of an assumed distribution from a known set of data, a popular criterion used for goodness is to maximize the likelihood function. Thus, the values of the parameters of the assumed distribution can be found by maximizing the likelihood function. MATLAB has several functions which can be used for fitting different types of assumed distributions for any data. The estimated parameters of the distribution corresponding to the maximization of the likelihood function are given by the program. In addition to the parameters, the MATLAB functions give the $100(1-\alpha)$ percent confidence intervals. The default value of $\alpha = 0.05$ is used in many of the functions. The user can specify other values of confidence levels (α) such as 0.01 that correspond to 99 percent confidence intervals around the parameter values given by the program. The following examples illustrate the procedure.

EXAMPLE 13.11

The data on the fracture toughness of a particular grade of steel $\left(X = K_{\mathrm{Ic}} \text{ in ksi } \sqrt{\text{in}}\right)$ are given below:

70.1, 71.7, 72.4, 73.2, 73.7, 73.9, 74.3, 74.8, 75.1, 75.4, 75.9, 76.2, 76.7, 77.3, 77.6, 78.2, 78.8, 79.2, 79.8, 80.3, 83.5, 89.2, 94.1, 95.4

Assuming that the data follows normal distribution, find the parameters of the distribution, mean μ and standard deviation σ, and the associated 95 percent confidence interval for each of the parameters using MATLAB.

Solution: The MATLAB command used and the results given by the program are shown below:

```
clear all; clc;
x = [70.1; 71.7; 72.9; 73.2; 73.7; 73.9; 74.3; 74.8; 75.1; 75.4; 75.9; 76.2; 76.7;...
    77.3; 77.6; 78.2; 78.3; 78.8; 79.2; 79.8; 80.3; 83.5; 89.2; 94.1; 95.4]
[mu, sigma, muci, sigmaci] = normfit (x)

    mu =

       78.2240

    sigma =

        6.3334

    muci =

       75.6097
       80.8383

    sigmaci =

        4.9453
        8.8107
```

Note: *mu* and *sigma* denote the estimated mean and standard deviation, respectively, while *muci* and *sigmaci* indicate the 95 percent confidence intervals for *mu* and *sigma*, respectively.

■

EXAMPLE 13.12

For the data given in Example 13.11, fit exponential distribution and find the parameter of the distribution (mean, μ) and the associated 99 percent confidence interval for the mean using MATLAB.

Solution: The MATLAB command used and results given by the program are shown below:

```
clear all; clc;
x = [70.1; 71.7; 72.9; 73.2; 73.7; 73.9; 74.3; 74.8; 75.1; 75.4; 75.9; 76.2; 76.7;...
    77.3; 77.6; 78.2; 78.3; 78.8; 79.2; 79.8; 80.3; 83.5; 89.2; 94.1; 95.4]
[muhat, muci] = expfit(x, 0.01)

    muhat =

        78.2240

    muci =

        49.2037
       139.7319
```

Note: The number, 0.01, used as the second argument in the command ***expfit*** denotes the value of α which corresponds to 99 percent confidence interval for *mu*.

■

EXAMPLE 13.13

For the data given in Example 13.11, fit Weibull distribution and find the parameters of the distribution (beta and eta) as well as the associated 99 percent confidence interval for each of the parameters using MATLAB.

Solution: The MATLAB command used and the results given by the program are given below:

```
clear all; clc;
x = [70.1; 71.7; 72.9; 73.2; 73.7; 73.9; 74.3; 74.8; 75.1; 75.4; 75.9; 76.2; 76.7;...
    77.3; 77.6; 78.2; 78.3; 78.8; 79.2; 79.8; 80.3; 83.5; 89.2; 94.1; 95.4]
[phat, pci] = wblfit (x, 0.01)

    phat =

       81.3395  10.7844

    pci =

       77.2976   7.5975
       85.5927  15.3082
```

Note: The first and second numbers under *phat* denote the estimated values of the parameters β and η, respectively, and the first and second columns under *pci* indicate the 99 percent confidence intervals on β and η, respectively.

■

13.8.2 Plotting of Data

MATLAB has functions that can be used to plot a given set of data on normal and Weibull probability papers. If the plot is a straight line on the probability paper, the data are assured to follow the distribution used for the plot. If all the data points do not fall on a straight line, it implies that the data do not follow the distribution used for the plot. The following examples illustrate the procedure.

EXAMPLE 13.14

Plot the data on the fracture toughness of a particular grade of steel given in Example 13.11 on normal probability paper.

Solution: The command used and the resulting plot are shown below:

```
clear all; clc;
x = [70.1; 71.7; 72.9; 73.2; 73.7; 73.9; 74.3; 74.8; 75.1; 75.4; 75.9; 76.2; 76.7;...
    77.3; 77.6; 78.2; 78.3; 78.8; 79.2; 79.8; 80.3; 83.5; 89.2; 94.1; 95.4]
y = normplot (x)
```

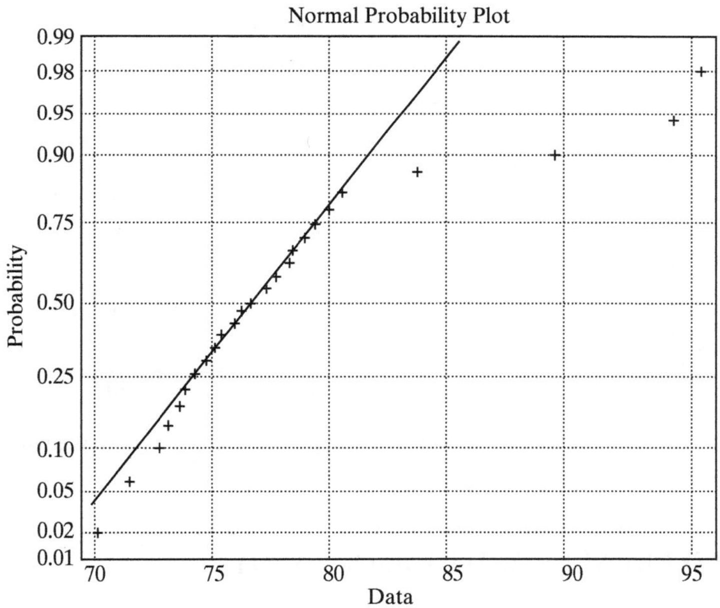

Note: Since all the data points do not fall on a straight line, the data do not follow normal distribution.

■

EXAMPLE 13.15

The data on the fracture toughness of steel welded joints ($X = K_{Ic}$ in ksi $\sqrt{\text{in}}$) are given below:

39.2, 45.9, 53.1, 54.7, 58.3, 61.8, 65.6, 67.2, 69.5, 72.9, 73.3, 74.4, 75.7, 79.4, 82.7, 83.8, 85.1, 87.2, 91.9, 92.6, 95.1, 99.9, 102.2, 106.7, 116.3

Plot the data on Weibull probability paper.

Solution: The command used and the resulting plot are shown below:

```
clear all; clc;
x2 = [39.2; 45.9; 53.1; 54.7; 58.3; 61.8; 65.6; 67.2; 69.5; 72.9; 73.3; 74.4; 75.7;...
     79.4; 82.7; 83.8; 85.1; 87.2; 91.9; 92.6; 95.1; 99.9; 102.2; 106.7; 116.3]
y = wblplot (x2)
```

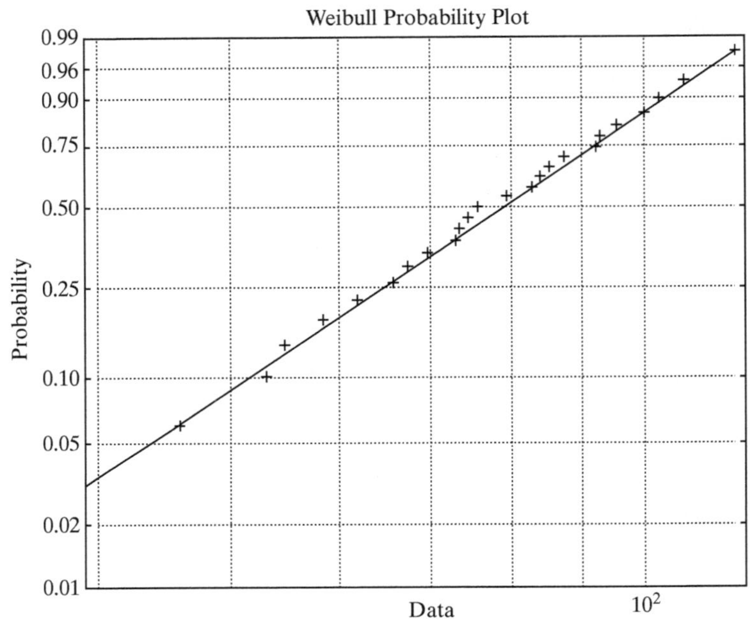

Note: Since all the data points fall on a straight line, the data follow Weibull distribution.

■

REFERENCES AND BIBLIOGRAPHY

13.1 P. A. Tobias and D. C. Trindade, *Applied Reliability*, Van Nostrand Reinhold Co., New York, 1986.

13.2 E. E. Lewis, *Introduction to Reliability Engineering*, John Wiley, New York, 1987.

13.3 J. R. King, *Probability Charts for Decision Making*, Industrial Press Inc., New York, 1971.

13.4 C. O. Smith, *Introduction to Reliability in Design*, McGraw-Hill, New York, 1976.

13.5 N. H. Roberts, *Mathematical Methods in Reliability Engineering*, McGraw-Hill, New York, 1964.

13.6 J. E. Shigley and C. R. Mischke, *Mechanical Engineering Design*, 5th ed., McGraw-Hill, New York, 1989.

13.7 "Automotive Electronics Reliability Handbook," Publication no. AE-9, Society of Automotive Engineers, Inc., Warrendale, PA., 1987.

13.8 A. Wald, *Sequential Analysis*, John Wiley, New York, 1947.

13.9 K. C. Kapur and L. R. Lamberson, *Reliability in Engineering Design*, John Wiley, New York, 1977.

13.10 E. B. Haugen, *Probabilistic Mechanical Design*, John Wiley, New York, 1980.

13.11 K. S. Trivedi, *Probability and Statistics with Reliability, Queuing and Computer Science Applications*, Prentice-Hall, Inc., Englewood Cliffs, NJ, 1982.

13.12 L. L. Lapin, *Probability and Statistics for Modern Engineering*, Brooks/Cole Publishing Co., Monterey, CA, 1983.

13.13 R. A. Heller, "The Weibull Distribution Did Not Apply to Its Founder," In S. Eggwertz and N. C. Lind, (eds.), *Probabilistic Methods in the Mechanics of Solids and Structures*, Springer-Verlag, Berlin, 1984, pp. xiii–xvi.

REVIEW QUESTIONS

13.1 Give brief answers to the following:

1. What type of reliability tests are conducted during the life cycle of a system?
2. State the objectives of a reliability test.
3. How many units or components are to be used in a reliability test?
4. Give three methods of conducting accelerated testing.
5. How is the "testing until partial failure" conducted?
6. How is acceleration achieved by using magnified loading?
7. What is the purpose of Arrhenius model?
8. What is the relationship between stress amplitude and the number of cycles in fatigue failure?
9. Describe the procedure used in sequential testing.
10. What is the significance of the average sampling number (ASN)?
11. What is meant by statistical inference?
12. What is the difference between parameter estimation and hypothesis testing?
13. What is the purpose of the maximum likelihood method?
14. Define the likelihood function.
15. How are the confidence level and the level of significance related?
16. What is confidence interval?
17. What is the significance of chi-square distribution?
18. State the principle used in the least-squares method.
19. What is linear rectification?
20. What is regression analysis?
21. How are the plotting positions defined for observed failure data?
22. What is the difference between mean rank and median rank?

13.2 Indicate whether each of the following statements is true or false:

1. Results of the tests conducted on prototype are used to modify/redesign the system.
2. Results of the tests conducted during manufacturing stage are used for redesigning the system.
3. Possible modifications of the design to improve reliability are considered based on the results of tests conducted during the operating stage.
4. Burn-in phase is same as the infant-mortality phase.
5. The usual sample size for testing subsystems is 4 to 20 units.

6. In the analysis of individual failure data, the cumulative distribution of the failure time at t_i (with i failures) is determined as $F(t_i) = \frac{i+1}{n}$.
7. With individual failure data, the reliability at time t_i can be found as $R(t_i) = \frac{n-i}{n}$.
8. In grouped failure data, only the time intervals in which groups of components fail are available.
9. It is not desirable to start with a larger sample size if failure time of only m units is to be tested.
10. The fractional time saved by using n units to observe the failure of m units $(m < n)$ is independent of the expected life of each of the units.
11. Accelerated testing using magnified loading is not suitable for fatigue testing.
12. Arrhenius model can be used for accelerated life testing with either increased temperature or increased stress.
13. In sudden-death testing, all remaining components are removed as soon as the first component fails in the group.
14. Sequential testing uses accept and reject lines.
15. Parameter estimation is different from point estimation.
16. Confidence intervals indicate the range in which the population mean lies.
17. Linear rectification transforms a linear equation to a nonlinear equation.

13.3 Fill in the blank with the proper word(s):

1. Reliability tests find whether a system can operate ______ over a specified period under prescribed operating conditions.
2. The usual sample size for testing individual components is ______ units.
3. The quantity $\frac{i}{n+1}$ is called the ______ rank in the analysis of individual failure times.
4. The aim of accelerated life testing is to ______ the time required for testing.
5. Accelerated life testing using magnified loading is useful only if the ______ between the normal and magnified load conditions is available.
6. Eyring model is more ______ than Arrhenius model for accelerated life testing.
7. The method of drawing conclusions about the population from the results of a sample is called statistical ______.
8. Hypothesis testing involves proving or ______ a hypothesis made about a parameter.
9. The difference between sample mean and population mean ______ as the sample size increases.

13.4 Select the most appropriate answer from the multiple choices given:

1. Better quality control methods can be devised to reduce defects based on the tests conducted during:
   a. Design stage **b.** Manufacturing stage **c.** Operating stage
2. Suitable warranty policies are developed based on the results of the tests conducted during:
   a. Design stage **b.** Manufacturing stage **c.** Operating stage

3. In general, high failure rates are observed during:
   a. Burn-in phase **b.** Operating-life phase **c.** Wear-out phase

4. Usual sample size for testing a complete system is:
   a. 1 to 10 units **b.** 20 to 100 units **c.** At least 100 units

5. In the analysis of individual failure data, a better formula for determining the cumulative distribution of failure time at t_i is:
   a. $\frac{1}{n}$ **b.** $\frac{i+1}{n}$ **c.** $\frac{i}{n+1}$

6. In the analysis of individual failure time, the quantity $\frac{i-0.3}{n+0.4}$ is called:
   a. Mean rank **b.** Median rank **c.** Mode rank

7. With grouped failure data, the reliability at time t_i, $R(t_i)$ can be estimated in terms of the number of surviving components (n_i) at t_i and the total number of components used in the test (n), as:
   a. $\frac{n-n_i}{n}$ **b.** $\frac{n_i}{n}$ **c.** $\frac{n_i+1}{n}$

8. Sudden-death testing is desirable if each component is
   a. Inexpensive **b.** Very expensive **c.** Moderately expensive

9. Sequential testing requires a sample of:
   a. One unit **b.** Ten units **c.** Sample size is not fixed

10. Average sampling number (ASN) is used in:
    a. Sudden-death testing
    b. Sequential testing
    c. Magnified-load testing

11. Parameter estimation, using n sample values, involves estimating:
    a. Single value of a parameter of the distribution
    b. Two values of a parameter of the distribution
    c. Single value for each of the multiple parameters of the distribution

12. The maximum-likelihood method is used for:
    a. Hypothesis testing
    b. Accelerated testing
    c. Parameter estimation

13. The sum of the level of significance and the confidence level is equal to:
    a. 0 **b.** 1 **c.** >1

13.5 Match the following tests with their purposes:

1. Tests at the end of design stage	**a.** To find the failure rate and reliability of the system
2. Tests during the manufacturing stage	**b.** To analyze failure modes and reliability of the system
3. Tests during the operating stage	**c.** To prove the achievement of reliability standards

13.6 Match the following tests with their objectives:

1. Longevity tests	**a.** To find the ability of the system to perform without failure for a specified minimum period
2. MTBF tests	**b.** To find the length of the useful life with constant failure rate
3. Operating life tests	**c.** To establish the margin of safety between extreme operating conditions
4. Reliability margin tests	**d.** To find the time between failures

13.7 Match the following in the context of accelerated life testing:

1. Sequential life testing	**a.** Speeds up chemical reaction time
2. Sudden-death testing	**b.** Aims to determine whether reliability goal is met
3. Testing with magnified loading	**c.** Aims at finding the average MTBF
4. Testing with enhanced temperature	**d.** Fatigue testing of materials
5. Testing until partial failure	**e.** Weibull distribution is used

PROBLEMS

13.1 The following failure times were observed for 15 hydraulic cylinders used in aircraft landing gears: 1719, 693, 1374, 2841, 1121, 1839, 947, 1746, 1972, 2477, 890, 1418, 1643, 2644, and 2165 hours. Plot the probability density, probability distribution, reliability, and hazard functions of the life of hydraulic cylinders.

13.2 The following failure data is collected for a group of 100 accelerometers:

Time Interval (100 Hours each)	Number of Failures Observed
1	21
2	13
3	9
4	7
5	6
6	6
7	5
8	7
9	11
10	15

Plot the probability density, probability distribution, reliability, and hazard functions of the failure time of the accelerometers.

13.3 Find the ratio of expected waiting times to observe the rth failure in samples of size n and r, respectively, by varying r from 1 to 20 and n from 1 to 25.

13.4 Find the magnification needed for the stress amplitude in the fatigue testing of AISI 1045 steel to reduce the testing time by a factor of 10. The *S–N* curve of this steel is given by $SN^{0.085} = 154.0$.

13.5 A sudden-death test is conducted by dividing randomly a group of 50 valves used in the ice-making unit of household refrigerators into groups of 5 each. The failure times observed for the first unit of each of the 10 groups are given by 192, 233, 274, 305, 359, 386, 411, 428, 457, and 585 hours. Find the median life of the population of the valves.

13.6 A lot of IC chips are subjected to a sequential test with values of $R_u = 0.95$, $R_l = 0.90$, $\alpha = 0.05$, and $\beta = 0.05$. The following observations are made sequentially about the condition of the chips: 40 good, 2 bad, 35 good, 1 bad, 30 good, and 2 bad. Plot the "accept" and "reject" lines and determine whether the lot is to be accepted or rejected. Also compare the number of chips tested with the ASN.

13.7 Find the equations of the "accept" and "reject" lines for a sequential test with $R_u = 0.99$, $R_l = 0.90$, $\alpha = 0.01$, and $\beta = 0.02$. Also estimate the number of units to be tested before making a decision about the acceptance/rejection of the lot.

13.8 The wave heights (x_i) at the site of an offshore platform are found to be 3.4, 3.7, 3.8, 4.1, 4.5, 4.8, 6.5, 5.1, 6.2, 5.3, 5.7, and 6.9 ft. Assuming Rayleigh's distribution

$$f_X(x) = \frac{x}{\alpha^2} e^{-1/2(x/\alpha)^2}; \quad x \geq 0$$

can be used to describe the wave heights, determine the maximum likelihood estimator of the parameter α.

13.9 The sample values $x_1, x_2, \ldots, x_n$ have been observed for a random variable X following binomial distribution

$$p_X(x_i) = p^{x_i}(1 - p)^{1-x_i}; \quad x_i = 0, 1$$

where p is a parameter. Find the maximum likelihood estimator of p.

13.10 The value of a parameter $\hat{\theta} = \hat{\theta}(x_1, x_2, \ldots, x_n)$ estimated from the observed values $x_1, x_2, \ldots, x_n$ of the random variable X is said to be an unbiased estimator of the parameter θ if $E[\hat{\theta}(x_1, x_2, \ldots, x_n)] = 0$. Prove that the variance of the random variable X computed using the formula

$$s_X^2 = \frac{1}{n} \sum_{i=1}^{n} (\overline{X} - x_i)^2$$

is a biased estimator of the population variance, σ^2.

13.11 Prove that the variance of the random variable X computed using the formula

$$s_X^2 = \frac{1}{n-1} \sum_{i=1}^{n} (\overline{X} - x_i)^2$$

is an unbiased estimator of the population variance σ^2.

13.12 If the value of a function $Y = Y(X_1, X_2, \ldots, X_n)$ is to be estimated from the measured values of $X_1, X_2, \ldots, X_n$, the confidence interval for Y is defined as

$$P\left[-z_{\alpha/2} < \frac{Y - \overline{Y}}{s_Y} < z_{\alpha/2}\right] = 1 - \alpha$$

where $\overline{Y} = Y(\overline{X}_1, \overline{X}_2, \ldots, \overline{X}_n)$,

$$s_{\overline{Y}}^2 \approx \sum_{i=1}^{n} \left(\frac{\partial Y}{\partial X_i} \bigg|_{\overline{X}_1, \overline{X}_2, \ldots, \overline{X}_n} \right)^2 s_{\overline{X}_i}^2$$

and $(\overline{X}_i, s_{X_i})$ are determined from the measured values of $X_i (i = 1, 2, \ldots, n)$. Thus, the $(1 - \alpha)$ confidence interval of Y can be determined as $(\overline{Y} - z_{\alpha/2}\, s_Y;\ \overline{Y} + z_{\alpha/2}\, s_Y)$. The dimensions of a casting, which is in the form of a rectangular prism, are measured independently six times and the following results are obtained:

Length (X_1): 18.2, 17.9, 18.3, 18.4, 18.1, and 18.0 in

Width (X_2): 11.9, 12.2, 12.0, 12.3, 11.8, and 12.1 in

Depth (X_3): 8.2, 8.1, 8.4, 8.0, 8.3, and 7.9 in

Determine the following:

a. The means and standard deviations of X_1, X_2, and X_3
b. The mean and standard deviation of the volume of the casting
c. The 95 percent confidence interval on the true volume of the casting

13.13 If 15 observations made for the yield strength of steel gave the sample mean as 29,800 lb/in^2 and the sample standard deviation as 1800 lb/in^2, determine the 99 percent confidence interval for the population standard deviation of the yield strength.

13.14 The standard deviation of the yield strength of steel is known to be 1500 lb/in^2. From a sample of 15 observations, the sample mean of the yield strength is found to be 29,800 lb/in^2: Find (a) the 95 percent confidence interval, and (b) the 99 percent confidence interval for the population mean of the yield strength of steel.

13.15 If 15 observations made for the yield strength of steel gave the sample mean as 29,800 lb/in^2 and the sample standard deviation as 1800 lb/in^2, determine the 99 percent confidence interval for the population mean of the yield strength.

13.16 The fatigue lives of 10 helical springs are found to be 1.1, 1.2, 1.4, 1.5, 1.7, 1.9, 2.0, 2.1, 2.4, and 2.8 million cycles. Determine the following: (a) The 95 percent confidence interval on the population mean of the fatigue life; (b) The 95 percent confidence interval on the population standard deviation of the fatigue life.

13.17 The temperatures measured at various distances along the thickness of the wall of a cold storage building are given as:

Distance, x (Inch)	Temperature, T (Degrees F)
0	90
4	69
8	58
12	47
16	40
20	34
24	30

Determine the following relationship between the temperature and the distance using least-squares technique:

a. Linear relationship: $T = a + bx$
b. Quadratic relationship: $T = a + bx + cx^2$

13.18 The average number of breakdowns experienced in different months in a machine shop due to the failure of various machines are given by 39.0, 14.5, 6.5, 75.0, 3.0, 19.5, 54.0, 31.5, 10.5, and 23.0. Plot the data on an exponential probability paper and determine the breakdown rate in the machine shop.

13.19 The tensile strengths of 12 welded joints have been observed to be 31,325, 31,556, 31,234, 31,998, 31,404, 31,263, 31,440, 31,785, 31,492, 31,357, 31,285, and 31,643 lb/in^2. Assuming that the tensile strength of welded joints follows exponential distribution, determine the parameters of the distribution graphically.

13.20 The crushing strengths of test specimens made of concrete used for a machine foundation are observed to be 18, 27, 24, 30, 22, 36, 41, 33, 29, and 34 $\text{k}(\text{lb/in}^2)$. Plot the data by assuming normal distribution to be valid for describing the crushing strength of concrete. Also, determine the parameters of the distribution from the graph.

13.21 Solve Problem 13.20 by assuming lognormal distribution for the crushing strength of concrete.

13.22 Solve Problem 13.20 by assuming a two-parameter Weibull distribution for the crushing strength of concrete.

13.23 Tests conducted to find the ultimate tensile strength of an alloy steel gave the following results: 63.3, 61.8, 67.4, 65.3, 59.7, 64.2, 60.9, 66.1, 62.6, and 63.9 $\text{k}(\text{lb/in}^2)$. Assuming that the ultimate tensile strength follows a two-parameter Weibull distribution, determine the parameters of the distribution.

13.24 Solve Problem 13.23 using a three-parameter Weibull distribution.

13.25 The times needed to restore power due to electrical outages by an electric power company are given by 1.0, 0.9, 2.4, 0.3, 0.7, 3.6, 2.0, 1.4, 0.4, and 1.6 hours. Plot the data on a lognormal paper and determine the mean value and standard deviation of the "restoring time" from the graph.

MATLAB Problems

13.26 Assuming that the data on the fracture toughness of steel welded joints given in Example 13.9 follows normal distribution, find the parameters of the distribution (mean and standard deviation) and the associated 95 percent confidence interval for each of the parameters using MATLAB.

13.27 Assuming that the data on the fracture toughness of steel welded joints given in Example 13.9 follows exponential distribution, find the parameter of the distribution (mean) and the associated 95 percent confidence interval of the parameter using MATLAB.

13.28 Assuming that the data on the fracture toughness of steel welded joints given in Example 13.9 follows Weibull distribution, find the parameters of the distribution (β and η) and the associated 95 percent confidence interval for each of the parameters using MATLAB.

13.29 Plot the data on the fracture toughness of steel welded joints given in Example 13.9 on normal probability paper.

13.30 Plot the data on the fracture toughness of a particular grade of steel given in Example 13.14 on Weibull probability paper.

CHAPTER 14

Quality Control and Reliability

What You Will Learn

In this chapter, you will learn

- The definition of quality
- The importance of controlling the dimensions of products from the point of view of reliability
- The important discrete probability distributions, such as binomial, hypergeometric, and Poisson distributions, used in statistical quality control
- The six sigma approach and its relationship to reliability, along with a discussion on the basic concepts of sigma level, yield, and defective parts per million (DPPM), as well as the implementation of the six sigma approach.
- The details of quality control charts—the p-chart, the $\overline{X}$-chart, the R-chart, and c-chart—with illustrative examples
- The basic ideas of the operating characteristic curve, producer's risk parameter, consumer's risk parameter, lot tolerance percent defective parameter, and acceptance quality level parameter
- The basic idea behind the Taguchi method and the concept of loss function
- The role of noise factors, experimental design, orthogonal arrays, and signal-to-noise(S/N) ratio in Taguchi method with illustrative examples

Walter Andrew Shewhart, the inventor of statistical quality control techniques, was born in New Canton, Illinois on March 18, 1891. He received Ph.D. in Physics from the University of California, Berkeley in 1917. He joined the Western Electric Company, which later became the Bell Telephone Laboratories, as a research engineer and statistician in 1918 and continued there until his retirement in 1956. He died on March 11, 1967 in Troy Hills, New Jersey. When the routine quality checks provided by inspectors were not good enough to have the required confidence in the instruments and appliances produced by Western Electric, Shewhart developed the first control chart in 1924 which has formed a basis for the widely used statistical quality control methods. He published the book, *Economic Control of Quality of Manufactured Product*, in 1931. In addition, after the founding of the Bell Telephone Laboratories in 1925, he published many important papers in the *Bell System Technical Journal*. He received many honors, medals, and an honorary D. Sc. Degree. The ASME presented him the Holley Medal in 1954 for his pioneering work on the application of statistical methods for the quality control of manufactured products.

14.1 Introduction

Quality. Quality can be defined in several ways. In simple terms, quality means fitness of a product or service for use. In a broader sense, quality means the ability of a product or service to satisfy a stated or implied need. Quality can also mean that the product or service is free from defects or deficiencies. According to the American Society for Quality (ASQ), quality is the totality of features and characteristics of a product or service that influence its ability to satisfy a given need. Improvement of quality implies elimination of waste in the form of scrap or rework (needed to make the component or system deliver the expected level of performance) and leads to lower costs, higher productivity, higher levels of customer satisfaction, more reputation for the product, and higher profits for the company.

Quality has two aspects—quality of design and quality of conformance. The quality of design is based on engineering decision regarding the degree of performance, function, or reliability of the component or system. It is accomplished through the design process to make the component or system achieve a specified level of reliability. The quality of conformance is based on the reduction of variability of performance. This means that each unit produced is defect-free and delivers the expected level of performance. Since it is impossible to inspect each and every product or unit produced in a production process, the product must be produced right the first time. This means that the manufacturing process must be stable or repeatable and must not cause any variation in the nominal or target specification such as a dimension. Statistical process control usually refers to the statistical techniques used in sampling production and monitoring the variability of the product from its specifications. It can be used to achieve both stability and reduction of variability of the production process.

Six Sigma Approach. Six sigma approach, a trademark of Motorola, was introduced by Motorola in the 1980s as a strategy to reduce the defects and product failure rate. The term *six sigma* was coined by William Smith, a reliability engineer at Motorola. Within few years of the introduction of the approach, six sigma became a major strategy to improve the reliability of products, systems, and organizations. It was used successfully at corporations such as General Electric, Motorola, and 3M. Initially, six sigma approach was developed to improve the manufacturing processes. It provides a well-defined quality target by specifying a maximum of 3.4 defects per million opportunities. Later, the concept has been used to improve any process, including businesses and organizations. The six sigma methodology and its use in improving the quality and reliability of products and systems through minimization of the number of defective units is discussed in this chapter.

Statistical Quality Control. The foundations of statistical quality control were laid in 1916 when Frazee, at Bell Telephone Laboratories, used statistical concepts in inspection and developed the well-known operating characteristic (OC) curve. Later, in 1924, Shewhart introduced the quality control (QC) charts. In 1940, the American Standards Association, at the request of the War Department, developed two standards involving the application of statistical quality control methods to manufactured products. Subsequently, the American Society for Quality Control (ASQC) was established in 1946. This society fosters the growth and development of quality control methods through the publication of the journals, such as *Quality Progress* and *Journal of Quality Technology*, and other activities.

In statistical quality control, tests are conducted on random samples of components or products and the results of the tests are plotted on graphs known as statistical control charts to determine whether the manufacturing process is under control. These tests are usually destructive to the components and are time-consuming and costly. Hence, it is important that efficient sampling strategies, based on sound statistical principles, be developed for controlling the quality of a product or process while using minimum number of components for inspection.

Since variability is inherent in all manufactured products, a compromise solution between reducing the variability of the quality characteristic and reducing the manufacturing cost is to be made in practice. In mass production, tests are conducted on random samples of manufactured components and systems to determine the quality and reliability of the products. Acceptance sampling plans involve testing of small samples to assess the acceptability of quality of the whole production lot. Statistical quality control techniques are used in sampling production and monitoring the variability of the product from its specification. It can be used to achieve both stability and reduction of variability of the production process. The operating characteristic curves are used to show the percentage of defective units in the lot versus the probability of acceptance of the lot including producer's and consumer's risks.

This chapter discusses the various types of sampling plans available for attribute data, the operating characteristic curves, and the choice of minimum sample sizes. The applications of various discrete probability distributions, including the binomial and Poisson distributions, are also discussed. In addition, the application of statistical process control charts for quality control is also outlined. The statistical quality control methods monitor a manufacturing process by testing samples drawn from the production line and verifying the quality levels produced. These methods can be considered as on-line quality control methods. The off-line quality control methods seek to improve the quality of a product or process by using robust design techniques during the development stages of the product or process.

Taguchi Method. The Taguchi method, which reduces the variation in the performance of a product or process, can be considered as an off-line quality control method. It can be used to achieve a robust design that is insensitive to external noises, manufacturing imperfections, and deterioration factors. This enables a consistent performance of the product or system, which leads to a higher reliability of the product or system. A brief discussion of Taguchi method is presented, along with illustrative examples, in this chapter.

14.2 Importance of Controlling Dimensions of Products

The importance of quality control, in terms of controlling, for example, the mean values or standard deviations of the dimensions of a machine part, from reliability point of view is illustrated with the following examples.

EXAMPLE 14.1

Pistons of IC engine are produced on a lathe in a manufacturing plant. It is known, from past production, that the dimensions of parts produced on the lathe have a standard deviation of 0.1 mm. Assuming that the diameters of pistons produced on the lathe follow normal distribution and the lathe is set to produce the mean value of piston diameters exactly at the middle of the specified lower and upper tolerance limits, determine the probability of producing a piston within the tolerance limits for the following specification of the diameter of the piston (D):

1. $D = 100^{+0.0}_{-0.1}$ mm, **2.** $D = 100^{+0.3}_{-0.3}$ mm, **3.** $D = 100^{+0.6}_{-0.6}$ mm

Solution:

1. The mean value of the diameter of the piston is given by

$$\overline{D} = \frac{100.0 + 99.9}{2} = 99.95 \text{ mm}$$

with standard deviation (σ_D) of 0.1 mm. The probability of producing a piston outside the specified tolerance limits can be determined as

$$\begin{aligned} P(D < 99.9) &+ P(D > 100.0) \\ &= P\left(\frac{D - \overline{D}}{\sigma_D} < \frac{99.9 - 99.95}{0.1}\right) + P\left(\frac{D - \overline{D}}{\sigma_D} > \frac{100.0 - 99.95}{0.1}\right) \\ &= P(Z < -0.5) + P(Z > 0.5) \\ &= 0.3085 + 0.3085 = 0.6170 \end{aligned}$$

It can be seen that the reliability of production (i.e., the probability of producing pistons within the acceptable limits) is only 0.3830.

2. Since the machine (lathe) is set to produce the mean value exactly at the midpoint of the tolerance range, we have

$$\overline{D} = 100.0 \text{ mm with } \sigma_D = 0.1 \text{ mm}$$

The probability of producing a piston outside the specified tolerance limits is given by

$$\begin{aligned} P(D < 99.7) &+ P(D > 100.3) \\ &= P\left(\frac{D - \overline{D}}{\sigma_D} < \frac{99.7 - 100.0}{0.1}\right) + P\left(\frac{D - \overline{D}}{\sigma_D} > \frac{100.3 - 100.0}{0.1}\right) \\ &= P(Z < -3.0) + P(Z > 3.0) \\ &= 0.00135 + 0.00135 = 0.0027 \end{aligned}$$

It can be observed that the reliability of production has improved to 0.9973 (this implies a rejection rate of 27 parts per 10,000 or 2700 units per million or 2700 ppm).

3. In this case, the probability of producing a piston outside the specified tolerance limits is given by

$$\begin{aligned} P(D < 99.4) &+ P(D > 100.6) \\ &= P\left(\frac{D - \overline{D}}{\sigma_D} < \frac{99.4 - 100.0}{0.1}\right) + P\left(\frac{D - \overline{D}}{\sigma_D} > \frac{100.6 - 100.0}{0.1}\right) \\ &= P(Z < -6.0) + P(Z > 6.0) \\ &= 0.000000001 + 0.000000001 = 0.000000002 \end{aligned}$$

It can be seen that the reliability of production has improved considerably to 0.999999998 (this implies a rejection rate of 2 units per billion or 2 ppb). Note that the specified tolerance limits correspond to *six-sigma quality* standard.

■

EXAMPLE 14.2

In Example 14.1, if the process is not centered in the tolerance range, but located at a distance of 1.0σ from the center of the tolerance range, then determine the probability of producing a piston within the specified tolerance limits for each of the three specifications for the diameter of the piston.

Solution:

1. For the specified tolerances on D of $100^{+0.0}_{-0.1}$ mm, the mean value of the diameter of the piston is

$$\overline{D} = \frac{100.0 + 99.9}{2} + 0.1 = 100.05 \text{ mm}$$

The probability of producing a piston outside the specified tolerance limits is given by

$$\begin{aligned} P(D < 99.9) + P(D > 100.0) & \\ &= P\left(\frac{D - \overline{D}}{\sigma_D} < \frac{99.9 - 100.05}{0.1}\right) + P\left(\frac{D - \overline{D}}{\sigma_D} > \frac{100.0 - 100.05}{0.1}\right) \\ &= P(Z < -1.5) + P(Z > -0.5) \\ &= (1 - 0.93319) + 0.69146 = 0.75827 \end{aligned}$$

2. In this case, the specified tolerances on D are $100^{+0.3}_{-0.3}$ mm with the mean value of the diameter of the piston as

$$\overline{D} = \frac{100.3 + 99.7}{2} + 0.1 = 100.1 \text{ mm}$$

The probability of producing a piston outside the specified tolerance limits will be

$$\begin{aligned} P(D < 99.7) + P(D > 100.3) & \\ &= P\left(\frac{D - \overline{D}}{\sigma_D} < \frac{99.7 - 100.1}{0.1}\right) + P\left(\frac{D - \overline{D}}{\sigma_D} > \frac{100.3 - 100.1}{0.1}\right) \\ &= P(Z < -4.0) + P(Z > 2.0) \\ &= (1 - 0.99996833) + (1 - 0.97725) = 0.02278167 \end{aligned}$$

3. The specified tolerances on D are given by $100^{+0.6}_{-0.6}$ mm. In this case, the mean value of the diameter of the piston is

$$\overline{D} = \frac{100.6 + 99.4}{2} + 0.1 = 100.1 \text{ mm}$$

Since $\sigma_D = 0.1$ mm, the probability of producing a piston outside the specified tolerance limits will be

$$
\begin{aligned}
P(D < 99.4) &+ P(D > 100.6) \\
&= P\left(\frac{D - \overline{D}}{\sigma_D} < \frac{99.4 - 100.1}{0.1}\right) + P\left(\frac{D - \overline{D}}{\sigma_D} > \frac{100.6 - 100.1}{0.1}\right) \\
&= P(Z < -7.0) + P(Z > 5.0) \\
&= 0 + (1 - 0.9999996981) = 0.0000003019
\end{aligned}
$$

It can be observed that the control of the variability (standard deviation) and the mean value of the diameter of pistons influence the reliability of the production process considerably.

■

14.3 Important Discrete Probability Distributions

Several discrete probability distributions find application in statistical quality control. We shall discuss the use of binomial, hypergeometric, and Poisson distributions in the following sections.

14.3.1 Binomial Distribution

As discussed in Chapter 3, the binomial distribution is applicable to the following situations:

1. There are only two outcomes in any trial, like success and failure.
2. There exists a fixed probability of success in each trial.
3. The outcome of each trial is independent of the outcomes of other trials.

The probability of realizing x successes in n trials is given by

$$P(X = x) = p_X(x) = \binom{n}{x} p^x (1 - p)^{(n-x)} \tag{14.1}$$

where p denotes the probability of success and $(1 - p)$ represents the probability of failure in each trial.

EXAMPLE 14.3

Fifty ball bearings are tested for 1500 hours under similar operating conditions. The probability of a ball bearing failing in 1500 hours is known to be 1 percent from previous experience. Assuming that all the bearings are selected from the same batch of production, find the following:

1. The probability of bearings surviving 1500 hours of operation.
2. The probability of at least one bearing failing in 1500 hours of operation.
3. The probability of exactly one bearing failing in 1500 hours of operation.

Solution:

1. The probability of all 50 bearings surviving 1500 hours of operation is same as the probability of zero bearings failing and can be found using Eq. (14.1) as:

$$P(X = 0) = \binom{50}{0}(0.01)^0(0.99)^{50} = 0.99^{50} = 0.6050$$

2. The probability of at least one bearing failing is given by

$$P(X > 0) = P(X \geq 1) = 1 - P(X = 0) = 1 - 0.6050 = 0.3950$$

3. The probability of exactly one bearing failing is given by

$$p_X(1) = \binom{50}{1}(0.01)^1(0.99)^{49} = 50(0.01)(0.99)^{49} = 50(0.01)(0.6111) = 0.3056$$

■

14.3.2 Hypergeometric Distribution

The hypergeometric distribution enables us to determine the probability of finding $X = x$ defective units in a sample of size n taken from a lot of N containing a total of m defective units as [14.1]:

$$P(X = x) = p_X(x) = \frac{\binom{n}{x}\binom{N-n}{m-x}}{\binom{N}{n}}; \quad x = 0, 1, 2, \ldots, n \tag{14.2}$$

If N is large compared to n, binomial distribution can be used to yield a good approximation to the hypergeometric distribution.

EXAMPLE 14.4

Ten castings are found to be defective in an inspection process. Out of these ten defective units, two castings have blowholes and eight have surface cracks. If three castings are selected at random from the 10 defective units, what is the probability that none of these will have blowholes?

Solution: We can find the required probability using the hypergeometric distribution with $N = 10$, $n = 3$, $m = 2$, and $x = 0$:

$$P(X = 0) = \frac{\binom{3}{0}\binom{10-3}{2-0}}{\binom{10}{3}} = \frac{\left(\frac{3!}{0!\,3!}\right)\left(\frac{7!}{2!\,5!}\right)}{\left(\frac{10!}{3!\,7!}\right)} = \frac{21}{120}$$

■

EXAMPLE 14.5

A production lot of 200 IC chips contains 5 defective units. Find the probability of not getting a defective chip in a sample of 10 drawn from the lot using (1) hypergeometric distribution and (2) binomial distribution.

Solution:

1. The hypergeometric distribution gives the probability of not getting a defective chip as

$$P(X = 0) = \frac{\binom{5}{0}\binom{200-5}{10-0}}{\binom{200}{10}} = \frac{190!\ 195!}{185!\ 200!} = 0.771716$$

2. Since $N = 200$ is large compared to $n = 5$, the binomial distribution can be used, with $p = \dfrac{5}{200} = 0.025$, to find the required probability as

$$P(X = 0) = \binom{10}{0}(0.025)^0(1 - 0.025)^{10} = (0.975)^{10} = 0.776330$$

This result can be seen to be very close to the result found in part(a).

■

14.3.3 Poisson Distribution

In many quality control applications, we need to distinguish between a defect and a defective product. In this connection, a defect is defined as a nonconformance to specifications and a defective product is defined as one with one or more defects. We can deal with defective units using binomial or hypergeometric distribution. To deal with components involving a number of defects per unit, including cases dealing with a number of failures or accidents per unit period, we can use Poisson distribution. The Poisson distribution is defined as [14.1]:

$$p_X(x) = P(X = x) = \frac{\lambda^x e^{-\lambda}}{x!}; \quad x = 0, 1, 2, \ldots \tag{14.3}$$

where λ is the mean value of X.

EXAMPLE 14.6

In the quality control of automobile brakes, a random sample of 40 units is tested periodically. If the process producing these brakes generates 3 percent defective units, determine the probability of finding no defective units in the sample tested.

Solution: The required probability can be found using Poisson distribution, with $\lambda = 40 \times 0.03 = 1.2$, as

$$P(X = 0) = \frac{(1.2)^0 e^{-1.2}}{0!} = 0.301194$$

■

14.3.4 Relationship Between Poisson and Exponential Distributions

The Poisson distribution is directly related to the exponential distribution. To see this, consider a system with an exponential life distribution with parameter λ (λ denotes the failure or repair rate). Assume that the system is repaired whenever it fails. The mean value of the number of failures (X) in time t is given by λt and the probability of having to make k repairs in time t is given by the Poisson distribution:

$$P(X = k) = \frac{(\lambda t)^k e^{-\lambda t}}{k!} \tag{14.4}$$

To verify this equation, consider a system which operates for time t without failure. The probability of having zero failures $(X = 0)$ is given by Eq. (14.4) as:

$$P(X = 0) = e^{-\lambda t} \tag{14.5}$$

Thus, the probability of realizing at least one failure is given by

$$P(X \geq 1) = 1 - P(X = 0) = 1 - e^{-\lambda t} \tag{14.6}$$

Equation (14.6) can be identified as the probability distribution function, $F(t)$, of the exponentially distributed life of the system. This shows that if the discrete random variable denoting the number of failures (X) follows Poisson distribution with parameter λt, then the probability of failure in time t is given by the continuous exponential distribution with parameter λ.

14.4 Six Sigma Approach and Reliability

The Greek letter sigma (σ) is commonly used to denote the standard deviation, which is a measure of the variability or spread of data about its mean value. It is well recognized that variability within a process leads to variations in the performance of the system and leads to customer dissatisfaction. Thus, the variability of a process about its mean value represents poor quality of the process. The term *six sigma*, a trademark of Motorola, was coined by William Smith, a reliability engineer at Motorola.

Six sigma approach is an extension of other quality improvement schemes such as Deming's statistical quality control technique. Similar to other management strategies used for quality improvement, the six sigma methodology is also focused on satisfying customer needs. The main difference between other quality strategies and six sigma approach lies in the disciplined quantitative approach used by six sigma for the improvement of the process. Six sigma approach consists of five phases, namely, Define, Measure, Analyze, Improve, and Control (DMAIC), for improving the process. The DMAIC cycle, as shown in Fig. 14.1, has several similarities with Deming's quality improvement process. Initially,

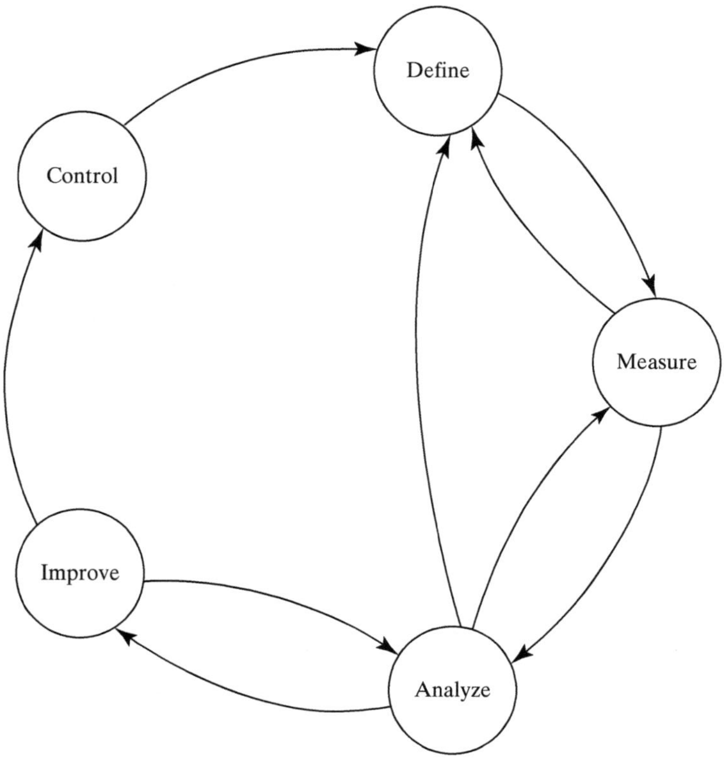

FIGURE 14.1 Five steps of the six sigma approach.

six sigma approach was developed to improve the manufacturing processes. It provides a well-defined quality target by specifying a maximum of 3.4 defects per million opportunities. Later, the concept has been used to improve any process, including businesses and organizations. The idea, in the context of manufacturing, is that it is not just enough to provide defect-free products to customers. An important characteristic of any product is that it should function with reliability and provide a lower total life time cost to the customer. A defective product can be identified, rather easily, during the manufacturing stage. However, a product with a poor reliability leads to a higher cost to the customer in the form of increased number of downtimes, maintenance costs, and logistics requirements.

In most commonly used statistical process techniques, the upper and lower control limits (UCL and LCL) or tolerances are assumed to be at $+3\sigma$ and -3σ deviations, respectively, from the mean. In the six sigma approach, the tolerance limits are set to be equal to $\mu - 6\sigma$ and $\mu + 6\sigma$. If the tolerances of a machine part are specified as UCL and LCL, respectively, the maximum permissible variation of the process, σ_{max}, is given by

$$\sigma_{max} = \frac{\text{UCL} - \text{LCL}}{6} \quad \text{in } 3\sigma \text{ process} \tag{14.7}$$

$$\sigma_{max} = \frac{\text{UCL} - \text{LCL}}{12} \quad \text{in } 6\sigma \text{ process} \tag{14.8}$$

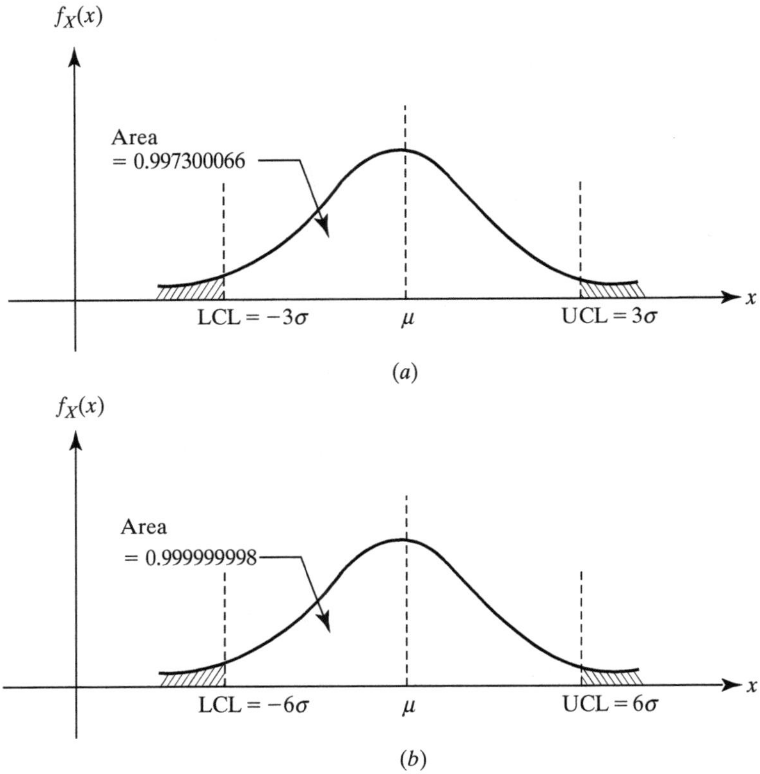

FIGURE 14.2 Comparison of 3σ and 6σ processes.

The comparison of the 3σ and 6σ processes is shown graphically in Figs. 14.2(*a*) and (*b*), respectively. It can be seen that the permissible variability in the 6σ process is much smaller (one-half) compared to the 3σ process and, as such, will have a significant impact on the number of defective parts per million (DPPM). The specification of a process as $\mu \pm 6\sigma$ results in less than one defective part per million and thus results in a near-perfect process, product, or transaction in service.

The following measures and terminology are commonly used in the six sigma approach:

1. Sigma level: The sigma level is a measure of the process defect rate and is commonly used to improve the process quality. The sigma level is defined as the z-value of the standard normal distribution (with zero shift in the process mean).
2. Yield (y): Yield is defined as the ratio of the number of defect-free parts to the total number of parts produced expressed as a percentage.

$$y = \frac{\text{Number of defect-free parts}}{\text{Total number of parts}} \times 100\% \tag{14.9}$$

$$y = \frac{\text{Total number of parts} - \text{Number of defective parts}}{\text{Total number of parts}} \times 100\% \tag{14.10}$$

TABLE 14.1 Values of yield (y) corresponding to DPPM

DPPM	y
1	99.9999%
10	99.999%
100	99.99%
1,000	99.9%
10,000	99%
20,000	98%
50,000	95%

3. Defective parts per million (DPPM): Defective parts per million is defined as the number of defective parts produced by the process (such as turning, milling, grinding, and casting) per million parts. Since, in practice, data on the number of defective parts in million parts is not available for most processes, the DPPM is estimated from a sample of N parts produced as follows:

$$\text{DPPM} = \frac{\text{Number of defective parts observed } (N_d)}{\text{Total number of parts produced } (N)} \times 10^6 \tag{14.11}$$

The defective parts per million is related to the yield (y) as

$$y = \left(1 - \frac{\text{DPPM}}{10^6}\right) \times 100\% \tag{14.12}$$

The yield corresponding to different values of defective parts per million are given in Table 14.1.

The six sigma process tries to quantify the process capability (C_p) and the process capability index (C_{pk}) as follows:

$$C_p = \frac{\text{UCL} - \text{LCL}}{6\sigma} \tag{14.13}$$

$$C_{pk} = \min\left\{\frac{\text{UCL} - \mu}{3\sigma}, \frac{\mu - \text{LCL}}{3\sigma}\right\} \tag{14.14}$$

where UCL and LCL denote, respectively, the upper and lower control limits or specifications of the process and μ and σ represent the mean and standard deviation of the process, respectively. The number of defective parts per million with the specification $\mu \pm k\sigma$ for different values of k are shown in Table 14.2.

In practice, it was observed that the process mean itself might drift by as much as $\pm 1.5\sigma$ over a long period of time as shown in Fig. 14.3. With this drift of the mean, the six sigma specification will correspond to the area under the standard normal distribution beyond 4.5σ, which is approximately equal to 0.0000034. Thus, 3.4 out of million parts fall outside the

TABLE 14.2 Number of defective parts per million with specification $\mu \pm k\sigma$ with no shift in the mean value μ

Value of k	Area Under the Standard Normal Distribution Curve Outside the Range $\mu \mp k\sigma$	DPPM
1	$2(0.1586552) = 0.3173104$	317,300
2	$2(0.0227500) = 0.0455000$	45,500
3	$2(0.001349) = 0.002698$	2,700
4	$2(0.00003167) = 0.00006334$	63
5	$2(0.000000285) = 0.00000057$	0.57
6	$2(0.000000001) = 0.000000002$	0.002

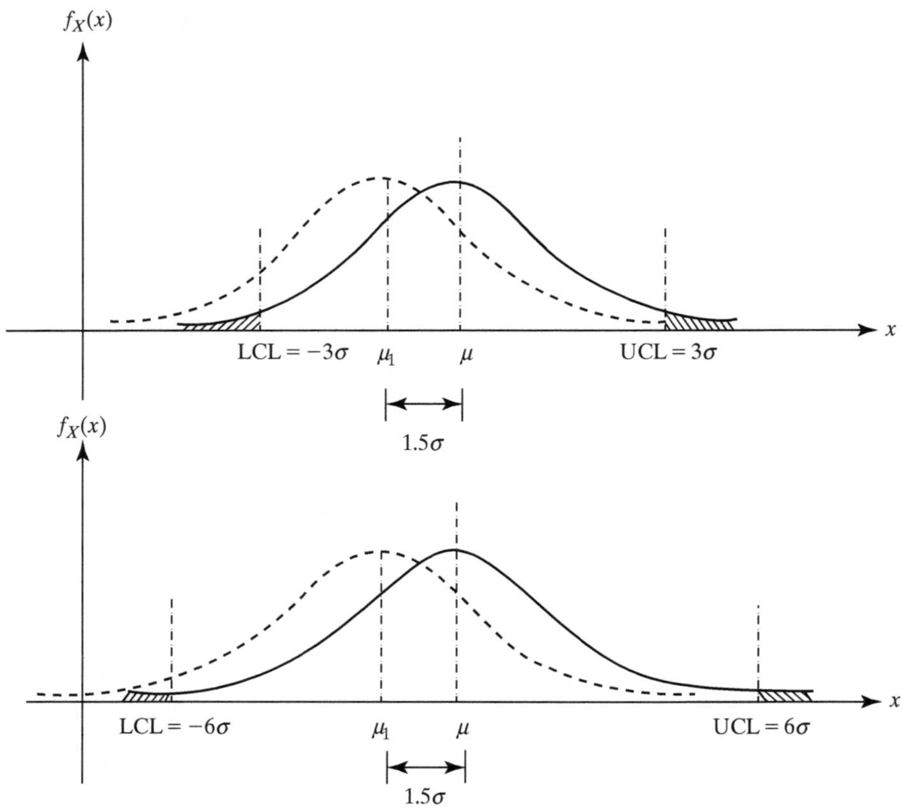

(*a*) Mean μ shifts to $\mu_1 = \mu - 1.5\sigma$

FIGURE 14.3 Effect of shifting of the mean.

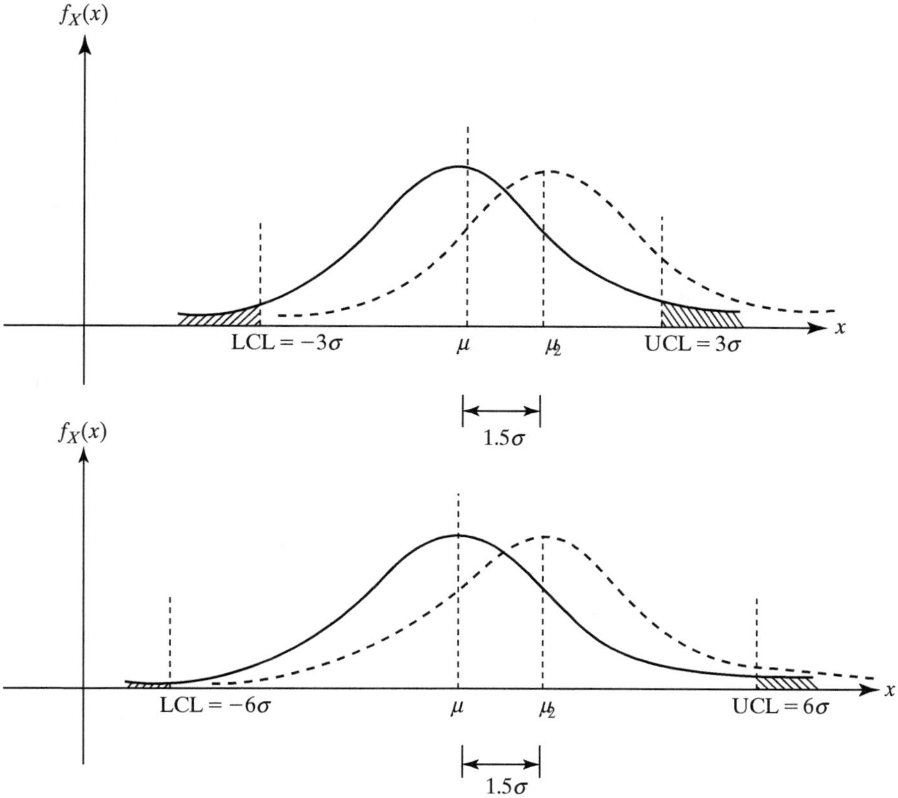

(*b*) Mean μ shifts to $\mu_2 = \mu + 1.5\sigma$

FIGURE 14.3 Effect of shifting of the mean. (*Continued*)

tolerance level. This implies that the DPPM will be 3.4 in a six sigma process (with a mean drift of $\pm 1.5\sigma$). In fact, when the mean shifts by $\pm 1.5\sigma$, the number of defective parts per million with the specification $\mu \pm k\sigma$ for different values of k are given in Table 14.3.

TABLE 14.3 Number of defective parts per million with specification $\mu \pm k\sigma$ with a shift of the mean value μ by $\pm 1.5\sigma$

Value of k	Area Under the Standard Normal Distribution Curve Outside the Range $\mu \mp k\sigma$	DPPM
−0.5	0.691462	697,672
0.5	0.308538	308,770
1.5	0.066807	66,811
2.5	0.006210	6,210
3.5	0.000233	233
4.5	0.0000034	3.4

EXAMPLE 14.7

The piston diameter of an automobile engine is specified as 3.000 ± 0.020 inches.

1. If the process capability index (C_p) is 1.0, what is the standard deviation of the diameter of the pistons produced on a lathe?
2. If the process capability index (C_p) is required to be 1.33, what standard deviation needs to be achieved for the diameters of the piston?

Solution:

1. Using the definition of the process capability index, C_p, given by Eq. (14.13), we find

$$C_p = 1.0 = \frac{\text{UCL} - \text{LCL}}{6\sigma} = \frac{3.02 - 2.98}{6\sigma} = \frac{0.04}{6\sigma}$$

or

$$\sigma = \frac{0.04}{6} = 0.006667 \text{ in}$$

2. Equation (14.13) gives

$$C_p = 1.33 = \frac{\text{UCL} - \text{LCL}}{6\sigma} = \frac{3.02 - 298}{6\sigma} = \frac{0.04}{6\sigma}$$

or

$$\sigma = \frac{1}{1.33} \times \frac{0.04}{6} = 0.005 \text{ in}$$

■

EXAMPLE 14.8

For the piston described in Example 14.7, assume that the process mean is centered within the tolerance range and the process capability index is $C_p = 1.33$. Find the number of oversized pistons produced on the lathe.

Solution: As shown in Example 14.7(b), the standard deviation of the piston diameter is $\sigma = 0.005$ in. If x denotes the diameter of the piston, the definition of the standard normal variate z yields (using UCL = 3.02, $\mu = 3.00$, and $\sigma = 0.005$)

$$z = \frac{x - \mu}{\sigma} = \frac{\text{UCL} - \mu}{\sigma} = \frac{3.02 - 3.00}{0.005} = 4.000$$

From the standard normal distribution tables, the area under the density function curve from $-\infty$ to $z = 4$ can be found as 0.999969. Thus, the number of oversized pistons produced on the lathe are given by $1 - 0.999969 = 0.000031$ or 0.0031 percent or 31 ppm (parts per million).

Note: Although the number of oversized pistons produced is determined in this example, in reality, there will be an equal number of undersized pistons produced due to the symmetry of the normal distribution. Thus, the total number of defective pistons (over- or undersized ones) produced on the lathe will be 0.0062 percent or 62 ppm.

To demonstrate the importance of six sigma approach in strength-based reliability problems, the following example is presented.

■

EXAMPLE 14.9

Consider a mechanical system in which the strength (S) and load (L) follow normal distributions with their mean values and variabilities given by

$$S = \mu_S \pm \Delta S \equiv 150 \pm 15$$
$$L = \mu_L \pm \Delta L \equiv 140 \pm 14$$

Assuming the variabilities to correspond to (a) $\pm 3\sigma$ and (b) $\pm 6\sigma$, find the reliabilities of the system.

Solution: The reliability of the system, with normally distributed strength and load, is given by Eq. (8.26):

$$R = \frac{1}{\sqrt{2\pi}} \int_{z=z_1}^{\infty} e^{-\frac{1}{2}z^2}\, dz \tag{E1}$$

where

$$z_1 = -\frac{\mu_S - \mu_L}{\sqrt{\sigma_S^2 + \sigma_L^2}} \tag{E2}$$

1. When variabilities correspond to $\pm 3\sigma$:
The standard deviations of strength and load are given by $\sigma_S = \frac{\Delta S}{3} = \frac{15}{3} = 5$ and by $\sigma_L = \frac{\Delta L}{3} = \frac{14}{3} = 4.6667$ and hence Eq. (E2) gives

$$z_1 = -\frac{150 - 140}{\sqrt{5^2 + 4.6667^2}} = -1.4621$$

This corresponds to a reliability of the system, $R = 0.9278$.

2. When variabilities correspond to $\pm 6\sigma$:
The standard deviations of strength and load are given by $\sigma_s = \frac{\Delta S}{6} = \frac{15}{6} = 2.5$ and by $\sigma_L = \frac{\Delta L}{6} = \frac{14}{6} = 2.3333$ and hence Eq. (E2) gives

$$z_1 = -\frac{150 - 140}{\sqrt{2.5^2 + 2.3333^2}} = -2.9242$$

This corresponds to a reliability of the system, $R = 0.9982$.

■

14.4.1 Implementation of the Six Sigma Approach

Six sigma method is a systematic and disciplined procedure that uses data and statistical analysis to measure and improve an organization's or company's operational performance. It identifies and eliminates defects in product development, manufacturing, or service-related activities. The goal of six sigma method is to eliminate variability, defects, and waste thereby increasing the profit and customer loyalty. There are five important steps in the implementation of the six sigma method—define, measure, analyze, improve, and control—as shown in Fig. 14.1.

In the first step, *define*, (1) the project team with a clear charter is formed, (2) the customers, their needs, and their expectations are identified, and (3) a high-level process map is created. In the second step, *measure*, the current sigma performance is determined. In the third step, *analyze*, the data and the process itself are analyzed by the team and the root cause of the current poor sigma performance of the process is determined. In the fourth step, *improve*, the team generates and selects a set of actions to improve the current sigma performance. In the fifth step, *control*, a set of tools and techniques are applied to the newly improved process so that the improved sigma performance of the process is retained over time.

The primary goal of any organization or business is customer satisfaction. Processes such as quality function deployment (QFD) may help in capturing what customers really want; but the customer's requirements and expectations may change with time. Six sigma methodology can help in designing the process to ensure a defect rate of no more than 3.4 defects per million opportunities. But achieving six sigma quality during the manufacturing stage alone cannot ensure success of an organization. Although a lesser number of defects is desirable, no customer would like to see the product fail within a short time, even if the product is still under warranty (due to the unavoidable downtime involved). The real effectiveness of an organization lies in the degree to which it maintains its effectiveness in production or manufacturing, reliability, and competitive edge in the market. The six sigma approach can be used in several ways in an organization: (a) as an operational strategy to reduce the number of defects during production, (b) as a management strategy to improve a business process or develop new business models, and (c) as a methodology to design and develop products and systems with smaller number of defects.

14.5 Acceptance Sampling

In most cases, it is costly, time-consuming, and, sometimes, physically impossible to inspect all the manufactured components. Examples of such cases include firing each missile to examine whether it functions normally, testing the breaking strength of each bolted joint, and testing each chocolate bar in a batch. Consider a lot consisting of a large number of exhaust fans. Let the fans with unbalanced blades be considered as defective units. Let the lot contain, say, 2 percent defective units. This will be revealed if the entire lot were tested; however, we would like to test only a small sample of fans to estimate the percentage defective units in the lot and determine if the lot has acceptable reliability. The procedure of using a small sample to assess the acceptability of quality of the whole production lot is known as "acceptance sampling" [14.2]. The most commonly used acceptance sampling plans are those provided by MIL-STD-105D and MIL-STD-414 [14.3, 14.4]. MIL-STD-105D provides three basic types of sampling plans: single, double, and multiple.

The single sampling plan is based on the inspection of a sample of size k. The production lot is considered acceptable if the number of defective units in the sample is less than or equal to an acceptance number, m. In addition, if the number of defective units exceeds a rejection number, n, the entire production lot is rejected. Both the acceptance and rejection numbers are given in MIL-STD-105D. In a double sampling plan, two sample sizes, with the corresponding acceptance and rejection numbers, are used. Usually the size of the first sample is smaller than that of the second sample. The procedure can be summarized as follows:

1. Inspect the first sample of size k_1 and find the number of defective units, N_1.
2. If N_1 is less than or equal to the first acceptance number (m_1), accept the production lot. If N_1 is greater than or equal to the first rejection number (n_1), reject the production lot.

3. If $n_1 > N_1 > m_1$, inspect the second sample of size k_2 and find the combined number of defective units, N_2.
4. If N_2 is less than or equal to the second acceptance number (m_2), accept the entire lot. If N_2 is greater than or equal to the second rejection number (n_2), reject the entire lot.

The multiple sampling plan can be considered to be an extension of the double sampling plan and is shown schematically in Fig. 14.4. A maximum of seven samples are permitted by MIL-STD-105D in a multiple sampling plan. Also note that in a multiple sampling plan, acceptance is not permitted in the first sample even if no defective units are detected.

14.5.1 Characteristics of Sampling Plans

The sampling plans have several advantages and disadvantages as indicated below:

1. The sampling plans are less costly compared to 100 percent inspection because only a small percentage of the total units are inspected. Also, monotony problem will be reduced resulting in a fewer inspection errors compared to the 100 percent inspection.
2. The sampling plans require less human power for inspection. There will be less handling damage during inspection compared to 100 percent inspection.
3. The sampling plans can be used even if destructive testing is involved.
4. The sampling plans encourage product improvement because entire lots are rejected instead of individual defective items.
5. In sampling plans, there is certain probability of rejecting a lot containing good items and accepting a lot having bad items.
6. The sampling plans do not provide as much information on the units as in the case of 100 percent inspection.

14.6 Process Capability

It is extremely important to select a proper manufacturing process that is capable of producing the required part or component within the specified tolerance range. A knowledge of the manufacturing process capability helps not only in setting the tolerances on the part, but also in selecting the most suitable supplier (out of several competing ones) who is most likely to satisfy the required quality specifications of the part. The process capability is measured in terms of the *process capability index*, C_p, which is defined as

$$C_p = \frac{\text{Acceptable part variation}}{\text{Machine or process variation}} = \frac{\text{Tolerance}}{\pm 3\sigma} = \frac{\text{USL} - \text{LSL}}{3\sigma - (-3\sigma)} = \frac{(\text{USL} - \text{LSL})}{6\sigma} \tag{14.15}$$

where USL and LSL denote, respectively, the upper and lower specification limits (limits on the tolerance) of the design parameter produced by the manufacturing process, which is assumed to be normally distributed and in a state of statistical control. Data from a control chart are used to determine how the process is performing. For a parameter such as the diameter of a piston, the mean and the variability of the population can be estimated from the control chart as μ and σ, respectively. In addition, the center of the tolerance range is assumed to coincide with the center of the range of the diameters produced by the machine

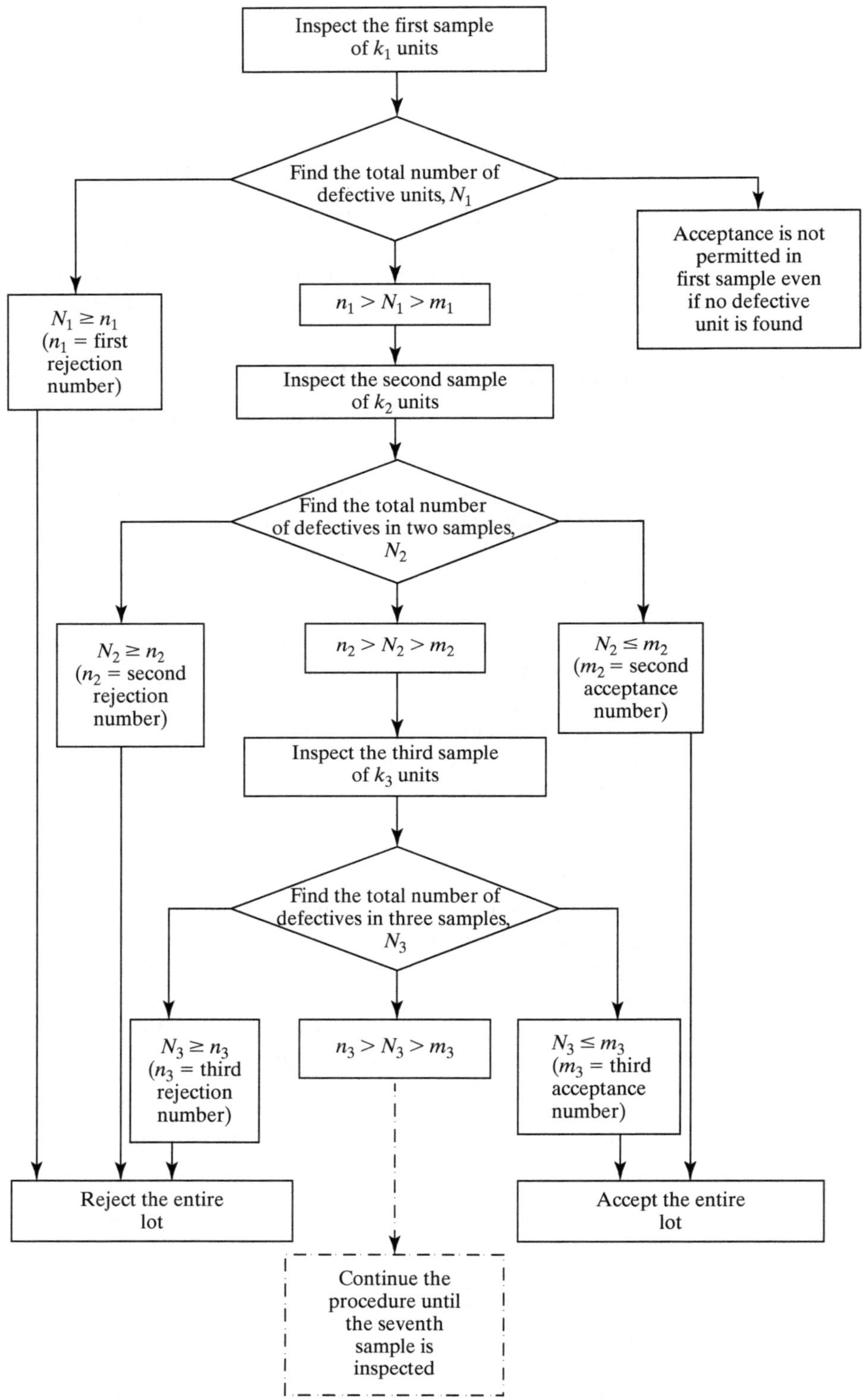

FIGURE 14.4 Multiple sampling plan.

in using Eq. (14.15). The conditions assumed are ideal, which result in the maximum (best) process capability without reducing the process standard deviation. These conditions are not usually satisfied unless careful adjustments are made to the machine.

The limits on machine variation are commonly set at $\pm 3\sigma$, which corresponds to 0.27 percent defective parts when $C_p = 1$. Figure 14.5 corresponds to $C_p \leq 1$ where the process variability or spread is larger than the acceptable part variation or tolerance range. This indicates that $C_p \leq 1$ and the process is not capable. In order to make the process capable, the variability in the process needs to be reduced or the tolerance needs to be relaxed. This corresponds to Fig. 14.6 in which $C_p = 1$. This situation indicates a marginal case, because any shift in the process mean, for example, to the right, will increase the number of defective parts. Figure 14.7 denotes the situation in which the process variability is much less than the tolerance range. This gives a considerable margin of safety for the movement of the process mean without going outside the band lying between the upper and lower specification limits.

In mass production, the percentage of defective units is critical and hence the value of C_p is required to be greater than 1.33. It can be noted that the process mean lies at a distance of 4 standard deviations from USL and LSL when $C_p = 1.33$. This is considered to be a good manufacturing process.

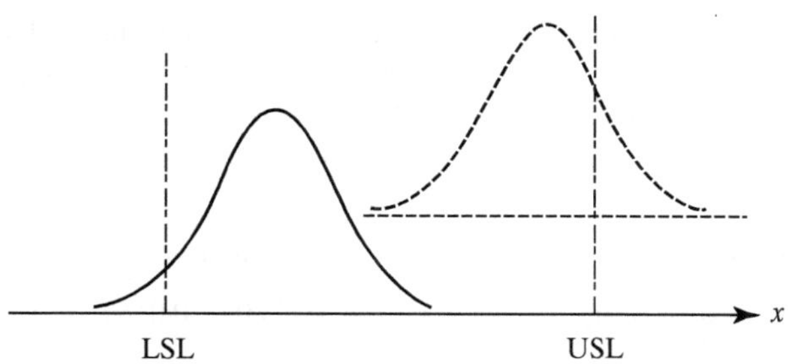

FIGURE 14.5 Process is not capable ($C_p < 1$).

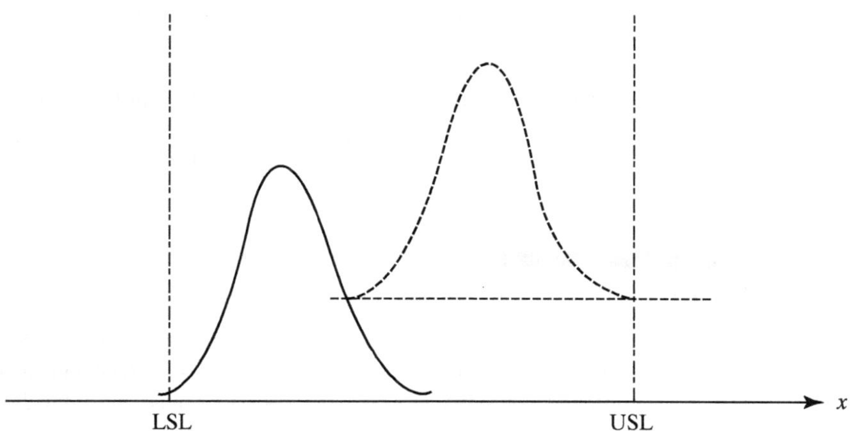

FIGURE 14.6 Process is marginally capable ($C_p = 1$).

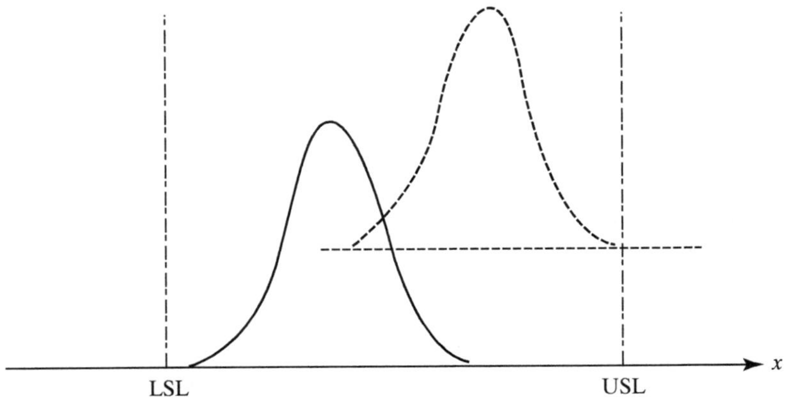

FIGURE 14.7 Process is capable ($C_p > 1$).

EXAMPLE 14.10

The pistons of an internal combustion engine are produced on a lathe with diameters specified as 3.0 ± 0.018 in. Determine the standard deviation of the diameter of the pistons produced if the process capability index (C_p) is (a) 1.0 and (b) 1.33.

Solution:

a. From the definition of the process capability index, we have

$$C_p = \frac{\text{USL} - \text{LSL}}{6\sigma} = \frac{(3.0 + 0.018) - (3.0 - 0.018)}{6\sigma} = \frac{0.036}{6\sigma} = 1.0$$

which gives $\sigma = 0.006$ in.

b. The definition of C_p can be used to obtain

$$C_p = \frac{\text{USL} - \text{LSL}}{6\sigma} = \frac{(3.0 + 0.018) - (3.0 - 0.018)}{6\sigma} = \frac{0.036}{6\sigma} = 1.33$$

from which the standard deviation of the diameter of the piston can be found as

$$\sigma = \frac{0.036}{6(1.33)} = 0.004511 \text{ in.}$$

■

EXAMPLE 14.11

If pistons of an internal combustion engine are produced on a lathe with diameters specified as 3.0 ± 0.018 inches with a process capability index of $C_p = 1.33$, determine the expected percentage of oversized pistons produced. Assume that the process mean is centered within the tolerance range.

Solution: Since the process capability index is known to be $C_p = 1.33$, the standard deviation of the diameter of the piston can be computed as $\sigma = 0.004511$ in (see Example 14.10(b)). If d is used to denote the diameter of the piston (following normal distribution), the standard normal variate z corresponding to the upper limit value of the diameter of the piston can be defined as

$$z = \frac{d - \mu}{\sigma} = \frac{\text{USL} - \mu}{\sigma} = \frac{3.018 - 3.0}{0.004511} = 3.9902$$

The value of the cumulative distribution function corresponding to $z = 3.9902$ (area under the probability density function curve from $z = -\infty$ to 3.9902) is given as 0.999971. This gives the expected number of the oversized pistons as $1.0 - 0.999971 = 0.000029$ or 0.0029 percent. Thus, the rejection rate would be 29 ppm (parts per million).

■

Shift of the Mean. In an ideal manufacturing process (as implied in Examples 14.10 and 14.11), the process mean lies at the middle of the upper and lower specification limits of the dimension of the part being produced. In practice, this is difficult to achieve and maintain. Even if the process starts off centered (i.e., with the mean at the center of the upper and lower specification limits), it has a tendency to move off the center with time due to tool wear and other variations of the process. If k denotes the distance of the process mean μ from the midpoint of the tolerance region, we have

$$k = \frac{\left| \dfrac{\text{USL} + \text{LSL}}{2} - \mu \right|}{\dfrac{\text{USL} - \text{LSL}}{2}} \tag{14.16}$$

with $0 \le k \le 1$. When the process mean is not centered, the process capability index, C_{pk}, can be defined as

$$C_{pk} = \text{minimum}\left\{ \frac{\text{USL} - \mu}{3\sigma}, \frac{\mu - \text{LSL}}{3\sigma} \right\} = \text{minimum}\{C_{pk_1}, C_{pk_2}\} \tag{14.17}$$

where

$$C_{pk_1} = \frac{\text{USL} - \mu}{3\sigma} \tag{14.18}$$

and

$$C_{pk_2} = \frac{\mu - \text{LSL}}{3\sigma} \tag{14.19}$$

Note that C_p and the proper value of C_{pk} are related as

$$C_{pk} = (1 - k)C_p \tag{14.20}$$

It can be observed that when $k = 0$, the mean is centered and $C_{pk} = C_p$. Since C_{pk} never exceeds C_p, the number of defective parts is usually larger than the one predicted by C_p. The percentages of acceptable parts and defective parts corresponding to different values of process standard deviations (sigmas) that can be accommodated within the tolerance range are shown in Table 14.2. The substantial increase in the number of defective parts that results from a shift of 1.5σ in the process mean (from the middle of the tolerance range) can also be seen from Table 14.3. A shift of $\pm 1.5\sigma$ in the process mean is considered typical in most of the common manufacturing processes.

EXAMPLE 14.12

If pistons of an internal combustion engine are produced on a lathe with diameter specified as 3.0 ± 0.018 inches with a process capability index of $C_p = 1.33$, determine the expected percentage of oversized pistons produced. Assume that the process mean has moved 1.5σ from the center of the tolerance range of the diameter of the pistons.

Solution: The standard deviation of the diameter of the pistons is given by 0.004511 in (see Example 14.10). The shift of the mean toward the upper specification limit is given by $k = 1.5(0.004511) = 0.0067665$ in. Thus, the new mean will be $\mu = 3.0 + 0.0067665 = 3.0067665$ in and hence Eqs. (14.18) and (14.19) give

$$C_{pk_1} = \frac{\text{USL} - \mu}{3\sigma} = \frac{3.018 - 3.0067665}{3(0.004511)} = 0.830082$$

$$C_{pk_2} = \frac{\mu - \text{LSL}}{3\sigma} = \frac{3.0067665 - 2.982}{3(0.004511)} = 1.830082$$

Note that the observation $C_{pk_1} \neq C_{pk_2}$ implies that the process mean is not centered. However, the process capability index of $C_{pk} = 1.830082$ indicates that the process is capable. The number of defective parts can be found by first computing the standard normal variates corresponding to the upper and lower specification limits as

$$z_{\text{USL}} = \frac{\text{USL} - \mu}{\sigma} = \frac{3.018 - 3.0067665}{0.004511} = 2.490246$$

$$z_{\text{LSL}} = \frac{\text{LSL} - \mu}{\sigma} = \frac{2.982 - 3.0067665}{0.004511} = -5.490246$$

The probability of diameters falling outside the tolerance range can be computed as

$$P(z \geq 2.490246) + P(z \leq -5.490246) = (1 - 0.993613) + (0.189896 \times 10^{-7}) = 0.00638702$$

This corresponds to a rejection rate of 6387 ppm.

It can be seen that the rejection rate increased substantially from a value of 29 ppm when the process was centered in the middle of the tolerance range to the current value of 6387 ppm when the process mean shifted by 1.5σ from the middle of the tolerance range.

■

14.7 Quality Control Charts

Acceptance sampling is done on the product after it is produced. The objective is to accept or reject the entire lot based on the results obtained from the sample. On the other hand, control charts are used to control the manufacturing process (while the production is being continued). Control charts represent important quality control tools for detecting the existence of assignable causes. Control charts are used to find the chance variation (also known as common cause variation) as well as the assignable variation caused by special causes in a manufacturing process. Chance variation occurs due to several noise factors associated with the process where each factor, by itself, will not be of much significance.

On the other hand, an assignable variation can be detected and controlled. Some of the causes of assignable variations include worn cutting tool, poorly trained operator, and so on.

Here the objective is to correct, if necessary, the production process while the product is being produced. The quality control charts were originally developed by Shewhart in 1924. The quality control chart method is a graphical technique to find whether a manufacturing process is in a state of statistical control or out of control. In this method, the process is considered to be out of statistical control whenever a sample value falls outside the upper and lower control limits. When a process is found to be out of statistical control, one can make an investigation to find the cause of its being out of control and then take necessary measures to bring back the process to be under control. The following types of quality control charts are considered in this section:

1. The p-chart
2. The $\overline{X}$-chart
3. The R-chart
4. The c-chart

14.7.1 The p-Chart

The p-charts are also known as control charts for attributes. To construct these charts, we classify the item inspected as either conforming or nonconforming to the quality specifications. The terms, *nondefective* and *defective* are also used to identify the two classes of products. Quality characteristics of this type are called "attributes." Typical examples are noncircular pistons in a day's production, nonfunctional integrated circuits in a production run, and so on. Since a product is classified either as defective or nondefective, the binomial distribution is used to establish the limits of the control chart. The mean value and standard deviation of binomial distribution are given by

$$\mu = \text{mean value} = \text{fractional defectives} = \frac{n}{mk} \tag{14.21}$$

and

$$\sigma = \text{standard deviation} = \left[\frac{\mu(1-\mu)}{m}\right]^{\frac{1}{2}} \tag{14.22}$$

where n = total number of defective units, k = number of samples considered, and m = size of each sample. The upper control limit (UCL) and the lower control limit (LCL) of the p-chart are given by

$$\text{UCL}_p = \mu + 3\sigma \tag{14.23}$$

$$\text{LCL}_p = \mu - 3\sigma \tag{14.24}$$

The fractions of defective units found in different samples are plotted on the p-chart to determine whether the manufacturing process is under control or not.

EXAMPLE 14.13

A manufacturer produces 20,000 gears over a period of 10 days. A random sample of 40 gears is tested every day to find the defective ones. The manufacturer wishes to determine whether the process is under control by establishing the control limits on the process. The inspection process revealed that the number of defective gears in the ten samples are 5, 3, 8, 5, 4, 10, 9, 6, 11, and 2. Construct the p-chart.

Solution: The fractions of defective components in the 10 samples are $\frac{5}{40} = 0.125$, $\frac{3}{40} = 0.075$, $\frac{8}{40} = 0.2$, $\frac{5}{40} = 0.125$, $\frac{4}{40} = 0.1$, $\frac{10}{40} = 0.25$, $\frac{9}{40} = 0.225$, $\frac{6}{40} = 0.15$, $\frac{11}{40} = 0.275$, and $\frac{2}{40} = 0.05$. Since the total number of defective gears is $5 + 3 + 8 + 5 + 4 + 10 + 9 + 6 + 11 + 2 = 63$, the mean and standard deviations are given by (with $n = 63$, $m = 40$, $k = 10$):

$$\mu = \frac{63}{40 \times 10} = 0.1575$$

$$\sigma = \left[\frac{0.1575(1.0 - 0.1575)}{40}\right]^{\frac{1}{2}} = 0.057596$$

The upper and lower control limits can be computed as

$$\begin{aligned} UCL_p &= \mu + 3\sigma = 0.1575 + 3(0.057596) = 0.330288 \\ LCL_p &= \mu - 3\sigma = 0.1575 - 3(0.057596) = -0.015288 \approx 0 \end{aligned}$$

Note that the lower control limit is to be taken as zero whenever the computed value is negative. The fractions of defective units found in the ten samples are plotted in Fig. 14.8 along with the upper and the lower control limits. This figure is known as the p-chart. Since all the sample fractions are within the control limits, the process is considered to be under control, that is, there is no abnormality in the process.

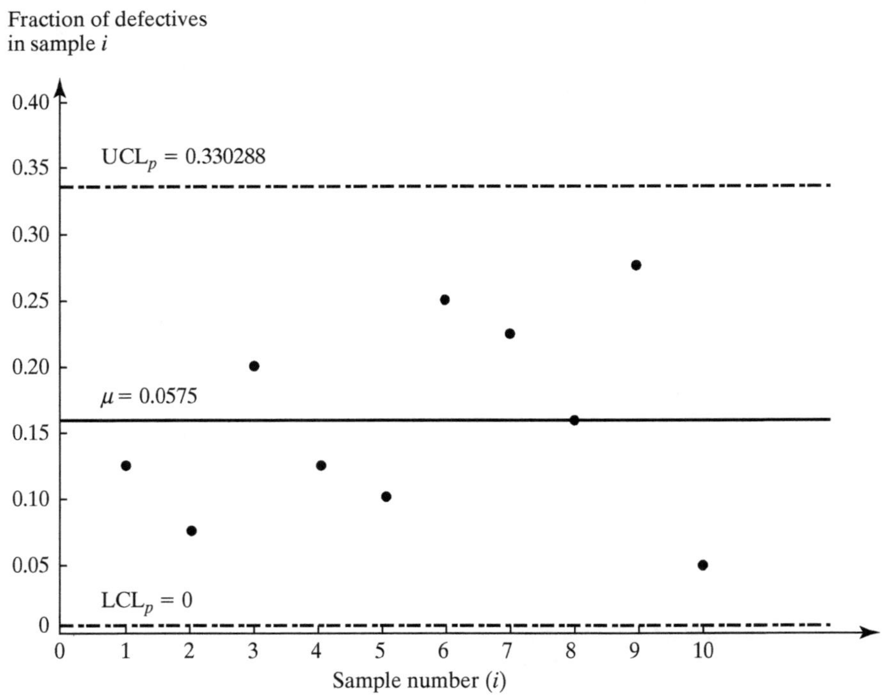

FIGURE 14.8 p-chart of Example 14.13.

■

14.7.2 The $\overline{X}$-Chart

The $\overline{X}$-charts are also known as the control charts for mean values. In these charts, the mean value of the quality characteristic of each sample is plotted and hence the $\overline{X}$-chart indicates whether the mean value of the process is under statistical control. If the control chart indicates that the process is not under control, the causes must be investigated and corrective measures taken to bring the process under control. Here the probability distribution of sample means, $\overline{X}$s, is assumed to follow approximately normal distribution if the sample size is five or more. The upper and lower control limits are given by

$$\text{UCL}_{\overline{X}} = \overline{\overline{X}} + 3\sigma \tag{14.25}$$

$$\text{LCL}_{\overline{X}} = \overline{\overline{X}} - 3\sigma \tag{14.26}$$

where

$$\overline{\overline{X}} = \text{mean value of the sample means} = \frac{1}{k}\sum_{i=1}^{k} \overline{X}_i \tag{14.27}$$

$$\sigma = \text{standard deviation of the sample means} = \left[\frac{\sum_{i=1}^{k}(\overline{X}_i - \overline{\overline{X}})^2}{k}\right]^{\frac{1}{2}} \tag{14.28}$$

k = number of samples, and $\overline{X}_i$ = mean value of the ith sample. Equations (14.27) and (14.28) are often rewritten as

$$\text{UCL}_{\overline{X}} = \overline{\overline{X}} + A_2\overline{r} \tag{14.29}$$

$$\text{LCL}_{\overline{X}} = \overline{\overline{X}} - A_2\overline{r} \tag{14.30}$$

where

$$A_2 = \frac{3}{d_2\sqrt{m}} \tag{14.31}$$

m is the sample size, d_2 is a factor dependent on the sample size (values are given in Table 14.4), and $\overline{r}$ is the mean value of sample ranges. The sample range is given by the difference between the largest and the smallest values in the sample.

TABLE 14.4 Values of d_2 and d_3 for $\overline{X}$- and R-charts

Sample Size (m)	d_2	d_3	Sample Size (m)	d_2	d_3
2	1.128	0.853	14	3.407	0.763
3	1.693	0.888	15	3.472	0.756
4	2.059	0.880	16	3.532	0.750
5	2.326	0.864	17	3.588	0.744

(*Continued*)

TABLE 14.4 Values of d_2 and d_3 for $\overline{X}$- and R-charts (*Continued*)

Sample Size (m)	d_2	d_3	Sample Size (m)	d_2	d_3
6	2.534	0.848	18	3.640	0.739
7	2.704	0.833	19	3.689	0.734
8	2.847	0.820	20	3.735	0.729
9	2.970	0.808	21	3.778	0.724
10	3.078	0.797	22	3.819	0.720
11	3.173	0.787	23	3.858	0.716
12	3.258	0.778	24	3.895	0.712
13	3.336	0.770	25	3.931	0.708

EXAMPLE 14.14

Piston rings for an automobile engine are produced by a forging process. Fifteen samples, each of size five, are taken and the internal diameters are measured in millimeters. The data are shown in Table 14.5. Establish the $\overline{X}$-chart of the process.

TABLE 14.5 Piston diameter data for control charts

Sample Number	Sample Values				
1	70.8	69.9	69.7	70.4	69.1
2	69.9	70.1	70.6	71.1	69.3
3	69.8	69.5	70.2	70.7	69.9
4	70.7	70.1	69.2	69.8	70.5
5	70.0	70.8	71.2	69.1	69.9
6	70.1	70.6	69.1	69.9	71.2
7	70.2	70.1	70.3	69.4	69.6
8	69.7	69.1	68.7	70.8	71.1
9	70.5	70.1	71.8	69.2	69.9
10	70.9	70.2	71.2	69.1	69.8
11	71.2	70.7	71.0	69.5	69.9
12	69.4	70.3	69.1	69.9	71.0
13	69.8	70.3	71.2	68.9	70.4
14	70.5	70.8	69.4	68.3	71.1
15	69.2	71.3	70.2	70.9	68.9

Solution: The sample range is computed as the difference between the largest and the smallest values in the sample. The sample ranges and the sample means are shown in Table 14.6. From the individual sample means, the mean value of the sample means is found as

$$\overline{\overline{X}} = \frac{1}{15} \sum_{i=1}^{15} \overline{X}_i = 70.1147$$

TABLE 14.6 Sample means and sample ranges of piston diameters

Sample Number	Sample Mean	Sample Range
1	69.98	1.7
2	70.20	1.8
3	70.02	1.2
4	70.06	1.5
5	70.20	2.1
6	70.18	2.1
7	70.02	1.4
8	69.88	2.4
9	70.30	2.6
10	70.24	2.1
11	70.46	1.7
12	69.94	1.9
13	70.12	2.3
14	70.02	2.8
15	70.10	2.4

The mean value of the ranges is determined as

$$\overline{r} = \frac{1}{15} \sum_{i=1}^{15} r_i = 2.0000$$

For the sample size of $m = 5$, we find that $d_2 = 2.326$ from Table 14.4. Thus Eq. (14.31) gives

$$A_2 = \frac{3}{2.326\sqrt{5}} = 0.5768$$

The upper and the lower control limits are given by

$$\text{UCL}_{\overline{X}} = \overline{\overline{X}} + A_2\overline{r} = 70.1147 + 0.5768(2.0000) = 71.2683$$
$$\text{LCL}_{\overline{X}} = \overline{\overline{X}} - A_2\overline{r} = 70.1147 - 0.5768(2.0000) = 68.9611$$

These upper and lower control limits, as well as the mean values of samples $(\overline{X}_i)$, are plotted in the $\overline{\text{X}}$-chart as shown in Fig. 14.9. It can be observed from Fig. 14.9 that all the sample mean values fall within the control limits implying that there is no abnormality in the process.

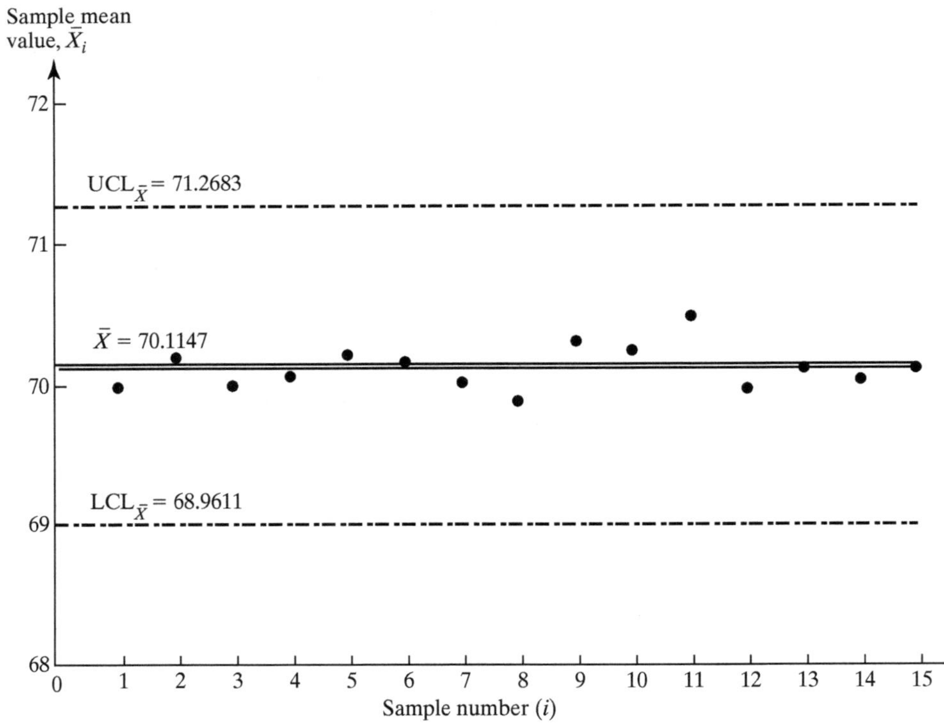

FIGURE 14.9 $\bar{X}$-chart of Example 14.14.

■

14.7.3 The R-Chart

The R-charts are also known as the control charts for ranges. Since the difference between the largest and the smallest values of the quality characteristic in a sample indicates the range, the R-chart is used to measure the variability of the process. The upper and the lower control limits are given by

$$\text{UCL}_r = D_3 \bar{r} \tag{14.32}$$

$$\text{LCL}_r = D_4 \bar{r} \tag{14.33}$$

where

$$D_3 = 1 + 3\frac{d_3}{d_2} \tag{14.34}$$

$$D_4 = 1 - 3\frac{d_3}{d_2} \tag{14.35}$$

where d_2 and d_3 are the constants given in Table 14.4 for different sample sizes.

EXAMPLE 14.15

Construct the R-chart for the data given in Example 14.14.

Solution: The sample ranges are shown in Table 14.6. The mean value of the sample ranges was computed in Example 14.14 as $\bar{r} = 2.0000$. For a sample size of five, Table 14.4 gives $d_2 = 2.326$ and $d_3 = 0.864$. The upper and the lower control limits of the R-chart can be computed as

$$UCL_r = \left[1 + \frac{3d_3}{d_2}\right]\bar{r} = \left[1 + \frac{3 \times 0.864}{2.326}\right]2.0000 = 4.2287$$

$$LCL_r = \left[1 - \frac{3d_3}{d_2}\right]\bar{r} = \left[1 - \frac{3 \times 0.864}{2.326}\right]2.0000 = -0.2287 \approx 0$$

Note that the lower control limit is taken as zero since the computed value is negative. The control limits and the sample ranges are plotted in Fig. 14.10. This R-chart shows that there is no abnormality in the production process.

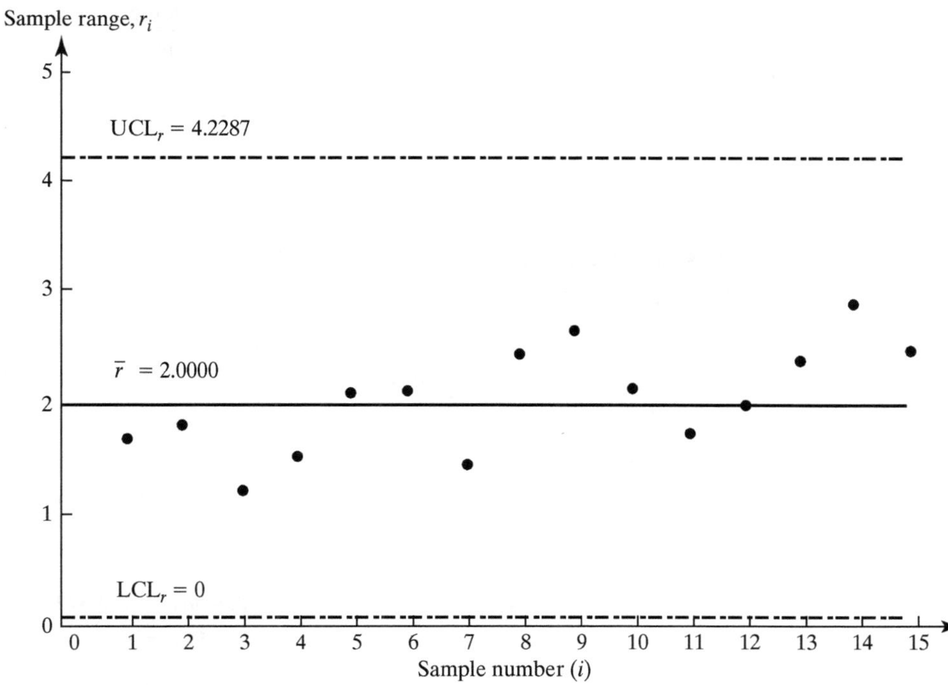

FIGURE 14.10 R-chart for Example 14.15.

■

14.7.4 The c-Chart

The c-charts are also known as the control charts for defects per unit. These charts are applied in situations where a number of defects can be observed in each item. The item may be a single component or an assembly of several components. Thus, the sample size is one in this case. Some example situations where the c-charts can be used are as follows:

1. The number of defective riveted joints in an airplane wing.
2. The number of defective welds in an automobile body.
3. The number of defects in a casting.

The number of defects in each unit is plotted in the c-chart and the upper and lower control limits are established as

$$\text{UCL}_c = \bar{c} + 3\sigma_c \tag{14.36}$$

$$\text{LCL}_c = \bar{c} - 3\sigma_c \tag{14.37}$$

where $\bar{c}$ and σ_c denote the mean and standard deviation of the number of defects per unit (c), respectively. The c-chart is based on the assumption that the number of defects in each unit follows Poisson distribution. Thus

$$\bar{c} = \frac{\text{Total number of defects}}{\text{Total number of units}} \tag{14.38}$$

and

$$\sigma_c = \sqrt{\bar{c}} \tag{14.39}$$

EXAMPLE 14.16

Twelve automobile gearboxes were examined for defects. The number of defects found in the gearboxes 1, 2, . . . , 9, and 10 were 4, 6, 19, 9, 3, 10, 8, 5, 7, 6, 2, and 5, respectively. Construct the c-chart.

Solution: The total number of defects found in the twelve gearboxes is $4 + 6 + 19 + 9 + 3 + 10 + 8 + 5 + 7 + 6 + 2 + 5 = 84$ and hence the mean value of the Poisson distribution is $\bar{c} = 84/12 = 7$ defects per gearbox. The standard deviation is $\sigma_c = \sqrt{7} = 2.6458$. Thus, the upper and lower control limits can be computed as

$$\text{UCL}_c = \bar{c} + 3\sigma_c = 7 + 3(2.6458) = 14.9374$$

$$\text{LCL}_c = \bar{c} - 3\sigma_c = 7 - 3(2.6458) = -0.9374 \approx 0$$

The c-chart is shown in Fig. 14.11 where the upper and lower control limits, as well as the number of defects in the various gearboxes, are plotted. This figure shows that one of the sample points (namely, the defects in gearbox number 3) falls outside the control limits and hence an investigation is necessary to find out the causes of the abnormal number of defects in gearbox number 3.

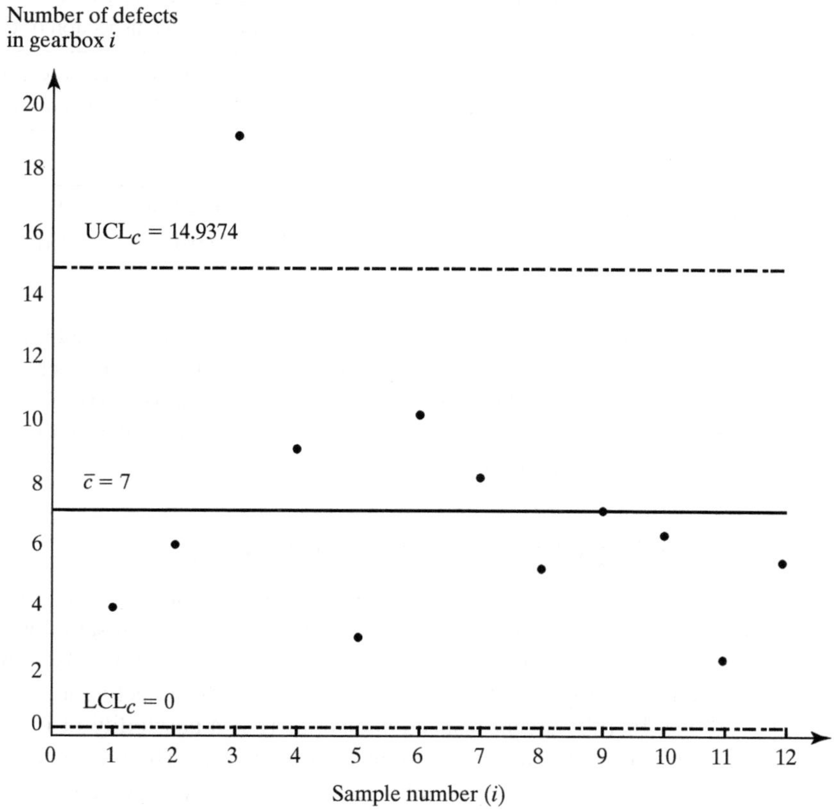

FIGURE 14.11 c-chart for Example 14.16.

■

14.8 Risks

As stated earlier, the decision made either to accept or reject a lot as a result of sampling involves certain risk. A simple approach might be to establish a maximum acceptable level of percent defectives, say, D% for the whole lot. Then we can select a convenient sample size, say 50 units, for inspection and set the criteria for the maximum number of defectives in the sample to provide a specified probability of accepting or rejecting the lot. For example, if the number of defective units in the sample is less than or equal to, say, d we will accept the lot and if the number of defectives exceed d, we will reject the lot. Since the percent defectives in a sample may not be same as the percent defectives in the lot, there are four possible outcomes as indicated in Table 14.7.

Thus, if the production lot has truly less than (greater than) D% defectives and we make a decision based on the sample results to accept (reject) the entire lot, we have made a correct decision. On the other hand, if the lot has less than D% defectives and if the sample shows more than d defectives, we end up rejecting the lot. Thus, we committed an error (called a Type-I error) and the probability of reaching this wrong decision is called the

TABLE 14.7 Acceptance and rejection of lot based on the inspection of the sample

		Actual Defectives in the Production Lot (Population)	
		$\leq D\%$	$> D\%$
Decision on the lot based on the inspection of the sample	Accept	Correct decision	Type-II error
	Reject	Type-I error	Correct decision

α–risk or producer's risk. Similarly, when we accept a lot based on sample results, when the lot has greater than D% defectives, we commit a Type-II error. The probability of reaching this wrong decision is known as the β–risk or consumer's risk. We shall now consider the incorporation of these two risks in the construction of the "operating characteristic curve" in the following section.

14.9 Operating Characteristic (OC) Curve

The OC curve is a graph between the percentage of defectives in a lot and the probability of accepting the lot by the sampling plan. In general, both sampling and 100 percent inspection processes may not guarantee the discovery of all the defective units in a lot. For example, in sampling, there is a small probability of rejecting a good lot and accepting a bad one. In 100 percent inspection also, there is a possibility of inspectors missing some defective units due to monotony and other reasons. The OC curve gives the probability with which a lot submitted with a certain percentage of defectives will be accepted or rejected. The following four parameters are often used in connection with a OC curve.

Producer's Risk Parameter (α): The producer's risk parameter, α, denotes the probability with which a lot will be rejected based on a sampling plan although it may meet the specified quality requirements. Usually a value of $\alpha = 0.05$ is used. This implies that there is a 95 percent chance of accepting a lot which meets the specified quality requirements.

Consumer's Risk Parameter (β): The consumer's risk parameter, β, denotes the probability with which a lot will be accepted based on a sampling plan although it may not meet the minimum specified quality requirements. Usually a value of $\beta = 0.1$ is used. This implies that there is a 90 percent chance of rejecting a lot which does not meet the lowest quality requirements.

Lot Tolerance Percent Defective Parameter (LTPD): The lot tolerance percent defective parameter (LTPD), in percentage defectives, represents the poorest level of quality that the consumer is willing to accept. Since this quality level is close to the minimum quality requirements, it has a low probability of acceptance, usually on the order of 10 percent.

Acceptance Quality Level Parameter (AQL): The acceptance quality level parameter (AQL) denotes the maximum percentage of defectives that can be considered satisfactory as the process average (for purposes of sampling inspection). Usually the lots with this quality level have a high probability of acceptance, on the order of 95 percent.

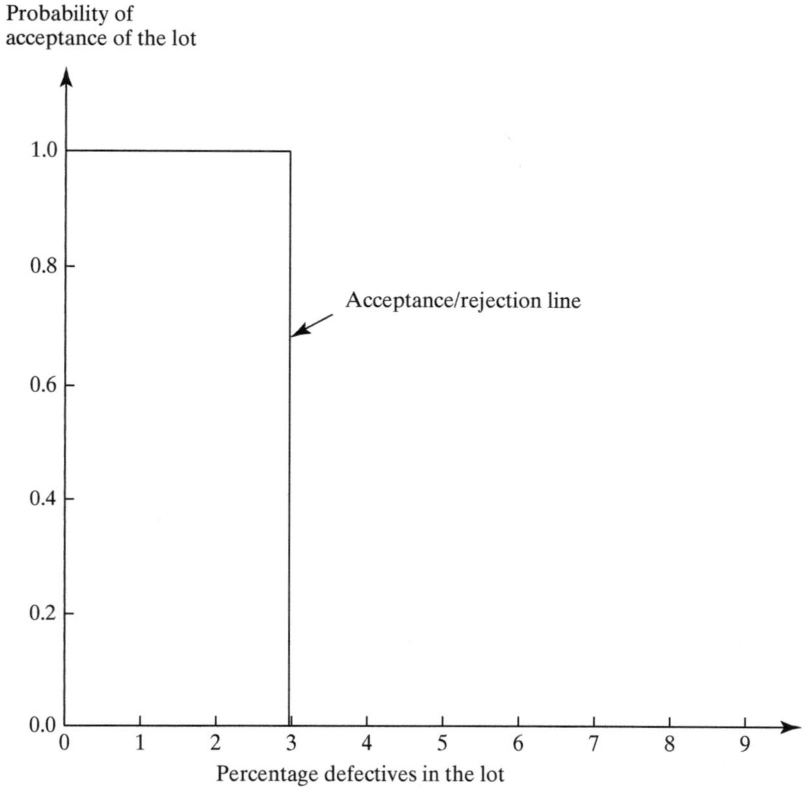

FIGURE 14.12 The ideal OC curve.

14.9.1 OC Curve

An ideal OC curve, which discriminates good lots and bad lots perfectly, is shown in Fig. 14.12. This curve shows that lots containing more than a specified percentage of defective units (taken as 3 percent in Fig. 14.12) would be rejected and lots containing less than the specified percentage of defective units (3 percent in Fig. 14.12) would be accepted. The ideal OC curve can be achieved only through 100 percent inspection (if it is done without any error). This curve, in practical cases, cannot be achieved. The shape of the actual OC curve depends on the sample size and the accept/reject criteria used in the sampling plan. A typical OC curve, in practice, with α, β, LTPD, and AQL is shown in Fig. 14.13.

14.9.2 Construction of OC Curve

The OC curve is constructed using the binomial distribution. The probability of observing exactly d defective units in a sample of size n (in a single sampling plan) drawn randomly from a lot containing p percent defective units is given by

$$P(d \text{ defectives}) = \frac{n!}{d!(n-d)!} p^d (1-p)^{n-d} \tag{14.40}$$

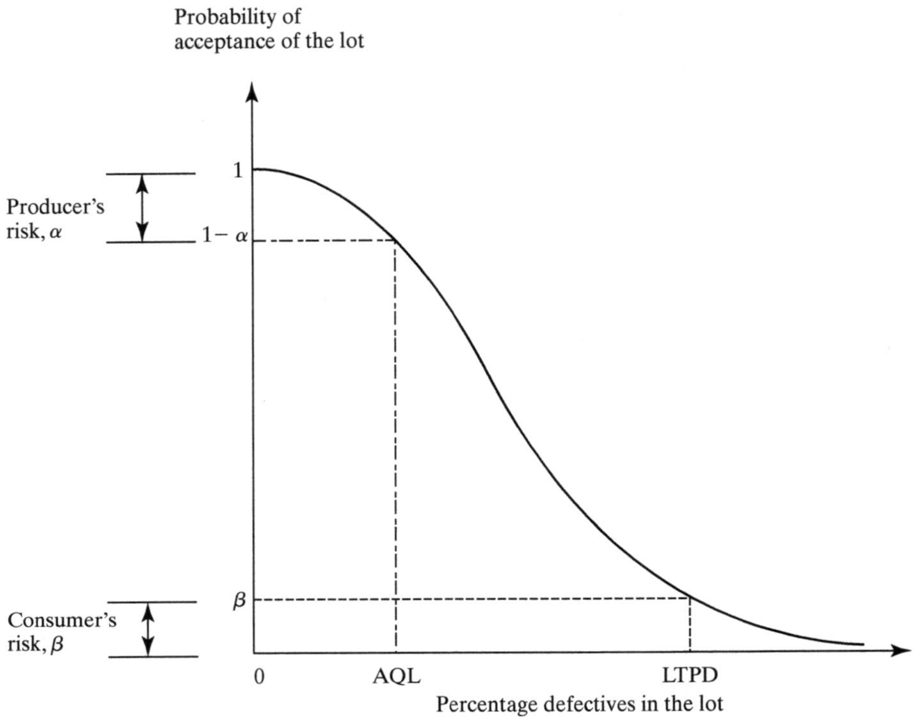

FIGURE 14.13 A typical practical OC curve.

If c denotes the acceptable number of defective units, then the probability of acceptance of the lot can be found as

$$P(\text{acceptance}) = P(d \leq c) = \sum_{d=0}^{c} \frac{n!}{d!(n-d)!} p^d (1-p)^{n-d} \tag{14.41}$$

The α-curve can be developed by evaluating Eq. (14.41) for various values of p.

EXAMPLE 14.17

In a manufacturing plant, it has been decided to use a sample of 25 components from a lot containing 250 components. Construct the operating characteristic curve if the percentage of defective units in the lot is expected to vary between 0 and 15. Assume that the lot will be accepted only if two or fewer components are found to be defective in the sample.

Solution: Here $n = 25$, $c = 2$, and p varies from 0 to 0.15. The calculations are done using Eqs. (14.40) and (14.41) and the results are shown in Table 14.8.

$$P(d=0) = \frac{25!}{0!\,25!}p^0(1-p)^{25} = p^0(1-p)^{25}$$

$$P(d=1) = \frac{25!}{1!\,24!}p^1(1-p)^{24} = 25p(1-p)^{24}$$

$$P(d=2) = \frac{25!}{2!\,23!}p^2(1-p)^{23} = 300p^2(1-p)^{23}$$

The resulting OC curve is shown in Fig. 14.14.

TABLE 14.8 Computations for the construction of the OC curve

Value of p	$P(d=0)$	$P(d=1)$	$P(d=2)$	$P(acceptance) = \sum_{i=0}^{c=2} P(d=i)$
0.00	1.000000	0.000000	0.000000	1.000000
0.01	0.777822	0.196420	0.023808	0.998050
0.02	0.603465	0.307890	0.075402	0.986757
0.03	0.466975	0.361063	0.134003	0.962041
0.04	0.360397	0.375413	0.187707	0.923516
0.05	0.277389	0.364986	0.230518	0.872893
0.06	0.212910	0.339750	0.260234	0.812895
0.07	0.162957	0.306640	0.276965	0.746562
0.08	0.124364	0.270357	0.282112	0.676834
0.09	0.094631	0.233979	0.277689	0.606299
0.10	0.071790	0.199416	0.265888	0.537094
0.11	0.054294	0.167762	0.248815	0.470870
0.12	0.040932	0.139542	0.228342	0.408816
0.13	0.030760	0.114907	0.206040	0.351706
0.14	0.023039	0.093763	0.183165	0.299966
0.15	0.017198	0.075873	0.160672	0.253742

■

14.9.3 Designing a Single Sampling Plan with a Specified OC Curve

A common procedure used for the design of an acceptance sampling plan requires the OC curve to pass through two designated points. Notice that one point is not enough to fully specify the sampling plan; however, two points are sufficient. In general, any two points can be specified. Usually, we wish to develop a sampling plan such that the probability of

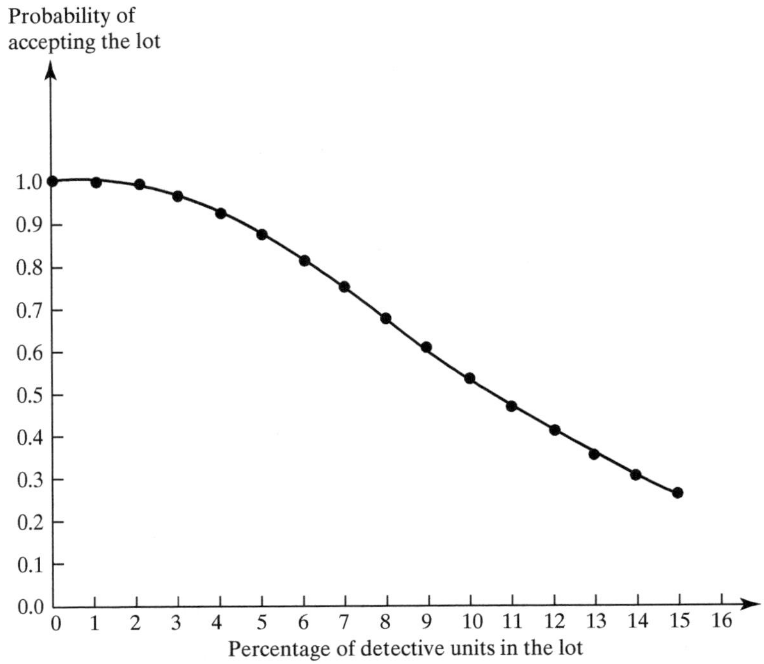

FIGURE 14.14 The OC curve of Example 14.17.

acceptance is $(1 - \alpha)$ for lots with p_1 fractional defectives, and the probability of acceptance is β for lots with p_2 fractional defectives. According to this scheme, the sample size n and the acceptance number c should satisfy the equations:

$$1 - \alpha = \sum_{d=0}^{c} \frac{n!}{d!(n-d)!} p_1^d (1 - p_1)^{n-d} \tag{14.42}$$

$$\beta = \sum_{d=0}^{c} \frac{n!}{d!(n-d)!} p_2^d (1 - p_2)^{n-d} \tag{14.43}$$

These equations, Eqs. (14.42) and (14.43), denote two points on the OC curve. These two simultaneous nonlinear equations are to be solved to find the values of n and c. Iterative and graphical solutions have been developed for solving these equations [14.5, 14.6].

14.10 Taguchi Method

14.10.1 Basic Concept

Taguchi method is based on the philosophy that high quality can be achieved economically by being "designed in," not by inspection and screening. The notion of "robust design" is used in Taguchi methods; this implies that the performance of a product must be consistent regardless of variations in material, manufacture, and use. These methods view the quality of a product as the loss imparted to society after the product is shipped. The rationale is that if a large number of customers are dissatisfied with a given product, the manufacturer will feel the loss, directly through increased service costs and refunds, and indirectly through a loss of market share in a highly competitive world. The performance characteristic of a product is the main quality characteristic that defines the product's performance which will satisfy the customer's needs. The ideal state of a performance characteristic is called its target value. A characteristic of a product can be either a qualitative entity, such as the sharpness of a TV picture, or a quantitative entity, such as the braking distance of a car. Thus, any deviation in the performance characteristic of a product from its specified target results in a loss of quality.

Traditionally, manufactured products are considered to be of good quality only if they meet certain specification limits. The most common method of inspection involves the go/no-go checks. According to the Taguchi's definition of quality, the variation of a product's quality characteristic about its target value should be as small as possible for a high-quality product.

14.10.2 Loss Function

In the Taguchi methods, the quality level of a product is defined as the total loss incurred by society due to the failure of the product to meet the expected performance and due to the harmful side effects of the product, including its operating cost. According to this concept, losses are incurred for any product from the time it is delivered to the customer to the end of its useful life or time of its retirement. Thus, every product will incur some quality loss; hence the smaller the loss, the more desirable the product. The quality loss is quantified in the form of a loss function in order to be able to compare alternative product designs and manufacturing processes. The loss function (or quality loss function), $L(y)$, is defined as

$$L(y) = k(y - \tau)^2 \tag{14.44}$$

where y is the quality or response characteristic of the product, τ is the target or desirable value of y of the product, and k is a constant, termed the quality loss coefficient. Equation (14.44) shows that the smaller the value of the loss function, the more desirable the product.

The loss function shown in Fig. 14.15(*a*) is applicable when the quality characteristic y is specified using a bilateral tolerance zone as $\tau \pm \Delta$. Thus, $\tau - \Delta$ and $\tau + \Delta$ represent the lower and upper specification limits of the quality characteristic y. In practical situations, the closer the performance characteristic (y) is to the target τ, the better the quality of the product. In other words, as the performance characteristic y deviates more from the target value τ, the more undesirable will be the product. Equation (14.44) indicates that the loss or undesirability of the product varies quadratically as its performance deviates from its target value.

It can be seen from Fig. 14.15(*a*) that the loss will be equal to A when the value of y falls at the lower or upper specification limit $(\tau - \Delta)$ or $(\tau + \Delta)$. When the value of y falls outside the tolerance range $(\tau - \Delta, \tau + \Delta)$, the product is rejected if it is in the manufacturing stage or repaired or replaced if the product is in service. For this case, the loss function becomes

$$L(\tau + \Delta) = A = k\{(\tau + \Delta) - \tau\}^2 = k\Delta^2$$

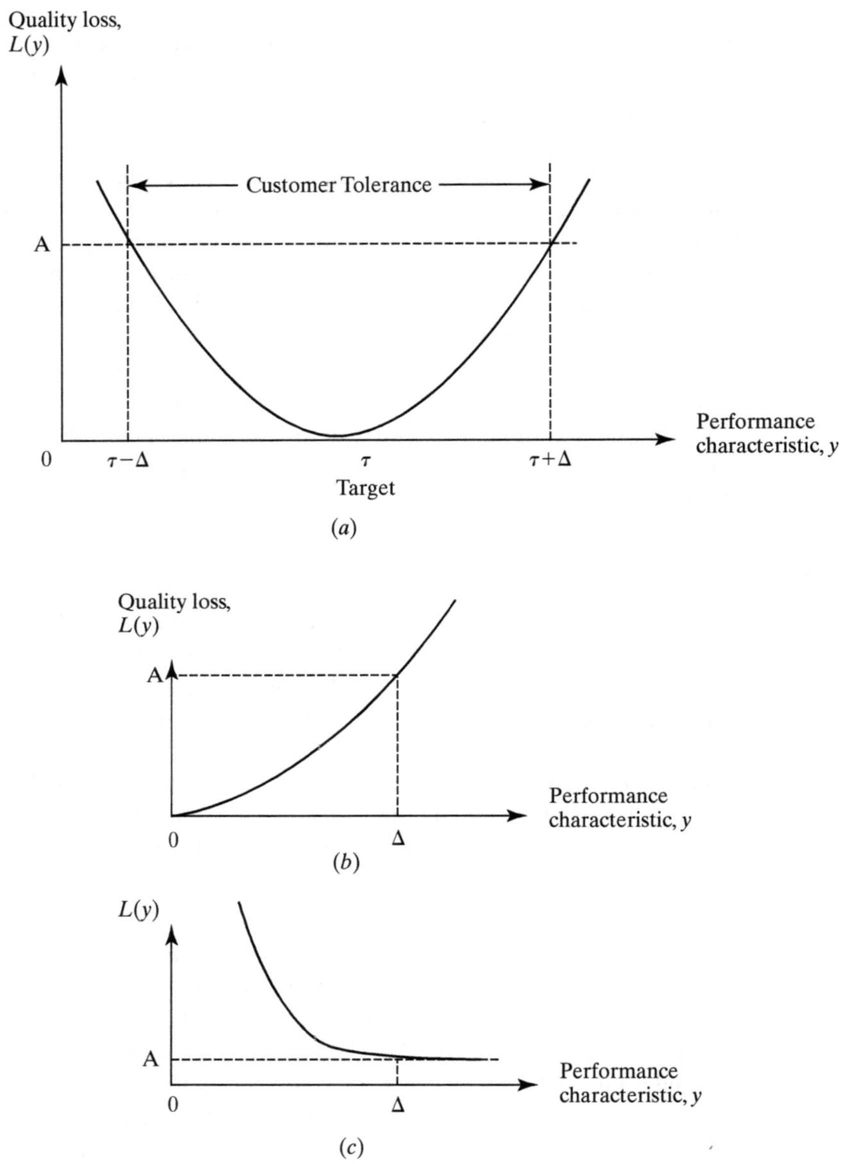

FIGURE 14.15 Quality loss function.

or

$$k = \frac{A}{\Delta^2} \tag{14.45}$$

Thus Eq. (14.44) can be rewritten as

$$L(y) = \frac{A}{\Delta^2}(y - \tau)^2 \tag{14.46}$$

Figure 14.15(*b*) shows the loss function when the quality characteristic y is required to be as small as possible (smaller the better). By proceeding as in the case of nominal the better, the loss function in the case of smaller the better can be expressed as

$$L(y) = \frac{A}{\Delta^2} y^2 \tag{14.47}$$

Finally, Fig. 14.15(*c*) shows the loss function when the quality characteristic y is required to be as large as possible (larger the better). The loss function in this case can be expressed as

$$L(y) = \frac{A\Delta^2}{y^2} \tag{14.48}$$

EXAMPLE 14.18

The cost of repair of a failed television set in a factory is 50 dollars per unit. Let color concentration be used as the performance characteristic, y, and let the customers consider the television set to be defective if y varies over the range $\tau \pm 10$ where τ is the target value of y. Find the quality loss of a customer whose television set has a color concentration of $\tau - 6$.

Solution: Equation (14.44) gives

$$k = \frac{50}{10^2} = 0.5 \text{ dollars per unit}$$

and

$$L(y) = 0.5(y - \tau)^2 \tag{E1}$$

This gives the quality loss of a customer whose television set has a color concentration of $\tau - 6$ as $L(\tau - 6) = 0.5(-6)^2 = 18$ dollars.

■

14.10.3 Noise Factors

As stated earlier, high quality can only be achieved economically by being designed in at an early stage of the product development. The design must also be robust and the product must perform consistently regardless of any variations in the environmental conditions, manufacturing imperfections, or deterioration factors. These variation factors are known as noise factors. The noise factors are grouped into three major types: outer or external noises, manufacturing imperfections, and deterioration. The outer noises are generally due to environmental conditions. The environment in which a product operates can affect the performance of that product. Examples of outer noise factors are temperature, humidity, vibration, supply voltage, and so on. For instance, an increase in the operating temperature can result in a severe reduction in copy quality of a sensitive copying machine. The manufacturing imperfections include the variations from unit to unit in a production line. Two similar products may not have exactly the same specified parameter, that is, dimensions, color, and so on. For example, the coil diameter of a spring may be specified as 2 inches on the design drawings, but after manufacture, some springs may have a coil diameter of 2.05 in. These variations in parameter values can result

TABLE 14.9 Noise factors [14.8, 14.9]

Outer Noises	Manufacturing Imperfections	Deterioration Factors
Load conditions	Variations from unit to unit in:	Wear
Temperature	–Dimensions	Fatigue/cracking
Humidity	–Mass/weight	Corrosion
Dust	–Purity of material	Permanent bending
Supply voltage	–Color	Aging
Vibrations	–Heat treatment	Fading of color
Human errors	Machine failure	Shrinkage
Ultraviolet light effects	Human errors	Drying out
Shocks	Tool wear	
Pressure conditions		

from differences in machining conditions, human errors during manufacturing, tool wear, and so forth. The deterioration factors occur during the service life of a product. When it is new, a product may function as conceived with all the performance characteristics at their desired values or targets. But as time passes, these characteristics may start to deviate from their target values resulting in the deterioration of the performance of the product. The deterioration factors include wear, fatigue, and corrosion. Table 14.9 gives a list of different noise factors which are applicable to most mechanical, electrical, and other types of products.

14.10.4 On-Line Versus Off-Line Quality Control

As discussed in the earlier sections, traditionally quality control is conducted during the actual manufacturing process. Finished products are inspected to check whether or not they meet the specified limits. Most inspections are based on statistical quality control procedures involving the use of control charts, process capability studies, and cause-and-effect diagrams. Since such control methods are carried in real time, they are called on-line control methods. The off-line control methods, on the other hand, deal with the implementation of quality measures as early as possible during the design stages of the product development cycle. Although the on-line quality control methods can reduce variations in quality characteristics due to manufacturing imperfections, they cannot reduce variations due to external noises or deterioration, both of which occur when a perfectly functioning product is put into service. For example, consider a car door. The nominal dimensions and tolerances of the door as well as those of the door frame are fixed at the design stage. If the dimensions of the door and the frame fall within their specified tolerance limits, then, after manufacture, the door may close just right. However, if some external factors such as metal expansion, hinge deflection, and rusting become significant later, then the door will not close properly. The off-line quality control can account for all three types of noises. The main purpose of off-line quality control is to improve product reliability and reduce lifetime costs which could be imparted to the customer in terms of service and repair charges, and to the manufacturer in terms of warranty costs and reputation.

14.10.5 Three-Step Design Approach

Taguchi suggested a three-step design approach for achieving off-line quality control: concept design, parameter design, and tolerance design [14.7].

Concept Design. In this stage, the base design and engineering concepts of the product or the process to be designed are established to produce a basic functional prototype design. The best design among many alternatives is selected and materials, components, and manufacturing modes are also chosen. The main focus of this stage is to create or redesign a product or a process. Since the optimum performance characteristics are not considered yet, the product may have crude characteristics. A fine tuning of these parameters is done in the next phase.

Parameter Design. To introduce the concept of parameter design, consider the example of a sensitive copying machine. We realize that any change in the surrounding temperature can result in a malfunction of the machine. One method of guarding the machine from damage is to control the temperature inside the copying room. But this solution is very costly since it requires additional equipment to be added to monitor the environment such as a thermostat, and an air-conditioner to maintain the room temperature at the desired level. Another way of solving the problem is to redesign the copying machine such that it can function under a wide range of temperatures. The method of designing a product to minimize its sensitivity to the noise factors is called parameter design. During parameter design, the tolerances are assumed to be large enough to reduce the manufacturing cost. Generally, the parameter design involves the use of experimental design procedures. Experimental design, in turn, may involve either actual experiments or computer simulations using mathematical models. So far, emphasis has been placed on experimental designs using orthogonal arrays and the analysis of variance (ANOVA) [14.9].

Tolerance Design. After the nominal values of the design parameters are determined, the tolerances are selected by conducting a trade-off study between the reduction in quality loss and the increase in manufacturing cost. Tolerance design is performed only after minimizing the sensitivity to noise factors in the parameter design stage. The computation of the tolerances is based on the cost of different grades of components and the variances found in the parameter design stage.

14.10.6 Experimental Design

The purpose of experimental design is to identify efficiently the optimal values of the design variables which will yield a minimum variation in the performance characteristic while keeping the cost as low as possible. Experimental design is a statistical procedure through which data are collected from a number of experiments in which different settings of the design variables and the noise factors are combined so as to probe the entire experimental space. The experimental space is defined as all possible combinations of the various settings of the design variables and the noise factors. Usually, the settings of a design variable or a noise factor are taken as the estimated largest, average, and the smallest values which that variable or factor can take.

It can be seen that the number of experiments becomes large as the number of design variables increases since each setting or combination of levels of the design variables and noise factors represents one experiment. For example, if there are six design variables/noise factors, each with three settings, the total number of experiments to be conducted will be $3^6 = 729$. Thus, the number of experiments can become very large which will be quite expensive and time-consuming to conduct. In addition, conducting too many experiments will defeat the very purpose of the experimental design procedure.

Orthogonal Arrays. Orthogonal arrays are used to reduce the number of experiments while allowing for the maximum amount of information to be obtained [14.9]. Orthogonal arrays are the modern version of the Graeco-Latin squares [14.11].

The orthogonal arrays have the pairwise property that every setting of a design parameter occurs with every setting of all other design parameters the same number of times. This balancing property minimizes the number of experiments to be conducted. For example, consider a problem with three factors or control parameters X_1, X_2, and X_3 influencing the performance of the design. If two levels (or settings) are considered for each control parameter, all possible combinations of the parameters can be represented by the 8 corners of the cube shown in Fig. 14.16 where the two levels of the parameter x_i are denoted

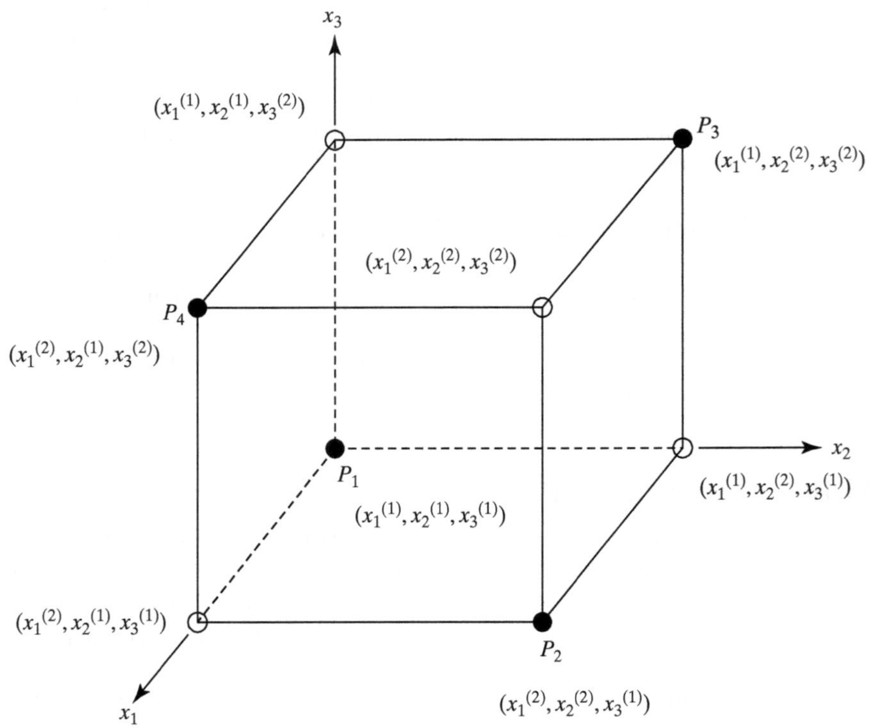

FIGURE 14.16 Possible combinations of three control parameters, each with two levels.

as $x_i^{(1)}$ and $x_i^{(2)}$. All the eight parameter combinations denoted by the corners of the cube constitute full factorial experiment. However, a fractional factorial experiment, involving only four corners (denoted as P_1, P_2, P_3, and P_4 in Fig. 14.16), can be used to obtain essentially the same information.

Tables 14.10 and 14.11 show the two-parameter orthogonal arrays L_4 and L_8 in which each row represents an experiment. Each column denotes the settings of the design variables and noise factors. The numerals 1 and 2 represent the first and the second setting, respectively. An L_8 array is the smallest array that can accommodate seven variables defined at two settings each. The orthogonal array L_8 thus reduces the number of experiments to be conducted from $2^7 = 128$ to eight. Table 14.12 shows a three-parameter orthogonal array, L_9, in which the numerals 1, 2, and 3 in the columns denote the first, second, and third levels or settings of the design variables. This orthogonal array reduces the number of experiments to be conducted from $3^4 = 81$ to 9. An extensive list of orthogonal arrays and the methods used in their selection are given in Ref. [14.9]. The use of an L_8 array is demonstrated in the following example.

TABLE 14.10 Orthogonal array L_4

	Column No.		
Trial No.	**1**	**2**	**3**
1	1	1	1
2	1	2	2
3	2	1	2
4	2	2	1

TABLE 14.11 Orthogonal array L_8

	Column No.						
Trial No.	**1**	**2**	**3**	**4**	**5**	**6**	**7**
1	1	1	1	1	1	1	1
2	1	1	1	2	2	2	2
3	1	2	2	1	1	2	2
4	1	2	2	2	2	1	1
5	2	1	2	1	2	1	2
6	2	1	2	2	1	2	1
7	2	2	1	1	2	2	1
8	2	2	1	2	1	1	2

TABLE 14.12 Orthogonal array L_9

	Column No.			
Trial No.	**1**	**2**	**3**	**4**
1	1	1	1	1
2	1	2	2	2
3	1	3	3	3
4	2	1	2	3
5	2	2	3	1
6	2	3	1	2
7	3	1	3	2
8	3	2	1	3
9	3	3	2	1

EXAMPLE 14.19

Consider the heat treatment of leaf springs used in trucks. The leaf spring assembly is first transported by a conveyor system through a high temperature furnace. Once the leaf spring comes out of the furnace, it is transferred to a forming machine where the spring is bent by passing it through a high pressure press for a short time. Finally, the spring is submerged and quenched in an oil tub and removed. The performance of the leaf spring is found to depend on five factors: (1) furnace temperature, (2) heating time, (3) transfer time, (4) hold down time (i.e., operating time of the press), and (5) quench oil temperature. These noise factors can be identified to be of manufacturing imperfection type.

In order to achieve the optimal performance of the leaf springs, the five factors are varied over their respective ranges until a combination, which reduces the characteristic variation, is determined. In order to carry out the design experiments, the five factors or design variables are designated as A, B, C, D, and E. For each of these factors, two settings are assumed as indicated in Table 14.13.

TABLE 14.13 Factor settings of leaf springs

Factor	First Setting	Second Setting
A: Furnace temperature (deg F)	$A_1 = 1700$	$A_2 = 1900$
B: Heating time (s)	$B_1 = 25$	$B_2 = 20$
C: Transfer time (s)	$C_1 = 12$	$C_2 = 10$
D: Hold down time (s)	2	3
E: Quench oil temperature (deg F)	100	150

For convenience, the orthogonal array L_8 is selected to conduct the experiments. Each experiment consists of producing 100 leaf springs at each setting and counting the number of defective units produced. The performance characteristic is taken as the percentage of defective leaf springs. The results from the eight settings are shown in Table 14.14. To interpret the entries of Table 14.14,

TABLE 14.14 Orthogonal array L_8 of Example 14.19

Experiment Number	Factors ABCDE	Furnace Temperature (deg F)	Heating Time (s)	Transfer Time (s)	Holding Time (s)	Quench Oil Temperature (deg F)	Number of Defective Leaf Springs
1	11111	1700	25	12	2	100	15
2	11122	1700	25	12	3	150	27
3	12211	1700	20	10	2	100	21
4	12222	1700	20	10	3	150	16
5	21212	1900	25	10	2	150	18
6	21221	1900	25	10	3	100	41
7	22112	1900	20	12	2	150	45
8	22121	1900	20	12	3	100	37

consider the third row, for example. This experiment is conducted with factors A, D, and E at their first settings, and the remaining factors at their second settings. Out of the 100 leaf springs produced under these settings, 21 are observed to be defective. Find the optimal settings of the five factors involved.

Solution: To study the effect of each setting of each factor in all of the eight experiments, the number of defective leaf springs in the first setting of each factor in all the experiments is compared to that of the second setting of the same factor. For example, when comparing A_1 and A_2, the total number of defectives in experiments 1, 2, 3, and 4 (which are conducted with setting A_1) and the total number of defectives in experiments 5, 6, 7, and 8 (which are conducted with setting A_2) are to be considered. This gives the result:

$$A_1 = 15 + 27 + 21 + 16 = 79; \quad A_2 = 18 + 41 + 45 + 37 = 141$$

Since the total of A_1 represents the number of defectives among $4 \times 100 = 400$ leaf springs, the fractional defectives would be $79/400 = 19.75$ percent. Similar results, obtained for the remaining factors, are shown in Table 14.15. The final step in the design of experiments procedure consists of comparing the individual settings of each factor. The settings with the lower percentage of defective leaf springs is to be selected for optimal performance. Thus, from the results shown in Table 14.15, the optimal settings of the various factors can be identified as A_1, B_1, C_2, D_1, and E_2. This result is expected to yield the lowest fractional defective leaf springs. A new experiment can be conducted at the optimal settings found to verify this result.

■

14.10.7 Signal-To-Noise Ratio

In order to quickly identify the control or design factors that yield reduced variation (or equivalently, improved quality) in the response, we can look at the amount of variation present in the response directly. However, in Taguchi approach, the repetitive data available from different experiments is used to define another quantity, known as the signal-to-noise

TABLE 14.15 Total and percent number of defective leaf springs (for different settings of the factors)

Setting of the Factor	Total Number of Defective Leaf Springs	Percent Defectives
A_1	15 + 27 + 21 + 16 = 79	19.75
A_2	18 + 41 + 45 + 37 = 141	35.25
B_1	15 + 27 + 18 + 41 = 101	25.25
B_2	21 + 16 + 45 + 37 = 119	29.75
C_1	15 + 27 + 45 + 37 = 124	31.00
C_2	21 + 16 + 18 + 41 = 96	24.00
D_1	15 + 21 + 18 + 45 = 99	24.75
D_2	27 + 16 + 41 + 37 = 121	30.25
E_1	15 + 21 + 41 + 37 = 114	28.50
E_2	27 + 16 + 18 + 45 = 106	26.50

(*S*/*N*) ratio, for identifying the design factors that yield the best quality. Three types of *S*/*N* ratios are used depending on the nature of the quality characteristic as follows:

1. Smaller the better

In some examples, it is desirable to have the quality characteristic, such as the stress induced during the operation of a machine, as small as possible for better quality. In this case, the S/N ratio is defined as

$$S/N = -10 \log_{10}\left(\frac{1}{n}\sum_{i=1}^{n} y_i^2\right) \tag{14.49}$$

where n is the number of repetitions or trials or experiments and y_i is the quality characteristic or response observed in ith experiment.

2. Nominal the better

In some situations, it is desirable to have the quality characteristic, such as the Brinell hardness number of steel, as close as possible to a specified or target value. In this case, the *S*/*N* ratio is defined in two possible ways as

By considering only variance:

$$S/N = -10 \log_{10}\left\{\frac{1}{n}\sum_{i=1}^{n} (y_i - \tau)^2\right\} \tag{14.50}$$

where τ is the target value.

By considering both mean and variance:

$$S/N = 10 \log_{10}\left(\frac{\mu}{\sigma}\right)^2 \tag{14.51}$$

where

$$\mu = \frac{1}{n}\sum_{i=1}^{n} y_i \tag{14.52}$$

and

$$\sigma = \left\{ \frac{1}{n-1} \sum_{i=1}^{n} (y_i - \mu)^2 \right\}^{\frac{1}{2}} \tag{14.53}$$

3. Larger the better
In some cases, it is desirable to have the quality characteristic, such as the stiffness of a beam or a machine component, as large as possible. In such a case, the *S/N* ratio is defined as

$$S/N = -10 \log_{10} \left(\frac{1}{n} \sum_{i=1}^{n} \frac{1}{y_i^2} \right) \tag{14.54}$$

14.10.8 Experimental Design in the Presence of Noise Factors

In order to achieve a robust design in the Taguchi approach, parameter design is used to select the proper levels of the design variables so as to reduce the variability or dampen the effect of noise on the product in the presence of uncontrollable noise factors. The parameter design of experiments is based on the use of a design parameter matrix (also called the inner array) and a noise matrix (also called the outer array) as shown, for a typical case, in Fig. 14.17. The design parameter matrix specifies the levels or settings of the design parameters. Figure 14.17 indicates four parameters x_1, x_2, x_3, and x_4

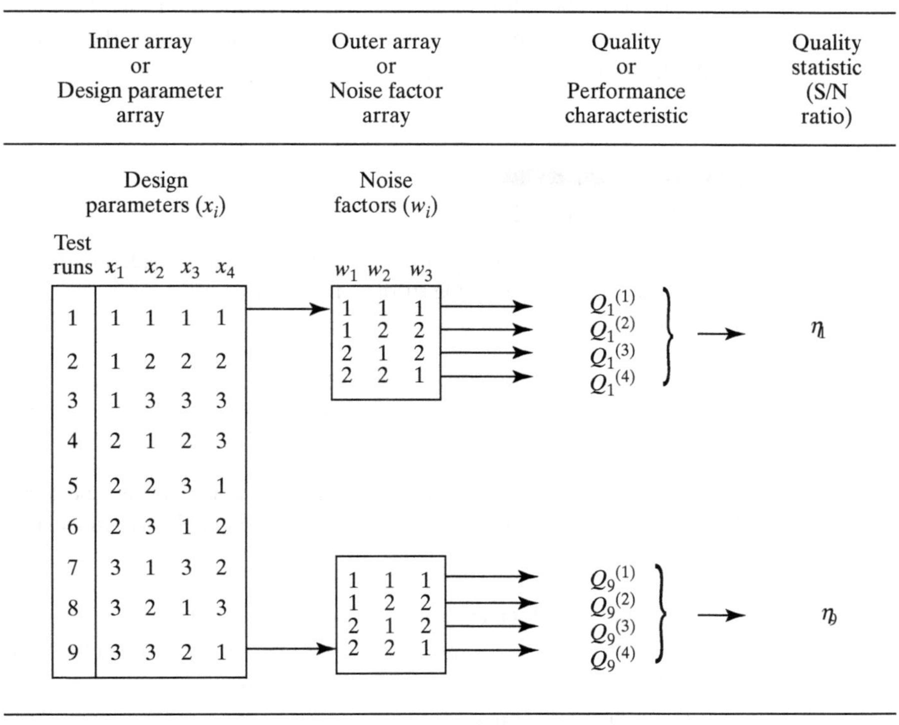

FIGURE 14.17 A typical Taguchi design of experiments for parameter design.

each consisting of three possible levels or settings during experimentation. With three levels for each of the four design parameters, there will be $3^4 = 81$ experiments possible, each with a different combination of levels of the design parameters. It can be observed that the design parameter matrix (orthogonal array) consists of only 9 experiments. Assuming that three of the design variables (among x_1, x_2, x_3, and x_4) are subject to variations or tolerances or noises w_1, w_2, and w_3, the noise matrix consists of three noise factors w_1, w_2, and w_3 each consisting of two possible levels or settings. Thus, there will be $2^3 = 8$ experiments possible, each with a different combination of levels of the noise factors. However, the noise matrix (orthogonal array) can be seen to consist of only 4 experiments.

To the design variable values or levels corresponding to each experiment (defined by different rows) of the design parameter matrix, the tolerances or noises corresponding to each trial of the noise matrix are added to determine the final values of the design variables to be used in the experimentation. Thus, for experiment 1, there would be four trials, one corresponding to each combination of noise factors (of a row) of the noise matrix. In each trial, the noises or tolerances specified by the corresponding row of the noise matrix are superposed to the respective design variable values or levels specified by the first row of the design parameter matrix. Similarly, for experiment 2, there would be another four trials. Thus, there would be 9 experiments, each with four trials, with a total of $9 \times 4 = 36$ trials or test conditions. In each test condition or trial, the performance or response of the system is found (by experimentation or simulation). Using the observed values of the performance or response of the system in each experiment, the mean value and the signal-to-noise ratio are computed. These results are used to determine the best levels or settings of the design parameters for the optimal performance (or response) of the system. The following example illustrates the procedure.

EXAMPLE 14.20

A helical spring, shown in Fig. 14.18, is to be designed considering the spring stiffness (k) as the quality characteristic. The spring stiffness is given by

$$k = \frac{d^4 G}{8D^3 N\left\{1 + \dfrac{d^2}{2D^2}\right\}} \tag{E1}$$

with d = wire diameter, G = shear modulus, D = coil diameter, and N = number of active turns denoting the design parameters. The ranges and the possible discrete values (settings) of the design parameters in the corresponding ranges are given by

d: 0.03−0.05 in (Discrete settings: 0.03 in, 0.04 in, 0.05 in)

G: $10 \times 10^6 - 13 \times 10^6$ psi (Discrete settings: 10×10^6 psi, 11.5×10^6 psi, 13×10^6 psi)

D: 0.4–0.6 in (Discrete settings: 0.4 in, 0.5 in, 0.6 in)

N: 8–12 (Discrete settings: 8, 10, 12)

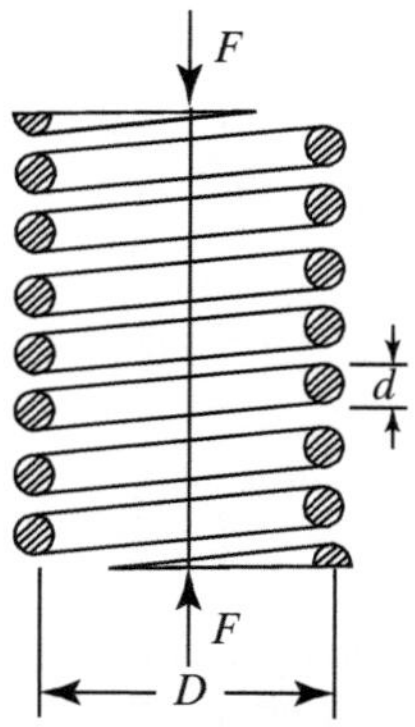

FIGURE 14.18 A helical spring.

The spring is made of steel which has a variability in shear modulus of $\pm 0.01 \times 10^6$ psi. The manufacturing equipment used in the production process is capable of achieving the following tolerances:

$$d: \pm 0.001 \text{ in}, D: \pm 0.01 \text{ in}, N: \pm 0$$

By using three settings for each of the design parameters, determine the optimal factor levels for the following type of the quality characteristic, k:

1. Larger the better
2. Nominal the better using the target value of k as 3.0 lb/in.

Use a $L_9(3^4)$ array for the inner array and a $L_4(2^3)$ array for the outer array.

Solution: In this case, the design variables are $x_1 = d$, $x_2 = G$, $x_3 = D$, and $x_4 = N$, and the noise factors correspond to $w_1 = d$, $w_2 = G$, and $w_3 = D$. The design variable values corresponding to the various experiments (rows) of the inner array (L_9) are shown in Table 14.16. Similarly, the tolerances on the design variables d, G, and D indicated by the various trials (rows) of the outer array (L_4) are shown in Table 14.17. As described earlier, for each experiment in the inner array, with

TABLE 14.16 L_9 array for parameters

EXPT #	*d*	*G*	*D*	*N*
1	0.03	1.00E + 07	0.4	8
2	0.03	1.15E + 07	0.5	10
3	0.03	1.30E + 07	0.6	12
4	0.04	1.00E + 07	0.5	12
5	0.04	1.15E + 07	0.6	8
6	0.04	1.30E + 07	0.4	10
7	0.05	1.00E + 07	0.6	10
8	0.05	1.15E + 07	0.4	12
9	0.05	1.30E + 07	0.5	8

TABLE 14.17 L_4 array for tolerances

EXPT #	*d* tol	*G* tol	*D* tol
1	−0.001	−1.00E + 04	−0.01
2	−0.001	1.00E + 04	0.01
3	0.001	−1.00E + 04	0.01
4	0.001	1.00E + 04	−0.01

design variable values shown in Table 14.16, the corresponding tolerances given by the outer array, with tolerances shown in Table 14.17, are superposed to determine the final values of the design variables to be used in the experimentation (a total of 36 tests) as shown in Table 14.18. For each test, the quality characteristic, namely the stiffness of the spring (k), is computed to obtain the values indicated in Table 14.19.

TABLE 14.18 Final values of design variables used in Taguchi experimentation

	Inner Array				Outer Array							
	Values of Design Variables				Values of Tolerances or Noises				Final Values of Design Variables to be Used in the Test			
Expt. No.	*d*	*G* (Unit: 10^6)	*D*	*N*	w_1 (for *d*)	w_2 (for *G*) Unit: 10^6	w_3 (for *D*)	Test No.	*d*	*G* (Unit: 10^6)	*D*	*N*
1	0.03	10.0	0.4	8	−0.001	−0.01	−0.01	1	0.029	9.99	0.39	8
					−0.001	0.01	0.01	2	0.029	10.01	0.41	8
					0.001	−0.01	0.01	3	0.031	9.99	0.41	8
					0.001	0.01	−0.01	4	0.031	10.01	0.39	8
2	0.03	11.5	0.5	10	−0.001	−0.01	−0.01	5	0.029	11.49	0.49	10
					−0.001	0.01	0.01	6	0.029	11.51	0.51	10
					0.001	−0.01	0.01	7	0.031	11.49	0.51	10
					0.001	0.01	−0.01	8	0.031	11.51	0.49	10
3	0.03	13.0	0.6	12	−0.001	−0.01	−0.01	9	0.029	12.99	0.59	12
					−0.001	0.01	0.01	10	0.029	13.01	0.61	12
					0.001	−0.01	0.01	11	0.031	12.99	0.61	12
					0.001	0.01	−0.01	12	0.031	13.01	0.59	12
4	0.04	10.0	0.5	12	−0.001	−0.01	−0.01	13	0.039	9.99	0.49	12
					−0.001	0.01	0.01	14	0.039	10.01	0.51	12
					0.001	−0.01	0.01	15	0.041	9.99	0.51	12
					0.001	0.01	−0.01	16	0.041	10.01	0.49	12

(Continued)

TABLE 14.18 Final values of design variables used in Taguchi experimentation *(Continued)*

	Inner Array: Values of Design Variables				Outer Array: Values of Tolerances or Noises				Final Values of Design Variables to be Used in the Test			
Expt. No.	d	G (Unit: 10^6)	D	N	w_1 (for d)	w_2 (for G) Unit: 10^6	w_3 (for D)	Test No.	d	G (Unit: 10^6)	D	N
5	0.04	11.5	0.6	8	−0.001	−0.01	−0.01	17	0.039	11.49	0.59	8
					−0.001	0.01	0.01	18	0.039	11.51	0.61	8
					0.001	−0.01	0.01	19	0.041	11.49	0.61	8
					0.001	0.01	−0.01	20	0.041	11.51	0.59	8
6	0.04	13.0	0.4	10	−0.001	−0.01	−0.01	21	0.039	12.99	0.39	10
					−0.001	0.01	0.01	22	0.039	13.01	0.41	10
					0.001	−0.01	0.01	23	0.041	12.99	0.41	10
					0.001	0.01	−0.01	24	0.041	13.01	0.39	10
7	0.05	10.0	0.6	10	−0.001	−0.01	−0.01	25	0.049	9.99	0.59	10
					−0.001	0.01	0.01	26	0.049	10.01	0.61	10
					0.001	−0.01	0.01	27	0.051	9.99	0.61	10
					0.001	0.01	−0.01	28	0.051	10.01	0.59	10
8	0.05	11.5	0.4	12	−0.001	−0.01	−0.01	29	0.049	11.49	0.39	12
					−0.001	0.01	0.01	30	0.049	11.51	0.41	12
					0.001	−0.01	0.01	31	0.051	11.49	0.41	12
					0.001	0.01	−0.01	32	0.051	11.51	0.39	12
9	0.05	13.0	0.5	8	−0.001	−0.01	−0.01	33	0.049	12.99	0.49	12
					−0.001	0.01	0.01	34	0.049	13.01	0.51	12
					0.001	−0.01	0.01	35	0.051	12.99	0.51	12
					0.001	0.01	−0.01	36	0.051	13.01	0.49	12

For example, for experiment 1 (first row of the inner array), we set the levels of each of the four design parameters at 1 so that their values will be $d = 0.03$, $G = 10 \times 10^6$, $D = 0.4$, and $N = 8$ (as shown in the first row of Table 14.16). But these values will be modified slightly due to the noise (outer) matrix. When the first row of the noise matrix or outer array (with levels of each of the noise factors, corresponding to the design parameters 1, 2, and 3 at 1) is considered, the values of the design parameters 1, 2, and 3 are subject to the tolerances -0.001, -0.01×10^6, and -0.01, respectively (as shown in the first row of Table 14.17). Hence the first trial of experiment 1 will be conducted with the following design parameter values: $d = 0.029$, $G = 9.99 \times 10^6$, $D = 0.39$, $N = 8$ (shown for test 1 in Table 14.18). When the same experiment 1 (first row of the inner matrix) is considered with the second row of the outer matrix (with levels of noise factors 1, 2, and 3

TABLE 14.19 "k" Values for each experiment

Expt #	Case 1	Case 2	Case 3	Case 4
1	1.85602866	1.60106697	2.08564839	2.42737703
2	0.86193369	0.76588797	0.99807856	1.12713415
3	0.46542485	0.42181114	0.54983798	0.60855206
4	2.03981612	1.81318774	2.20962215	2.49570003
5	2.01787692	1.82926896	2.23000634	2.46846734
6	6.3011194	5.43417738	6.62425373	7.70434325
7	3.49308057	3.16767053	3.70895424	4.10628663
8	11.5405083	9.95740825	11.6581853	13.5579308
9	9.89598881	8.7936915	10.2999067	11.6263604

corresponding to the design parameters 1, 2, and 3 at 1, 2, and 2, respectively), the values of the design parameters 1, 2, and 3 are subject to the tolerances -0.001, $+0.01 \times 10^6$, and $+0.01$, respectively (row 2 in Table 14.17). Hence the second trial of experiment 1 will be conducted with the following design parameter values: $d = 0.029$, $G = 10.01 \times 10^6$, $D = 0.41$, $N = 8$ (shown for test 2 in Table 14.18). Similarly, trials 3 and 4 of experiment 1 are conducted by considering, respectively, rows 3 and 4 of the outer matrix to determine the values of the first three design parameters. The procedure is repeated and four trials (each corresponding to different rows of the outer matrix) of each of the experiments 2, 3, ..., 9 are conducted. Thus, a total of $9 \times 4 = 36$ values of the quality characteristic (stiffness of the spring, k) are determined (computed in the present case) to obtain the values indicated in Table 14.19.

1. *Larger the better*
   For each of the experiments 1 through 9, the mean squared deviation (MSD) and the corresponding signal-to-noise ratio (S/N or η) are computed as

$$\text{MSD} = \frac{1}{4}\left\{\frac{1}{k_1^2} + \frac{1}{k_2^2} + \frac{1}{k_3^2} + \frac{1}{k_4^2}\right\} \tag{E2}$$

where k_i denotes the value of the stiffness of the spring (quality characteristic) found in case i, $i = 1, 2, 3, 4$, and

$$\eta = -10 \log_{10} \text{MSD} \tag{E3}$$

The resulting MSD and η values are shown in Table 14.20. The value of η corresponding to level 1 of the design variable $d(x_1)$ is given by the average of the η values given by experiments 1, 2, and 3 (since the level of x_1 is 1 in experiments 1, 2, and 3 in L_9) as:

$$\eta_{\text{level 1 of } d} = \frac{1}{3}(5.68636458 - 0.82946419 - 6.08652154) = -0.40987371 \tag{E4}$$

Similarly, the value of η corresponding to levels 2 and 3 of d are given by the averages of the η values given by experiments 4, 5, 6 and 7, 8, 9, respectively, as:

$$\eta_{\text{level 2 of } d} = \frac{1}{3}(6.43079676 + 6.43191958 + 16.0780385) = 9.64691826 \tag{E5}$$

TABLE 14.20 Larger the better observations

Expt #	MSD	η
1	0.26999986	5.68636458
2	1.21044878	−0.82946419
3	4.06117921	−6.08652154
4	0.22746801	6.43079676
5	0.22740921	6.43191958
6	0.02467153	16.0780385
7	0.07840408	11.0566134
8	0.007598	21.1930087
9	0.00999179	20.0035661

and

$$\eta_{\text{level 3 of } d} = \frac{1}{3}(11.0566134 + 21.1930087 + 20.0035661) = 17.417729 \tag{E6}$$

The values of η corresponding to the levels 1, 2, and 3 of the other design parameters G, D, and N can be computed in a similar manner. The results are shown in Table 14.21. It can be seen that the optimal levels of the design variables d, G, D, and N are given, respectively, by 3, 3, 1, and 1 (corresponding to the largest values of η in Table 14.21). Thus the optimum values of the design variables are given by

$$d^* = 0.05 \text{ in}, G^* = 13 \times 10^6 \text{ psi}, D^* = 0.4 \text{ in}, N^* = 8 \tag{E7}$$

Since the optimal levels of the design variables do not correspond to any of the levels in L_9, the optimum value of the quality characteristic (k^*) can be determined by conducting another experiment (or computation/simulation) using the optimum values of the design variables. In the present case, the optimum value of design variables, given in Eq. (E7), are substituted in the expression of k, Eq. (E1), to obtain

$$k^* = 19.68265 \text{ lb/in} \tag{E8}$$

TABLE 14.21 Larger the better optimum levels

Factor	1	2	3
d	−0.40987371	9.64691826	***17.417729***
G	7.72459159	8.93182135	***9.998361***
D	***14.319137***	8.53496622	3.80067049
N	***10.707283***	8.7683959	7.17909462

2. *Nominal the better with the target value of $k = 3$*
For each of the experiments 1 through 9, the mean squared deviation and the corresponding signal-to-noise ratio are computed as

$$\text{MSD} = \frac{1}{4}\{(k_1 - 3)^2 + (k_2 - 3)^2 + (k_3 - 3)^2 + (k_4 - 3)^2\} \tag{E9}$$

where k_i denotes the value of the stiffness of the spring (quality characteristic) found in case i, $i = 1, 2, 3, 4$, and

$$\eta = -10 \log_{10} \text{MSD} \tag{E10}$$

The resulting MSD and η values are shown in Table 14.22. The value of η corresponding to level 1 of the design variable d (x_1) is given by the average of the η values given by experiments 1, 2, and 3 (since the level of x_1 is 1 in experiments 1, 2, and 3 in L_9) as:

$$\eta_{\text{level 1 of } d} = \frac{1}{3}(-0.44306476 - 6.30374476 - 7.92276904) = -4.88985952 \tag{E11}$$

Similarly, the value of η corresponding to levels 2 and 3 of d are given by the averages of the η values given by experiments 4, 5, 6 and 7, 8, 9, respectively, as:

$$\eta_{\text{level 2 of } d} = \frac{1}{3}(0.95623691 + 0.95474585 - 11.1468327) = -3.0786167 \tag{E12}$$

and

$$\eta_{\text{level 3 of } d} = \frac{1}{3}(3.01523618 - 18.8618025 - 17.1772095) = -11.0079253 \tag{E13}$$

The values of η corresponding to the levels 1, 2, and 3 of the other design parameters G, D, and N can be computed in a similar manner. The results are shown in Table 14.23. It can be seen

TABLE 14.22 Nominal the better observations

Expt #	Target Value	MSD	η
1	3	1.10740499	−0.44306476
2	3	4.26947501	−6.30374476
3	3	6.19836153	−7.92276904
4	3	0.802373	0.95623691
5	3	0.80264853	0.95474585
6	3	13.0221673	−11.1468327
7	3	0.49943202	3.01523618
8	3	76.9449719	−18.8618025
9	3	52.2060637	−17.1772095

TABLE 14.23 Nominal the better optimum levels

Factor	1	2	3
d	−4.88985952	***−3.0786167***	−11.0079253
G	***1.1761361***	−8.07026712	−12.0822704
D	−10.1505666	−7.50823911	***−1.3175957***
N	−5.55517613	***−4.8117804***	−8.60944486

that the optimal levels of the design variables d, G, D, and N are given, respectively, by 2, 1, 3, and 2 (corresponding to the largest values of η in Table 14.23). Thus, the optimum values of the design variables are given by

$$d^* = 0.04 \text{ in}, G^* = 10 \times 10^6 \text{ psi}, D^* = 0.6 \text{ in}, N^* = 10 \tag{E14}$$

Since the optimal levels of the design variables do not correspond to any of the levels in L_9, the optimum value of the quality characteristic (k^*) can be determined by conducting another experiment (or computation/simulation) using the optimum values of the design variables. In the present case, the optimum value of design variables, given in Eq. (E14), are substituted in the expression of k, Eq. (E1), to obtain

$$k^* = 1.4782 \text{ lb/in} \tag{E15}$$

■

14.11 Numerical Solutions Using MATLAB

MATLAB's *Statistics Toolbox* supports three types of control charts: X-bar charts, S charts, and experimentally weighted moving average (EWMA) charts. The plotting of X-bar chart using MATLAB is demonstrated through the following example.

EXAMPLE 14.21

Plot the X-bar chart for the data given in Table 14.5, labeled as *data1*, given below using MATLAB.

Solution: The MATLAB command used to plot the X-bar chart, along with the plot, is shown below.

```
clear all; clc ; close all

data1 = [70.8 69.9 69.7 70.4  69.1; 69.9 70.1 70.6 71.1  69.3; 69.8 69.5 70.2 70.7  69.9;
    70.7 70.1 69.2 69.8 70.5; 70.0  70.8 71.2 69.1 69.9; 70.1  70.6 69.1 69.9 71.2;
    70.2 70.1 70.8 69.4 69.6; 69.7  69.1 68.7 70.8 71.1; 70.5  70.1 71.8 69.2 69.9;
    70.9 70.2 71.2 69.2 69.8; 71.2  70.7 71.0 69.5 69.9; 69.4  70.3 69.1 69.9 71.0;
    69.8 70.3 71.2 68.9 70.4; 70.5  70.8 69.4 68.3 71.1; 69.2  71.3 70.2 70.9 68.9]

xbarplot(data1)
```

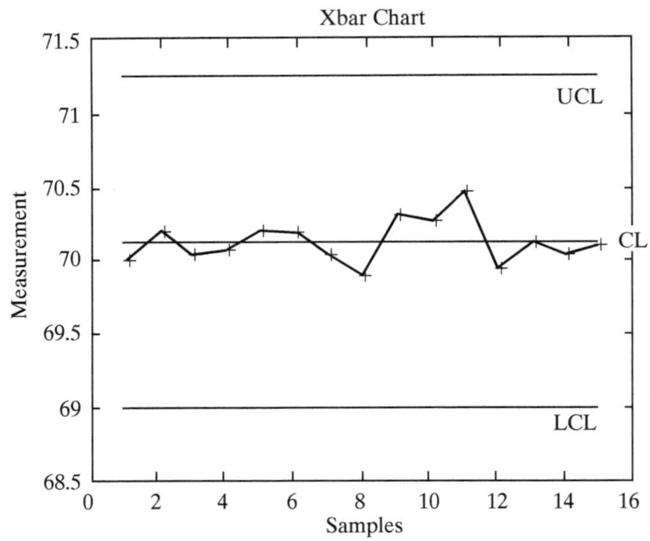

■

REFERENCES AND BIBLIOGRAPHY

14.1 L. L. Lapin, "Probability and Statistics for Modern Engineering," Brooks/Cole Publishing Co., Monterey, CA, 1983.

14.2 S. Halpern, "The Assurance Sciences: An Introduction to Quality Control and Reliability," Prentice-Hall, Englewood Cliffs, NJ, 1978.

14.3 MIL-STD-105D, Military Standard, "Sampling Procedures and Tables for Inspection by Attributes," April 29, 1963.

14.4 MIL-STD-414, Military Standard, "Sampling Procedures and Tables for Inspection by Variables for Percent Defectives," June 11, 1957.

14.5 P. A. Tobias and D. C. Trindale, "Applied Reliability," Van Nostrand Reinhold, New York, 1986.

14.6 D. Del Mar and G. W. Sheldon, "Introduction to Quality Control," West Publishing Co., St. Paul, MN, 1988.

14.7 M. S. Phadke, "Quality Engineering Using Robust Design," Prentice-Hall, Englewood Cliffs, NJ, 1989.

14.8 R. N. Kackar, "Off-Line Quality Control, Parameter Design, and the Taguchi Method," *Journal of Quality Technology*, Vol. 17, No. 4, pp. 176–209, 1985.

14.9 G. Taguchi, *Introduction to Quality Engineering*, Asian Productivity Organization, Tokyo, Japan, and Kraus International Publications, White Plains, NY, 1986.

14.10 U. Dinesh Kumar, J. Crocker, T. Chitra, and H. Saranga, *Reliability and Six Sigma*, Springer Science-Business Media, Inc., New York, 2006.

14.11 G. Eckes, *Six Sigma for Everyone*, John Wiley, Hoboken, NJ, 2003.

14.12 P. J. Ross, *Taguchi Techniques for Quality Engineering*, 2nd ed., McGraw-Hill, New York, 1996.

14.13 J. X. Wang, *Engineering Robust Design with Six Sigma*, Prentice Hall, Upper Saddle River, NJ, 2005.

14.14 G Taguchi, *System of Experimental Design*, Volumes 1 and 2, UNIPUB/Krauss International Publications and American Supplier Institute Press, New York, 1987.

REVIEW QUESTIONS

14.1 Give brief answers to the following:

1. Define the single, double, and multiple sampling plans.
2. Define the terms: *acceptance quality level* (AQL) and *lot tolerance percent defectives* (LTPD).
3. What are the roles of AQL and LTPD in acceptance sampling?
4. What is the significance of the operating characteristic (OC) curve?
5. Why is the 100 percent inspection plan not necessarily the most accurate sampling plan?
6. What is acceptance sampling? What are its advantages?
7. Explain the difference between an ideal OC curve and a practical OC curve.
8. Which probability distribution is used in the construction of OC curves?
9. What is statistical quality control? Why is it important?
10. What is the difference between producer's risk and consumer's risk?
11. Which military document is commonly used for acceptance sampling?
12. Explain the difference between a $\overline{X}$-chart and a R-chart.
13. What are quality control charts?
14. What is a p-chart?
15. What is the significance of the c-chart?
16. Explain the significance of upper and lower control limits in quality control charts.
17. How is a single sampling plan designed with a specified OC curve?
18. What is Taguchi method?
19. What is the difference between on-line and off-line quality control methods?
20. What is the significance of loss function?
21. What are noise factors and how are they related to the quality of a product?
22. What is robust design?
23. Explain the terms: *concept design*, *parameter design*, and *tolerance design*.
24. What is experimental design?
25. What are orthogonal arrays and how are they used in experimental design?

14.2 Indicate whether each of the following statements is true or false:

1. In mass production, the value of C_p is required to be greater than 1.33.
2. Poisson distribution for discrete random variables and exponential distribution for continuous random variables are directly related.
3. In six sigma process, the process capability (C_p) is given by

$$C_p = \frac{\text{UCL} - \text{LCL}}{12\sigma}$$

4. In statistical quality control, tests are conducted on random samples of components.
5. Statistical quality control techniques are used to achieve both stability and reduction of variability of the production process.
6. Taguchi method can be considered as an on-line quality control method.
7. If the sample size n is large compared to the lot size N, binomial distribution can be used as a good approximation to the hypergeometric distribution.
8. The sample means are assumed to follow normal distribution in $\overline{X}$-chart.
9. In Taguchi method, the quality level of a product is defined as the loss to society due to the poor performance of the product.
10. Taguchi method can be considered as a robust design procedure.

14.3 Fill in the blank with the proper word(s):

1. Quality is the ability of a product or service to satisfy a stated or implied ______.
2. Quality of conformance can be achieved by reducing the ______ of performance.
3. In acceptance sampling plans, tests are conducted on small samples to assess the acceptability of ______ of the whole production lot.
4. The quality target in the six sigma approach is to achieve a maximum of ______ defects per million opportunities.
5. The tolerance limits in six sigma process are given by ______ from the mean.
6. The specification of a process as $\mu \pm 6\sigma$ results in less than ______ defect per million.
7. Yield is the ratio of the number of ______ parts to the total number of parts produced.
8. Acceptance sampling uses a ______ sample to assess the acceptability of the whole production lot.
9. Sampling plans encourage product improvement because entire lot is ______ otherwise.
10. Tool wear causes the process ______ to move off the center of specification limits.
11. The p-chart is based on ______ distribution.
12. The graph between the percent of defectives in a lot and the probability of accepting the lot is called the ______ characteristic curve.

14.4 Select the most appropriate answer from the multiple choices given:

1. Six sigma approach is a trademark of:
   a. General Electric **b.** Motorola **c.** 3M
2. For components involving a number of defects per unit or number of failures per unit period, we can use:
   a. Binomial distribution
   b. Hypergeometric distribution
   c. Poisson distribution
3. The number of basic types of sampling plans provided by the military standard MIL-ST-105D is:
   a. 2 **b.** 3 **c.** 4
4. When the process variability is larger than the acceptable part variation, the process capability (Cp) will be:
   a. <1 **b.** 1 **c.** >1

5. Quality control charts were developed by:
   a. Shewhart **b.** Frazee **c.** Taguchi
6. In c-chart, the number of defects in each unit is assumed to follow:
   a. Binomial distribution
   b. Poisson distribution
   c. Normal distribution
7. The following is an outer noise:
   a. Tool wear **b.** Humidity **c.** Shrinkage
8. Orthogonal arrays permit:
   a. Use of small number of experiments
   b. To speed up the experiments
   c. Study of more characteristics of the product
9. Which quality control technique is applied after completion of production?
   a. Taguchi method **b.** Quality control charts **c.** Acceptance sampling

14.5 Match the following:

1. Control chart for ranges	**a.** p-chart
2. Control chart for defects per unit	**b.** $\overline{X}$-chart
3. Control chart for mean values	**c.** R-chart
4. Control chart for attributes	**d.** c-chart

PROBLEMS

14.1 A manufacturer of shock absorbers tested 40 units for 3000 hours. If the probability of failure of a shock absorber in 3000 hours of operation is known to be 2 percent from previous experience, determine the following:

a. The probability of realizing no failures.
b. The probability of realizing exactly two failures.
c. The probability of realizing at least two failures.
d. The probability of realizing at most two failures.

14.2 A robot used for spot welding is known to produce 10 percent defective welds. The robot is used in an assembly plant to weld the body panels of automobiles each requiring 25 spot welds.

a. If 4 randomly selected welds are inspected on each body panel, find the probability that all the four welds are defective given that there are four defective welds on the body panel.
b. If the population (25 spot welds) is considered to be very large, find the solution of part(a) using binomial distribution.

14.3 The number of defective welds produced by a spot welding machine is known to be 10 percent. Assuming Poisson distribution for the defects, determine the probability that a set of 400 spot welds made by the robot and picked at random will contain only (i) one defective weld, and (ii) two defective welds.

14.4 The strain gages produced by a manufacturer contain 5 percent defective items. If the strain gages are supplied in lots of 15 items, determine the probability that a lot will contain 5 or more percent defective items assuming (i) binomial distribution, and (ii) Poisson approximation to the binomial distribution.

14.5 A manufacturer produces IC chips which contain 0.2 percent defective units. Find the probability of finding 2 or more defective units among a randomly selected sample of 100 chips.

14.6 A batch of 200 forged crankshafts is known to contain 30 defective units. When a random sample of 40 crankshafts is inspected, 5 defective units are found. Find the probability of occurrence of this result using:

a. Binomial approximation to the hypergeometric distribution.
b. Poisson approximation to the hypergeometric distribution.

14.7 A production line of ball bearings contains 2 percent defectives. Find the probability of finding 3 or less defectives in a sample lot of 50 drawn randomly from the production line.

14.8 The results of inspection of 10 sample lots of 50 units each are given below:

Sample lot number	1	2	3	4	5	6	7	8	9	10
Number of defects	10	4	9	15	6	11	7	5	8	12

Plot the attribute control chart (p-chart).

14.9 A manufacturer of resistors tested ten sample lots, each containing five resistors, and obtained the following values of resistances in ohms. Construct the R-chart.

Resistor Number \ Lot Number	1	2	3	4	5	6	7	8	9	10
1	188	179	202	178	224	199	183	175	194	184
2	190	206	183	220	189	218	179	182	189	214
3	222	184	194	203	177	185	224	190	218	221
4	175	195	178	195	180	174	188	216	198	176
5	197	217	213	201	192	190	192	220	213	193

14.10 Construct the $\overline{X}$-chart for the data given in Problem 14.9.

14.11 The number of surface defects observed in ten castings are 3, 5, 8, 12, 9, 7, 18, 8, 6, and 11. Construct a c-chart and determine whether the casting process is under control.

14.12 A spring manufacturer tests a random sample of 50 springs from each batch containing 500 springs. If the number of nonconforming springs in ten batches are found to be 8, 4, 7, 6, 5, 9, 11, 3, 7, and 10, construct the corresponding p-chart.

14.13 A sample of 30 parts is drawn from a lot of 150 strain gages. If the percentage of defective units in the lot is expected to vary between 0 and 20, construct the operating characteristic curve. Assume that the lot will be accepted only if two or fewer components are found to be defective in the sample.

14.14 Construct a family of two OC curves for the following conditions:

Sample size: 25 units, Range of percent defectives: 0 to 25

Acceptance criterion: (i) 0 defectives, (ii) 4 defectives or less.

Identify the AQL and LTPD points (probability of acceptance is 95 percent for AQL and 10 percent for LTPD).

14.15 A single sampling plan with $n = 150$ and $c = 2$ is being used for receiving inspection where the vendor ships the product in lots of size $N = 2{,}000$. Construct the OC curve for this plan.

14.16 Fifteen automobile bodies were inspected for defective welds. The number of defects observed in the fifteen units was 6, 3, 10, 5, 9, 7, 4, 8, 11, 5, 6, 9, 7, 5, and 10. Determine the control limits for the c-chart and find whether the process is under control.

14.17 Consider the problem of design of a helical spring discussed in Example 14.20. Considering the quality characteristic as the natural frequency of vibration of the spring (instead of the stiffness of the spring), determine the optimal factor levels of the design variables for the following cases:

a. Larger the better
b. Smaller the better
c. Nominal the better with a target value of the natural frequency as 300.0 Hertz

The natural frequency of vibration of a helical spring (f) is given by

$$f = \frac{1}{2}\sqrt{\frac{kg}{W}} \text{ Hertz}$$

where g = acceleration due to gravity = 386.4 in/sec ^{2}, k is the spring stiffness given by

$$k = \frac{d^4G}{8D^3N\left\{1 + \dfrac{d^2}{2D^2}\right\}}$$

and W is the weight of the spring given by

$$W = \frac{\pi^2 d^2 DN\rho}{4}$$

with ρ = weight density of the material = 0.283 lb_f/in^3.

14.18 Consider the problem of design of a helical spring discussed in Example 14.20. Considering the quality characteristic as the shear stress induced in the spring (instead of the stiffness of the spring), determine the optimal factor levels of the design variables for the following cases:

a. Smaller the better
b. Nominal the better with a target value of the induced stress as 30,000.0 psi

The shear stress induced in the spring under an axial load F is given by

$$\sigma = \frac{8FD}{\pi d^3}\left(1 + \frac{d}{2D}\right)$$

where F = axial load = 4 lb_f.

14.19 A helical spring, shown in Fig. 14.18, is to be designed considering the spring stiffness (k) as the quality characteristic. The spring stiffness is given by

$$k = \frac{d^4 G}{8D^3 N\left\{1 + \dfrac{d^2}{2D^2}\right\}} \tag{E1}$$

with d = wire diameter, G = shear modulus, D = coil diameter, and N = number of active turns denoting the design parameters. The ranges and the possible discrete values (settings) of the design parameters in the corresponding ranges are given by

d: 0.03–0.05 in (Discrete settings: 0.03 in, 0.05 in)

G: $10 \times 10^6 - 13 \times 10^6$ psi (Discrete settings: 10×10^6 psi, 13×10^6 psi)

D: 0.4–0.6 in (Discrete settings: 0.4 in, 0.6 in)

N: 8–12 (Discrete settings: 8, 12)

The spring is made of a steel, which has a variability in shear modulus of $\pm 0.01 \times 10^6$ psi. The manufacturing equipment used in the production process is capable of achieving the following tolerances:

$$d: \pm 0.001 \text{ in}, D: \pm 0.01 \text{ in}, N: \pm 0$$

By using two settings for each of the design parameters, determine the optimal factor levels for the following type of the quality characteristic, k:

a. Larger the better

b. Nominal the better using the target value of k as 3.0 lb/in.

Use a $L_8(2^4)$ array for the inner array and a $L_4(2^3)$ array for the outer array.

14.20 The color variation of an extruded part is found to depend on seven factors, each with two levels, as shown below:

Factor	Factor Name	Level 1	Level 2
A	Back pressure	1000 psi	2000 psi
B	Formulation	Type 1	Type 2
C	Gum space	0.06	0.07
D	Screw speed	50 rpm	100 rpm
E	Vulcanizing temperature	400°F	425°F
F	Vulcanizing time	5 minutes	10 minutes
G	Heat profile	Type 1	Type 2

An L_8 array is used to conduct the experiments by repeating each experiment five times to account for the noise factors. The S/N ratios computed in various experiments are given below:

Expt. Number	1	2	3	4	5	6	7	8
S/N Ratio (dB)	−7.19	−2.92	−8.71	−10.26	−10.01	−8.92	−11.93	−8.69

The overall influence of each level of the various control factors on the *S/N* ratio is shown below:

Factor	A	B	C	D	E	F	G
S/N ratio, Level 1	−7.27	−7.26	−7.68	−9.46	−8.38	−9.04	−9.57
S/N ratio, Level 2	−9.89	−9.90	−9.47	−7.70	−8.78	−8.12	−7.58

a. Find the optimal levels of the various control factors.

b. Find the optimal *S/N* ratio predicted by the Taguchi method (i.e., the *S/N* ratio corresponding to the solution of part (a)).

14.21 A Taguchi experiment was designed to investigate the influence of five factors on the heat treatment of leaf springs. The factors and their settings or levels are given in the following table:

Factor	Description of the Factor	First Setting	Second Setting
A	Furnace temperature (deg F)	$A_1 = 1500$	$A_2 = 2000$
B	Heating time (s)	$B_1 = 30$	$B_2 = 40$
C	Transfer time (s)	$C_1 = 16$	$C_2 = 12$
D	Hold down time (s)	$D_1 = 1$	$D_2 = 2$
E	Quench oil temperature (deg F)	200	250

The L_8 orthogonal array was used for the design and the quality characteristic was the percentage number of defective springs. The results are shown below.

Expt. Number	Factor Levels A	B	C	D	E	Furnace Temperature (deg F)	Heating Time (s)	Transfer Time (s)	Holding Time (s)	Quench Oil Temp (deg F)	*S/N* Ratio (η)
1	1	1	1	1	1	1600	30	15	4	200	–23.5
2	1	1	1	2	2	1600	30	15	6	300	–28.6
3	1	2	2	1	1	1600	24	12	4	200	–26.4
4	1	2	2	2	2	1600	24	12	6	300	–24.1
5	2	1	2	1	2	1800	30	12	4	300	–25.1
6	2	1	2	2	1	1800	30	12	6	200	–32.3
7	2	2	1	1	2	1800	24	15	4	300	–33.1
8	2	2	1	2	1	1800	24	15	6	200	–31.4

Find the optimal settings of the factors and the corresponding optimal percentage number of defective springs. State the assumptions made.

MATLAB Problem

14.22 Plot the X-bar chart for the data given in Problem 14.9 using MATLAB.

C H A P T E R 1 5

Maintainability and Availability

What You Will Learn

In this chapter, you will learn

- Importance of a suitable maintenance program for repairable systems to restore them to operating condition after failure
- The two types of maintenance programs—preventive and corrective maintenances—that are used in practice and their objectives
- The aspects of periodic inspections and replacement of worn-out parts in preventive maintenance, which is conducted even before a failure
- The aspects of bringing the failed system to an operating condition as soon as possible in the corrective maintenance, which is undertaken after the failure of a component or system
- The equations for the reliability and mean time to failure of a maintained system and their application to systems whose failure time follows exponential, normal, and Weibull distributions
- The influence of imperfect maintenance
- The optimal replacement strategy for minimum cost and the aspect of spare parts requirement
- The availability analysis using a Markovian approach and its application to systems with single components, series systems, and parallel systems

Andrei Andreevich Markov (or Markoff) was a Russian mathematician who was born on June 14, 1856, and died on May 20, 1922. He graduated from St. Petersburg University in 1878 with a gold medal for his thesis on the solution of differential equations. He was a student of Chebyshev, and taught at St. Petersburg University, becoming a professor there in 1893. He developed the chain theory that has become popular in weather prediction, social sciences, biological sciences, molecular and statistical physics, quantum theory, and genetics. A chain is associated with a process, which changes from stage to stage. The chain theory provides the probability of being in the next stage once the probability of being in the current stage is known. Markov chains are applied to random walks, which are a succession of walks along a particular path. Entire books are written on Markov chains and separate courses are offered in colleges and universities on the subject. Markov extended many of the works of Chebyshev and made many contributions to statistics. He explained sampling theory within the framework of probability theory [15.17, 15.18].

15.1 Introduction

The previous chapters were concerned with the problem of assessing the reliability of a component or a system without considering the aspects of maintenance and repair. This chapter deals with maintainable and repairable systems that can be restored to service either at regular intervals of time or after a failure. Both the aspects of maintainability (preventive as well as corrective maintainability) and availability are discussed. In general, the maintainability of a system provides a measure of the repairability of a system when it fails while the availability of a system provides a measure of the readiness of a system for use at any instant of time.

The availability analysis of a system requires a knowledge of the following aspects: (1) the system configuration describing how the components are functionally connected, (2) the failure process of the components, (3) the method of operation and the definition of failure of the system, and (4) the repair or maintenance policy. A variety of failure and repair time distributions can be used in the availability analysis. Lie, Hwang, and Tillman [15.1] gave a detailed classification of references on availability with regard to both failure and repair time distributions. The most frequently used distributions for failure process (or failure times) include exponential, Weibull, and normal distributions. For the repair process (or repair times), the most prominent ones are exponential, lognormal, gamma, Poisson, and uniform distributions. For example, if the components of a system that fail frequently have a relatively short repair times compared to those components that fail infrequently, the repair times can be assumed to follow exponential distribution. On the other hand, if every component of the system has the same failure rate and the same repair time, the repair times can be assumed to follow uniform distribution.

In this chapter, we assume, mostly, that the failure times as well as repair times of individual components follow exponential distribution. This permits us to use Markov models for the analysis.

15.2 Maintainability

15.2.1 Overview

The purpose of maintenance is to restore a deteriorating or failed system to the normal operating state. Maintainability $M(t)$ is defined as the probability of repairing a failed component or system in a specified period of time. Two types of maintenance are possible: preventive maintenance and corrective maintenance. In *preventive maintenance,* the system is periodically inspected, some components are replaced, lubrication systems are checked, and adjustments are made before the system fails. Preventive maintenance is intended to eliminate costly repairs involved during the corrective maintenance stage when the system fails. On the other hand, *corrective maintenance* is used after the system fails. While the aim of preventive maintenance is to increase the reliability and prolong the life of the system by overcoming the effects of aging, fatigue, and wear, the objective of corrective maintenance is to bring the system from the failed state to the operating state as soon as possible to increase its availability. Preventive maintenance, by its nature, can be scheduled and controlled to minimize the cost, while corrective maintenance cannot be controlled. The type of maintenance used depends on the type of system, cost involved, and the risk associated with the failure. For example, the type of maintenance used for an automobile engine is totally different from the one used for an aircraft engine.

15.2.2 Preventive Maintenance

Reliability of the System. Let the reliability of the system with no maintenance be given by $R(t)$ and let preventive maintenance be used at times t_0, $2t_0$, $3t_0$, ...[1] We assume that each time a maintenance action is taken; the system is restored to "as good as new" condition as indicated in Fig. 15.1 [15.4]. The reliability of the maintained system, $R_m(t)$, during $0 \le t \le t_0$ can be stated as

$$R_m(t) = R(t); \quad 0 \le t \le t_0 \tag{15.1}$$

since no maintenance action is taken until $t = t_0$. At any time during the next interval of time, $t_0 \le t \le 2t_0$, the reliability of the maintained system is given by the product of the probability of the original system operating at time $t = t_0$ and the probability of the system, restored to new condition at t_0, operating at $t = t - t_0$

$$R_m(t) = R(t_0)R(t - t_0); \quad t_0 \le t \le 2t_0 \tag{15.2}$$

Next, the reliability of the maintained system during $2t_0 \le t \le 3t_0$ can be obtained as the product of the probability of the system operating at $t = 2t_0$ and the probability of the system, restored to new condition at $2t_0$, operating at time $t = t - 2t_0$

$$\begin{aligned} R_m(t) &= R(t_0)\, R(t - t_0)\,|_{t=2t_0} R(t - 2t_0) \\ &= R^2(t_0)\, R(t - 2t_0); \quad 2t_0 \le t \le 3t_0 \end{aligned} \tag{15.3}$$

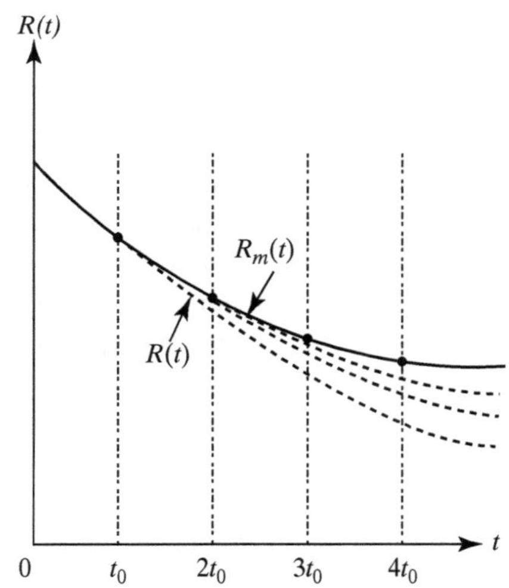

FIGURE 15.1 Reliability of a system with and without preventive maintenance.

[1]Note that t denotes the actual operating time of the system and excludes the time periods during which the system is shut off for maintenance action.

By proceeding in a similar manner, the reliability of the maintained system during $it_0 \le t \le (i+1)t_0$ can be expressed as

$$R_m(t) = R^i(t_0)R(t - it_0); \quad it_0 \le t \le (i+1)t_0; \quad i = 0, 1, 2, \ldots \tag{15.4}$$

Mean Time to Failure. The mean time to failure (see Eq. 6.28) of a maintained system ($MTTF_m$) is given by

$$\begin{aligned} MTTF_m &= \int_0^\infty R_m(t)\,dt = \int_0^{t_0} R_m(t)\,dt + \int_{t_0}^{2t_0} R_m(t)\,dt + \cdots + \int_{it_0}^{(i+1)t_0} R_m(t)\,dt + \cdots \\ &= \sum_{i=0}^{\infty} \int_{it_0}^{(i+1)t_0} R_m(t)\,dt \end{aligned} \tag{15.5}$$

By substituting Eq. (15.4) into Eq. (15.5) and noting that $R(t_0)$ is independent of t, we obtain

$$MTTF_m = \sum_{i=1}^{\infty} R^i(t_0) \int_{t=it_0}^{(i+1)t_0} R(t - it_0)\,dt \tag{15.6}$$

Defining $\tau = t - it_0$, Eq. (15.6) can be rewritten as

$$MTTF_m = \sum_{i=1}^{\infty} R^i(t_0) \int_{\tau=0}^{t_0} R(\tau)\,d\tau \tag{15.7}$$

In view of the equivalence [15.9]

$$\sum_{i=1}^{\infty} R^i(t_0) = \frac{1}{1 - R(t_0)} \tag{15.8}$$

Equation (15.7) gives

$$MTTF_m = \frac{1}{1 - R(t_0)} \int_0^{t_0} R(\tau)\,d\tau \tag{15.9}$$

EXAMPLE 15.1

Find the reliability of a system without and with preventive maintenance performed at regular intervals of time t_0 when the failure time distribution of the original system is given by

1. $f_T(t) = \lambda e^{-\lambda t}$ (exponential)
2. $f_T(t) = \dfrac{1}{\sqrt{2\pi}\sigma} \exp\left\{-\dfrac{1}{2}\left(\dfrac{t-\mu}{\sigma}\right)^2\right\}$ (normal)
3. $f_T(t) = \dfrac{\beta}{\eta}\left(\dfrac{t}{\eta}\right)^{\beta-1} \exp\left\{-\left(\dfrac{t}{\eta}\right)^\beta\right\}$ (Weibull)

Solution:

Without Repair. The reliability of the system without repair (see Section 6.7) can be expressed as follows

1. $R(t) = e^{-\lambda t}$
2. $R(t) = 1 - \Phi\left(\frac{t-\mu}{\sigma}\right)$
3. $R(t) = \exp\left\{-\left(\frac{t}{\eta}\right)^{\beta}\right\}$

With Repair. When preventive maintenance is performed at time intervals t_0, the reliability of the system is given by Eq. (15.4)

$$R_m(t) = R^i(t_0)R(t - it_0); \quad it_0 \le t \le (i+1)t_0; \quad i = 0, 1, 2, \ldots \tag{E1}$$

1. Here $R(t) = e^{-\lambda t}$ and Eq. (E1) gives

$$R_m(t) = (e^{-\lambda t_0})^i e^{-\lambda(t-it_0)} = e^{-i\lambda t_0}e^{-\lambda t}e^{i\lambda t_0} = e^{-\lambda t} \text{ for any } t$$

$$\therefore R_m(t) = R(t) = e^{-\lambda t}; \quad t \ge 0$$

2. In this case, $R(t) = 1 - \Phi(\{t-\mu\}/\sigma)$, and hence

$$R_m(t) = R^i(t_0)R(t - it_0)$$

$$= \left[1 - \Phi\left(\frac{t_0-\mu}{\sigma}\right)\right]^i\left[1 - \Phi\left(\frac{t - it_0 - \mu}{\sigma}\right)\right]; \quad it_0 \le t \le (i+1)t_0; \quad i = 0, 1, 2, \ldots$$

3. Since the reliability function of the system without maintenance (see Section 6.7.4) is given by

$$R(t) = \exp\left\{-\left(\frac{t}{\eta}\right)^{\beta}\right\}$$

the reliability of the maintained system becomes

$$R_m(t) = R^i(t_0)R(t - it_0) = \left(\exp\left\{-\left(\frac{t_0}{\eta}\right)^{\beta}\right\}\right)^i \exp\left\{-\left(\frac{t - it_0}{\eta}\right)^{\beta}\right\}$$

$$= e^{-i\left(\frac{t_0}{\eta}\right)^b} e^{-\left(\frac{t-it_0}{\eta}\right)^{\beta}}$$

$$= e^{-\left\{i\left(\frac{t_0}{\eta}\right)^{\beta} + \left(\frac{t-it_0}{\eta}\right)^{\beta}\right\}}; \quad it_0 \le t \le (i+1)t_0; \quad i = 0, 1, 2, \ldots$$

■

EXAMPLE 15.2

Derive the condition under which preventive maintenance is beneficial to the system for the following cases:

1. When the failure time is governed by exponential distribution
2. When the failure time is governed by Weibull distribution

Solution: The ratio of reliabilities of the system with and without preventive maintenance is given by

$$\frac{R_m(t)}{R(t)} = \frac{R^i(t_0)R(t - it_0)}{R(t)}; \quad it_0 \le t \le (i+1)t_0; \quad i = 0, 1, 2, \ldots \tag{E1}$$

For specificness, we set $t = it_0$ so that Eq. (E1) becomes

$$\frac{R_m(it_0)}{R(it_0)} = \frac{R^i(t_0)R(0)}{R(it_0)} = \frac{(R(t_0))^i}{R(it_0)}; \quad i = 0, 1, 2, \ldots \tag{E2}$$

Case 1: For exponential failure time distribution, the reliability function is given by

$$R(t) = e^{-\lambda t} \tag{E3}$$

and Eq. (E2) reduces to

$$\frac{R_m(it_0)}{R(it_0)} = \frac{(e^{-\lambda t_0})^i}{e^{-i\lambda t_0}} = 1 \tag{E4}$$

This shows that preventive maintenance is not beneficial in the case of exponential failure time distribution, which has a constant failure rate.

Case 2: For Weibull failure time distribution, the reliability function (see Section 6.7.4) is given by

$$R(t) = \exp\left\{-\left(\frac{t}{\eta}\right)^{\beta}\right\} \tag{E5}$$

and Eq. (E2) leads to

$$\frac{R_m(it_0)}{R(it_0)} = \frac{\left[\exp\left\{-\left(\frac{t_0}{\eta}\right)^{\beta}\right\}\right]^i}{\exp\left\{-\left(\frac{it_0}{\eta}\right)^{\beta}\right\}} = \exp\left\{-i\left(\frac{t_0}{\eta}\right)^{\beta} + i^{\beta}\left(\frac{t_0}{\eta}\right)^{\beta}\right\} \tag{E6}$$

This indicates that $R_m(it_0) > R(it_0)$ if $-i + i^{\beta} > 0$, or $i^{\beta-1} > 1$. This inequality will be satisfied only when $\beta > 1$ (i.e., when the failure rate increases). This indicates that preventive maintenance is useful over the wear-out or aging period when β increases with time.

■

EXAMPLE 15.3

A system is scheduled for preventive maintenance at times $t_0, 2t_0, 3t_0, \ldots$. Find the mean and standard deviation of the number of times preventive maintenance is performed before the occurrence of a system failure.

Solution: The probability of the system undergoing exactly i preventive maintenances, P_i, can be assumed to be same as the probability of the maintained system failing at time $t = (i+1)t_0$. This gives, using Eq. (15.4):

$$P_i = [R(t_0)]^i[1 - R(t_0)]; \quad i = 0, 1, 2, \ldots$$

The expected value of i can be found as

$$E[i] = \sum_{i=1}^{\infty} iP_i = \sum_{i=1}^{\infty} i[R(t_0)]^i[1 - R(t_0)] = R(t_0)[1 - R(t_0)]\sum_{i=1}^{\infty} i[R(t_0)]^{i-1}$$

$$= \frac{R(t_0)[1 - R(t_0)]}{[1 - R(t_0)]^2} = \frac{R(t_0)}{1 - R(t_0)}$$

The standard deviation of i can be evaluated as

$$\sigma_i^2 = E[(i - E[i])^2] = E[i^2] - (E[i])^2$$

$$= \sum_{i=0}^{\infty} i^2 P_i - (E[i])^2 = \frac{R(t_0)}{[1 - R(t_0)]^2}$$

■

15.2.3 Imperfect Maintenance

It has been assumed in the previous section that maintenance is performed perfectly so that the system can be considered to be "as-good-as-new" after the maintenance action. However, there is a possibility of performing imperfect maintenance due to human error. In such cases, the system will fail immediately after the preventive maintenance action. If the probability of performing imperfect maintenance is denoted as p, the reliability of the system is to be multiplied by $(1 - p)$ each time a preventive maintenance action is taken. Thus, the reliability of a maintained system can be expressed, by modifying Eq. (15.4), as

$$R_m(t) = (1 - p)^i R^i(t_0) R(t - it_0); \quad it_0 \le t \le (i + 1)t_0; \quad i = 0, 1, 2, \ldots \quad (15.10)$$

In order to find whether the replacement of wearing parts during preventive maintenance is worthwhile compared to the reduced reliability caused by faulty maintenance, we consider the ratio $R_m(t)/R(t)$ at $t = it_0$

$$\frac{R_m(it_0)}{R(it_0)} = \frac{(1 - p)^i R^i(t_0) R(0)}{R(it_0)} = \frac{(1 - p)^i R^i(t_0)}{R(it_0)}; \quad i = 0, 1, 2, \ldots \quad (15.11)$$

since $R(0) = 1$.

EXAMPLE 15.4

Determine the condition under which faulty preventive maintenance is useful when the failure time of the original (unmaintained) system follows: (1) exponential distribution and (2) Weibull distribution.

Solution:

1. When the failure time of the original (unmaintained) system follows exponential distribution,

$$R(t) = e^{-\lambda t} \quad \text{(E1)}$$

and Eq. (15.11) gives

$$\frac{R_m(it_0)}{R(it_0)} = (1 - p)^i \frac{e^{-i\lambda t_0}}{e^{-i\lambda t_0}} = (1 - p)^i \tag{E2}$$

Since the ratio given by Eq. (E2) is less than 1 for all practical values of p, the faulty preventive maintenance is not desirable.

2. When the failure time of the original system follows Weibull distribution, we have

$$R(t) = \exp\left\{-\left(\frac{t}{\eta}\right)^\beta\right\} \tag{E3}$$

and Eq. (15.11) gives

$$\frac{R_m(it_0)}{R(it_0)} = \frac{(1 - p)^i \left(\exp\left\{-\left(\frac{t_0}{\eta}\right)^\beta\right\}\right)^i}{\exp\left\{-\left(\frac{it_0}{\eta}\right)^\beta\right\}}$$

$$= (1 - p)^i \exp\left\{\left(\frac{it_0}{\eta}\right)^\beta - i\left(\frac{t_0}{\eta}\right)^\beta\right\}; \quad i = 0, 1, 2, \ldots \tag{E4}$$

For small values of p, we can use the approximation

$$(1 - p)^i = e^{-ip} \tag{E5}$$

and Eq. (E4) can be expressed as

$$\frac{R_m(it_0)}{R(it_0)} \approx \exp\left\{-ip - i\left(\frac{t_o}{\eta}\right)^\beta + \left(\frac{it_0}{\eta}\right)^\beta\right\}; \quad i = 0, 1, 2, \ldots \tag{E6}$$

It was observed in Example 15.2 that ideal preventive maintenance will be beneficial only when $\beta > 1$ (i.e., when failure rate increases, as in the case of wear-out period). Equation (E6) shows that faulty preventive maintenance will be useful, for $\beta > 1$, only when

$$-ip - i\left(\frac{t_0}{\eta}\right)^\beta + \left(\frac{it_0}{\eta}\right)^\beta > 0$$

that is,

$$p < \left(i^{\beta - 1} - 1\right)\left(\frac{t_0}{\eta}\right)^\beta \tag{E7}$$

■

15.2.4 Repair-time Distributions

Some of the distributions used for the repair times include exponential, normal, lognormal, and Weibull distributions. A typical distribution of repair times is shown in Fig. 15.2. The maintainability of the system, in different cases, can be computed as follows.

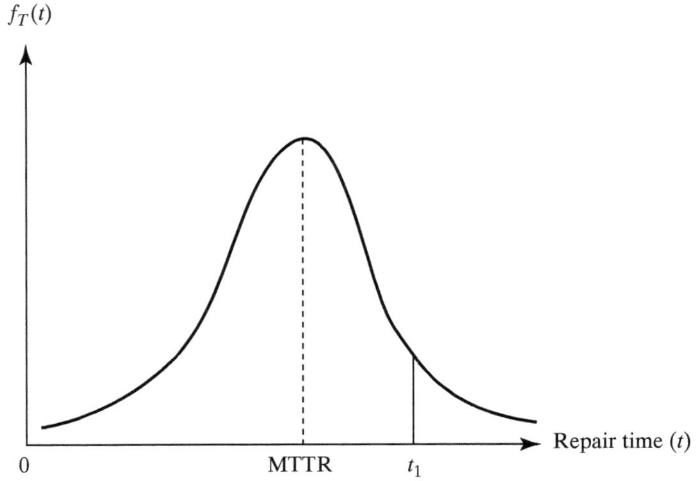

FIGURE 15.2 Probability density function of repair time.

When Time to Repair Follows Exponential Distribution. The probability density function of repair time (T) is given by

$$f_T(t) = \mu e^{-\mu t} \tag{15.12}$$

where $\mu = 1/\text{MTTR}$ is the repair rate and MTTR = mean time to repair. If $t_1, t_2, \ldots, t_n$ denote the observed repair times, MTTR can be computed as

$$\text{MTTR} = \frac{1}{n}\sum_{i=1}^{n} t_i \tag{15.13}$$

In this case, the maintainability function, $M(t)$, can be determined as

$$M(t) = P[T \le t] = \int_0^t f_T(\tau)\cdot d\tau = 1 - e^{-\mu t} \tag{15.14}$$

When Time to Repair Follows Lognormal Distribution. Let the observed repair times be $t_1, t_2, \ldots, t_n$. Then the sample mean $(\overline{X})$ and variance s^2 of $X = \ln t$ are given by

$$\overline{X} = \frac{1}{n}\sum_{i=1}^{n} \ln t_i \tag{15.15}$$

$$s^2 = \frac{1}{n-1}\sum_{i=1}^{n} (\ln t_i - \overline{X})^2 \tag{15.16}$$

If T_0 denotes the permissible repair time, the maintainability will be equal to the probability that the repair time is less than T_0. This can be found from the standard normal tables by finding the value of the standard normal variate z_1 corresponding to T_0 as

$$z_1 = \frac{\ln T_0 - \overline{X}}{s} \tag{15.17}$$

EXAMPLE 15.5

The total number of failures of a numerically controlled milling machine in a year is 84. The total number of maintenance hours spent in correcting all these failures is 578. Compute the probability of repairing a failure in (1) 1 hour, (2) 5 hours, and (3) 20 hours assuming an exponential repair time.

Solution: The data gives $\mu = 84/578 = 0.1453$ maintenance actions per hour and MTTR $= 1/\mu = 578/84 = 6.8810$ hours (average time for one repair). Thus the maintainability or probability of repair can be determined as follows:

1. $M(1) = 1 - e^{-\mu t} = 1 - e^{-0.1453(1)} = 1 - 0.8648 = 0.1352$
2. $M(5) = 1 - e^{-0.1453(5)} = 1 - 0.4836 = 0.5164$
3. $M(20) = 1 - e^{-0.1453(20)} = 1 - 0.0547 = 0.9453$

These results indicate that a failure has only 13.52 percent chance of being repaired in 1 hour but a 94.53 percent chance of being repaired in 20 hours.

■

EXAMPLE 15.6

The repair times t_i for a mainframe computer system are observed to be 1.3, 1.5, 1.7, 1.8, 2.2, 2.6, 3.0, 3.1, 3.3, and 3.9 hours. Assuming lognormal distribution for the repair times, determine the following:

1. Maintainability of the system for an allowed downtime of 5 hours.
2. Downtime required to achieve a maintainability of 0.99.

Solution: Since the repair time T follows lognormal distribution, the mean of $X = \ln t$ is given by

$$\begin{aligned}\overline{X} &= \frac{1}{10}\sum_{i=1}^{10} \ln t_i = \frac{1}{10}(\ln 1.3 + \ln 1.5 + \cdots + \ln 3.9)\\ &= \frac{1}{10}[0.2624 + 0.4055 + 0.5306 + 0.5878 + 0.7885 + 0.9555\\ &\qquad + 1.0986 + 1.1314 + 1.1939 + 1.3610] = 0.83152\end{aligned}$$

The standard deviation of $\ln t$ can be computed as

$$\begin{aligned}s^2 &= \frac{1}{9}\sum_{i=1}^{10}(\ln t_i - \overline{X})^2 = \frac{1}{9}[(0.2624 - 0.83152)^2 + (0.4055 - 0.83152)^2\\ &\quad + (0.5306 - 0.83152)^2 + (0.5878 - 0.83152)^2 + (0.7885 - 0.83152)^2\\ &\quad + (0.9555 - 0.83152)^2 + (1.0986 - 0.83152)^2 + (1.1314 - 0.83152)^2\\ &\quad + (1.1939 - 0.83152)^2 + (1.3610 - 0.83152)^2] = 0.1384\end{aligned}$$

This gives $s = 0.3720$.

1. The maintainability of the system corresponding to a downtime of 5 hours is given by

$$M(5) = \Phi\left(\frac{\ln 5 - \overline{X}}{s}\right) = \Phi\left(\frac{1.6094 - 0.83152}{0.3720}\right) = \Phi(2.0911) = 0.9817$$

2. To achieve a maintainability of 0.99, the required downtime t can be determined from the relation

$$M(t) = 0.99 = \Phi\left(\frac{\ln t - \overline{X}}{s}\right) = \Phi\left(\frac{\ln t - 0.83152}{0.3720}\right)$$

From standard normal tables, we obtain

$$\frac{\ln t - 0.83152}{0.3720} = 2.327$$

that is,

$$\ln t = 1.69176 \quad \text{or} \quad t = 5.4584 \text{ hours}$$

■

EXAMPLE 15.7

Solve Example 15.6 assuming that the repair times follow Weibull distribution.

Solution: The 10 sample repair times (t_i) can be used to find the parameters of Weibull distribution as (using the procedure outlined in Section 13.7.7) $\eta = 2.75$ and $\beta = 2.93$ so that the maintainability $M(t)$ is given by

$$M(t) = 1 - \exp\left\{-\left(\frac{t}{\eta}\right)^{\beta}\right\}$$

1. The maintainability corresponding to a downtime of 5 hours is given by

$$M(5) = 1 - \exp\left\{-\left(\frac{5}{\eta}\right)^{\beta}\right\} = 1 - \exp\left\{-\left(\frac{5}{2.75}\right)^{2.93}\right\}$$
$$= 1 - \exp(-5.764177) = 0.996862$$

2. The downtime required, t_0, to achieve a maintainability of 0.99 is given by

$$t_0 = \eta\,[-\ln(1 - 0.99)]^{\frac{1}{\beta}} = 2.75[-\ln 0.01]^{\frac{1}{2.93}} = 4.631231 \text{ hours}$$

■

15.2.5 Unrepaired Failures

The number of unrepaired failures, that is, the number of failures that are each not likely to be repaired in a specified time T_0 will be of interest to the management. If λ denotes the failure rate, the expected number of failures during time T_0 will be λT_0. If the repair time follows exponential distribution with a repair rate of μ, the maintainability function is given by

$$M(t) = 1 - e^{-\mu t} \tag{15.18}$$

Thus the probability of not being able to repair one system failure in time t_r is given by $e^{-\mu t_r}$. Hence the number of unrepaired failures in time T_0, N_{ut}, will be equal to

$$N_{ut} = \lambda T_0 e^{-\mu t_r} \tag{15.19}$$

EXAMPLE 15.8

The air-conditioning unit used in a computer installation had an average of 0.1 failure per month. In a system life of 12 months, how many failures are likely not to get repaired if a repair time of 2 hours is allowed for each repair. Assume the average number of repairs per hour as 0.5.

Solution: The expected number of failures during the life, T_0, is given by $\lambda T_0 = 0.1(12) = 1.2$ failures. The number of failures which are not likely to get repaired is given by Eq. (15.19)

$$N_{ut} = \lambda T_0 e^{-\mu t_r} = 1.2\, e^{-0.5(2)} = 1.2(0.3679) = 0.4415$$

Thus, 0.4415 failures every 12 months, on the average, will take longer than 2 hours to repair.

■

15.2.6 Optimal Replacement Strategy

As seen in the previous sections, preventive maintenance improves the reliability of a system during the wear-out period. In some systems, a component is automatically replaced after an operational time of t_0. This strategy is known as age replacement. Changing the oil and oil filter in an automobile after every t_0 months is an example of age replacement. In some cases, a set or batch of components is replaced regularly at times $t_0, 2t_0, 3t_0, \ldots$, and also whenever a failure occurs. This strategy, known as batch replacement, is used whenever a large number of inexpensive components are involved such as a set of lightbulbs in a factory. In all the cases, the cost of improving the reliability through preventive maintenance should be compared with the cost of restoring a failed system through repair. Let a component be replaced at regular times $t_0, 2t_0, 3t_0, \ldots,$. Let the cost of replacing a component through preventive maintenance be c_m and the cost of replacing a failed component through corrective maintenance be c_r. In most cases, $c_r > c_m$, since the failure of a system involves not only the cost of replacing the failed component but also the costs associated with several additional factors. For example, if the transmission of an automobile breaks down during a long-distance trip, several additional costs, such as the costs of towing and extra lodging, are involved besides the cost of replacing the transmission.

Let the component be replaced n_m times through preventive maintenance (unfailed components replaced) and n_r times through corrective maintenance (failed component replaced) during a sufficiently long operating time t. Then the total number of times the component is replaced, n, is given by

$$n = n_m + n_r \tag{15.20}$$

and the total cost of replacement c by

$$c = c_m n_m + c_r n_r \tag{15.21}$$

Since the total operating time t is very large compared to the mean time between replacements (MTBR), we have

$$n = \frac{t}{\text{MTBR}} \tag{15.22}$$

where MTBR, similar to Eq. (6.28), is given by

$$\text{MTBR} = \int_0^\infty \widetilde{R}(\tau)\, d\tau \tag{15.23}$$

and $\widetilde{R}(\tau)$ is the reliability of the component which is automatically replaced after a time t_0

$$\widetilde{R}(\tau) = \begin{cases} R(\tau) & \text{for} \quad \tau \le t_0 \\ 0 & \text{for} \quad \tau > t_0 \end{cases} \tag{15.24}$$

with $R(\tau)$ denoting the reliability of the original (unmaintained) component. Equations (15.24) and (15.23) yield

$$\text{MTBR} = \int_0^{t_0} \widetilde{R}(\tau)\, d\tau + \int_{t_0}^\infty \widetilde{R}(\tau)\, d\tau = \int_0^{t_0} R(\tau)\, d\tau \tag{15.25}$$

Since the fraction of components that survive until the next preventive maintenance is given by $R(t_0)$, the number of surviving components, n_m, that are replaced through preventive maintenance, are given by

$$n_m = R(t_0) n \tag{15.26}$$

and similarly, the number of failed components n_r, which are replaced through corrective maintenance, are given by

$$n_r = (1 - R(t_0)) n \tag{15.27}$$

By substituting Eqs. (15.22), (15.25), (15.26), and (15.27) into Eq. (15.21), we obtain

$$c = \frac{c_m R(t_0) t}{\int_0^{t_0} R(\tau)\, d\tau} + \frac{c_r (1 - R(t_0)) t}{\int_0^{t_0} R(\tau)\, d\tau} \tag{15.28}$$

To minimize the total cost of maintenance by selecting a proper value of the preventive maintenance interval t_0 we set

$$\frac{dc}{dt_0} = 0 \tag{15.29}$$

where c is given by Eq. (15.28) and solve for t_0. Equation (15.29) will be a nonlinear equation in most practical cases (see Problems 15.8 and 15.9).

15.2.7 Spare Parts Requirement

When the failure times of components of complex systems follow exponential distribution, the number of failures in a specified time follows Poisson distribution [15.2]. Thus, the probability of having n failures in time t is given by

$$P(n) = \frac{(\lambda t)^n}{n!} e^{-\lambda t} \tag{15.30}$$

where $\lambda = 1/\text{MTBF}$ and MTBF = mean time between failures. The probability of having less than or equal to r failures in time t is given by

$$P(n \leq r) = \sum_{i=0}^{r} P(i) \tag{15.31}$$

The value given by Eq. (15.31) is also the probability of having adequate replacement parts available if r spare parts are stocked at the beginning of the period. Equation (15.31) can also be used to find the number of spare parts needed as follows.

If a desired cumulative probability is given for a particular item, the number of spare parts needed can be determined by summing the values of the individual probabilities of failure until this sum is equal to or greater than the desired cumulative probability. A simpler, but approximate, formula for determining the number of spare parts in a mission time T_0 is given by

$$\text{Number of spare parts} = \lambda T_0 + z_{1-\alpha}\sqrt{\lambda T_0} \tag{15.32}$$

where λT_0 and $\sqrt{\lambda T_0}$ denote the mean and standard deviation of the number of failures in time T_0, according to Poisson distribution, and $z_{1-\alpha}$ is the value of standard normal variate corresponding to the confidence level α.

EXAMPLE 15.9

An electric surveillance system consists of three types of components. There are 20 components of type A, 130 components of type B, and 100 components of type C in the system. The failure rates (number of failures per hour) of types A, B, and C components are given by 0.002, 0.0001, and 0.0005, respectively. Find the number of spare parts required for each type of component to cover a mission period of 30 days with 99 percent probability.

Solution: The mission time is equal to $T_0 = 30 \times 24 = 720$ hours and the standard normal variate corresponding to 99 percent probability is given by $z_{1-\alpha} = 2.33$. Thus, the number of spare parts required can be determined as follows:

Type A. $n_A = \lambda_A T_0 + 2.33\sqrt{\lambda_A T_0} = (0.002)(720) + 2.33\sqrt{(0.002)(720)} = 4.2360$ per unit

Type B. $n_B = \lambda_B T_0 + 2.33\sqrt{\lambda_B T_0} = (0.0001)(720) + 2.33\sqrt{(0.0001)(720)} = 0.6972$ per unit

Type C. $n_C = \lambda_C T_0 + 2.33\sqrt{\lambda_C T_0} = (0.0005)(720) + 2.33\sqrt{(0.0005)(720)} = 1.758$ per unit

Thus, the total number of spares of type A, B, and C required is given by $4.2360(20) = 84.72$, $0.6972(130) = 90.636$, and $1.758(100) = 175.8$, respectively.

■

15.3 Availability

The term *availability* is used to indicate the probability of a system or equipment being in operating condition at any time *t*, given that it was in operating condition at $t = 0$. In order to be in operating condition at time *t*, the system must not have failed or, if it had failed during the period *t*, it must have been repaired. Thus, availability includes both the aspects of reliability and maintainability.

15.3.1 Definitions [15.1, 15.3]

Availability can be defined in several ways as indicated by the following figure.

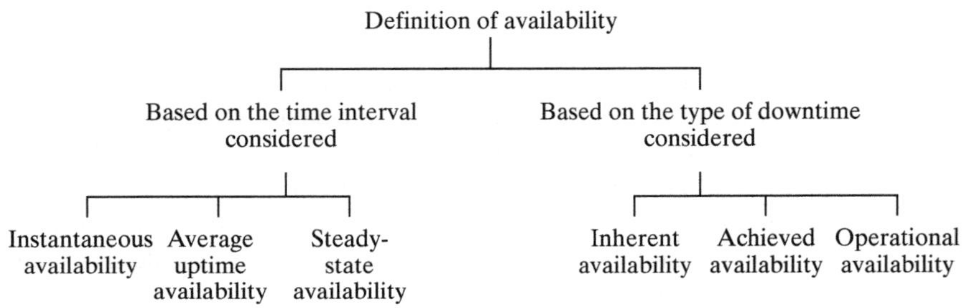

1. *Instantaneous availability, A(t)*: It is the probability that the system is operational at any arbitrary time *t*. It is given by the expected up-time of the system

$$A(t) = E[z(t)] \tag{15.33}$$

where $z(t)$ is an indicator variable defined as

$$z(t) = \begin{cases} 0; & \text{if the system is in operating state at time } t \\ 1; & \text{if the system is in failed state at time } t \end{cases} \tag{15.34}$$

In view of (15.34), Eq. (15.33) can be rewritten as

$$A(t) = (1)P[z(t) = 0] + (0)P[z(t) = 1] = P[z(t) = 0] \tag{15.35}$$

2. *Average uptime availability, A(T)*: It is defined as the proportion of time during which the system is available for use in a specified interval (0, *T*):

$$A(T) = \frac{1}{T}\int_0^T A(t)\,dt \tag{15.36}$$

3. *Steady-state availability*, $A(\infty)$: It is defined as the probability that the system is operational when the time interval considered is very large:

$$A(\infty) = \lim_{T\to\infty} A(T) = \lim_{T\to\infty}\frac{1}{T}\int_0^T A(t)\,dt \tag{15.37}$$

4. *Inherent availability*, A_i: It is defined as the proportion of time during which the system is operational, by considering only corrective maintenance downtime and excluding ready time, preventive maintenance downtime, logistics (supply) time, and waiting downtime:

$$A_i = \frac{\text{MTBF}}{\text{MTBF} + \text{MTTR}} \tag{15.38}$$

where MTBF = mean time between failures and MTTR = mean time to repair.

5. *Achieved availability*, A_a: It is defined as the proportion of time during which the system is operational by considering both corrective and preventive maintenance downtimes and excluding ready time, logistics (supply) time, and waiting downtime

$$A_a = \frac{\text{MTBM}}{\text{MTBM} + M} \tag{15.39}$$

where MTBM = mean time between maintenances and M = mean maintenance downtime due to breakdown and preventive maintenance actions.

6. *Operational availability*, A_o: It is defined as the proportion of time during which the system is operational, by considering ready time, logistics (supply) time, and waiting time along with corrective and preventive maintenance downtimes

$$A_o = \frac{\text{MTBF} + \text{ready time}}{\text{MTBF} + \text{ready time} + \text{MDT}} \tag{15.40}$$

where ready time = (operational cycle − MTBF − MDT) and MDT = mean downtime = M + delay time due to supply and administrative factors.

15.3.2 Availability Analysis

Several different approaches have been used in the literature for the availability analysis of systems [15.1]. If prior information is available about the system parameters, a Bayesian approach can be used to obtain additional information about the system. The Bayesian approach in the formulation of availability models was used by several authors [15.10 – 15.12]. Brender [15.10] carried a statistical assessment of system availability by considering alternating sequence of independent exponentially distributed operational and repair intervals, with the failure time and repair time parameters described by distinct gamma distributions. The Monte Carlo simulation approach was suggested for the availability analysis of extremely complex systems and/or systems requiring costly experimentation [15.13].

The Markovian approach has been frequently used for the availability analysis using exponential distributions for failure times and repair times [15.14, 15.15]. Due to the mathematical complexities involved, relatively less work has been done using failure time and repair time distributions other than the exponential type. The Markovian approach is described in the following sections.

15.3.3 Development of the Model

The following assumptions are made in developing a model for the availability analysis of a system:

1. At any given time, the system is either in the operating state or in the failed state.
2. The state of the system changes as time progresses.
3. The transition of the system from one state to the other takes place instantaneously.
4. The failure and repair rates are constant.

The transition of the system from one state to the other is illustrated in Fig. 15.3. The probability of failure of the system and the probability of being returned to the operating state play important roles in the availability analysis. One of the best-known state-space analysis techniques, namely, the Markov analysis, is presented in this section for the availability analysis of a system.

For a system consisting of one component, an indicator variable $z(t)$ can be defined as indicated in Eq. (15.34). If the system has a constant failure rate λ and a constant repair rate μ, the probabilities of the four possible events can be found as follows:

$P[z(t+dt) = 0|_{z(t)=1}]$ = probability of the system being in operating state at time $t + dt$ given that it was in failed state at time t

$= \mu\, dt$

$P[z(t+dt) = 0|_{z(t)=0}]$ = probability of the system being in operating state at time $t + dt$ given that it was in operating state at time t

$= 1 - \lambda\, dt$

$P[z(t+dt) = 1|_{z(t)=0}]$ = probability of the system being in failed state at time $t + dt$ given that it was in operating state at time t

$= \lambda\, dt$

$P[z(t+dt) = 1|_{z(t)=1}]$ = probability of the system being in failed state at time $t + dt$ given that it was in failed state at time t

$= 1 - \mu\, dt$

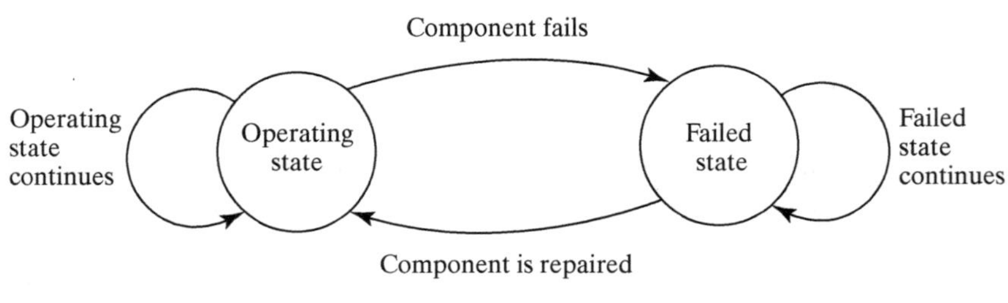

FIGURE 15.3 Transition diagram for the states of the system.

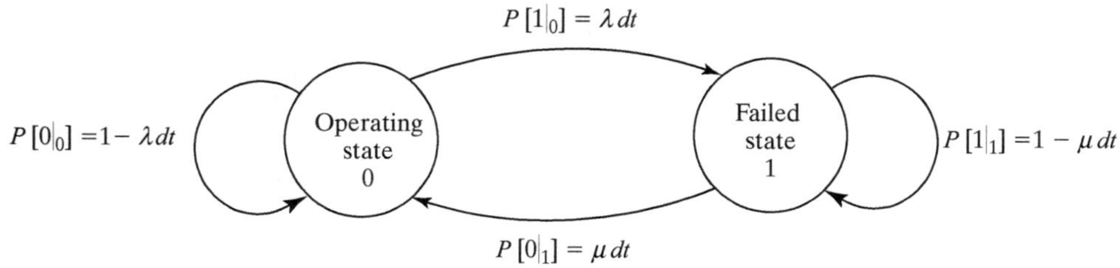

FIGURE 15.4 Markov transition diagram.

All the four transition probabilities are shown in the Markov diagram of Fig. 15.4. The transition probabilities are also shown in the form of a transition (or stochastic) matrix $[P]$ as

$$[P] = \begin{matrix} & 0 & 1 \\ 0 & (1-\lambda) & \lambda \\ 1 & \mu & (1-\mu) \end{matrix} \tag{15.41}$$

where the ijth entry denotes the probability that the system will be in state i at time $t + dt$ given that it was in state j at time t.

15.3.4 Systems with a Single Component

For a system consisting of one component, the probability of its being in operating state (state 0) at time $t + dt$, $P_0(t + dt)$, can be determined by adding the probability that it was in state 0 at time t and did not fail in $(t, t + dt)$ to the probability that it was in state 1 at time t and was brought to state 0 in $(t, t + dt)$. Thus,

$$P_0(t + dt) = P_0(t)(1 - \lambda\, dt) + P_1(t)\mu\, dt \tag{15.42}$$

Similarly, the probability of the system being in state 1 at time $t + dt$, $P_1(t + dt)$, is given by the sum of the probability that it was in state 0 at time t and failed in $(t, t + dt)$ and the probability that it was in state 1 at time t and the repair was not completed in $(t, t + dt)$. Thus,

$$P_1(t + dt) = P_0(t)\lambda\, dt + P_1(t)(1 - \mu\, dt) \tag{15.43}$$

Since the derivative, $dP_i(t)/dt$, can be defined as

$$\frac{dP_i(t)}{dt} = \lim_{dt \to 0} \frac{P_i(t + dt) - P_i(t)}{dt} \tag{15.44}$$

Equations (15.42) and (15.43) can be rewritten as

$$\frac{dP_0(t)}{dt} = -\lambda P_0(t) + \mu P_1(t) \tag{15.45}$$

$$\frac{dP_1(t)}{dt} = +\lambda P_0(t) - \mu P_1(t) \tag{15.46}$$

If the system was in operating state at time $t = 0$, the initial conditions are given by

$$P_0(0) = 1, P_1(0) = 0 \tag{15.47}$$

The solution of the simultaneous differential equations, Eqs. (15.45) and (15.46), with the initial conditions of Eq. (15.47) can be obtained as [15.3]

$$P_0(t) = A(t) = \frac{\mu}{\lambda + \mu} + \frac{\lambda}{\lambda + \mu} e^{-(\lambda+\mu)t} \tag{15.48}$$

$$P_1(t) = 1 - A(t) = \frac{\lambda}{\lambda + \mu} - \frac{\lambda}{\lambda + \mu} e^{-(\lambda+\mu)t} \tag{15.49}$$

On the other hand, when the system was in failed state at time $t = 0$, the initial conditions are

$$P_0(0) = 0, P_1(0) = 1 \tag{15.50}$$

The solution of Eqs. (15.45) and (15.46), with the initial conditions of Eq. (15.50), can be found as [15.3]

$$P_0(t) = A(t) = \frac{\mu}{\lambda + \mu} - \frac{\lambda}{\lambda + \mu} e^{-(\lambda+\mu)t} \tag{15.51}$$

$$P_1(t) = 1 - A(t) = \frac{\lambda}{\lambda + \mu} + \frac{\lambda}{\lambda + \mu} e^{-(\lambda+\mu)t} \tag{15.52}$$

It can be seen that as t becomes large, the solutions of Eqs. (15.48) and (15.51) converge to

$$A(\infty) = \frac{\mu}{\lambda + \mu} \tag{15.53}$$

Similarly, as $t \to \infty$, the solutions of Eqs. (15.49) and (15.52) converge to

$$1 - A(\infty) = \frac{\lambda}{\lambda + \mu} \tag{15.54}$$

Equations (15.53) and (15.54) indicate that the state of the system, after a long time, becomes independent of its starting (or initial) state.

The steady-state solution can also be found by setting derivatives of P_i equal to zero since the derivative of the steady-state value, which is a constant, is zero. By setting $dP_0/dt = 0$ and $dP_1/dt = 0$ in Eqs. (15.45) and (15.46), we obtain

$$-\lambda P_0 + \mu P_1 = 0 \tag{15.55}$$

$$\lambda P_0 - \mu P_1 = 0 \tag{15.56}$$

Since the sum of the probabilities of mutually exclusive events is one, we have

$$P_0 + P_1 = 0 \tag{15.57}$$

Equations (15.55) to (15.57) yield the steady-state solutions as

$$P_0 = A(\infty) = \frac{\mu}{\lambda + \mu} \tag{15.58}$$

$$P_1 = 1 - A(\infty) = \frac{\lambda}{\lambda + \mu} \tag{15.59}$$

In some cases, we will be interested in finding the average uptime of the system over a specific period of time T. This can be found by integrating $A(t)$ over the time interval T and dividing by the total time:

$$A(T) = \frac{1}{T}\int_0^T A(t)\,dt \tag{15.60}$$

If the system was in operating state at $t = 0, A(t)$ is given by Eq. (15.48) and Eq. (15.60) yields

$$A(T) = \frac{\mu}{\lambda + \mu} + \frac{\lambda}{(\lambda + \mu)^2 T} - \frac{\lambda}{(\lambda + \mu)^2 T} e^{-(\lambda+\mu)T} \tag{15.61}$$

Similarly, if the system was in failed state at $t = 0, A(t)$ is given by Eq. (15.51) and (15.60) yields

$$A(T) = \frac{\mu}{\lambda + \mu} - \frac{\lambda}{(\lambda + \mu)^2 T} + \frac{\lambda}{(\lambda + \mu)^2 T} e^{-(\lambda+\mu)T} \tag{15.62}$$

The steady-state or long-term availability of the system can be determined by letting $T \to \infty$ in Eqs. (15.61) and (15.62). This gives

$$A(\infty) = \frac{\mu}{\lambda + \mu} \tag{15.63}$$

Equation (15.63) can also be expressed as

$$A(\infty) = \frac{\text{MTTF}}{\text{MTTF} + \text{MTTR}} \tag{15.64}$$

where MTTF $= 1/\lambda =$ mean time to failure and MTTR $= 1/\mu =$ mean time to repair.

EXAMPLE 15.10

A system has a mean time between failures of 150 hours and a mean time to repair of 20 hours. What is the steady-state availability of the system?

Solution: The steady-state availability of the system can be found from Eq. (15.64) as

$$A(\infty) = \frac{150}{150 + 20} = 0.8824$$

■

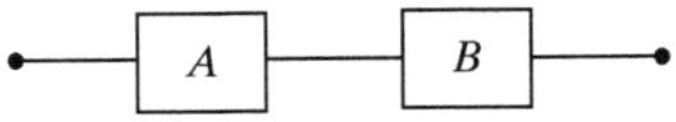

FIGURE 15.5 A series system.

15.3.5 Series Systems

Consider a series system with two components A and B as shown in Fig. 15.5. Let the failure rate be λ and the repair rate be μ for each component. The system will be in any of the following three possible states at any time t:

1. Both components are in operating condition (state 0)
2. One component is in operating condition and the other under repair (state 1)
3. Both components are under repair (state 2)

Since both components must be in operating condition for the system to be in operating condition, the availability of the system is given by the probability of the system being in state 0 at time t, that is, $A(t) = P_0(t)$. The system will be down in states 1 and 2.

One Repair Person. The availability analysis of a multicomponent system depends on the number of repair persons available for servicing the failed components. First we consider the case with one repair person. The resulting transition probabilities are summarized in Table 15.1.

From this table, the transition matrix $[P]$ can be identified as [15.3]

$$[P] = \begin{array}{c} \\ 0 \\ 1 \\ 2 \end{array}\begin{array}{c} \begin{array}{ccc} 0 & 1 & 2 \end{array} \\ \begin{bmatrix} (1-2\lambda) & 2\lambda & 0 \\ \mu & (1-\lambda-\mu) & \lambda \\ 0 & \mu & (1-\mu) \end{bmatrix} \end{array} \tag{15.65}$$

Thus the probability of the system being in state 0 or 1 or 2 at time $t + dt$ can be written as:

$$P_0(t+dt) = P_0(t)(1-2\lambda\, dt) + P_1(t)\mu\, dt \tag{15.66}$$

$$P_1(t+dt) = P_0(t)\,2\lambda\, dt + P_1(t)(1-\lambda\, dt - \mu\, dt) + P_2(t)\mu\, dt \tag{15.67}$$

$$P_2(t+dt) = P_1(t)\lambda\, dt + P_2(t)(1-\mu\, dt) \tag{15.68}$$

These equations can be rewritten, using the definition given in Eq. (15.44), as

$$\frac{dP_0(t)}{dt} = -2\lambda P_0(t) + \mu P_1(t) \tag{15.69}$$

$$\frac{dP_1(t)}{dt} = 2\lambda P_0(t) - (\lambda+\mu)P_1(t) + \mu P_2(t) \tag{15.70}$$

$$\frac{dP_2(t)}{dt} = \lambda P_1(t) - \mu P_2(t) \tag{15.71}$$

TABLE 15.1 Transition probabilities of a two-component series system with one repair person

Serial Number	State of the System at Time t	State of the System at Time $t + dt$	Transition of Components	Transition Probability
1	0 (operating)	0 (operating)	Both components remain in operating condition	$(1 - \lambda\, dt)^2 \approx 1 - 2\lambda\, dt$
2	0 (operating)	1 (one component failed)	Either of the components fails and the other remains in operating condition	$2\lambda\, dt(1 - \lambda\, dt) \approx 2\lambda\, dt$
3	0 (operating)	2 (two components failed)	Both components fail	$(\lambda\, dt)^2 \approx 0$
4	1 (one component failed)	0 (operating)	One remains in operating condition and repair is complete for failed one.	$\mu\, dt$
5	1 (one component failed)	1 (one component failed)	One remains in operating condition and repair is not complete for failed one.	$(1 - \mu\, dt)(1 - \lambda\, dt) \approx 1 - (\lambda + \mu)\, dt$
6	1 (one component failed)	2 (two components failed)	One component fails and repair of the other is not complete.	$\lambda\, dt(1 - \mu\, dt) \approx \lambda\, dt$
7	2 (two components failed)	0 (operating)	Repair of both components completed	$(\mu\, dt)^2 \approx 0$
8	2 (two components failed)	1 (one component failed)	Only one component repaired	$\mu\, dt$
9	2 (two components failed)	2 (two components failed)	Repair of two components not completed (only one component is being repaired)	$1 - \mu\, dt$

These equations can be solved for any specified initial conditions to find P_0, P_1, and P_2 as functions of time. For example, if the system was in operating state at time $t = 0$, the initial conditions are given by

$$P_0(0) = 1; \quad P_1(0) = 0; \quad P_2(0) = 0 \tag{15.72}$$

In most cases, the steady-state solution of Eqs. (15.69) to (15.71) will be of interest. For this, we set $dP_0/dt = 0$, $dP_1/dt = 0$, and $dP_2/dt = 0$ in Eqs. (15.69) to (15.71) and use the condition $P_0 + P_1 + P_2 = 1$:

$$-2\lambda P_0 + \mu P_1 = 0 \tag{15.73}$$

$$2\lambda P_0 - (\lambda + \mu)P_1 + \mu P_2 = 0 \tag{15.74}$$

$$\lambda P_1 - \mu P_2 = 0 \tag{15.75}$$

$$P_0 + P_1 + P_2 = 1 \tag{15.76}$$

The solution of Eqs. (15.73) to (15.76) gives

$$P_0 = \frac{\mu^2}{\mu^2 + 2\lambda\mu + 2\lambda^2} \tag{15.77}$$

$$P_1 = \frac{2\lambda\mu}{\mu^2 + 2\lambda\mu + 2\lambda^2} \tag{15.78}$$

$$P_2 = \frac{2\lambda^2}{\mu^2 + 2\lambda\mu + 2\lambda^2} \tag{15.79}$$

Thus the steady-state availability of the system is given by

$$A(\infty) = P_0 = \frac{\mu^2}{\mu^2 + 2\lambda\mu + 2\lambda^2} \tag{15.80}$$

Two Repair Persons. In this case, each repair person can work on a different component. If the system was in state 2 at time t, it can return to state 1 at time $t + dt$ if repair of one of the components is complete, while that of the other is incomplete. Since the repair of either of the two components can be completed by the time $t + dt$, the probability is $2\mu\, dt(1 - \mu\, dt) \approx 2\mu\, dt$. On the other hand, if the system was in state 2 at time t, it continues to remain in state 2 if the repair of both the components is incomplete. The probability of this event is equal to $(1 - \mu\, dt)^2 \approx 1 - 2\mu\, dt$. Thus, the transition matrix, [P], of Eq. (15.65) will be modified as

$$[P] = \begin{array}{c} \\ 0 \\ 1 \\ 2 \end{array} \begin{array}{c} \begin{array}{ccc} 0 & 1 & 2 \end{array} \\ \begin{bmatrix} (1 - 2\lambda) & 2\lambda & 0 \\ \mu & (1 - \lambda - \mu) & \lambda \\ 0 & 2\mu & (1 - 2\mu) \end{bmatrix} \end{array} \tag{15.81}$$

This matrix can be used to derive the steady-state equations as

$$-2\lambda P_0 + \mu P_1 = 0 \tag{15.82}$$

$$2\lambda P_0 - (\lambda + \mu)P_1 + 2\mu P_2 = 0 \tag{15.83}$$

$$\lambda P_1 - 2\mu P_2 = 0 \tag{15.84}$$

$$P_0 + P_1 + P_2 = 1 \tag{15.85}$$

The solution of Eqs. (15.82) to (15.85) is given by

$$P_0 = \frac{\mu^2}{\mu^2 + 2\lambda\mu + \lambda^2} \tag{15.86}$$

$$P_1 = \frac{2\lambda\mu}{\mu^2 + 2\lambda\mu + \lambda^2} \tag{15.87}$$

$$P_2 = \frac{\lambda^2}{\mu^2 + 2\lambda\mu + \lambda^2} \tag{15.88}$$

Thus, the steady-state availability of the system will be

$$A(\infty) = P_0 = \frac{\mu^2}{\mu^2 + 2\lambda\mu + \lambda^2} \tag{15.89}$$

15.3.6 Parallel Systems

Consider a parallel system consisting of two components A and B as shown in Fig. 15.6. For this system, there are three possible states: state 0 (both components operating), state 1 (one component operating and the other under repair), and state 2 (both components under repair).

One Repair Person. If there is only a single repair person to service the two components, the transition matrix can be derived as:

$$[P] = \begin{matrix} & 0 & 1 & 2 \\ 0 & (1 - 2\lambda) & 2\lambda & 0 \\ 1 & \mu & (1 - \lambda - \mu) & \lambda \\ 2 & 0 & \mu & (1 - \mu) \end{matrix} \tag{15.90}$$

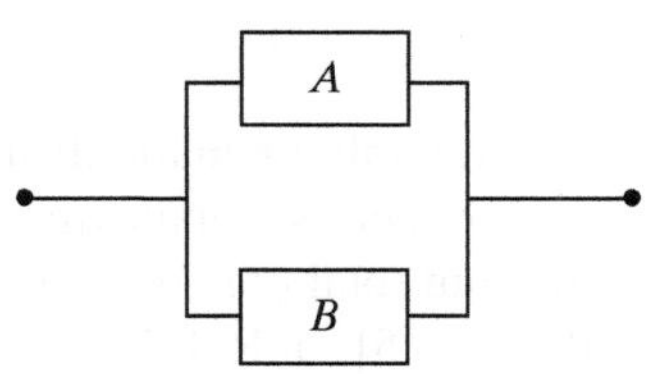

FIGURE 15.6 A parallel system.

This gives the steady-state equations as

$$-2\lambda P_0 + \mu P_1 = 0 \tag{15.91}$$

$$2\lambda P_0 - (\lambda + \mu)P_1 + \mu P_2 = 0 \tag{15.92}$$

$$\lambda P_1 - \mu P_2 = 0 \tag{15.93}$$

$$P_0 + P_1 + P_2 = 1 \tag{15.94}$$

The solution of these equations is given by Eqs. (15.77) to (15.79). Since states 0 and 1 constitute the operation of the system, the steady-state availability of the system can be found as

$$A(\infty) = P_0 + P_1 = \frac{\mu^2 + 2\lambda\mu}{\mu^2 + 2\lambda\mu + 2\lambda^2} \tag{15.95}$$

Two Repair Persons. If there are two repair persons, one repair person can be assigned to each component and the transition matrix can be derived as

$$[P] = \begin{array}{c} \\ 0 \\ 1 \\ 2 \end{array} \begin{array}{c} \begin{array}{ccc} 0 & 1 & 2 \end{array} \\ \begin{bmatrix} (1 - 2\lambda) & 2\lambda & 0 \\ \mu & (1 - \lambda - \mu) & \lambda \\ 0 & 2\mu & (1 - 2\mu) \end{bmatrix} \end{array} \tag{15.96}$$

From this matrix, the steady-state equations can be obtained as

$$-2\lambda P_0 + \mu P_1 = 0 \tag{15.97}$$

$$2\lambda P_0 - (\lambda + \mu)P_1 + 2\mu P_2 = 0 \tag{15.98}$$

$$\lambda P_1 - 2\mu P_2 = 0 \tag{15.99}$$

$$P_0 + P_1 + P_2 = 1 \tag{15.100}$$

The solution of these equations is given by Eqs. (15.86) to (15.88). The steady-state availability of the system can be determined as

$$A(\infty) = P_0 + P_1 = \frac{\mu^2 + 2\lambda\mu}{\mu^2 + 2\lambda\mu + 2\lambda^2} \tag{15.101}$$

15.4 Optimization Approaches

Several optimization methods have been used for solving maintainability and availability problems. The allocation of availability parameters (repair times and failure rates) to various components of a system for minimum total cost was formulated and solved using the Lagrange multiplier method in Ref. [15.5]. In Ref. [15.6], the optimal values of MTBF, MTTR, and the number of redundant units to achieve a specified availability for minimum cost were found using a dynamic programming approach. The steady-state availability of a repairable system with cold standbys and nonzero replacement time is maximized under constraints of total cost and total weight in Ref. [15.7]. The provision of spare parts significantly influences the operating costs, maintainability, and availability of a system. The optimal determination of spare parts can be carried for maximum availability or reliability, and minimum spare part and operating costs or weight of spare parts. Different models have been developed for the optimal spare part determination and some of the methods are being routinely used by airlines, rental car agencies, and so on [15.8].

15.5 Numerical Solutions Using MATLAB and Excel

EXAMPLE 15.11

The failure time of a machine follows normal distribution with a mean value of $\mu = 20$ months and a standard deviation of $\sigma = 4$ months. The machine undergoes preventive maintenance at regular intervals of 12 months. Find the reliability of the machine at $t = 30$ months without and with maintenance using MATLAB.

Solution: The reliability of the machine at time t with no maintenance is given by

$$R_{\text{no maintenance}}(t) = 1 - \Phi\left(\frac{t-\mu}{\sigma}\right) \quad \text{(E1)}$$

Noting that the reliability of the machine at time t (when undergoing preventive maintenance at intervals of time t_0) is given by Eq. (E1) of Example 15.1, we have for $t_0 = 12$ and $t = 30$ months, $it_0 \le t \le (i+1)t_0;\ i = 2$

$$\begin{aligned} R_m(t) &= R^i(t_0)\, R(t - it_0) = \left[1 - \Phi\left(\frac{t_0 - \mu}{\sigma}\right)\right]^i \left[1 - \Phi\left(\frac{t - it_0 - \mu}{\sigma}\right)\right] \\ &= \left[1 - \Phi\left(\frac{12-20}{4}\right)\right]^2 \left[1 - \Phi\left(\frac{30 - 24 - 20}{4}\right)\right] = [1 - \Phi(-2)]^2[1 - \Phi(-3.5)] \end{aligned} \quad \text{(E2)}$$

The MATLAB program to find the reliabilities indicated in Eqs. (E1) and (E2) and the results are shown below.

```
clear all ; clc ;

mu = 20         ;
sd = 4          ;
time0 = 12      ;
time = 30       ;

etime = 6       ;
i = 2           ;

z10 = (time - mu)/sd
```

```
Relno = 1 - normcdf(z10)
z1 = (time0 - mu)/sd
z2 = (time - i*time0 - mu)/sd

Prob1 = normcdf(z1)
Prob2 = normcdf(z2)

Relm = ((1 - Prob1)^2)*(1 - Prob2)

z10 =

   2.5000

Relno =

   0.0062

z1 =

   -2

z2 =

   -3.5000

Prob1 =

   0.0228

Prob2 =

   2.3263e-004

Relm =

   0.9548
```

■

EXAMPLE 15.12

The failure time of a machine follows two-parameter Weibull distribution with parameters $\beta = 2.0$ and $\eta = 50$ months. The machine undergoes preventive maintenance at regular intervals of 12 months. Find the reliability of the machine at $t = 30$ months without and with maintenance using Excel.

Solution: The reliability of the machine at time t with no maintenance is given by

$$R_{\text{no maintenance}}(t) = \exp\left\{-\left(\frac{t}{\eta}\right)^{\beta}\right\} \tag{E1}$$

Noting that the reliability of the machine at time t (when undergoing preventive maintenance at intervals of time t_0) is given by Eq. (E1) of Example 15.1, we have for $t_0 = 12$ and $t = 30$ months,

$$it_0 \le t \le (i+1)t_0; \; i = 2$$

$$\begin{aligned} R_m(t) = R^i(t_0)\, R(t - it_0) &= \left[\exp\left\{-\left(\frac{t_0}{\eta}\right)^{\beta}\right\}\right]^i \left[\exp\left\{-\left(\frac{t - it_0}{\eta}\right)^{\beta}\right\}\right] \\ &= \left[\exp\left\{-\left(\frac{12}{50}\right)^{2}\right\}\right]^2 \left[\exp\left\{-\left(\frac{30 - 24}{50}\right)^{2}\right\}\right] \end{aligned} \tag{E2}$$

The Excel program to find the reliabilities indicated in Eqs. (E1) and (E2) and the results follow.

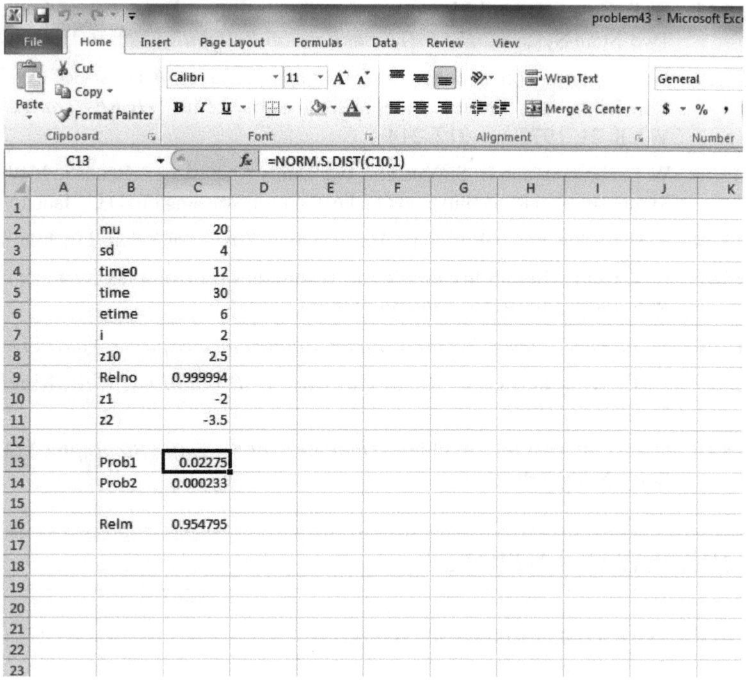

REFERENCES AND BIBLIOGRAPHY

15.1 C. H. Lie, C. L. Hwang, and F. A. Tillman, "Availability of Maintained Systems: A State-of-the-art Survey," *AIIE Transactions,* Vol. 9, 1977, pp. 247–259.

15.2 M. O. Locks, "Reliability, Maintainability, and Availability Assessment," Hayden Book Co., Rochelle Park, NJ, 1973.

15.3 G. H. Sandler, *System Reliability Engineering,* Prentice-Hall, Englewood Cliffs, NJ, 1963.

15.4 E. E. Lewis, *Introduction to Reliability Engineering,* John Wiley, New York, 1987.

15.5 R. J. McNichols and G. H. Messer, Jr., "A Cost-Based Availability Allocation Algorithm," *IEEE Transactions on Reliability,* Vol. R-20, 1971, pp. 178–182.

15.6 B. K. Lambert, A. G. Walvekar, and J. P. Hirmas, "Optimal Redundancy and Availability Allocation in Multistage Systems," *IEEE Transactions on Reliability,* Vol. R-20, 1971, pp. 182–185.

15.7 M. Sasaki, S. Kaburaki, and S. Yanagi, "System Availability and Optimum Spare Units," *IEEE Transactions on Reliability,* Vol. R-26, 1977, pp. 182–188.

15.8 E. G. Frankel, *Systems Reliability and Risk Analysis,* Martinus Nijhoff Publishers, The Hague, The Netherlands, 1984.

15.9 S. M. Selby, *Standard Mathematical Tables,* 17th ed., The Chemical Rubber Co., Cleveland OH, 1969.

15.10 D. M. Brender, "The Prediction and Measurement of System Availability: A Bayesian Treatment," *IEEE Transactions on Reliability,* Vol. R-17, 1968, pp. 127–138.

15.11 D. P. Gover, Jr. and M. Mazumdar, "Some Bayes Estimates of Long-Run Availability in a Two-State System," *IEEE Transactions on Reliability,* Vol. R-18, 1969, pp. 184–189.

15.12 W. E. Thompson and M. D. Springer, "A Bayes Analysis of Availability for a System Consisting of Several Independent Subsystems," *IEEE Transactions on Reliability,* Vol. R-21, 1972, pp. 212–214.

15.13 W. E. Faragher and H. S. Watson, "Availability Analysis—A Realistic Methodology," 10th National Symposium on Reliability and Quality Control, Washington, D.C., January 1964, pp. 365–378.

15.14 R. E. Barlow and F. Proschan, *Mathematical Theory of Reliability,* John Wiley, New York, 1965.

15.15 L. T. Hunt, "Reliability Prediction Techniques for Complex Systems," *IEEE Transactions on Reliability,* Vol. R-15, 1966, pp. 58–69.

15.16 M. A. Moss, *Designing for Minimal Maintenance Expense,* Marcel Dekker, New York, 1985.

15.17 A. C. King and C. B. Read, *Pathways to Probability,* Holt, Rinehart & Winston, Inc., New York, 1963.

15.18 C. C. Gillespie (ed.-in-chief), *Dictionary of Scientific Biography,* Vol. 9, Charles Scribner's Sons, New York, 1980.

REVIEW QUESTIONS

15.1 Give brief answers to the following:

1. Define the terms *maintainability* and *availability.*
2. What is the difference between preventive maintenance and corrective maintenance?
3. What is MTTR?
4. How is the reliability of a maintained system defined?
5. Define instantaneous availability of a system.
6. What is average uptime availability?
7. How is steady-state availability defined?
8. Explain the difference between inherent, achieved, and operational availabilities.
9. What is a transition matrix?
10. What is Markov analysis?
11. What is the effect of imperfect maintenance on the reliability of a system?
12. State the factors that influence the optimal replacement strategy.
13. How is the number of spare parts determined for a single-component system?

15.2 Indicate whether each of the following statements is true or false:

1. The purpose of maintenance is to restore a deteriorating or failed system to normal operating state.
2. If mean time to repair is given by MTTR, the repair rate (μ) is given by $\mu = \dfrac{1}{MTTR}$.
3. Markovian approach is not useful for the availability analysis.
4. Markovian transition diagram and transition matrix provide different information.

15.3 Fill in the blank with the proper word(s):

1. Maintainability provides a measure of the ______ of the system.
2. Availability provides a measure of the ______ of the system for use at any instant of time.

3. Maintainability is defined as the probability of ______ a failed component or system in a specified period of time.
4. The preventive maintenance is intended to eliminate future costly ______.
5. The purpose of corrective maintenance is to bring the failed system to the ______ state to increase its availability.
6. The purpose of preventive maintenance is to increase reliability and ______ the life of the system.

15.4 Select the most appropriate answer from the multiple choices given:

1. In preventive maintenance, the system is inspected and some components are replaced:
   a. Before any failure
   b. After complete failure
   c. After partial failure.

2. The corrective maintenance is used:
   a. After failure
   b. Much before failure
   c. Just before failure

3. When the failure times of components of a system follow exponential distribution, the number of failures in a specified time follows:
   a. Normal distribution
   b. Poisson distribution
   c. Binomial distribution

4. The long-term (steady state) availability, $A(\infty)$, is given in terms of repair rate (μ) and failure rate (λ) as:
   a. $\dfrac{\mu}{\lambda + \mu}$ **b.** $\dfrac{\lambda}{\lambda + \mu}$ **c.** $\dfrac{\lambda + \mu}{\mu}$

15.5 Match the following availabilities:

1. Instantaneous availability	**a.** Proportion of time during which the system is operational by considering ready time, logistics time, and waiting time along with corrective and preventive maintenance downtimes
2. Average uptime availability	**b.** Proportion of time during which the system is operational by considering only the corrective and preventive maintenance downtimes
3. Steady-state availability	**c.** Proportion of time during which the system is available for use in a specified interval
4. Inherent availability	**d.** Probability that the system is operational when the time interval considered is very large
5. Achieved availability	**e.** Proportion of time during which the system is operational by considering only the corrective maintenance downtimes
6. Operational availability	**f.** Proportion of time during which the system is operational at any arbitrary time

PROBLEMS

15.1 The repair time data for a boiler in a thermal power plant are given by 4.2, 4.5, 4.9, 5.0, 5.2, 5.6, 5.8, 6.1, 6.3, 6.4, 6.7, 6.8, 7.0, 7.3, and 7.5 hours. Find

a. the maintainability of the boiler for an allowed downtime of 6 hours.
b. the downtime required to achieve a maintainability of 0.95.
c. the required MTTR when the allowable downtime is 6 hours for a maintainability of 0.95. Assume that the repair times follow exponential distribution.

15.2 Solve Problem 15.1 assuming that the repair times follow normal distribution.

15.3 Solve Problem 15.1 assuming that the repair times follow lognormal distribution.

15.4 Solve Problem 15.1 assuming that the repair times follow Weibull distribution.

15.5 The times required to restore power outages by an electric power company are 0.5, 0.7, 0.8, 1.0, 1.2, 1.3, 1.6, 1.7, 1.9, 2.1, 2.2, 2.6, 2.9, 3.4, and 4.2 hours. Determine the following:

a. The maintainability of the power supply if the outage is permitted for 5 hours.
b. The duration of the outage corresponding to a maintainability of 0.99.

Solve the problem assuming (i) normal distribution and (ii) lognormal distribution for the outage restoring times.

15.6 Find the reliability of a system for which preventive maintenance is performed after every 500 hours of operation when the failure time T in hours of the original system is given by the following distribution:

a. $f_T(t) = 0.001\, e^{-0.001\, t}$

b. $f_T(t) = \dfrac{1}{\sqrt{2\,\pi}\,(50)} \exp\left\{ -\dfrac{1}{2}\left(\dfrac{t - 1000}{50}\right)^2 \right\}$

15.7 A system has a mean time between failures of 200 hours, mean time to repair of 10 hours, mean ready time of 5 hours, and a mean downtime of 15 hours. Find the inherent, achieved, and operational availabilities of the system.

15.8 Find the time interval, t_0, needed for replacing the components in preventive maintenance if the reliability of the original (unmaintained) components is given by

$$R(t) = \exp\left\{ -\left(\frac{t}{100}\right)^2 \right\}; \quad t \text{ in hours}$$

when c_m and c_r are given by 100 and 1000 dollars, respectively.

15.9 Find the time interval, t_0, needed for replacing the components in preventive maintenance if the reliability of the original (unmaintained) components is given by

$$R(t) = \exp\left\{ -\left(\frac{t}{50}\right)^4 \right\}; \quad t \text{ in hours}$$

where c_m and c_r are given by 100 and 5000 dollars, respectively.

15.10 A system consists of three types of components with 10 units of type A, 20 units of type B, and 30 units of type C. The mean times before failure of type A, B, and C components are, respectively, 2000, 1500, and 3000 hours. Find the number of spare parts required over a mission time of 5000 hours for a confidence of (a) 90 percent (b) 95 percent, and (c) 99 percent.

15.11 Find the mean time to failure of a system for which preventive maintenance is performed at times $t_0, 2t_0, 3t_0, \ldots$, when the failure time follows (a) exponential distribution, $f_T(t) = \lambda e^{-\lambda t}$, (b) Weibull distribution,

$$f_T(t) = \frac{\beta}{\eta}\left(\frac{t}{\eta}\right)^{\beta-1} \exp\left\{-\left(\frac{t}{\eta}\right)^{\beta}\right\}$$

15.12 Find the condition under which the ideal preventive maintenance would be beneficial for a system for which the failure time distribution is given by

$$f_T(t) = k^2 t e^{-kt}$$

15.13 The number of failures in a thermal power station in a year is 65. The total time spent in restoring the system is 220 hours. Find the probability of repairing a failure in (a) 1 hour, (b) 2 hours, and (c) 10 hours.

15.14 A welding robot in an assembly shop is found to fail once in every 100 hours of operation and a repair takes, on the average, 1 hour. Determine the number of failures that are likely not to get repaired over a period of 1 year if a repair time of 1 hour is allowed for each repair. Assume that the robot is operated 8 hours per day for 350 days in a year.

15.15 A system has a mean time between failures of 200 hours and a mean time to repair of 30 hours. Find the steady-state availability of the system.

15.16 A series system consists of two identical components each with a failure rate of 0.005 per hour and a repair rate of 0.1 per hour. Find the steady-state availability of the system if the number of repair persons is (a) one and (b) two.

15.17 A parallel system consists of two identical components each with a failure rate of 0.005 per hour and a repair rate of 0.1 per hour. Find the steady-state availability of the system if the number of repair persons is (a) one and (b) two.

15.18 A turbine is known to sustain two types of failures: bearing failures and blade failures. The bearing failure times follow exponential distribution with a failure rate of 0.0005 per hour and the blade failure times follow Weibull distribution with

$$f_T(t) = \frac{1}{100}\left(\frac{t}{200}\right)\exp\left\{-\left(\frac{t}{200}\right)^2\right\}; \quad t \text{ in hours}$$

a. Find the reliability of the unmaintained system after 10,000 hours of operation of the turbine.

b. If the reliability of the turbine is to be increased by 20 percent at the end of 10,000 hour operating period by replacing the turbine blades at times $t_0, 2t_0, 3t_0, \ldots$, find the value of t_0.

15.19 The failure rate of a system is given by $h(t) = 0.001\sqrt{t}$; t in hours. If the design life of the system is 10,000 hours, determine the following:

a. Reliability of the unmaintained system at the end of the design life.

b. Reliability of the system assuming perfect maintenance after every 1000 hours of operation.

c. Reliability of the system assuming imperfect preventive maintenance after every 1000 hours of operation with a 1 percent probability of failure of the system immediately after the maintenance.

15.20 The MTTF and MTTR of a spot welding robot are 60 hours and 3 hours, respectively. Find the availability of the robot. If the MTTR is reduced to 1 hour, find the smallest value of MTTF which gives at least the same value of availability.

15.21 The times at which a multicomponent system failed and the times at which the system is restored through repair over a mission period of 1000 hours are shown in the following table.

Time at Which Failure Occurred (Hours)	Time at Which System Restored (Hours)
74.2	77.5
102.1	104.2
193.4	194.1
245.5	247.8
381.0	381.9
415.3	418.2
464.6	467.1
499.8	502.9
580.2	583.4
633.1	635.1
678.9	680.3
711.7	714.2
788.5	791.6
891.0	893.3
982.4	985.8

Find the following: (a) average availability of the system, (b) average failure and repair rates of the system, and (c) availability of the system using the failure and repair rates found in part (b).

15.22 Plot the availability functions given by Eqs. (15.48), (15.51), and (15.63) for a system having a failure rate of 0.05 per hour and a repair rate of 0.5 per hour.

MATLAB and Excel Problems

15.23 The failure time of a machine follows exponential distribution with a failure rate of 0.0001 failures per hour. The machine undergoes preventive maintenance at regular intervals of $t_0 = 5000$ hours. Find the reliability of the machine without and with maintenance after 12,500 hours of operation using MATLAB.

15.24 Solve Example 15.6 using MATLAB.

15.25 Solve Problem 15.23 using Excel.

15.26 Solve Example 15.6 using Excel.

CHAPTER 16

Warranties

What You Will Learn

In this chapter, you will learn

- That a warranty is a contractual agreement that requires a manufacturer to rectify a failed unit, either through replacement or repair, if the failure occurs during the specified warranty period
- That the warranty period may be specified in terms of calendar time or usage, or both
- The various types of warranties and the costs associated with them
- The warranty costs, with a consideration of the time value of money and the future changes in the price of the product
- The warranty analysis with a consideration of multiple failures of the product/system during the warranty period using renewal theory
- The determination of optimal warranty period
- The two-dimensional warranty policies

Agner Krarup Erlang, was a Danish mathematician, statistician, and engineer born on January 1, 1878, in Lonborg, Jutland, who invented the fields of queueing theory and traffic engineering. Erlang passed the entrance examination of the University of Copenhagen in 1896 with distinction and majored in mathematics. He received his master's degree in 1901 and taught at several schools during the following seven years. Erlang joined the Copenhagen Telephone Company, an offshoot of the International Bell Telephone Company, in 1908. It was while working with the Copenhagen Telephone Company that Erlang used the theory of probabilities to come up with a formula, now known as Erlang's formula, to calculate the fraction of callers attempting to call someone outside the town that must wait because all of the lines are in use. In addition, he used mathematics to assess the number of telephone operators needed to handle a given volume of telephone calls. At that time, most telephone exchanges used human operators. Although Erlang's model is a simple one, the mathematics underlying today's complex telephone network is still based on his work. In 1909, Erlang published his work "The Theory of Probabilities and Telephone Conversations," in which Poisson distribution was applied to random telephone traffic. In 1917, he published "Solution of Some Problems in the Theory of Probabilities of Significance in Automatic Telephone Exchanges" which contained his classic formulas for loss and waiting time. These works brought him international recognition. Erlang continued to work with Copenhagen Telephone Company until his death on February 3, 1929. The British Post Office accepted his formula as the basis of calculating circuit facilities. The unit of communication activity was named Erlang in recognition of his contributions. The probability distribution that was based on his work is now known as Erlang distribution [16.19].

16.1 Introduction

Manufacturers produce new products and systems at an ever-increasing rate due to the availability of new materials and design procedures, technological changes, and global competition. Some of these products and systems are very complex to operate and maintain and their reliabilities are often a concern to consumers and users. In most cases, the manufacturer offers a warranty at the time of the sale to assure the buyer that the product or system will perform satisfactorily. In the case of failure of the product or system during the warranty period, the manufacturer will either repair or replace the unit or refunds a part or whole of the sale price. If the system is complex involving many parts, the manufacturer will repair or replace the failed parts and restores the system to operating condition.

For new manufacturers, providing better warranties to their products or systems helps them to establish themselves in the market and, with established reputation, can improve their future sales. For consumers also, warranty helps in their decision-making process in the buying of products or systems. Particularly, when several products or systems that perform basically the same function are available in the market and their prices are similar, consumers' preference will be to buy products or systems produced by the manufacturers with established reputation or those that provide larger warranty periods. Manufacturers who traditionally did not provide warranties for some of their products and systems are also forced to provide some type of warranty to their products and systems either due to global competition or due to mandatory requirements. For example, no warranties were required for weapon systems and military equipment until 1985. In 1985, the Defense Procurement Reform Act was enacted requiring defense contractors to provide written warranties for specific types of systems and assure possible remedial actions in case of failures. Thus, warranties play an important role for both the manufacturer and the consumer in terms of the sale and use of consumer as well as military products and systems.

A warranty can be considered as an agreement or contract between the manufacturer and the consumer under which the manufacturer of the product, system, or service agrees to repair or replace the product or system in the case of the failure of the product or system, or provide the service again in case the service provided originally does not meet the customer's expectations during a specified time, known as the warranty period. The warranty period may be specified in terms of calendar time or usage, or both. In the case of calendar time, the units may be hours, days, months, or years and in the case of usage, the units may be miles or hours of operation or the number of times the product or system is used (e.g., the miles driven in the case of an automobile engine, the number of months in the case of an automobile battery, the number of pages printed in the case of a printer, and the number of copies made in the case of a Xerox machine). While warranties of some products, such as computers, do not restrict the amount of usage, the warranties of other products, such as automobiles, impose restrictions on the amount of use. For example, it is common for automobiles to carry a warranty such as "for five years or 50,000 miles, whichever comes first." The warranties of the type given to computers are termed one-dimensional warranties, while those given to automobiles are called two-dimensional warranties.

Warranties have become a critical and integral part of modern industries. Although they originated because of the legal context of product liability (an aspect that has arisen from a shift in philosophy from *caveat emptor* to *caveat venditor*), they have now become a prominent means not only for conveying information on the quality of a product but also for increasing the market share of the product [16.1]. Furthermore, in some cases, local, state, and federal governments mandate and regulate warranties. In most warranties, the rectification of the failed unit is done at no cost to the customer. However, under limited warranties, the cost of rectification is shared with the customer. The cost of warranty (also called the basic warranty) is built into the selling price of the unit. In some cases, the manufacturer offers an extended warranty on a product at a premium price. The customer can either accept or reject the extended warranty. If the customer buys the extended warranty, the cost of rectification is borne by the manufacturer (or insurer). Usually, the extended warranty becomes active after the expiration of the initial warranty period, also known as the basic warranty.

Since the cost of the basic warranty is to be built into the selling price of the product, the manufacturer needs to estimate the amount of money that should be put into reserve to meet the warranty claims, which would start from the date of the sale of the product and continues for some years after the unit is sold. In some industries, such as the automobile industry, the total warranty costs might run into millions of dollars, which makes the problem of accurate forecasting of the warranty costs to be extremely important. In general, manufacturers are risk neutral and can make the level of quality of their products to be either high or low. Since consumers cannot observe the quality of the product at the time of purchase, they tend to purchase products that carry longer warranties because the longer the warranty period of a product, the greater the assurance of its good quality and the better the protection to the customer. Thus, the customers assume that a product with a longer warranty period will have a better quality. Thus, the manufacturers are indirectly forced to produce good quality products. At the same time, consumers are risk averse implying that a longer warranty period will reduce their incentive and motivation to take care of the product. This is considered to be a double-sided moral hazard, which along with the competitive actions by manufacturers, leads to a conflicting and game theoretic considerations in setting up an optimal warranty period for any product.

This chapter presents the various types of warranties and the costs associated with them. The determination of optimal warranties corresponding to different warranty models with different failure time distributions is also considered. The following assumptions are made in the present analysis:

1. When a product fails, the customer makes a claim for the repair/replacement of the product immediately.
2. The failure of the product occurs under normal use and the claim made by the customer is valid. In other words, failure of the product resulting from the misuse or operation in a way not covered by the warranty is not assumed.
3. The manufacturer has the necessary spare parts and facilities to repair or enough inventory of the product to replace a failed product.
4. The manufacturer repairs or replaces the failed product immediately so that the time to repair or replace can be assumed to be zero.

16.2 Types of Warranties

Basically, two types of warranties are commonly used in consumer products and systems. These are (1) the free replacement warranty (FRW) and (2) the pro-rata warranty (PRW). In the free replacement warranty, if a product or system fails during the warranty period, it is repaired or replaced at no cost to the consumer. Warranties can also be classified, based on the number of criteria used to determine the length of warranty eligibility, as one dimensional, two dimensional, and multidimensional warranties. In one dimensional warranties, only one criterion or parameter is used. For example, the warranty of a television may be given for a period of one year (time is the only parameter used). In the case of an automobile engine, the warranty may be given up to 50,000 miles of operation of the automobile (driving distance is the only parameter used). In two dimensional warranties, two criteria or parameters are used to establish the eligibility for warranty. For example, an automobile transmission system may be given for a period of 5 years or 50,000 miles, whichever occurs first. Here two parameters—time and driving distance—are used to establish the warranty. The costs to the manufacturer associated with different types of warranty are given in the following sections.

16.3 Warranty Costs Based on a Single Failure During the Warranty Period

In this section, we consider the expected cost of warranty to the manufacturer according to different warranty models. A simple method, by considering only a single failure of the product during the warranty period, is presented to estimate the warranty costs. The method is approximate because more than one failure of the product can occur during the warranty period. The inclusion of possible multiple failures of the product during the warranty period is considered in Section 16.6.

16.3.1 Free Replacement Warranty

In this case, the manufacturer absorbs the cost associated with the repair or replacement of the product if it fails during the warranty period. The free replacement warranty (FRW) is usually provided to simple and inexpensive products such as simple electronic components, one-time use or disposable products, and flashbulbs. The FRW policy is also given to repairable products such as lawn mowers, televisions, ovens, and refrigerators which, upon failure, can be repaired to bring them to a state of as-good-as-new operating condition. Let the manufacturing cost of the product be c_1. Thus the cost of replacement of the product to the manufacturer will be c_1 if the product fails at time t, $0 \leq t \leq t_w$, where t_w is the warranty time. Thus, the cost per one unit of the product, $c_w(t)$, can be expressed as

$$c_w(t) = \begin{cases} c_1, & 0 \leq t \leq t_w \\ 0, & t > t_w \end{cases} \tag{16.1}$$

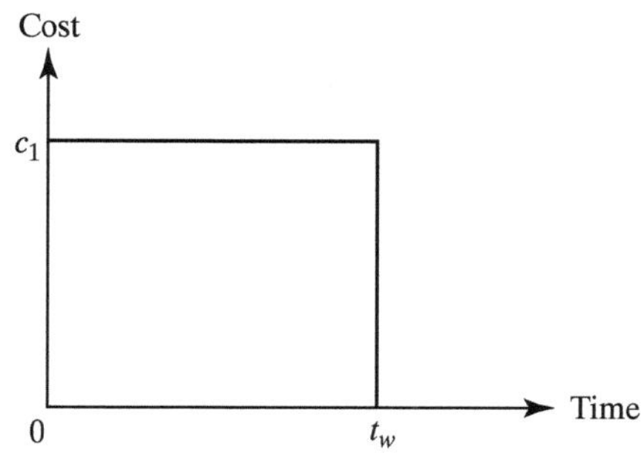

FIGURE 16.1 Cost of warranty in free replacement warranty (FRW) policy.

Equation (16.1) is shown in Fig. 16.1. The probability of failure of the product during the interval $(t, t + dt)$ is given by $f(t)\,dt$ where $f(t) = f_T(t)$ is the probability density function of the failure time T of the product. Thus, the expected cost to the manufacturer due to failure of the product is

$$E[c_w(t)] = \bar{c}_w = \int_{t=0}^{t_w} c_1 f(t)\,dt = c_1\,F(t)\big|_0^{t_w} = c_1\,F(t_w) \tag{16.2}$$

where $F(t) = F_T(t)$ is the cumulative probability distribution function of T. For example, if the failure time of the product (T) follows exponential distribution, $F(t)$ is given by

$$F(t) = 1 - e^{-\lambda t} \tag{16.3}$$

where λ is the constant failure rate, and the expected warranty cost to the manufacturer per product is

$$E[c_w(t)] = \bar{c}_w = c_1\,F(t_w) = c_1(1 - e^{-\lambda t_w}) = c_1\left(1 - e^{-\frac{t_w}{\mu}}\right) \tag{16.4}$$

where μ is the mean life of the product (MTTF) with

$$\mu = \frac{1}{\lambda} \tag{16.5}$$

If the manufacturer produces the product in a lot of size N (i.e., the number of products in the lot is N), the total warranty reserve fund required is given by

$$C_w = N\,\bar{c}_w = N\,c_1\left(1 - e^{-\frac{t_w}{\mu}}\right) \tag{16.6}$$

The ratio of the warranty reserve cost per unit of the product and the product cost, $\dfrac{C_w}{N\,c_1}$, given by

$$\frac{C_w}{N\,c_1} = \left(1 - e^{-\frac{t_w}{\mu}}\right) \tag{16.7}$$

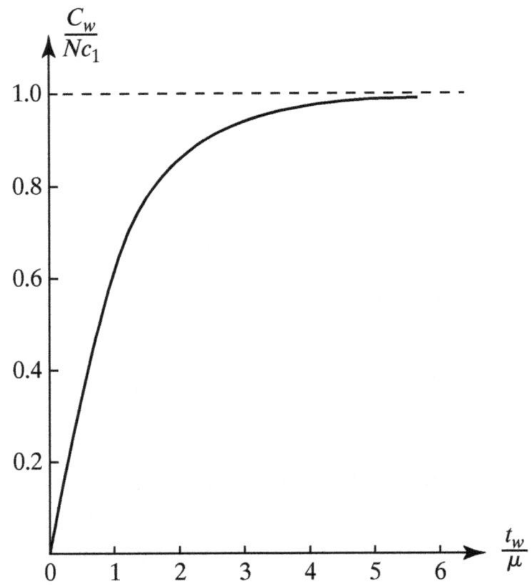

FIGURE 16.2 Variation of warranty reserve cost with warranty period in FRP.

can be seen to increase with the ratio of the warranty period and the MTTF, $\frac{t_w}{\mu}$, as shown in Fig. 16.2.

EXAMPLE 16.1

A manufacturing firm offers a one-year free replacement warranty on its treadmills. The failure time of the treadmills was determined experimentally to follow exponential distribution with a failure rate of $\lambda = 0.005$ failures per month. The firm's cost of production of one treadmill is \$250. If the firm produces 10,000 treadmills in a batch, determine the amount of warranty reserve fund to be kept aside by the firm for the whole batch of treadmills.

Solution: With the available values of $c_1 = \$250$, $t_w = 1$ year $= 12$ months, $\lambda = \frac{1}{\mu} = 0.005$ or $\mu = 200$ months, and $N = 10{,}000$ units, Eq. (16.6) can be used to find amount of the warranty reserve fund as

$$C_w = N c_1 \left(1 - e^{-\frac{t_w}{\mu}}\right)$$
$$= (10000)(250)\left(1 - e^{-\frac{12}{200}}\right) = (2.5 \times 10^6)(1 - 0.94176) = \$145{,}600$$

■

16.3.2 Pro-rata Warranty

In this warranty, we consider a nonrepairable product, which implies that any failure during the warranty period involves the replacement of the failed product by a new unit. The consumer is given a rebate which decreases linearly with time. The replacement cost to the manufacturer will be equal to the total manufacturing cost of the product, including

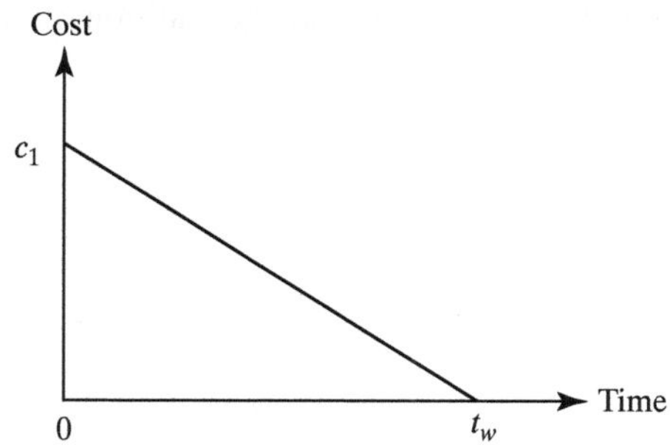

FIGURE 16.3 Cost of warranty in pro-rata warranty (PRW) policy.

the warranty cost, per unit minus the rebate given to the consumer. Thus, the cost of a unit at time t to the manufacturer can be expressed as (see Fig. 16.3):

$$c_w(t) = \begin{cases} c_1\left(1 - \dfrac{t}{t_w}\right), & 0 \leq t \leq t_w \\ 0, & t > t_w \end{cases} \tag{16.8}$$

This type of warranty is given to nonrepairable products such as automobile tires and batteries. It is assumed that the customer pays the complement portion of the cost, $c_1\dfrac{t}{t_w}$. The expected warranty cost for one unit of the product is given by

$$E[c_w(t)] = \bar{c}_w = \int_0^{t_w} c_1\left(1 - \frac{t}{t_w}\right) f(t)\, dt = c_1 F(t_w) - \frac{c_1}{t_w}\int_0^{t_w} t f(t)\, dt \tag{16.9}$$

If the failure time of the product follows exponential distribution, $f(t)$ is given by

$$f(t) = \lambda e^{-\lambda t} \tag{16.10}$$

and $F(t)$ by Eq. (16.3). Thus, Eq. (16.9) becomes

$$\begin{aligned} E[c_w(t)] = \bar{c}_w &= c_1(1 - e^{-\lambda t_w}) - \frac{c_1}{t_w}\int_0^{t_w} \lambda t\, e^{-\lambda t}\, dt \\ &= c_1(1 - e^{-\lambda t_w}) - \frac{c_1}{\lambda t_w}[1-(1 + \lambda t_w)e^{-\lambda t_w}] \\ &= c_1\left(1 - \frac{\mu}{t_w}\left\{1 - e^{-\frac{t_w}{\mu}}\right\}\right) \end{aligned} \tag{16.11}$$

where $\mu = \dfrac{1}{\lambda}$ is the expected or mean life, Eq. (16.11) can be rewritten as

$$\frac{\bar{c}_w}{c_1} = 1 - \frac{\mu}{t_w}\left(1 - e^{-\frac{t_w}{\mu}}\right) \tag{16.12}$$

If $\overline{c}$ denotes the cost of one unit of the product before the addition of the warranty cost, we have

$$c_1 = \overline{c} + \overline{c}_w$$

or

$$c_1 = \frac{\overline{c}}{1 - \dfrac{\overline{c}_w}{c_1}} \tag{16.13}$$

Equations (16.12) and (16.13) yield

$$c_1 = \frac{\overline{c}}{\dfrac{\mu}{t_w}\left(1 - e^{-\frac{t_w}{\mu}}\right)} \tag{16.14}$$

If the manufacturer produces the product in a lot of size N, the expected warranty reserve fund required $\overline{C}_w$, is given by

$$\overline{C}_w = N\,\overline{c}_w = N\,c_1\left\{1 - \frac{\mu}{t_w}\left(1 - e^{-\frac{t_w}{\mu}}\right)\right\} \tag{16.15}$$

The ratio of the warranty reserve cost per unit of the product and the product cost, $\dfrac{\overline{C}_w}{c_1}$, given by Eq. (16.12), increases with the ratio of the warranty period and the MTTF, $\dfrac{t_w}{\mu}$, as shown in Fig. 16.4.

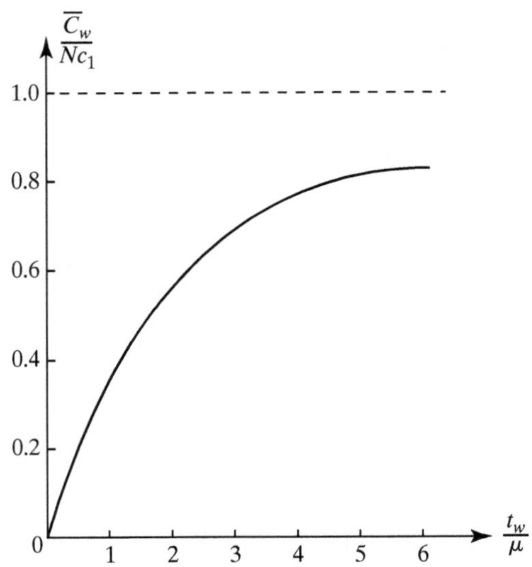

FIGURE 16.4 Variation of warranty reserve cost with warranty period in PRW.

EXAMPLE 16.2

A manufacturing firm offers a one-year pro-rata warranty on its treadmills. The failure time of the treadmills was determined experimentally to follow exponential distribution with a failure rate of $\lambda = 0.005$ failures/month. The firm's cost of production of one treadmill, including the warranty cost, is \$250. If the firm produces 10,000 treadmills in a batch, determine the following:

1. Cost of the treadmill to the manufacturer without the warranty cost.
2. Amount of warranty reserve fund to be kept aside by the firm for the whole batch of treadmills.

Solution:

1. With the available values of $c_1 = \$250$, $t_w = 1$ year $= 12$ months, $\lambda = \frac{1}{\mu} = 0.005$ or $\mu =$ 200 months, and $N = 10{,}000$ units, Eq. (16.12) can be used to find

$$\frac{\overline{c}_w}{c_1} = 1 - \frac{\mu}{t_w}\left(1 - e^{-\frac{t_w}{\mu}}\right)$$

$$= 1 - \frac{200}{12}\left(1 - e^{-\frac{12}{200}}\right) = 1 - 16.66667(1 - 0.94176) = 0.029333$$

Thus the expected warranty cost of a treadmill is

$$\overline{c}_w = (0.0229333)c_1 = (0.029333)(250) = \$7.33$$

2. Thus the amount of warranty reserve fund needed by the firm $(\overline{C}_w)$ is

$$\overline{C}_w = N\,\overline{c}_w = 10000(7.33) = \$73{,}300$$

■

16.3.3 Combined Free Replacement Warranty and Pro-rata Warranty (FRW/PRW) Policy

In this case, the manufacturer covers the full cost of repair/replacement if the product fails at any time t during the initial warranty period, $0 \le t \le t_{w1}$, and provides a pro-rata coverage for failures occurring at any time t during the next warranty period, $t_{w1} < t \le t_{w2}$. If the initial cost of the product is c_1, the warranty cost $c_w(t)$ to the manufacturer is given by (Fig. 16.5):

$$c_w(t) = \begin{cases} c_1; & 0 < t \le t_{w1} \\ c_1\left(\dfrac{t_{w2} - t}{t_{w2} - t_{w1}}\right); & t_{w1} < t \le t_{w2} \\ 0; & t > t_{w2} \end{cases} \tag{16.16}$$

The expected warranty cost per unit of the product can be found as

$$\overline{c}_w = c_1 \int_0^{t_{w1}} f(t)\,dt + \frac{c_1}{(t_{w2} - t_{w1})} \int_{t_{w1}}^{t_{w2}} (t_{w2} - t)f(t)\,dt \tag{16.17}$$

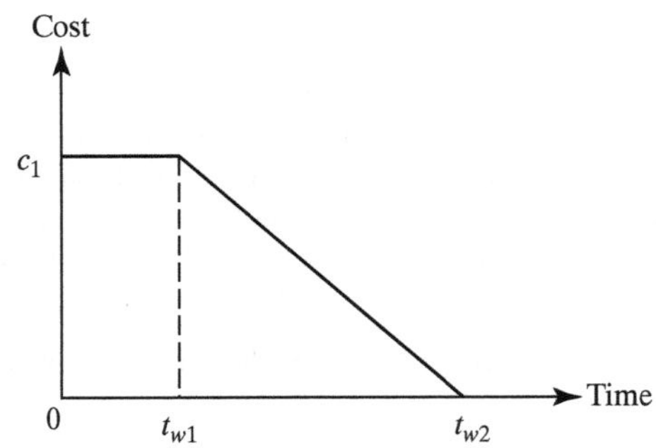

FIGURE 16.5 Cost of warranty in combined FRW/PRW policy.

From the definition of the probability distribution function of the failure time, $F(t)$, Eq. (16.17) can be expressed as

$$\bar{c}_w = c_1 F(t_{w1}) + \frac{c_1\, t_{w2}}{(t_{w2} - t_{w1})}\left\{F(t_{w2}) - F(t_{w1})\right\} - \frac{c_1}{(t_{w2} - t_{w1})}\int_{t_{w1}}^{t_{w2}} t f(t)\, dt \tag{16.18}$$

Using integration by parts, the integral in Eq. (16.18) can be evaluated as

$$\int_{t_{w1}}^{t_{w2}} t f(t)\, dt = t\,F(t)\big|_{t_{w1}}^{t_{w2}} - \int_{t_{w1}}^{t_{w2}} f(t)\, dt = t_{w2} F(t_{w2}) - t_{w1}\, F(t_{w1}) - \int_{t_{w1}}^{t_{w2}} F(t)\, dt \tag{16.19}$$

In view of Eq. (16.19), Eq. (16.18) becomes

$$\bar{c}_w = c_1 F(t_{w1}) + \frac{c_1\, t_{w2}}{(t_{w2} - t_{w1})}\left\{F(t_{w2}) - F(t_{w1})\right\}$$

$$- \frac{c_1}{(t_{w2} - t_{w1})} t_{w2}\, F(t_{w2}) + \frac{c_1}{(t_{w2} - t_{w1})} t_{w1}\, F(t_{w1}) + \frac{c_1}{(t_{w2} - t_{w1})}\int_{t_{w1}}^{t_{w2}} F(t)\, dt$$

which can be simplified to obtain

$$\bar{c}_w = \frac{c_1}{(t_{w2} - t_{w1})}\int_{t_{w1}}^{t_{w2}} F(t)\, dt \tag{16.20}$$

If the failure time of the product follows exponential distribution, Eq. (16.20) gives

$$\bar{c}_w = \frac{c_1}{(t_{w2} - t_{w1})}\int_{t_{w1}}^{t_{w2}} \left[1 - e^{-\frac{t}{\mu}}\right] dt = \frac{c_1}{t_{w2} - t_{w1}}\left[(t_{w2} - t_{w1}) + \mu\left(e^{-\frac{t_{w2}}{\mu}} - e^{-\frac{t_{w1}}{\mu}}\right)\right] \tag{16.21}$$

EXAMPLE 16.3

A manufacturing firm offers a three-months free replacement warranty followed by a nine-month pro-rata warranty on its treadmills. The failure time of the treadmills was determined experimentally to follow exponential distribution with a failure rate of $\lambda = 0.005$ failures/month. The firm's cost of production of a treadmill, including the warrant cost, is \$250. If the firm produces 10,000 treadmills in a batch, determine the amount of warranty reserve fund to be kept aside by the firm for the whole batch of treadmills.

Solution: With the available values of $c_1 = \$250$, $t_{w1} = 3$ months, $t_{w2} = 12$ months, $\lambda = \frac{1}{\mu} = 0.005$ or $\mu = 200$ months, and $N = 10{,}000$ units, Eq. (16.21) can be used to find

$$\begin{aligned}\overline{c}_w &= \frac{c_1}{t_{w2} - t_{w1}}\left[(t_{w2} - t_{w1}) + \mu\left(e^{-\frac{t_{w2}}{\mu}} - e^{-\frac{t_{w1}}{\mu}}\right)\right] \\ &= \frac{250}{12 - 3}\left[(12 - 3) + 200\left(e^{-\frac{12}{200}} - e^{-\frac{3}{200}}\right)\right] = \$9.44\end{aligned}$$

Thus, the amount of warranty reserve fund needed (expected value) by the firm is

$$\overline{C}_w = N\,\overline{c}_w = 10{,}000(9.44) = \$94{,}400$$

■

16.3.4 FRW Policy Equivalent to a FRW/PRW Policy

In some cases, the manufacturer might want to give an option to the customer to choose either a combined FRW/PRW policy for a total period of t_{w2} or an equivalent FRW policy for a shorter period t_{we} with $t_{w1} < t_{we} < t_{w2}$. This could be a marketing strategy of the manufacturer to increase the sales of the product. The equivalent FRW policy is expected to cost the same as the combined FRW/PRW policy to the manufacturer. To find the warranty period, t_{we}, of the equivalent FRW/PRW policy, we equate Eq. (16.2) with $t_w = t_{we}$ and (16.20):

$$c_1 F(t_{we}) = \frac{c_1}{t_{w2} - t_{w1}} \int_{t_{w1}}^{t_{w2}} F(t)\, dt \tag{16.22}$$

or

$$F(t_{we}) = \frac{1}{t_{w2} - t_{w1}} \int_{t_{w1}}^{t_{w2}} F(t)\, dt \tag{16.23}$$

By using the inverse of the probability distribution function of the failure time T, Eq. (16.23) can be solved to obtain

$$t_{we} = F^{-1}\left\{\frac{1}{t_{w2} - t_{w1}} \int_{t_{w1}}^{t_{w2}} F(t)\, dt\right\} \tag{16.24}$$

If the failure time follows exponential distribution,

$$F(t) = 1 - e^{-\lambda t} \tag{16.25}$$

and hence

$$\begin{aligned}\int_{t_{w1}}^{t_{w2}} F(t)\, dt &= \int_{t_{w2}}^{t_{w2}} \{1 - e^{-\lambda t}\}\, dt = t\Big|_{t_{w1}}^{t_{w2}} - \left(-\frac{e^{-\lambda t}}{\lambda}\right)\Bigg|_{t_{w1}}^{t_{w2}} \\ &= (t_{w2} - t_{w1}) + \frac{1}{\lambda}\left(e^{-\lambda t_{w2}} - e^{-\lambda t_{w1}}\right)\end{aligned} \tag{16.26}$$

Hence Eq. (16.24) can be rewritten in the form

$$F(t_{we}) = 1 - e^{-\lambda t_{we}} = \left(1 + \frac{e^{-\lambda t_{w2}} - e^{-\lambda t_{w1}}}{\lambda(t_{w2} - t_{w1})}\right) \tag{16.27}$$

Equation (16.27) yields

$$e^{-\lambda t_{we}} = \frac{e^{-\lambda t_{w1}} - e^{-\lambda t_{w2}}}{\lambda(t_{w2} - t_{w1})} \tag{16.28}$$

or

$$-\lambda t_{we} = \ln\left\{\frac{e^{-\lambda t_{w1}} - e^{-\lambda t_{w2}}}{\lambda(t_{w2} - t_{w1})}\right\} \tag{16.29}$$

or

$$t_{we} = \frac{1}{\lambda}\ln\left\{\frac{\lambda(t_{w2} - t_{w1})}{e^{-\lambda t_{w1}} - e^{-\lambda t_{w2}}}\right\} \tag{16.30}$$

EXAMPLE 16.4

For the problem stated in Example 16.3, find the length of the FRW policy that is equivalent to the combined free replacement and pro-rata warranty. The relevant data are $\lambda = 0.005$ failures/month, $t_{w1} = 3$ months, and $t_{w2} = 12$ months.

Solution: Using the known data $\lambda = 0.005$ failures/month, $t_{w1} = 3$ months, and $t_{w2} = 12$ months, Eq. (16.30) can be used to find the length of the equivalent FRW policy as

$$t_{we} = \frac{1}{\lambda}\ln\left\{\frac{\lambda(t_{w2} - t_{w1})}{e^{-\lambda t_{w1}} - e^{-\lambda t_{w2}}}\right\} = \frac{1}{0.005}\ln\left\{\frac{0.005(12 - 3)}{e^{-(0.005)(3)} - e^{-(0.005)(12)}}\right\}$$

$$= 200\ln\left\{\frac{0.045}{0.98511 - 0.94176}\right\} = 200\ln(1.03806) = 7.47116 \text{ months}$$

■

16.3.5 Lump-sum Payment Type of Warranty

In some cases, the manufacturer may like to pay a fixed or lump-sum amount to the customer if the product or system fails before the warranty period. This type of policy is particularly attractive if the administrative costs of managing other types of warranty are too expensive. In order to determine the warranty reserve fund, we assume the lump-sum amount (L) paid to the customer for failure of the product within the warranty period to be proportional to the cost of the product (c_1) as

$$L = ac_1 \tag{16.31}$$

where a is a constant $(0 < a \leq 1)$. Assuming the probability of failure of the product in time t_w to be given by (exponential distribution)

$$F(t) = 1 - e^{-\frac{t_w}{\mu}} \tag{16.32}$$

the warranty reserve fund for one unit of the product can be expressed as

$$\overline{c}_w = a c_1 \left(1 - e^{-\frac{t_w}{\mu}}\right) \tag{16.33}$$

In case the lump-sum amount is made equal to the pro-rata amount considered in Section 16.3.2, we have

$$a c_1 \left(1 - e^{-\frac{t_w}{\mu}}\right) = c_1 \left\{1 - \frac{\mu}{t_w}\left(1 - e^{-\frac{t_w}{\mu}}\right)\right\} \tag{16.34}$$

or

$$a = \frac{1}{1 - e^{-\frac{t_w}{\mu}}} - \frac{\mu}{t_w} \tag{16.35}$$

If the total number of units sold or the lot size is N, the total warranty reserve fund (expected value) will be

$$\overline{C}_w = N a c_1 \left(1 - e^{-\frac{t_w}{\mu}}\right) \tag{16.36}$$

EXAMPLE 16.5

The manufacturing firm described in Example 16.2 considers to offer a lump-sum warranty plan equivalent to the pro-rata plan by offering a lump-sum amount (a percentage of the initial price) to its customers whose treadmills fail during the warranty period. Find the percentage of the price to be refunded to customers whose treadmills fail before the expiry of the warranty period.

Solution: With the known data $t_w = 1$ year $= 12$ months, $c_1 = \$250$, $\lambda = \frac{1}{\mu} = 0.005$ or $\mu = 200$ months, and $N = 10{,}000$ units, Eq. (16.35) can be used to find

$$\begin{aligned} a &= \frac{1}{1 - e^{-\frac{t_w}{\mu}}} - \frac{\mu}{t_w} \\ &= \frac{1}{1 - e^{-\frac{12}{200}}} - \frac{200}{12} = \frac{1}{1 - 0.94176} - 16.66667 = 0.50366 \end{aligned}$$

This implies that the manufacturer should pay 50.4 percent of the initial price to customers whose treadmills fail before the expiration of the warranty period. From Example 16.2, the total cost of the pro-rata warranty to the manufacturer is given by

$$c_{\text{total}} = c_1 = \overline{c} + \overline{c}_w = \$250.00$$

and hence the lump-sum payment is equal to

$$L = c_{\text{total}}\, a = 250.00(0.50366) = \$125.915$$

The total warranty reserve fund needed (expected value) is given by

$$\overline{C}_{we} = NL = 10000(125.915) = \$1{,}259{,}150.00$$

■

16.4 Warranty Costs Considering the Time Value of Money

Depending on the economic conditions, when the interest rates are high, the time value of money can be considered in computing the warranty costs. In this case, the expected replacement cost of the product will be smaller for the manufacturer. A consideration of the time value of money becomes important particularly if the product has a high repair (or replacement) cost and a lengthy warranty period. The reduction in the cost of the product at time t after it is sold at time $t = 0$ is assumed to be of the form $e^{-\delta t}$ where δ is called the discounting factor with $0 < \delta < 1$. Thus, the unit cost of the product at time t is given by $c_1 e^{-\delta t}$, where c_1 is the cost of the product at $t = 0$.

16.4.1 FRW Policy

The unit cost of the product at time t is given by

$$c_w = \begin{cases} c_1 e^{-\delta t}; & 0 < t \le t_w \\ 0; & t > t_w \end{cases} \tag{16.37}$$

The expected warranty cost per product can be determined as

$$\overline{c}_w = E\left[c_w(t)\right] = c_1 \int_0^{t_w} e^{-\delta t} f(t)\, dt \tag{16.38}$$

If the failure time of the product, T, follows exponential distribution, Eq. (16.38) can be expressed as

$$\begin{aligned} \overline{c}_w &= c_1 \int_0^{t_w} e^{-\delta t}\left(\lambda e^{-\lambda t}\right) dt = c_1 \lambda \int_0^{t_w} e^{-(\delta+\lambda)\,t}\, dt \\ &= \frac{c_1 \lambda}{\lambda + \delta}\left\{1 - e^{-(\lambda+\delta)\,t_w}\right\} \end{aligned} \tag{16.39}$$

EXAMPLE 16.6

A manufacturing firm offers a one-year free replacement warranty on its treadmills. The failure time of the treadmills follows exponential distribution with a failure rate of $\lambda = 0.005$ failures per month. The firm's cost of production of one treadmill is \$250. By considering the time value of money with a discounting factor of $\delta = 0.05$, determine the expected amount of warranty payment to be made by the manufacturer per unit.

Solution: With the available values of $c_1 = \$250$, $t_w = 1$ year $= 12$ months, $\lambda = \dfrac{1}{\mu} = 0.005$ or $\mu = 200$ months, and $\delta = 0.05$, Eq. (16.39) can be used to find the expected warranty payment per each failed treadmill as

$$\begin{aligned} \overline{c}_w &= \frac{c_1 \lambda}{\lambda + \delta}\left\{1 - e^{-(\lambda+\delta)\,t_w}\right\} = \frac{(250)(0.005)}{0.005 + 0.05}\left\{1 - e^{-(0.005+0.05)(12)}\right\} \\ &= (22.7273)(1 - 0.5168) = \$10.9818 \end{aligned}$$

■

16.4.2 PRW Policy

In this case, the unit cost of the product at time t is given by

$$c_w(t) = \begin{cases} c_1 e^{-\delta t}\left(1 - \dfrac{t}{t_w}\right); & 0 < t \le t_w \\ 0; & t > t_w \end{cases} \tag{16.40}$$

The expected unit warranty cost can be determined as

$$\begin{aligned} \overline{c}_w = E\left[c_w(t)\right] &= c_1 \int_0^{t_w} \left(1 - \frac{t}{t_w}\right) e^{-\delta t} f(t)\, dt \\ &= c_1 \int_0^{t_w} e^{-\delta t} f(t)\, dt - \frac{c_1}{t_w} \int_0^{t_w} t\, e^{-\delta t} f(t)\, dt \end{aligned} \tag{16.41}$$

The second term on the right-hand side of Eq. (16.41) can be rewritten, using integration by parts, as

$$\frac{c_1}{t_w} \int_0^{t_w} t\, e^{-\delta t} f(t)\, dt = \frac{c_1}{t_w}\left[t \int_0^t e^{-\delta \tau} f(\tau)\, d\tau\right]_0^{t_w} - \frac{c_1}{t_w} \int_0^{t_w} dt \int_0^t e^{-\delta \tau} f(\tau)\, d\tau \tag{16.42}$$

In view of Eq. (16.42), Eq. (16.41) becomes

$$\overline{c}_w = \frac{c_1}{t_w} \int_0^{t_w} dt \int_0^t e^{-\delta \tau} f(\tau)\, d\tau \tag{16.43}$$

If the failure time of the product (T) follows exponential distribution, Eq. (16.43) can be expressed as

$$\overline{c}_w = \frac{c_1}{t_w} \int_{t=0}^{t_w} dt \int_{\tau=0}^{t} e^{-\delta \tau}\left(\lambda e^{-\lambda \tau}\right) d\tau \tag{16.44}$$

The integrations of Eq. (16.44) can be evaluated as

$$\begin{aligned} \overline{c}_w &= \frac{c_1}{t_w} \int_{t=0}^{t_w} dt \int_{\tau=0}^{t} \lambda e^{-(\delta+\lambda)\tau}\, d\tau = \frac{c_1}{t_w} \int_0^{t_w} \lambda \left[\frac{e^{-(\delta+\lambda)\tau}}{\delta + \lambda}\right]_0^t dt \\ &= \frac{c_1}{t_w} \int_0^{t_w} \frac{\lambda\left\{1 - e^{-(\delta+\lambda)t}\right\}}{\delta + \lambda}\, dt = \frac{c_1 \lambda}{t_w(\delta + \lambda)} \int_0^{t_w} \left(1 - e^{-(\delta+\lambda)t}\right) dt \\ &= \frac{c_1 \lambda}{\delta + \lambda} - \frac{c_1 \lambda}{t_w(\delta + \lambda)^2}\left\{1 - e^{-(\delta+\lambda)t_w}\right\} \end{aligned} \tag{16.45}$$

EXAMPLE 16.7

A manufacturing firm offers a one-year pro-rata warranty on its treadmills. The failure time of the treadmills follows exponential distribution with a failure rate of $\lambda = 0.005$ failures per month. The firm's cost of production of one treadmill is \$250. By considering the time value of money with a discounting factor of $\delta = 0.05$, determine the expected amount of warranty payment to be made by the manufacturer per unit.

Solution: With the available values of $c_1 = \$250$, $t_w = 1$ year $= 12$ months, $\lambda = \frac{1}{\mu} = 0.005$ or $\mu = 200$ months, and $\delta = 0.05$, Eq. (16.45) can be used to find the expected warranty payment per each failed treadmill as

$$\bar{c}_w = \frac{c_1 \lambda}{\delta + \lambda} - \frac{c_1 \lambda}{t_w(\delta + \lambda)^2} \left\{1 - e^{-(\delta+\lambda)\, t_w}\right\}$$

$$= \frac{(250)\,(0.005)}{(0.05 + 0.005)} - \frac{(250)\,(0.005)}{(12)\,(0.05 + 0.005)^2} \left\{1 - e^{-(0.05+0.005)\,(12)}\right\}$$

$$= 22.7273 - \frac{1.25}{0.0363}\left(1 - e^{-0.66}\right) = 22.7273 - 16.6373 = \$6.0899$$

■

16.5 Warranty Reserve Fund Considering the Time Value of Money and Future Changes in the Price of the Product

In this section, we consider the time value of money and future changes in the price of the product in estimating the warranty reserve fund to be set up by the manufacturer for a product produced in a lot size of N products [16.7]. Warranty costs are incurred in the future when a product under warranty fails and the consumer is given a rebate. Because the rebate costs are incurred at a future time, the actual reserve fund to be set up at present should be discounted by considering the time value of money. In addition, the changes in the price of the product in future (due to inflation) should also be taken into account. Assuming that all consumers experiencing a product failure within the warranty period make a claim, the expected cost of the warranty claims, in the case of pro-rata policy, in the time interval t to $t + dt$ is given by:

$$dC_w = c_w(t)\; dE[n(t)] = c_1\left(1 - \frac{t}{t_w}\right)\frac{N}{\mu}\, e^{-\frac{t}{\mu}}\, dt \tag{16.46}$$

where c_1 denotes the unit price of the product including the warranty cost and $dE[n(t)]$ is the expected number of product failures during the time increment dt. This cost, dC_w, will occur in the future after the product is sold. Hence, the present value of the amount given by Eq. (16.46) is reduced by the opportunity (for the manufacturer) to invest and earn interest until the time of payment. In addition, the possible changes in the general price level of the product may cause the real value of the rebate cost to be larger or smaller than their nominal values in future periods. If θ denotes the rate of return (i.e., interest per one time period) that could be earned through investment of warranty reserve fund, and ϕ indicates the expected change in the price of the product (similar to interest rate, but ϕ could be positive or negative), the real present value of the warranty claims during the interval t to $t + dt$, $dC_w{}^*$, is given by

$$dC_w{}^* = \frac{dC_w}{(1 + \theta)^t(1 + \phi)^t} \tag{16.47}$$

where t is the number of time periods and $(1 + \theta)^t(1 + \phi)^t$ denotes the interest and price change factor over t time periods.

Since θ and ϕ are usually small compared to unity, we can use the approximation

$$(1 + \theta)^t(1 + \phi)^t \approx (1 + \theta + \phi)^t \tag{16.48}$$

and rewrite Eq. (16.47) as

$$dC_w{}^* = \frac{dC_w}{(1 + \theta + \phi)^t} = c_1\left(1 - \frac{t}{t_w}\right)(1 + \theta + \phi)^{-t}\left(\frac{N}{\mu}\right)e^{-\frac{t}{\mu}}\,dt \tag{16.49}$$

The real present value of the costs associated with the warranty claims on a product lot of size N units can be found by adding all (integrating) the incremental costs, $dC_w{}^*$, from $t = 0$ to $t = t_w$:

$$\begin{aligned} C_w{}^* &= \frac{Nc_1}{\mu}\int_{t=0}^{t_w}\left(1 - \frac{t}{t_w}\right)\left[(1 + \theta + \phi)e^{-\frac{1}{\mu}}\right]^{-t} dt \\ &= \frac{Nc_1}{1 + \mu\ln(1 + \theta + \phi)}\left\{1 - \frac{\mu}{t_w}\frac{1}{1 + \mu\ln(1 + \theta + \phi)}\left[1 - (1 + \theta + \phi)^{-t_w}e^{-\frac{t_w}{\mu}}\right]\right\} \end{aligned} \tag{16.50}$$

Thus, the warranty cost assigned by the manufacturer per unit of the product (present value) is

$$c_w{}^* = \frac{C_w{}^*}{N} \tag{16.51}$$

The actual unit price of the product to the manufacturer will be

$$c_1{}^* = \bar{c} + c_w{}^* = \bar{c} + \frac{C_w{}^*}{N} \tag{16.52}$$

where $\bar{c}$ denotes the cost of manufacturing a unit of the product without including the warranty cost with

$$\begin{aligned} \frac{c_w{}^*}{c_1{}^*} &= \frac{C_w{}^*}{Nc_1{}^*} \\ &= \frac{1}{1 + \mu\ln(1 + \theta + \phi)}\left\{1 - \frac{\mu}{t_w}\frac{1}{1 + \mu\ln(1 + \theta + \phi)}\left[1 - (1 + \theta + \phi)^{-t_w}e^{-\frac{t_w}{\mu}}\right]\right\} \end{aligned} \tag{16.53}$$

Table 16.1 gives the values of $\frac{c_w{}^*}{c_1{}^*}$ for $\theta + \phi = 0.1$ and $\mu = 20$ time periods (such as months) for different values of $\frac{t_w}{\mu}$. As one would expect, for any specific value of $\frac{t_w}{\mu}$, the values of $\frac{c_w{}^*}{c_1{}^*}$ are less compared to the values of $\frac{\bar{c}_w}{c_1}$ given by Eq. (16.12) (where the present worth of warranty cost and the price change of the product are not considered).

TABLE 16.1 Variation of $\dfrac{c_w^*}{c_1^*}$ with the ratio $\dfrac{t_w}{\mu}$

Value of $\dfrac{t_w}{\mu}$	0.05	0.10	0.20	0.50	1.0	2.0	3.0	4.0
Value of $\dfrac{c_w^*}{c_1^*}$	0.024	0.045	0.083	0.163	0.232	0.285	0.305	0.315

Note:
Since $C_w^* < C_w$ with $c_w^* < c_1$ and hence $c_1^* < c_1$. This result has important implications for the manufacturer depending on the price elasticity of demand. For example, if the demand for the manufacturer's product primarily depends on the price, the manufacturer can reduce the unit price from c_1 to c_1^* and increase its sales and hence the revenues. In addition, in terms of the aggregate economy, if a future price increase (due to inflation) is expected (with $\phi > 0$), the manufacturer in fact can reduce its current price and try to get a bigger market share for its product. On the other hand, the manufacturer should be very careful in reducing the current price of the product in anticipation of the expected increase in the price of the product in future due to inflation. If the overall economy does not experience the anticipated inflation in future, the manufacturer will suffer a loss.

EXAMPLE 16.8

The manufacturing firm described in Example 16.2 plans to invest the warranty reserve fund to earn an annual interest rate of 6 percent and to increase the price of its treadmills by 4 percent in the following year. Find the corresponding warranty reserve fund and the price of the treadmill.

Solution: The following data are known in this case: $t_w = 1$ year $= 12$ months, $\bar{c} = \$250$, $\lambda = \dfrac{1}{\mu} = 0.005$ or $\mu = 200$ months, and $N = 10{,}000$ units, $\theta = 0.06$, and $\phi = 0.04$. With the interest rate and percentage price increase known, we have

$$1 + \theta + \phi = 1 + 0.06 + 0.04 = 1.10$$

Equation (16.53) gives

$$\frac{c_w^*}{c_1^*} = (1 + 200 \ln 1.10)^{-1}\left\{1 - \frac{200}{12}(1 + 200 \ln 1.10)^{-1}\left[1 - (1.10)^{-12}e^{-\frac{12}{200}}\right]\right\}$$

$$= (1 + 19.0620)^{-1}\left\{1 - 16.6667(1 + 19.0620)^{-1}\left[1 - \frac{0.9418}{3.1384}\right]\right\} = 0.02086$$

and Eq. (16.52) yields

$$c_1^* = \frac{\bar{c}}{1 - \dfrac{c_w^*}{c_1^*}} = \frac{250}{1 - 0.02086} = \$255.33$$

Thus, the required reserve fund will be

$$C_w^* = N\left(\frac{c_w^*}{c_1^*}\right)c_1^* = 10000(0.02086)\,(255.33) = \$52{,}261.02$$

■

16.6 Warranty Analysis Considering Multiple Failures During the Warranty Period

In Sections 16.2 and 16.3, we considered the probability of only a single failure during the warranty period of the product. However, in practice, it is possible to have multiple failures of the product during the warranty period. For example, in the free replacement warranty, let the product sold fail at time X_1 before the expiration of the warranty period $(X_1 < t_w)$. If the product is nonrepairable, it is immediately replaced by a new product by the manufacturer. If this new product also fails after a time X_2, before the expiration of the warranty period $(X_1 + X_2 < t_w)$, it too is replaced by another new product by the manufacturer. This process is continued until the total operating time reaches the warranty period t_w. In this case, the warranty is said to be nonrenewing, implying that the warranty period for a replaced product (say, product i) is ordinarily not t_w, but $t_w - S_i$, provided $S_i < t_w$ where $S_i = \sum_{j=1}^{i} X_j$ with X_j denoting the life of the jth product. The warranty ends at time t_w, and the product in service at that time is no longer covered by the warranty. The process is shown in Fig. 16.6. It is evident that multiple failures occur during the warranty period in the case of nonrenewing warranty.

In another type of free replacement warranty, called renewing warranty, the warranty period begins a new after each replacement under warranty in addition to the one after the initial purchase. This means that when the original product fails during the warranty period $(0, t_w)$, it is replaced by a new product by the manufacturer. The replaced item is again covered by a new full warranty for a period t_w. When the replaced product fails during its own warranty period t_w, it is again replaced by a new product, which is again covered by a full warranty for a period t_w. The process is shown in Fig. 16.7. It is evident that only one failure occurs during the warranty period of any product in the renewing warranty.

The expected warranty cost of a single failure based on free replacement policy can be found as

$$E[c_w(t_w)] = c_1[1 + F(t = t_w)] \tag{16.54}$$

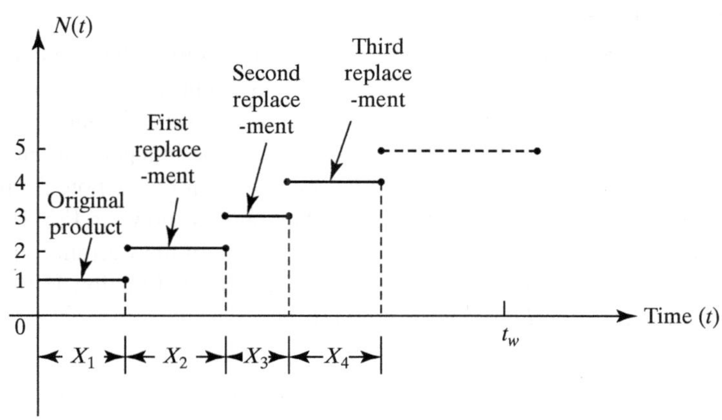

FIGURE 16.6 Occurrence of events in a nonrenewing FRW $(N = \text{number of products})$.

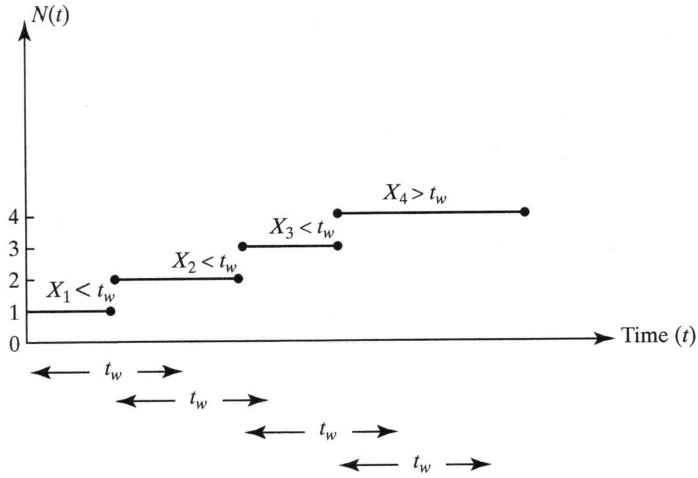

FIGURE 16.7 Occurrence of events in a renewing FRW.

Because of the possibility of multiple failures within the warranty period, we introduce a new random variable $N = N(t_w)$ to denote the number of products failed (original and the replacements). The random variable $N(t_w)$ can also be thought of as the first N such that $S_{N+1} = \sum_{i=1}^{N+1} X_i \geq t_w$. When the lives of the products, X_i, are independent random variables with identical distributions, the expected value of $N(t_w)$ is given by $E[N(t_w)] = M(t_w)$, where $M(t)$ is the renewal function associated with the failure time distribution function $F(t)$. Thus, the expected total cost of warranty per unit under the free replacement policy can be expressed as

$$E[c_w(t_w)] = E[c_1 + c_1 N(t_w)] = c_1[1 + M(t_w)] \tag{16.55}$$

The determination of the renewal function $M(t)$ is discussed in the following subsection.

16.6.1 Renewal Process

A renewal process is a stochastic process which counts the number of events that have occurred up to some point in time. In order for a stochastic process to be a renewal process, each time we count an event, the system must reset itself, that is, the time between the occurrence of events (also called renewals) must be independent and identically distributed random variables. For example, $X_1, X_2, \ldots, X_n$ might denote the times elapsed before the first, second, . . . , nth machine or product fails as shown in Fig. 16.6. As stated earlier, the expected number of failures of the product during the warranty time, $E[N(t_w)]$, is the value of the renewal function at t_w, $M(t_w)$. The renewal function $M(t)$ corresponding to the distribution function $F(t)$ is defined by [16.1]:

$$M(t) = F(t) + \int_0^t M(t - \tau) f(\tau)\, d\tau \tag{16.56}$$

The renewal function $M(t)$ can be derived in closed form only for few probability distribution functions $F(t)$. The exponential distribution function is one such distribution for which $M(t)$ can be derived analytically. For most other distributions, analytical expressions are not available for the renewal function, $M(t)$. Hence, it is necessary to use either some approximations or tables of renewal function [16.1]. The following two examples illustrate the analytical derivation of the renewal function.

16.6.2 Computation and Use of Renewal Functions

1. Renewal function corresponding to exponential distribution:

EXAMPLE 16.9

Find the renewal function corresponding to the exponential failure time distribution.

Solution: For exponential distribution, the probability distribution and density functions are given by

$$F(t) = 1 - e^{-\lambda t} \tag{E1}$$

$$f(t) = \lambda e^{-\lambda t} \tag{E2}$$

The Laplace transforms of Eqs. (16.56), (E1), and (E2) can be expressed as [16.15]

$$\overline{M}(s)[1 - \overline{f}(s)] = \overline{F}(s)] \tag{E3}$$

$$\overline{F}(s) = \frac{\lambda}{s(\lambda + s)} \tag{E4}$$

$$\overline{f}(s) = \frac{\lambda}{\lambda + s} \tag{E5}$$

Equation (E3) can be rewritten as

$$\overline{M}(s) = \frac{\overline{F}(s)}{1 - \overline{f}(s)} \tag{E6}$$

Substitution of Eqs. (E4) and (E5) in Eq. (E6) gives

$$\overline{M}(s) = \frac{\lambda}{s(\lambda + s)} \frac{1}{1 - \left(\dfrac{\lambda}{\lambda + s}\right)} = \frac{\lambda}{s^2} \tag{E7}$$

On taking the inverse Laplace transform of Eq. (E7), we obtain the renewal function as

$$M(t) = \lambda t \tag{E8}$$

The following example illustrates the computation of the expected warranty costs of products considering the possibility of multiple failures during the warranty period using the renewal function.

■

EXAMPLE 16.10

The manufacturer's cost of a high-definition (HD) television is $c_1 = \$600$. The manufacturer sells the HD televisions with a one-year free replacement warranty. If the failure time of the televisions follows exponential distribution with an average life of 3 years, determine the expected warranty cost to the manufacturer.

Solution: Since the average life of a HD television (μ) is 3 years, the failure rate will be $\lambda = 1/\mu = 1/3 = 0.3333$ failures per year. The expected cost of warranty per unit to the manufacturer can be obtained from Eq. (16.55) as:

$$E[c_w(t_w)] = E[c_1 N(t_w)] = c_1[M(t_w)] \tag{E1}$$

Because the renewal function $M(t)$ corresponding to exponential failure time distribution is given by Eq. (E8) of Example 16.9, Eq. (E1) gives

$$\overline{c}_w = E[c_w(t_w)] = c_1 M(t_w) = c_1 \lambda t_w = (600)(0.3333)(1) = \$200 \tag{E2}$$

Thus, the expected cost of warranty to the manufacturer per television is \$200.

■

2. *Renewal function corresponding to two-stage Erlang distribution:*

EXAMPLE 16.11

Find the renewal function when the failure time follows the two-stage Erlang distribution with

$$F(t) = 1 - (1 + \lambda t)\, e^{-\lambda t} \tag{E1}$$

Solution: The probability density function of the two-stage Erlang distribution corresponding to Eq. (E1) is given by

$$f(t) = \frac{dF(t)}{dt} = \lambda^2\, t\, e^{-\lambda t} \tag{E2}$$

The Laplace transforms of Eqs. (E2) and (E1) yield

$$\overline{f}(s) = \frac{\lambda^2}{(\lambda + s)^2} \tag{E3}$$

$$\overline{F}(s) = \frac{\overline{f}(s)}{s} = \frac{\lambda^2}{s(\lambda + s)^2} \tag{E4}$$

In view of Eqs. (E3) and (E4), the Laplace transform of the renewal function, given by Eq. (E6) of Example 16.9, can be expressed as

$$\overline{M}(s) = \frac{\overline{F}(s)}{1 - \overline{f}(s)} = \frac{\overline{f}(s)}{s[1 - \overline{f}(s)]} = \frac{\lambda^2}{(\lambda + s)^2} \frac{1}{s\left[1 - \dfrac{\lambda^2}{(\lambda + s)^2}\right]} = \frac{\lambda^2}{s^2(s + 2\lambda)} \tag{E5}$$

The right-hand side of Eq. (E5) can be expanded in partial fractions as

$$\frac{\lambda^2}{s^2(s + 2\lambda)} = \frac{a_1}{s} + \frac{a_2}{s^2} + \frac{a_3}{(s + 2\lambda)} \tag{E6}$$

Equation (E6) yields the values of the constants a_1, a_2, and a_3 as

$$a_1 = -\frac{1}{4}, \quad a_2 = \frac{\lambda}{2}, \quad a_3 = \frac{1}{4} \tag{E7}$$

Thus $\overline{M}(s)$ of Eq. (E5) can be expressed as

$$\overline{M}(s) = -\frac{1}{4s} + \frac{\lambda}{2s^2} + \frac{1}{4(s + 2\lambda)} \tag{E8}$$

The inverse Laplace transform of Eq. (E8) gives the renewal function corresponding to the two-stage Erlang distribution as

$$M(t) = -\frac{1}{4} + \frac{\lambda t}{2} + \frac{1}{4} e^{-2\lambda t} \tag{E9}$$

The following example illustrates the computation of the expected warranty costs of products considering the possibility of multiple failures during the warranty period using the renewal function when the failure time follows the two-stage Erlang distribution.

■

EXAMPLE 16.12

The manufacturer's cost of a high-definition television is $c_1 = \$600$. The manufacturer sells the HD televisions with a one-year free replacement warranty. The failure time of the televisions follows a two-stage Erlang distribution with

$$F(t) = 1 - (1 + \lambda t)\, e^{-\lambda t} \tag{E1}$$

If the average life of the television is 3 years, determine the expected warranty cost to the manufacturer.

Solution: The expected cost of warranty per unit to the manufacturer can be obtained from Eq. (16.55) as

$$E[c_w(t_w)] = E[c_1 N(t_w)] = c_1 M(t_w) \tag{E1}$$

Because the renewal function $M(t)$ corresponding to two-stage Erlang failure time distribution is given by Eq. (E9) of Example 16.11, Eq. (E1) becomes

$$\begin{aligned} \bar{c}_w &= E[c_w(t_w)] = c_1 M(t_w) = c_1\left(-\frac{1}{4} + \frac{\lambda t_w}{2} + \frac{1}{4} e^{-2\lambda t_w}\right) \\ &= (600)\left(-\frac{1}{4} + \frac{0.3333(1)}{2} + \frac{1}{4} e^{-2(0.3333)(1)}\right) = \$27.03 \end{aligned} \tag{E2}$$

Thus, the expected cost of warranty to the manufacturer per television is \$27.03.

■

3. Renewal function corresponding to Weibull distribution:

Consider the two-parameter Weibull distribution:

$$F(t) = 1 - e^{-(\lambda t)^\beta}, \quad t \geq 0 \tag{16.57}$$

where $\beta > 0$ and $\lambda > 0$ are the parameters of the distribution. In this case, a closed-form expression cannot be derived for the renewal function, $M(t)$. However, the following methods can be used to find the approximate values of the renewal function for any specified data.

Analytical approximations:

A series expansion of $M(t)$ is given by Leadbetter [16.18]. For this, first the failure time distribution given by Eq. (16.57) is expressed in an infinite series as

$$F(t) = 1 - e^{-t^\beta} = \sum_{i=1}^{\infty} \frac{(-1)^{i-1}}{\Gamma(1 + i\beta)} C_i \, (t^\beta)^i \tag{16.58}$$

where the coefficients C_i are given by

$$C_i = \frac{\Gamma(1 + i\beta)}{i!} \tag{16.59}$$

Then the renewal function $M(t)$ is also expressed in an infinite series as

$$M(t) = \sum_{i=1}^{\infty} \frac{(-1)^{i-1}}{\Gamma(1 + i\beta)} A_i \, (t^\beta)^i \tag{16.60}$$

with A_i's given by

$$A_1 = C_1, \quad A_2 = C_2 - A_1 C_1, \quad A_j = C_j - \sum_{i=1}^{j-1} A_i C_{j-1}; \quad j > 1 \tag{16.61}$$

When $t < 1$, the series in Eq. (16.60) can be truncated at some finite value. Of course, the accuracy of $M(t)$ can be improved by considering a larger number of terms. For large values of t, the renewal function $M(t)$ can be approximated as [16.1, 16.2]

$$M(t) \approx \frac{t}{\mu} + \frac{\sigma^2}{2\mu^2} - \frac{1}{2} \tag{16.62}$$

where μ is the mean value and σ is the standard deviation of the failure time T corresponding to Eq. (16.57):

$$\mu = \frac{1}{\lambda} \Gamma\left(1 + \frac{1}{\beta}\right) \tag{16.63}$$

$$\sigma = \frac{1}{\lambda} \left[\Gamma\left(1 + \frac{2}{\beta}\right) - \Gamma^2\left(1 + \frac{1}{\beta}\right)\right]^{\frac{1}{2}} \tag{16.64}$$

The error associated with using Eq. (16.60) decreases for larger values of t. When $t \to \infty$, the renewal function converges to

$$\lim_{t \to \infty} \frac{M(t)}{t} = \frac{1}{\mu} \tag{16.65}$$

so that $M(t) \approx \dfrac{t}{\mu}$ as $t \to \infty$.

Numerical approximation and use of tables:

The renewal function corresponding to Weibull distribution is computed numerically and the results are tabulated by Baxter et al. [16.17]. For any value of the parameter β (with $\lambda = 1$), the value of the renewal function $M(t)$ can be found from the tables for any specific value of t. If $\lambda \neq 1$, the value of λt is to be used in place of t to find the value of $M(t)$ from the tables.

The following example illustrates the computation of the expected warranty costs of products considering the possibility of multiple failures during the warranty period when the failure time follows Weibull distribution using approximate values of the renewal function.

EXAMPLE 16.13

The manufacturer's cost of a high-definition television is $c_1 = \$600$. The manufacturer sells the HD televisions with a one-year free replacement warranty. The failure time of the televisions follows Weibull distribution with

$$F(t) = 1 - e^{-(\lambda t)^\beta}; \quad t \geq 0 \tag{E1}$$

If the average life of the television is 3 years, determine the expected warranty cost to the manufacturer for $\beta = 2$.

Solution: For $\beta = 2$ ($\beta = 1$ corresponds to exponential distribution), the mean and standard deviation of the failure time (T) are given by (from Eqs. (16.63) and (16.64))

$$\mu = \frac{1}{\lambda}\Gamma(1.5), \quad \sigma = \frac{1}{\lambda}\left[\Gamma(2) - \Gamma^2(1.5)\right]^{\frac{1}{2}} \tag{E2}$$

From the formulas of gamma function, we find that $\Gamma(1.5) = 0.88623$ and $\Gamma(2) = 1.0$ and hence

$$\mu = \frac{0.88623}{\lambda}, \quad \sigma = \frac{0.46324}{\lambda} \tag{E3}$$

In order to have $\mu = 3$ years, λ must be equal to $\lambda = \dfrac{0.88623}{3} = 0.29541$ and hence the value of σ becomes $\dfrac{0.46324}{0.29541} = 1.56812$. Since the warranty time is $t = t_w = 1$ year, t is not large and hence Eq. (16.62) cannot be used. The value of the renewal function corresponding to $\beta = 2$, $\lambda = 0.29541$, and $t = t_w = 1$ is found from the table of renewal function [16.1, 16.17]. Because $\lambda \neq 1$, the value of $\lambda t_w = (0.29541)(1) = 0.29541$ is used for t to obtain the value of $M(t_w)$ as 0.0874. Thus, the expected warranty cost per television is given by

$$\bar{c}_w = E[c_w(t_w)] = c_1 M(\lambda t_w) = (600)(0.0874) = \$52.44 \tag{E4}$$

■

16.7 Optimum Warranty Period

Once the product is perceived to be of reasonably good quality by the customers, the manufacturer can establish the warranty period to minimize the expected total warranty cost. The total warranty cost to the manufacturer, C_{total}, is given by the sum of the expected warranty cost $(\bar{c}_w(t_w))$, the expected cost benefit of warranty $(B_w(t_w))$, and the fixed cost to administer the warranty program (A_w):

$$C_{\text{total}}(t_w) = \bar{c}_w(t_w) + B_w(t_w) + A_w \tag{16.66}$$

In general, the expected warranty cost to the manufacturer increases with an increase in the warranty period, the expected cost benefit decreases with the warranty period, the fixed cost of administration does not depend on the warranty period, and the total warranty cost will vary nonlinearly with the warranty period as indicated in Fig. 16.8 with a minimum at $t_w{}^*$, known as the optimum warranty period. For simplicity, the expected cost benefit of warranty to the manufacturer can be taken as a quadratic function of the warranty period as

$$B_w(t_w) = k_1 + k_2\, t_w + k_3\, t_w^2 \tag{16.67}$$

where k_1, k_2, and k_3 are known constants. The fixed administrative costs to manage the warranty program, A_w, are also known to the manufacturer. Because these costs are constant, they do not have any effect on the optimum warranty period.

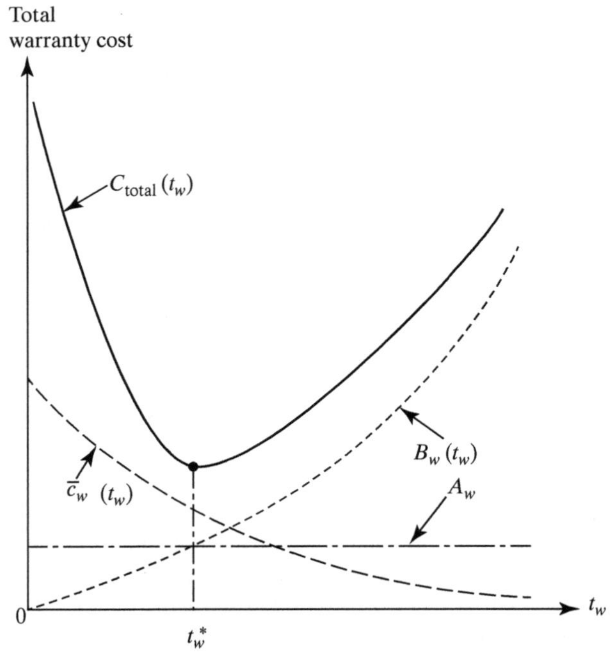

FIGURE 16.8 Variation of total warranty cost showing the optimum warranty period.

Once the equation for the total warranty cost is known, the optimum warranty period can be determined using a suitable method of optimization [16.14]. As there are no constraints, the optimum warranty cost can be found by solving the equation:

$$\frac{dC_{\text{total}}}{dt_w} = \frac{d\bar{c}_w}{dt_w} + \frac{dB_w}{dt_w} = 0 \tag{16.68}$$

The following examples illustrate the procedure for finding the optimum warranty period.

EXAMPLE 16.14

The cost of a television to the manufacturer is \$600. The failure rate of the television is known to be 0.2 per year. The cost of warranty to the manufacturer for the failure of each television during the useful life of the television is estimated to be \$600. The cost to the manufacturer to market the television without warranty is \$1600 and this cost is expected to decrease according to the relation

$$B_w(t_w) = \$(1600 - 1000t_w + 250t_w^2) \tag{E1}$$

Assuming that the failure time of the television follows exponential distribution, find the optimal warranty period for the television.

Solution: For exponential failure time distribution, the number of failures expected during the warranty period is given by the renewal function

$$M(t_w) = \lambda t_w \tag{E2}$$

Hence, the cost of warranty to the manufacturer is given by

$$\bar{c}_w = c_1 M(t_w) = c_1 \lambda t_w \tag{E3}$$

Using Eq. (E1) for B_w (with $k_1 = 1600$, $k_2 = -1000$, $k_3 = 250$) and Eq. (E3) for $\bar{c}_w$ with $c_1 = \$600$ in Eq. (16.68), we obtain

$$\frac{dC_{\text{total}}}{dt_w} = \frac{d}{dt_w}(c_1 \lambda t_w) + \frac{d}{dt_w}(1600 - 1000t_w + 250t_w^2) = 0$$

that is,

$$c_1 \lambda - 1000 + 500\, t_w = 0$$

or

$$t_w{}^* = \frac{1000 - c_1 \lambda}{500} = \frac{1000 - (600)(0.2)}{500} = 1.76 \text{ years}$$

Note: The value of the second derivative of the total warranty cost is given by

$$\frac{d^2 C_{\text{total}}}{dt_w^2} = 500 > 0$$

This shows that the optimum warranty period, $t_w{}^* = 1.76$ years, indeed corresponds to the minimum warranty cost to the manufacturer.

■

EXAMPLE 16.15

If the failure time of the television considered in Example 16.14 follows a two-stage Erlang distribution with

$$F(t) = 1 - (1 + \lambda t)\, e^{-\lambda t} \tag{E1}$$

and $\lambda = 0.2$ per year, determine the optimum warranty period to minimize the total warranty cost to the manufacturer. Assume the expected cost-benefit function to be same as the one given in Example 16.14.

Solution: For the two-stage Erlang failure time distribution, the number of failures expected during the warranty period is given by the renewal function

$$M(t_w) = -\frac{1}{4} + \frac{\lambda t_w}{2} + \frac{1}{4} e^{-2\,\lambda t_w} \tag{E2}$$

By using $\lambda = 0.2$, the cost of warranty to the manufacturer can be expressed as

$$\overline{c}_w = c_1 M(t_w) = c_1\left(-\frac{1}{4} + \frac{t_w}{10} + \frac{1}{4} e^{-0.4\, t_w}\right) \tag{E3}$$

Using $\lambda = 0.2$ per year, $c_1 = \$600$, Eq. (E1) of Example 16.14 for B_w (with $k_1 = 1600$, $k_2 = -1000$, $k_3 = 250$), and Eq. (E3) for $\overline{c}_w$, Eq. (16.68) gives

$$\frac{dC_{\text{total}}}{dt_w} = \frac{d}{dt_w}\left\{600\left(-0.25 + 0.1\, t_w + 0.25\, e^{-0.4\, t_w}\right)\right\} + \frac{d}{dt_w}\left(1600 - 1000\, t_w + 250\, t_w^2\right) = 0$$

that is,

$$60 + (-0.4)(150)\, e^{-0.4\, t_w} - 1000 + 500\, t_w = 0$$

that is,

$$60\, e^{-0.4\, t} = (500\, t_w - 940) \tag{E4}$$

The nature of the left- and right-hand sides of Eq. (E4) are shown graphically in Fig. 16.9. Using some trial and error values of t_w, the following results are obtained:

t_w* (trial value of $t = t_w$)	$-0.4\, t_w$	$60\, e^{-0.4\, t_w}$	$(500\, t_w - 940)$	Is this a solution of Eq. (E4)?
1.0	−0.4	40.2192	−440	No
2.0	−0.8	26.9597	60	No
1.9	−0.76	28.0600	10	No
1.89	−0.756	28.1724	5	No
1.935	−0.774	27.6699	27.5	Yes

From these results, the optimum warranty period can be seen to be $t_w^* = 1.935$ years.

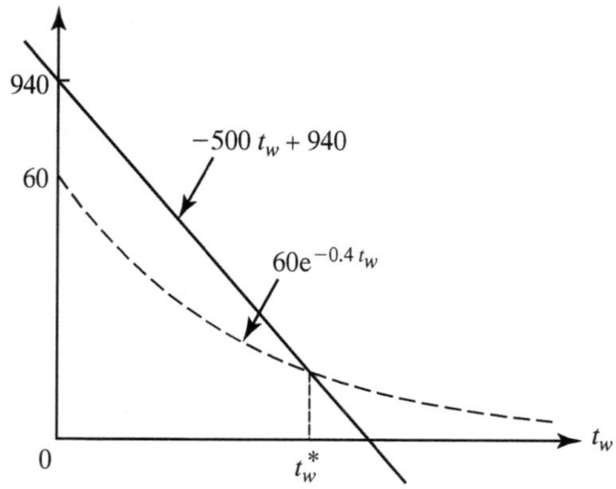

FIGURE 16.9 Graphs of left- and right-hand side expressions of Eq. (E.4).

Note: The value of the second derivative of the total warranty cost is given by

$$\frac{d^2 C_{\text{total}}}{dt_w^2} = 24\, e^{-0.4\, t_w^*} + 500 = 11.0679 + 500 = 511.0679 > 0$$

This shows that the optimum warranty period, $t_w^* = 1.935$ years, indeed corresponds to the minimum warranty cost to the manufacturer.

■

EXAMPLE 16.16

If the mean life of the television considered in Example 16.14 follows the Weibull distribution given by

$$F(t) = 1 - e^{-t^2}, \quad t \geq 0 \tag{E1}$$

determine the optimum warranty period to minimize the total warranty cost to the manufacturer. Assume the expected cost benefit function to be same as the one given by Eq. (E1) of Example 16.14.

Solution: For $\beta = 2$, the mean and standard deviation of the failure time (T) are given by (from Eqs. (16.63) and (16.64))

$$\mu = \frac{1}{\lambda}\Gamma(1.5), \quad \sigma = \frac{1}{\lambda}\left[\Gamma(2) - \Gamma^2(1.5)\right]^{\frac{1}{2}} \tag{E2}$$

From the formulas of gamma function, we find that $\Gamma(1.5) = 0.88623$ and $\Gamma(2) = 1.0$ and hence

$$\mu = \frac{0.88623}{\lambda}, \quad \sigma = \frac{0.46324}{\lambda} \tag{E3}$$

In order to have $\mu = 5$ years, λ must be equal to $\lambda = \dfrac{0.88623}{5} = 0.177246$ and hence the value of σ becomes $\dfrac{0.46324}{0.177246} = 2.61354$. Since the warranty time is $t = t_w = 1$ year, t is not large and hence Eq. (16.62) cannot be used. The value of the renewal function corresponding to $\beta = 2$, $\lambda = 0.177246$, and $t = t_w = 1$ is found from the table of renewal function [16.1, 16.17]. Because $\lambda \neq 1$, the value of $\lambda t_w = (0.177246)(1) = 0.177246$ is used for t to obtain the value of $M(t_w)$, using interpolation between the values of renewal function for $t = 0.15$ and $t = 0.20$, as 0.031588. Thus, the expected warranty cost per television is given by

$$\bar{c}_w = E[c_w(t_w)] = c_1 M(\lambda t_w) \approx (600)(0.031588) = \$18.9528 \tag{E4}$$

Using $c_1 = \$600$, Eq. (E4) for $\bar{c}_w$ and Eq. (E1) of Example 16.14 for B_w (with $k_1 = 1600$, $k_2 = -1000$, $k_3 = 250$), Eq. (16.68) yields

$$\frac{dC_{\text{total}}}{dt_w} \approx \frac{d}{dt_w}\{18.9528\} + \frac{d}{dt_w}(1600 - 1000\, t_w + 250\, t_w^2) = 0$$

or

$$-1000 + 500\, t_w{}^* \approx 0$$

or

$$t_w{}^* \approx 2.0 \text{ years}$$

Note: The value of the second derivative of the total warranty cost is given by

$$\frac{d^2 C_{\text{total}}}{dt_w^2} = 500 > 0$$

This shows that the optimum warranty period, $t_w{}^* \approx 2.0$ years, indeed corresponds to the minimum warranty cost to the manufacturer.

■

16.8 Two-dimensional Warranties

Consider the most common type of two-dimensional warranty such as the one offered for automobiles—for example, five years or 50,000 miles, whichever occurs first. This can be represented graphically by a rectangular region as shown in Fig. 16.10(a) where the horizontal axis denotes the time (T) in years and the vertical axis denotes the usage (U) in thousands of miles. A two-dimensional warranty can represent any of the following three characteristics:

1. Just warranty: It is a warranty that favors neither the manufacturer nor the consumer.
2. Consumer-friendly warranty: It is a warranty that favors the consumer.
3. Manufacturer-friendly warranty: It is a warranty that favors the manufacturer.

It is to be noted that since the warranties are specified by the manufacturer, most warranties tend to be manufacturer-friendly. However, if the competition among the manufacturers increases, manufacturers tend to make the warranties to be consumer-friendly to increase their market share. The diagonal of the rectangle denoting this warranty is the line joining the points $(0, 0)$ and (T, U). An advantage of this warranty is that it is simple to

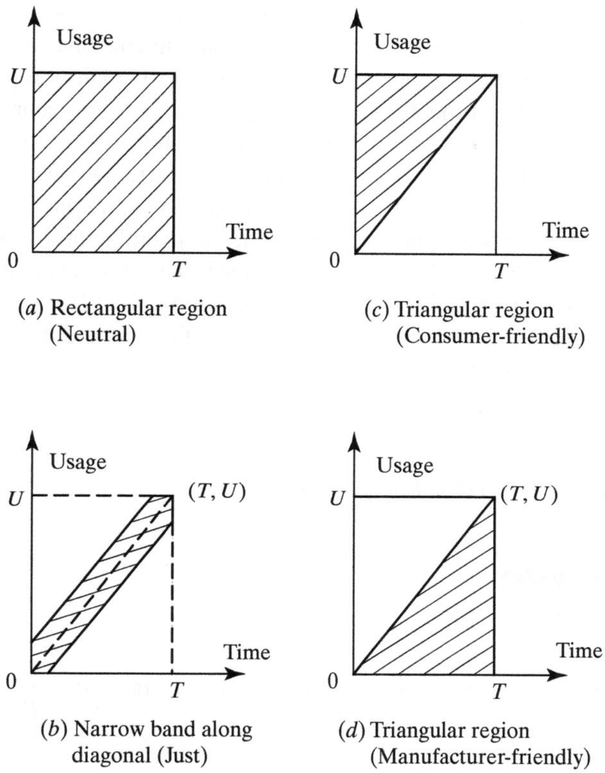

FIGURE 16.10 Warranty regions in two-dimensional warranties.

implement. The disadvantage is that it tends to encourage above-average use during the initial period of purchase, which poses a risk to the manufacturer. However, these warranties protect the consumer from product failures during the *infant mortality period*, which occurs during the initial period of use. Thus, a narrow band around the diagonal connecting the points $(0, 0)$ and (T, U), as shown in Fig. 16.10(*b*), appears to represent a just warranty. The warranty shown in Fig. 16.10(*b*) becomes more just as the band becomes narrower. The reason is that the diagonal represents what would be considered a normal usage of the product. In fact, the upper triangle of the rectangular warranty region, shown in Fig. 16.10(*c*), can be considered to be consumer-friendly while the lower triangular region, shown in Fig. 16.10(*d*), can be considered to be manufacturer-friendly. Thus, the whole region shown in Fig. 16.10(*a*) can be considered to represent neutral warranty since the usage rate is above the norm for 50 percent of the consumers and below the norm for nearly 50 percent of the consumers. Irrespective of the pros and cons, the rectangular warranty region represents the one corresponding to the most commonly used type of two-dimensional warranty.

The estimation of the cost of warranty in two-dimensional warranty can be made by using the models similar to those considered for one-dimensional warranty. For example, in the free replacement policy, the manufacturer agrees to repair or replace the failed product free of charge up to a time T or up to a usage U, whichever occurs first, from the time of the initial purchase. The warranty region is a rectangle $(0, T) \times (0, U)$ as shown in Fig. 16.10(a). If the usage is heavy, for example, the warranty can expire much before the time T. Similarly, if the usage is light, the warranty can expire well before the usage limit of U. If the product fails at time t and usage u, the manufacturer replaces or repairs the product only if $t < T$ and $u < U$. In this case, the replaced or repaired product will have a warranty for the time period $(T - t)$ and usage amount for $(U - u)$. The expected warranty cost to the manufacturer can be found using either a one-dimensional approach or a two-dimensional approach as outlined in Ref. [16.1].

16.9 Numerical Solutions Using MATLAB

EXAMPLE 16.17

Plot the variation of $\frac{c_w*}{c_1*}$ with $\frac{t_w}{\mu}$ given by Eq. (16.53) in the range $0 \leq \frac{t_w}{\mu} \leq 5$ using MATLAB. Assume $\mu = 20$ and $\theta + \phi = 0.1$.

Solution: The MATLAB program developed and the resulting plot of Eq. (16.53) are shown below:

```
clear all ; clc ;
mu = 20      ;
tha_phi = 0.1    ;
for i = 1:49
    x(i) = 0.01 + i * 0.01 ;
    r(i) = define_r(x(i),mu,tha_phi)  ;
end

x
r
for j = 1 : 45
    x(j+49) = 0.5 + 0.1 * j         ;
    r(j+49) = define_r(x(j+49),mu,tha_phi)  ;
end

x(49)
x(50)
plot(x,r,'-')
end

function r = define_r(x,mu,tha_phi)
tw = mu*x  ;
r = (1/(1+mu*log(1+tha_phi))) * ...
    (1 - (1/x)*(1/(1 + mu * log(1+tha_phi))) * (1- (exp(-x)/((1+tha_phi)
^(20*x)))))   ;
end
```

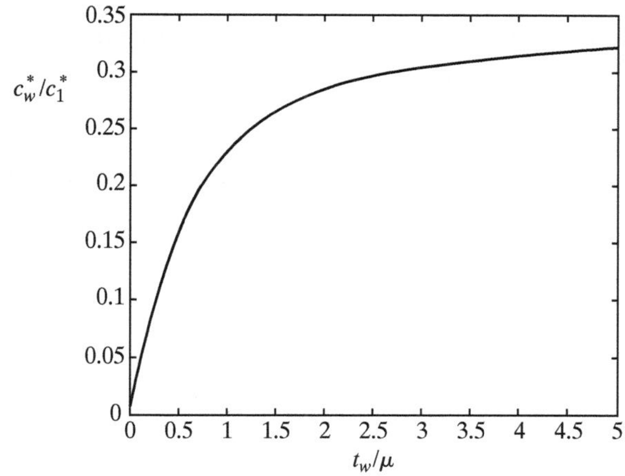

■

EXAMPLE 16.18

Using MATLAB, find the optimum warranty period of a television to minimize the total warranty cost to the manufacturer when the failure time of the television follows a two-stage Erlang distribution. Use the data given in Example 16.15.

Solution: The total warranty cost to the manufacturer can be determined as the sum of the expected warranty cost $\bar{c}_w(t_w)$ and the expected cost benefit of warranty, $B_w(t_w)$, given by Eq. (E3) of Example 16.15 and Eq. (E1) of Example 16.14, respectively:

$$C_{\text{total}} = 600(-0.25 + 0.1\, t_w + 0.25\, e^{-0.4\, t_w}) + (1600 - 1000\, t_w + 250\, t_w^2) \qquad \text{(E1)}$$

The minimum value of C_{total} is found using the unconstrained minimization program ***fminunc*** (from the Optimization Toolbox of MATLAB). The program written for the minimization of C_{total} and the results of optimization are given below:

```
clear all ; clc ;
x0 = 0.1
options=optimset('Display','iter') ;
[x,fval,flag] = fminunc(@obj_fun,x0,options)
end

function f = obj_fun(x)
f = 1600 - 1000 * x +250 * x^2 + 600 *(-0.25 + 0.1*x + 0.25*exp(-0.4
*x)) ;

end

x0 =

    0.1000

Warning: Gradient must be provided for trust-region method;
  using line-search method instead.
> In fminunc at 281
  In problem35 at 9
```

```
                                                           First-order
Iteration   Func-count        f(x)        Step-size        optimality
    0            2         1502.62                                948
    1            4         815.105      0.00105524                429
    2            6         636.354               1               4.82
    3            8         636.331               1             0.0195
    4           10         636.331               1                  0
Optimization terminated: relative infinity-norm of gradient less than options.TolFun.

x =

    1.9353

fval =

  636.3309

flag =

     1

>>
```

■

REFERENCES AND BIBLIOGRAPHY

16.1 W. R. Blischke and D. N. P. Murthy, *Warranty Cost Analysis*, Marcel Dekker, New York, 1994.

16.2 B. D. Bunday, *Statistical Methods in Reliability Theory and Practice,* Ellis Horwood, London, 1991.

16.3 M. U. Thomas, *Reliability and Warranties: Methods for Product Development and Quality Improvement,* CRC Taylor & Francis, Boca Raton, FL, 2006.

16.4 M. U. Thomas, "Optimum Warranty Policies for Nonrepairable Items," *IEEE Transactions on Reliability,* Vol. R-32, No. 3, 1983.

16.5 P. H. Ritchken, "Warranty Policies for Non-repairable Items under Risk Aversion," *IEEE Transactions on Reliability,* Vol. R-34, No. 2, 1985.

16.6 K. S. Park, "Optimal Use of Product Warranties," *IEEE Transactions on Reliability,* Vol. R-34, No. 5, 1985.

16.7 H. N. Amato and E. E. Anderson, "Determination of Warranty Reserves: An Extension," *Management Science,* Vol. 22, No. 12, 1976 pp. 1391–1394.

16.8 W. R. Blischke and D. N. P. Murthy, *Reliability: Modeling, Prediction, and Optimization,* Wiley, New York, 2000.

16.9 W. W. Menke, "Determination of Warranty Reserves," *Management Science,* Vol. 15, No. 10, 1969 pp. B542 – B549.

16.10 J. W. Mamer, "Discounted and Per Unit Costs of Product Warranty," *Management Science,* Vol. 33, No. 7, 1987 pp. 916–930.

16.11 N. D. Singpurwalla and S. Wilson, "The Warranty Problem: Its Statistical and Game Theoretic Aspects," *SIAM Review,* Vol. 35, No. 1, 1993 pp. 17–42.

16.12 M. U. Thomas and S. S. Rao, "Warranty Economic Decision Models: A Summary and Some Suggested Directions for Future Research," Operations Research, Vol. 47, No. 6, 1999 pp. 807–820.

16.13 H.-G. Kim and B. M. Rao, "Expected Warranty Cost of Two-attribute Free-replacement Warranties based on a Bivariate Exponential Distribution," *Computers & Industrial Engineering,* Vol. 38, 2000, pp. 425–434.

16.14 Singiresu S. Rao, *Engineering Optimization: Theory and Practice,* 4th ed., John Wiley, Hoboken, NJ, 2009.

16.15 I. Cochin and W. Cadwallender, *Analysis and Design of Dynamic Systems,* 3rd ed., Addison-Wesley, Reading, MA, 1997.

16.16 P. H. Ritchken and D. Fuh, "Optimal Replacement Policies for Irreparable Warrantied Items," *IEEE Transactions on Reliability,* Vol. R-35, No. 5, 1986.

16.17 L. A. Baxter, E. M. Scheuer, D. J. McConalogue, and W. R. Blischke, "On the Tabulation of the Renewal Function," *Technometrics,* Vol. 24, pp. 151–156, 1982.

16.18 M. R. Leadbetter, "On Series Expansion for the Renewal Moments," *Biometrica,* Vol. 50, pp. 75–80, 1963.

16.19 E. Brockmeyer, H. I. Hatstrom, and Arns Jensen, *The Life and Works of A. K. Erlang,* The Copenhagen Telephone Company, Copenhagen, 1948.

REVIEW QUESTIONS

16.1 Give brief answers to the following:

1. Give two reasons why warranty is needed for the manufacturer.
2. Give two reasons why warranty is needed for the customer.
3. What is a one-dimensional warranty?
4. State two assumptions made in estimating the warranty costs.
5. What is a renewal process?
6. Give two examples of products or units that are usually sold with a two-dimensional warranty.
7. What is a free replacement warranty?
8. What is a pro-rata warranty?
9. What is a combined FRW/PRW policy?
10. Although the warranty cost to the manufacturer is not reduced, what is the motivation to the manufacturer to offer a FRW policy equivalent to a FRW/PRW policy?

16.2 Indicate whether each of the following statements is true or false:

1. Warranty period is always specified in terms of calendar time.
2. For some products, warranty is required by law.
3. Warranties originated because of the legal context of product liability.
4. The combined FRW/PRW policy does not require any payment by the customer in the case of a failure of the product.
5. The lump-sum payment type of warranty is provided by the manufacturer primarily to attract more customers.
6. Warranty reserve funds set aside by the manufacturer initially is usually smaller than the expected amount to be paid.
7. Closed-form expressions are available for the renewal function corresponding to Weibull distribution.

8. The renewal function corresponding to a two-stage Erlang distribution is available in closed form.
9. The pro-rata warranty is better to the customer compared to a free replacement warranty.
10. The free replacement warranty is applicable to only one-dimensional warranties.

16.3 Fill in the blank with proper word(s):

1. A warranty can be considered as a ______ between the manufacturer and the customer of the product.
2. Usually, the automobile batteries are sold with a ______ dimensional warranty.
3. For simplicity, warranty costs can be estimated by assuming only ______ failure of the product during the warranty period.
4. The cost of warranty is usually included in the ______ price of the product.
5. For a product having a mean life of five years, with the failure time distribution given by exponential distribution, the failure rate is ______.
6. The customer is expected to pay a fraction of the replacement cost in the case of ______ warranty policy.
7. The computation of expected warranty costs considering the possibility of multiple failures during the warranty period requires the use of ______ function.
8. The warranty reserve fund is set up by incorporating the interest rates (time value of money) and expected future changes in the ______ of the product.
9. A renewal process is a ______ process which counts the number of events that occur in a period of time.
10. The determination of the optimal warranty period requires the warranty costs, administration costs, and the ______ of the warranty program.

16.4 Select the most appropriate answer from the multiple choices given:

1. A renewal process is a random process involving:
   a. Rare occurrences
   b. Extreme values
   c. Repeated occurrences

2. The free replacement warranty policy is usually provided for:
   a. Multicomponent systems
   b. Nonrepairable products
   c. Repairable products

3. The estimation of the warranty costs of a product requires:
   a. Failure time distribution function
   b. Failure rate
   c. Mean time to failure

4. The reason for a manufacturer to offer a free replacement policy equivalent to a FRW/PRW policy is:
   a. To increase the sales of the product
   b. To reduce the warranty cost
   c. To increase the reliability of the product

5. The lump-sum payment type of warranty is offered by the manufacturer to:
   a. Attract more customers
   b. Reduce administrative costs of warranty
   c. To be able to sell products of inferior quality

6. The renewal function, $M(t)$, corresponding to exponential failure time distribution is:

   a. λt b. $-\frac{1}{4} + \frac{\lambda t}{2} + \frac{1}{4}e^{-2\lambda t}$ c. λ

16.5 Match the following:

Expression	Denotes expected cost of warranty in the case of
1. $c_1\,(1 - e^{-\lambda t_w})$	a. Weibull distribution with one failure
2. $c_1\,(1 - \{1 + \lambda t\}\, e^{-\lambda t_w})$	b. Erlang distribution with multiple failures
3. $c_1\,\lambda t_w$	c. Weibull distribution with multiple failures
4. $c_1\left(-\frac{1}{4} + \frac{\lambda t_w}{2} + \frac{1}{4}e^{-2\lambda t_w}\right)$	d. Exponential distribution with multiple failures
5. $c_1(1 - e^{-(\lambda t_w)^{\beta}})$	e. Exponential distribution with one failure
6. $c_1\left(\frac{t_w}{\mu} + \frac{\sigma^2}{2\,\mu^2} - \frac{1}{2}\right)$	f. Erlang distribution with one failure

PROBLEMS

16.1 A manufacturer sells lawn mowers with a six-month free replacement warranty. The failure time of the lawn mowers is known to follow exponential distribution with a failure rate of 0.02 per month. The manufacturer's cost of production of a lawn mower is \$200. If the lawnmowers are produced in a lot of 10,000 units, determine the total amount of warranty reserve fund to be kept aside by the manufacturer.

16.2 A manufacturer offers a six-month pro-rata warranty on its lawn mowers. The failure time of the lawn mowers is known to follow exponential distribution with a failure rate of 0.02 per month. The manufacturer's cost of production of a lawn mower is \$200, including the warranty cost. If the lawn mowers are produced in a lot of 10,000 units, determine the following:

a. Cost of the lawn mower to the manufacturer without the warranty cost.
b. Amount of warranty reserve fund to be kept aside by the firm for the whole batch of lawn mowers.

16.3 A manufacturing firm offers a one-month free replacement warranty followed by a five-month pro-rata warranty on its lawn mowers. The failure time of the lawn mowers follows exponential distribution with a failure rate of $\lambda = 0.02$ per month. The manufacturer's cost of production of a lawn mower, including the warrant cost, is \$200. If the lawn mowers are produced in a batch of 10,000 units, determine the amount of warranty reserve fund to be kept aside by the manufacturer for the whole batch of lawn mowers.

16.4 In Problem 16.3, find the length of the free replacement warranty policy that is equivalent to the combined free replacement and pro-rata warranty. The relevant data are $\lambda = 0.02$ failures/month, $t_{w1} = 1$ month and $t_{w2} = 6$ months.

16.5 In Problem 16.2, the manufacturer considers to offer a lump-sum warranty plan equivalent to the pro-rata plan by offering a lump-sum amount (a percentage of the initial price) to its customers whose lawn mowers fail during the warranty period. Find the percentage of the price to be refunded to customers whose lawn mowers fail before the expiration of the warranty period.

16.6 A manufacturing firm offers a six-month free replacement warranty on its lawn mowers. The failure time of the lawn mowers follows exponential distribution with a failure rate of $\lambda = 0.02$ failures per month. The firm's cost of production of one lawn mower is \$200. By considering the time value of money with a discounting factor of $\delta = 0.05$, determine the expected amount of warranty payment to be made by the manufacturer per unit.

16.7 A manufacturing firm offers a one-year pro-rata warranty on its lawn mowers. The failure time of the lawn mowers follows exponential distribution with a failure rate of $\lambda = 0.02$ failures per month. The firm's cost of production of a lawn mower is \$200. By considering the time value of money with a discounting factor of $\delta = 0.05$, determine the expected amount of warranty payment to be made by the manufacturer per unit.

16.8 The failure time of an electrical appliance follows exponential distribution with a failure rate of 0.02 per month. The manufacturer sells the appliance with a free replacement warranty for a period of one year. If the unit cost of the appliance to the manufacturer is \$100, find the following:

a. Expected unit warranty cost to the manufacturer.
b. Variance of the unit warranty cost to the manufacturer[@].

16.9 The failure time of an electrical appliance follows the following probability density function:

$$f(t) = k_1\,\lambda_1\,e^{-\lambda_1 t} + k_2\,\lambda_2\,e^{-\lambda_2 t} \tag{E1}$$

where $k_1 = 0.6$, $k_2 = 0.4$, $\lambda_1 = 0.01$ failure per month, and $\lambda_2 = 0.02$ failure per month. Equation (E1) is a combined exponential distribution, which indicates that the whole group of appliances is composed of those produced in two different batches; 60 percent from the first batch with the failure rate λ_1 and 40 percent from the second batch with the failure rate λ_2. If the cost of manufacturing one electrical appliance is \$150, find the unit warranty cost corresponding to a warranty period of 12 months.

16.10 The failure time distribution corresponding to Gamma distribution is given by

$$f(x) = \begin{cases} \dfrac{1}{\Gamma(\beta)}\lambda(\lambda x)^{\beta-1}\,e^{-\lambda x}; & x > 0 \\ 0; & x \le 0 \end{cases}$$

The renewal function corresponding to Gamma distribution, for the parameters $\beta = 2$ and $\lambda = 0.5$, is given by

$$M(t) = 0.25\,(t - 1 + e^{-t})$$

[@] See Ref. [16.5] for the expression of variance of the unit warranty cost.

In a television manufacturing company, the manufacturer's cost of a television is $c_1 = \$600$. The manufacturer sells the television with a one-year free replacement warranty. The failure time of the television follows Gamma distribution with $\beta = 2$ and $\lambda = 0.5$, If the average life of the television is 4 years, find the expected warranty cost to the manufacturer.

16.11 The manufacturer's cost of an electrical appliance is $c_1 = \$200$. The manufacturer sells the appliance with a one-year free replacement warranty. If the failure time of the appliance follows exponential distribution with an average life of 4 years, determine the expected warranty cost to the manufacturer by considering the possibility of multiple failures during the warranty period.

16.12 The manufacturer's cost of an electrical appliance is $c_1 = \$200$. The manufacturer sells the appliance with a one-year free replacement warranty. The failure time of the appliance follows a two-stage Erlang distribution:

$$F(t) = 1 - (1 + \lambda t)e^{-\lambda t}$$

If the average life of the appliance is 4 years, determine the expected warranty cost to the manufacturer by considering the possibility of multiple failures during the warranty period.

16.13 The manufacturer's cost of an electrical appliance is $c_1 = \$200$. The manufacturer sells the appliance with a one-year free replacement warranty. The failure time of the appliance follows Weibull distribution with

$$F(t) = 1 - e^{-(\lambda t)^{\beta}}; \quad t \geq 0$$

If the average life of the television is 4 years, determine the expected warranty cost to the manufacturer for $\beta = 2$ by considering the possibility of multiple failures during the warranty period.

16.14 The cost of an electrical appliance to the manufacturer is \$200. The failure rate of the appliance is known to be 0.25 per year. The cost of warranty to the manufacturer for the failure of each unit during the useful life of the appliance is estimated to be \$200. The cost to the manufacturer to market the appliance without warranty is \$1000 and this cost is expected to decrease according to the relation

$$B_w(t_w) = \$\left(1000 - 500\, t_w + 100\, t_w^2\right)$$

Assuming that the failure time of the appliance follows exponential distribution, find the optimal warranty period for the appliance.

16.15 If the failure time of the electrical appliance considered in Problem 16.13 follows a two stage Erlang distribution with

$$F(t) = 1 - (1 + \lambda t)\, e^{-\lambda t}$$

Using $\lambda = 0.25$ failures per year, determine the optimum warranty period to minimize the total warranty cost to the manufacturer. Assume the expected cost-benefit function to be the same as the one given in Problem 16.14.

16.16 If the mean life of the electrical appliance considered in Problem 16.13 follows the Weibull distribution

$$F(t) = 1 - e^{-t^2}, t \geq 0$$

determine the optimum warranty period to minimize the total warranty cost to the manufacturer. Assume the expected cost-benefit function to be the same as the one given in Problem 16.14.

16.17 The hazard rate of an electronic device is given by $\lambda(t) = 0.0005\, t$ per hour of operation. The manufacturer's cost of the device, not including the warranty cost, is \$100. Because of the competition from other manufacturers, the manufacturer desires to limit the selling price of the device to \$120 after including the warranty cost. If the device is produced in a lot of 5000 units, determine the following:

a. Warranty period for the device.
b. Warranty reserve fund to be kept aside by the manufacturer for the whole lot.

16.18 The failure rate of a product is given by

$$\lambda(t) = \alpha\, t^{\beta}$$

where α and β are constants. If the manufacturer's cost of the product is c_1 per unit, determine the warranty cost if the product is sold with a free replacement warranty for a period of t_w.

16.19 An electrical appliance, having a failure rate of 0.25 per year, is sold with a one-year free replacement warranty. The average warranty cost to the manufacturer is \$100 per claim. If the manufacturer decides to reduce the warranty costs by 25 percent by improving the quality of the appliance (thereby reducing the failure rate), determine the new failure rate to be achieved.

16.20 A particular model of an automobile tire is sold by a tire manufacturer with a three-month free replacement warranty followed by a nine-month pro-rata warranty. If the tires have a mean life of 36 months and the manufacturer's cost is \$60 per tire, determine the unit warranty cost to the manufacturer.

16.21 Plot the variations of a and $\dfrac{\overline{C}_w}{N\, c_1}$ with respect to $\dfrac{t_w}{\mu}$ in Eqs. (16.35) and (16.36).

16.22 Plot the graph between $\dfrac{c_w{}^*}{c_1{}^*}$ and $\dfrac{t_w}{\mu}$ in Eq. (16.53) for the specific values of $\theta + \phi = 0.1$ and $\mu = 20$ time periods.

MATLAB Problems

16.23 Use MATLAB to find the expected warranty cost per unit of the product under the combined free replacement warranty and pro-rata warranty, given by Eq. (16.21), when the failure rate (λ) changes from 0.001 to 0.010 failures per month. Other data: $c_1 = 250$, $t_{w1} = 3$ months, and $t_{w2} = 12$ months.

16.24 Use MATLAB to find the expected warranty cost per unit of the product under the combined free replacement warranty and pro-rata warranty, given by Eq. (16.21), when the free replacement and pro-rata warranty periods vary as $(t_{w1}, t_{w2}) = (1, 11)$ months, (2, 10) months, . . . , (11, 1) months. Other data: $\lambda = 0.005$ failures/month and $c_1 = 250$.

16.25 Use MATLAB to find the FRW policy period equivalent to a FRW/PRW policy period, given by Eq. (16.30), when the failure rate (λ) varies from 0.001 to 0.010 failures per month. Other data: $t_{w1} = 3$ months and $t_{w2} = 12$ months.

APPENDIX A

Standard Normal Distribution Function

Karl Pearson, born on 27 March, 1857, in London, England, is credited for establishing the discipline of mathematical statistics. He studied mathematics at the University of Cambridge during 1876–1879 and went to Germany to study physics at the University of Heidelberg. Later, he went to Berlin to study Roman law and German literature. He joined the University College London in 1883 as a Professor and was appointed as the Goldsmid Chair of Applied Mathematics and Mechanics in 1884. In 1911, Pearson founded the world's first university statistics department at University College, London. He received several awards in a wide range of disciplines, including the Darwin medal in 1989, the honorary degree LLD from the University of St. Andrews in 1911, and D.Sc. from University of London in 1911. He was elected as Fellow of the Royal Society in 1896. He founded the journal *Biometrika*, dedicated to the development of statistical theory, in 1901 and the journal *Annals of Eugenics* (now *Annals of Human Genetics*) in 1925. His contributions to statistics include the concept of correlation coefficient, method of moments, Pearson's system of continuous curves, which forms the basis for the continuous probability distributions, Chi distance, P-value, foundations of statistical hypothesis testing theory and statistical decision theory, and Pearson's chi-square test. He died on 27 April, 1936, in Surrey, England.

Values of $\Phi(z)$ where $\Phi(z) = \frac{1}{\sqrt{2\pi}} \int_{-\infty}^{z} e^{-x^2/2}\,dx$

z	.00	.01	.02	.03	.04	.05	.06	.07	.08	.09
.0	.50000	.50399	.50798	.51197	.51595	.51994	.52392	.52790	.53188	.53586
.1	.53983	.54380	.54776	.55172	.55567	.55962	.56356	.56749	.57142	.57534
.2	.57926	.58317	.58706	.59095	.59483	.59871	.60257	.60642	.61026	.61409
.3	.61791	.62172	.62552	.62930	.63307	.63683	.64058	.64431	.64803	.65173
.4	.65542	.65910	.66276	.66640	.67003	.67364	.67724	.68082	.68439	.68793
.5	.69146	.69497	.69847	.70194	.70540	.70884	.71226	.71566	.71904	.72240
.6	.72575	.72907	.73237	.73565	.73891	.74215	.74537	.74857	.75175	.75490
.7	.75804	.76115	.76424	.76730	.77035	.77337	.77637	.77935	.78230	.78524
.8	.78814	.79103	.79389	.79673	.79955	.80234	.80510	.80785	.81057	.81327
.9	.81594	.81859	.82121	.82381	.82639	.82894	.83147	.83398	.83646	.83891

(*Continued*)

Values of $\Phi(z)$ where $\Phi(z) = \frac{1}{\sqrt{2\pi}} \int_{-\infty}^{z} e^{-x^2/2}\, dx$ *(Continued)*

z	.00	.01	.02	.03	.04	.05	.06	.07	.08	.09
1.0	.84134	.84375	.84614	.84849	.85083	.85314	.85543	.85769	.85992	.86214
1.1	.86433	.86650	.86864	.87076	.87286	.87493	.87698	.87900	.88100	.88298
1.2	.88493	.88686	.88877	.89065	.89251	.89435	.89616	.89796	.89973	.90147
1.3	.90320	.90490	.90658	.90824	.90988	.91149	.91309	.91466	.91621	.91774
1.4	.91924	.92073	.92220	.92364	.92507	.92647	.92785	.92922	.93056	.93189
1.5	.93319	.93448	.93574	.93699	.93822	.93943	.94062	.94179	.94295	.94408
1.6	.94520	.94630	.94738	.94845	.94950	.95053	.95154	.95254	.95352	.95449
1.7	.95543	.95637	.95728	.95818	.95907	.95994	.96080	.96164	.96246	.96327
1.8	.96407	.96485	.96562	.96638	.96712	.96784	.96856	.96926	.96995	.97062
1.9	.97128	.97193	.97257	.97320	.97381	.97441	.97500	.97558	.97615	.97670
2.0	.97725	.97778	.97831	.97882	.97932	.97982	.98030	.98077	.98124	.98169
2.1	.98214	.98257	.98300	.98341	.98382	.98422	.98461	.98500	.98537	.98574
2.2	.98610	.98645	.98679	.98713	.98745	.98778	.98809	.98840	.98870	.98899
2.3	.98928	.98956	.98983	$.9^{2}0097$	$.9^{2}0358$	$.9^{2}0613$	$.9^{2}0863$	$.9^{2}1106$	$.9^{2}1344$	$.9^{2}1576$
2.4	$.9^{2}1802$	$.9^{2}2024$	$.9^{2}2240$	$.9^{2}2451$	$.9^{2}2656$	$.9^{2}2857$	$.9^{2}3053$	$.9^{2}3244$	$.9^{2}3431$	$.9^{2}3613$
2.5	$.9^{2}3790$	$.9^{2}3963$	$.9^{2}4132$	$.9^{2}4297$	$.9^{2}4457$	$.9^{2}4614$	$.9^{2}4766$	$.9^{2}4915$	$.9^{2}5060$	$.9^{2}5201$
2.6	$.9^{2}5339$	$.9^{2}5473$	$.9^{2}5604$	$.9^{2}5731$	$.9^{2}5855$	$.9^{2}5975$	$.9^{2}6093$	$.9^{2}6207$	$.9^{2}6319$	$.9^{2}6427$
2.7	$.9^{2}6533$	$.9^{2}6636$	$.9^{2}6736$	$.9^{2}6833$	$.9^{2}6928$	$.9^{2}7020$	$.9^{2}7110$	$.9^{2}7197$	$.9^{2}7282$	$.9^{2}7365$
2.8	$.9^{2}7445$	$.9^{2}7523$	$.9^{2}7599$	$.9^{2}7673$	$.9^{2}7744$	$.9^{2}7814$	$.9^{2}7882$	$.9^{2}7948$	$.9^{2}8012$	$.9^{2}8074$
2.9	$.9^{2}8134$	$.9^{2}8193$	$.9^{2}8250$	$.9^{2}8305$	$.9^{2}8359$	$.9^{2}8411$	$.9^{2}8462$	$.9^{2}8511$	$.9^{2}8559$	$.9^{2}8605$
3.0	$.9^{2}8650$	$.9^{2}8694$	$.9^{2}8736$	$.9^{2}8777$	$.9^{2}8817$	$.9^{2}8856$	$.9^{2}8893$	$.9^{2}8930$	$.9^{2}8965$	$.9^{2}8999$
3.1	$.9^{3}0324$	$.9^{3}0646$	$.9^{3}0957$	$.9^{3}1260$	$.9^{3}1553$	$.9^{3}1836$	$.9^{3}2112$	$.9^{3}2378$	$.9^{3}2636$	$.9^{3}2886$
3.2	$.9^{3}3129$	$.9^{3}3363$	$.9^{3}3590$	$.9^{3}3810$	$.9^{3}4024$	$.9^{3}4230$	$.9^{3}4429$	$.6^{3}4623$	$.9^{3}4810$	$.9^{3}4991$
3.3	$.9^{3}5166$	$.9^{3}5335$	$.9^{3}5499$	$.9^{3}5658$	$.9^{3}5811$	$.9^{3}5959$	$.9^{3}6103$	$.9^{3}6242$	$.9^{3}6376$	$.9^{3}6505$
3.4	$.9^{3}6631$	$.9^{3}6752$	$.9^{3}6869$	$.9^{3}6982$	$.9^{3}7091$	$.9^{3}7197$	$.9^{3}7299$	$.9^{3}7398$	$.9^{3}7493$	$.9^{3}7585$
3.5	$.9^{3}7674$	$.9^{3}7759$	$.9^{3}7842$	$.9^{3}7922$	$.9^{3}7999$	$.9^{3}8074$	$.9^{3}8146$	$.9^{3}8215$	$.9^{3}8282$	$.9^{3}8347$
3.6	$.9^{3}8409$	$.9^{3}8469$	$.9^{3}8527$	$.9^{3}8583$	$.9^{3}8637$	$.9^{3}8689$	$.9^{3}8739$	$.9^{3}8787$	$.9^{3}8834$	$.9^{3}8879$
3.7	$.9^{3}8922$	$.9^{3}8964$	$.9^{4}0039$	$.9^{4}0426$	$.9^{4}0799$	$.9^{4}1158$	$.9^{4}1504$	$.9^{4}1838$	$.9^{4}2159$	$.9^{4}2568$
3.8	$.9^{4}2765$	$.9^{4}3052$	$.9^{4}3327$	$.9^{4}3593$	$.9^{4}3848$	$.9^{4}4094$	$.9^{4}4331$	$.9^{4}4558$	$.9^{4}4777$	$.9^{4}4988$
3.9	$.9^{4}5190$	$.9^{4}5385$	$.9^{4}5573$	$.9^{4}5753$	$.9^{4}5926$	$.9^{4}6092$	$.9^{4}6253$	$.9^{4}6406$	$.9^{4}6554$	$.9^{4}6696$

(Continued)

Values of $\Phi(z)$ where $\Phi(z) = \dfrac{1}{\sqrt{2\pi}} \displaystyle\int_{-\infty}^{z} e^{-x^2/2}\, dx$ *(Continued)*

z	.00	.01	.02	.03	.04	.05	.06	.07	.08	.09
4.0	$.9^{4}6833$	$.9^{4}6964$	$.9^{4}7090$	$.9^{4}7211$	$.9^{4}7327$	$.9^{4}7439$	$.9^{4}7546$	$.9^{4}7649$	$.9^{4}7748$	$.9^{4}7843$
4.1	$.9^{4}7934$	$.9^{4}8022$	$.9^{4}8106$	$.9^{4}8186$	$.9^{4}8263$	$.9^{4}8338$	$.9^{4}8409$	$.9^{4}8477$	$.9^{4}8542$	$.9^{4}8605$
4.2	$.9^{4}8665$	$.9^{4}8723$	$.9^{4}8778$	$.9^{4}8832$	$.9^{4}8882$	$.9^{4}8931$	$.9^{4}8978$	$.9^{5}0226$	$.9^{5}0655$	$.9^{5}1066$
4.3	$.9^{5}1460$	$.9^{5}1837$	$.9^{5}2199$	$.9^{5}2545$	$.9^{5}2876$	$.9^{5}3193$	$.9^{5}3497$	$.9^{5}3788$	$.9^{5}4066$	$.9^{5}4332$
4.4	$.9^{5}4587$	$.9^{5}4831$	$.9^{5}5065$	$.9^{5}5288$	$.9^{5}5502$	$.9^{5}5706$	$.9^{5}5902$	$.9^{5}6089$	$.9^{5}6268$	$.9^{5}6439$
4.5	$.9^{5}6602$	$.9^{5}6759$	$.9^{5}6908$	$.9^{5}7051$	$.9^{5}7187$	$.9^{5}7318$	$.9^{5}7442$	$.9^{5}7561$	$.9^{5}7675$	$.9^{5}7784$
4.6	$.9^{5}7888$	$.9^{5}7987$	$.9^{5}8081$	$.9^{5}8172$	$.9^{5}8258$	$.9^{5}8340$	$.9^{5}8419$	$.9^{5}8494$	$.9^{5}8566$	$.9^{5}8634$
4.7	$.9^{5}8699$	$.9^{5}8761$	$.9^{5}8821$	$.9^{5}8877$	$.9^{5}8931$	$.9^{5}8983$	$.9^{6}0320$	$.9^{6}0789$	$.9^{6}1235$	$.9^{6}1661$
4.8	$.9^{6}2067$	$.9^{6}2453$	$.9^{6}2822$	$.9^{6}3173$	$.9^{6}3508$	$.9^{6}3827$	$.9^{6}4131$	$.9^{6}4420$	$.9^{6}4696$	$.9^{6}4958$
4.9	$.9^{6}5208$	$.9^{6}5446$	$.9^{6}5673$	$.9^{6}5889$	$.9^{6}6094$	$.9^{6}6289$	$.9^{6}6475$	$.9^{6}6652$	$.9^{6}6821$	$.9^{6}6981$

$\Phi(z)$	z
0.60	0.253
0.70	0.524
0.80	0.842
0.90	1.282
0.95	1.645
0.99	2.326
0.999	3.090
0.9999	3.719
0.99999	4.27
0.999999	4.75
0.9999999	5.20
0.99999999	5.61
0.999999999	6.00
0.9999999999	6.36

Note: $\Phi(-z) = 1 - \Phi(z)$; $\Phi(3) = 0.9^{2}8650 = 0.998650$, etc.

APPENDIX B

Values of $t_{\alpha, n}$ for Specific Values of α and n of t-Distribution

Probability density function: $f_T(t) = \dfrac{\Gamma\left(\dfrac{n+1}{2}\right)}{\sqrt{n\pi}\,\Gamma\left(\dfrac{n}{2}\right)}\left(1 + \dfrac{t^2}{n}\right)^{-\frac{n+1}{2}}$

$P(t > t_{\alpha, n}) = \alpha; \quad t_{1-\alpha, n} = -t_{\alpha, n}$

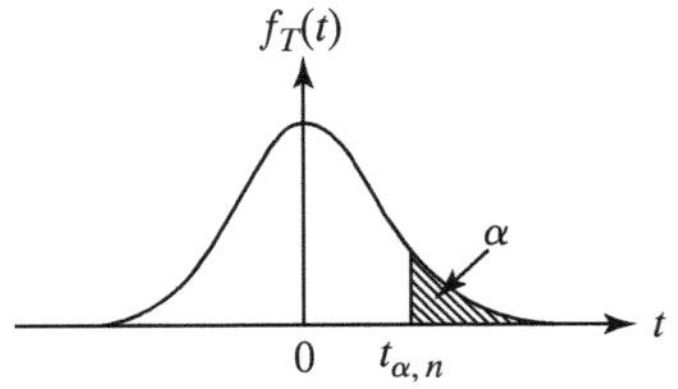

n \ α	0.200	0.150	0.100	0.050	0.025	0.010	0.005	0.0025	0.0010	0.0005
1	1.376	1.963	3.078	6.314	12.706	31.820	63.657	127.321	318.309	636.619
2	1.061	1.386	1.886	2.920	4.303	6.964	9.925	14.089	22.327	31.599
3	0.978	1.250	1.638	2.353	3.182	4.541	5.841	7.453	10.214	12.924
4	0.941	1.189	1.533	2.132	2.776	3.747	4.604	5.597	7.173	8.610
5	0.919	1.156	1.476	2.015	2.570	3.365	4.032	4.773	5.893	6.869
6	0.906	1.134	1.440	1.943	2.447	3.143	3.707	4.317	5.208	5.959
7	0.896	1.119	1.415	1.894	2.365	2.998	3.499	4.029	4.785	5.408
8	0.889	1.108	1.397	1.859	2.306	2.896	3.355	3.832	4.501	5.041
9	0.883	1.100	1.383	1.833	2.262	2.821	3.250	3.690	4.297	4.781
10	0.879	1.093	1.372	1.812	2.228	2.764	3.169	3.581	4.144	4.587
11	0.875	1.088	1.363	1.796	2.201	2.718	3.106	3.497	4.025	4.437
12	0.873	1.083	1.356	1.782	2.179	2.681	3.054	3.428	3.930	4.318
13	0.870	1.079	1.350	1.771	2.160	2.650	3.012	3.372	3.852	4.221
14	0.868	1.076	1.345	1.761	2.145	2.624	2.977	3.326	3.787	4.140

(Continued)

(Continued)

n \ α	**0.200**	**0.150**	**0.100**	**0.050**	**0.025**	**0.010**	**0.005**	**0.0025**	**0.0010**	**0.0005**
15	0.866	1.073	1.341	1.753	2.131	2.602	2.947	3.286	3.733	4.073
16	0.865	1.071	1.337	1.746	2.120	2.583	2.921	3.252	3.686	4.015
17	0.863	1.069	1.333	1.740	2.110	2.567	2.898	3.222	3.646	3.965
18	0.862	1.067	1.330	1.734	2.101	2.552	2.878	3.196	3.610	3.922
19	0.861	1.065	1.328	1.729	2.093	2.539	2.861	3.174	3.579	3.883
20	0.860	1.064	1.325	1.725	2.086	2.528	2.845	3.153	3.552	3.849
21	0.859	1.063	1.323	1.721	2.080	2.518	2.831	3.135	3.527	3.819
22	0.858	1.061	1.321	1.717	2.074	2.508	2.819	3.119	3.505	3.792
23	0.857	1.060	1.319	1.714	2.069	2.500	2.807	3.104	3.485	3.768
24	0.857	1.059	1.318	1.711	2.064	2.492	2.797	3.090	3.467	3.745
25	0.856	1.058	1.316	1.708	2.059	2.485	2.787	3.078	3.450	3.725
26	0.856	1.057	1.315	1.706	2.055	2.479	2.779	3.067	3.435	3.707
27	0.855	1.057	1.314	1.703	2.052	2.473	2.771	3.056	3.421	3.689
28	0.855	1.056	1.312	1.701	2.048	2.467	2.763	3.047	3.408	3.674
29	0.854	1.055	1.311	1.699	2.045	2.462	2.756	3.038	3.396	3.659
30	0.854	1.055	1.310	1.697	2.042	2.457	2.750	3.030	3.385	3.646
40	0.851	1.050	1.303	1.684	2.021	2.423	2.704	2.971	3.307	3.551
50	0.849	1.047	1.299	1.676	2.008	2.403	2.678	2.937	3.261	3.496
60	0.848	1.045	1.296	1.671	2.000	2.390	2.660	2.914	3.232	3.460
70	0.847	1.044	1.294	1.667	1.994	2.381	2.648	2.899	3.211	3.435
80	0.846	1.043	1.292	1.664	1.990	2.374	2.639	2.887	3.195	3.416
90	0.846	1.042	1.291	1.662	1.987	2.368	2.631	2.878	3.183	3.402
100	0.845	1.042	1.290	1.660	1.984	2.364	2.626	2.871	3.174	3.390
1000	0.842	1.037	1.282	1.646	1.962	2.330	2.581	2.813	3.098	3.300
∞	0.842	1.036	1.281	1.645	1.960	2.326	2.576	2.807	3.090	3.290

APPENDIX C

Values of $\chi^2_{n,\alpha}$ Corresponding to Specific Values of α and n of χ^2-Distribution

$$\alpha = P[X \geq \chi^2] = 1 - F_X(\chi^2) = \int_{\chi^2}^{\infty} \frac{1}{2^{\frac{n}{2}}\,\Gamma\left(\frac{n}{2}\right)} x^{\frac{n-2}{2}} e^{-x/2}\,dx$$

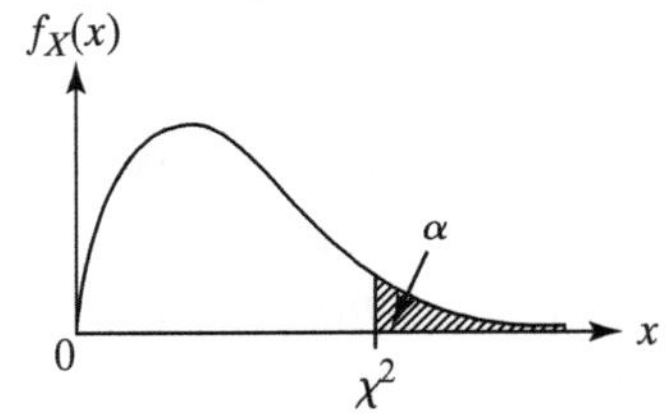

n \ α	**0.999**	**0.995**	**0.990**	**0.975**	**0.950**	**0.900**
1	0.0^516	0.0^439	0.0^316	0.0^398	0.0039	0.0158
2	0.0020	0.0100	0.0201	0.0506	0.103	0.211
3	0.0243	0.0717	0.115	0.216	0.352	0.584
4	0.0908	0.207	0.297	0.484	0.711	1.064
5	0.210	0.412	0.554	0.831	1.145	1.610
6	0.381	0.676	0.872	1.237	1.635	2.204
7	0.598	0.989	1.239	1.690	2.167	2.833
8	0.857	1.344	1.646	2.180	2.733	3.489
9	1.152	1.735	2.088	2.700	3.325	4.168
10	1.479	2.156	2.558	3.247	3.940	4.865
11	1.834	2.603	3.053	3.816	4.575	5.578
12	2.214	3.074	3.571	4.404	5.226	6.304
13	2.617	3.565	4.107	5.009	5.892	7.041
14	3.041	4.075	4.660	5.629	6.571	7.789

(Continued)

(*Continued*)

n \ α	**0.999**	**0.995**	**0.990**	**0.975**	**0.950**	**0.900**
15	3.483	4.601	5.229	6.262	7.261	8.547
16	3.942	5.142	5.812	6.908	7.962	9.312
17	4.416	5.697	6.408	7.564	8.672	10.085
18	4.905	6.265	7.015	8.231	9.390	10.865
19	5.407	6.844	7.633	8.906	10.117	11.651
20	5.921	7.434	8.260	9.591	10.851	12.443
25	8.649	10.520	11.524	13.120	14.611	16.473
30	11.588	13.787	14.953	16.791	18.493	20.599
35	14.688	17.192	18.509	20.569	22.465	24.797
40	17.916	20.706	22.164	24.433	26.509	29.050
45	21.251	24.311	25.901	28.366	30.612	33.350
50	24.674	27.991	29.707	32.357	34.764	37.689
60	31.738	35.534	37.485	40.482	43.188	46.459
70	39.036	43.275	45.442	48.758	51.739	55.329
80	46.520	51.172	53.540	57.153	60.391	64.278
90	54.155	59.196	61.754	65.647	69.126	73.291
100	61.918	67.328	70.065	74.222	77.929	82.358

n \ α	**0.250**	**0.100**	**0.050**	**0.025**	**0.010**	**0.001**
1	1.323	2.705	3.841	5.024	6.635	10.828
2	2.773	4.605	5.991	7.378	9.210	13.815
3	4.108	6.251	7.815	9.348	11.345	16.266
4	5.385	7.779	9.488	11.143	13.277	18.467
5	6.626	9.236	11.070	12.832	15.086	20.515
6	7.841	10.645	12.592	14.449	16.812	22.458
7	9.037	12.017	14.067	16.013	18.475	24.322
8	10.219	13.362	15.507	17.534	20.090	26.124
9	11.389	14.684	16.919	19.023	21.666	27.877

(*Continued*)

(Continued)

n \ α	**0.250**	**0.100**	**0.050**	**0.025**	**0.010**	**0.001**
10	12.549	15.987	18.307	20.483	23.209	29.588
11	13.701	17.275	19.675	21.920	24.725	31.264
12	14.845	18.549	21.026	23.337	26.217	32.909
13	15.984	19.812	22.362	24.736	27.688	34.528
14	17.117	21.064	23.685	26.119	29.141	36.123
15	18.245	22.307	24.996	27.488	30.578	37.697
16	19.369	23.542	26.296	28.845	32.000	39.252
17	20.489	24.769	27.587	30.191	33.409	40.790
18	21.605	25.989	28.869	31.526	34.805	42.312
19	22.718	27.204	30.143	32.852	36.191	43.820
20	23.828	28.412	31.410	34.170	37.566	45.315
25	29.339	34.382	37.652	40.646	44.314	52.620
30	34.800	40.256	43.773	46.979	50.892	59.703
35	40.223	46.059	49.802	53.203	57.342	66.619
40	45.616	51.805	55.758	59.342	63.691	73.402
45	50.985	57.505	61.656	65.410	69.957	80.077
50	56.334	63.167	67.505	71.420	76.154	86.661
60	66.981	74.397	79.082	83.298	88.379	99.607
70	77.577	85.527	90.531	95.023	100.425	112.317
80	88.130	96.578	101.879	106.629	112.329	124.839
90	98.650	107.565	113.145	118.136	124.116	137.208
100	109.141	118.498	124.342	129.561	135.807	149.4491

APPENDIX D

Product Liability

D.1 Basic Concept

In legal terms, product liability describes an action (such as a lawsuit) in which the plaintiff (the injured party) seeks to recover damages for personal injury or loss of property from the defendent (the seller or the manufacturer) when it is alleged that the damage was caused by a defective product [D.1–D.4]. A product defect may be either a manufacturing (or production) defect or a design defect. A manufacturing defect may arise due to an error in the production process. In such a case, the product may not meet the manufacturer's own standards and may escape from detection due to the use of inadequate quality control procedures. A classic example is the soda pop bottle that exploded, either due to an imperfection in the glass structure of the bottle or due to an inadvertent overcarbonization. The design defect may arise when inferior standards are used by the manufacturer in producing the product. Thus, the manufacturer and the designer become responsible for the reliable and safe performance of a product.

D.2 Definitions

Some of the basic terminologies employed in product liability are indicated in the following list:

1. *Tort.* Any wrongful act, damage, or injury for which the aggrieved can seek redress by legal action.
2. *Privity.* A direct contractual relationship between two persons or parties.
3. *Plaintiff.* The person (organization) who (which) initiates legal action for redress of a loss.
4. *Negligence.* The failure to take reasonable amount of care, which results in an injury or property damage to another person or party.
5. *Warranty.* An assurance by the seller to the purchaser that the product will be as promised.
6. *Implied warranty.* An automatic warranty, implied by law, that a product is suitable for a specific purpose and is reasonably safe for use.
7. *Express warranty.* A statement made by a seller, either orally or in writing, that the product sold is suitable for a specific use.

D.3 Theories of Product Liability

The following three legal principles form a basis in settling most product liability suits in favor of the consumer:

1. Negligence
2. Breach of express warranty
3. Breach of implied warranty

Negligence. *Negligence* can be described as the "conduct which falls below the standard established by law for the protection of others against unreasonably great risk of harm." This principle became the keystone of tort law. It can be seen that this law is flexible in that the "standard" would always be current, appropriate to the situation at hand, and whatever the court declares it to be. Hence, the manufacturer should produce the product based on the state-of-the-art technology known to them at the time of manufacture so that they will not be "negligent" in taking "reasonable care" in producing a safe product. The state-of-the-art technology may be accomplished through the following means:

In-house training and seminars
Professional society meetings
Continuing education courses
Scientific and business periodicals

A proof of such endeavors by a company might prove its progressive attitude in any future product liability litigation involving its products.

Breach of Express Warranty. *Breach of express warranty* occurs when the claims made by the manufacturer about his product are not met, and this results in an injury or damage. The claims made by the manufacturer need not be in writing. Some examples of express warranties according to the Uniform Commercial Code, adopted by most states, are as follows:

1. Catalog descriptions of products offered for sale by mail-order businesses
2. Photographs, display material, and manuals accompanying a product
3. Any model or sample that is made part of the basis of bargain
4. Any affirmation of fact or promise made by the seller to the buyer, which then becomes part of the basis of bargain.

Breach of Implied Warranty. *Implied warranty* is one which is implied by the law rather than given explicitly by the seller. The Uniform Commercial Code, which has been adopted by all the states except Louisiana, forms a basis for liability suits under breach of implied warranty. The implied warranty may be in regard to merchantability or fitness of a product. *Merchantability* indicates that the mere act of offering a product for sale implies that the product is safe and has at least the minimum qualities one would expect of such a product. In this case, the plaintiff need not prove that the entire population of products is unsafe; he needs to prove that only the particular one, which may be called *the worst-case* *product,* was unsafe. The *fitness-type* of implied warranty arises when a manufacturer or his representative makes a wrong recommendation about the suitability of a product for a particular use. An example is that of a paint store clerk recommending to the customer a particular type of lacquer thinner to remove paint from a floor. When a customer died due to an explosion during the use of the thinner, the store had to pay damages on the basis of breach of warranty of fitness.

D.4 Prevention of Product Liability

In the last two decades, the number of product liability suits increased dramatically to the extent of causing what some people label it as a crisis or an epidemic. Under the doctrine of strict liability, the manufacturers have become increasingly responsible for their products and designs; the reason is that the producer cannot assume that his product will be used correctly, safely, and solely for the purpose for which it is intended. For example, a person used a rotary lawn mower to trim the hedges, got injured, filed a lawsuit against the manufacturer and recovered the damages. To prevent product liability suits, the manufacturer should follow the old saying, "an ounce of prevention is worth a pound of cure," and follow the guidelines which are given.

Guidelines to Prevent Design Defects

1. Evaluate every design from safety point of view.
2. Carefully scrutinize the functioning of the product during prototype development and initial testing phases.
3. Many problems and/or defects will be discovered by the consumers. Correct these problems and/or defects by modifying the design in a suitable manner.
4. Evaluate the risks involved and balance the design between product utility (benefits) and potential risks.
5. Conduct a thorough mathematical analysis, which should include kinematic, kinetic, stress, weight, and reliability analyses, during the design stage to insure the safety of the product.
6. Use proper safety and reliability factors during design.
7. Use materials with known properties.
8. Design the product to be safe against all foreseeable misuses and abuses.
9. Use well-established standards such as those recommended by ANSI, ASTM, ASME, SAE, and Military Specifications in the design.

Guidelines to Prevent Manufacturing Defects

1. Use proper process control, quality control, and inspection techniques to reduce manufacturing defects.
2. Use statistical sampling techniques to evaluate the adherence of production employees to design and manufacturing specifications.

3. If the potential risk of the product in causing injuries is high, consider using 100 percent inspection instead of statistical sampling.
4. Document all inspection, quality control, and testing activities and report the results to the product design and development department. This not only aids in making the product safer and cheaper but also in the manufacturer's defense in the event of a lawsuit.
5. Use proper labels and warnings about the use of the product.
6. Build all safety features and devices as part of the basic product instead of making them available as optional equipment.

REFERENCES AND BIBLIOGRAPHY

D.1 V. J. Colangelo and P. A. Thornton, *Engineering Aspects of Product Liability,* American Society of Metals, Metals Park, OH, 1981.

D.2 C. E. Witherell, *How to Avoid Products Liability Lawsuits and Damages,* Noyes Publications, Park Ridge, NJ, 1985.

D.3 D. G. Owen and M. J. Davis, *Products Liability and Safety* (University Casebooks), 6th Ed., Foundation Press, Spring Valley, WI, 2010.

D.4 J. F. Thorpe and W. H. Middendorf, *What Every Engineer Should Know About Product Liability,* Marcel Dekker, Inc., New York, 1979.

Answers to Selected Problems

Chapter 1

1.1 (c) Mean $= 29.576 \times 10^6$ psi, standard deviation $= 2.194 \times 10^6$ psi

1.4 Mean $= 5.79$ kpsi, standard deviation $= 1.73$ kpsi

1.5 (c) Density function is bimodal **(d)** Mean $= 38.015$ kpsi, standard deviation $= 3.961$ kpsi

1.7 (c) Mean $= 1704.3$ cfs, standard deviation $= 105.8$ cfs

1.9 (c) Mean $= 9.375$, standard deviation $= 0.347$ **1.12** (c) Mean $= 4732$ cycles, standard deviation $= 578$ cycles

Chapter 2

2.3 **(a)** 658,008 **(b)** 23,310 **2.5** 20

2.7 **(a)** $\overline{A} = \{x|_{x \le 10.5}\}$ **(b)** $\overline{B} = \{x|_{x \ge 15.0}\}$ **(d)** $A \cap B = \{x|_{10.5 < x < 15.0}\}$

2.9 **(a)** 0.20 **(b)** 0.55 **(c)** 0.45 **(d)** 0.30 **2.11** 0.08418 **2.13** **(a)** 0 **(b)** 0.9 **(c)** 0 **(d)** 0.985

2.16 **(a)** $\frac{188}{265}$ **(b)** $\frac{219}{265}$ **(c)** $\frac{147}{219}$ **(d)** $\frac{147}{188}$ **2.18** **(a)** 9 **(b)** 15 **(c)** 6 **2.21** **(a)** $\frac{1}{15}$ **(b)** $\frac{8}{15}$ **(c)** $\frac{9}{15}$

2.23 **(a)** 0.6 **(b)** 0.026 **(c)** 0.308 **(d)** 0.598 **2.25** $P(A) = 0.4, P(B) = 0.6$ **2.27** 0.032

2.29 0.04828 **2.31** 0.972 **2.33** **(a)** 0.865 **(b)** 0.4722 **2.35** 0.9652 **2.37** 27 percent

2.39 0.02262 **2.41** **(a)** Not independent, **(b)** independent **2.44** **(a)** 0.01485 **(b)** 0.9999

Chapter 3

3.2 **(a)** $\overline{X} = 2.91, s = 1.55$, skewness coefficient $= 0.0535$ **(b)** 0.35 **(c)** 0.75

3.4 $E(X^2) = 4473, E(X^3) = 364{,}270, \sigma_X = 25.3257$, skewness coefficient $= 0.4917$

3.6 $p_X(x) = \begin{cases} 0.1, & x = -5 \\ 0.7, & x = 1 \\ 0.2, & x = 4 \end{cases}$ **3.8** $\overline{X} = 15.26, s_X = 1.6530$ **3.10** **(a)** 0.17456 **(b)** 0.09626

3.12 **(a)** 0.6065 **(b)** 0.0821 **(c)** 0.28735 **3.15** 0.006 percent

3.18 **(a)** 6 **(b)** 8 **(c)** 6.3246 **(d)** 2.1602 **(e)** skewness coefficient $= -0.4761$ **3.20** $5e^{-5x},\ x > 0$

3.22 **(a)** $\mu_X = 2.0202\ \text{cm}^3, \sigma_X = 9.8494\ \text{cm}^3$ **(b)** \$0.91

3.27 (57172, 62828), (56838, 63162), (56349, 63651), (55528, 64472), (53675, 66325), (−140000, 260000)

3.32 **(a)** $f_X(x) = \frac{1}{102}\left(x + \frac{9}{2}x^2\right), f_Y(y) = \frac{1}{102}\left(\frac{15}{2} + 68y\right)$ **(b)** $\mu_X = 3.018, \mu_Y = 1.5515$

3.37 **(a)** $F_X(x) = \begin{cases} 0; & x < a \\ \frac{x-a}{b-a}; & a \le x \le b \\ 1; & x \ge b \end{cases}$ **3.41** **(i)** $a = -11.4, b = 12.6, c = -1.5$ **3.43** $\mu - \sqrt{3}\sigma,\ \mu + \sqrt{3}\sigma$

3.45 **(a)** 2.074 **(b)** 0.1695 **(c)** mode = 0, median = 2 **3.51** Yes

3.56 **(a)** $f_X(x) = \dfrac{2}{\pi a^2}\sqrt{a^2 - x^2}$, $f_Y(y) = \dfrac{2}{\pi a^2}\sqrt{a^2 - y^2}$ **(b)** $\dfrac{1}{16}$ **(c)** 0.315 **3.58** **(a)** 0.4876 **(b)** 0.58

Chapter 4

4.3 $$F_{X_n}(x) = \begin{cases} \left(\dfrac{x^2}{20000} - \dfrac{3}{50}x + 18\right)^n; & 600 \le x \le 700 \\ \left(-\dfrac{x^2}{20000} + \dfrac{2}{25}x - 31\right)^n; & 700 \le x \le 800 \end{cases}$$

4.4 $$F_{X_1}(x) = \begin{cases} 1 - \left(\dfrac{x^2}{20000} + \dfrac{3}{50}x - 17\right)^n; & 600 \le x \le 700 \\ 1 - \left(\dfrac{x^2}{20000} - \dfrac{2}{25}x + 32\right)^n; & 700 \le x \le 800 \end{cases}$$

4.7 $\mu_Z = 145.9$ ksi, $\sigma_Z = 10.52$ ksi **4.9** 44.5097 m^3/s **4.10** **(a)** 0.0119 **(b)** 0.01623 **4.14** 0.0005, 0.009953

4.16 Type I: $\ln \ln \dfrac{1}{1 - F_Z(z)} = az - a\omega$, type II: $-\ln \ln \dfrac{1}{1 - F_Z(z)} = m \ln z - m \ln \omega$,

type III: $\ln \ln \dfrac{1}{1 - F_Z(z)} = m \ln(z - \varepsilon) - m \ln(v - \varepsilon)$

4.18 **1.** $\mu = 50.75$, $\sigma = 8.25$ (from the graph)
2. Type I: $F_Y(y) = e^{-e^{-0.151(y-46.417)}}$, type II: $F_Y(y) = e^{-\left(\frac{46.010}{y}\right)^{7.514}}$, type III: $F_Y(y) = e^{-\left(\frac{74.492-y}{27.842}\right)^{2.8424}}$

4.19 **1.** $\mu = 58.0$, $\sigma = 10.5$ (from the graph)
2. Type I: $\mu = 58.51$, $\sigma = 10.66$, type II: Not applicable, type III: $\mu = 58.84$, $\sigma = 9.73$

4.23 **1.** $\overline{X} = 1.02$, $s_X = 0.3140$, probability of damage $= P_f = 0.0009043$
2. Type I: $P_f = 0.01936$
3. Type II: $P_f = 0.04737$
4. Type III: $P_f = 0.32683$

Chapter 5

5.1 $$f_Y(y) = \begin{cases} 0; & y < 19.7 \\ 0.4348; & 19.7 \le y \le 22.0 \\ 0; & y > 22.0 \end{cases}$$

5.3 $f_X(x) = \dfrac{1}{\pi\sqrt{1 - x^2}}$; $-1 \le x \le 1$

5.5 $$f_Y(y) = \begin{cases} ye^{-y}; & y \ge 0 \\ 0 & \text{elsewhere} \end{cases}$$

5.8 $$f_Y(y) = \frac{7.9788}{\sqrt{\pi y}}\left[\exp\left\{-200\left(\sqrt{\frac{4y}{\pi}} - 2\right)^2\right\} + \exp\left\{-200\left(-\sqrt{\frac{4y}{\pi}} - 2\right)^2\right\}\right]$$

5.10 $\overline{\eta} = 0.59194$ **5.12** $\bar{l} = 247.3736$ cm **5.16** $f_I(i) = 2.2507 \exp\{-100.0019(i - 0.05)^2\}$; i = interference

5.19 $f_Y(y)$: $\dfrac{y}{2}$ for $0 \le y \le 1$; $\dfrac{1}{2}$ for $1 < y \le 2$; $\dfrac{3 - y}{2}$ for $2 < y \le 3$; 0 elsewhere with $Y = X_1 + X_2$

5.21 $\Delta\overline{p} = 20 \times 10^6$ $kg/m\text{-}s^2$ **5.24** For independent X and Y: $\mu_Z = \mu_X\mu_Y$, $\sigma_Z = \sqrt{\mu_X^2\sigma_Y^2 + \mu_Y^2\sigma_X^2 + \sigma_X^2\sigma_Y^2}$

5.26 $f_Y(y) = \dfrac{\lambda}{\sqrt{y}}e^{-\lambda\sqrt{y}}$; $\lambda > 0$, $y > 0$ **5.28** $f_Y(y) = \dfrac{1}{8\sqrt{y}}$; $1 \le y \le 25$

5.32 Take logarithms of $Y = X_1 X_2$ and $Y = \dfrac{X_1}{X_2}$ and use the properties of normal variates.

Chapter 6

6.4 $h(t) = 0.00223$ at 200 hours, 0.0008379 at 600 hours, 0.0042262 at 1000 hours

6.6 **(a)** $h(t) = 2t$ **(b)** $f_T(t) = 2te^{-t^2}$

6.8 **(i)** $R_i(t) = e^{-10^{-4}t}$ **(ii)** $R(t) = e^{-4\times 10^{-4}t}$ **(iii)** $R(t) = 1 - (1 - e^{-10^{-4}t})^4$

6.9 $h(t) = \dfrac{2}{t_2 - t}, R(t) = \dfrac{2}{(t_2 - t_1)^2}(t_2 - t)^2$

6.26 **1.** Component 5 involves least increase in its failure rate with a new failure rate of 0.0003618.
2. Component 2 involves least increase in cost.

6.28 **(a)** 0.006036 **(b)** 0.000005 **6.30** Three components of type B **6.32** 4

6.34 $R(t) = 1 - (1 - \exp(-10^{-3t})^n$

6.38 **1.** Component 1 with new reliability of $\approx$ 1.0
2. Component 1 with new reliability $\approx$ 1.0

6.40 $R_{\text{system}} = R_A R_C + R_A R_D + R_B R_D + R_B R_E - R_A R_C R_D - R_A R_B R_D - R_B R_D R_E - R_A R_B R_C R_E + R_A R_B R_C R_D R_E$

6.42 $R_{\text{system}} = R_A R_E R_F + R_A R_C R_D + R_B R_E R_F - R_A R_C R_D R_E R_F - R_A R_B R_E R_F$

6.44 $R_{\text{system}} = R_A R_E R_F + R_A R_C R_D + R_B R_E R_F - R_A R_C R_D R_E R_F - R_A R_B R_E R_F$

6.45 At 100 hours: $R_{\text{system}} = 0.999999951$ **6.48** 0.82237

6.50 $\lambda_i = 0.00030379, 0.003645, 0.0001215, 0.000506, 0.000162, 0.00010126$ for $i = 1$ to 6.

Chapter 7

7.1 0.5 **7.2** 0.999999999 **7.3** 0.5 **7.4** 0.017 **7.5** 0.003649 **7.6** 0.09342 **7.7** 0.15386

7.8 0.15866 **7.9** 0.26639 **7.11** 0.39939

7.13 **1.** Mean = 22.4522 cm/s, standard deviation = 15.7839 cm/s
2. Mean = 141.0717 cm/s^2, standard deviation = 99.1734 cm/s^2
3. 0.66749

7.15 $S_u = 4$ in, $S_a = 0.4$ g **7.16** $S_u = 0.5$ in, $S_v = 1$ in/sec, $S_a = 1$ g

Chapter 8

8.1 0.2174 **8.3** 0.1801 **8.5** 1.0 **8.7** 0.6535 **8.9** 0.8251 **8.12** 0.54975 **8.15** 0.3936

8.17 0.8461 **8.19** 0.5940 **8.22** 0.04356 **8.24** 0.7193 **8.25** 0.77097 **8.30** 0.95 **8.32** 0.393

8.37 $\beta_{0.05} = -1.645, \beta_{0.55} = 0.125, \beta_{0.85} = 1.033$ **8.42** $\beta = \dfrac{\mu_{\ln Y}}{\sigma_{\ln Y}}$ where $Y = \dfrac{S}{L}$ **8.44** 1.1894

8.48 **(a)** 0.32 **(b)** 0.5504 **(c)** $0.32 \leq P_f \geq 0.6856$

Chapter 9

9.4 0.001524 in **9.7** 0.768943 in **9.9** 1.125 **9.10** 0.71226 **9.11** 0.65173

9.12 2.8248 in **9.13** 2.5825 in

Chapter 10

10.1 $x_i = \dfrac{1}{1 - u_i}$ or $\dfrac{1}{u_i}$ **10.2** $x_i = k\left(1 - \sqrt{1 - u_i}\right)$ **10.4** $x_i = \exp(\mu_{\ln X} + z\sigma_{\ln X})$

10.6 **(a)** $d_r = w_1, D = 1.5\,w_1 + w_2$ **(b)** $\mu_p = 1.628$ MPa, $\sigma_p = 0.222711$ MPa

10.11 $\bar{p} = 56534816.3$ N/m^2, $\sigma_p = 14985703.8$ N/m^2 **10.13** $\bar{p}_{\max} = 4.36196 \times 10^5$ psi, $\sigma_{p_{\max}} = 2.3383 \times 10^4$ psi

10.14 $\bar{p}_{\max} = 27573.29$ psi, $\sigma_{p_{\max}} = 1221.70$ psi

Chapter 11

11.1 **1.** Find $n_1, n_2, \ldots, n_N$ to maximize $R_s = \prod_{j=1}^{N} [1 - (1 - R_j)^{n_j}]$ subject to $\sum_{j=1}^{N} c_j n_j \le c_{\max}$, $\sum_{j=1}^{N} w_j n_j \le w_{\max}$, and $n_i \ge 0, i = 1, 2, \ldots, N$

11.2 **1.** Find $n_1, n_2, \ldots, n_N$ to maximize $R_s = 1 - \prod_{j=1}^{N} [1 - (R_j)^{n_j}]$ subject to $\sum_{j=1}^{N} c_j n_j \le c_{\max}$, $\sum_{j=1}^{N} w_j n_j \le w_{\max}$, and $n_i \ge 0, i = 1, 2, \ldots, N$

11.8 $\lambda_i^* = (2a_i)^{\frac{1}{3}} \left\{ \frac{1}{b} \sum_{j=1}^{n} \frac{a_j}{(2a_j)^{\frac{2}{3}}} \right\}^{\frac{1}{2}}; \quad i = 1, 2, \ldots, n$ **11.12** $x^* = 2, f^* = -11$

11.13 $\begin{Bmatrix} x_1^* \\ x_2^* \end{Bmatrix} = \begin{Bmatrix} 17/3 \\ 7/3 \end{Bmatrix}, f^* = 58/3$

Chapter 12

12.27 $R_{\text{system}} = 0.9498$ **12.28** $R_{\text{system}} = 0.9227$ **12.29** **(a)** $\lambda_{\text{series}} = 506 \times 10^{-7}$, **(b)** $R_{\text{series}} = 0.9506$

12.30 $R_{\text{press}} = 0.7982$

Chapter 13

13.1 At $t_i = 2644$ hours: $f(t_i) = 3.3841 \times 10^{-4}$, $R(t_i) = 0.0667$, $h(t_i) = 50.7361 \times 10^{-4}$ **13.4** 1.0607

13.6 Reject, Number of IC chips to be tested as per ASN = 110

13.7 Accept line: $2.3026f - 0.0953\,s = 4.5850$, reject line: $2.3026f - 0.0953\,s = -3.902$, number of units from ASN: 54

13.17 **1.** $T(x) = 81.2857 - 2.3928x$,
2. $T(x) = 88.3095 - 4.5x + 0.0878x^2$

Chapter 14

14.1 **(a)** 0.4457, **(b)** 0.1448, **(c)** 0.1905, **(d)** 0.9543

14.3 **(i)** 1.6993×10^{-16}, **(ii)** 3.3986×10^{-15} **14.5** 0.00112 **14.7** 0.982243 **14.9** No abnormality

14.11 Process is under control **14.12** Process is under control **14.16** $UCL_c = 14.9371$, $LCL_c = 0$, process is under control

Chapter 15

15.1 **(a)** $M(6) = 0.6351$, **(b)** $t = 17.8317$ hours, **(c)** Required MTTR = 2.0028 hours

15.2 **(a)** $M(6) = 0.5179$, **(b)** $t = 7.6468$ hours, **(c)** Required MTTR = 4.3065 hours

15.13 **(a)** 0.2558, **(b)** 0.4461, **(c)** 0.9479 **15.15** $A(\infty) = 0.8696$

15.16 **(a)** $A(\infty) = 0.9050$, **(b)** $A(\infty) = 0.9070$

15.17 **(a)** $A(\infty) = 0.9955$, **(b)** $A(\infty) = 0.9977$

15.20 $A(\infty) = 0.9524$, $(\text{MTTF})_{\text{new}} = 20$ hours

Chapter 16

16.1 \$226,159 **16.2** **(a)** \$188.47, **(b)** \$115, 300 **16.3** \$134,426.53 **16.4** 3.4790 months

16.5 51 percent of initial price **16.6** \$19.60 **16.11** \$50 **16.12** \$5.33 **16.13** \$9.76

Index